A COMPREHENSIVE GUIDE TO THE DESIGN AND MANUFACTURE OF PRINTED BOARD ASSEMBLIES

VOLUME 1

A COMPREHENSIVE GUIDE TO THE DESIGN AND MANUFACTURE OF PRINTED BOARD ASSEMBLIES
— VOLUME 1 —

Edited
by
William MacLeod Ross

ELECTROCHEMICAL PUBLICATIONS LTD
1996

ELECTROCHEMICAL PUBLICATIONS LTD
Asahi House, Church Road, Port Erin,
Isle of Man, British Isles

©
Electrochemical Publications Ltd
1996

ISBN 0 901150 32 0

All rights reserved. No part of this publication may be reproduced, stored in a retrieval system, or transmitted, in any form or by any means, electronic, mechanical, photocopying, recording, or otherwise, without the prior written authorisation of the publishers.

DISCLAIMER

We believe the information provided in this work is reliable and useful, but it is furnished without warranty of any kind from the publisher, editor or authors. Readers should make their own determination of the suitability or completeness of any of the material included for specific purposes and adopt any safety, health, and other precautions as may be deemed necessary by the user. No license under any patent or other proprietary right is granted or to be inferred from the provision of the information herein.

Typeset by Electrochemical Publications Ltd, Port Erin, Isle of Man, British Isles
Printed by Arrowsmith, Bristol, England

ACKNOWLEDGEMENTS

In compiling a book of this description, help, advice and expert knowledge are available, and freely given, from many quarters. Friends, acquaintances and strangers have been of enormous assistance to me, as an editor and as an author. To mention each by name is impossible in the space available and I hesitate to name a few for fear of offending the many.

A considerable debt is owed to the many companies who have provided expert advice and many of the illustrations. When practicable, due acknowledgement has been made in the text for such contributions. It must be acknowledged here that all proprietary product names used in the text are registered trademarks or tradenames of their respective companies. Many of these are in common usage in the electronics industry and will be familiar to readers.

The authors have responded magnificently throughout the period of compiling this book. While it is obvious that there would be no book without the efforts of the authors, it must be acknowledged that the path to publication has not been an easy one and the manner in which my co-authors accepted the trials and tribulations which beset us and reacted to them was of enormous help to the beleaguered editor. Please accept my grateful thanks for all your efforts.

It is almost needless to pay my respects to the publisher and his staff. Despite staff movements, each and every one of them who was, even in part, responsible for editing, redrawing figures or correcting my omissions and mistakes, performed their tasks and assisted me very well. Their unfailing encouragement and attention to detail during the proof stages proved invaluable. Thank you all for helping to achieve the final success!

Finally, but far from least, I must tender my heartfelt thanks to my wife. Over a number of years her support and encouragement have been my main props. Without her unswerving devotion, I should not have been able to carry on to this conclusion.

W. MacLeod Ross

PREFACE

This handbook is the result of a desire to compile the best and most comprehensive guide to the design and manufacture of printed board assemblies available.

The first book to recognise that there was a need for a text which addressed not only printed boards but also other electronic components, and the design and manufacture of electronic assemblies from them, was written in the late seventies and first published in English in 1981. In the late eighties, the author of that book, Giovanni Leonida, set out to consider an update of the text and started to compile material appropriate to a revision of the Handbook of Printed Circuit Design, Manufacture, Components and Assembly.

Changing circumstances meant that Dr Leonida could not complete the revision and the present work was born.

It was recognised that, while some aspects of electronic components and assemblies had changed by only a small amount, giant strides had been made in the technology of others. Consideration had to be given to covering new materials, new methods of using them, and newer techniques in design, manufacture and assembly of printed boards and their assemblies. After considerable deliberation, a list of essential subjects was drawn up and put into a logical sequence. From the intended revision, a totally new book was created.

It rapidly became clear that it was extremely unlikely that any one person could be found having the depth and breadth of knowledge (and availability of time!) adequate to cover all subjects envisaged necessary. Accordingly, much time was expended in finding authors with expert knowledge of their subject matter and the ability to impart that knowledge in an understandable and readable manner.

The results of our authors' efforts are presented in two volumes. The first of these deals with components and assembly, including design and layout of boards. The second volume covers board manufacture, the environment, quality assurance and financial considerations. (Originally, it was intended that the book would be published in one volume. It became obvious, however, that the result would be an over-large and heavy tome and the present arrangement of two volumes resulted.) In both volumes, the authors selected are authorities in their particular fields and many will be familiar to readers as contributors to national and international journals, conferences and symposia.

Trouble has been taken in the editing of each chapter to produce reasonable uniformity of style and presentation without losing the individuality of the writer. In some cases it may be considered that undue repetition has been permitted. Closer examination should show that such text is *additive* rather than *repetitive*. It contributes something new to the subject, often showing a different viewpoint or an alternative approach. In certain instances, repetition is essential to the sense and content of the subject.

Cross-referencing in each volume between chapters (and between volumes) has been carried out with a view to assisting the reader to find the text contributory to aiding understanding and knowledge of the subject being pursued. This will also be assisted by the comprehensive list of contents and the indices.

Every effort has been made to ensure that the text is as up-to-date as is possible. In a field which has developed, and is still developing, at a steadily increasing pace for many years, to talk of being up-to-date is tempting providence. What is at the forefront of the technology today will be outmoded and 'old hat' tomorrow. Every day the frontiers of electronic science and technology move farther out, extending to a horizon which is itself steadily moving away from the observer. For that reason, while specific examples of, e.g., component types and of mounting equipment are given as illustrations, no attempt has been made to provide the sort of information which can be readily gleaned from standards and specifications, suppliers' catalogues, from trade associations or from trade fairs and conferences. Rather we have attempted to guide the reader on how to find and how to use such information to achieve the best advantage.

A particular problem of the age in which we live arises from the reluctance with which persons in different lands co-operate one with the other. While governments and international standardisation bodies promote the metric system of units for all types of measurement, at a less exalted level individuals tend to use the unit they know and/or like best. For example, there are those to whom the 'mil' is the area of a circle of 0.001 inch diameter. To others the area of the 0.001 in. diameter circle is called the 'circular mil' and the unqualified 'mil' is a linear measure equal to 0.001 inch. There are other persons, too lazy or slipshod to pronounce 'millimetre', who give the name 'mil' to 1 millimetre. 'Micron', depending upon who is talking about it, can mean 'microinch' or micrometre'. Throughout both volumes of this book every effort has been made to ensure that units employed are clear and unambiguous. In many cases, particularly where the original units are imperial units, a conversion, e.g., into metric units, has been provided. In some instances where dual dimensions are given, the conversion is not a true value but rather an approximation that an engineer thinking in that system of units would think of naturally.

The text has been carefully edited to minimise the appearance of equations and lengthy calculations in an attempt to produce a practical, readable and instructive book that does not require a degree in mathematics to understand. It is true that Volume 1, Chapter 11, does have some mathematical formulae. These may be considered inescapable in a chapter devoted to thermal management of printed board assemblies. However, Mr Dean does minimise the need for calculation and provides a series of very helpful tables in addition to giving advice on computer-assisted methods of calculation. The result is a text which is eminently understandable and easy to put into practice.

Finally, while a section is set aside in Volume 2 to cover quality assurance and control and inspection and test, much information is given on inspection and test in individual chapters. For example, considerable detail is given about on-line inspection and test of assemblies in the chapters covering assembly techniques.

It is confidently expected that the knowledge contained in these pages will assist designers and all persons involved in the production and use of printed boards and their assemblies.

W. MacLeod Ross
Bishop's Stortford, UK
March 1996

CONTENTS

ACKNOWLEDGEMENTS		v
PREFACE		vi

CHAPTER ONE

Introduction — 1

1.1	Electronics	1
1.2	Conduction — A Simple View	2
1.2.1	Superconductivity	3

CHAPTER TWO

Historical Background — 5

2.1	Introduction	5

Part 1 — Printed Boards — 5

2.2	Early days	5
2.2.1	Prior Art	5
2.2.2	The Birth of the Printed Circuit	6
2.2.3	The First Printed Circuit Symposium	7
2.2.4	Module Development	7
2.3	Further Development	8
2.3.1	Plated-through Hole Boards	8
2.3.2	Multilayer Boards	8
2.3.3	Additive Processes	8
2.3.4	Discrete Wired Boards	9
2.3.5	Flexible Circuits	10
2.3.6	Other PB Developments	10
2.4	Materials and Processes	10
2.4.1	Laminates	10
2.4.2	Resists and Etchants	11
2.4.3	Machining	12
2.4.4	Hole Pretreatment, Metallising and Plating	12
2.4.5	Finishes	14
2.4.6	CAD	14
2.5	In Conclusion	15
2.5.1	Printed Boards and the Environment	16
2.5.2	The Future of Printed Circuits	17

Part 2 — Electronic Components — 17

2.6	General	17
2.7	1701-1800	17
2.8	1801-1900	18
2.8.1	1801-1830	18
2.8.2	1831-1860	18
2.8.3	1861-1880	19
2.8.4	1881-1900	19
2.9	1901- Present day	20
2.9.1	1901-1920	20
2.9.2	1921-1940	21
2.9.3	1941-1960	21
2.9.4	1961-1980	22
2.9.5	1981- Present day	24

Part 3 — Soldering — 26

2.10	From Ancient Craft to Modern Science	26

CHAPTER THREE

Conventional Components — 31

3.1	General	31
3.2	Classification	31
3.3	Resistors	32
3.3.1	Packages	32
3.3.2	Characteristics	33
3.3.3	Manufacturing Technology	35
3.3.4	Identification	37
3.3.5	Variable Resistors	40
3.4	Capacitors	40
3.4.1	Packages	40
3.4.2	Characteristics	42
3.4.3	Manufacturing Technology	43
3.4.4	Electrolytic Capacitors	46
3.4.5	Identification	49
3.4.6	Variable Capacitors	49
3.5	Semiconductor Diodes	50
3.5.1	Packages	50
3.5.2	Characteristics	51
3.5.3	Technology	56
3.5.4	Identification	58
3.5.5	Rectifiers	58
3.6	Transistors	58
3.6.1	Packages	62
3.6.2	Characteristics	63
3.6.3	Manufacturing Technology	65
3.6.4	Identification	67
3.7	Silicon Controlled Rectifiers and Thyristors	67
3.8	Optoelectronic Devices	69
3.8.1	Packages	73
3.9	Monolithic Integrated Circuits	74
3.9.1	Classification	74
3.9.1.1	Technology	74
3.9.1.2	Operating Mode	77

3.9.1.3	Function	78
3.9.1.4	Level of Integration	78
3.9.2	Manufacturing Technology	79
3.9.3	Analogue ICs	80
3.9.4	Digital ICs	82
3.9.4.1	Elementary Circuits	82
3.9.4.2	Memories	84
3.9.4.3	Microprocessors	86
3.9.4.4	ASICs	86
3.9.5	Packages	87
3.10	Hybrid Integrated Circuits	93
3.10.1	Thin Film Hybrids	94
3.10.2	Thick Film Hybrids	97
3.10.3	Packages	97

CHAPTER FOUR

Surface Mounted Components 102

4.1	General	102
4.2	Classification	103
4.3	Passive Surface Mounting Components	105
4.3.1	MELF Components	106
4.3.1.1	Resistors	107
4.3.1.2	Jumpers	108
4.3.1.3	Positive Temperature Coefficient (PTC) Resistors	108
4.3.1.4	Capacitors	108
4.3.1.5	Inductors	109
4.3.2	Mini-MELF Components	109
4.3.3	Other Cylindrical Components	110
4.3.4	Chip Components	112
4.3.4.1	Resistors	112
4.3.4.2	Jumpers	115
4.3.4.3	Thermistors	115
4.3.4.4	Ceramic Chip Capacitors	115
4.3.4.5	Microchip Capacitors	118
4.3.4.6	Chip Inductors	119
4.3.5	Other Flat Components	119
4.3.5.1	Metal Film Square Chip Resistors	119
4.3.5.2	High Power Resistors	120
4.3.5.3	Metallised Polyester Capacitors	120
4.3.5.4	Wet Type Aluminium Electrolytic Capacitors	120
4.3.5.5.	Solid Aluminium Electrolytic Capacitors	123
4.3.5.6	Tantalum Electrolytic Capacitors	124
4.3.5.7	Inductors	125
4.3.5.8	Oscillators and Filters	127
4.3.6	Miscellaneous Passive Components	128
4.3.6.1	Aluminium Electrolytic Capacitors	128
4.3.6.2	Potentiometers	129
4.3.6.3	Trimmer Capacitors	129
4.3.6.4	Connectors	130
4.4	Active Surface Mounting Components	132
4.4.1	Diodes	133
4.4.2	SOT-23	135
4.4.3	SOT-89	137
4.4.4	SOT-143	137

4.4.5	SOT-223	137
4.4.6	Adapted Packages	138
4.4.7	Small Outline Integrated Circuits — SOICs	139
4.4.8	Very Small Outline Packages — VSOs	141
4.4.9	Flatpacks and Quadpacks — QFPs	141
4.4.10	Plastic Flat Packs — PFPs	143
4.4.11	Plastic Leaded Chip Carrier — PLCC	144
4.4.12	Small Outline J-leaded Package — SOJ	147
4.4.13	Leadless Ceramic Chip Carrier — LCCC	148
4.4.14	Leaded Ceramic Chip Carrier — LDCC	150
4.4.15	Other Complex Packages	150
4.4.16	Tape Automated Bonding — TAB	152
4.4.17	Chip-on-board (COB) and Multichip Modules (MCMs)	155
4.4.17.1	Chip-on-board	155
4.4.17.2	Multichip Modules	156
4.5	Supply Packaging	157
4.5.1	Bulk	157
4.5.2	Magazine	158
4.5.3	Rail and Tube	158
4.5.4	Tray or Palette	159
4.5.5	Adhesive Tape	159
4.5.6	Tape-on-reel	160
4.6	Advantages of SMT	161
4.6.1	Design Freedom	162
4.6.2	Size and Weight	165
4.6.3	Reliability	166
4.6.4	Electrical Characteristics	166
4.6.5	Effect of SMT on Automation	167
4.6.6	Cost	167
4.7	Disadvantages of SMT	168
4.8	The Future of Surface Mounting	169

CHAPTER FIVE

Manual Assembly 171

5.1	General	171
5.1.1	Manpower and Manual Assembly	172
5.2	Low Volume Assembly	173
5.3	Mass Production	178
5.3.1	Without Pre-cut and Formed Components	178
5.3.2	Automatic Positioning Tables	179
5.3.3	Using Pre-cut and Formed Components	180
5.4	Lead Preforming	180
5.4.1	Types of Preformed Shape — Axial Components	181
5.4.2	Manual Preforming — Axial Components	186
5.4.3	Preforming Equipment for Axial Components	186
5.4.4	Preforming Non-axial Leaded Components	190
5.5	Production Aids	194
5.5.1	Layout Memorisation	195
5.5.2	Stage Assembly	196
5.5.3	Sequence Assembly	197
5.5.4	Component Sequencing	199
5.6	Visual Aids	200
5.6.1	Screen Printing	200
5.6.2	Slide Projection	202
5.6.3	Lamp Displays	202
5.6.4	LED Displays	203

5.6.5	Optical Fibre Displays	204
5.6.6	Laser Scanning Systems	204
5.7	Assembly Inspection	206
5.7.1	Visual Inspection	206
5.7.2	Equipment Assisted Inspection	207

CHAPTER SIX

Automatic Assembly — Conventional Components 209

6.1	Introduction	209
6.1.1	Economics of Automation	209
6.1.2	The 'Mechanical Horse'	210
6.2	Early Automation	212
6.2.1	Semi-automatic Equipment	212
6.2.2	The Transfer Line	214
6.3	Component Packaging	215
6.3.1	Axial Component Taping	215
6.3.2	Radial Component Taping	218
6.3.3	DIP Components	221
6.3.4	Odd Components	222
6.4	Machine Types	222
6.4.1	Dedicated Inserters	223
6.4.2	Off-line Sequencing	224
6.4.2.1	Axial Component Sequencing	224
6.4.2.2	Sequence Verifiers	228
6.4.3	In-line Sequencing	228
6.4.3.1	Chain Sequencing	228
6.4.3.2	Air Delivery	229
6.4.3.3	Shuttle Delivery	230
6.4.3.4	Gravity Chute	230
6.4.3.5	Sort and Place	232
6.4.3.6	Pick-and-place	232
6.4.4	X-Y Tables and Rotary Tables	232
6.4.5	Positioning Aids — Optical Verifiers	234
6.4.6	Cutting and Clinching	235
6.4.7	Insertion Verification On-line	236
6.4.8	Control Unit	237
6.5	Axial Component Insertion Machines	238
6.5.1	Inserters versus Sequence-inserters	239
6.5.2	Fixed versus Variable Centre Distance	241
6.5.3	Insertion Programs	242
6.5.4	Insertion Cycle	245
6.5.5	Typical Axial Inserters	247
6.6	Radial Component Insertion Machines	248
6.6.1	Machine A	248
6.6.2	Machine B	249
6.6.3	Machine C	249
6.7	Dual Inserters	250
6.8	DIP Insertion Machines	250
6.8.1	Component Feeding	252
6.8.2	Insertion Cycle	253
6.8.3	Typical DIP Sequencer-inserters	254
6.8.3.1	Example A	254
6.8.3.2	Example B	254
6.8.3.3	Example C	254
6.8.3.4	Example D	255
6.8.3.5	Equipment Manufacturers	255

xiv *A Comprehensive Guide to the Design and Manufacture of Printed Board Assemblies*

6.9	Odd Component Insertion Machines	255
6.9.1	Dedicated Insertion Machines for Odd Components	256
6.9.2	General Purpose Robots	257
6.10	Assembly Inspection	259
6.10.1	Visual Inspection	260
6.10.2	Inspection Equipment	261
6.11	Board Handling	262

CHAPTER SEVEN

Automatic Assembly — Surface Mount Components 266

7.1	Types of Surface Mount Assembly (SMA)	266
7.1.1	Single-sided SMA	267
7.1.2	Double-sided SMA	269
7.1.3	Mixprint — SMDs One Side Only	269
7.1.4	Full Mixprint	272
7.2	Surface Mounting Adhesives	272
7.2.1	Requirements for Adhesives	274
7.2.2	Application Methods	276
7.2.2.1	Pin Transfer	276
7.2.2.2	Syringe	280
7.2.2.3	Screen Printing	281
7.2.3	Adhesive Dot Criteria	281
7.2.4	Storage of Boards	283
7.2.5	Adhesive Curing	283
7.3	Solder Pastes	285
7.4	Conductive Adhesives	286
7.5	Component Mounting	287
7.5.1	Evolution of Pick-and-place Equipment from Manual Assembly	288
7.5.2	Automatic Mounting	289
7.5.3	An Automatic Mounting Machine	291
7.5.4	PB Loading and Positioning	292
7.5.5	Component Feeding	293
7.5.5.1	Rails and Tubes	295
7.5.5.2	Vibratory Bowl	295
7.5.5.3	Vibrating Conveyor	296
7.5.5.4	Hopper	296
7.5.5.5	Tape on Reel	297
7.5.5.6	Others	297
7.5.6	Application Heads	297
7.5.6.1	Picking Up	298
7.5.6.2	Centring	299
7.5.6.3	Rotation	299
7.5.6.4	Testing	300
7.5.6.5	Placement Control	301
7.5.6.6	Teach-in Options	302
7.5.6.7	Bad Circuit Detector	303
7.5.6.8	Optical Recognition Equipment	303
7.5.6.9	Turret Heads	303
7.5.6.10	Adhesive Applicators	304
7.5.7	Control	304
7.5.8	Software Packages	307
7.6	Mounting Equipment	308
7.6.1	Table-top	308
7.6.2	Pick-and-place	309
7.6.2.1	Examples of Sequential Pick-and-place Machines	310
7.6.3	Multiple Arm Pick-and-place	315

7.6.4	Parallel Equipment	316
7.6.5	Parallel-sequential Equipment	321
7.7	Ancillary Equipment	322
7.7.1	Curing Ovens	322
7.7.2	General Purpose Robots	324
7.8	Integrated Assembly Lines	324
7.9	Equipment Selection	326

CHAPTER EIGHT

General Principles of Design and Layout (of Printed Board Assemblies)   328

8.1	General	328
8.2	Basic Definitions	329
8.3	Classification of PBs	334
8.4	Initial Assessment	340
8.5	Tentative Dimensioning	342
8.5.1	Vertical Mounting	343
8.6	Working Out the Volume	344
8.7	Multiple Board Assembly	345
8.7.1	Advantages of a Single Board Solution	345
8.7.2	Advantages of a Multiple Board Solution	345
8.7.3	Point-to-point Wiring	346
8.7.4	Soldered Backplanes	348
8.7.5	Book Connection	349
8.7.6	Motherboard	350
8.7.7	Soldered Add-ons	351
8.7.8	Sandwich	351
8.7.9	Others	352
8.8	Design of Boards	352
8.8.1	Partition of the Assembly	352
8.8.1.1	Number of Connections	353
8.8.1.2	Assembly Technology	353
8.8.1.3	Making a Layout	353
8.8.1.4	Testability and Reparability	354
8.8.1.5	Maintainability	354
8.8.1.6	Environment	355
8.8.1.7	Logistics	355
8.8.1.8	Costs	356
8.8.2	Gross and Net Area	356
8.8.3	PIH Assembly	359
8.8.4	SMC Assembly	364
8.8.5	Mixprint Assemblies	370
8.8.6	Heavy Components	371
8.8.7	Other Points	372
8.9	Conductor Dimensioning	373
8.10	Further Reading	382

CHAPTER NINE

Layout of Printed Circuit Boards   384

9.1	Introduction	384
9.1.1	Manual Layout	384
9.1.2	Amateur's Layout	387
9.1.3	Manual Drawing of Artworks	387
9.1.4	Master Drawing	393
9.1.5	Component Map	395

9.2	Automated Production of Films	396
9.2.1	Digitising	396
9.2.2	Editing	397
9.2.3	Photoplotting	398
9.3	General Principles	400
9.3.1	Why Use a Grid?	401
9.3.2	Grid Selection	402
9.3.3	Grid Sizes in Common Use	404
9.3.4	A Bad Example	409
9.3.5	A Good Example	413
9.4	PIH (Pin-in-hole) Components	414
9.4.1	Axial Components	415
9.4.2	Other Discrete Components	418
9.4.3	Integrated Circuits	419
9.5	Surface Mounted Components	421
9.5.1	Is a Grid Essential?	423
9.5.2	Minimum Distance	423
9.5.3	Dimensioning of Lands	425
9.5.3.1	Wave Soldering	425
9.5.3.2	Reflow Soldering	428
9.5.4	Footprints	428
9.5.4.1	Chip SMD Footprints	429
9.5.4.2	MELF Component Footprints	429
9.5.4.3	Other Discrete SMC Footprints	430
9.5.4.4	SM Integrated Circuit Footprints	431
9.6	Multiple Boards	432
9.7	Conclusion	433

CHAPTER TEN

CAD/CAM 439

10.1	Introduction	439
10.2	Input Formats	439
10.2.1	Pin List	439
10.2.2	Schematic Entry	440
10.2.2.1	Style of Schematic Entry	443
10.2.2.2	Hierarchical Design	443
10.2.2.3	Libraries	443
10.2.3	No Format Supplied	444
10.3	Simulation	444
10.3.1	PAL, PLA and ASIC	445
10.4	Printed Board Layout Requirements	446
10.4.1	Capability Required	446
10.4.1.1	Co-ordinate System	446
10.4.1.2	Resolution	446
10.4.1.3	Layers	446
10.4.1.4	Track/Trace Widths	447
10.4.1.5	Pad Shapes	447
10.4.2	Placement	447
10.4.2.1	Autoplacement	448
10.4.3	Routing	449
10.4.3.1	Manual Routing	449
10.4.3.2	Autoroute Styles	449
10.4.4	Design Checking	452

10.4.5	Modifications	454
10.4.6	Schematic Annotation	455
10.4.7	Automatic Test	455
10.4.8	Mechanical Drawing	456
10.5	Outputs	456
10.5.1	Schematics	456
10.5.2	Layouts	457
10.5.3	Photoplotting	457
10.5.3.1	Pen Plotters	458
10.5.3.2	Matrix Plotters	459
10.5.3.3	Laser Printers	460
10.6	Computer Aided Manufacturing (CAM)	460
10.7	Interfaces	461
10.8	The Processing Platform	462
10.8.1	The Personal Computer (PC)	462
10.8.2	Memory Needs	463
10.8.3	Data Transfer	464
10.8.4	Machine Architecture	464
10.8.5	Graphics	465
10.8.6	Networks and Workstations	465
10.8.7	Processor Performance	465
10.9	Man-machine Interaction	466
10.10	Choosing a CAD System	467

CHAPTER ELEVEN

Thermal Management Aspects 469

11.1	Introduction	469
11.1.1	Thermal Design	469
11.1.1.1	The Influence of Components and Cooling Methods on Temperature	470
11.1.1.2	The Rôle of Temperature Prediction	471
11.1.1.3	How Far Thermal Design?	471
11.1.2	Format of this Chapter	471
11.2	Heat Transfer Mechanisms and Temperature Effects	472
11.2.1	Mechanisms of Heat Transfer	473
11.2.1.1	Conduction	473
11.2.1.2	Convection	473
11.2.1.3	Radiation	473
11.2.1.4	Phase-change Heat Absorption	474
11.2.2	Difficulties Involved in Evaluating the Effects of Temperature on Reliability and Performance	474
11.2.3	Approach to Thermal Design	475
11.2.4	Designing for Reliability and Performance	476
11.2.4.1	Will the System Operate?	476
11.2.4.2	Can the System be Improved?	477
11.2.4.3	Relationship Between Temperature and Performance	477
11.3	Printed Board Construction and Examples	478
11.3.1	Types of Printed Board	478
11.3.1.1	A Combination of Circuit Board Types	479
11.3.2	Circuit Board Thermal Properties	480
11.3.2.1	The Thermal Conductivity of Basic Epoxide Printed Boards	480
11.3.2.2	Effective Thermal Conductivity of Printed Circuit Boards Having a High Thermal Conductivity Core	482
11.3.2.3	Surface Heat Loss and Gain from Boards Having a High Thermal Conductivity Core	483

11.3.2.4	Poor Thermal Conductivity Boards Having Designed Heat Pathways Bonded to Their Surface	483
11.3.3	Circuit Board Examples for Illustrating the Estimation of Effective Thermal Properties	485
11.4	'First Look' Methods of Circuit Board Thermal Design	488
11.4.1	'First Look' Temperature Prediction for Circuit Boards Without Surface Heat Loss	489
11.4.1.1	Concentration of Heat Generation at the Mean Distance from the Heat Sink	489
11.4.1.2	Uniform Distribution of Heat Generation with No Surface Heat Loss	490
11.4.1.3	Effect of Component Size	491
11.4.1.4	Estimate of Component Area Temperatures	492
11.4.2	'First Look' Temperature Prediction for Surface-cooled Circuit Boards	493
11.4.2.1	Simplification of Temperature Profile Approach	494
11.4.2.2	Exact Evaluation of Uniformly Heated Circuit Board with Surface Heat Loss	498
11.4.2.3	Effect of Component Size Under Surface Heat Loss Conditions	500
11.4.2.4	Estimate of Component Area Temperatures for Surface Cooling	502
11.4.3	General Features of Temperature Prediction Using 'First Look' Methods	502
11.4.3.1	Area of Sparse Power Dissipation	503
11.4.3.2	Poor Thermal Conductivity Circuit Boards	504
11.4.4	'First Look' Temperature Prediction for Circuit Boards Cooled by a Rear-mounted Heat Sink	506
11.4.5	'First Look' Temperature Values for Components	509
11.5	More Accurate Methods of Estimating the Temperatures Reached in Circuit Boards	510
11.5.1	Balanced Circuits	510
11.5.1.1	Balanced Circuits Mounted on Conduction-only Cooled Circuit Boards	512
11.5.1.2	Balanced Circuits Mounted on Surface-cooled Circuit Boards	513
11.5.2	Excess Temperature Estimation	514
11.5.2.1	A Graphical Method of Temperature Prediction for Conduction-cooled Circuit Boards	514
11.5.2.2	Temperature Estimates for Surface-cooled Circuit Boards	518
11.5.3	Estimates of Temperatures Arising on Circuit Boards Cooled by a Rear-mounted Heat Sink	518
11.5.4	Temperature Values for Components	519
11.6	Desk-computer-supported Circuit Board Thermal Design	519
11.6.1	Defects of Computer Packages when used for Thermal Design	519
11.6.2	Two Reliable Computer-assisted Methods	521
11.6.2.1	A Method for Edge-cooled Circuit Boards	521
11.6.2.2	A Method for Rear-cooled Circuit Boards	525
11.7	Convection Cooling	525
11.7.1	The Mechanisms of Convection	525
11.7.1.1	Design Features of Air-cooled Systems	526
11.7.2	Accuracy of Heat Transfer Coefficients	527
11.7.2.1	Influence of Changes in Air Temperature and Edge-connector Efficiency	529
11.7.3	Packages with an Attached Heat Sink	529
11.7.3.1	The Use of Die-cast Zinc Heat Sinks	529
11.7.4	Cooling Within the Cabinet	530
11.8	Other Aspects of Thermal Design	534
11.8.1	Thermal Expansion Mismatch	534
11.8.1.1	Soft Solder Joint Embrittlement	534
11.8.1.2	Component Features	534
11.8.2	Liquid Cooling	535

11.8.2.1	Convection Cooling Using Liquids	535
11.8.2.2	Phase-change Cooling by Nucleate Boiling	535
11.8.3	Heat Pipes	535
11.9	Postscript	536

CHAPTER TWELVE

Soldering　538

12.1	Fundamentals of Soldering	538
12.2	Making a Solder Joint	539
12.3	Cost Breakdown (The Economics of Soldering)	539
12.3.1	Manufacturing Costs	539
12.3.2	Materials Cost versus Failure Cost	540
12.3.3	Machine Cost	541
12.3.4	Quality Costs	541
12.4	Problem Areas in Soldering	542
12.5	Prerequisites for a Sound, Reliable Solder Joint	543
12.5.1	The Design Phase	544
12.5.1.1	Placement of the Solder Joint	544
12.5.1.2	The Joint Design (Dimensions, Geometry and Tolerances)	544
12.5.1.3	Thermal Problems	545
12.5.1.4	Repair	545
12.5.1.5	Demands Placed on the Solder Joint in Manufacturing, Storage, Transport and Operation	
12.5.1.6	Properties of the Materials Used	546
12.5.1.7	Choice of Solder and Flux	546
12.5.1.8	Choice of Soldering Method	547
12.5.2	Preproduction Phase	547
12.5.2.1	Equipment and Workshop	547
12.5.2.2	Personnel	548
12.5.2.3	Preparation for Soldering	548
12.5.2.4	Preproduction Checks	548
12.5.2.5	Storage of Material	548
12.5.3	Production Phase	548
12.6	Metallurgy	549
12.6.1	Soldering	549
12.6.2	Dissolution of Metals	549
12.6.3	Solidification of Solder	552
12.6.4	Intermetallic Phases	552
12.6.5	Diffusion	552
12.7	Wetting	555
12.8	Thermal Considerations	558
12.9	Solderability	560
12.9.1	General	560
12.9.2	What is Solderability?	560
12.9.3	The Advantages of Solderability Testing	561
12.9.4	Preservation of Solderability	561
12.9.5	Solderability Testing Methods	563
12.9.5.1	The Wetting Balance Method	564
12.9.5.2	The Scanning Method	565
12.9.5.3	The Workshop Method	567
12.9.5.4	The Solder Globule Method	568
12.9.5.5	The Spread Test	569
12.9.5.6	Testing of Printed Boards	570
12.9.5.7	Artificial Ageing Methods	572

12.9.5.8	Analysing Solderability Results	574
12.9.5.9	Solderability Testing *versus* Actual Performance	576
12.9.6	Correcting Bad Solderability	576
12.10	Quality and Reliability	577
12.10.1	Education, Information, Training	579
12.10.1.1	Management	579
12.10.1.2	Designers	579
12.10.1.3	Soldering Operators and Inspectors	579
12.10.2	Statistical Considerations	580
12.10.2.1	The Problem of the Great Number of Solder Joints	580
12.10.2.2.	How Exact is a Measured Value?	581
12.10.2.3	SPC — Statistical Process Control	582
12.10.3	Inspection of Solder Joints	586
12.10.3.1	Visual Inspection	586
12.10.3.2	Automated Optical Inspection (AOI)	587
12.10.3.3	X-ray Inspection	589
12.10.3.4	Ultrasonic Inspection	589
12.10.3.5	Thermal Inspection	589
12.10.4	Soldering Defects	590
12.10.4.1	What is a Soldering Defect?	590
12.10.4.2	Soldering Defects and Failure Rate	591
12.10.4.3	Solder-filled Through-plated Holes or not?	592
12.10.4.4	Defect Terminology	594
12.10.5	Expert Systems in Soldering	598
12.11	Mounting Methods	600
12.12	Classification of Assemblies	600
12.13	Health and Safety	601
12.13.1	Solders	601
12.13.2	Fluxes	602
12.3.3	Soldering Equipment	602
12.14	Terms and Definitions	603
12.14.1	General Definitions Related to Soldering	603
12.14.2	Terms Related to Soldering	604
12.14.3	Time-temperature Terms	604
12.14.4	Terms Related to Joint Form and Size	606
12.14.5	Terms Related to Solder Joints	607
12.14.6	Materials to be Joined	607
12.14.7	Flux Terms	608
12.14.8	Solder Terms	609
12.14.9	Soldering Aid Material	610
12.14.10	Soldering Methods and Processes	610
12.14.11	Quality Terms	610
12.14.12	Soldering Defects	611

CHAPTER THIRTEEN

Materials Used in Soldering   614

13.1	General	614
13.2	Base Materials	614
13.3	Surface Treatments	614
13.3.1	Methods of Surface Treatment	615
13.3.2	Mechanism of Bonding	615
13.3.3	Layer Structure	615
13.3.3.1	The Bond	616
13.3.3.2	Diffusion Barriers	617
13.3.3.3	Protective Layers	617
13.3.3.4	Materials Build-up for Soldering	617

13.3.4	Solderable Coatings	618
13.3.4.1	Electrolytic Tinning	618
13.3.4.2	Hot Tinning	619
13.3.4.3	Nickel	620
13.3.4.4	Gold	620
13.3.4.5	Silver, Silver-palladium Plating	628
13.3.4.6	Problems in Plating	629
13.3.4.7	Points to be Checked	629
13.4	Solder for Electronic Purposes	629
13.4.1	General	629
13.4.2	Properties of Solder	630
13.4.3	Requirements Specified for Solders	630
13.4.4	Low Melting Point Solders	631
13.4.5	High Melting Point Solders	631
13.4.6	Step Soldering	632
13.4.7	Impurities in Solder	632
13.4.7.1	Copper	635
13.4.7.2	Gold	636
13.4.7.3	Silver	636
13.4.7.4	Iron	636
14.7.4.5	Nickel	637
13.4.7.6	Bismuth	637
13.4.7.7	Antimony	637
13.4.7.8	Arsenic	637
13.4.7.9	Aluminium, Zinc and Cadmium	637
13.4.7.10	Oxygen	638
13.4.7.11	Sulphur	638
13.4.7.12	Phosphorus	639
13.4.7.13	Dross	639
13.5	Fluxes	639
13.5.1	Definition of Fluxes	639
13.5.2	General	640
13.5.3	Classification of and Tests on Fluxes	640
13.5.4	Resin Fluxes	644
13.5.5	Colophony (Rosin)	644
13.5.6	Activated Rosin Fluxes	644
13.5.7	Halogen-containing Fluxes and Corrosion	645
13.5.8	Water-soluble Fluxes	645
13.5.9	Gaseous Fluxes and Soldering Without Fluxes	646
13.5.10	Low Solid Content Flux, No Residue Flux	647
13.5.11	Cleaning Off Fluxes and Flux Residues	647
13.6	Solder Pastes	648
13.6.1	General	648
13.6.2	Solder Balling	648
13.6.3	Viscosity	649
13.6.4	Slump	649
13.6.5	Corrosion of Residues	649
13.6.6	Assessment of Shape and Size of Solder Powder Particles	649
13.6.7	Storage of Solder Pastes	650
13.7	Properties of Solder Influencing the Solder Joint	650
13.7.1	Cracks in the Solder Joint	650
13.7.2	Mechanisms of Solder Joint Failure	651
13.7.3	The Different Types of Strength	653
13.7.4	Fatigue	656
13.7.4.1	Lifetime Prediction	658
13.7.4.2	Testing of Fatigue Properties	662
13.7.4.3	Measures to Increase the Lifetime of a Solder Joint	662

13.7.5	Tin Pest	665
13.7.6	Whiskers	666

CHAPTER FOURTEEN

Manual Soldering — 668

14.1	Manual Soldering	668
14.2	The Soldering Iron	668
14.2.1	How Does a Soldering Iron Work?	669
14.2.2	The Soldering Iron Tip	671
14.2.3	The Wear of the Soldering Iron Tip	673
14.2.4	Heat Flow from the Iron to the Workpiece	673
14.2.5	Measuring the Tip Temperature	675
14.2.6	The Rating of the Soldering Iron	676
14.2.7	Tip Temperature and Heat Output from the Tip	676
14.2.8	Criteria for a Soldering Iron	677
14.2.9	Soldering Iron Support	679
14.2.10	'Tin Surplus' Collector	679
14.3	Soldering Printed Board Assemblies	680
14.3.1	Boards, Components and Solder Requirements	680
14.3.2	Operator and Working Conditions	680
14.3.3	Soldering Conditions	680
14.3.4	Making the Joint	681
14.4	Rework and Repair	681
14.4.1	Repair of Conductors on the Board	682
14.4.2	Repair of Solder Joints	682

CHAPTER FIFTEEN

Mass Soldering — 685

15.1	General Considerations	685
15.1.1	Heat Supply	686
15.1.2	Temperature Profiles	686
15.1.3	Reflow Soldering	687
15.1.4	Controlled Atmosphere Soldering	688
15.2	The Mass Soldering Machine	690
15.2.1	The Soldering Fixture	691
15.2.2	The Conveyor	691
15.2.3	The Fluxing Station	692
15.2.3.1	Brush Fluxing and Dip Fluxing	693
15.2.3.2	Rotary Brush Fluxing	693
15.2.3.3	Wave Fluxing	693
15.2.3.4	Foam Fluxing	694
15.2.3.5	Spray Fluxing	694
15.2.3.6	Density Control	695
15.2.4	The Preheating Station	696
15.2.5	The Soldering Station	698
15.2.5.1	Solder Replacement	699
15.2.6	The Cleaning Station	700
15.2.7	Process Control Aids	700
15.2.7.1	Temperature Indication	700
15.2.7.2	Soldering Process Testing	700
15.2.8	The Maintenance of a Soldering Machine	701
15.2.9	Buying a Soldering Machine	703
15.3	Dip Soldering	704
15.4	Drag Soldering	704
15.5	Wave Soldering	705

15.5.1	The Wave Form and the Nozzles	708
15.5.2	Wave Soldering Machines for SMT	709
15.5.2.1	The Double Wave Soldering Machine	709
15.5.2.2	The Pulsed Wave Soldering Machine	709
15.5.3	Oil in the Wave	710
15.6	Infra-red Soldering	711
15.7	Convection and Forced Convection Soldering	711
15.8	Vapour Phase Soldering or Condensation Soldering	712
15.9	Laser	715
15.9.1	Types of Laser used for Soldering	715
15.9.2	Pulsed and CW Lasers	716
15.9.3	Controlled Laser for Soldering	716
15.9.4	The Heat Flow in Metals Radiated by Laser	716
15.9.5	The Laser Solder Joint	717
15.9.6	Maximising the Advantages of Laser Soldering	718
15.9.7	Areas for Laser Soldering	719
15.9.8	The Economics of Laser Soldering	720
15.10	Light Soldering	720
15.11	Hot Bar Soldering	721
15.12	Hotplate Soldering	722
15.13	Belt Soldering	722
15.14	Hot Gas Soldering	722
15.15	Furnace Soldering	723
15.16	Robot Soldering	723
15.17	Ultrasonic (US) Soldering	724
15.18	High Frequency (HF) Soldering	724

CHAPTER SIXTEEN

Microjoining Methods 726

16.1	Introduction	726
16.2	The Principles of Metallurgical Joining Methods	727
16.2.1	Soldering Methods	728
16.2.2	Welding Methods	728
16.2.2.1	Resistance Welding	729
16.2.2.2	Diffusion Welding	729
16.2.2.3	Ultrasonic (US) Welding	731
16.2.3	The Equipment	731
16.2.4	The Materials	732
16.2.4.1	Materials to be Joined	732
16.2.4.2	Chips, Dice or Die	732
16.2.4.3	Electrodes and Thermodes	733
16.2.5	Process Control	734
16.3	Different Joining Methods	734
16.3.1	Wire Bonding	734
16.3.2	Tape Automated Bonding (TAB)	735
16.3.3	Flip-chip	738
16.3.4	Beam Lead	739
16.3.5	Isothermal Soldering	740
16.3.6	Wire Wrap	740
16.3.7	Explosive Welding	741
16.3.8	Adhesives	741
16.3.9	Chip-on-board (COB) and Multichip Module (MCM)	742
16.3.10	Board Wiring Techniques	742
16.3.10.1	Impulse Bonded Wiring	742
16.3.10.2	Stitchwire	743
16.3.10.3	Multiwire and Microwire	743

Section 1
Origins of an Industry

Chapter 1

INTRODUCTION

W. MACLEOD ROSS
Welnorth Ltd, Bishop's Stortford, UK

1.1 ELECTRONICS

The precise time at which 'electronics', the *scientific study and application of the movement of electrons*, became household terminology is unknown. Black box electronic systems were commonplace at the beginning of World War Two, when scientists and other 'boffin' types began to appear in the armed forces with mysterious devices which could detect enemy aircraft, shipping and submarines. However, electronic equipments were found in the home and in industry many years previously. The foundation stone of modern electronic circuits, the semiconductor, was seen in the first crystal radio sets but it took many years before real attempts were made to understand and appreciate fully their unique properties.

Electronic assemblies and systems now take many forms and humanity as a whole depends upon the ingenuity of the electronics industry in the home, at work and at play. The social impact is far-reaching and not only due to things such as television, computerised office systems, video games, personal computers and spacecraft. Activities which were totally undreamt of, even by the most fertile imaginations, in the early part of the twentieth century are commonplace in modern societies as the advent of the twenty-first century appears on the horizon.

This is all due to a range of equipments covering a size spectrum from miniature hearing aids and micro-miniature medical and surgical devices, to the most complex data processing and storage equipment. If we adopt the simplest view, we can say that the only thing these electronic marvels have in common is that they perform their functions by the manipulation of electrons. Each equipment consists of an electronic circuit, or numbers of circuits, which by transmitting, receiving, modifying and/or storing the energy derived from the movement of electrons produces a pulse of energy, a visual and/or audible signal, a motion or just a memory.

It can be readily appreciated by most that all electronic circuits are made up from 'active' devices, which give gain control and are typified by transistors, valves and items such as ferromagnetic pot cores, and from 'passive' elements such as capacitors, resistors and conductors. These component parts are manufactured from three broad classes of material — insulators, conductors and semiconductors. This classification is not absolute and there is no hard and fast division between the three types of material.

1.2 CONDUCTION — A SIMPLE VIEW

It may be considered that conduction normally takes place in one of two ways, electrolytic conduction or conduction by electrons. Electrolytic or ionic conduction is always associated with the movement of ions and chemical electrode reactions. Contamination of an insulator by polar or ionisable matter will allow conduction in the presence of an applied potential and will almost certainly produce corrosion and/or migration of metals. The best insulating materials are those which are essentially non-polar. A low dielectric constant is also desirable since a high dielectric constant can increase ionic conduction by reducing the energy needed to produce ionisation. The inadvertent introduction of polar material at any stage of manufacture (or use) can negate the circuit designer's best efforts.

The form of conduction more important to electronics is conduction by electrons. While the atom is described best by using quantum mechanics, the simple view sees a central, positively charged nucleus around which rotate negatively charged electrons. In a neutral atom, the number of electrons equals the number of protons in the nucleus. The electrons must normally occupy specific orbits or shells and only a specific number of electrons can occupy a given shell. The electrons in the outermost shell determine the valency of the atom and, in certain circumstances, can be dislodged from the valency shell and become free. A free electron will carry one unit of *negative* current by moving under an applied potential. The *positive* counter-current is carried by the 'hole' left in the valence shell by the removal of the electron.

The number of free electrons or the energy needed to free an electron determines whether the material is an insulator, a semiconductor or a conductor. The energy required to create free electrons may be viewed as the difference between two energy bands, the conduction band and the valence band. In most metals, these bands will overlap and the metal will conduct at low electrical potentials. In an insulator, there will be a finite, high difference in the band energies and a high voltage will be needed to make an electron jump the 'forbidden gap'. Putting this differently, an insulator will be characterised by a scarcity of free electrons and they will conduct only minute currents when subjected to high voltages. Insulators must be viewed as conductors of very high resistivity. While the current may be neglected for power equipment, it can be of critical importance in electronic equipment. As implied by the name, a semiconductor is intermediate between insulators and conductors. In a semiconductor the forbidden gap is small and, at normal temperatures, the thermal energy available is sufficient to enable some of the electrons to enter the conduction band and conduction is possible.

It is important to remember that the difference between conductors, semiconductors and insulators is only one of degree. Large inputs of energy, light, thermal radiation, gamma radiation etc., can transfer electrons to the conduction band and reduce resistivity. The presence of foreign atoms in a crystal or of dislocations in a crystal lattice will alter the conductivity of a metal. The inclusion of impurities in insulators or semiconductors can reduce the energy needed to produce free electrons and hence conduction. This is readily illustrated by the deliberate addition of particular impurities or *dopants* to an intrinsic semiconductor — one part of a suitable dopant per million parts of pure semiconductor is sufficient to increase the conductance 70 times. Band distortion can be used to explain why copper or silver deposited from apparently identical solutions can vary by a factor greater than ten. The different results are certainly due to either the levels of trace impurities, or variation in plating conditions and the effects of pre- and post-plating treatments, or a combination of all of these.

1.2.1 Superconductivity

No discussion of conduction and resistivity is complete without mention of the phenomenon called *superconductivity*. This phenomenon was first noted in mercury by the Dutch physicist Kamerlingh Onnes in 1911 at the University of Leiden. Onnes had liquefied helium (at -269°C, 4.16 kelvin [K]) in 1908 and was carrying out a systematic study into the electrical resistivity of metals at very low temperatures.[1] He observed a sudden drop in the electrical resistance of mercury at 4.16 K and could find no remaining resistance below that temperature.

The metal had become *superconducting*. Superconductivity is defined as the property of zero electrical resistance together with the expulsion of magnetic fields from the body in question. The expulsion of magnetic fields means that, whereas if a piece of metal in its normal state is placed in a magnetic field the field exists within the metal, if the metal is cooled below a critical temperature, dependent on the metal and usually below -260°C (13.16 K), the magnetic field can no longer be detected within the metal, being forced to flow around it. This was found by the German physicist Walther Meissner in 1933 and is termed the *Meissner effect*. A true superconductor must exhibit both zero electrical resistance and the Meissner effect. Provided the temperature is held below the critical temperature, a current established in a closed circuit of a superconducting material will continue indefinitely.

Superconductors are used in large magnets and devices such as the Josephson junction, employed in rings of superconducting material for measuring very small magnetic fields, can be used for measurement of brain activity. However, uses are restricted because of the need to cool the metal to, and hold it at, very low temperatures.

If a 'high-temperature' superconducting material can be developed, costs can be reduced and the uses of superconductivity extended. (A high temperature is usually defined as any temperature higher than about -250°C (213 K).) Up to the early seventies the highest known critical temperature was 18 K for Nb_3Sn, an intermetallic compound of niobium and tin, the remarkable current-carrying properties of which were reported by Kunzler *et al.* in 1961.[2] In 1986 Müller and Bednorz observed superconductivity at temperatures above 23 K for a ceramic based on copper oxide, barium and lanthanum. Later still, other ceramics were found to be superconducting at temperatures of the order of 93 kelvin. As yet, however, the brittle nature of these materials, coupled with other problems, prevents commercial development.

REFERENCES

1 Dummer, G. W. A., 'Materials for Conductive and Resistive Functions', Hayden Book Company, NY.
2 Kunzler, J. E., Beuhler, E., Hsu, F. S. L. and Wernick, J. H., 'Superconductivity in Nb_3Sn at High Current Density in a Magnetic Field of 88 kgauss', *Physical Review Letters*, **Vol. 6**, p. 89 (1961).

Chapter 2

HISTORICAL BACKGROUND

2.1 INTRODUCTION

The history of components and circuits is the story of electronics. Initially, the printed wiring board was developed in order to mount and interconnect components, now the development of the printed circuit and the development of an electronic component are inextricably linked. It all started with the Leyden jar in 1745/46 although it was some 50 years before the voltaic cell was developed and very nearly 90 years before Michael Faraday carried out his experiments on electromagnetic induction, establishing the principles for the generation of electricity.

Thereafter the rate of experimentation and discovery accelerated, leading to the invention of radio transmitters and receivers in 1896. This was a major spur for the research and development into components which led to the modern electronics industry.

It will not be possible to give credit to every inventor or invention in what is after all an introduction to a practical textbook. It is hoped that enough will be said to whet the reader's appetite for the history of a living, modern industry and guide them to further reading.

Since the electronics industry could not have made the progress which has been seen over the last 50-60 years without the invention of the printed circuit, it is pertinent to start with the history of printed wiring and circuits.

PART 1 — PRINTED BOARDS

MARY POLE-BAKER
PCB Design Consultant, Southwood, Farnborough, UK

2.2 EARLY DAYS

2.2.1 Prior Art

Although the major inventions giving rise to the modern printed circuit industry were patented during the early 1940s, the first patents connected with printed circuit techniques coincided with the start of the electronics industry in the radio field.

Early methods included a method suggested by Rhuyssenaers in 1926, to manufacture conducting networks by spraying molten metal through a stencil on to an insulating panel,[1] and H. G. Arlt, in 1937, patented a process for spraying metal patterns interconnected by conductive rivets and bolts.[2] Charles Ducas in the United States proposed the electrolytic deposition of metal to form conductors upon the surface of an insulating material containing a printed pattern and in his patent of 1925[3] he also suggested that connections between conductor lines on different levels could be made by metal filled holes. In 1927, Parolini suggested dusting an insulating panel containing an inked circuit pattern with metallic powder and solidifying the powder by electrolysis.[4]

These ideas were used in research by John Sargrove and resulted in his development of automatic production equipment for the manufacture of radio sets in the early 1940s,[5] based on a method of spraying metal on to moulded plates. John Sargrove's method never became an important factor in printed circuit manufacture, but his ideas on the automatic production of electronic equipment made a great impact.

2.2.2 The Birth of the Printed Circuit

The person generally referred to as 'the father of printed circuits' is Dr Paul Eisler. His early experiments were with printed metals and powders applied directly to the substrate but none of these methods satisfied the crucial aspects of being universally applicable and yet still compatible with specialisation.

Printing was not considered by the British Government to be an essential industry during the Second World War and consequently a great part of the productive capacity of the industry lay idle. Dr Eisler envisaged that the unused technical resources of the industry could be used to solve the problem of the urgent need to provide increasing quantities of reliable electronic devices in as short a time as possible.

The first printed circuit apparatus, a radio set, was made by Dr Eisler and his assistant, G. Parker, in 1942, using a foil technique. The method involved backing a copper foil with insulating paper laminate, silk-screening the metal circuit pattern, then etching the foil to remove the unwanted copper.[6] Forty-eight printed circuit radio sets were made and distributed in 1943 but, although a great deal of interest was shown by the engineers and military personnel who attended demonstrations, the British Government rejected the idea for use in military equipment and no industrial firms could be persuaded to give it a trial.

Dr Eisler was bitterly disappointed but, although the British Government had rejected the idea of printed circuits during the war years, the Americans had applied the idea in a modified form.[7] The US Signal Corps Engineering Labs, with the American National Bureau of Standards (NBS — now the American National Institute for Science and Technology, NIST) and the Harry Diamond Ordnance Fuze Labs, had developed a method of printing silver circuit paths on a steatite base and produced a miniature electronic device as a proximity fuse, housed in the cap of an aircraft shell. These were being produced at the rate of five thousand per day ready for operational use when the war ended.[8]

2.2.3 The First Printed Circuit Symposium

In 1947, the United States Aeronautical Board and the NBS sponsored the first technical symposium, which was held in Washington. The technical publication of the symposium listed twenty-six individual and distinct methods of producing printed circuits and described the first commercial application, a hearing aid, which had already been placed on the market. The twenty-six methods were subdivided into six different classes: painting, spraying, chemical deposition, vacuum processes, die-stamping and dusting.[9]

The symposium was arranged because, after the story of the proximity fuse became known, the NBS had received a huge demand for the technical information on printed circuit techniques from other Government departments, from industry and from commercial firms.

An important technique in the manufacture of hybrid circuits developed from the printing of metallic paints, usually containing highly conductive silver, and resistive paints loaded with carbon powders on to a base using conventional printing methods. The substrates were mainly restricted to ceramics, it being necessary to heat the material to high temperatures.

Vacuum techniques were not developed for printed circuits because of the expense of processing in a vacuum, coupled with size and material limitations, but evaporation techniques are used today for making thin film circuits.

Millions of antennae for use in radios were made by die-stamping using a method patented by A. W. Franklin in 1946 but, although this was, and is, an important mechanical method of producing relatively simple circuits, it never became widely used because of the difficulty of making complex patterns in mass production situations.[8]

From the confusion of ideas at the symposium, only the technique of electrodeposition eventually became of major use in the manufacture of printed boards.

2.2.4 Module Development

Under the threat of the Korean War in 1950, the NBS was funded by the Bureau of Aeronautics for project 'Tinkertoy', modules made from layers of small ceramic printed wafers on an etched foil-clad laminate chassis. The wafers were held together by riser wires which also connected the wafers electrically, and were soldered to the foil conductors on the base.[6] Several thousand sonobuoys were produced, but very little interest was aroused in industry.

The module idea was developed further by the US Army as the 'Micromodule Program'.[10] These modules were described as a combination of film metal oxide resistors, multilayer thin film capacitors, ferrite core inductors, silicon diodes, prototype round transistors and quartz crystals, three-tenths of an inch (7.62 mm) square and one thousandth of an inch (0.025 mm) thick — the predecessor of the integrated circuit and the multichip module.

The US Army Signal Corps Engineering Laboratories (SCEL) considered that printed wiring had to be inexpensive, possess adequate electrical and mechanical properties, be able to withstand the heat of mass soldering and have good solderability. It concentrated therefore on Dr Eisler's etched foil technique. This was the only method which seemed to satisfy its requirements for printed wiring and lend itself to mass production, an extremely important factor in the rapid production of large quantities of military devices. By the early 1950s the etched foil technique had become the principal method for the production of printed circuits.

British commercial enterprises were slower to make use of the new technology than Government bodies and by 1955 British radio and television manufacturers were just starting to use printed circuits in their equipment, although in the United States the use of printed circuit boards was already well established in the commercial field.

2.3 FURTHER DEVELOPMENT

2.3.1 Plated-through Hole Boards

Once the initial interest had been aroused, printed circuit techniques developed very rapidly. There were continuous demands to fit more circuitry on a board and soon boards were being formed with patterns on both sides and this raised the problem of how best to connect the two patterns. At first connections were made by eyelets or connecting pins and the first firm to use a plated connection between two sides of a board in production was Motorola in 1953.[8]

Motorola's process started with an unclad board, such as paper phenolic, with the hole pattern already punched. The board was dip coated in adhesive to cover the surface and the hole walls. A silver coating was deposited by spraying and, after a priming treatment, a reverse resist pattern was applied. The board was then copper plated down the holes and on the unprotected silvered surface and finally the ink resist and the silver coating under it were removed.

2.3.2 Multilayer Boards

The American Tinkertoy and Micromodule programmes had shown that layers of etched circuits could be built up to provide a complete unit, and high densities could be achieved. The Hazeltine Corporation in America perfected a technique in which three or more layers with coincident holes were bonded together with insulating sheets and the layers connected through by using the existing plated through hole technique. Their 'Multiplanar' process was disclosed in 1961,[11] and is the method on which all existing specifications are based.

2.3.3 Additive Processes

Electroless plating processes date back to 1861, when Justis von Liebig invented a process for silvering mirrors, but the first practical process of depositing metal from an inherently unstable solution on to a non-metallic surface, without the use of electricity, was invented by Dr Abner Brenner of NBS in 1946.

The American Photocircuits Corporation began research into additive procedures in 1955 in an effort to produce printed boards of equal quality but more cheaply than by print and etch methods. At that time there were no really stable copper baths. Copper deposits were either powdery or brittle, adhesion was poor, and selective deposition practically impossible. After nine years of research, Photocircuits announced their CC-4 process in 1964,[12] claiming the advantages of copper being deposited evenly over the board and down the holes, extremely good pattern definition, and reduced material costs. Automatic controllers were devised by Grundig in 1969 for maintaining solution stability.

PD-R, a competitive additive process, was developed by Philips Research in 1969 based on the use of a special laminate coated with an adhesive containing titanium dioxide, a light-sensitive compound which was exposed to ultra-violet light.[13] Despite intense interest, this process achieved very little commercial success because of problems with incomplete metal removal and the necessity of buying special laminates.

2.3.4 Discrete Wired Boards

During the latter part of the 1960s, Photocircuits were searching for a technique which would combine in one system the best features of discrete wiring, plated-through hole circuitry and computer technology and this resulted in the Multiwire process introduced in 1970.[14] Instead of copper surface tracking, insulated copper wires are laid on to a laminate surface. The insulation means that there is no restriction on wire crossovers so that a much denser interconnection pattern can be achieved on each layer. Interconnections are formed by drilling the circuit holes, so cutting the wires. The wire insulation is chemically etched around the exposed copper surface, then electroless and electrolytic processes are used to plate around the wire ends and down the holes. Multiwire has become a successful alternative to multilayer systems, although growth of the technology has been smaller in Europe than in the United States and Japan.

Microwire was developed from Multiwire at the end of the 1970s specifically for use with surface mount devices.[15] The board is built up from a copper-Invar core which has a thermal coefficient of expansion much nearer to that of ceramic surface mount devices than conventional laminates. A fine wire of 0.0025 in. (0.064 mm) diameter copper core is used, the insulation being removed at connection points by laser drilling before plating.

Wire wrapping is a very popular method for high density interconnection of electronic components, particularly for prototype production where design flexibility is important. It was developed by Bell Telephone Laboratories in the late 1940s to provide highly reliable long-life wire connections. Insulated wire (usually copper) is wrapped around a terminal pin using a special tool. This terminal pin has two sharp edges which scrape the surface film off the wire to make a clean metal-to-metal contact. The connection becomes stronger as it ages owing to solid state diffusion. The main disadvantage of wire-wrap is weight.[16]

2.3.5 Flexible Circuits

Flexible circuits were made first by Paul Eisler by his etched foil method, using paper or plastic film base materials.[6] He made flat strip circuit patterns, then folded or rolled them to reduce the area. The circuits were produced on thin flexible laminates and the pattern encapsulated with another layer of flexible material. In 1960, V. Dahlgreen invented a new technique for bonding foil to thermoplastic film, which involved treating the copper chemically to produce an oxidised surface to which the plastic film could be bonded. The most popular base materials in the early 1960s were various flexible polymers such as polyhalocarbons or polyesters but there were a number of processing problems with these materials. It was not until polyimide film laminates became available at the end of the 1960s that flexible circuits were accepted as suitable for high reliability applications.[17]

2.3.6 Other PB Developments

Some newer developments are moving away from conventional printed board technology. One such development is polymer thick film; successive dielectric, resistive and conducting layers can be printed on to a single-sided PB, effectively transforming it into a multilayer interconnection system with significant improvement in packing density.[18]

Various moulded and 3-D circuits have been developed, such as the Konec process using no wet chemistry from Union Carbide with Chromenics; the Photoform process from ICI with PCK using glass reinforced polyethersulphone and 100% electroless plating;[19] and the Mold-N-Plate process (also from PCK) with Smiths Corona creating circuit patterns by moulding, then additive plating.

Much work has been done, and is continuing, on the use of optical fibres for high speed transmission.

2.4 MATERIALS AND PROCESSES

The development of the materials used for manufacture has been inseparable from the development of the processes. Laminates and resist materials are now vastly different from those available when Dr Eisler approached the manufacturers of Bakelite and Formica for a supply of suitable laminates and used printer's ink as a resist.

2.4.1 Laminates

Laminates could be said to have started with the discovery of phenolic resins by Dr Leo Baekeland in 1909, but the Formica company was the first to plasticise phenolic laminates and much of its black phenolic sheet was used in the 1920s when home radio manufacture became popular. The Owens-Corning Glass Corporation began the manufacture of glass fibre yarn in 1938 and, in the same year, the first epoxide resin was produced by Pierre Costain. During 1940-41 researchers produced fibreglass reinforced with polyester resins and by 1942 the United States had replaced all its electrical terminal boards with fibreglass melamine and asbestos melamine laminates.

When Dr Eisler began making printed circuits by the etched foil method, no laminates were commercially available with copper cladding. After his etched foil patents were filed in 1943, he approached the manufacturers of Bakelite and Formica about providing copper-clad laminates, and these were soon commercially available from the United States although not from British manufacturers until 1953.[20]

Epoxide resins were first introduced in adhesives in 1947,[21] and by 1951 fire retardant resins had been introduced.[22] The first patents on addition polyimide resins were registered in the mid 1950s, but production on polyimide base resins only commenced on a pilot scale in 1968.[23] Increasing demands for better dimensional stability, better control of impedance, and higher T_g have produced results in new resin and reinforcement developments, although glass epoxy laminate is still the industry standard to which everything else is compared. (The use of paper phenolic laminate is essentially restricted to the domestic market, despite the steady improvement in its properties over the years.) Polyimide film has replaced practically every other material for flexibles in recent years, despite its disadvantages of high moisture absorption and high cost, because of its excellent working temperature range.

Various metal core materials (including the extremely expensive titanium, molybdenum and nickel-iron alloys such as Invar) and Kevlar fibre reinforced polyimide have been developed for surface mounted devices which need mechanical and thermal properties which differ from those obtainable from the usual laminates.[24]

2.4.2 Resists and Etchants

Dr Eisler's original patent claimed the application of acid resisting ink to copper-clad substrates by any of the recognised printing methods and, until the second half of the 1960s, the main methods used for patterning were those of lithography and screen printing. By the late 1950s thermoplastic inks had become available, strippable in wet steam.[6] Later resin based inks were solvent strippable and, by the end of the 1960s, resist inks were produced which could be stripped in common solvents.

Kodak was one of the first companies to market a photoresist based on polyester resin in the 1950s.[6] Dry film resists were introduced in 1968[25] and by 1980 these had largely replaced liquid resists. In the late 1980s direct imaging on to dry film resists using laser exposure systems was already popular.[26]

When only simple single-sided or double-sided boards were made without plated-through holes, the two main etchants used were ferric chloride and ammonium persulphate. The latter became widely used as printed boards began to be used in substantial numbers. Cupric chloride was found to be an excellent etchant for bulk etching purposes, and although it is incompatible with solder resists it is more generally used than ferric chloride for print and etch boards. Peroxide-sulphuric acid etchants have been known for some time and improvements have been made with the addition of stabilisers. About 1970, alkaline etchants, based on solutions containing ammonium hydroxide, were introduced and are still very popular.

2.4.3 Machining

Originally most printed boards were hot punched because of the brittle nature of the available laminates. When epoxy glass laminates were introduced, it became preferable to drill holes instead of punching them, although punching remains useful for many applications.

At first general engineering drilling machines were used for hole production but gradually more specialised machines and drills became available. Tape controlled drill machines were introduced during the 1960s, at first with a single spindle, then with multiple spindles.[27] Later machine innovations included the replacement of electric motors and feed systems by hydraulic or pneumatic ones.[28] Drill speeds were increased from 10,000 rpm in 1965 to around 72,000 rpm in 1970. Software controlled acceleration, high-feed and peck drilling now help to make small 0.3 mm (0.012 in.) holes and 0.1 mm (0.004 in.) vias possible.

The first drill bits used were a miniaturised version of the standard engineering twist drills made from steel but, with the introduction of laminates containing glass fibre cloth, the abrasive wear was so great that even with the use of flash chrome plating the drill life was only about fifty holes. Drills made entirely of tungsten carbide appeared in the 1960s. About the same time the turboshank drill with a relatively large, constant size shank was introduced to enable automatic drill changes on the new tape controlled machines. Further developments have included 'back taper' of the drill, and thin-webbed drill bits which provide a larger flute cavity enabling faster swarf removal.[29] Microdrill drills are made of ultrafine grain cemented carbide.[30]

Since the late 1980s, there has been growing use of excimer lasers, pulsed gas lasers with ultra-violet wavelengths, which enable the removal of small quantities of organic material with precision.

In 1990 Hitachi patented a multifunction step-feed drill or peck drill for drilling high aspect ratio holes in multilayer panels.[31] By 1991 Excellon engineers were reporting that 20% of their business was connected with drilling holes less than 0.018 in. (0.46 mm). Their latest machines at that time had independent Z-axis and depth control, designed to drill holes down to 0.004 in. (0.10 mm). Trudrill had direct drive Z-axis spindles utilising peck drilling, capable of drilling all sizes from 0.004 in. to 0.25 in. (0.10 mm to 6.35 mm); one machine could drill 0.003 in. and 0.002 in. (0.076 mm and 0.051 mm) holes.[32]

Back-up and entry boards used to be selected from the cheapest materials available, such as fibreboard and low-cost phenolic/paper board, but these cheap materials contain contaminants which can cause smearing problems. The use of copper- or aluminium-clad material and improvements in tools and techniques reduced smearing considerably.[33] Back-up materials such as aluminium foil on specialised cellulose, or on special wood core, improved phenolic/paper-based materials with precise resin control, and phenolic/paper laminated to cellulose core, have been used for small hole drilling.

2.4.4 Hole Pretreatment, Metallising and Plating

With the development of the rigid multilayer printed board in the mid 1960s, resin smear on the hole walls became a major problem. The earliest processes tried for the removal of resin smear utilised organic solvents including chloroform, methylene chloride and dimethyl formamide but the reaction was very difficult to control and there were toxic hazards. Concentrated sulphuric acid was also tried but, in early use, the epoxide resin in the laminate was dissolved at the same time as the smear, leaving exposed glass fibres which could result in very poor quality holes. The rapid reaction of the sulphuric acid also produced exposed copper layer ends in the hole, resulting in stronger plated holes and, because of this discovery, concentrated sulphuric acid is still used as a major technique of smear removal, now combined with other chemicals to remove offending glass fibres.

A later development in desmearing uses alkaline potassium permanganate. Although this is a relatively slow process, a number of boards can be processed simultaneously, and the general degradation of the hole wall is reduced. Potassium permanganate is also effective in smear removal techniques for polyimide resins, the only other effective chemical for these being chromic acid. Plasma techniques using a mixture of oxygen and fluorine based gases at low pressure with a high applied voltage have also been developed. This plasma attacks epoxide resin at a consistent rate, and can also be used for desmearing polyimide resin materials, using pure oxygen.[34]

With all plated-through hole processes it is necessary to provide a method of promoting adhesion of the metal to the insulating surface. Before the adoption of metallising processes based on chemical reduction, extensive use was made of conductive silver, copper and carbon paints, and colloidal graphite but they were difficult to apply to plastics, had low adhesion and variable conductivity. By the mid sixties it had become clear that a complex compound of palladium and tin salts with hydrochloric acid produced the best results for adhesion. (It is pertinent to note here that a commercially and technically successful hole pretreatment for metallising based on a colloidal solution of graphite was developed in America in the late eighties and first marketed in Europe in 1990.)

Cyanide based solutions were originally tried for copper plating but were found to attack the adhesives used in laminates. Fluoroborate copper solutions and alkaline pyrophosphate copper[35] were the two preferred plating solutions used in the 1960s, the latter being the first to be used on a large scale for printed board manufacture. By the end of the 1970s, new 'high throw' acid copper sulphate solutions used with phosphorised copper anodes were being selected in preference to the pyrophosphate solution.

Anode bags became necessary about the time that the advantages of solution agitation were recognised, since the film formed on the anode would not stay in place. The first bags were made of cotton, but this had poor wet strength. Nylon bags were tried. These had excellent wet strength but relatively poor sieving action. For modern copper and tin-lead plating, polypropylene anode bags are used.[36]

Tin-lead plating began to be used as a solderable coating for circuit boards at a very early stage. The use of fluoroboric acid tin-lead solutions was first suggested by Groff in 1920, and all commercial plating solutions are based on this acid. Groff used skin or bone glue as an additive to give a smooth, dense deposit but in the 1950s this was replaced by peptone when it was found that the amount of tin in the deposit was affected by the glue.[37] In 1960

Rothschild and Sanders improved the throwing power by reducing the total metal content and increasing the percentage of fluoroboric acid,[38] and in the 1970s, new tin-lead solutions were introduced using proprietary additives, easier to analyse and control.[39]

2.4.5 Finishes

Circuit boards were originally left with a plain copper finish for immediate assembly,[6] otherwise a lacquer finish was used which gave limited protection. The use of metal etch resists to perform the secondary function of providing a solderable finish was introduced during the 1950s, and by 1970 it was quite normal to use a plated metal, usually tin-lead, with bright tin as an alternative.[40] Precious metal resists are also used for some boards, usually gold, often with a nickel or tin-nickel undercoat. Silver is now used only when overplated with gold.[41]

Tin-lead has been the most popular finish since the late 1950s, because of its excellent solderable finish and long shelf life. Applications in the early days were made by dip soldering but automatic methods of application were available by 1957. The rival process of flow or wave soldering provides a virtually oxide-free surface (Rudolf Strauss et al. applied for the first flow soldering machine patent in 1955).[42] Later machines used new wave shapes, enabling conveyor speeds to be increased.[43] During the late 1980s increasing environmental pressures to reduce the level of CFCs in the atmosphere prompted the development of a wave soldering machine using a nitrogen blanket which reduces the need for cleaning. Low residue fluxes have also been developed.

As plated-through hole and multilayer boards became commonplace during the 1960s, the further process of flow melting became necessary for a smooth finish with electrodeposited tin-lead finishes. Several methods of reflow soldering were developed such as forced heated air and infra-red fusing. Asymptotic heating, or immersion heating in hot oil, was developed by the European Space Agency. Automatic hot air levelling machines were developed in North America in the early 1970s by Electrovert. These can be used for additive as well as subtractive boards and for salvage operations on poorly tin-lead plated boards. Reflow soldering by vapour phase reflow, using the latent heat of vaporisation to melt tin-lead, was introduced in 1974[44] and was the preferred method for SMDs by the 1980s.

2.4.6 CAD

The increasing complexity of circuit board designs would not have been possible without the development of computer-aided design (CAD).

Before CAD was widely available, board layouts were often prepared using tape on vellum at 4:1 scale, from which a precision camera produced a 1:1 negative as a manufacturing tool (see Section 4). CAD became available in the mid 1960s and since then it has increased the accuracy of tooling and the board packaging density by about 1000% and has aided the development of new packaging techniques such as MCMs.

Early CAD systems were blind digitisers, using punched cards to provide computer input. The output from the computer was then used to drive a photoplotter. The first mechanical-optical photoplotter was designed by Joseph Gerber; Gerber Scientific photoplotters dominated the market for many years until the advent of laser photoplotters during the 1980s.

The concept of auto-interactive CAD was developed by Eric Wolfendale, an English engineer; his prototype system was demonstrated in 1971 on an ICL computer. Racal Electronics founded Racal-Redac Incorporated which produced its first commercial product using a Digital Equipment Corporation PDP-15. In the early 1980s, workstation hardware rapidly replaced mini-computers, offering automation of the complete design cycle. In 1983-84 the IBM-XT PC, capable of supporting a reduced-scale CAD system, was able to design about 70% of boards at an affordable price. Even small firms were able to have access to CAE and CAD through use of PCs. In 1987, PC-based software with full workstation capability became available at less than $1000.

Autorouters originated in the 1970s. The first commercially available autorouter was probably the German developed Calay autorouter but, as densities increased, the Calay router faded from the scene. In 1987-88 the German Oliver Bartels developed a rip-up and re-route autorouter and, in the late 1980s, Massteck of Littleton, Massachusetts, introduced a new concept, the push-and-shove router. At the same time Walter Katz produced a router with variable grid as a function of density, one of the best for MCMs.

The early history of CAD for printed boards and circuits was covered by an article by H. G. Marsh in early 1993.[45]

2.5   IN CONCLUSION

The invention of the transistor in the late 1940s, the first small scale integrated circuit in the middle sixties and the very large scale integrated circuits of the 1980s have all had their influence on printed circuit technology. As component size has decreased, the electronics industry has demanded greater packaging densities, and this has stimulated research into printed circuit processes and the products associated with them.

Plated-through hole techniques evolved because of the need to have a connection from side to side on double-sided boards, and the research into the electroless plating associated with the holes led to the development of additive technology. Multilayer boards were a logical development of double-sided boards, and this led to research into the production of better holes by desmearing and drilling techniques, and also improved plating processes.

One of the most promising technologies to address the problems of high density and performance is that of multichip modules (MCMs). Since Kyocera in Japan was credited with introducing multilayer MCM techniques in the early 1980s, they have been used widely. After 1990, interconnect densities increased rapidly.

MCMs use bare dice which are attached and wire-bonded to a substrate containing traces that interconnect dice signal lines and bring signals to the input/output pins of the module. Since signal lines are generally shorter and smaller than those on PBs, they incur fewer parasitic effects from loading, transmission line effects and crosstalk. MCMs are complex hybrids; they may contain one or more ASICs, or multiple-die packages; they may be made using thick and/or thin film technology, the substrates being made from a wide range of materials. Unpackaged IC dice can be connected to the MCMs using wire bonded or solder 'bump' connections.

As will be seen in the chapters which follow, packaging is moving steadily from DIPs to smaller outline cases, TAB, multichip and direct substrate mounting, requiring finer tracking.

High speed devices need controlled track lengths and impedance, with tight tolerances on the conductor width and thickness requiring high track densities and spacings. Typical track widths for 'fine-line' circuits were 0.012 in. (0.30 mm) in the mid sixties, 0.008 in. (0.20 mm) in the mid seventies and 0.004 in. (0.10 mm) in the mid eighties. Now a fine-line circuit will have tracks of the order of 0.003 in. (0.076 mm), with 0.002 in. (0.05 mm) confidently expected to be achieved in full scale production.[46] The ability to achieve finer tracking and high densities has been due to precision photo-imaging techniques, the use of dry film resists and higher quality board materials. Increases are also attributable to improvements in plating processes, such as the improved throwing power of tin-lead plating solutions.

Rigid multilayer boards now dominate the market, with surface mounting technology gaining steadily at the expense of pin-in-hole technology. An increasing market share will be taken by the further development of technologies such as flexible circuits, moulded circuits, polymer thick film and, in particular, multichip modules.[47]

2.5.1 Printed Boards and the Environment

During the latter part of the 20th century, a growing awareness of the detrimental effect on the environment of many of the waste products generated by industry gave rise to an increasing number of regulations governing waste disposal in Europe and in the USA. The Montreal Protocol of 1987 aimed at the reduction of CFCs in the atmosphere and the Du Pont company led industry by announcing a complete phase-out of their $750 million a year business producing CFCs.

Between 1972 and 1991 the European Commission issued more than 150 environmental directives, including the Toxic and Dangerous Wastes Directive in 1978, which also covered disposal of substances such as lead and copper. In the UK, various Acts of Parliament were passed from 1972 onwards, controlling waste disposal and control of pollution; one of the major ones was the 1990 Environmental Protection Act. Pressure from such regulations and from environmental groups had considerable effect on the circuit board industry, causing changes not only in waste disposal, but also to the substances and processes used for manufacture.

The industry had already begun to move away from the use of solvent-developing photoimaging materials during the late 1970s. In 1990 alternative cleaning strategies were being introduced, using aqueous and semi-aqueous agents; by 1994 the semi-aqueous system had almost been abandoned in favour of fully aqueous systems. Other approaches were the 'no-clean' method of soldering using solder pastes without rosin based fluxes or the use of a nitrogen blanket during wave soldering.[48]

Strict regulations governing worker exposure to formaldehyde made manufacturers examine alternative plating strategies; the first direct metallisation installation in the USA was at Photocircuits in 1986.[49]

In 1994 it was reported that, to minimise lead wastes, alternatives to tin-lead solder were being investigated, based on tin but with the addition of elements such as bismuth, antimony, indium, silver, copper and zinc. There was also an increasing use of gold over nickel for surface finishing although the primary reason was not environmental, see Volume 2, Chapter 6.[50]

2.5.2 The Future of Printed Circuits

The growth of printed circuit technology has been concurrent with the growth of the electronics industry, and has developed very rapidly since its obscure beginnings in 1943. Printed circuitry is a multi-million pound industry with little sign that its rate of growth will slacken in the near future.

PART 2 — ELECTRONIC COMPONENTS

W. MACLEOD ROSS

Welnorth Ltd, Bishop's Stortford, UK

2.6 GENERAL

It is not practicable in a text of this nature to mention all inventions and discoveries which have contributed to the development of modern components for electronics. Many full length books have been written on the subject. The author of this text has, therefore, omitted much of the history of electrical and electronic discoveries and inventions and concentrated on those things which were steps forward in electrical/electronic component technology or spurred on component development. For that reason, such things as the works of Pascal, Leibnitz and Babbage on the early development of computers are not mentioned. Those seeking information on that subject will find that much information has been published elsewhere, see, for example, Goldstine.[51]

Such was the rate of development of component technology, particularly after the invention of the solid state transistor, that examples only are given to indicate how the technology was continuing to advance.

2.7 1701-1800

It is generally accepted that the first electronic component was developed in Germany by von Kleist in late 1745 and, independently, by von Muschenbrock at the University of Leyden in January 1746, with the construction of the first man-made capacitor, the Leyden jar.[52] The construction and appearance differed markedly from a modern capacitor, but the principle of operation had been established.

In 1772, the first moulded iron cores were made in Britain by G. Knight and in 1780 Luigi Galvani began his studies of animal electricity but, possibly, the next truly notable invention was Alessandro Volta's voltaic pile, described by him in 1800 in a letter to the President of the Royal Society. This was constructed from copper and zinc discs separated by cloth moistened with an electrolyte. This was the first 'dry battery'. This was later improved upon by the use of paper discs, tinned on one side with manganese dioxide on the other; when stacked, this construction produced 0.75 V between 1 in. (25.4 mm) diameter discs.

2.8   1801-1900

2.8.1   1801-1830

In the early years of the nineteenth century much work was done on dry batteries by various experimenters in Europe. One of these, J. W. Ritter (who discovered ultra-violet rays just after Herschel's discovery of infra-red), invented the first accumulator or secondary pile.[53] He used only one metal interleaved with absorbent material moistened with an electrolyte which did not chemically attack the metal. When charged from a voltaic pile the charge was retained. (It should be noted, however, that the storage cell battery, in which the electrode reactions have to be reversible, did not appear until Planté's discovery of the lead-acid system in 1860.[54])

In the 1820s, Oersted discovered electromagnetism, Seebeck reported on thermoelectricity and Ohm formulated the law which bears his name.

2.8.2   1831-1860

In 1831, Michael Faraday discovered electromagnetic induction and described a multi-winding transformer and a transformer experiment in his diary. Faraday went on to enunciate his laws of electrolysis, demonstrating that chemical affinity was in fact electrical force acting at the molecular level.

The 1830s also saw patents by Cooke and Wheatstone, and by Davy, on the design and use of electromagnetic relays in telegraphy. Possibly less well known is the fact that in the late thirties Wheatstone invented a sound magnification system which he called a 'microphone'. In 1837-38 Samuel Morse developed and demonstrated equipment for telegraphy and a binary code for use with it which is still employed today. That was the first effective demonstration of telegraphy, the first form of telecommunication.[55] Morse's invention filled a great and growing need for speedy communication and its use spread rapidly.

Having become aware of Ohm's work establishing the relationship between potential difference (*electromotive force* — EMF), resistance and current, Charles Wheatstone developed his famous bridge, and the first rheostat, in 1843. This was an invaluable contribution to the science of electrical testing at a time when galvanometers were notoriously unstable and there was no accurate method for calibration.[56] Gaugain (No! Not the painter — he was *Gauguin*!) developed a cold cathode discharge tube (a cathode ray tube or CRT) in 1855 and is credited with being the first to observe that a glow tube having its two electrodes of different size is capable of rectifying the oscillating current produced by an induction coil. Nevertheless, such discharge tubes were used only for lighting and display purposes for many years. This and other early discharge tube developments were described by Cobine.[57]

2.8.3 1861-1880

The Leclanché cell[58] and the standard Clark cell[59] were invented in 1868 and 1870 respectively. Other notable component developments around that time were the mica capacitor (M. Bauer)[60] and the rolled paper capacitor (D. G. Fitzgerald).[61] Surprisingly, although Bauer reported fully on the excellent properties of mica as a capacitor dielectric in 1874, mica was not utilised commercially until some 40 years later when the 1914-1918 War brought demands for a capacitor dielectric more robust than glass. The better dielectric properties gave an added bonus by permitting significant size reduction for a given performance.

On February 14, 1876, Alexander Graham Bell filed the now famous or infamous (depending on whether you *love* it or *loathe* it!) patent for the first telephone.[62] Whether one loves the telephone or loathes it, it has to be admitted that the growth of the electronic component industry owes a great deal to the growth of telecommunications.

That the phonograph followed the telephone so quickly is not a coincidence. The invention of the telephone drew attention to the difficulties of speech reproduction and T. A. Edison was particularly interested since he was partially deaf. His patent application for the phonograph, USA Patent 200,521, was filed December 24, 1877. To Edison, also, goes the credit for being the first to design a microphone transducer using granules of carbonised hard coal.[63] In the same year E. W. Siemens invented the first moving coil type of loudspeaker[64] and the way was paved for yet another colossus of the electronics industry.

2.8.4 1881-1900

Three gentlemen from Budapest, O. T. Blathy, M. Deri and C. Zipernowski, are credited with the first patent for a power transformer, the patent being granted in 1885.[65] Max Deri appears again in 1885, this time described as from Austria, as the author of a patent for the use of transformers in an electricity distribution system.[66] 1885 was also the year in which British Patent No. 8076/1885 was issued to C. S. Bradley to cover a moulded carbon composition resistor.[67] The resistor was blended from a mixture of carbon and rubber, heated and moulded to shape before being vulcanised to form a hard body.

In 1887, Emile Berliner, the father of disc recording and sound reproduction, applied for a patent for his 'gramophone'. The first model, produced in 1886, had a flat disc record as opposed to the cylinder employed by Edison (and shown on Berliner's own patent). Berliner used lamp black as his recording medium and combined this with an etching process to permit transfer of the original to copper or nickel. By this means he created a master pattern and made possible the mass production of records. The system he developed by the end of the century stood as the industry standard for many years.[68]

1891 saw the advent of the Weston cadmium cell, the patent for which, USA Patent 494,827, was granted to Dr Weston in 1893. The Weston cell had a much better temperature coefficient of EMF than did the Clark cell

(40 µV/°C as opposed to 1200 µV/°C) and, after further improvements by various workers from 1893 on, was allocated the value of 1.0183 volts at 20°C.[69]

Following the proof by Heinrich Hertz of the existence of radio waves, Guglielmo Marconi took out the first patent for wireless telegraphy on June 2, 1896. This led to the first ship-to-shore transmission in the following year and to the classical transatlantic transmission in 1901. It also led in due course to the development of a new industry and a need for new types of component.

Returning to significant events in the 1890s, in 1897 J. J. Thomson showed that cathode rays were fast-moving particles of mass almost 2000 times less than that of a hydrogen atom. The world owes a debt of gratitude to J. G. Stoney since Thomson initially called the particle 'corpuscle' but then adopted the word 'electron' which had been invented by Stoney for the charge on a hydrogen atom. (The assonance of 'corpuscular component' does not compensate for the risk of confusion with the blood supply.) Also in 1897, Gambrell and Harris patented a carbon film type of resistor.[70] The following year another brilliant invention, magnetic recording, was the subject of a patent application filed by the Dane, Valdemar Poulsen. The principles invented by Poulsen of drawing a steel wire or tape past an electromagnet through which is passing the current from a microphone still apply to all magnetic recorders. Two major advantages of magnetic recording are (a) the ability to play recordings many times without degradation of quality of sound and (b) the ability to re-use the same wire or tape many times by demagnetising to 'clean' the tape before using for the new recording.

In the year 1900, L. Lombardi filed a patent for ceramic capacitors[71] and two new types of battery, the nickel-iron cell and the nickel-cadmium cell,[72] were invented in the USA and Sweden respectively.

2.9   1901- PRESENT DAY

2.9.1   1901-1920

In 1904, J. A. Fleming filed his patent for a two-electrode tube or thermionic valve (a diode) for rectifying high frequency alternating currents. This was followed in 1906 by L. de Forest's patent application for a three-electrode (triode) valve for use as an amplifier or oscillator. Regarded by many as a radio pioneer and the 'grandfather of television', Dr de Forest is also credited by some with the invention of the four-electrode valve or tetrode. This is debatable since H. J. Round has been named as the person who brought the tetrode into practical use in 1926.[73]

An early paper by Swann on the subject of sputtered platinum thin film resistors and their properties illustrates how little, in some respects at least, ideas have changed.[74] During and just after the First World War, few totally new components were developed although much work was done on improving production techniques. The first electromagnetic filter appeared in 1915 and in 1917 Czochralski developed the crystal pulling technique for the production of single crystal silicon.

2.9.2 1921-1940

The cracked carbon or pyrolytic carbon film resistor probably appeared first in a patent assigned to the Siemens and Halske organisation in 1925. While considerable skill and control are needed to produce consistent results, a highly stable resistive coating can be achieved. Again in Germany, in 1926, Lowe patented another method of making fixed resistors. The method involved applying a spray of platinum resinate to an insulating base material and heating the film to convert it to platinum metal.[75]

The pentode valve, invented by Tellegen and Holst at the Philips Company in Holland in 1928,[76] was developed in order to suppress secondary emission in a tetrode, especially where it was desired to operate with high screen and anode potentials. The Tellegen and Holst solution inserted a suppressor grid, maintained at the filament potential, between the screen grid and the anode. By the outbreak of World War Two, the pentode valve had become extremely popular for high and low frequency applications.

The insulated gate field effect transistor, see Chapter 3, might well have appeared in the 1930s, at least two patents being issued by German workers which described devices remarkably similar in operation (but not in form or material) to the familiar MOSFET. Work on liquid crystals started in earnest at the Marconi Research Laboratories in the UK, and ICI (Imperial Chemical Industries Limited) developed polyethylene.

2.9.3 1941-1960

During the years of the Second World War much work was done in developing circuitry for different applications but component development appeared to be concentrated on improving existing designs and materials, for example, to produce more rugged versions of valves and resistive and conductive pastes for thick film circuits.

Semiconducting materials had been known for some time but had not been used in a knowledgeable manner. This changed dramatically in 1948 when the Bell Telephone Laboratory announced the invention of the solid state transistor.[77] This resulted from Bell setting up a team in 1946 under William Shockley to study solid state physics and paved the way for the semiconductor industry and integrated circuits. In the next few years, Bell pioneered ion implantation,[78] Darlington pairs, so named after the inventor, S. Darlington, thermocompression bonding and wire wrapping and, in 1954, the silicon solar cell for converting sunlight into electricity. In 1956, Bell Telephone Laboratories were the first to produce transistors based on the results of studies of the diffusion of selected impurities into single crystal germanium and silicon. Frosch, also at the Bell Laboratories, followed this in 1957 by reporting the development of thermally grown oxide films as maskants for silicon. (Also in 1956, McLean and Power announced the development of the first solid electrolyte capacitor.[79])

The first commercial field effect transistor was produced in France by S. Teszner in 1958, the year in which L. Esaki of Japan invented the tunnel diode.[80] The following year saw the invention of the planar transistor by J. A. Hoerni (Fairchild). Because the *p-n* junction was protected with a layer of boron- and phosphorus-diffused silicon oxide, dust and other foreign material could not contaminate it, resulting in a more reliable component. Shortly after this discovery, Fairchild applied the technique to the fabrication of the new 'integrated circuits'. (The first integrated circuit patent, US Patent Number 3,138,743, had been filed in the USA by J. S. Kilby in February 1959.)

1960 was the year in which the Bell Telephone Laboratories announced their new method of making transistors by epitaxial crystal growth from the gas phase. *Epitaxy* is defined as the growth of a thin layer on a single-crystal substrate that determines the lattice structure of the layer. This means that it is possible to grow very thin regions of controlled purity, for example, of phosphorus-doped silicon (n-silicon) on a boron-doped (p-silicon) base, to give a single crystal, defect-free, p-n junction.

2.9.4   1961-1980

RCA developed equipment and methods for epitaxial growth of gallium arsenide and germanium from a liquid phase for which advantages over vapour phase growth were claimed. This work was reported in 1961 but the major development from the RCA Electronic Research Laboratory around this time has to be the development of the first MOS (metal-oxide-semiconductor) integrated circuit. The circuit, a logic block of 16 MOS transistors in a silicon chip 0.050 in. x 0.050 in. (1.27 mm x 1.27 mm), was reported to a conference on electron devices in October 1962.[81] In the same year (1962), Holonyak at General Electric, Syracuse, NY, invented the first gallium arsenide phosphide LED *(light emitting diode)* and the first semiconductor laser to operate in the visible spectrum. Another notable invention in '62 was the flatpack, described in Chapter 3, 3.9.5. It was invented by Yung Tao, Texas Instruments.

In 1963, J. B. Gunn reported his work on the Gunn diode oscillator[82], silicon-on-sapphire became important with the advent of practical MOS technology, Mattox of Sandia Laboratories, New Mexico, invented ion plating, and Rowan and Sittig of Bell Telephone Laboratories filed a number of patents on surface acoustic wave devices. The Bell Laboratories were back in the news in 1964 with the IMPATT (impact avalanche transit time) diode, a diode which generates microwaves when a DC voltage is applied to it, and the invention of the beam lead, see Chapter 16, 16.3.4. RCA developed the overlay transistor for the US Army as a UHF power transistor to replace the valve output stages then employed in military radio transmitters.

Flip-chip bonding, see Chapters 4, 4.4.16, and 16, 16.3.3, was invented and patented by Wiessenstern and Wingrove in 1966.[83] While it is still being used some thirty years later attempts are still being made to realise its full potential.

In 1964, G. W. A. Dummer of the UK Royal Radar Establishment (later the Royal Signals and Radar Establishment) speculated on the replacement of electric current and transmission wires in telecommunications by light and glass fibres in a talk he gave to the British Association for the Advancement of Science. There is no doubt that others were also considering the potential use of fibre optics but it was not until Kao and Hockham published their paper on the subject in 1966[84] that development work began in earnest and the full benefits of optical fibre transmission began to be realised. In March 1966, General Electric (USA) patented the use of a passivation layer of silicon nitride over silicon dioxide as a means of preventing instability problems in semiconductors. Later in the sixties methods of producing CMOS (*complementary metal oxide silicon*), where both p- and n-channel transistors are combined on the same chip, were developed. Up to that time it had been very difficult to fabricate both n- and p-channels in the same substrate. The TRAPATT diode was developed in 1967. The TRAPATT (trapped plasma avalanche transit time) diode allowed the realisation of high efficiency solid state microwave oscillators and amplifiers to be achieved.

In the following year, Robert Noyce, of Intel fame, suggested the use of aluminium for forming the base and emitter contacts of transistors, an apparently simple suggestion but of great importance in creating practical devices. He was later given credit for proposing nickel as a good electrical contact for devices. The late 1960s are also notable for investigations into and development of bipolar processes for device fabrication. Several processes were developed which were compatible with circuits operating at frequencies in excess of 1.5 GHz and all had the advantage of using less surface area than the processes previously employed. Semiconductor memory devices were also developed and rapidly replaced the earlier magnetic arrays which occupied a great deal more space, were more expensive to fabricate and slower in operation.[85]

In 1970, Boyle and Smith of the Bell Telephone Laboratories reported on the structure and operation of the *charge coupled device* (CCD), described in Chapter 3, 3.9.1.1. The most familiar application of the CCD is in television cameras. The floppy disk first appeared in 1970 as a read-only device (rather like the compact disks of the mid-1980s) for fault-finding in an IBM computer. After further development, it rapidly became an industry standard. In 1971 Intel developed the FAMOS (floating-gate avalanche-injection metal oxide semiconductor) integrated circuit as a memory device. It showed many advantages over MNOS (*metal nitride on silicon*) memories. In the following year Intel produced the first microcomputer, now successfully used in a multitude of applications and becoming smaller, faster and more powerful every year. In addition, the company produced a 1024 bit RAM (random access memory), demonstrating for the first time that over 1000 bits of read-write could be supplied in low cost MOS structure on a single chip.

By the end of 1975 a 16-bit single-chip microprocessor had been produced (by the National Semiconductor Corporation in the USA), a laser had been combined with light guides, modulators and filters in a single crystal microcircuit about 0.006 in. x 0.015 in. (0.15 mm x 0.38 mm) in area (by the Bell Laboratories, of course) and Fairchild Semiconductor produced a 4096 bit I^{2}L (integrated injection logic) RAM with an access time of 100 nanoseconds, double the speed of a 4 kilobit NMOS DRAM (*nitride metal-oxide silicon dynamic random access memory*).

In 1976, the Intel Corporation produced a 16 kilobit RAM and then the 'super component', a one-board computer with programmable input/output.[86] This was followed in June 1978 by the original 8086 device, a 16-bit device running at an original clock frequency of 5 MHz, which Intel called a complete one-chip microcomputer. The 8086 started a whole family of microprocessors and led to the proliferation of the PC. The 80186 and 80286 chips appeared in 1980. Like the 8086, these were 16-bit devices. By then, 64 kilobit DRAMS were available and, also in 1980, the first UV-erasable EPROM (*erasable programmable read-only memory*), see Chapter 3, 3.9.4.2, was developed.

2.9.5   1981- Present Day

IBM launched the personal computer, the PC, in 1981, an event which was to affect the development of, and need for, electronic components significantly. The success of the PC made it an unofficial world standard and encouraged a variety of clones, many offering improved performance over the IBM machine for a lower price. Japan had already developed a large and technically expert semiconductor industry but the PC clone industry encouraged other countries in the Far East, such as Taiwan and Korea, to enter the electronics industry.

As the PC revolution took effect, so there was a drive towards faster and more powerful ICs and bigger and better memories. There was also a distinct movement towards the development and use of packages specifically designed for surface mounting for a number of reasons such as cost reduction, simplification of handling and assembly, and improvements in packing density.[87] In 1982, Inmos (later part of Thorn EMI) launched a 64 kilobit DRAM and Fujitsu announced a 128 kilobit EPROM. Continual improvement in technology was making nonsense of earlier beliefs in the limits achievable. As an illustration of this, in the late seventies a 'state of the art' *application specific integrated circuit* (ASIC) had 100 gates, in 1980 1,000 gates, in 1985 10,000 gates. Some ten years later, the number of gates was rising to 1,000,000 with 10,000,000 gates expected by the end of the century.[88]

In 1982-83, Philips and Sony jointly sounded the death knell of the gramophone disc by announcing the birth of the compact disc, a 120 mm (4.72 in.) plastic disc on which digitally encoded sound was recorded, and in 1984 Apple launched the Apple Mac alternative to the IBM PC. This was the first personal computer to use 32-bit technology. This allows longer and more complex instructions to be processed and was made possible by the Motorola Corporation's development of the 68020 chip. Intel Corporation then made the 80386 chip available (in 1985). While using the same basic structures as earlier Intel devices, the 386 included 32-bit registers which could be controlled by software to 'split' down the middle for use as 16-bit registers. This allowed the 80386 to run the earlier 8086 and 80286 code. The original 80386 chip had 275,000 transistors (VLSI, *very large scale integration* — see Chapter 3) and a clock speed of 12 or 16 MHz.

Hitachi announced the first supertwist LCD (*liquid crystal display*) and CDROM (or CD-ROM, i.e., *compact disc read-only memory*) in 1986. The CDROM is capable of holding more than 500 Mb and by the mid 1990s was widely used to store software of all types. This was followed in 1987 by the announcement from Mitsubishi of the 4 megabit RAM and, from Motorola, the 68000 processor chip. Motorola followed the 68000 with the 88000 family in 1988. These RISC (reduced instruction set computer) devices ran at a clock speed of 20 MHz. Not to be outdone, the Intel Corporation released the i486 in 1989. This device had 1,200,000 transistors (an early example of ULSI, *ultra large scale integration*, see Chapter 3) and, essentially, was an implementation on one chip of the 386, its maths coprocessor, the 387, the 385 memory management unit and a large unitary cache. The original clock speeds were 25 and 33 MHz. In the same year, IBM scientists claimed that they had developed the fastest p-n-p transistor. Digital circuits built with the transistor switched on and off 25 *billion* times per second.

Motorola updated the 68000 microprocessor family by introducing the 68040 MPU in 1990. The architecture of the new device utilised many lessons learned from the design of the RISC 88000 processors to produce high processing power.[89] An agreement between Motorola and Toshiba for exchange of technology and expertise led to Toshiba producing a 64-pin QFP, see Chapter 4, 4.4.9, having the smallest package size for 68000 devices. When announcing this development, Toshiba reported development of a MOSFET with a breakdown voltage of 1,500 V and a laser diode 'with the bright red-light output normally only available from bulky power-hungry gas lasers'.[90] Still in 1990, Alupower-Chloride developed a primary cell which runs partly on air as a standby cell for a telecommunications or computer system. The battery operates by the electrochemical oxidation of aluminium and was claimed to generate 6-10 times the power of a similar size lead-acid battery.[91] Also by the end of 1990, Hitachi had developed a neural computer on a chip and demonstrated it at the Hitachi Technology Fair held in New York. The 1,152 neural circuits can, when combined, run 2.3 billion learning operations per second. Using neural networks instead of conventional chips maximises the learning capability of a computer.

In January 1991, it was reported in the electronics press that Edinburgh University had invented a complete camera on a chip. The chip was fabricated with CMOS technology and comprised a photodiode array to detect light and charge sense amplifiers to produce an electric signal. The chip was simply made into a camera by gluing a lens on to the top of the photodiode array. Another example of American and Japanese collaboration was particularly interesting as it illustrated the increasing importance of matching the package to the component (and *vice versa*). Texas Instruments and Hitachi, working together on 16 Mbit DRAM development, jointly designed a special J-lead package to reduce the variation in lead capacitance which all conventional small outline packages exhibit. The bonding pads were placed centrally on the long axis of the chip so that the leads from the pads to the pins could be of approximately equal length. The technology, termed *lead-on-chip with centre bond* (LOCCB), increased the amount of silicon the package could hold, reduced on-chip noise and reduced thermal and mechanical stress on the semiconductor material.[92]

1991 was also the year in which Texas Instruments reported the invention of the first multi-material monolithic optoelectronic integrated circuit. The method employed was developed in the TI Research Laboratories in Dallas, USA, in 1988 and is a unique co-integration process called 'gallium arsenide on silicon'. This allowed the researchers to combine silicon CMOS logic circuitry with an array of 8 GaAs infra-red LEDs on the same chip. The advantages of using optical techniques to replace conventional copper conductors are higher transmission speed, lower power requirements and reduction in board size.[93]

In 1992, NEC announced a new manufacturing technique for producing higher yields of carbon *nanotubes*. In this rare form of carbon the carbon atoms are arranged in concentric sheets of hexagons organised in a helical pattern. It is mooted by several researchers that carbon nanotubes could be the basis of new types of semiconductors once the properties are better understood. Dowty Batteries in the UK disclosed that development work by the company in conjunction with the University of St Andrews had resulted in a rechargeable lithium manganese dioxide cell. The new battery has advantages as a non-toxic high power density battery for industrial and military use.

In 1993, Texas Instruments and Hitachi revealed their joint development of a 64 Mbit DRAM and IBM, together with Analog Devices, announced the intention to develop silicon-germanium integrated circuits. Several other researchers, in the USA and Britain, revealed that they too had been working on SiGe as an alternative to the more costly gallium arsenide process.[94] In 1994 the Intel Corporation announced that it was starting to build its third 8 inch wafer fabrication facility for completion in 1997. The facility, in Arizona, will produce the next generations of Intel microprocessors using 0.25 µm technology. It was revealed that the existing Intel fabrication facilities could more than cope with the demand for 486 and the 100 MHz Pentium processors.

Making predictions in the electronics industry can be a dangerous thing. In the 1960s, well-respected engineers were confidently predicting that the printed board would be superseded by other methods of interconnection by the end of the decade. There is still no sign of its demise in the 1990s. At the start of the 1980s, it was confidently predicted that one micrometre was the limit for optical technology. We all know how wrong that prediction was.

One thing can be forecast with some degree of confidence. Electronic and electro-optics technologies will continue to expand for many years to come.

PART 3 — SOLDERING

(Based on material supplied by Gert Becker, Consultant, Sköndal, Sweden)

2.10   FROM ANCIENT CRAFT TO MODERN SCIENCE

In the introduction 'Soldering Spans the Centuries' to the first edition of his book on solders and soldering[95] Howard Manko pointed out that, while soldering is the oldest metallurgical joining method known, it is as essential to the space age as is the computer. Thirty years later, that statement is still true. Soldering is still vital in electronics.

There is no doubt that soldering is a very ancient craft although when and where it started is not known. Tin, used in the most common solders, is not found in many places but was known in Mesopotamia 3,200 years before the birth of Christ and in Europe about 2,000 BC.[96] By around 800 (AD) the art of soldering was generally known in Europe, the Near East and parts of Asia.

The first known book on soldering was published in 1760 in Berlin by Johann Georg Friedrich Klein for the purpose of training his instrument makers in his workshop in St Petersburg.[97] In that book Klein covered subjects such as soldering on gold, copper, brass, iron and other metals, how to make fluxes and solders for different purposes, and cleaning after soldering. In addition, he dealt with the issues of strength and lifetime of solders and included many illustrations of the tools used.

The first known solder joint made for electrical purposes can be found in the British Science Museum in London on a nineteenth century electrical machine. The use of solder for joining has been accepted as a natural thing to do at every stage of the development of electronics. Electricity was employed as a source of energy for soldering in 1890, even though the tools used were very different from the electrical soldering iron as we know it today.[98]

Mass soldering in electronics was really first made possible by the patent by Strauss and his colleagues[42] on the wave soldering principle. Other processes such as drag soldering, vapour phase soldering, laser soldering and others followed, see Chapter 15.

Fluxes for electronic soldering were based in the beginning on naturally occurring resin materials. Occasionally, and only when it was possible to clean thoroughly after soldering, the flux would be based on organic or inorganic acid compounds. During the Second World War there was a shortage of natural resin in Germany which led to the application of synthetic resins for fluxing. The recognition of the need for improved solderability promoted the search for better fluxes. Many fluxes, and their residues, needed to be cleaned off the board assemblies after the joints were formed and, when the damage caused by cleaners such as CFCs to the ozone layer became apparent, efforts were concentrated on development of low solid content and 'no-clean' fluxes.

Flux development is still going on and will continue while soldering is still a major method for making electronic joints. With continual emphasis on improved producibility, increased productivity, better reliability and the trend to smaller and smaller joints, the solders and methods which have been commonly used for many years are changing. Many modern electronic assemblies could not be produced without solder pastes and solders containing materials such as bismuth and/or indium. New materials are constantly being developed and used in more and more applications.

Materials used in soldering are covered in Chapter 13 of this book.

REFERENCES

1 British Patent No. 349706, Rhuyssanaers (1929).
2 United States Patent No. 2066511, Arlt (1937).
3 United States Patent No. 1563731, Ducas (1925).
4 British Patent No. 269729, Parolini (1926).
5 Sargrove, J. A. 'New Methods of Radio Production', *Journal of Brit. IRE.*, January, February (1947).
6 Eisler, Dr P., 'Technology of Printed Circuits', Haywood, London (1959).
7 Eisler, Dr P., 'Printed Circuits and their Future Prospects within the Industry', *British Printer*, July-August (1954).
8 Schnorr, D. P., 'Printed Wiring, History of a Technology'.
9 Various authors, 'New Advances in Printed Circuits', Proceedings of the First Technical Symposium on Printed Circuits, Washington, October 1947, Miscellaneous Publication 192 of the NBS (USA) (1948).
10 'Micromodular System', US Army Reports (1958 and 1959).

11 British Patent No. 911718, Hazeltine Corporation (1961).
12 Swigget, R. L. and Schneble, F. W., '"CC-4'. A Reliable and Economical Additive Printed Circuit Process', National Electronics Conference, October (1964).
13 Jansen, C. J., Jonger, H. and Postma, L., 'PD-R, A New Modification of PD Photoplating', IMF Symposium, PC Group (1969).
14 Burr, R. P. and Keogh, R. J., 'A New Simplified Approach to High Density Connections', Nepcon West, February (1970).
15 Messner, G. and Burr, R. P., 'New Multiwire Meets the Challenge of Interconnecting Chip Carriers', *Electronics*, December (1979).
16 Staller, J., 'Point to Point Wiring Automation', in 'Automation in Electronics' (1970).
17 Cannizzaro, 'The Development of Polyimide Multilayer Boards Containing Integral Flexible Circuitry', InterNepcon (1970).
18 *New Electronics*, January (1987).
19 Thompson, M. K., 'Polyethersulphone — a Thermoplastic PCB Substrate', *Circuit World*, **Vol. 9**, No. 3, pp. 41-45 (1983).
20 Eisler, Dr P., 'Printed Circuits — Some General Principles and Application of the Foil Technique', *Journal of Brit. IRE*, November (1953).
21 Lubin, G., 'Handbook of Fibreglass and Advanced Plastic Composites', Polymer Technology Services, Van Nostrand Reinhold (1969).
22 Brody, M., 'Laminates — PCB Manufacturing Begins Here', *Printed Circuit Fabrication*, December (1982).
23 Marnott, P. J., 'Polyimide and Related Resins as Dielectric Materials', InterNepcon (1973).
24 *Surface Mount Technology*, April (1989).
25 United States Patent No's 3469982 and 3622334, Du Pont (1968).
26 *Electronic Production*, July (1989).
27 'Drilling Printed Circuit Cards under Tape Control', *Machinery* (1965).
28 'NC Drilling System — State of the Art', *Circuits Manufacturing*, January (1972).
29 Cox, D., 'Know the Drill', *Circuit World*, **Vol. 7**, No. 3, pp. 55-59 (1981).
30 *Printed Circuit Fabrication*, February (1990).
31 United States Patent No. 4872787, Hitachi (1990).
32 *Printed Circuit Fabrication*, January (1991).
33 Block, J., 'Drill Heat Problems in PCBs', *Electronic Production*, February (1980).
34 Duffeck, E. F., 'Trends in Manufacturing Printed Circuits', *Plating and Surface Finishing*, January (1979).
35 Schlaback, T. D. and Rider, D. K., 'Printed and Integrated Circuitry', McGraw-Hill (1963).
36 *Product Finishing*, September (1966).
37 Gerstberger, H., 'Electroplated Tin and Tin-Lead Coatings', InterNepcon (1973).
38 Rothschild, B. and Sanders D., 'High Throwing Power Solder Plating Solutions', *Plating*, December (1969).
39 Luke, D. A., 'Tin-Lead Plating and its Problems', IMF PCG Symposium (1980).
40 Tucker, Dr W. B., 'A Survey of Resists and Protective Coatings', Institute of Printed Circuits, Chicago, September (1970).
41 Woods, B., 'Printed Circuit Resins — A Review', *Transactions of the IMF, PC Supplement* (1968).
42 British Patent No. 798701, Barnes, A. F. C., Elliott, V. B. and Strauss, R. S., (1956).
43 Woodgate, R. W., 'Thermal Effects on Machine Soldered Joints', *Electronic Packaging & Production*, June (1982).
44 Chu, T. Y., Mallendorf, J. C. and Pfahl, R. C. Jr., 'A New Mass Soldering Process — Condensation Soldering', InterNepcon (1974).
45 Marsh, H. G., 'A Genealogy of PCB CAD', *PC Design*, January (1993).
46 Britton, P. L. and Chopko, D. A., 'Conductor Line Imaging — Dry Film or Liquid Photoimaging Technology?', Printed Circuit World Convention, UK, June (1990).
47 Pearne, N., 'Interconnection Strategies of the Future', Printed Circuit World Convention, UK, June (1990).
48 *Electronic Production*, January-February (1991).

49 *Printed Circuit Fabrication*, February (1994).
50 *Electronic Production*, February (1994).
51 Goldstine, H. H., 'The Computer from Pascal to von Neumann', Princeton University Press (1972).
52 Darnell, P. S., 'History, Present Status and Future Developments of Electronic Components', *IRE Transactions on Component Parts*, pp. 127-8, September (1958).
53 Motteley, 'Biographical History of Electricity and Magnetism', p. 381, Charles Griffin, London (1922).
54 Morehouse, C. K., Glicksman, R. and Lozier, G. S., 'Batteries', Proceedings of the IRE, pp. 1474-1475, August (1958).
55 Everitt, W. L., 'Telecommunications — the Resource not depleted by Use. A Historical and Philosophical Résumé', *Proceedings of the IEEE*, **Vol. 64**, No. 9, p. 1293, September (1976).
56 Bowers, B. P., 'Wheatstone's Contribution to Electrical Engineering', *Electronics and Power*, p. 295, May (1976).
57 Cobine, J. D., 'The Development of Gas Discharge Tubes', Proceedings of the IRE, May (1962).
58 Wheeler, N. D., 'Survey of Electrochemical Batteries', *Electro-Technology*, p. 68, June (1963).
59 'The Clark Cell', Proceedings of the Royal Society, **Vol. XX**, p. 444 (1872).
60 Coursey, P. R., 'Electrical Capacitors in our Everyday Life', *ERA Journal*, No. 6, p. 10, January (1959).
61 Darnell, P. S., 'History, Present Status and Future Developments of Electronic Components', *IRE Transactions on Component Parts*, p. 124, September (1958).
62 Flood, J. E., 'Alexander Graham Bell and the Invention of the Telephone', *Electronics and Power*, p. 159, March (1976).
63 Bauer, B. B., 'A Century of Microphones', Proceedings of the IRE, p. 721, May (1962).
64 Olson, H. F., 'Loudspeakers', Proceedings of the IRE, p. 730, May (1962).
65 Carr, L. H. A. and Wood, J. C., 'Patents for Engineers', pp. 89-90, Chapman and Hall, London (1959).
66 Carr, L. H. A. and Wood, J. C., 'Patents for Engineers', p. 91, Chapman and Hall, London (1959).
67 Coursey, P. R., 'Fixed Resistors for use in Communication Equipment', Proceedings of the IEE, **Vol. 96**, Pt. III, p. 169 (1949).
68 Bachman, W. S., Bauer, B. B. and Goldmark, P. C., 'Disk Recording and Reproduction', Proceedings of the IRE, pp. 738-739, May (1962).
69 'Standard Cells by Muirhead', *Muirhead Technique*, **Vol. 18**, No. 3, p. 19, July (1964).
70 Coursey, P. R., 'Fixed Resistors for use in Communication Equipment', Proceedings of the IEE, **Vol. 96**, Pt III, p. 170 (1949).
71 British Patent No. 9133, Lombardi, L., (1900).
72 Morehouse, C. K., Glicksman, R. and Lozier, G. S., 'Batteries', Proceedings of the IRE, p. 1478, August (1958).
73 Appleton, Sir E., 'Thermionic Devices from the Development of the Triode up to 1939', IEE Publication Thermionic Valves 1904-1954, pp. 22-23, IEE London (1955).
74 Swann, W. F. G., *Philosophical Magazine*, **Vol. 28**, p. 467 (1914).
75 Marsten, J., 'Resistors — a Survey of the Evolution of the Field', Proceedings of the IRE, p. 922, May (1962).
76 Appleton, Sir E., 'Thermionic Devices from the Development of the Triode up to 1939', IEE Publication Thermionic Valves 1904-1954, pp. 23-24, IEE London (1955).
77 Kelly, M., 'The First Five Years of the Transistor', *Bell Telephone Magazine*, Summer (1953).
78 Brown, W. C. and MacRae, A. U., 'Ion Implantation', *Bell Laboratories Research*, p. 389, November (1975).
79 McLean, D. A. and Power, F. S., 'Tantalum Solid Electrolyte Capacitor', Proceedings of the IRE, p. 872, July (1956).

80 Esaki, L., 'New Phenomenon in Narrow Germanium p-n Junctions', *Physics Review*, **Vol. 109**, p. 603 (1958).
81 Socolovsky, A., 'The First MOS', *The Electronic Engineer*, p. 56, February (1970).
82 Gunn, J. B., 'Microwave Oscillations of Current in III-V Semiconductors', *Solid State Communications*, **Vol. 1**, p. 88 (1963).
83 United States Patent No. 3,256,465, Weissenstern, M. and Wingrove, G. A. S., (1966).
84 Kao, K. C. and Hockham, G. A., 'Dielectric-fibre Surface Waveguides for Optical Frequencies', Proceedings of the IEE, **Vol. 113**, pp. 1151-8 (1966).
85 Chang, J. J., 'Nonvolatile Semiconductor Memory Devices', Proceedings of the IEEE, **Vol. 64**, p. 1039, July (1976).
86 Garrow, R., Johnson, J. and Maerz, M., 'The 'Super Component': the One-board Computer with Programmable I/O', *Electronics*, p. 77, February 5 (1976).
87 Messner, G. and Lassen, C., 'Substrates for Surface Mounted Components', *Circuit World*, **Vol. 8**, No. 1, pp. 10-14 (1981).
88 'ASICs, Your Pathway to Tomorrow', Toshiba Technical Data (1994).
89 'Motorola's MC68040 is Launched', *Motorola Semiconductor News*, No. 73, July (1990).
90 *Toshiba Semiconductor News*, Issue 6, Autumn (1900).
91 'Running on Fresh Air', *Electronics Weekly*, p. 14, November 7 (1990).
92 'Special Package for 16 Mbit chips', *Electronics Weekly*, p. 3, February 27 (1991).
93 'Chips using Light could break Speed Bottlenecks', *Electro Optics*, p. 9, June (1991).
94 Parry, S., 'SiGe looks to plug GaAs leak', *Electronics Weekly*, p. 14, January 12 (1994).
95 Manko, H. H., 'Solders and Soldering', McGraw-Hill (1964).
96 'Werkstoffhandbuch Nichteisenmetalle', VDI-Verlag, Düsseldorf (1960).
97 Klein, J. G. F., 'Ausführliche Beschreibung der Metallothe und Löthungen', Berlin (1760). Original edition reprinted in Leipzig for the Deutscher Verlag für Schweisstechnik GmbH, Düsseldorf, Germany (1987).
98 de Fodor, E., 'Schweissung und Löthung', A. Hartleben's Verlag, Wien, Pest Leipzig (1892). Reprinted by the Deutscher Verlag für Schweisstechnik GmbH, Düsseldorf, Germany (1980).

Section 2
Electronic Components

Chapter 3

CONVENTIONAL COMPONENTS

GIOVANNI LEONIDA
Milan, Italy

W. MACLEOD ROSS
Welnorth Ltd, Bishop's Stortford, UK

3.1 GENERAL

If the printed circuit board is the foundation, electronic components are the building blocks used by the electronic designer to realise his circuit.

There are more than one hundred thousand types of component and new components appear almost daily. To know the characteristics of all of these is impossible and the engineer must continually refer to standard reference data and to suppliers' literature.[1]

While manufacturers of electronic assemblies do not need to be experts in components, as the circuit designer or components specialist will make all decisions on new types, new sources, and replacements for obsolete components, they do need some basic information. Typically they need to know:

  (i) how to identify components;
 (ii) outline dimensions;
(iii) precautions needed during handling, preforming, assembly, soldering, and so on.

In this chapter and the next the most common types of electronic components are described. Major attention is given to those parameters of primary importance to production engineering and assembly.

3.2 CLASSIFICATION

Two broad categories of component are defined:

  (i) *passive*, e.g., resistors, capacitors, inductors, connectors;
 (ii) *active*, e.g., diodes, transistors, SCR, integrated circuits.

Another classification is into *discrete components* which represent an elementary function of the circuit — such as a resistor, capacitor, diode, transistor, etc., — and *integrated components*, such as an integrated circuit (IC) which is functionally equivalent to, i.e., will perform the same tasks as, a circuit made of many discrete components enclosed in a single package.

For assembly, components may be divided into:

(i) *conventional or traditional*, i.e., to be assembled on one side of the printed board, with their leads passing through mounting holes on the board to be soldered on the other side; for this reason the component must have long leads;

(ii) *surface mounted*, i.e., to be soldered on the side to which they are assembled; components can be mounted on both sides of the board and have only very short leads or no leads at all.

Surface mounted components may be called *leadless* while conventional components may be referred to as *leaded*. Both terms may be incorrect and should not be used except to describe a particular component.

This chapter deals with the conventional or traditional components. Chapter 4 deals with surface mounted components or devices, SMCs or SMDs.

3.3 RESISTORS

Resistors are the most frequently used components in electronic circuits. A resistor exhibits a controlled value of resistance across its two terminations. The resistance value is fixed (except for drifts) for *fixed* resistors and is adjustable within a specified range for *variable* resistors.

Common resistors have a resistance value which is largely independent of the voltage across them. They are said to be *linear* because the current I which flows across them is directly proportional to the applied voltage V in accordance with Ohm's law:

$$I = V/R$$

where R is the resistance of the resistor.

Some circuit applications require *non-linear* resistors or VDRs (*voltage dependent resistors*) the resistance of which decreases when the applied voltage increases. For these the current I depends upon the voltage according to the formula

$$I = V^n/r$$

where r depends on V and n is a constant whose value is normally 2 to 6. These unusual resistors will be ignored as will *thermistors*,[2] i.e., resistors whose resistance is temperature dependent. The latter include two basic types, NTC (*negative temperature coefficient*) where resistance decreases with increase in temperature, and PTC (*positive temperature coefficient*) where resistance increases with increase in temperature. Electrical symbols for resistors are shown in Figure 3.1.

3.3.1 Packages

Commonly, fixed resistors resemble small cylinders with leads connected to the centre of the terminations and aligned on the long axis of the cylinder,

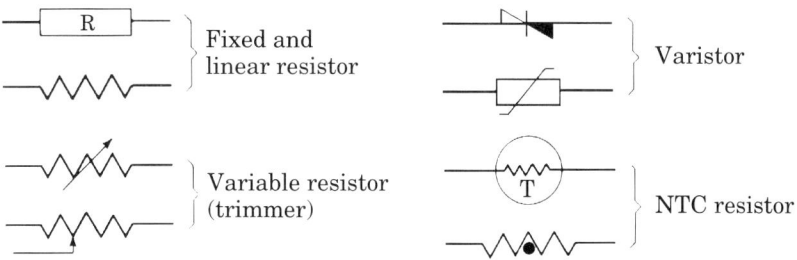

Fig. 3.1 Symbols for resistors.

Figure 3.2. Less usually they are cylindrical with radial leads. Arrays or networks of resistors are often enclosed in a DIP or DIL package, 3.9.5.

Variable resistors, often termed *trimmers* or *pots,* are available in a large number of packages, some examples of which are shown in Figure 3.3.

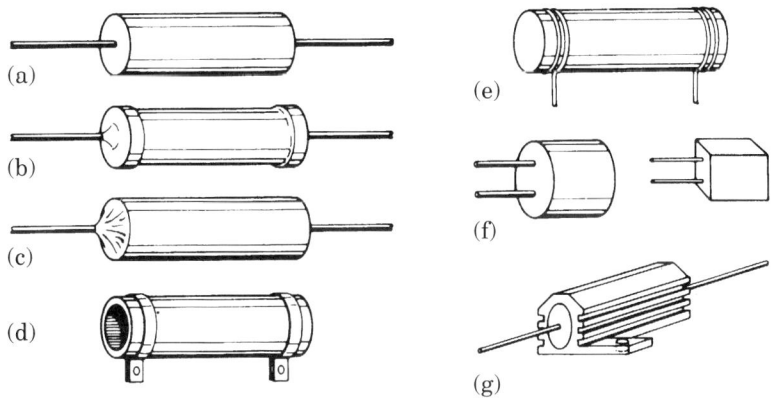

Fig. 3.2 Common packages for fixed resistors: (a), (b) and (c) cylindrical with axial leads (or axial); (d) and (e) cylindrical with radial leads; (f) radial; and (g) high power package, with axial leads and copper body for increased heat dissipation.

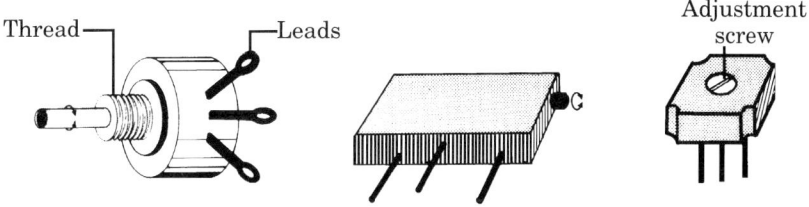

Fig. 3.3 Three typical packages for variable resistors.

3.3.2 Characteristics

A resistor is defined by a set of parameters, changes in which may restrict interchangeability in many applications. The main parameters are:

Resistance: This is the nominal value of resistance between the two leads of the component at 25°C. It is measured in ohms or multiples thereof such as kilo-ohm (kΩ) and mega-ohm, usually shortened to megohm (MΩ). One kΩ = 1,000 ohms; one MΩ = 1,000,000 ohms. Most resistors have a nominal value which is one value of the standard decade, Table 3.1, or a multiple thereof by 10^n, where n is an integer.

Table 3.1
Standard Decade for Resistors

1.00	1.47	2.15	3.16	4.64	6.81
1.02	1.50	2.21	3.24	4.75	6.98
1.05	1.54	2.26	3.32	4.87	7.14
1.07	1.58	2.31	3.40	4.99	7.32
1.10	1.62	2.37	3.48	5.11	7.50
1.13	1.65	2.43	3.56	5.23	7.68
1.15	1.69	2.49	3.65	5.36	7.86
1.18	1.74	2.55	3.74	5.49	8.05
1.21	1.78	2.61	3.83	5.62	8.25
1.24	1.82	2.67	3.92	5.76	8.45
1.27	1.87	2.74	4.02	5.90	8.66
1.30	1.91	2.80	4.12	6.04	8.87
1.33	1.96	2.87	4.22	6.19	9.08
1.37	2.00	2.94	4.32	6.34	9.30
1.40	2.05	3.01	4.42	6.49	9.53
1.43	2.10	3.09	4.53	6.65	9.76

Tolerance: The resistor tolerance is the maximum deviation of the actual resistance value from the nominal value. For example, in a batch of resistors of 130 ohms with a 10% tolerance the resistance of each should lie between 117 and 143 ohms.

The combination of two factors, tolerance and stability, is referred to as precision. Very high precision resistors have a tolerance better than 0.5%, high precision resistors 0.5 to 2%, medium precision resistors 5 to 10%, and low precision resistors have a tolerance of 20%.

Power Rating: This is the maximum power, measured in watts or fractions of a watt, that a resistor can dissipate continuously at a temperature of 70°C (158°F). Above 70°C the nominal power rating must be reduced according to a derating factor which is specified by the manufacturer. The power dissipated by a resistor in a dc circuit is $VI = I^2R = V^2/R$. The actual power dissipated is often much less than the power rating. Most common resistors have a power rating of ½, ¼ or ⅛ watt.

Temperature Coefficient (TCR): TCR is the measure of the change in resistance value with temperature of a resistor. It is usually expressed in parts per million of the nominal value per degree Celsius (ppm/°C). The TCR of the most commonly used resistors is in the range 25 to 500 ppm/°C.

Maximum Voltage (dc): This, sometimes called *Rated Continuous Working Voltage* (RCWV), is the maximum dc voltage which can be safely applied to a resistor for an indefinite period of time. The RCWV of most resistors of resistance 100 ohms or more is at least 1,000 volts.

Parasitic Effects: A resistor can be considered as a perfect resistor with an inductor in series and a capacitor in parallel to both the resistor and the inductor. The value of parasitic inductance and capacitance depends upon the structure and dimensions of the resistor. Usually circuit engineers are more concerned about inductance than capacitance although, in modern 'impedance controlled' printed board assemblies, both must be evaluated and controlled.

Stability or Drift: Stability, or time stability, is a measure of how much the value of resistance of a resistor changes in time under load. It is normally expressed as a percentage change after 1,000 hours' operation at 70°C.

Noise: This is a measure of the small fluctuations of voltage applied across a resistor at constant current. In ac it is defined as small random deviations from Ohm's law. It is due to electron movement and depends primarily on the method of construction. For a given resistor type it increases at higher values of resistance and at higher values of frequency.

3.3.3 Manufacturing Technology

The characteristics of resistors are profoundly affected by the method of manufacture. Leaving aside printed resistors produced, for example, by thick film polymer technology, see Volume 2, Chapter 12, fixed resistors can be divided into three manufacturing categories. These are composition, wire wound, and film.

Composition resistors have a controlled conductivity core comprising a mixture of a conductor such as carbon or graphite and an insulating binder. Leads extend into the core for some length to make good electrical contact, Figure 3.4. The core is encapsulated by a resin, normally by transfer moulding, to insulate it and protect it from mechanical and environmental damage.

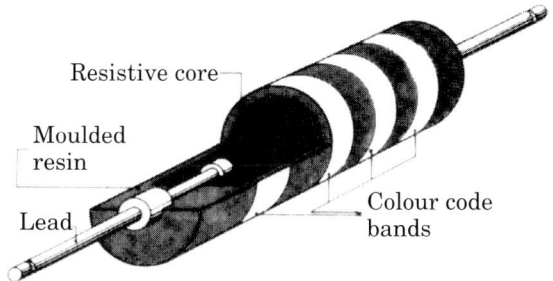

Fig. 3.4 Section of a carbon composition resistor. The nail head end of the lead holds it in position before transfer moulding of the resin.

The finished resistors have resistance values in the range of 100 ohms to 100 MΩ with normal tolerances of 5%, 10% and 20%; power ratings are in the range of ¹/₈ to 2 watts. These resistors have a high TCR of 500 ppm/°C or more and low stability but are nevertheless very popular because of their low price.

Wire wound resistors are made of a cylindrical insulating core, usually ceramic, e.g., steatite, around which a controlled value high resistance wire is wound. The body is then coated with a vitreous enamel or inserted into a high thermal conductivity case, e.g., a metal or high alumina ceramic, with a quartz or other insulating powder infill of high thermal conductivity between case and body.

Such resistors have a TCR of about 20 ppm/°C and high power ratings, up to hundreds of watts if adequate heat dissipation is provided. Tolerance is 5 to 20% for common types while tolerance for high precision wound types can be as low as 0.5%.

A wide range of resistance values is available, from a fraction of an ohm to 100 kilo-ohms. As the conductive element is wound helically, like a spring, wire wound resistors have intrinsically high parasitic inductance. This can be reduced considerably by techniques such as winding half of the wire clockwise and half counter-clockwise, or winding a doubled wire, both ends of which emerge from the same end of the core.

Film resistors, Figure 3.5, have a central cylindrical core of insulating material, e.g., glass or ceramic, with a thin film of low conductivity material on the surface. The film is produced with a wide tolerance value and each component has to be trimmed to adjust the resistance to the required value. Adjustment of such resistors is only possible upwards, i.e., by increasing the resistance value. For example, to produce resistors of value between 100 and 1000 ohms, the metallised cores could have an initial resistance between 64 and 96 ohms. This would then be trimmed to the precise value required. This is done by special machines which cut the film along a spiral. Trimming is stopped automatically when the preset resistance value is reached.

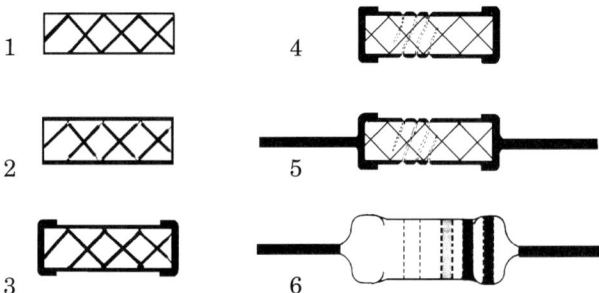

Fig. 3.5 Manufacturing steps of a film resistor. A ceramic core (1) is coated with a resistive film (2) and end caps are fitted to it (3). After trimming to adjust the resistance value (4), two leads are soldered or welded to the caps (5) and the assembly is coated with liquid resin (6). Alternatively, step (5) may be followed by a transfer moulded sealer.

These resistors have critical features: (a) the attachment of caps to the conductive film; (b) attachment of leads to caps or body; and (c) eventual cracking of the core due to maladjustment of the trimming machine. Properties of film resistors depend largely on the conductive film employed. Common materials used are:

metals: vacuum coated nickel-chromium alloy of thickness 0.1 μm or less;
metal glaze: a coating of metallic powder such as silver and palladium is fired on to the surface of the core;
tin oxide: heating the core in contact with tin oxide produces a thin conductive film;
cermets: the core is coated with a metal/glass ink,[3] such as used in the production of thick film microcircuits, and fired at high temperature;
carbon: carbon black is mixed with a resinous binder and coated on to the surface.

Film resistors are available in a wide range of resistance values, from fractions of an ohm to many megohms. Power ratings span from 1/8 watt to several watts and tolerances are 5% or better. The best types, e.g., metal film, have TCRs as low as 20 ppm/°C while cermets or carbon films have TCRs as high as 250 ppm/°C.

Resistor networks, Figure 3.6, can be included in the classification of film resistors as they are usually made by either thin or thick film hybrid microcircuit techniques, see 3.10.

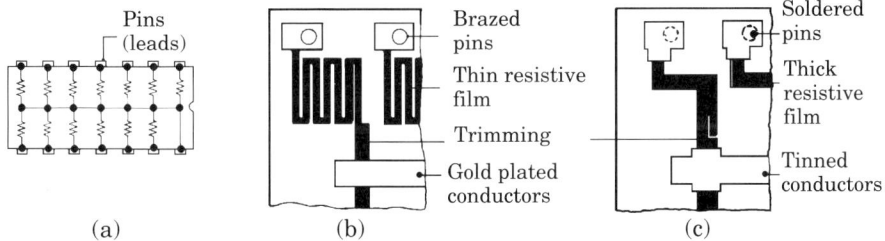

Fig. 3.6 A 13-resistor network microcircuit: (a) package outline and connections; (b) typical thin film construction of the network; and (c) typical thick film construction of the network. Value is adjusted if necessary by a cut in the film.

3.3.4 Identification

Standardised marking schemes have been devised for resistors using colour bands, shape conventions and alphanumeric designations. Colours to be used for coding and identification of resistors (and capacitors) for use in electronic equipment are given in such internationally agreed specifications as the International Electrotechnical Commission's publication IEC 425. International agreement has been reached in most cases, based largely on EIA standards issued by the Electronic Industries Association in the USA.[4] Colour coding on resistors is subject to discolouration at high dissipation temperatures and the colour bands do not adhere well to some case materials or can be attacked by solvents employed in cleaning after assembly. Nevertheless, colour banding is a very common means of marking small resistors.

The colour and position indicate the value of resistance and tolerance as shown in Figure 3.7 and Table 3.2. The two major codes use four bands and, for greater precision, five bands. A five band standard, e.g., DIN 41429 or EIA RS-196, is justified when the tolerance is 5% or better. In the four band standard the tolerance band may be omitted if the tolerance is ± 20%. In both standards one or more bands can be left off if they are the same colour as the body of the component.

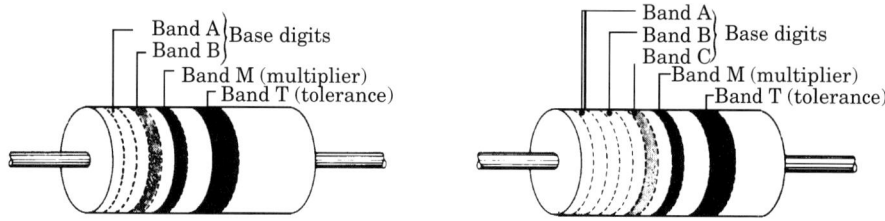

Fig. 3.7 Colour codes for axial resistors. When the components are orientated as shown, colour bands are read from left to right.

Table 3.2

Colour Coding for Axial Resistors

Colour	Bands		
	A, B & C	M	T
Black	0	0	-
Brown	1	1	1%
Red	2	2	2%
Orange	3	3	-
Yellow	4	4	-
Green	5	5	0.5%
Blue	6	6	-
Violet	7	7	-
Grey	8	8	-
White	9	9	-
Gold	-	- 1	5%
Silver	-	- 2	10%
No colour	*	*	20%

*Same colour as the resistor body

The following examples illustrate the use of Figure 3.7 and Table 3.2:

Four Band Standard

The first band, band A, is the one closest to a lead. If the bands are not shifted towards one lead, the last band, band T, must be broader and spaced further apart than the others. The bands must be read in the order A to T, orientating band A on the left hand side.

The value of resistance in ohms is given by the first two bands, A and B, multiplied by the power of 10 indicated by the third band M. The fourth band gives the tolerance. It should be noted that the use of the four band standard necessitates rounding off to two significant digits so that 97.6 (Table 3.1) becomes 98.

If A, B, M and T are the values from Table 3.2, the resistance value is given by the formula:

$$(10A + B)10^m \pm T\%$$

Examples:

White-grey-orange-red $= 98 \times 10^3 \pm 2 = 98 \text{ k}\Omega \pm 2\%$

Yellow-green-brown $= 45 \times 10^1 = 450 \text{ }\Omega \pm 20\%$

Brown-black-green-silver $= 10 \times 10^5 \pm 10 = 1 \text{ M}\Omega \pm 10\%$

Five Band Standard

After having orientated the resistor as described, three significant digits are read, bands A, B and C, instead of two, plus the multiplier M and the tolerance T.

The value of resistance is given by the formula:

$$(100A + 10B + C)10^m \pm T\%$$

Examples:

White-violet-blue-red-red $= 976 \times 10^2 \pm 2 = 97.6 \text{ k}\Omega \pm 2\%$

Yellow-green-orange-black-gold $= 453 \times 10^0 \pm 5 = 453\Omega \pm 5\%$

Brown-black-black-yellow-green $= 100 \times 10^4 \pm 0.5 = 1 \text{ M}\Omega \pm 0.5\%$

Red-grey-black-silver-brown $= 280 \times 10^{-2} \pm 1 = 2.8 \text{ }\Omega \pm 1\%$

Large resistors are often coded alphanumerically. These abbreviated markings are easily translated, for example:

7.5 K - 0.5 W -5% $= 7.5 \text{ k}\Omega \pm 5\%, \frac{1}{2}$ watt

3.4M/10/1 $= 3.4 \text{ M}\Omega \pm 10\%$, 1 watt

In the second example ambiguity may exist between tolerance and power rating. However, a 10 watt resistor is much larger than a 1 watt resistor and comparison with other resistors of the same type removes any doubt.

Colour coded resistors have no indication of power rating (except with specific codes). However, for a class of resistors of a given supplier, power rating increases with increase in dimensions independently of the resistance value. For example, metal film resistors from one supplier have a maximum body length L and a maximum body diameter D which change with power rating as follows: for $1/8$ watt L = 0.20 in. and D = 0.10 in.; for $1/4$ watt L = 0.30 in. and D = 0.125 in.; for $1/2$ watt L = 0.40 in. and D = 0.156 in.

Reference must be made to specifications and suppliers' data sheets for all other characteristics.

3.3.5 Variable Resistors

These are available in a wide variety of styles. The conductive element can be a wound metal wire, a carbon or cermet film, or (for the least expensive types) a metal coated plastic or a conductive plastic. These components have three leads, the third being connected to a wiper which contacts the conductive element at a point which can be adjusted by a suitable control.

While for fixed resistors the two leads are freely interchangeable, on a variable resistor the lead connected to the wiper cannot be exchanged with either of the other two.

3.4 CAPACITORS

A capacitor is a passive component which is used to store electrical charges. It is employed to block direct current without blocking an alternating current signal, as a filter, to accumulate energy, and so on. The simplest capacitor has two facing, parallel, plates, the *electrodes,* separated by an insulating material, the *dielectric.* A lead is connected to each plate.

The charge Q in coulombs which can be stored in a capacitor is related to the voltage V across it by the formula

$$Q = CV$$

where C is the *capacitance* of the capacitor. The value of C is directly proportional to the facing area of the electrodes and to the relative dielectric constant of the dielectric, and it is inversely proportional to the thickness of the dielectric, i.e., the distance between the electrodes.

Capacitors can be fixed or variable, the latter allowing adjustment of capacitance by, for example, a tuning control. Electrical symbols for capacitors are shown in Figure 3.8. In the following text only fixed capacitors are discussed unless otherwise stated.

3.4.1 Packages

Capacitors are available in a bewildering variety of packages, differing in both shape and dimensions. There are a multitude of national, European and international specifications for capacitors, particularly for governing dimensions and major characteristics for use with printed circuits, but overall standardisation is lacking and each major supplier lists many types in his catalogue. Some of the more common shapes are shown in Figures 3.9 and 3.10.

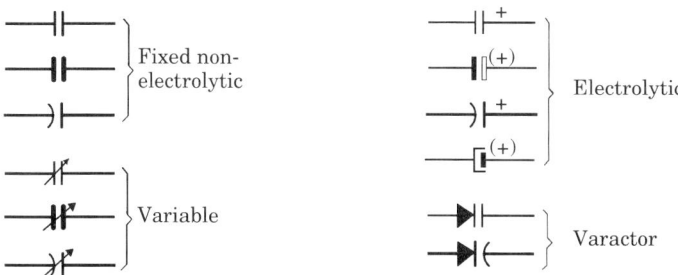

Fig. 3.8 Symbols for capacitors. The polarity mark '+' may be omitted where it is shown in brackets.

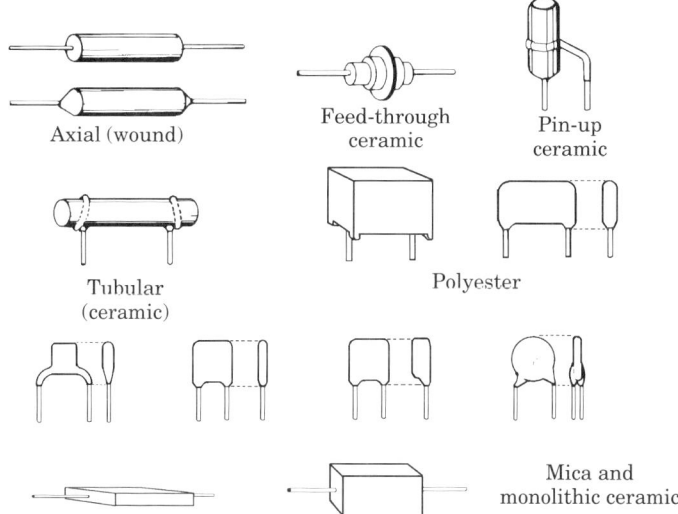

Fig. 3.9 Some typical packages of fixed non-electrolytic capacitors. There is an enormous variety of shapes and sizes.

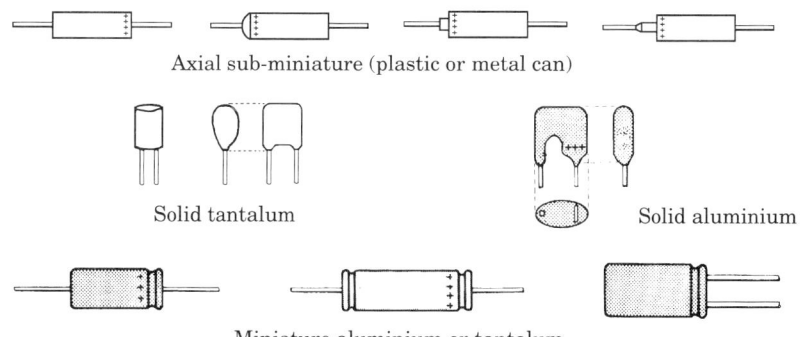

Fig. 3.10 Typical packages for small electrolytic capacitors. The polarity mark '+' may be omitted on non-symmetrical packages. Metal cans may be insulated by a heat-shrunk sleeve.

3.4.2 Characteristics

The parameters which most commonly restrict interchangeability are:

Capacitance: This is the nominal value of C. A capacitor has a capacitance of one farad (F) if its potential is raised one volt when it receives a charge of one coulomb. The farad is too large a unit for convenience and sub-multiples are usually employed. The most common sub-multiples are the microfarad or $\mu F = 10^{-6}$ F and the picofarad or $pF = 10^{-12}$ F. Less often used units are the millifarad or $1\ mF = 10^{-3}$ F and the nanofarad or $nf = 10^{-9}$. The nanofarad is sometimes called a millimicrofarad (mμF) and the picofarad a micromicrofarad ($\mu\mu$F). Capacitance is temperature dependent and, if not otherwise specified, is measured at 25°C (77°F).

Tolerance: This is the maximum allowable deviation of the actual value from the nominal value.

Working Voltage: This is the maximum voltage which can be applied continuously across the capacitor. The temperature of test and the type of voltage, ac or dc, must be stated. The value is normally quoted as Vdc or Vac.

Breakdown Voltage: This is the minimum voltage which causes permanent damage to the dielectric. It is usually at least twice the working voltage.

Temperature Coefficient: This gives the change in capacitance value with change in temperature. It is quoted in parts per million per degree Celsius (ppm/°C).

DC Leakage/Insulation Resistance (IR): All dielectrics used in the manufacture of capacitors will allow some dc flow. This current is the leakage current and at specified measurement voltage and temperature can be expressed as the resistance in the capacitor. For small paper, film, mica and ceramic capacitors the IR is about 10^5 MΩ. For electrolytic capacitors, which have very high capacitance and very thin dielectric leading to a low relative resistance, dc leakage current is normally specified. As resistance is inversely proportional to temperature, the converse is true for leakage current.

Loss Angle: If a sinusoidal ac voltage is applied across a perfect capacitor, the current flowing leads the voltage by 90°. In practice, the current leads the voltage by a lesser phase angle, the complement of which is called the loss angle, ∂ (delta). The *power factor* is a measure of the loss in a capacitor as is the *dissipation factor* and either can be used depending on the type of capacitor. These are related to the loss angle thus

$$\text{Power factor} = \sin \partial$$

$$\text{Dissipation factor} = \tan \partial$$

For small values of ∂ the tangent and the sine are effectively equal, which has led to some interchangeability of the terms in the industry.

3.4.3 Manufacturing Technology

The need to obtain high values of capacitance in small volumes and to control other parameters has enforced the use of many materials, both for electrodes and for dielectrics. An outline of the technology of the most common non-electrolytic capacitors follows. (Electrolytic capacitors are described in 3.4.4.)

Aluminium-paper

The electrodes are two long strips of thin aluminium foil; two strips of waxed paper are used, one as the dielectric and one to insulate each turn when the sandwich of the four strips is rolled to a compact diameter. After lead attachment, the roll is packaged in a can, usually aluminium, Figure 3.11, or potted, dipped, or moulded in a resin, dependent on the finished product requirement.

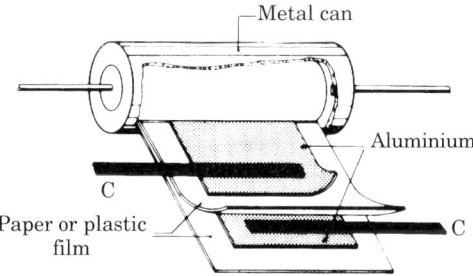

Fig. 3.11 Structure of an aluminium-paper or plastic film capacitor. Two metallic strips 'C' connect the capacitor plates to the leads.

These capacitors still find use because of low cost, but are being replaced by plastic-film capacitors which have better capacity-to-volume ratios (often termed *volumetric efficiency*) and significantly higher resistance values than paper.

Plastic Film

These are produced in similar fashion to the aluminium-paper types. The paper is replaced by a very thin plastic film which increases the volumetric efficiency to the order of 0.1-1 μF/cc^3. Miniature and ultra-miniature capacitors have a very high volumetric efficiency and are about the size of a one watt resistor.

Commonly used plastic film materials are polystyrene, polycarbonate, and polyester resins. Polystyrene capacitors are cheap and have good stability but their use is confined to below 85°C (185°F). Polycarbonate capacitors have excellent low dielectric loss and can be used up to 140°C (285°F). Polyester capacitors have electrical properties similar to polycarbonate but are cheaper and only withstand 125°C (255°F).

Metallised Dielectric Capacitors

Metals, usually zinc or aluminium, can be vapour deposited in vacuum on to paper or plastics films to give a layer about 1000 Å thick. (1 Å = 1 Ångström unit = 10^{-10} metres.) The most common combination is aluminised Mylar (Du Pont Electronics) or Melinex (ICI). Two strips of the metallised film are then rolled in an extended-foil configuration and the rolled ends sprayed with finely divided molten metal to which the capacitor leads are soldered or welded. This construction gives high volumetric efficiency and self-healing capability. Self-healing is a property where a momentary short, because of pinholes or high dielectric voltage stress, is eliminated by burning away of the metal at the short.

Volumetric efficiency has limitations. At 200 V a metallised capacitor is 0.75 of the volume of a metal foil capacitor, i.e., the volume factor is 0.75. At 600 V the volume factor increases to 0.8 and above 600 V the metallised capacitor offers little size advantage. However, if the voltage is limited to about 125 V a metallised polyester film capacitor can have a capacitance-to-volume ratio which is 2 to 4 times that of a metal foil/polyester film capacitor.

This method of fabrication, shown in Figure 3.12, gives considerable scope for mechanisation of the process and in consequence allows relatively low prices.

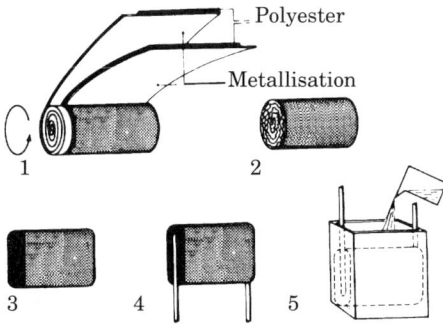

Fig. 3.12 Manufacturing steps for a metallised polyester capacitor: (1) two strips of metallised plastic film are wound together; (2) the roll is pressed at both ends to bend the overhanging metal film and then laterally to make it oval; (3) both ends are coated with a tin alloy at high temperature; (4) leads are electrically welded to the ends; (5) the assembly is placed in a plastic can which is then filled with liquid resin. Alternatively, the roll is not pressed laterally and axial leads are welded to make an axial package.

Mica and Silver-mica

An excellent dielectric for capacitors, mica has been used for many years. Formerly mica capacitors were made by stacking alternate layers of mica and tin-lead alloy foils. The conductive foils were stacked off-centre so that they could be soldered to two leads, alternately on either side of the stack, to produce an interdigitated structure.

Modern mica capacitors have a similar structure but the electrodes are produced by metallising the mica with silver or another (cheaper) metal. The stack is then fired to produce a compact or monolithic structure. Two sides of the stack are metallised for lead attachment, Figure 3.13.

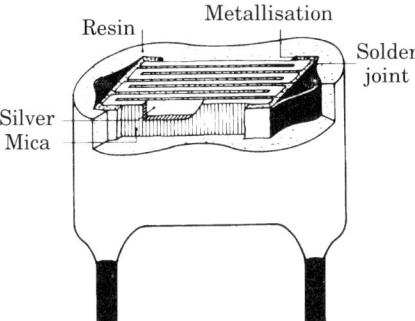

Fig. 3.13 Section of a silver-mica capacitor. The interdigitated structure of plates provides a high specific capacitance. Silver may be replaced by cheaper metals.

Ceramics

Ceramic capacitors have a ceramic dielectric, usually in the form of a tube, disc or other shape, which is coated with a thin metallic film to make the electrodes, Figure 3.14. As with mica an interdigitated structure may be used to increase the capacitance/volume.

Some types of ceramics reduce the TCE of the capacitor to almost zero. Small unpackaged ceramic capacitors, known as *chip capacitors,* are available for hybrid microcircuits and SMT assemblies, see Chapter 4, 4.3.4.4 and 4.3.4.5.

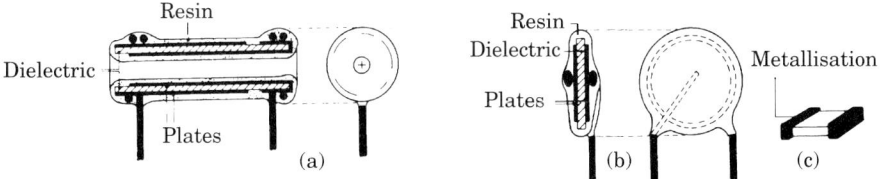

Fig. 3.14 Three typical constructions for ceramic capacitors: (a) cylindrical with radial leads; (b) disc type; and (c) chip capacitor, with no lead and metallic end coating, to be assembled on hybrid microcircuits.

Glass

Glass capacitors are made up from stacks of thin gold or aluminium foils with glass wafer dielectrics in similar fashion to monolithic mica capacitors.

Stability is excellent (as for mica capacitors) and, in addition, they tolerate high humidity and temperature.

3.4.4 Electrolytic Capacitors

Many low frequency filtering, coupling, decoupling and some bypass applications require high capacitance values. Electrolytic capacitors are unique in having a high capacitance x voltage, CV, to volume ratio and cost least per microfarad of any capacitor type in use. Electrolytic capacitors differ markedly from types previously discussed in construction and operating parameters. The high CV to volume ratio results from using as dielectric a thin layer of oxide grown directly on the surface of one electrode, Figure 3.15. Such capacitors are polarised and must be connected so that the anode, the oxide-coated electrode, is positive with respect to the cathode, which has no oxide coating. Semipolarised types with a thinner oxide coating on the cathode to minimise the effects of reverse voltage must be similarly connected. Non-polarised types have equal thicknesses of oxide on both electrodes and may be connected in either direction.

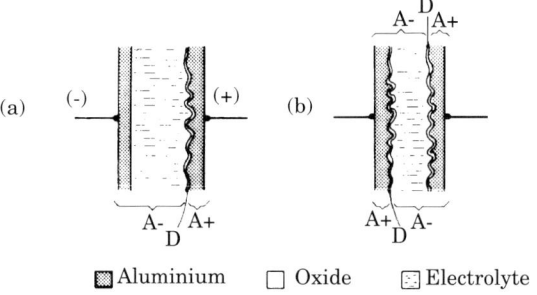

Fig. 3.15 Aluminium electrolytic capacitor. Diagram (a) refers to a polarised type: the positive plate 'A+' is a roughened aluminium foil, the negative one 'A-' is the electrolyte and the other foil; a thin layer of oxide 'D' is the dielectric. Diagram (b) refers to a non-polarised type: both plates are roughened and oxide coated. Positive plate, negative plate and dielectric may be as shown at the top or bottom of the diagram.

Polarised capacitors are the most commonly used in electronic circuits. They are mostly used in dc circuits where the applied voltage is unidirectional with a specified maximum ac component. Semipolarised capacitors are used where a specified dc potential, lower than the rated voltage, may be applied in the reverse direction for extended periods of time. Non-polarised capacitors are used in certain dc and ac circuits.

Two metals have been found most suitable for electrolytic capacitor manufacture, aluminium and tantalum. Both form thin, about 0.1 μm, adherent oxide layers having low porosity which is readily sealed. Other metals, notably niobium, titanium and zirconium, can be anodised to form dielectric films but are of limited use.

The earliest form of electrolytic capacitor is illustrated in Figure 3.16. It is constructed from a rolled sandwich, similar to the aluminium-paper capacitor, except that the paper is unwaxed and one of the metal foils is anodically oxidised. The sandwich is immersed in a liquid electrolyte contained in a hermetically sealed can fitted with a pressure relief valve. Typical electrolytes are aqueous solutions of ammonium borate, or ethylene glycol/boric acid/ammonium nitrate, and various proprietary solutions. The paper becomes saturated with the electrolyte and, with the un-oxidised metal foil, becomes the cathode; the oxidised foil is the anode and its oxide layer the dielectric. To increase surface area and hence capacitance, the anode foil is usually roughened before anodising.

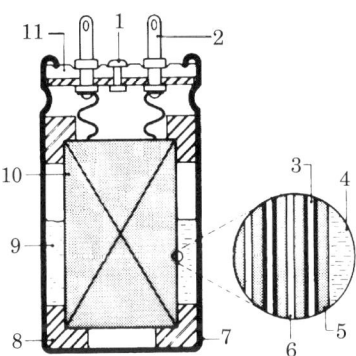

Fig. 3.16 Vertical section of a wet aluminium electrolytic capacitor. 1 — safety pressure release valve; 2 — terminal; 3 — positive plate (oxidised); 4 — liquid electrolyte; 5 — paper (impregnated with liquid); 6 — negative plate; 7 — aluminium cold drawn can; 8 — spacer; 9 — excess electrolyte; 10 — rolled sandwich; 11 — rubber gasket.

The presence of a liquid meant that the capacitor had to be kept upright and new electrolytes were introduced in the form of gels and pastes. Such electrolytics are termed *dry*. In appearance they resemble the aluminium-paper non-electrolytics.

Tantalum electrolytics require a more advanced technology due to the difficulty of working the metal. Tantalum metal is produced in powder form which is sintered and vacuum melted to give a more compact metal which can be cold rolled. It cannot be soldered or welded except by special techniques. Tantalum capacitors can be divided into three types, foil, wet anode and solid anode. Foil tantalums use solid tantalum foil; wet and solid anode types use sintered tantalum.

While tantalum foil capacitors are similar to aluminium capacitors, the lower thickness of tantalum oxide gives a higher capacitance to volume ratio and a lower maximum working voltage. (Tantalum capacitors have a maximum dc working voltage of about 150 volts, for all types, while aluminium electrolytics have a maximum dc working voltage of about 600 volts.) The foil tantalum electrolytics can be made with wet, dry or paste type electrolytes. The problem of sealing in the liquid electrolyte is solved by using high temperature glass-to-metal seals and, since tantalum capacitors are usually very small, the can may be hermetically sealed without provision for pressure relief. The components can therefore operate in any position.

Wet sintered anode tantalum capacitors are illustrated in Figure 3.17. The anode is a small sintered pellet which is oxidised resulting in a very high surface-to-volume ratio. A length of solid tantalum wire in contact with the sintered pellet is electrically welded to the external anode lead. The liquid electrolyte is the cathode and the metal can in contact with the electrolyte is silver lined. The complexity results in a high price but the capacitance to volume ratio is very high. A 50 microfarad capacitor (at a working voltage of 25 volts) is available in a can only 0.15 inch (3.81 mm) diameter and 0.4 inch (10.16 mm) long.

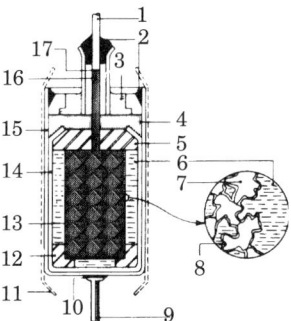

Fig. 3.17 Section of a wet sintered anode tantalum capacitor. Despite its complexity, its diameter may be as low as 9/64 in. (3.6 mm). 1 – tinned-nickel lead (positive); 2 – solder seals; 3 – glass seal; 4 – cast resin; 5 – spacer; 6 – liquid electrolyte; 7 – tantalum particle (enlarged); 8 – tantalum oxide film; 9 – tinned-nickel lead (negative); 10 – electrical weld; 11 – insulating sleeve (optional); 12 – spacer; 13 – sintered tantalum pellet (anode); 14 – silver can (internal); 15 – tinned-brass can (external); 16 – solid tantalum leads; 17 – electrical weld.

Solid sintered anode tantalum capacitors have a similar pellet anode but the voids between the sintered particles are filled with a solid electrolyte, for example, manganese dioxide. The pellet is then covered with a graphite layer to make good electrical contact over a large area. The anode lead is a solid tantalum wire which is welded to the external lead. The cathode contacts the metal can, which may be zinc, through the graphite layer. If a non-metallic pack is used instead of a metal can, connection to the graphite layer may be made by a tinned copper lead.

The use of a dry electrolyte eliminates all sealing problems. These capacitors are available in plastics (dip coated) packages as well as metal in a variety of shapes such as disc, drop, and so on.

On polarised and semipolarised capacitors the positive lead (anode) must be indicated by a '+' mark. A '-' mark *may* be seen close to the negative electrode (cathode). Connection of polarised capacitors to a reverse voltage can cause permanent damage to the oxide layer by electrolytic reactions. Much heat is generated and gas may be produced. In the worst case the can will explode. The same effects can be produced if the maximum rated working voltage is grossly exceeded or by overheating.

3.4.5 Identification

Capacitors are often marked with full details of capacitance, tolerance and working voltage; for electrolytic types polarity and working temperature range are usually added.

On small packages markings can be abbreviated and require decoding. Common markings are of the type '1.5u/10/25DC', which means '1.5 microfarad ± 10%, 25 volts dc' or '10K/5/25V-' which translates to '10,000 pF ± 5%, 25 volts dc'. For graphic reasons microfarad (μF) is often written 'UF' or 'uF' or, quite improperly, 'M'; picofarad can be abbreviated to 'p' or omitted.

Additionally, alphanumeric coding is often used for both military and industrial style capacitors. Designations such as 'DNS 226 D 050 M R' and 'CL66 B J 1R7 K P G' indicate capacitance and other characteristics such as type, tolerance, working voltage, temperature coefficient and package outline. It is necessary to refer to the manufacturer's catalogue to decipher them. Close attention must be given to capacitance value. In the examples cited it is indicated by '226' and by '1R7'. In the first instance, the third digit is the power of ten of a multiplier and 226 means $22 \times 10^6 = 22 \times 10^6$ pF = $22 \mu F$. Thus 221 means 220 pF, 222 means 2,200 pF, and so on; however, 229 means 2.2 pF because the digit 9 means 10^{-1} and not 10^9.

In the second example, which is relevant to the same coding, the capacitance value is given in microfarad and the letter 'R' is used as a decimal point so that 1R7 means 1.7 μF.

Colour coding is also employed but there are too many standards, both for the number of colour bands and their meanings, to give details here. As an example, there are at least five standards for mica capacitors: (a) nine dots (6 + 3 on the two sides); (b) six dots; (c) five dots (on one side); (d) five dots (3 + 2); and (e) three dots. It is necessary to refer to the manufacturer's catalogue or to the standard referenced in it.

3.4.6 Variable Capacitors

Small variable capacitors are frequently called *trimmer* or *tuning capacitors*. Adjustment of the capacitance value is usually achieved by changing the facing area of the electrodes. Most common types have two sets of metal plates, one fixed and one moving, so that the moving set can penetrate into the fixed set to a controlled depth without any plate of the moving set contacting a plate of the fixed set. All plates in a set are electrically connected to one another. Air is often used as the dielectric but in some types the plates are very thin and a plastics film is interposed between each pair of plates to prevent contact.

The *varactor* is an unusual type of variable capacitor. Capacitance depends upon the voltage across the terminals and decreases, in non-linear fashion, when the voltage increases. A varactor is basically a reverse biased diode, see 3.5 . The electrodes are the *p* and *n* regions while the depletion layer is the dielectric. As voltage increases the depletion layer increases in thickness and the capacitance decreases in value.

3.5 SEMICONDUCTOR DIODES

A diode is an active component through which current flows more easily in one direction than in the other. The resistance which it opposes to the current flow is dependent on its direction and changes as a complex function of current intensity.

The characteristic of a diode is normally defined by a diagram which shows current intensity as a function of the applied voltage.

The two leads of a semiconductor diode are called the anode and the cathode, based on the analogy of the vacuum diode (the so-called thermionic tube or valve).

The current flows more easily when the anode is positive with respect to the cathode (*forward bias*) than when the cathode is positive with respect to the anode (*reverse bias* or *back bias*). For this reason it is usual to draw the I_R - V_R characteristic, Figure 3.18, to a different scale from that of the I_F - V_F. Subscript R means reverse bias, subscript F means forward bias. These symbols are standard, see JEDEC publication 77A. JEDEC stands for the Joint Electron Device Engineering Council which operates within the USA Electronic Industries Association.

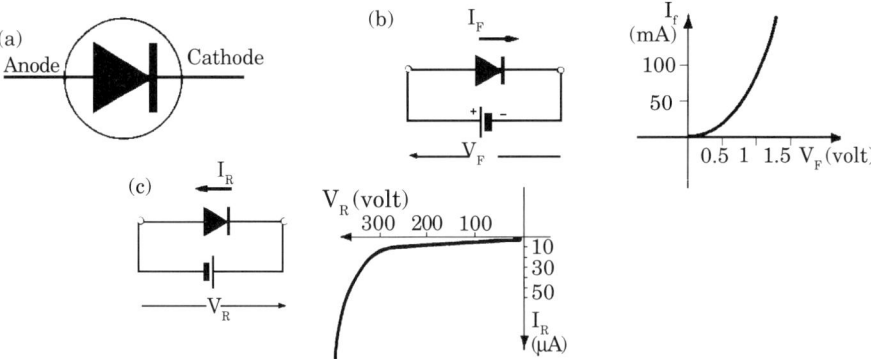

Fig. 3.18 Diode (a) symbol (the circle may be omitted); (b) voltage/current diagram when forward biased; and (c) voltage/current diagram when reverse biased (note the different scale).

3.5.1 Packages

Low and medium power diodes are usually available in axial packages; high power diodes are obtainable in a large variety of packages of many shapes and sizes. Many packages incorporate screw threads for mounting on to a PCB or a heat sink.

The JEDEC standard designation for diode packages has the format DO-xx, where xx stands for 1 to 3 digits plus, if needed, 1 or 2 letters. For semiconductor package outlines, reference can be made to JEDEC publication 12-F 'Registered Outlines and Gauges for Semiconductor Devices'. The British Standards Institution has BS 3934, equivalent to IEC publication 191-1 etc. In addition to standard packages (JEDEC, IEC and National standards) there are a large number of non-standard packages of many shapes and sizes. Many of these are similar to the standards.

Suppliers' data sheets normally specify a standard outline for the component or include a dimensioned drawing. However, despite statements such as 'all JEDEC DO-xx dimensions and notes apply', or 'in accordance with JEDEC DO-xx', or 'similar to IEC 191-xx', some suppliers' data sheet drawings differ from the referenced standard's outline dimensions. All dimensions quoted must always be checked against the best available information before placing reliance on them. The most used packages are DO-4, DO-5, DO-7, DO-8, DO-14 and DO-35.

Diodes can also be provided in three lead packages like transistors, see 3.6.1, or in two lead versions of the three lead ones. Diode arrays or networks, a step on the way to total integration, are available in packages derived for integrated circuits, see 3.9.4. Chip diodes (unpackaged) are available for surface mounting, e.g., for hybrid circuit manufacture.

It must be remembered that standards for packages refer only to the package outline and each manufacturer can adapt one of these to his needs within the specified outline. Threads, where provided, may be non-metric. Metric equivalents are given in Table 3.3.

3.5.2 Characteristics

The most important characteristics of diodes depend upon the application. For further detail on diode characteristics see, for example, European specification CECC 50 000.[5] According to this, diodes can be divided into many classes, the most common of which are now discussed. Common symbols and typical current/voltage diagrams are given in Figure 3.19.

Signal Diodes

These are general purpose diodes, suitable for applications involving low currents and a wide range of voltage, from a few volts to 50 kV. Switching diodes change in a very short time from a conducting to a non-conducting state (and *vice versa*) when voltage reverses. Basic symbols and definitions for these are given in Table 3.4.

Rectifiers

Similar to signal diodes, these are more suitable for high currents. The maximum voltage can be as high as 200 kV but, as the maximum voltage increases, so the maximum current decreases. See also 3.5.5.

Video Detectors

These are signal diodes which are suited to the detection of video signals in television sets. They are specially classed because of the unusual application.

Schottky Diodes

These differ from previous types in passing a larger current when forward biased at a given voltage. Their high speed recovery suits them to uhf switches and detectors.

Table 3.3

Machine Screws Most Frequently Used in Electronics (non-metric screws)

Gauge	Maximum Diameter (in.)	Maximum Diameter (mm)	Minimum Diameter (in.) UNF	Minimum Diameter (in.) UNC	Minimum Diameter (mm) UNF	Minimum Diameter (mm) UNC	Thread Threads per inch UNF	Thread Threads per inch UNC	Thread Pitch (mm) UNF	Thread Pitch (mm) UNC	Suggested Clearance Hole Diameter (in.)	Suggested Clearance Hole Diameter (mm)
1	0.0730	1.854	0.0550	0.0527	1.397	1.338	72	64	0.353	0.397	0.0781 ± 0.0081	1.984 ± 0.206
2	0.0860	2.184	0.0657	0.0628	1.669	1.595	64	56	0.397	0.454	0.0935 ± 0.0075	2.375 ± 0.190
3	0.0990	2.515	0.0758	0.0719	1.925	1.826	56	48	0.454	0.529	0.1040 ± 0.0050	2.642 ± 0.127
4	0.1120	2.845	0.0849	0.0795	2.156	2.019	48	40	0.529	0.635	0.1200 ± 0.0080	3.048 ± 0.203
5	0.1250	3.175	0.0955	0.0925	2.426	2.349	44	40	0.577	0.635	0.1360 ± 0.0110	3.454 ± 0.279
6	0.1380	3.505	0.1055	0.0974	2.680	2.474	40	32	0.635	0.794	0.1440 ± 0.0060	3.658 ± 0.152
8	0.1640	4.166	0.1279	0.1234	3.249	3.134	36	32	0.706	0.794	0.1730 ± 0.0090	4.394 ± 0.229
10	0.1900	4.826	0.1494	0.1359	3.795	3.452	32	24	0.794	1.058	0.1990 ± 0.0090	5.055 ± 0.229
12	0.2160	5.486	0.1696	0.1619	4.308	4.112	28	24	0.907	1.058	0.2280 ± 0.0120	5.791 ± 0.305
1/4"	0.2500	6.350	0.2036	0.1850	5.171	4.699	28	20	0.907	1.270	0.2610 ± 0.0110	6.629 ± 0.279
5/16"	0.3125	7.937	0.2584	0.2403	6.563	6.104	24	18	1.058	1.411	0.3281 ± 0.0156	8.334 ± 0.396
3/8"	0.3750	9.525	0.3209	0.2938	8.151	7.463	24	16	1.058	1.587	0.3906 ± 0.0156	9.921 ± 0.396
1/2"	0.5000	12.700	0.4351	0.4001	11.052	10.166	20	13	1.270	1.954	0.5156 ± 0.0156	13.096 ± 0.396

Example: 10-32-UNF-2A is read as follows: 10 - gauge number; 32 - threads per inch; UNF - Unified Fine (UNC - Unified Coarse); 2 - class of fit; A - external thread (B - internal thread).

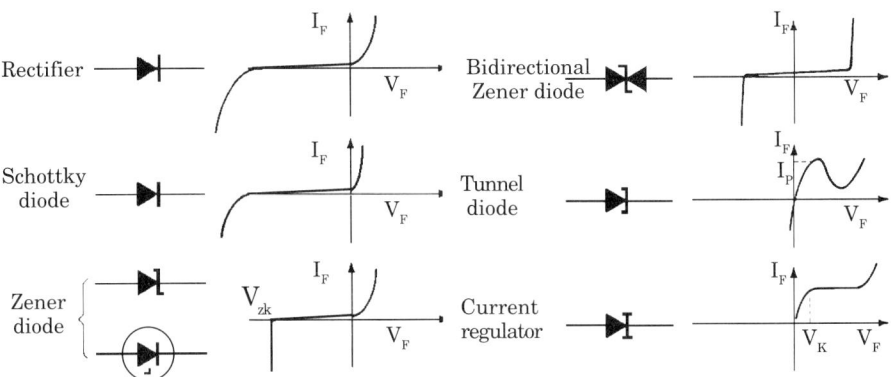

Fig. 3.19 Common symbols and typical current/voltage diagrams for common types of diode. The symbol may be enclosed in a circle.

Zener Diodes

The Zener diode is unique in being designed specifically to work in a reverse biased condition. While strictly speaking the term Zener breakdown is applied to a breakdown mechanism characterised by a negative temperature coefficient, i.e., the breakdown voltage decreases as the *pn*-junction temperature increases, in practice all diodes operated in the reverse breakdown region are called Zener diodes. When the absolute value of reverse voltage V_R exceeds a given value, a small increase of V_R causes a large increase in the reverse current due to the Zener effect.

The Zener effect is obtained by designing a junction with an extremely narrow depletion region and occurs when the electrical field strength across the junction becomes 10^6 volt/cm or more. Breakdown on reverse bias occurs with all diodes but is usually destructive while for Zener diodes it is largely reversible.

Zeners are principally used for voltage regulation, Figure 3.20, or as reference diodes. Terms characteristic of Zener diodes are given in Table 3.5.

Double-anode Regulator

The user can view this device as equivalent to two opposed diodes, the anode of each being connected to the cathode of the other. In fact it consists of two diodes with single cathodes and a common anode. Such devices are typically used for clipping or chopping ac waveforms.

Tunnel Diodes

The characteristic of the tunnel diode in the forward bias region shows a length of the I_F-V_F curve in which current decreases with increase in voltage as if the device has negative resistance. This is due to a construction similar to that of Zener diodes, that is an extremely narrow depletion layer. This permits minority carriers to flow or 'tunnel' through the barrier. Tunnel

Table 3.4

Most Common Abbreviations, Terms and Definitions for
Semiconductor Rectifier Diodes
(Based upon JEDEC Publication 77-A)

Abbreviation	Term	Definition (short)
V_F	dc forward voltage	Voltage (dc) across a directly biased diode (anode positive with respect to the cathode)
I_F	dc forward current	Current flowing in a diode when directly biased by V_F
V_R	dc reverse voltage	Voltage (dc) across a reverse biased diode (anode negative with respect to the cathode)
I_R	dc reverse current (leakage current)	Current flowing in a diode when reverse biased by V_R
V_{FM}	Forward peak voltage	Maximum (peak) voltage across a directly biased diode
I_{FM}	Forward peak current	Maximum (peak) current in a directly biased diode; it is measured in ac over a short time (waveform and time interval must be specified)
I_O	Average rectified forward current	Average value of the current flowing in a diode used for rectifying a 60 Hz ac current (180° conduction angle averaged over a full cycle)
P_F	Forward power dissipation	Power dissipated (dc) by a directly biased diode
P_R	Reverse power dissipation	Power dissipated (dc) by a reverse biased diode
t_{rr}	Forward recovery time	Time required for the diode current (or for the voltage across it) to switch from a reverse bias condition to a forward bias condition (current values and test circuit to be specified)

diodes, sometimes called Esaki diodes after their inventor, operate up to 50 GHz and are used in uhf and microwave oscillators, amplifiers and switches.

Current Regulators

In the forward bias region, their characteristic shows a length in which current is virtually independent of the applied voltage.

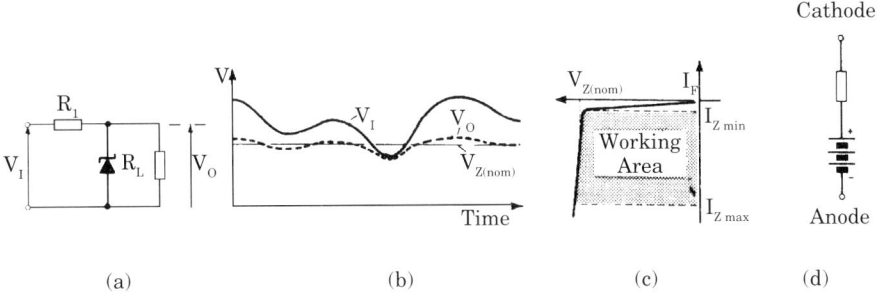

Fig. 3.20 Application of a Zener diode to voltage regulation: (a) typical circuit, where R_L is the load; (b) qualitative input voltage (V_i) and regulated output voltage (V_o); (c) working area of the diode, which is limited at the top by the knee of the curve and at the bottom by the maximum power rating; and (d) equivalent diagram for a Zener diode.

Table 3.5

Most Common Abbreviations, Terms and Definitions for Zener Diodes
(Based upon JEDEC Publication 77-A)

Abbreviation	Term	Definition (short)
V_F, I_F	—	(see Table 3.4)
V_R, I_R	—	(see Table 3.4)
I_Z	dc regulator current	Current (dc) flowing in a Zener diode when it is reverse biased and operates in its breakdown region
I_{ZK}	dc near breakdown knee current	Minimum (dc) reverse current I_R to activate the breakdown of the diode (a peculiar point of its characteristic must be specified)
I_{ZM}	dc maximum rated current	Maximum (dc) current the diode can withstand in its breakdown region; it is based upon the maximum power it can dissipate
V_Z	dc regulator voltage	Voltage across a reverse biased diode at the current I_Z (to be specified); the nominal voltage of a voltage regulator (if not specified otherwise)
V_{ZM}	dc voltage at maximum rated current	Reverse (dc) voltage applied to the diode when the current flow is I_{ZM}
Z_Z	Regulator impedance	Small signal impedance of a reverse biased diode in its breakdown region when the current flow is I_Z (to be specified)
α_{VZ}	Temperature coefficient	Temperature coefficient of the voltage regulator; usually it is given in percentage of change of V per °C

Varactor Diodes

Varactor diodes are described in 3.4.6, Variable Capacitors.

Others

Photodiodes and light activated diodes are optoelectronic devices, see 3.8. Silicon controlled rectifiers are described in 3.7.

3.5.3 Technology

Long established, but still used, contact diodes are based on the rectifying effect of a non-ohmic contact of a fine wire (of gold, tungsten, platinum alloy, etc.) on a semiconductor chip.

Junction diodes are now the most common in electronics. These are manufactured by starting from a very pure single crystal of a semiconductor material such as silicon or germanium. The crystal is cut into thin slices and then 'doped', usually by vapour diffusion, into 'p' regions and 'n' regions. The concentration of dopant atoms is normally of the order of one atom of dopant to 100 million atoms of semiconductor. When the concentrations are greatly increased the p and n regions are called p^+ and n^+ respectively.

A p-doped region is diffused with an element such as boron or aluminium. As the dopant is trivalent and germanium and silicon are quadrivalent, the p region will contain free sites or holes from which electrons are missing. These holes can be considered as positive charges which can move under an applied potential. The n-doped region is obtained by diffusing the base material with an element such as phosphorus or arsenic. The dopant in this region is pentavalent so that there is an excess of electrons, or free negative charges, which can move under the influence of an applied potential. Many diode chips are produced on one slice of the semiconductor material and are separated into the individual chips after the diffusion processes and removal of masks, etc.

The p region forms the anode of the diode; the n region forms the cathode of the diode. The boundary between the two regions is called a *p-n* junction or simply a *pn junction*. In a thin layer across the junction, the combination of free electrons and free holes creates a *depletion layer* in which the density of free charges is close to zero and the electric potential is discontinuous. If an external voltage is applied with the negative terminal connected to the p side of the junction, the depletion layer grows wider and current flow is effectively cut off; if, however, the positive terminal is connected to the p side of the junction, the forward bias causes the layer to become narrower and current flows readily across the junction. Depending on the polarity of the applied voltage, the junction either blocks or passes current, that is, it rectifies. Figure 3.21 shows this effect schematically.

The chip is assembled into a package which is selected according to the power to be dissipated, the equipment into which it will go, and the intended use. The two basic types of package are hermetically sealed packages (which include metal cans and the so-called glass diodes) and plastics packages. The metal canned devices, such as the JEDEC DO-1, DO-4, and DO-8, are strong but expensive. Glass and plastics packs are less robust and must be handled with care. The lead anchorages in plastics packs are susceptible to overstressing which, by cracking the plastic, can allow ingress of moisture which will reduce component life.

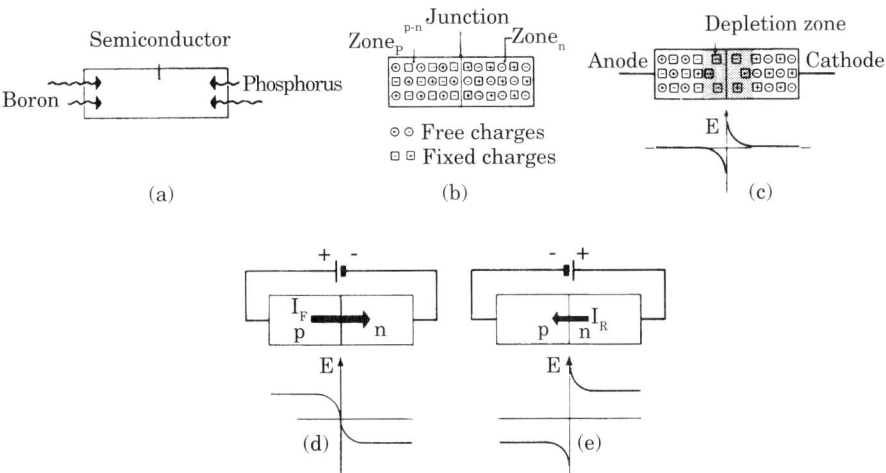

Fig. 3.21 Qualitative explanation of the *p-n* junction of a diode: (a) a junction is produced by doping the semiconductor; (b) free charges and fixed charges are generated by the dopant atoms; (c) reaction between free charges generates a depletion layer around the junction and a discontinuity of potential E in the semiconductor; (d) when the diode is biased directly, the discontinuity is reduced and disappears; and (e) when the diode is biased in reverse, the discontinuity is enhanced. This explains qualitatively why the forward current I_F is much larger than the reverse current I_R.

Plastics packaged diodes which generate a lot of heat, such as some Zener types, dissipate most of the heat by their leads. These are usually of heavier gauge than is needed for the currents used. The diode must be mounted proud of the printed board and enough lead left between the body of the device and the solder joint. Ideally, the leads should be soldered to large copper areas of the board, possibly on the component side, to improve thermal dissipation, Figure 3.22.

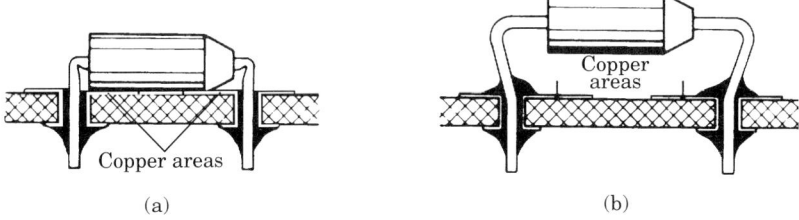

Fig. 3.22 In the assembly of plastics-packaged diodes which produce a lot of heat, care must be taken to avoid the mistakes shown in (a). First, the leads have been indented during forming and may crack and break in service. Secondly, the leads are too short and do not allow good heat dissipation. Thirdly, the component is in actual contact with the PB so that it may be damaged by the heat emitted. The ideal assembly is shown in (b). The leads are of good length and have been correctly formed with a stand-off. In addition, copper areas have been provided on the board to act as a heat sink.

3.5.4 Identification

Most diodes are marked alphanumerically by the manufacturer, the marking identifying them clearly. There are many different diodes on sale and reference to manufacturers' data sheets, or other sources of data, is necessary. Each manufacturer can identify a new diode as he wishes. If, however, he wishes to register the device with the CENELEC Electronic Components Committee (CECC) or with another generally accepted body he must follow the rules of that body for identification. For example, if the component is registered with the JEDEC then a marking such as 1NXXXX will be assigned in accordance with JEDEC standard RS-370-A where XXXX stands for 2-4 digits and, if needed, 1-2 additional letters. Devices originating from different sources will be largely interchangeable if they bear the same 1NXXXX designation. National and international bodies approving and/or registering such components publish lists of registered components with updates at regular intervals.

While there are standard colour codes for diodes, e.g., JEDEC Standard RS-236-B, they are seldom used. It can be difficult to identify the polarity of the diode from such marking. In these cases the polarity may be checked by measuring the dc resistance which is much lower when the positive probe is connected to the anode than *vice versa*.

3.5.5 Rectifiers

Diodes expressly designed for rectifying high ac currents are generally larger than those used in electronic circuits. They will often be fitted with a finned body to promote thermal dissipation. For large currents, selenium rectifiers, based on the rectifying effect of a contact between selenium and cadmium or its alloys, or copper oxide rectifiers, based on the effect of a contact between copper and copper oxide, are still used, generally in low cost entertainment applications. For most purposes, silicon rectifiers have replaced these polycrystalline materials, being the most widely used power semiconductor with a range up to some 1500 A and 1000 V_{rms}. Bridge rectifiers can be made by combining two, four or more rectifiers in a single package.

There are hundreds of different rectifying devices listed in the major suppliers' catalogues. Despite this, the majority of voltage and current combinations can be covered with approximately 150 silicon devices.

3.6 TRANSISTORS

The transistor is the basis of all solid state electronics, either as a single component or as an element of integrated circuits. It is an active device which may be employed to amplify a current or a voltage. It is a three terminal device; in the most common type, the bipolar junction transistor (BJT), the terminals are called the *base* (B), *collector* (C) and *emitter* (E). Although applications for the transistor are almost unlimited, in most it is used as an amplifier or a switch. As an amplifier, the transistor is operated in a region of its characteristics where the output is directly proportional to the input. As a switch the transistor is operated between the cut-off region, 'off' state, and the saturation region, 'on' state. There are thousands of transistor types available, with older types being withdrawn almost as quickly as new types appear.

BJTs are based on the *p-n* junction. A semiconductor chip is doped to give two *p* regions which surround a thin *n* region or two *n* regions surrounding a thin *p* region. The two types are called *p-n-p* (or PNP or pnp) and *n-p-n* (or NPN or npn) and are shown in Figure 3.23.

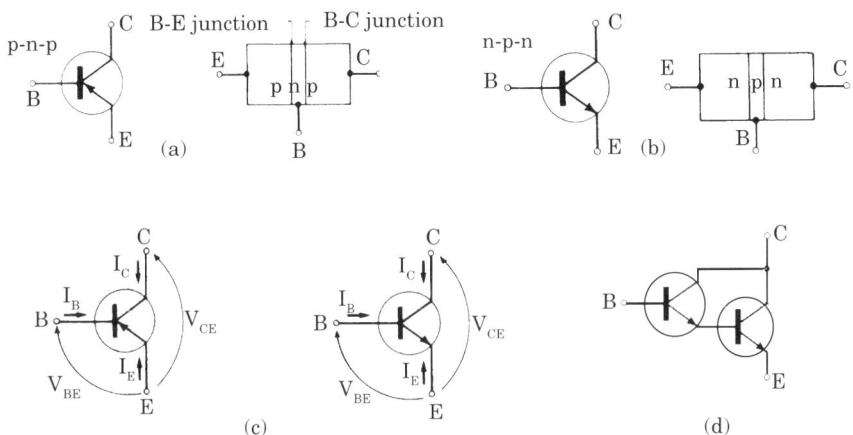

Fig. 3.23 (a) and (b) Usual symbols (the circle may be omitted) and structural diagrams for bipolar *PNP* and *NPN* transistors; (c) usual symbols for major electrical parameters; and (d) Darlington connection of two transistors.

Figure 3.24 shows the common-emitter connection in which the collector current I_C depends primarily on the base current I_B. Within a specific range, a small change of I_B causes a much larger change in I_C. This is shown by the graph of I_C as a function of VCE (collector-to-emitter voltage) for different values of I_B. This connection is very popular for *n-p-n* transistors used in amplifier and in switch modes. The *p-n-p* junction transistor behaves similarly, except for the reversal of polarities. The BJT operates because the two *p-n* junctions are very close together. With the base open, current flow is inhibited. With the base connected, the current flow between C and E is inhibited or enhanced depending on the sign and intensity of the current through the base terminal. For further explanation of this subject reference should be made to text books on basic electronics.

The field effect transistor (FET) was developed later than the power transistor and, like that device, is growing in usage in integrated form. They are unipolar devices, using charge carriers of only one polarity, in contrast to the BJT which uses both polarities of charge carriers. The FET takes its name from the fact that current flow in the device is controlled by varying an electric field. It is a three terminal device which, as a voltage amplifier, is closer to vacuum tubes than to the BJT.

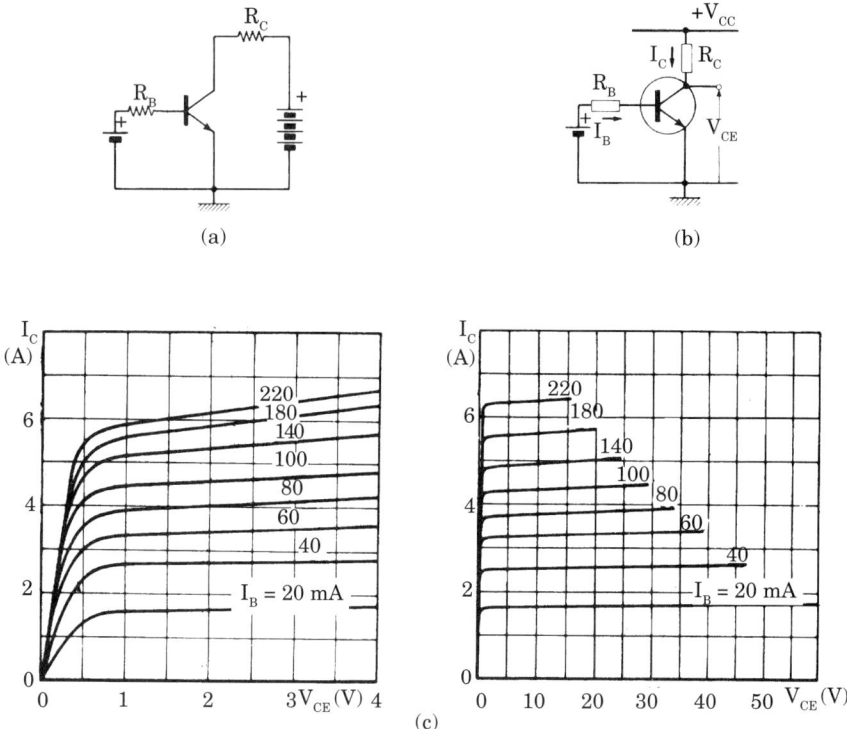

Fig. 3.24 Characteristic curves of a bipolar transistor (BU-102 by SGS-ATES); (a) and (b) usual emitter test circuits; and (c) output characteristic curve drawn with two different scales for V_{CE}.

There are two basic types of FET: the junction FET (JFET) and the insulated gate FET (IGFET). The most common insulated gate FET is the metal-oxide semiconductor FET or MOSFET. While what follows is based on silicon technology, FETs and other transistors are commonly manufactured from other materials, e.g., gallium arsenide (GaAs), when particular properties are needed.

The JFET in its simplest form starts as a 'bar' of doped silicon. The doping can give an n channel type or a p channel type as required. A terminal is connected to each end of this channel; the terminal through which current is introduced is the *source* (S) and the terminal at the other end is the *drain* (D). As shown in Figure 3.25, two areas of the opposite polarity, i.e., p type for an n channel device and n for a p channel device, are doped into the chip and connected in parallel to form the *gate* (G) terminal. (It should be noted that the schematic structure shown in Figure 3.25 is misleading in that normally a single-ended structure, diffused from one side only, is used. In this case, and in the case of MOSFETs where the construction is different, a fourth lead may be present. This is connected to the undoped substrate.) The gates are used to control the current between source and drain by changing the width of the depletion regions surrounding the junctions.

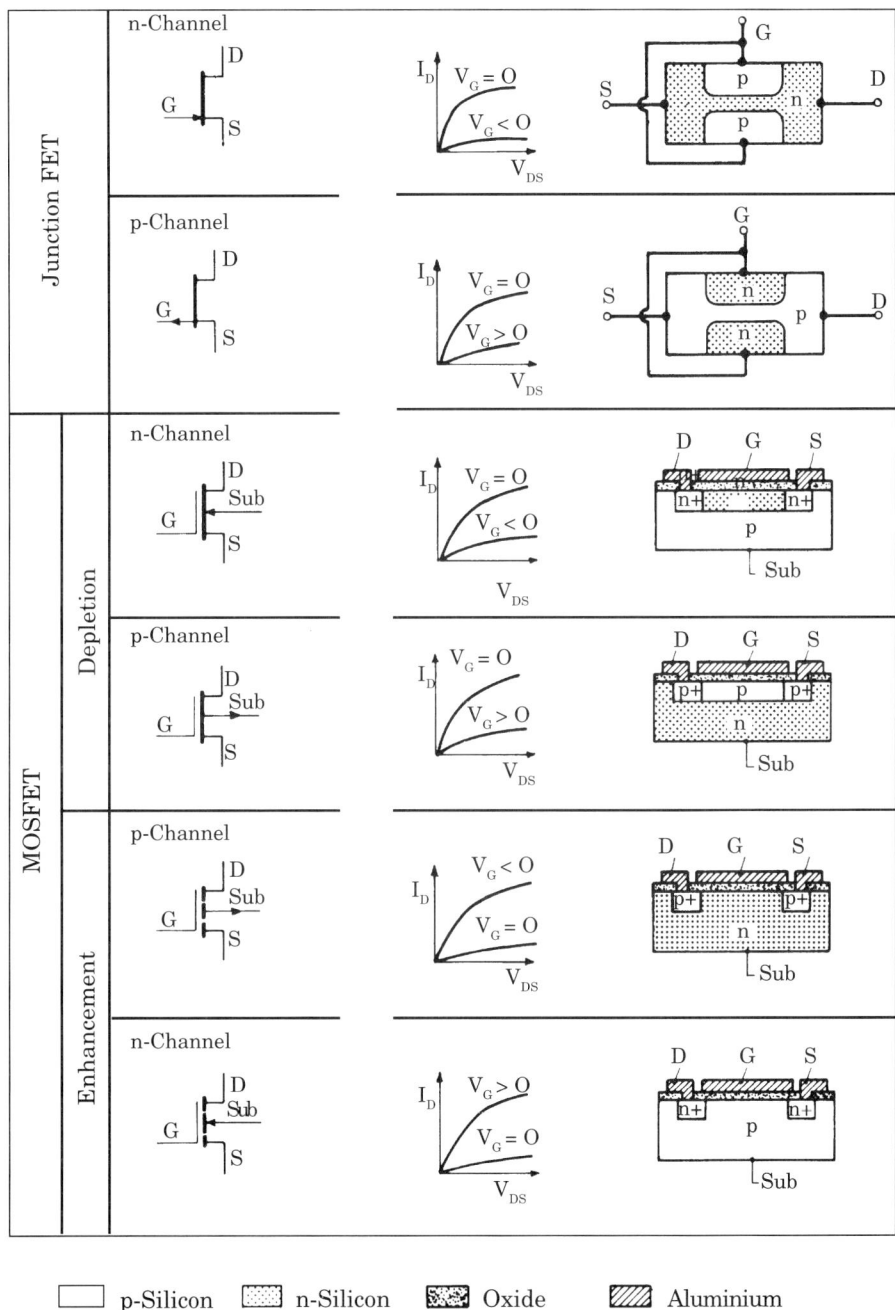

Fig. 3.25 Summary of FETs. For each type shown, it gives the usual symbol (without the circle), a qualitative I_D-V_{DS} diagram (in absolute values) and a typical structure.

When a voltage V_{DS} is applied between D and S the current will flow if both gates are open. A voltage V_G applied to the gate controls the current. In the n channel device a negative V_G induces positive charges into the channel which combine with free electrons increasing the width of the depletion region. As the resistivity of a depleted region is very high, the lower the gate voltage V_G, the less the current flow between S and D at constant V_{DS}. Similarly, a p channel JFET is ON when the gate is open and current flow is controlled by a positive voltage applied to the gate.

MOSFETs can be divided into two types, depletion and enhancement. Each type can be made with n channel or p channel. Depletion MOSFETs are ON when the gate is open, i.e., they conduct between S and D. As with JFETs, the operation of the devices is controlled by the voltage applied to the gate (V_G). In this case the gate is working as the electrode of a capacitor, the other electrode being the semiconductor. The material, usually silicon dioxide, insulating the gate from the semiconductor acts as the dielectric. An n channel depletion MOSFET has two heavily doped n regions, i.e., n^+ regions, connected by the n doped channel. The source and drain are connected to the n^+ regions such that the current flows between them. The more negative the gate voltage becomes, the more restricted the n channel and the lower the current, Figure 3.25. The p channel MOSFET operates similarly except that a positive voltage must be applied to the gate to increase the depletion zone and hence the resistance to current flow, S to D.

In *enhancement* type MOSFETs the channel is created by induction when a voltage is applied to the gate. For a p channel type a negative voltage applied to the gate induces positive charges into the semiconductor and creates a p channel, the depth of which increases with an increase in the absolute value of V_G, to allow current to flow, S to D. In the n channel type a positive voltage is applied to the gate to create a conductive path between S and D. Both types of MOSFET are OFF when $V_G = 0$.

There is a strong trend in the discrete component industry towards multiple devices. Examples found include diode arrays, which have already been mentioned, dual and quad transistors with matching of particular characteristics, Darlington pairs, and complementary *npn-pnp* devices.

Darlington pairs are effectively two transistors directly coupled to give a high gain device and are available in gains of up to 150,000. Power Darlingtons offer 5 to 10 amps available current for servos, motor drives, converters and relay drives.

3.6.1 Packages

Over 600 packages are listed in manufacturers' literature for transistor devices. Standards exist for these just as for diodes, the JEDEC code being TO-XXX where XXX stands for 1-3 digits plus, if needed, 1-2 letters. Metal cans, such as the TO-5, TO-3 and TO-18, have been in use for many years. For assembly, they are robust, the major weakness being the glass-to-metal seal for the leads, but are difficult to insert automatically because of the shape and poor standardisation.

The relatively high cost of metal cans has led to their replacement by plastics packages. Metallic packs for low and medium power devices are now being replaced by plastics packages. For high power devices, the stud or bolt type, and flat type cans once common are superseded by plastics packages with metal tabs to improve heat dissipation. The most commonly used packs are the TO-3, TO-5 (or TO-39), TO-18, TO-126, TO-202 and TO-220.

3.6.2 Characteristics

To describe fully the characteristics of a transistor requires a set of diagrams and specification of the relevant test conditions. Manufacturers' data sheets usually give some basic parameter values, as in Table 3.6 for an n-p-n BJT, or an extended set of diagrams. In some catalogues, an appendix is given to explain terms, definitions and test conditions and it is sensible to read these thoroughly since, although these should be standardised, there are still variations which may cause problems.

Table 3.6
Most Common Abbreviations, Terms and Definitions for Bipolar Junction Transistors
(Based Upon JEDEC Publication 77-A)

Abbreviation	Term	Definition (brief)
C_{cb}	Collector to base capacitance	Capacitance between collector and base terminal when the C-B junction is reverse biased and the B-E junction is not biased
C_{ibo}	Common base open circuit input capacitance	Input capacitance, i.e., the capacitance between B and E with the common base connection and the collector open
C_{ieo}	Common emitter open circuit input capacitance	Same as above, but with common emitter connection
C_{obo}	Common base open circuit output capacitance	Output capacitance, i.e., the capacitance between C and B with the common base connection and the emitter open
C_{oeo}	Common emitter open circuit output capacitance	Same as above, but with common emitter connection
f_T	Transition frequency	Frequency at which the forward current gain is unity (with common emitter connection)
h_{FE}	Static forward current transfer ratio (common emitter)	The ratio of the dc output current (collector current) to the dc input current (base current) with the common emitter connection
h_{FB}	Static forward, etc., (common base)	The ratio of the dc output current to the dc input current with common base connection

Cont.

Abbreviation	Term	Definition (brief)
h_{fe}	Small signal forward current transfer ratio (common emitter)	Ratio of the ac output current (collector current) to the small signal ac input current (base current) with the output short circuited to ac and the common emitter connection
h_{fb}	Small signal, etc., (common base)	Similar to the above definition, but with a common base connection
I_B, I_C, I_E	dc current (terminal B,C,E)	dc current flowing into the transistor through its base, its collector or its emitter
I_{BEV}	Base cut-off current	dc base current when the B-E junction is reverse biased (and for a value of V_{CE} to be specified)
I_{CBO}	Collector cut-off current (base open)	dc collector current when C is reverse biased with respect to B (i.e., $V_{CE} > 0$ for NPN transistors and $V_{CE} < 0$ for PNP ones) and the base is open
I_{CEV}	Collector cut-off current	Same as above, for a given value of V_{BE}
I_{EBO}	Emitter cut-off current (collector open)	dc emitter current when the E-B junction is reverse biased and the collector is open
V_{BB}, V_{CC}, V_{EE}	dc supply voltage	dc supply voltages applied to the circuit to which the terminals B, C and E are connected
V_{BE}	Base to emitter voltage	Base voltage with respect to the emitter
V_{BC}, V_{CB}, V_{CE}, etc.,	—	Voltage of a terminal (first subscript) with respect to another terminal (second subscript)
V_{BE}(sat)	dc saturation voltage (base to emitter)	Voltage V_{BE} when the transistor is in the saturation region of its characteristic (i.e., it can be compared to a closed switch between C and E)
V_{CE}(sat)	dc saturation voltage (collector to emitter)	Voltage V_{CE} in same condition as above
$V_{(BR)CBO}$	Collector to base breakdown voltage	Voltage V_{CB} which causes the breakdown of the reverse biased C-B junction with open emitter
$V_{(BR)CEO}$	Collector to emitter breakdown voltage	Voltage V_{CE} which causes the breakdown between C and E (with $V_{CE} > 0$ for NPN and $V_{CE} < 0$ for PNP), with open base

Cont.

Abbreviation	Term	Definition (brief)
$V_{(BR)CER}$	Collector to emitter breakdown voltage	Same as previous entry, with a resistance (to be specified) between B and E
$V_{(BR)EBO}$	Emitter to base breakdown voltage	Voltage V_{EB} which causes the breakdown of the reverse biased B-E junction, with open collector
V_{CBO}, V_{CEO}, V_{CER}, V_{EBO}	—	Same as $V_{(BR)CBO}$, $V_{(BR)CEO}$, etc., when the voltages between the two terminals (C and B, C and E, etc.) are not sufficient to cause the breakdown

3.6.3 Manufacturing Technology

Most transistors are manufactured using the planar technology illustrated in Figure 3.26. This technique is applicable to silicon only and is widely used for transistors since it gives good performance and high reliability while permitting large batch production. A single transistor, called a chip, dice or

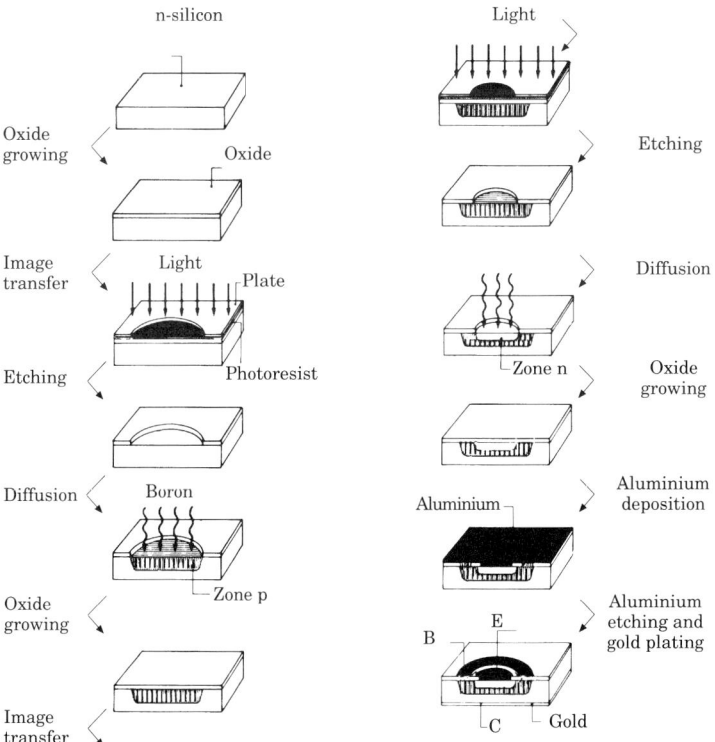

Fig. 3.26 Manufacturing steps for an *NPN* bipolar transistor made by the planar technique. Many hundreds of similar transistors are produced together on a single silicon slice (wafer).

die, is usually 0.13-0.25 mm (0.005-0.01 inch) thick and, depending on its power, 0.25-2.5 mm (0.01-0.1 inch) square. Chip production starts from a silicon single crystal slice or wafer of the required thickness. The slice is cut from a rod of 76-200 mm (3-8 inch) diameter and some hundreds or thousands of devices can be made on the same wafer in a single sequence of manufacturing operations. After parting from the wafer, the chip or die is assembled into a package and electrically connected to the package leads. This is illustrated in Figure 3.27.

Attachment of chip to package and connection of the terminal pads to the package leads is still a time-consuming process and the cost of the chip is usually less than half the cost of the completed component. Fixing the chip to its mounting in the can, *die attach,* can be by adhesive bonding or by eutectic bonding. The latter is achieved by heating the chip in contact with a gold plated area of the mount at about 400°C (750°F) when a silicon-gold eutectic alloy is formed.

Connection of the chip to the package leads is normally by bonding a thin wire, from 0.018 mm to 0.10 mm (0.0007 to 0.004 inch) diameter, of aluminium, gold or an aluminium-silicon alloy. Several methods can be used.

Ball bonding can only be used with gold wire. When the wire is cut by a hydrogen flame, a small ball is formed. The ball is pressed against the chip pad to form a joint with the aid of pressure and heat.

Stitch bonding is another form of thermocompression bonding. No ball is formed, a length of wire being bonded by heat and pressure to the pad. Both gold and aluminium can be used.

Ultrasonic bonding requires no heat. The wire is pressed against the pad by a head which vibrates at ultrasonic speed. The wiping action of the vibrations removes any oxide films present and a joint is formed by the applied pressure.

These and other chip mounting methods such as beam lead and flip-chip are described in more detail in Chapter 16. Packages for high power devices

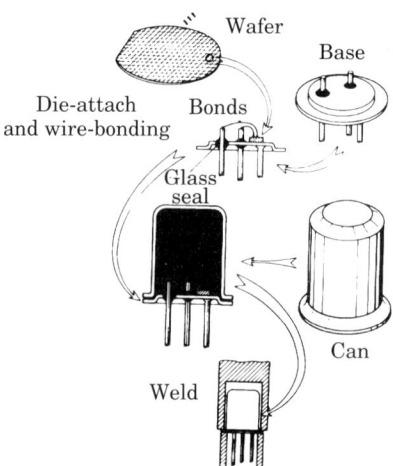

Fig. 3.27 Packaging of a transistor in a metal can. As this operation is still partly manual and only one transistor can be packaged at a time, it accounts for most of the cost of the packaged transistor.

may avoid wire bonding by using direct contact of the package leads to the chip pads, as shown in Figure 3.28.

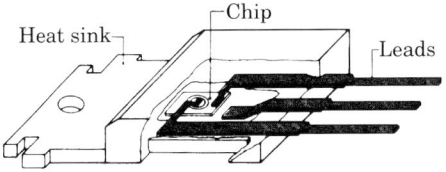

Fig. 3.28 Section of a TO-220 package with direct contact of the leads to the chip.

3.6.4 Identification

Transistors are identified in similar manner to diodes. The JEDEC standard code for registered types is 2N XXX. All components having the same code should be interchangeable but it is advisable to check equivalence when boundary conditions are being used for some parameters. Before replacing a transistor with a similar one from another supplier, the circuit designer or component engineer should be asked to check all characteristics thoroughly for all circuits in which it will be used.

3.7 SILICON CONTROLLED RECTIFIERS AND THYRISTORS

The term thyristor applies to a family of semiconductor devices known as *pnpn* devices. The SCR, silicon controlled rectifier, is by far the most widely used member of the family, largely because of its versatility. Thyristors have wide use in solid state power control and power switching circuits which replace electromechanical relays.

The SCR, for which the IEC official name is '*Reverse blocking triode thyristor*', has three terminals, the anode (A), cathode (K) and the gate (G). When the anode is negative with respect to the cathode, the device is reverse biased and behaves as a diode. When forward biased, anode positive with respect to the cathode, it is controlled by the voltage V_G applied to the gate (measured with respect to the cathode). If $V_G = 0$ no current flows until the anode to cathode voltage V_{AK} is raised to a low positive value. When it reaches a threshold value, the SCR switches abruptly to a conduction state and the voltage V_{AK} drops to a low value which is largely independent of the current, see Figure 3.29. At this point the SCR goes on conducting until V_{AK} is reduced so that the current drops below a specific value which is called the holding current.

When forward biased and V_{AK} is less than that required for spontaneous triggering of conduction, the device can be switched from a non-conducting state to a conducting state by a positive pulse of the gate voltage V_G. Again the triggered SCR remains in the conducting state until the current drops below the holding current value.

Because of the mode of operation and the very low voltage drop across AK when the device is conducting, the SCR can be used as an electronic relay which is OFF until triggered. Other applications include phase controls, inverters, choppers and pulse modulators.

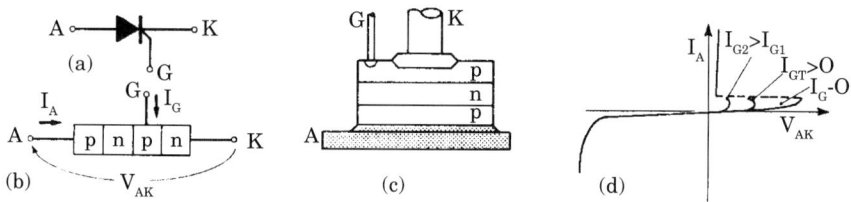

Fig. 3.29 Silicon controlled rectifier (SCR): (a) symbol (without circle); (b) schematic diagram of the *pn* junctions; (c) structure; and (d) typical current/voltage diagram.

The Triac, official IEC name '*bidirectional triode thyristor*', derives its name from 'tri-AC', an abbreviation of *Triode AC Semiconductor Switch*, i.e., three electrodes switch for ac. The name thyristor was coined from thyratron and transistor, the first being a vacuum tube now largely replaced by the SCR or Triac. The Triac is similar to the SCR but operates as a forward biased SCR both when forward and reverse biased, Figure 3.30. The three terminals of a Triac are anode 1, anode 2 and gate (A1, A2 and G) or terminal 1, terminal 2 and gate (T1, T2 and G).

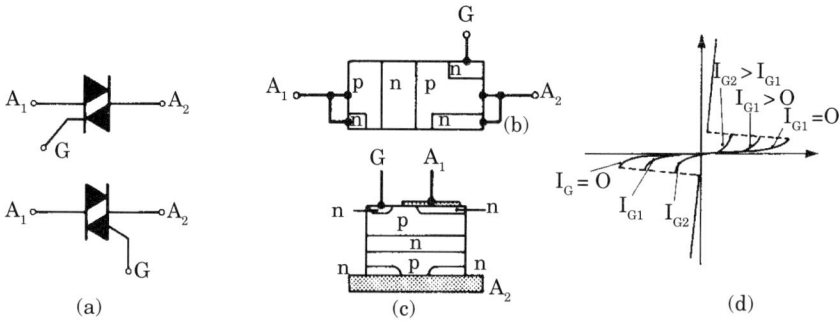

Fig. 3.30 TRIAC: (a) symbols (without circle); (b) schematic diagram of the *p-n* junctions; (c) structure; and (d) typical current/voltage diagram.

Both for SCRs and Triacs the currents involved can be very large and, despite the low internal voltage drop, the power to be dissipated is often very high.

Many transistor packages, e.g., TO-3, TO-5, TO-39, TO-59, TO-61, TO-63, TO-66 and especially TO-126, TO-202 and TO-220, are used for SCRs and Triacs and special packages have had to be developed. The latter include the TO-49, TO-118, the TO-65, similar to the DO-5 with an additional terminal, and the TO-200, similar to the DO-200 with the addition of a side terminal. For medium and high power the most common packages have a plastics body with a heat dissipating metal tab, such as the TO-202 and TO-220. These are cheap and suited to automatic manufacture.

Internal connections of these devices are not standardised for all packages and all suppliers. The most common pin assignments are shown in Table 3.7. (Distinguishing between the two anode terminals of a Triac is of minor importance because of the symmetrical characteristics.)

Table 3.7

More Common Pin Assignments for Three Terminal Packages

	Terminal	Transistor (BJT)	SCR	TRIAC
TO-18, TO-5	1	E	K	T1
(and similar)	2	B	G	G
	3(†)	C	A	T2
TO-59	1	E	K	T1
(and similar)	2	B	G	G
	3(‡)	C	A	T2
TO-3	1	B	G	G
(and similar)	2	E	K	T1
	3(††)	C	A	T2
TO-220	1	B	K	T1
(and similar)	2(‡‡)	C	A	T2
	3	E	G	G

(†) Usually in contact with the can
(‡) Usually in contact with the body and the thread
(††) Component body
(‡‡) Usually in contact with the metal tab

3.8 OPTOELECTRONIC DEVICES

Optoelectronic devices detect, respond to or emit electromagnetic radiation in the visible, infra-red and/or ultraviolet regions of the spectrum. The more common types of optoelectronic device, symbols for which are shown in Figure 3.31, are described briefly in the following paragraphs.

Photodiodes

Solid-state *pn* junctions possess a number of useful properties other than those associated with rectification. Light-sensitive voltaic or conductive diodes have been optimised for generative or conductive modulation so that they can be used as sensors or detectors. Photovoltaic diodes work with no external bias and provide an output current proportional to the intensity of the incident radiant energy. Photoconductive diodes need external, reverse bias. The current flow through them is very low in the absence of light, 'dark current', and increases proportionally to the intensity of incident radiant energy.

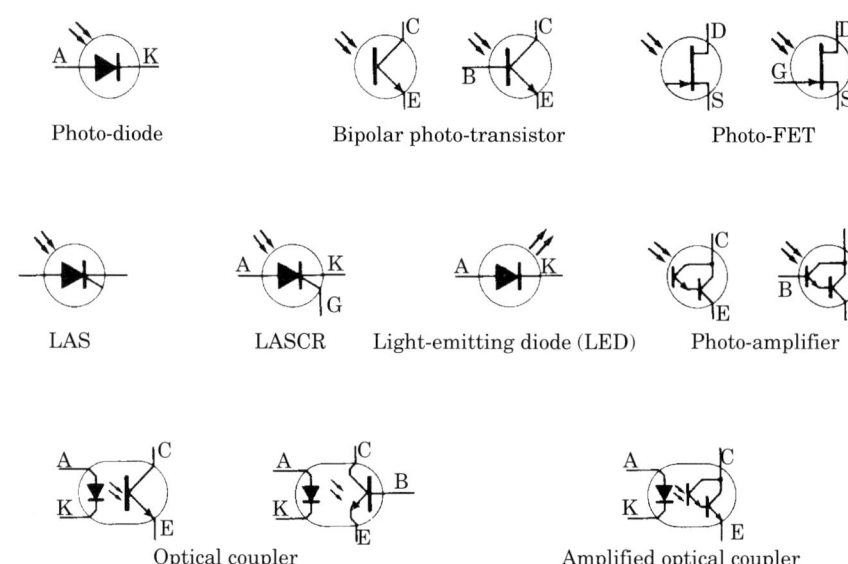

Fig. 3.31 Symbols of common optoelectronic devices. The circle is usually shown.

Phototransistors

Phototransistors work as photoconductive diodes but provide larger output current because they amplify internally the current generated by incident radiant energy. Different semiconductors are employed to enhance properties needed. The simplest phototransistor is an *npn* BJT which is activated by incident light energy instead of base current. Most transistor chips are light sensitive. A simple phototransistor can be built by carefully removing the upper part of an *npn* BJT and replacing it with transparent resin. If the device is connected with a common emitter, photoconduction will appear between collector and emitter. If an external connection to the base is present, the transistor will operate both by incident light and by base current.

FET phototransistors conduct between source and drain in accordance with the light intensity reaching the gate; an external connection to the gate is optional.

LAS and LASCR

An LAS is defined by IEC as a *reverse blocking diode thyristor*. It is of similar construction to a four-layer (Shockley) diode but is triggered by infrared and visible light radiation. The LASCR is defined by IEC as a reverse blocking triode thyristor and is triggered either by light activation or by a gate signal as is a standard SCR. Both types are used in static switches, photoelectric controls and trigger circuits. The LAS is particularly useful as a trigger for high voltage SCR applications and the LASCR in position monitors and limit switches.

Light Emitting Diode (LED)

The LED or solid state lamp emits visible or infra-red light when forward biased. This is the result of the combination of electrons and holes in the *p* region when the electrons flow from the *n* region under the influence of a current from anode to cathode. In particular semiconductors it produces an emission of luminous energy. Gallium arsenide is used for infra-red light, wavelength about 7-10000 Å, gallium arsenide phosphide, GaAsP, on phosphide for amber light, about 5900 Å, gallium phosphide, GaP, for green light, about 5650 Å, and GaAsP for visible red light, about 6500 Å. Other colours can be produced by filters or coatings. The rate of development progress may be judged from the following. In 1989 it was proudly announced that a blue LED was at last available although the yield was very low, the reliability poor and the cost high. In early 1994, a Japanese company announced that it was starting production of a blue light emitting diode based on potassium nitride. The light emitted was 100 times brighter than the brightest blue LED then existing. The LED produced light of wavelength 450 nm at an intensity of one hundred million candela.[6]

Double LEDs in three lead packages are also available. If, e.g., one diode emits in the red and the other in the yellow, by flickering the two a green light results.

Solid State Laser

Semiconductor laser materials, i.e., those solid state laser materials in which the method of excitation is by the *pn* junction, are generally based on $Ga(As_xP_{1-x})$ or GaAs or indium combinations with gallium, arsenic, phosphorus or antimony. The possible materials grow almost daily. They emit coherent, polarised radiations having the same characteristics as laser light and are sometimes termed laser diodes.

Solid state lasers have applications in telemetry, bar code readers, printers, proximity switches and signal transmission (through optical fibres). The most popular application is the reading of compact discs and CD-ROMs.

Photodarlington

The active element is a chip having a bipolar junction phototransistor connected to an *npn* transistor which amplifies its output. This gives increased output current.

Optical Coupler (Optocoupler, optoisolator or optically coupled isolator)

In this device an LED, usually an infra-red emitter, is assembled in the same pack with a phototransistor. A signal sent to the LED produces radiation which drives the phototransistor to give an output proportional to the input signal. Amplified optical couplers use a photodarlington instead of a phototransistor.

Solid State Displays

LEDs can be arranged in such a way that selective activation generates a symbol which can be readily recognised. Numeric displays generate only digits and possibly some signs and symbols while alphanumeric displays can show both letters and digits, plus possible other symbols. The common numeric display employs seven segment diodes (or seven pairs of diodes, as shown in Figure 3.32). A typical alphanumeric display uses 35 point diodes arranged in a 7 x 5 matrix, illustrated in Figure 3.33.

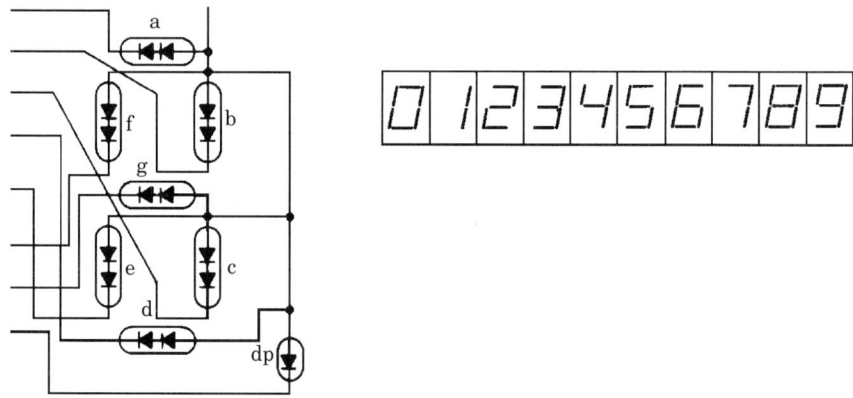

Fig. 3.32 Arrangement of LEDs in a 7-segment display and set of numerical characters that may be displayed. Two diodes in series are used for each single segment (from a to g); a single diode is used for the decimal point dp. (Courtesy of Texas Instruments)

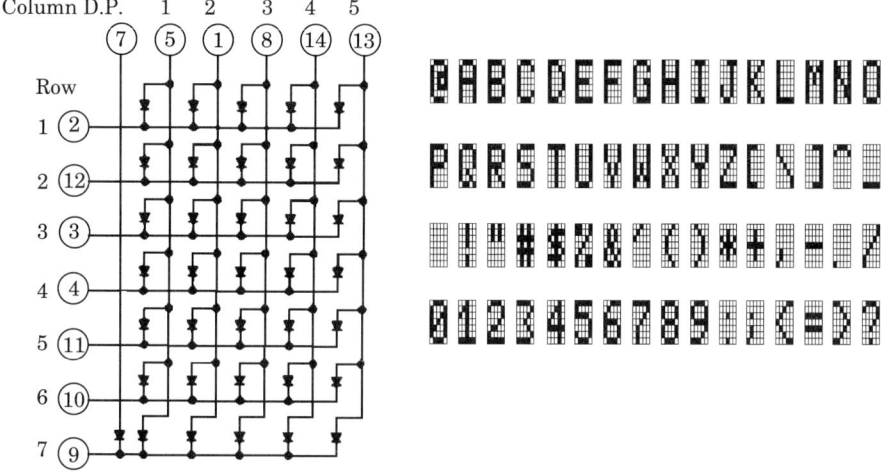

Fig. 3.33 Matrix of 7 x 5 LEDs for displaying alphanumeric characters. An integrated circuit attached to the matrix receives a 7-bit US ASCII code, decodes it and drives the 5 columns (plus one for the decimal point) and the 7 rows. According to the input code, any one of the characters shown on the right may be displayed. (Courtesy of Texas Instruments)

Packages for displays can hold either a single character or an array of characters. Some may include an IC chip to provide the logic to drive the LEDs.

Liquid crystal displays (LCDs) are an alternative in many applications for LEDs. These do not generate light; they reflect or transmit it. A liquid crystal display is made from two flat glass strips in close proximity to each other, between which there is a liquid. The liquid becomes opaque or transparent according to the value and intensity of an electrostatic field acting upon it. Both glass strips must have a conductive surface; one can be completely conductive, the other must be conductive only in seven areas corresponding to the seven segments. The conducting material on one of the strips must be transparent, for example, tin oxide.

Both LEDs and LCDs are widely used for electronic watches, calculators, instruments, and so on. LCDs have very low power consumption (less than 1 milliwatt per square centimetre) and may be always left ON when powered by small batteries such as those used for wrist watches.

3.8.1 Packages

Photodiodes and phototransistors are often supplied in TO-5 and TO-8 type packages, modified by the insertion of a small lens in the upper part. Specially designed packages, Figure 3.34, are frequently used, some being very small. Both plastic and metal bodies are used. Numerous different types have been developed, many of which are used by a single supplier, so that package standardisation is poor.

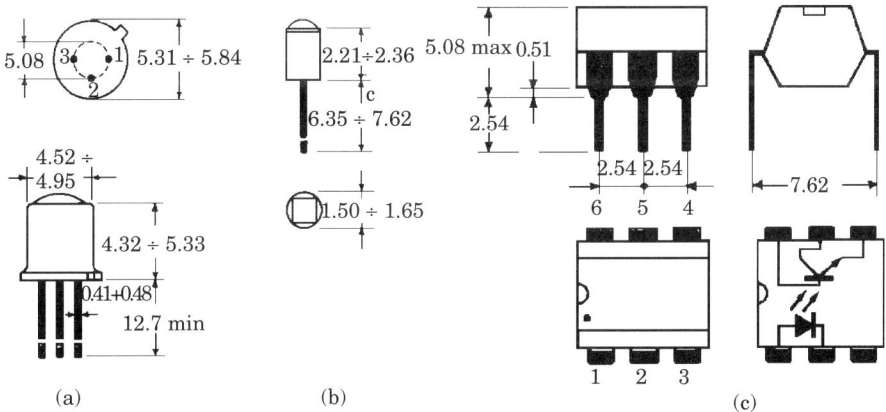

Fig. 3.34 Typical packages for optoelectronic devices: (a) is derived from a TO-18 by replacing its top with a lens; (b) is developed for these devices; and (c) is a package derived from the TO-116 which contains an optocoupler. Dimensions are in millimetres.

LEDs are usually encapsulated in plastic transparent packages, the body also acting as the lens; the solid angle in which the light is emitted depends on its shape.

Laser diodes are relatively more expensive and used in environmentally more demanding applications so that metal packs are preferred. The most common is derived from a TO-5 or TO-39 by providing a small transparent window on the top.

Hermetically sealed packages, e.g., TO-99, are used for optical couplers. A cheap six-pin DIP derived from the TO-116, 3.9.5, is also offered.

The seven segment or matrix displays are packaged in TO-100 packages, modified by inclusion of a lens, or in special DIPs similar to the TO-116.

3.9 MONOLITHIC INTEGRATED CIRCUITS

The term 'monolithic' — etymologically 'from a single block of stone' — refers here to a device which is made from a single chip of semiconductor or, by extension, of a single material. It is used where there may be confusion with hybrid ICs which are made from two or more elements. If nothing is specified, any IC is assumed to be monolithic. Monolithic integrated circuits or ICs, also written ic, account for over 50% of the cost of consumer and professional assemblies. Although the cost of a given IC is continually decreasing, the usage of ICs is steadily increasing, their market growing by some 20% per annum while the growth of the total market for electronics is in the range of 10 to 15% per annum.

A monolithic IC contains many more active components than the same circuit made with discrete components. This is because, while it is possible to make resistors, e.g., with a small strip of doped silicon, and capacitors, using aluminium and silicon for plates and silicon dioxide as dielectric, they are space consuming and precision is not always adequate. For this reason the complexity of an IC is still sometimes expressed in 'equivalent transistors', that is the number of discrete transistors needed to build a functionally equivalent circuit. This does not necessarily mean the cost of the IC is proportional to the 'equivalent transistor' count. The cost depends on other factors such as the development cost, the interconnection complexity, production quantities and yields, and competition.

To describe all families of ICs would take hundreds of pages and is not the intention of this book. ICs change rapidly as the technology advances and to remain up to date in the subject is a task beyond all except the most dedicated. The following paragraphs are therefore devoted exclusively to a general description of the most important families, for the benefit of the non-specialist.

3.9.1 Classification

A single comprehensive classification for the multitudes of integrated circuit types seems impossible to find. Certainly, all ICs fall into one of two major classes, namely analogue and digital, but within these broad classifications there are many different devices which may be further classed, by use, by manufacturing technology or by function.

While no single classification is exhaustive, most of them are necessary. The major types are discussed in the following text, in which there will be overlapping as the same device belongs in different categories from different standpoints.

3.9.1.1 TECHNOLOGY

Depending on the way in which their basic element functions, ICs may be divided into the following classes:

Bipolar: The basic element is a bipolar junction transistor. This is the oldest technology and is used for the majority of analogue devices. In digital circuits it has the advantage of high speed and high output power. However, it has rather high power consumption and the density of components on the chip is relatively low, Figure 3.35.

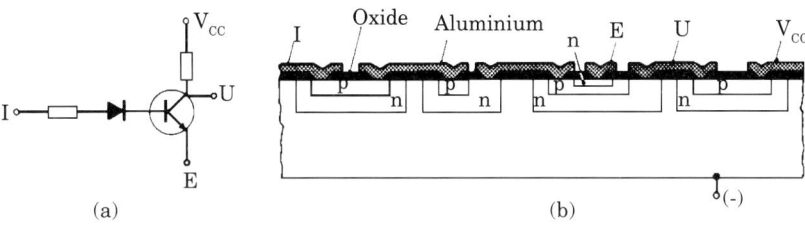

Fig. 3.35 Bipolar integrated circuit: (a) electrical diagram of a part of a circuit; and (b) how this may be arranged on a silicon slice. Note that all elements are insulated from each other by means of N diffusions and are connected to each other by aluminium strips on the surface.

MOS: The basic element is a field effect transistor. The class may be further divided into sub-classes depending on the polarity of the channel (n channel or p), on the material used for the gate (aluminium, silicon, or silicon nitride or oxide) and the mode of operation (deplction or enhancement).

MOS technology, Figure 3.36, uses smaller devices and the manufacturing process is simpler than bipolar technology. For a given chip size it is possible to have a higher level of integration, i.e., more devices, and better yield, hence lower cost. As it is possible to manufacture very large chips with a reasonable yield, MOS is a good technology for the highest levels of integration.

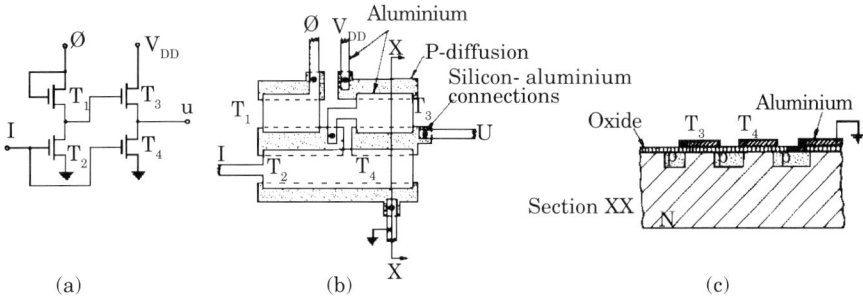

Fig. 3.36 MOS integrated circuit: (a) electrical diagram of a part of a circuit (a push-pull drive); and (b) typical arrangement of it on a silicon slice. Note that all connected terminals have been produced by a single p diffusion or a single aluminium strip. This technique provides a higher density of components than the bipolar technique.

BI-MOS: In this both bipolar and MOS technologies are used on the same chip to combine the speed of bipolar with the device density of MOS.

CMOS (or *complementary MOS,* sometimes written *C-MOS*): In this variant of MOS technology, both p channel and n channel transistors are combined on the same chip to make complementary stages and further reduce the low power consumption of the MOS, Figure 3.37. In this technique one of the complementary transistors is always turned off so that no direct current can flow from the power supply to the substrate. Power dissipation is limited to that absorbed by the parasitic capacitance of the devices. CMOS devices can operate at fairly high speed, show good noise immunity, and voltage supply is not critical. It is the standard technology for battery operated items such as pocket calculators and watches, and for items which must always be in the 'ON' state.

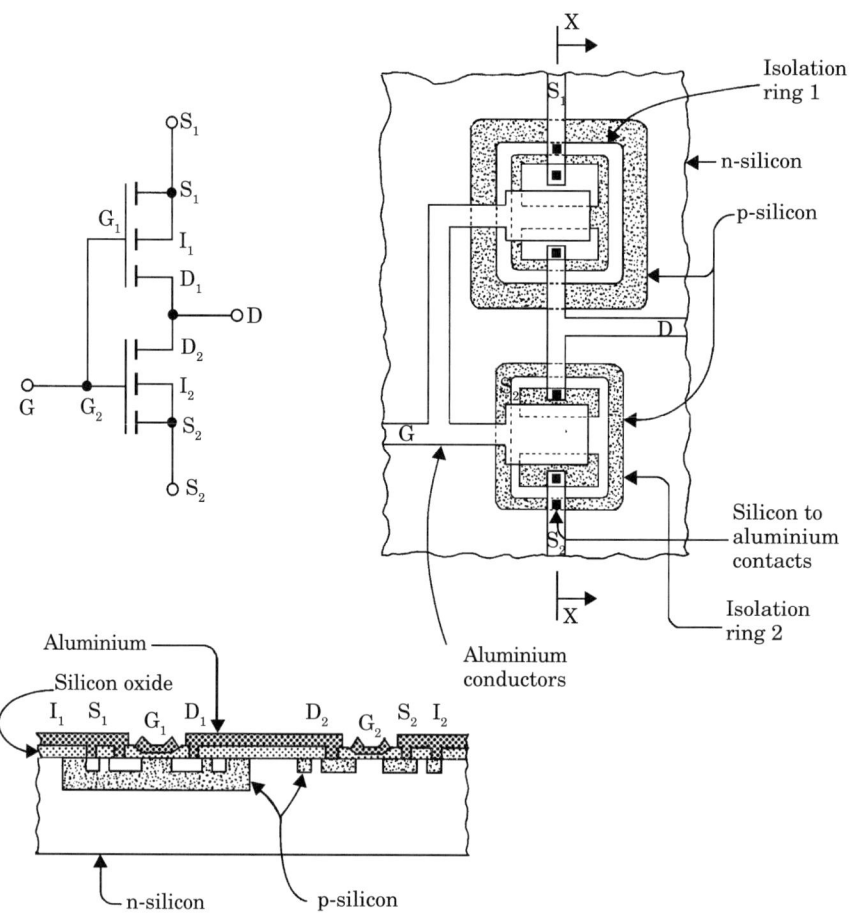

Fig. 3.37 Structure of a complementary MOS device, CMOS.

CCD (Charge Coupled Devices): The CCD is based on a completely different principle from the foregoing devices. The basic element is a chain of closely spaced thin aluminium electrodes (gates) deposited on a thin layer of silicon dioxide grown over a doped silicon area, Figure 3.38. A signal applied to the first gate induces charges in the doped silicon underneath. The induced charges may be moved, by controlled clock signals, step by step along the chain to its end, just as in a shift register, described later. Loss of charges along the chain may be restored by an MOS transistor amplifier. The operating speed can be very high.

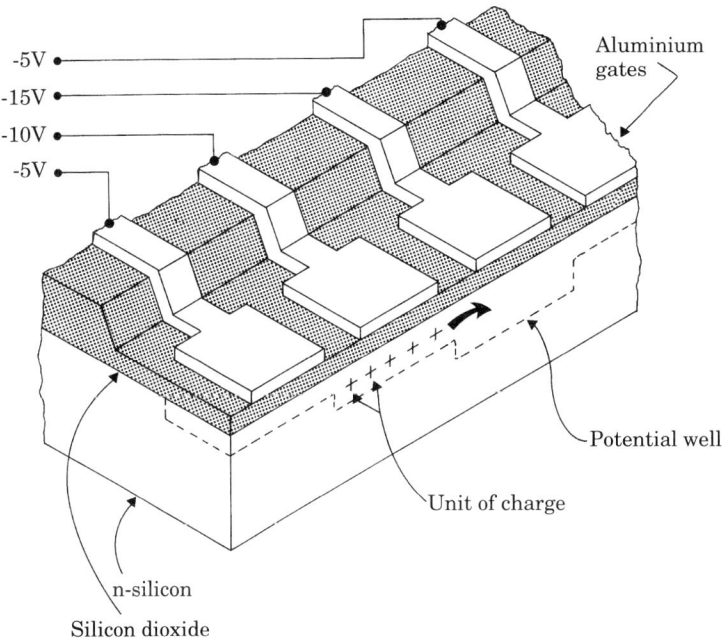

Fig. 3.38 Basic structure of a charge coupled device, CCD.

CCDs may be used as memories, both analogue and digital. The latter type gives a very high density of bits per unit area (and thus a low cost per bit) but information is not accessible bit-wise. Non-volatile CCD memories are possible by adding silicon nitride capacitors (NMOS technology), but the best known use of CCDs is for picture storage in TV cameras, in which they have replaced the bulky and fragile vacuum tubes. Devices with a resolution of over 1 million pixels, e.g., 1280x980, have been mass marketed since as early as 1987.

Other technologies, such as silicon-germanium (SiGe) HBT (heterojunction bipolar transistor) technology, publicised in 1994 as a successor to CMOS for high speed circuits,[7] have been or are being employed in relatively small quantities but are not considered further in this text.

3.9.1.2 OPERATING MODE

All electronic circuits are either *analogue* or *digital* and the same classification is applied to ICs.

Analogue circuits deal with continuously variable values of voltage or current which represent continuously variable physical quantities, such as temperature, pressure, stress and distance. Digital circuits work only by switching between two states, that is 'off' or 'on'. Anything may be represented by either state, for example, 'voltage' and 'no voltage', or — conventionally — by one binary digit or bit (either '1' or '0'). A digital circuit may deal also with analogue quantities, provided that these have been transformed into digital format, i.e., have been digitised. (The existence of three-state devices, which operate between three states, does not change the classification.)

It is not always simple to classify devices in this way. An IC can include both digital and analogue functions, making assignment questionable. The common practice is to restrict the definition 'digital' to fully digital ICs, and to consider all others to be analogue.

Linear ICs are a sub-class of analogue ICs, in which mode they operate. At least one of their outputs is directly, or inversely, proportional to one of their inputs.

3.9.1.3 FUNCTION

ICs, digital or analogue, may be placed in one of two broad classes: (a) general purpose and (b) application specific. The general purpose device is used widely in many different circuits; the application specific device is designed for a single application or a single complex function.

Among digital ICs the following, non-exhaustive, classes have outstanding importance:

Logic Circuits: These are devices which implement the operators of Boolean algebra and some other basic functions; Boolean operators — from the English mathematician George Boole (1815-64) who did extensive work on mathematical analysis of logic — are the basis of digital electronics; each uses an elementary circuit, which performs a specific operation and may be realised in different ways; most common are:

the NOT or inverter	output is '1' when input is '0' and *vice versa*
the NOR	same function as a NOT followed by a NOT
the OR	output is '1' when at least one input is '1'
the AND	output is '1' when all inputs are '1'
the NAND	= AND + NOT).

Memories: Devices which can store information (bits).

Microprocessors: Devices which include complex computing functions (i.e., several logic circuits and memory).

With analogue ICs the major distinction is related to the circuit or equipment in which they are used, e.g., radio, TV, telecommunications, controls, and so on, except for linear types.

3.9.1.4 LEVEL OF INTEGRATION

According to the number of devices per single chip, the 'level of integration' or transistor equivalent, ICs may be divided into:

SSI   (small scale integration):   up to 50 transistors;
MSI   (medium scale integration):   from 50 to 1,000 transistors;
LSI   (large scale integration):   from 1,000 to 50,000 transistors;
VLSI (very large scale integration):   from 50,000 to 1 million transistors;
ULSI (ultra large scale integration): over 1 million transistors.

Although there is agreement on the definition, different companies may refer to different limits for one class (or more). As microprocessors are expected to have a transistor count in excess of 100 million by the turn of the century, classes need to be re-defined to have any meaning. As stated earlier, the significance of the level of integration has been lost as most manufacturers are able to produce all levels and there is no close correlation with cost.

3.9.2 Manufacturing Technology

The basic process is planar technology, essentially as for producing transistors, but much modified. The process normally starts from a thin slice (wafer) of single crystal silicon, which is doped where necessary by using silicon dioxide as a resist and very precise photoimaging techniques.

Many devices, from about one hundred to one thousand and more, are manufactured in parallel on a single wafer which is then divided into single chips. Each of these has conductive pads, normally on the borders, for the inputs and outputs, the remainder generally being protected by a passivation layer. Tested good chips are then assembled in a package which has a central area to hold the chip and as many conductors as needed, made up from an etched and plated *leadframe* or deposited on to an insulated substrate. The chip is attached to the package and the leads are bonded as described for transistors. Alternative connection methods, such as beam leads (conductors projecting from the chip), have lost favour with the advent of precision automatic bonding machines.

The major challenge in IC manufacture will always be in terms of miniaturisation. Small chips mean more per wafer with higher parallelism in all production steps up to packaging. Also, the best single crystal will include some imperfections, and small chips will give higher yield than large ones. On the other hand, more functions on the same chip should mean a higher price for around the same production cost but the space occupied by internal connections will increase dramatically and can occupy over 50% of the chip area.

Emphasis in IC development is given to (i) reducing the area occupied by each single device, (ii) reducing the width of internal conductors, (iii) reducing the dimensions of interconnection pads and (iv) increasing the size of wafers and chips.

With reference to (i), in 1987 ICs with transistors as small as 9 square microns were being marketed. In theory this means that more than 100,000 devices could be accommodated on a chip 1 millimetre square. As a human hair has a diameter of about 100 µm (0.1 mm) and thus an area of about 8,000 square microns, its section could accommodate some 1,000 transistors. Seven years later, transistors could be made with an area of only 1.22 square microns!

The benefit of reducing the area of the elementary device cannot be realised unless the space taken up by connections between them is also reduced. Two methods are adopted in parallel, multilayer connection and reduction of conductor width. In the early eighties, devices with four connection layers were available, two in metal and two in silicon. By 1987 this had increased to six, three in metal and three in silicon. This was then believed to be the maximum for mass production — but it will be foolish to imply that something is impossible in microelectronics! Reduction in conductor width, or the *design rule,* is a case in point.

Up to 1980, 3 µm was the lower width limit. In 1986 1 µm width was achieved and, in 1987, companies started to market devices with a 0.75 µm design rule, breaking the so-called *sub-micron barrier*. This was both a psychological and physical barrier. For years, chip designers accepted that 1 µm would be the practical limit for the current technology. Visible light has a wavelength in the range of 0.4 to 0.8 µm and sub-micron objects cannot be seen, even by the most powerful optical microscope. The micron was assumed to be the limit for optical photoimaging. Methods to replace optical photoimaging, such as electron beam exposure, direct ion implantation and X-ray lithography, have been developed, yet the sub-micron dimension was achieved by refining the traditional technology. The major semiconductor manufacturers were predicting in 1992-1993 that 0.35 µm would be the practice by 1995, 0.25 µm by 1997 and that, by using X-ray technology, the design rule will be 0.1 to 0.15 µm by the turn of the century.

In parallel with these developments, the dimensions of pads have been reduced from 250 µm x 250 µm to below 100 µm x 100 µm. The key factor here is the improved precision of automatic bonding machines.

Chip dimensions have also increased. However, the current silicon refining technique does not permit good yields on chips larger than 100 to 150 mm^2 and attempts to produce a one-wafer chip have not been commercially successful.

All these improvements have resulted in a dramatic increase in the integration level. For instance, the capacity of memories is being multiplied by 4 each two to three years with the Gbit DRAM expected by the year 2000 and the price per bit is decreasing even faster.

Several new problems are foreseen. The cost of developing a new ULSI circuit is now in the region of US $10,000,000 and is sharply increasing. Sub-micron geometries need absolute cleanliness and very high degrees of automation to avoid contamination. The necessary investment in plant and equipment costs many millions of US dollars.

The new frontier, *gigascale integration,* which is expected to provide the so-called super-chips, one billion components per chip, is so capital intensive that only 5 to 10% of the VLSI manufacturers will be able to reach it and more collaboration between such manufacturers can be expected, see Chapter 2, 2.9.5.

3.9.3 Analogue ICs

Analogue ICs find application in all electronic equipments, either to replace old, well known circuits or to provide new functions. The following are the most common, although the list is not exhaustive.

Amplifiers

Amplifiers — the typical linear ICs — are available in an enormous number of types. A general purpose operational amplifier (μA 741) was among the earliest ICs and is still used. There are now many variants: dual, quad (short for quadruple and refers to devices with either 4 parts or 4 functions), low noise, high slew rate, high power, high frequency, etc. Similar considerations apply to *differential amplifiers*. Application specific amplifiers have been developed for home and in-car entertainment equipment, including stereo, hi-fi and high power versions.

Converters

An analogue to digital converter, or A/D converter, is a device which converts an analogue signal into a digital one. It has one input and many outputs (4, 6, 8, 10, 12, 16 or more) and may be called a 4-bit A/D converter, a 6-bit A/D converter and so on. If the analogue quantity, say voltage, is applied to the input, the outputs, read in the right order, will give a binary number representing the value of the voltage. The *digital to analogue* converter, D/A converter, performs the converse function, converting digital signals into analogue, e.g., voltage or current.

Note: A binary number is a set of bits (binary digits). The maximum decimal number which can be represented by N bits is 2^{N-1}. If a converter has four outputs, the value can span from 0000 to 1111, in decimal figures, from 0 to 15. Any value of the analogue quantity will be given by one of 16 values from 0 to 15. Thus a 16 bit A/D converter gives a more accurate conversion than a 12 bit one, and so on.

Comparators

This device gives an output proportional to the difference between two quantities, usually voltage, applied to two inputs. Dual and quad versions are available.

Control Circuits

This broad class includes voltage regulators, current regulators, switch mode power supply controllers, motor speed controllers, and so on. Often they involve high power or high voltage.

Peripheral Interfaces

These ICs are used to drive peripherals such as LCD, LED or fluorescent displays, computer peripherals, modems, internal communication links, tape decks, etc.

Communication Circuits

This class includes oscillators, mixers, encoders, decoders, modulators, demodulators and their combinations. Specific ICs are also supplied for equipment such as telephone sets and mobile telephones.

Radio and Audio Circuits

ICs developed for this field include tuners, FM and AM/FM receivers, medium frequency amplifier/detectors, stereo decoders, noise suppressors, Dolby processors, and frequency synthesizers, as well as complex circuits incorporating several functions.

Video Circuits

The very large market for TV has resulted in specific devices being developed which cover almost every function in the set. In addition to the types mentioned above, ICs have been developed for video control, vertical deflection, sound processing, synchronisation, and remote control transmitters and receivers. Video tape recorders and cameras include high performance devices for frequency modulation, video processing and VHS chrominance, among others.

Voice Synthesizers

An unusual type of IC, the output applied to a loudspeaker reproduces the human voice. Cheaper devices are used mainly for toys, the best ones being used for public address systems in airports, railway stations, and so on.

3.9.4 Digital ICs

Digital circuits process and store information, numbers, text, instructions and commands, in the form of binary numbers. Typical applications are in computers, calculators and electronic watches. The most common types are described in the following paragraphs.

3.9.4.1 ELEMENTARY CIRCUITS

In the early days of ICs these were the only circuits available. As higher levels of integration are more economic, elementary circuits are now used only in special situations, for example, in prototyping, or where very high speed or current is needed. These are the basic blocks from which any IC, however complex, is usually constructed.

Gates

A gate is a simple circuit which performs an elementary logic function in implementing an operator of Boolean algebra. A gate may be a NOT, an OR, a NOR, an AND, a NAND or an XOR. (An XOR, or 'exclusive OR', operator has two inputs and one output; the output will be at logic level '1' when one input is '0' and the other is '1'.) Gates were the first logic ICs and are the simplest. They come in 14-pin packages, so that each device contains 2, 3, 4 or 6 gates depending on the number of inputs on each. (Even a small chip can contain thousands of gates, as happens with LSI and VLSI.) See also gate arrays, 3.9.4.4.

Flip-flops

A flip-flop is the simplest memory element which can be used to store a bit, i.e., an elementary unit of information. The flip-flop is constituted of two or more NAND or NOR gates, properly connected such that the inputs can be used to set an output to logic 0 or to logic 1. Once the output has been set to a given state, it will remain in that state until a suitable trigger is applied to one of the inputs. Many flip-flops can be interconnected to produce counters, shift registers and memories.

Astable Multivibrators

The output of this circuit is a rectangular type waveform of constant frequency. It is termed *astable* (free running) to differentiate it from the flip-flop which is a bistable multivibrator. The astable multivibrator is often called the clock, as it provides a synchronisation signal for all circuits in a digital machine.

Counters

A counter gives an output which is a rectangular type waveform of frequency 1/N of the frequency of a rectangular type waveform applied to its input. N is an integer, usually 2, 4, 8, 10 or 16. It is a 'divide by N' circuit and can be used to divide the frequency of a clock by N.

Other types of counters can be used to count the pulses applied to their input, either by increasing the binary number present on their output, an up-counter or forward-counter, or by decreasing it, a down-counter.

Encoders

An encoder translates information into a proper code. For instance, each time a key is pressed on a digital computer, the keyboard encoder generates a set of N bits which corresponds to the activated key. Decoder and transcoder circuits are also available.

Multiplexers

A typical application for a multiplexer is to serialise information which is available in parallel or to select one line out of N, where N is an integer. Similar but opposite functions are carried out by demultiplexers.

Adders

These circuits add two or more binary digits in accordance with the binary system rules. The basic circuit is the *half adder*, which outputs the sum of the bits applied to its two inputs and, as another output, the relevant carry over (if there is one). A full adder for as many bits (in parallel) as needed can be made using more half adders and some gates. In numerical computers the adder also performs subtraction, multiplication and division. It is the heart of the *arithmetic logic unit*, ALU, which is in turn the heart of the *central processing unit*, CPU, of the computer.

Shift Registers

A set of N flip-flops connected in cascade is an N-bit shift register, the simplest version of which is the serial shift register. Each time a bit is applied to the input, it is stored in the first flip-flop. The bit previously held by the first flip-flop is shifted to the next flip-flop and so on for all N cells. Each time a new bit is entered, all previous information is shifted, hence the name of the device.

Other common versions are the serial-in/parallel-out (the N bits are stored serially and read out in parallel), the parallel-in/serial-out (the bits are stored in parallel and read out serially) and the parallel-in/parallel-out types.

3.9.4.2 MEMORIES

If the microprocessor is the 'king' of electronics, the memory must be 'queen'.

The amazing progress of electronics is based on the availability of these two components with ever increasing performance and constantly decreasing prices.

The capacity of a memory is measured in bits (i.e., the elementary units of information 0 or 1). As this is a very small unit, capacity is normally quoted in kilobits, Kb or K, megabits, Mb or M, and gigabits, Gb or G. In this application, the prefixes 'K', 'M' and 'G' do not mean 10^3, 10^6 and 10^9 because reference is made to the powers of 2. That is, 1 Kb means 1×2^{10} = 1,024 bits, 1 Mb = 2^{20} = 1,048,576 bits, and 1 Gb = 2^{30} = 1,073,741,824 bits. Note that 1 Mb = 1,024 Kb and 1 Gb = 1,024 Mb.

In 1966 the cost of storing a single bit was about 1 US dollar; in 1976 some 1000 bits could be stored for the same cost; in 1986 1 US dollar was enough for about 100,000 bits. The space required in the electronic assembly was decreasing even faster because of the higher levels of integration. A 1 Mb memory does not occupy more space than a 256 Kb one if both are realised on a single chip. Moreover, with surface mount packages it may take up even less space.

Acronyms for different types of memories seem to multiply daily. Some of the most common are described here.

ROMs

A ROM is a *r*ead *o*nly *m*emory and is used for permanent storage of information. The machine into which it is built can read the ROM as many times as needed, but the information, usually instructions, cannot be altered. It is an intrinsically non-volatile memory because the content will not disappear, even when it is not powered.

The content, usually user specified, is permanently implanted into the chip during manufacture, and includes all information the final equipment needs for performing its basic tasks. ROMs may have a very large capacity and a low cost/bit.

PROMs

A PROM is a *p*rogrammable *r*ead *o*nly *m*emory, that is a blank ROM whose content can be written, permanently, or programmed by the user. The PROM has the same function as the ROM. It is more expensive but offers much greater flexibility as it allows for last minute changes of design and for customising an equipment, e.g., putting messages in different national languages.

EPROMs

The problem with a PROM is that, if there is something wrong with the content, the device must be discarded. There is, therefore, a need for an *E*rasable PROM, an EPROM, whose content may be cancelled and rewritten. Two primary erasing methods are used: UV-EPROMs are cancelled by exposing the chip to ultra-violet light through a window in the package; EEPROMs, *E*lectrically *E*rasable PROMs, may be cancelled electrically.

RAMs

A *r*andom *a*ccess *m*emory, or RAM, is a device which stores information which can be both written and read, as many times as necessary, by the machine during its normal operation. In computers, it is used to store the input data, programs, intermediate and final results, etc., as well as internal information the machine generates for its own use.

'Random access' means that any part of the memory may be reached (to store or read information) directly through an address. The amount of RAM built into a computer is normally measured in bytes. The byte has a value of 8 bits and represents the number of bits needed for a machine command, or a symbol, letter or two decimal figures, etc. Its multiples are kilobyte, megabyte and gigabyte, referred to as K, M and G. Thus, the same letter is used for bits and bytes, depending on whether the reference is to a component or to a machine. Again, the scale is non-decimal, i.e., 1 kilobyte = 1,024 bytes, 1 megabyte = 1,024 K, and 1 gigabyte = 1,024 M. In the smallest computers a large RAM is incorrectly assumed to be synonymous with high computing power.

The elementary circuit of a RAM, which stores a single bit, is called a *cell* and is duplicated on the chip as many times as necessary to give the required capacity. Cell design is unique to each manufacturer, but they can be divided into two basic types which give the two basic RAM families. These are the *static* RAM (s-RAM or SRAM) and the *dynamic* RAM (d-RAM or DRAM). The cell of the SRAM is essentially a flip-flop. Once information has been written it remains there, unless changed intentionally by overwriting or the power fails.

The cell of the DRAM behaves like a capacitor, in which the charge slowly decays, and after a period of time the information is lost. To keep the information the memory must be refreshed, usually many times per second. The relevant circuitry is often built into the chip and, during the refreshing cycle, read/write operations are inhibited to avoid errors. The cell of a DRAM can be much smaller than that of a SRAM and, for the same chip size, the

DRAM can have more capacity, of the order of 4:1. Indeed, a 256 K SRAM needs about the same level of technology as a 1 M DRAM.

The same ratio is reflected in the price per bit and the major reasons for using the SRAM are technological. They provide much higher speed than DRAMs, they do not need interrupts for refreshing, and their access time, i.e., the time to store or retrieve information, is smaller.

Both SRAMs and DRAMs are volatile, that is, the stored information is lost when the power is turned off. However, there are small DRAMs available which can store the content for some months, like CCD memories.

3.9.4.3 MICROPROCESSORS

A microprocessor is a highly integrated component, with the ability to perform all the basic functions of a small computer. It contains all circuits needed to input data, store them (in an internal or external memory), process them (add, subtract, multiply, divide, compare, transcode, etc.), storing intermediate results as necessary, and sort them. The chip also contains control circuits for all these tasks. It operates from simple coded instructions, called microcode, which are specific to a device or a family of devices.

The microprocessor is the central processing unit of all digital equipments, including computers. In such equipment there may be a main microprocessor and several ancillary ones, each devoted to specific tasks. They are difficult to classify as they are all unique (except that one may be derived from another). One characteristic is roughly related to complexity and performance. This is the number of bits which they can handle in parallel, by which they may be classed as 4-, 8-, 16- or 32-bit microprocessors.

Early microprocessors needed several external components (ROM, RAM, clock, input/output circuits, bus drive, etc.) to work. Now most of these are on the same chip which may include even more functions. Application specific devices are available, including VLSI and ULSI circuits which incorporate a microprocessor.

3.9.4.4 ASICs

The term ASIC stands for *Application Specific IC* and refers to a digital device specifically developed for a particular application or customer. One example of these are Teletext ICs, for use in TV sets to select, filter and store Teletext signals, received with the video signal in normal transmissions, prior to conversion into characters and delivery to the screen.

Any customer who cannot solve a particular problem using standard logic devices has several alternatives. A *standard cell device* is a custom-made IC produced to meet the customer's needs by combining standard elements or cells, developed by the IC manufacturer, and held in a standard cell library. The IC user must have access to the cell library and to the support software needed for design. While development cost will be relatively low and the timescale to realisation quite short (of the order of a few months), the cost per device can be higher than for a full custom IC.

Gate arrays have anything from a few hundred to many hundred thousand gates and can be used to develop a design with the flexibility available from a single gate. Having resolved the interconnection pattern, the circuit can then be realised on a single chip by the IC manufacturer. The result is a semi-custom IC at low development cost, which can be obtained in weeks rather than months.

The term *Programmable Logic Device* (PLD) refers to a family of devices, the elements of which are only partially connected by the IC manufacturer. The intending user may interconnect them as preferred, rather like writing a PROM. There are several types of these, the most common being:

FPGA Field Programmable Gate Array;

FPLA Field Programmable Logic Array;

FPLS Field Programmable Logic Sequencer.

The relative cost of these devices is quite high compared with standard ICs or other custom made circuits. They can, however, be programmed in hours or days by a user with the necessary equipment and are extensively used for prototyping and small-scale production.

If all else fails, an IC tailored to the application will have to be developed. Such 'full-custom' ICs are expensive to develop and the process may take months or even years to satisfactory completion. If the potential market is very large, a successful design can be sold in sufficient quantity to recover design costs at a price much lower per device than that of any of the other solutions given above.

The use of ASICs restores to the machine designer some of the design freedom lost by the enforced use of 'standard' logic circuits.

3.9.5 Packages

Packages for ICs fall into three main categories: metal case TO type, DIP type, and flat package or flatpack.

For a component requiring only three pins, the TO metal case packages used for transistors may be employed. For components with more leads, the most used metal packages are the TO-99, the TO-100, Figure 3.39, and a few others such as:

Fig. 3.39 A TO-100 package assembled and soldered on a PB. Note that leads are bent irregularly, yet at the proper distance from the body. Mounting of such packages requires more labour than mounting DIPs.

TO-73   similar to TO-5, but with 12 leads;
TO-74   similar to TO-5, but with 10 leads;
TO-77   similar to TO-5, but with 8 leads;
TO-78   similar to TO-99, but with no central insulating seating plane;
TO-101 similar to TO-100, but with 12 leads.

All the pins on these cases are in a circle. The pins are numbered sequentially counter-clockwise, with the component viewed from the bottom, so that the pin closest to the orientation tab has the highest number.

Plastic dual-in-line packages (DIPs or DILs) have all leads in two parallel lines. Developed in the 1960s, for nearly twenty years DIPs accounted for some 80% of all IC packs because of their low cost and ease of assembly. The TO-116, Figure 3.40, is considered the forerunner of the DIP, although it originally had a ceramic body.

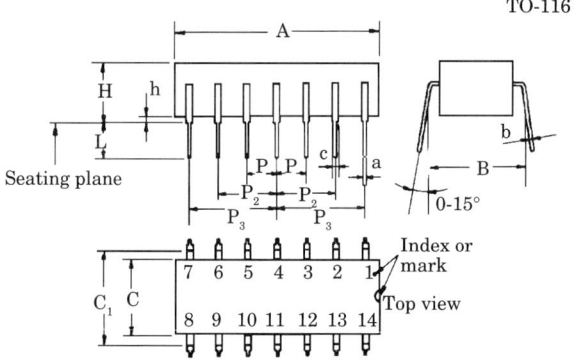

	millimetres		mils	
	min	max	min	max
A	16-76	19-94	660	785
B[1]	7-37	7-87	290	310
C	5-59	7-11	220	280
C_1	-	8-26	-	325
H	-	5-08	-	200
L	2-54	-	100	-
P[2]	2-29	2-79	90	100
P_2[2]	4-83	5-33	190	210
P_3[2]	7-37	7-87	290	310
a[3]	0-381	0-584	15	23
b	0-203	0-381	8	15
c	0-76	1-78	30	70
h[4]	0-51	-	20	-

(1) Total width when installed.
(2) Measured at the seating plane or up to 0.030 in. (0.762 mm) under it.
(3) Lead transition from dimension 'c' to dimension 'a' is optional on both sides of the seating plane.
(4) Clearance between the installed component and the PB.

Fig. 3.40 Package TO-116 (dimensions in millimetres are derived from those in inches). Note that most dimensions have wide tolerances.

Ceramic TO-116 bodies were made by a few suppliers and were quite similar to one another. When plastic bodies were produced, each IC manufacturer made his own transfer moulding dies with a 'personal' outline within the JEDEC standard outline dimensions, which have wide tolerances. This resulted in a number of outlines which are interchangeable on a PCB but not for automatic assembly, see Chapter 6.

Many versions have been developed from the original 14 pin package. An important one is the plastic bodied 16 pin DIP, having the same overall body dimensions as the TO-116 body, the four corner leads having been partially cut away. The 14 or 16 pin package named *quad-in-line*, QIL, has leads which are bent alternately inwards and outwards, Figure 3.41. This increases the mounting hole centre distance between adjacent pins from 2.54 mm to 3.59 mm (0.1 in. to 0.1424 in.), Figure 3.42, simplifying the design and manufacture of consumer PCBs. DIPs have also been produced with 6, 8, 10, 18, 20 and 24 pins. Some have metal fins to be soldered to the printed board or connected to a heat sink to improve heat dissipation, Figure 3.43.

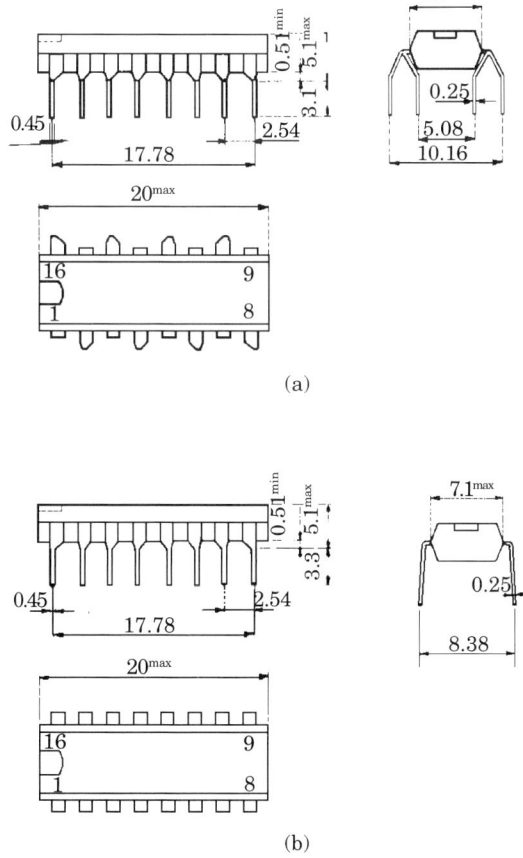

Fig. 3.41 Examples of packages derived from the TO-116: (a) 16 pin DIP; and (b) staggered 16 pin DIP. Dimensions are in millimetres. (Courtesy of SGS-ATES)

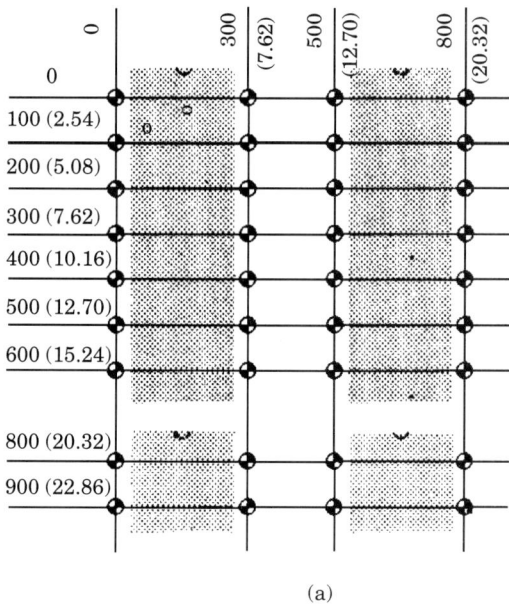

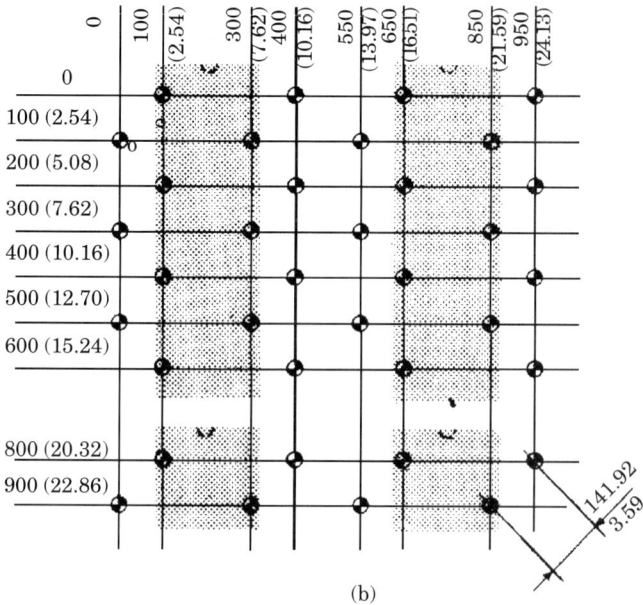

Fig. 3.42 Typical insertion hole pattern for 14 pin DIPs: (a) standard package; and (b) staggered pin package. Dimensions are in US mil and (in brackets) in millimetres. Suggested hole diameter is from 32 to 43 mil (0.8 to 1.1 mm).

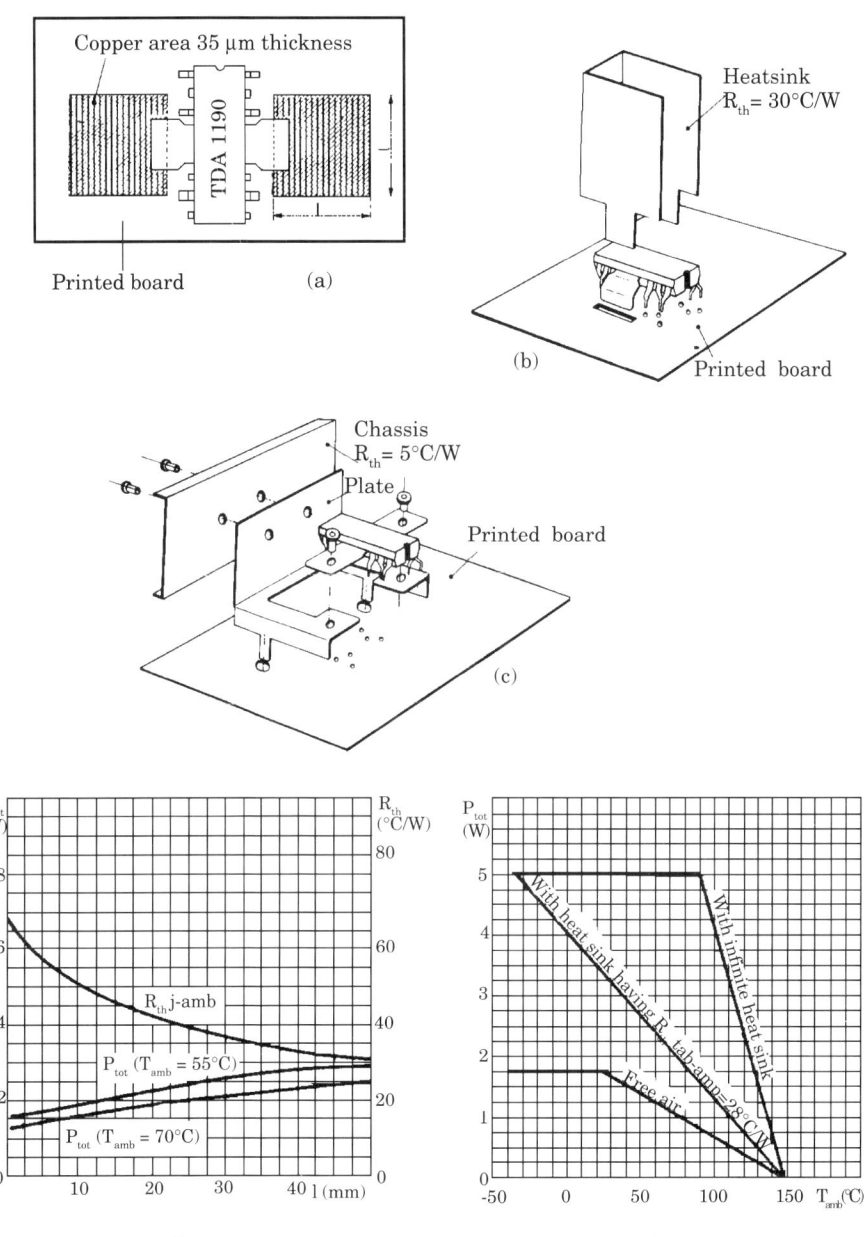

Fig. 3.43 Typical techniques for assembly of a DIP with built-in heat conduction tabs: (a) tabs are soldered to two copper areas on the PB; (b) and (c) tabs are used to fasten the component to a heatsink or to the chassis; (d) maximum dissipation allowed with assembly (a) as a function of dimension 'l'; and (e) maximum dissipation as a function of ambient temperature (T_{amb}) and thermal resistance of the heatsink. (Courtesy of SGS-ATES)

As chip size increased, larger packages were needed. A larger bodied version, 15.24 mm (0.6 in.) between the two rows of pins instead of 7.62 mm (0.3 in.), was initially produced in an alumina ceramic and then in plastic. A few versions appeared with a 10.16 mm (0.4 in.) wide body.

As the number of pins increases, the DIL becomes very long. A 40 pin package has a body length of about 52 mm (>2 in.) while a 48 pin package is around 62 mm long, almost 2.5 inches. Apart from occupying too much space on the printed board, the length of conductors on the leadframe is detrimental to high speed performance.

By reducing the pitch of the leads from 2.54 mm to 1.27 mm (0.1 in. to 0.05 in.) and staggering the leads to maintain the mounting hole centres at 2.54 mm, the length can be reduced, Figure 3.44. The resulting QILs, or 'QUIPs', have had limited success.

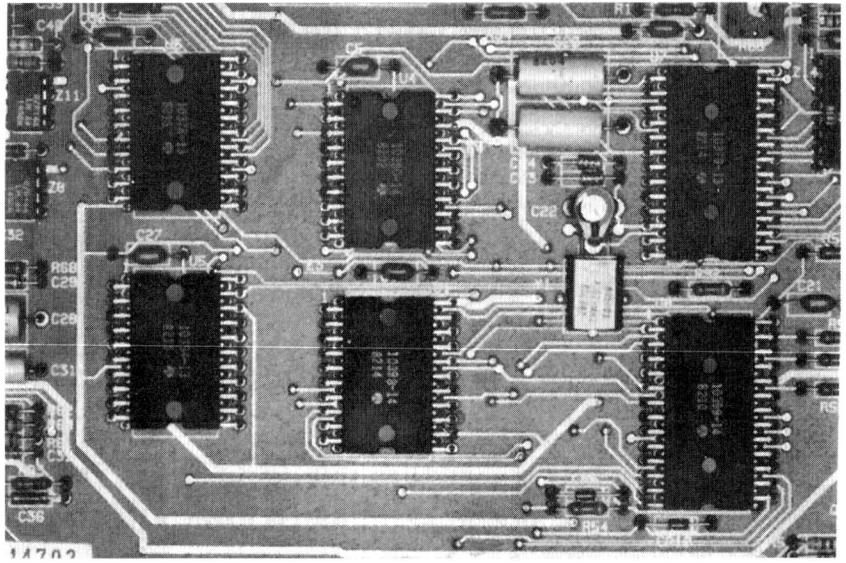

Fig. 3.44 Quad-in-line packages (QILs) in a telecommunications modem.

Flat packages, or flatpacks, are smaller than the DIPs and are assembled with the thin strip leads soldered or welded to the surface of the PB. A flatpack with long leads could, with some difficulty, be mounted as a QIL. Because it was not so easily adapted to mass assembly methods then available, the flatpack tended to be used only where weight and high packing densities were prime considerations, e.g., in the military field. A plastic bodied version has been adapted to modern surface mounting technology, see Chapter 4.

Devices needing a very large number of pins, say, 144 or 200, cannot be packaged in any standard case. The most common package, the pin grid array, is a flat case with all the leads arranged in rows on the bottom, Figure 3.45. Leads must be sturdy and precisely positioned or assembly is impossible. Types are available with leads on 2.54 mm pitch or on 1.27 mm pitch or even less. For conventional boards, the 2.54 mm pitch is preferred. For high pin count devices, surface mounting technology can offer more workable, and cheaper, solutions, see Chapter 4. In this context, it is relevant to note that, at an NEC Semiconductor conference 'Integration 2000', held in Birmingham, UK, 1992, one of the Japanese engineers showed a prototype 1,000-pin TAB packaged gate array. It is difficult to imagine in what other manner the package could be fabricated.

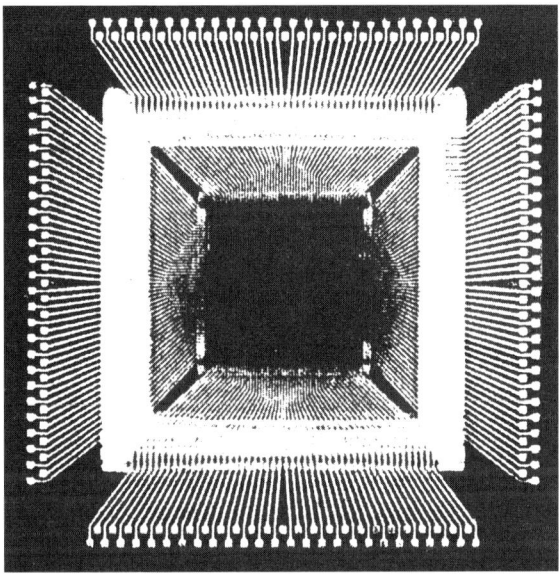

Fig. 3.45 Typical contact arrangement of pad array device, with pin grid array for comparison. (Courtesy of Motorola)

3.10 HYBRID INTEGRATED CIRCUITS

Certain components, such as inductors, cannot be realised by monolithic IC technology. Others, such as resistors and capacitors, can be incorporated in the chip but the resulting devices have a limited range of value and are of low precision.

The demand for well-defined and reproducible functional modules which can incorporate any type of component (including monolithic ICs!) is satisfied by hybrid integrated circuits, referred to simply as *hybrids*.

Hybrids achieve very high component density and are used to improve the packing density of an assembly. The advent of surface mounting assemblies has reduced this advantage, but other advantages of hybrids remain, for example, the availability of pre-adjusted circuits, and simplified assembly and test. They can be produced effectively even in small quantities and can, with limited investment, be produced in-house by the intending user.

The term 'hybrid' shows that these circuits are not limited to a single technology and may incorporate many. The two main classes refer to the manner of producing the conductive, and resistive, elements. These are *thin film* and *thick film*. To these may be added the present pinnacle of hybrid technology, the *multichip module* (MCM). However, as the latter modules are more truly treated as surface mounting modular components, they are dealt with in Chapter 4. It will suffice here to say that the IPC definitions adopted back in 1990[8] show that MCM type D modules use thin film techniques, MCM type C modules employ thick film technology, and the type L modules are based on the use of laminate structures, i.e., printed boards, for interconnection.

3.10.1 Thin Film Hybrids

There have been many definitions of 'thin film'. One such is that the circuit elements are less than 5 μm thick. To a production engineer, the most important fact is possibly that they are produced by a mixture of vacuum technology and photomechanical/etching techniques. To the design engineer, the primary reason for thin film technology is the degree of precision available. For example, let us consider the humble resistor. A resistor can be produced in single crystal silicon with resistivity in the range 100-300 ohms/square. The normal precision as produced will be of the order of 20% and it is not practicable to trim the resistor. A thin film resistor produced using nichrome can be made with resistivity in the range of 10 to 1000 ohms/square. The precision as produced can be better than 5%, and this can be improved by trimming to within 0.01% or better.

The production process starts with a thin ceramic substrate, usually high grade alumina, which is of a suitable size to allow many circuits to be made in parallel. After careful cleaning the substrate is placed in a vacuum evaporator where final cleaning is carried out by ion bombardment. The resistive layer, usually nichrome, is then applied either by evaporation or by cathode sputtering to a thickness adequate for all resistance values required. If conductors only are needed, a few nanometres of nichrome will be deposited first to act as a bonding coat. A thin layer of copper is then deposited. Alternative resistive materials, e.g., tantalum nitride, are sometimes used.

The conductors are formed by coating with a photosensitive resist, photoimaging, electroplating with copper and/or gold to give the required thickness, and then removing the photoresist and etching away the vacuum deposited copper to expose the resistive layer. The process is then repeated to etch the resistive layer selectively so that it remains only where a resistor is required. (The basic processes employed in these sequences are essentially similar to the processes used for printed board fabrication, described in detail in Volume 2, Section 1.) As the specific resistance of the resistive layer is defined by the material and the thickness deposited, the different resistor values are controlled by the width and length of each resistor, Figure 3.46.

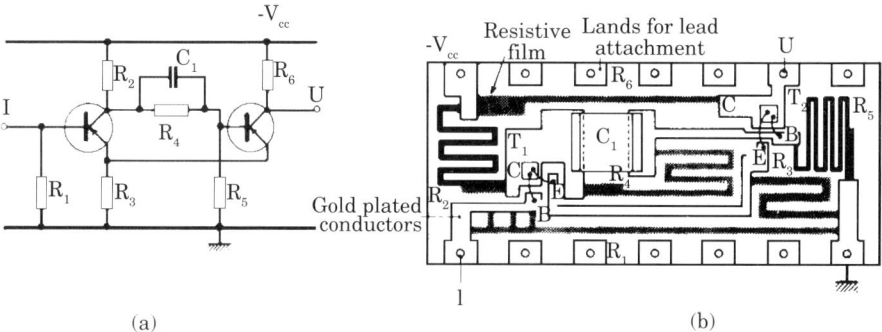

Fig. 3.46 Electrical diagram (a) and arrangement (b) for a thin film hybrid microcircuit ready to be enclosed in a 14 pin DIP package. The same circuit is used as an example for PCB design with discrete components in Chapter 9.

Resistors are normally produced to be 10% or 5% *below* the nominal value and can be laser trimmed upwards in value to within 0.01% of the required resistance, Figure 3.47. Trimming is carried out with continuous measurement of resistance so that it may be stopped when the required value is reached.

With a tantalum nitride resistive film, capacitors may be made directly on the substrate by selectively oxidising the capacitor areas, to form the dielectric, and then covering the dielectric with a second metal layer. Other dielectrics, ceramics, oxides and insulators can be deposited, e.g., by radio frequency sputtering, and, if desired, a multilayer thin film circuit can be constructed by sequential deposition.

It is also possible to fabricate diodes and transistors directly on the substrate by thin film techniques.

Normally, the thin film technique is employed to fabricate only high precision resistors, capacitors and inductors of limited range, and conductors, all other components being added after the substrate has been divided into individual circuits. Diodes, transistors and ICs may be added as unpackaged chips, attaching them to a gold plated land with a conductive resin or by eutectic bonding. It is pertinent to note that eutectic bonding can cause problems when used on hybrids since, if several devices are on the hybrid, some devices will have to stay at a high temperature for quite a long time. This has caused much trouble in the past by giving noticeable device degradation. The use of an adhesive processed at lower temperatures will preserve device characteristics. It should be noted that, for attachment of ICs, the adhesive need not be conducting — in fact there are cases where it is possible to route conductors under the ICs.[9] Connections between lands

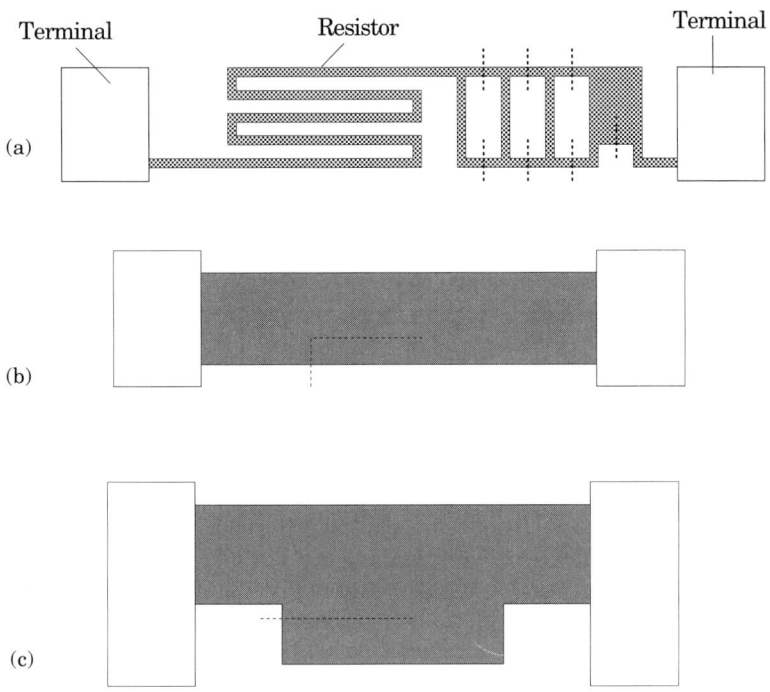

Fig. 3.47 Trimming resistors. The dotted lines show some of the ways in which cuts may be made to increase the resistance.

and the chip pads are usually made by the methods described earlier for transistors, although the 'bump' technique, described in Chapter 4, 4.4.16, has been used. When bonding is complete, the assembled chips and bonds will be protected by a drop of suitable encapsulant or conformal coating material. The bonding operations and encapsulation must be carried out in an ultra-clean room, to prevent contamination of the chip surfaces.

Alternatively, active components in small packages like the SOT (*Small Outline Transistor*) and SO (*Small Outline*) types can be connected to the conductors by reflow soldering, see Chapter 15. The same technique is used for other passive components — capacitors, inductors, special resistors, etc. — which will usually be in surface mount packages, see Chapter 4. Leads are added and soldered, unless a metal package is to be used. The substrate itself may be used as a leadless ceramic carrier like those for surface mounting. (Hybrids were, in fact, the first type of surface mount assemblies.)

As stated previously, it is possible to make multilayer thin film circuits, but the cost penalty is high. The great majority of thin film hybrids have only one conductor layer. The devices are still expensive but very accurate and stable resistors can be made in this way with high overall reliability. Because it is possible to produce very accurately controlled line widths, thin film circuits have excellent high frequency performance and are extensively used in low-loss type circuits.

3.10.2 Thick Film Hybrids

Thick film technology provides a less expensive and more flexible method of making hybrid circuits.[3] Thick film hybrids are almost invariably produced by screen printing methods, see Volume 2, Chapter 3, on ceramic substrates. The major ceramic used is alumina, the surface finish of which does not have to be so fine as for thin film. The inks employed have been developed specifically for this technology and are not to be confused with the inks used for polymer thick film PBs, see Volume 2, Chapter 12.

Inks for thick film hybrid circuits are mixtures of metal and/or oxide powders, particle shape and size being very important, glass powders, and resins and solvents, which act as the carriers and binders until the ink has been printed ready for firing. For conductors, while copper or nickel-containing inks and some other metals or alloys used, will have to be fired (to eliminate solvents and resinous binders) in an inert atmosphere, noble metals, e.g., gold, silver and the platinum metals, may be fired in air. Palladium-silver is the most commonly used metal.

Resistors are fabricated with inks based on materials such as ruthenium oxide, bismuth ruthenate and tin oxide. Atmosphere control during firing can be critical. After firing, the resistors must be trimmed, usually by laser. Precision is not as good as with thin film resistors but, in large volume production, better than 0.1% is now routinely achieved. Providing the firing atmosphere is controlled as necessary, inks may be fired more than once, allowing the production of layers of resistive material of different thickness and hence different specific resistance. Thick film resistors can be produced to cover a range from $0.1\ \Omega$ to $1000\ M\Omega$.

The ability to fire and refire makes it relatively easy to screen and fire an insulating layer, followed by another layer of conductors. While doing this, capacitors may be produced, albeit within a limited range of values. Inks are available with high dielectric constants although at the highest values (up to 2000) the loss factor is also high and it is difficult, if not impossible, to control the spread of values. An ink with a dielectric constant up to about 100 can produce a low value capacitor — up to 50 pF/mm^2 of print area. It is normally preferred to use add-on chip capacitors, as for thin film circuits.

When conductors, resistors, dielectrics, and so on have been printed and fired, the substrate will be split into the individual circuits. Any required add-on items are assembled, Figure 3.48, and testing and packaging complete the circuit.

While being more flexible and cheaper to produce, thick film hybrids do not have the high reliability of thin film and are seldom applicable at very high frequencies.

3.10.3 Packages

High reliability thin film hybrids are often packaged in metal cans, either standard or specially made, Figure 3.49. If suitable, with pins on a DIL layout, Figure 3.50, they may be encapsulated in a moulded package. Thick film hybrids may be packaged similarly, although less expensive solutions may be employed. The so-called *single-in-line* (SIL) package, which is not standardised, is often used, Figure 3.51. As the SIL is made by dipping the hybrid into a resin or putting it into a small box and filling with resin, the body tolerances can be quite large, Figure 3.52. Transfer moulding, which would give more reproducible dimensions, is seldom used because of the small production volumes involved.

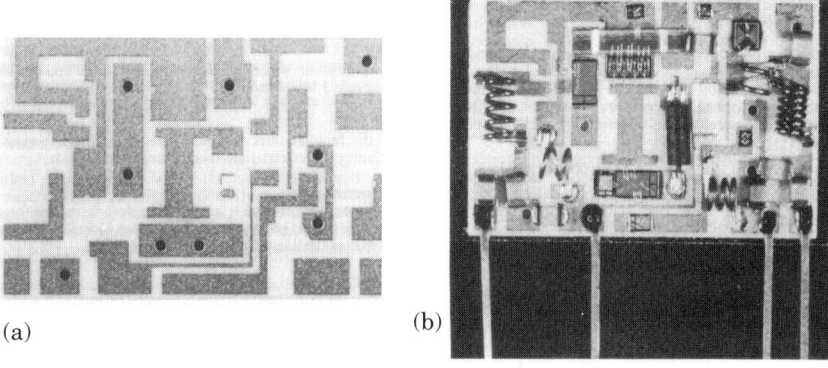

Fig. 3.48 The conductor pattern (a) and the completed assembly (b) for an r.f. application using thick film technology.

Fig. 3.49 (a) Complex LSI hybrid with SSI and MSI TTL chips; (b) High reliability metal encapsulations. (Courtesy of RIFA AB, Sweden)

Fig. 3.50 A thin film hybrid circuit. (Courtesy of Philips and MBLE)

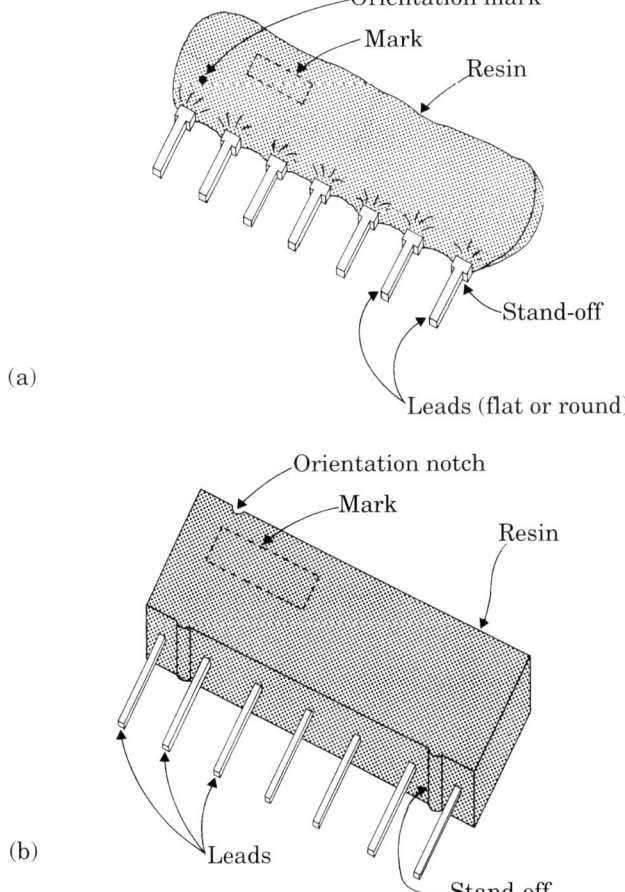

Fig. 3.51 Single-in-line (SIL) packages for thin film hybrids. (a) SIL package (resin dipped); (b) SIL package (potted or injection moulded).

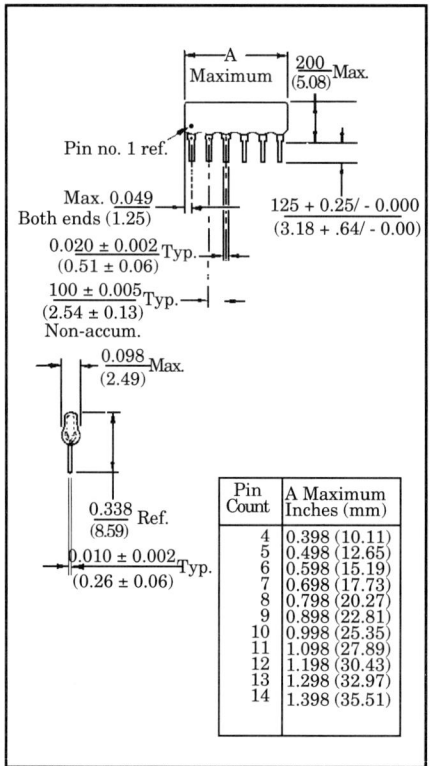
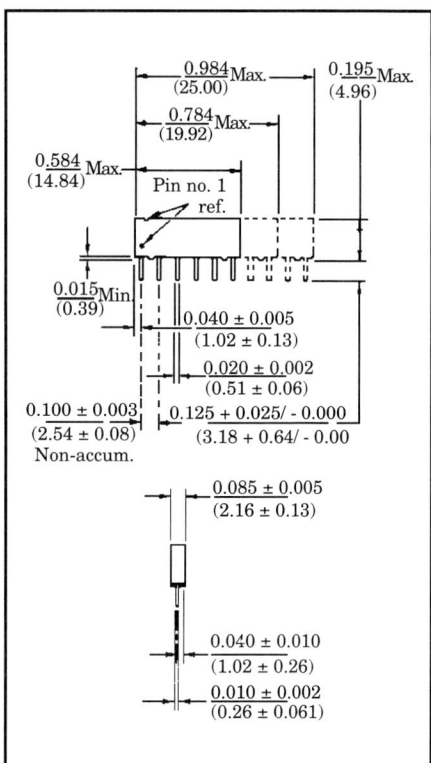

(a) resin dipped package   (b) potted or moulded package

Fig. 3.52 Dimensions of SIL packages.

REFERENCES

1. For example the DATA books (published by Data Books Inc., 2 Lincoln Avenue, Orange, New Jersey 07050, USA) which describe over 150,000 solid state components. The rapid evolution of components means that frequent updating of these books is essential.
2. See 'Thermistors' by E. D. Macklen, Electrochemical Publications Ltd, Asahi House, Church Road, Port Erin, Isle of Man, British Isles (1979).
3. For further reading see:
   Holmes, P. J. and Loasby, R. G., 'Handbook of Thick Film Technology', Electrochemical Publications Ltd (1976). Pitt, K. E. G., 'An Introduction to Thick Film Component Technology', Mackintosh (1981). Haskard, M. R. and Pitt, K. E. G., 'Thick Film Technology & Applications', Electrochemical Publications Ltd, Asahi House, Church Road, Port Erin, Isle of Man, British Isles (to be published 1996).
4. Electronic Industries Association, Engineering Department, Standard Sales, 2001 Pennsylvania Avenue, NW, Washington, DC 20006, USA.
5. CECC 50 000. 'Harmonized system of quality assessment for electronic components. Generic specification: discrete semiconductor devices'.
6. 'Blue LED 100 times brighter', *Electronics Weekly*, January 19 (1994).

7   Parry, S., 'SiGe looks to plug GaAs leak', *Electronics Weekly*, January 12 (1994).
8   Geschwind, G. and Clary, R. M., 'Multichip Modules: an Overview', *Printed Circuit Fabrication*, November (1990).
9   Private communication, Griffiths, G., Cammax Precima Ltd, Mildenhall, UK, April (1994).

Chapter 4

SURFACE MOUNTED COMPONENTS

GIOVANNI LEONIDA
Milan, Italy

W. MACLEOD ROSS
Welnorth Ltd, Bishop's Stortford, UK

4.1 GENERAL

IEC Publication 194, Third Edition, 1988, defines surface mounting as "Electrical connection of components on the surface of a conductive pattern without utilising component holes". This was amended in the revision circulated in 1994 to "The electrical connection of components to the surface of a conductive pattern that does not utilise component holes".[1] Whatever the definition, the technique is not new, having been employed for many years in the production of thin film and thick film hybrid circuits. In 1960, IBM introduced the 360 series computers which were based on thick film hybrid modules, utilising the then available passive and active surface mounting devices such as leadless or chip capacitors, some resistors, chip diodes and transistors. Such devices were more expensive than conventional leaded components and other uses of them were generally restricted to military and other relatively low volume applications, where the higher cost was less important than space saving and reliability.

With the increased complexity of the monolithic IC, new functions were incorporated on the chip, the area needed by the IC was reduced, and the board area occupied by discrete passive components became relatively more important. The volume of the final assembly was dictated by a limited number of discrete components and by the need to interconnect active and passive devices. Simultaneously, the general growth of electronics in the sixties and seventies had stimulated component manufacturers to increase the production of leadless components so that the cost per unit was close to that of conventional leaded devices. At the end of the seventies, the variety of available components was large enough for them to be introduced into some products where space was creating a premium on cost.

Wider usage brought further reductions in price and spurred the development of good automatic mounting or 'pick-and-place' machines, specifically designed for use with surface mounting components. In the early eighties, surface mounting technology began to grow and has continued to grow during the nineties.

Surface mounting technology must not be dismissed as simply an extension of the conventional or traditional PCB technology. While it is resolving many problems associated with the functional realisation of electronic circuits, it has brought new problems to the fore. The resolution of these has spurred designers, material and equipment suppliers, and production engineers to new heights of achievement.

4.2 CLASSIFICATION

The term *leadless component* is often misused as a *surface mounting component* or *surface mounted component*, an SMC, may have leads. Strictly speaking, the device is 'surface mounting' before being assembled and 'surface mounted' after it is assembled to the board. *Surface mounted device*, SMD, is also used although the term is registered to the North American Phillips Corporation. The less grammatical terms of *'surface mount'* component, device, assembly or technology have been widely adopted. A printed board having SMCs on one or both sides is called a surface mount board and, when assembled, an SMA or *surface mount assembly*.

An SMC is a component intended to be connected to the surface conductors of a PB without the use of mounting holes. SMCs may therefore be attached to one or both sides of a PB and may or may not be mixed with conventional components, designed for mounting in holes, Figure 4.1. The four types of printed board assembly (PBA) are:

(a) *traditional*, with leaded components mounted through holes in the board;
(b) *single-sided SMA*, i.e., SMCs only, with all components on one side, typically the primary side of the PB;
(c) *double-sided SMA*, i.e., SMCs only, with components on both primary and secondary sides of the PB;
(d) *mixed technologies* (*mixprint* or *mixed print*), i.e., both SMCs and conventional components with components located on both primary and secondary sides of the PB. (Usually the conventional components are mounted on one side only.)

It should be noted that terms such as 'solder side' and 'component side' are meaningless in surface mounting. Terminology which accommodates both leaded and surface mounting components is:
    Primary side — coincident with design layer 1.
    Secondary side — side furthermost from layer 1.
For fuller definitions, see Reference 1.

SMT (*surface mounting technology*) offers the designer and packaging engineer much more choice in tailoring the assembly to the specific need. As mixed technology boards offer two alternatives, there are five possible solutions instead of only one.

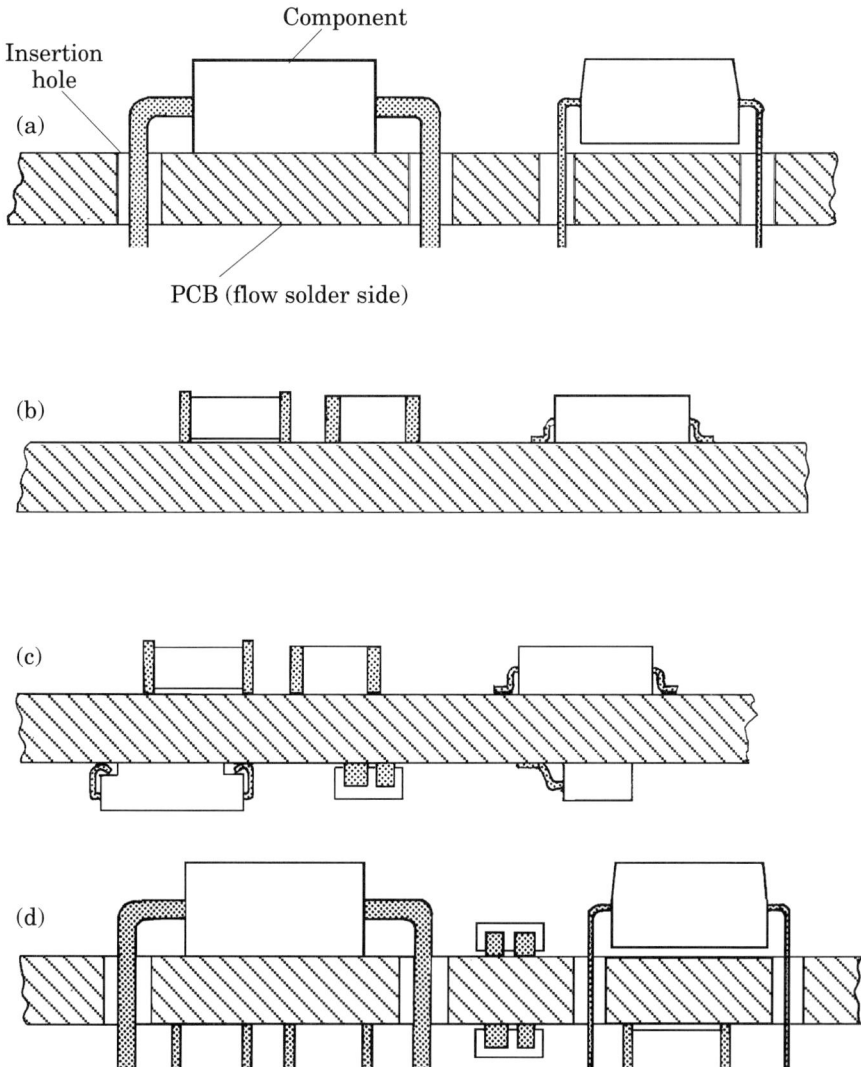

Fig. 4.1 Types of assembly. (a) Traditional or conventional assembly; (b) surface mounted components, one side only; (c) surface mounted components on both sides; (d) mixed technologies with SMCs on both sides.

To classify SMCs properly, the first characteristic to examine is the leads. These devices may have no lead, the electrical connection being made to two or more solderable surfaces, normally at the ends of the component. Leaded devices may be of several types. They may have flattened leads, normally formed outwards in an inverted 'gull-wing' shape; alternatively, the SMC leads may be formed under the body of the device in the shape of a 'J' (the foot of the J may or may not fit in a recess in the device body), Figure 4.2. Other shapes or forms are used and in some cases the SMC has mixed connections, e.g., one bent lead and one solderable surface. It should be noted that conventional leaded devices can be converted readily into a surface mounting format. The type of lead is very important for ICs as ease of assembly, soldering and reliability are all affected by the type selected.

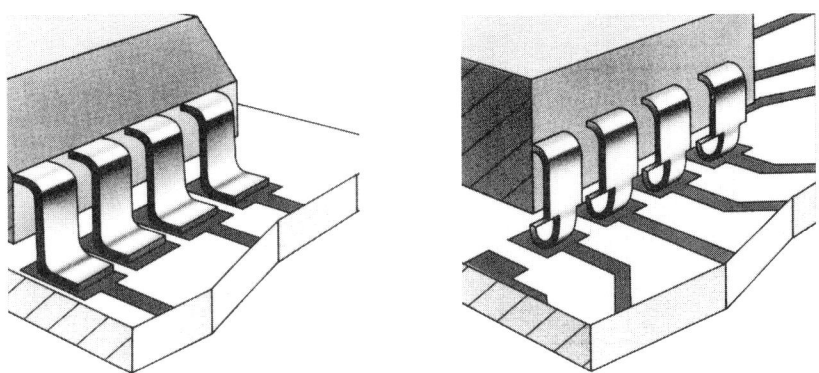

Fig. 4.2 An SOIC having gull-wing leads and a PLCC having 'J' leads.

Whatever the lead form, all SMC package connections have a common requirement: they must be readily solderable and maintain good solderability over long storage periods. Manufacturing SMAs with components of poor solderability is a very tough job!

Most discrete SMCs have no leads and are divided into two categories by body shape. The two shapes are cylindrical (the body is roughly a cylinder), and rectangular (the body is essentially a parallelepiped).

A MELF (*Metal Electrode Face Bonding*) is typical of cylindrical components, the Mini-MELF and Micro-MELF being smaller versions. These components were originally developed by the Taiyo Yuden Company, Japan, and are now available from several sources. (The JEDEC SOD-80 outline is sometimes called a Mini-MELF.)

The rectangular devices are often called *chip components*. The term is strictly only applicable to devices manufactured all together and divided into chips at the end of the major manufacturing process. To avoid confusion with semiconductor chips, the term 'chip' should be used only in the form 'chip capacitor' or 'chip resistor', etc.

There are other package shapes and ICs, in particular, have specific packages which are covered later in this chapter.

4.3 PASSIVE SURFACE MOUNTING COMPONENTS

Most packages for passive SMCs fall into one of the types described hereafter. Differences may exist in the packaging between one supplier and another and data sheets for a specific SMD must be checked before finalising a layout. Dimensions given in the following paragraphs must be taken as nominal or typical values, not as standard values. New components and new

4.3.1 MELF Components

The concept of MELF components is to eliminate leads and, consequently, the cost, space, difficulty of handling and potential failure associated with them. If we go back to the film resistor described in Chapter 3, Figure 3.5, we see that prior to lead attachment they are functionally complete, except for soldering or welding the leads, coating and marking. At that stage they resemble closely a MELF resistor, which is manufactured by exactly the same process.

The conventional resistor is completed by making a joint (i) between the resistive film and the end cap, (ii) the end cap and the lead, (iii) the lead and solder and (iv) the solder and the PB conductor. As there is a lead at each end of the resistor the jointing operations must be repeated, giving each resistor a total of eight joints in the chain between the resistive element and the PB. Each of these is a potential source of unreliability. On the other hand, the cap of the MELF is soldered directly to the PB conductor, replacing the cap-to-lead and lead-to-solder joints by a single joint between the cap and the solder, Figure 4.3. Space is saved by removing the leads and the reliability of the assembly is increased by reducing the number of joints.

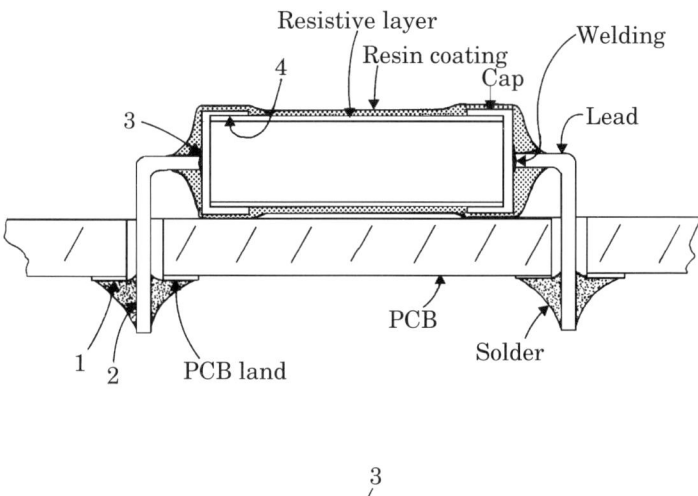

Fig. 4.3 The conventional component, top, has 4 solder joints/end; the MELF equivalent has only 3.

The shorter distance between the functional element of the SMC and the PB pads and the smaller overall component length may also increase the performance of the circuit in terms of speed. However, the high packing densities achievable can create problems in dissipating heat generated by the components and may require special types of PB, see Chapter 11 and Volume 2, Section 2, in particular Chapter 10.

When MELFs are employed in mixed technology boards they are normally placed on the secondary side of the board, the conventional components being on the primary side. This means that they can be wave soldered at the same time as the conventional leads. As the MELFs will be in direct contact with solder at about 240-260°C (464-500°F) the bodies must be ceramic, the resistive film being carbon or metal. Any coating on the bodies must be heat resistant, possibly with a second layer to improve moisture resistance. The coating must not be allowed to contaminate the end caps, which will be plated, usually with tin-lead alloy to give good and durable solderability.

While the MELF is cheap and takes up minimum space on the board, the cylindrical shape can cause problems on assembly due to their tendency to roll. Even a highly tacky solder paste will not always secure the component. A sure solution is to use a spot of adhesive which adds time and cost. For this reason many authorities advise that, unless used in a situation which demands the use of adhesive for all SMCs, the MELF should be replaced with rectangular devices. Although the prime cost may be doubled, the real cost of one reworked MELF resistor joint can be equivalent to the purchase price saving on 500 or more resistors.

4.3.1.1 RESISTORS

The MELF resistor is smaller and lighter (typically, 1000 MELFs weigh only 60-70 grams, about 2 ounces) than the conventional resistor and, when soldered to a PB, are much more resistant to shock and vibration. A MELF resistor is typically 5.9 mm (0.232 in.) long and has a maximum diameter (adjacent to the caps) of 2.2 mm (0.087 in.). The diameter of the body between the caps is usually less, about 2.1 mm, Figure 4.4. The tolerances on dimensions are quite large and the actual outline size can vary significantly from supplier to supplier.

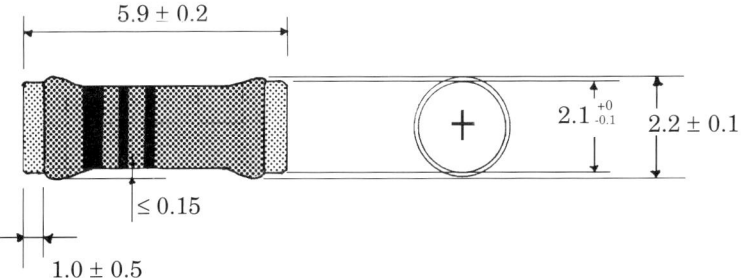

Fig. 4.4 A MELF resistor. All dimensions are in millimetres.

Most manufacturers use colour bands for identification of value, the colours having the same significance as for conventional components, see Chapter 3, Table 3.2. The first band to be read is that closest to a cap. It is often wider than the others to aid recognition. Carbon film and metal film resistors are available in MELF packages, the metal film being more precise and more stable but also more expensive. An extensive range of resistance values is obtainable, with one or more values of tolerance. The small size restricts power rating to $1/4$ watt as the casing will not dissipate more. Resistors of $1/8$ watt rating are packaged in Mini-MELF cases, see later text.

As there are no leads to dissipate heat, much of the heat generated passes to the substrate through the solder joints. The temperature rise of the MELF is therefore dependent on the thermal conductivity of the substrate. If the MELF is mounted on alumina, the rise will be only half that reached if it is mounted on epoxide/glass or paper/phenolic laminate.

On an alumina substrate a $1/4$ watt MELF resistor may be loaded to dissipate 0.35 watt. The temperature rise over ambient will be about 20°C, which is acceptable; on an epoxide laminate, 0.35 watt would produce a rise of about 45°C.

4.3.1.2 JUMPERS

Large numbers of MELFs are used on single-sided PCBs to realise particular SMAs. In such assemblies there may be a need for *jumpers* or links to connect all components. To save space, it can be useful to put some of the jumpers on the secondary (soldering) side of the PCB.

This can be done using MELF *zero-ohm* resistors. A zero-ohm resistor has such a low resistance, of the order of 0.015 Ω, that it may be neglected in most circuits. The allowable current for these devices may be as high as 5 amps.

4.3.1.3 POSITIVE TEMPERATURE COEFFICIENT (PTC) RESISTORS

PTC resistors are available in MELF packages from a few suppliers. They are basically metal film resistors with a special nickel film on a ceramic core of very high thermal conductivity. Several types can be obtained, with positive temperature coefficients ranging from 0.1 to 0.4% per°C and linear change in resistance over a wide temperature range, e.g., -30 to +130°C.

4.3.1.4 CAPACITORS

Ceramic bodied MELF capacitors have roughly the same dimensions as resistors, Figure 4.4. Colour coding is used for values and some manufacturers colour the body to indicate voltage rating, e.g., pink = 16 V or 35 V and pale green = 50 V.

Construction is similar to conventional tubular capacitors, Figure 3.14, except that each of the two plates projects slightly on one end and is metallised to make the solderable contact. Inductance is lower than for the leaded version due to the larger area of contact between each plate and its termination. The construction is capable of withstanding fairly high temperatures providing care is taken in selecting an encapsulating resin.

A wide range of values is available. Tolerance on the nominal capacitance is normally +5% or +10%. The change in value of capacitance with change in temperature may be as low as 0.5% in going from -20°C to +80°C but does depend on the dielectric.

4.3.1.5 INDUCTORS

Progress in ferrite technology has resulted in cores which can be exposed to soldering temperatures without losing their ferromagnetic properties. This has led to inductors becoming available in MELF configuration.

4.3.2 Mini-MELF Components

The Mini-MELF is typically 3.5 mm (0.14 in.) long, with a maximum diameter of 1.45 mm (0.06 in.), as shown in Figure 4.5. There is a relatively large difference in outline size from one supplier to another. The Mini-MELF was introduced because, in some applications, the MELF was too large. The small outline (and low cost) allows the Mini-MELF components to compete more favourably with chip components.

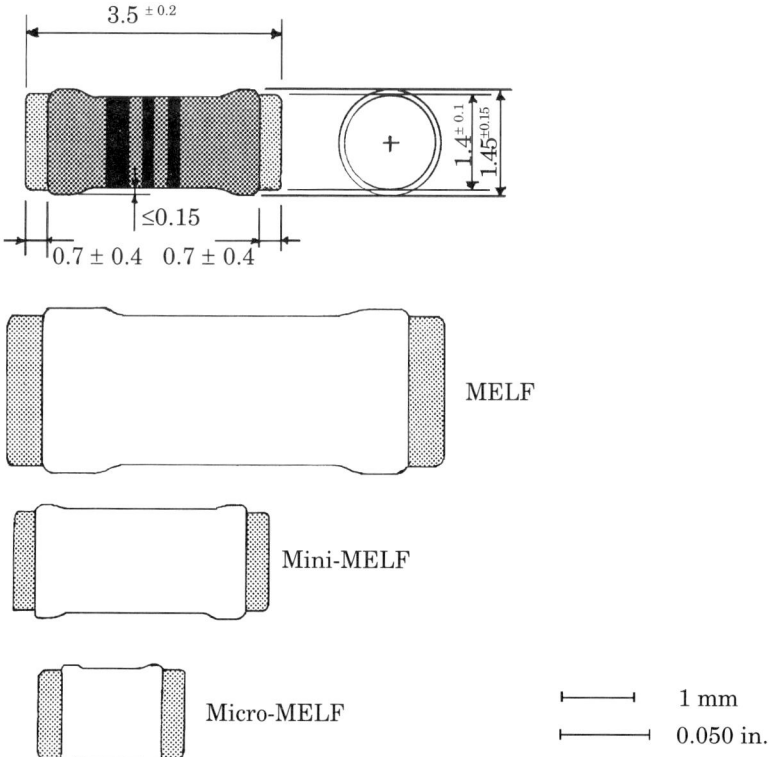

Fig. 4.5 The Mini-MELF compared with the MELF and Micro-MELF.

This family of components includes all those described in MELF configuration. In addition, new families of components are being packaged in this form, including some active devices. Power rating for resistors is restricted to $1/8$ watt, although this may be increased provided that the heat generated can be removed.

The Mini-MELF is so small that assembly by hand is almost impossible. Fortunately, most modern automatic and semi-automatic assembly machines will deal with the even smaller packages now available, see Chapter 7.

4.3.3 Other Cylindrical Components

There are other components which resemble the MELF or Mini-MELF but have different shapes or dimensions. The majority of these can be assembled using the same equipment.

The *Cerachip*, a single layer tubular ceramic capacitor made by Taiyo Yuden, was originally produced in Mini-MELF size to be compatible in assembly with Mini-MELF components, Figure 4.6. They are now available in three sizes as shown in the figure. The major feature (due to the single layer structure) is very low internal inductance and hence better high frequency performance.

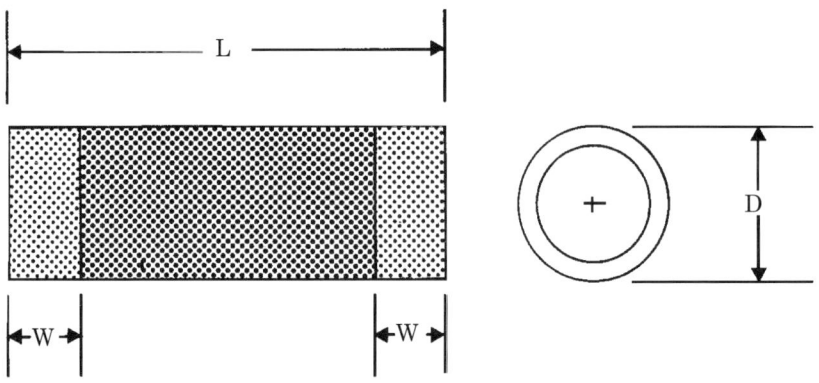

Dimensions (mm)

Type	L	D	W
033	1.6 + 0.15 − 0.1	1.0 ± 0.075	0.3 ± 0.15
053	2.0 + 0.15 − 0.1	1.25 ± 0.1	0.3 + 0.2 − 0.1
103	3.4 ± 0.2	1.5 + 0.1 − 0.15	0.6 + 0.3 − 0.2

Fig. 4.6 Cerachip ceramic capacitors.

In 1987, the continuing pressure for miniaturisation led to the introduction by Taiyo Yuden of a range of small resistors, from 4.7 ohm to 1 Megohm, in a package called a Micro-MELF. As shown in Figure 4.5, these are much smaller than the Mini-MELF. The diameter is 1.25 mm (1.35 maximum to 1.20 minimum) and the length is only 2.0 + 0.1 mm, approximately 0.080 in. As with the MELF, the diameter may be less in the middle of the component. These components had the same power rating ($1/16$ watt) as a thick film chip resistor, which has the same dimensions (2.0 mm length and 1.25 mm thick) although a parallelepiped. With proper design of the PCB, the two types of component may be interchangeable, mechanically and electrically. This is one of the few situations in which cylindrical and chip components can be mounted as alternatives to each other. Normally the designer must decide in advance which shall be used in each position (although both types may be present on the same board).

By 1994, Taiyo Yuden offered a carbon film resistor in the 'Micro-MELF' package with a power rating of $1/8$ watt and a metal film resistor with a power rating of $1/16$ watt in a cylindrical package only 1 ± 0.075 mm (0.039 ± 0.003 in.) long and 0.65 ± 0.075 mm (0.025 ± 0.003 in.) diameter.

The *Ceralock* series of ceramic resonators is a diverse series of extremely small resonators covering a range of frequencies from about 1.8 MHz to 34 MHz in a variety of package shapes and sizes. The Type CSAC MGC, illustrated in Figure 4.7, covers the range from 1.80 to 6.00 MHz and is in a

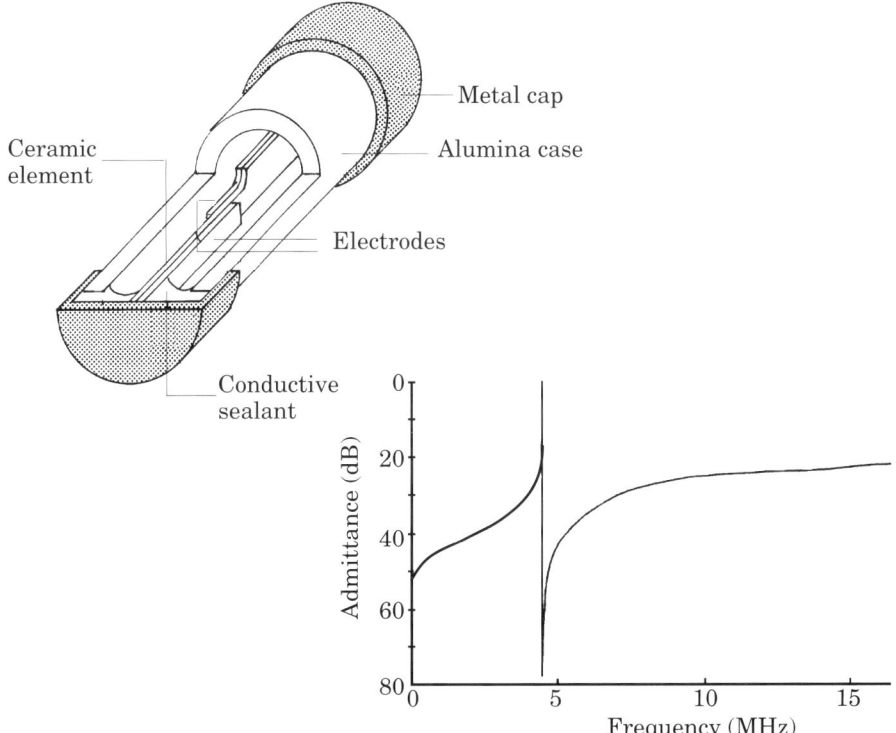

Fig. 4.7 The Ceralock Type CSAC MGC ceramic resonator.

tubular case 7.0 ± 0.2 mm (0.276 ± 0.008 in.) long with a maximum diameter, at the end caps, of 2.8 ± 0.1 mm (0.110 ± 0.004 in.). The middle 4 mm (0.16 in.) is 2.6 mm (0.1 in.) in diameter. The outline is, therefore, larger than that of a MELF, but it can be mounted by the same equipment. Other types of Ceralock are rectangular and look like chip components although the name of 'chip Ceralock' is a misnomer. These devices are used as reference oscillators to replace quartz crystals, for instance, in generating the clock of a microcomputer. Together with small size and absence of leads, they offer good performance at low cost. Ceralock resonators are manufactured by the Murata Manufacturing Company Limited.

4.3.4 Chip Components

Chip components were developed specifically for SMAs and are conceptually quite similar to MELFs as far as connection to a substrate is concerned. They were developed originally for the hybrid IC market and many of the devices described are common to hybrid circuits and surface mounted PCBs.

The chip component family is quite small, being confined to resistors, multilayer ceramic capacitors, and some inductors.

4.3.4.1 RESISTORS

A common type of chip resistor is the 'size 1206'. The figure should be read as 12-06 and gives the length and width (thickness is not specified) in hundredths of an inch. That is, the 1206 is 0.12 in. long and 0.06 in. wide. This is a standard size for a resistor having a power dissipation of $1/8$ watt. Another common outline is the 'size 0805', which is the standard for $1/16$ watt power rating. The actual dimensions for both sizes are given in Figure 4.8. The weight of these devices is, naturally, very small. 1,000 of the size 1206 chips weigh about 9 grams (0.02 lb) and 1,000 of the size 0805 weigh about 4 grams (0.009 lb).

This method of specifying size applies to resistor and capacitor chip components and was developed by the American EIA. All manufacturers now specify dimensions in metric terms while still using the EIA code as a general reference. The EIA code may, however, eventually be superseded by the metric code used mainly by Far Eastern component manufacturers. The equivalents are:

Metric Code		mm (in.) x mm (in.)		EIA Code
1005	=	1.0 (0.040) x 0.5 (0.20)	=	0402
1608	=	1.6 (0.060) x 0.8 (0.030)	=	0603
2012	=	2.0 (0.080) x 1.2 (0.050)	=	0805
3216	=	3.2 (0.120) x 1.6 (0.060)	=	1206
3225	=	3.2 (0.120) x 2.5 (0.100)	=	1210
4532	=	4.5 (0.180) x 3.2 (0.120)	=	1812
5650	=	5.6 (0.220) x 5.0 (0.200)	=	2220

The shape is not a perfect parallelepiped and, again, sizes and tolerances of the packages differ from supplier to supplier. If space is critical, careful

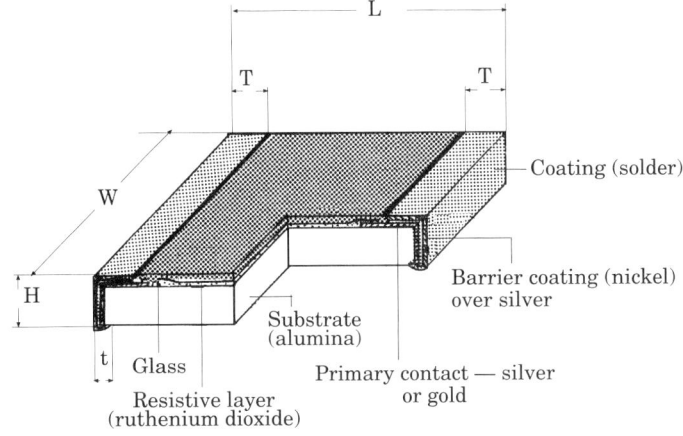

		L	W	H	T	t
Size 1206	1/8 watt	3.2 ± 0.15	1.6 ± 0.15	0.6 ± 0.10	0.5 ± 0.25	0.35 max
Size 0805	1/16 watt	2.0 ± 0.20	1.25 ± 0.20	0.5 ± 0.10	0.40 ± 0.25	0.35 max

Fig. 4.8 Thick-film chip resistors.

checks must be made on the values given in the data sheets of current and potential suppliers. Resistors are widely available with values ranging from about 10 Ω to 2.21 MΩ. The tolerance can be from ≤1% to 453 Temperature coefficient ranges from as high as 500 ppm/°C for a 20% tolerance resistor to less than 100 ppm/°C for a 1% tolerance resistor. Chip resistors (and MELFs) are supplied to withstand immersion in molten solder for 5-8 seconds at 260°C (500°F). Not all manufacturers guarantee this (as not all resistors are wave soldered) so the specific supplier's data sheet must be checked. (To ensure that such factors are not overlooked, it will be sensible to consider each component in conjunction with a specification such as CECC 00 802.[2])

The manufacture of chip resistors, outlined in Figure 4.9, starts from a sheet of alumina ceramic, usually 96% Al_2O_3. This is scribed as shown in the X and Y axes to facilitate breaking subsequently into chips. A conductive paste, silver, gold or a precious metal alloy, is screened across the scribed lines in one direction only to provide contacts for the resistive elements and fired to a solid film.

A resistive paste is screen printed within the areas bounded by the scribed lines to overlap the conductive strips and fired to a solid mass. (Ruthenium dioxide is the most commonly used resistive component.) After trimming the resistors to the required value, a glass is applied and fired to protect the resistive elements against mechanical and environmental damage. The glass extends on to the conductor to ensure that the resistive element is fully protected.

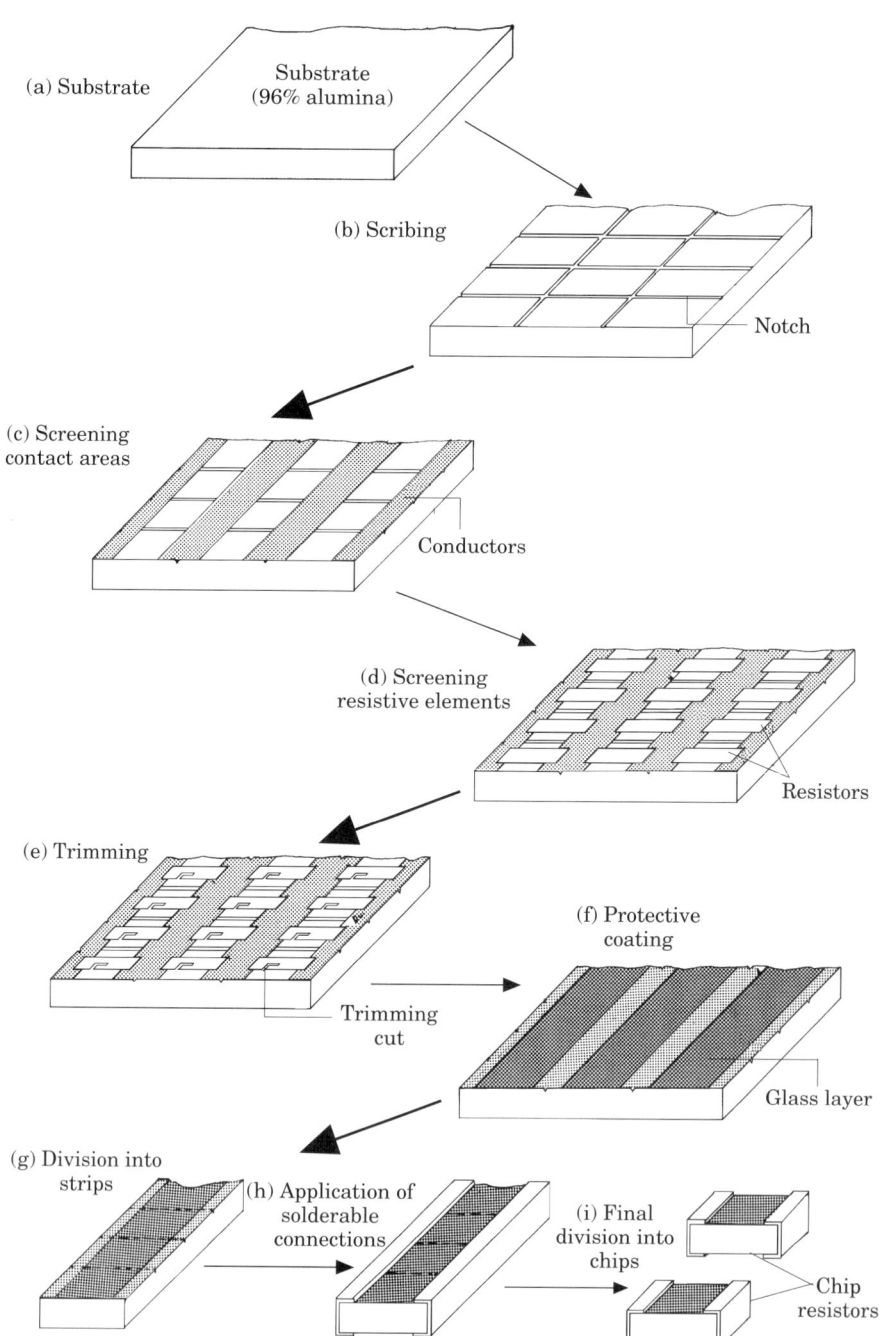

Fig. 4.9 Manufacture of chip resistors. (a) Substrate; (b) scribing; (c) screening contact areas; (d) screening resistive elements; (e) trimming; (f) protective coating; (g) division into strips; (h) application of solderable connections; (i) final division into chips.

The substrate is now broken into strips and coated with a solderable alloy. This is normally achieved by dipping in a silver or silver-palladium alloy conductive ink and firing, followed by deposition of a layer of nickel (to prevent leaching of silver by molten solder) and, finally, a layer of tin, tin-lead or tin-lead-silver, to give a good solderable coating. (Such coatings are also applied to the terminations of MELFs and to chip capacitors.)

The component is mounted with the alumina surface facing the PB.

4.3.4.2 JUMPERS

The jumper or zero-ohm chip resistor is produced in the same way, the resistive layer being replaced with a conductive layer, normally as used for the contact areas. Dimensions and shape are as shown in Figure 4.8. Chip jumpers are not used in the same numbers as MELF/Mini-MELF types. Chips are most often used on double-sided and multilayer substrates and external links are seldom required.

4.3.4.3 THERMISTORS

Both negative and positive temperature coefficient (NTC and PTC) resistors (or *thermistors*) may be obtained in chip form. The NTC type is available in 1206 outline with a power rating up to 0.5 watt (using the board as a heat sink). Other outlines are available, the 0603 and the 0805 being popular. NTC resistors are used to compensate for thermal drift in other components.

PTC resistors are more widely used and are supplied in various outlines, e.g. 0805, 1210, 1812 and 2220. Respective dimensions are (L x W x H in mm): 2.0 x 1.25 x 1.0, 3.15 x 2.55 x 1.7, 4.55 x 3.15 x 1.7, and 5.7 x 4.85 x 1.7.

PTC resistors find uses as temperature sensors to indicate maximum temperature; thresholds of 110, 120 and 130°C (230, 248 and 266°F) with a tolerance of ±5% are available. Other uses are in overload protection, delayed switching, surge current protection and current stabilisation. The rated current at 80 volts is 55 or 65 mA but the maximum switched current may go up to 600 mA.

4.3.4.4 CERAMIC CHIP CAPACITORS

The predominant type of chip capacitor is the multilayer capacitor because it has the most useful range of values. While similar to a chip resistor in shape and size, they may be easily differentiated by colour. They are available in a range of sizes, standardised by common usage, although tolerances can vary significantly from one manufacturer to another. Figure 4.10 illustrates the general construction and gives dimensions for the most common sizes. It is important to note that thickness changes with capacitance value since, as shown in Figure 4.10, these capacitors have a true interdigitated multilayer structure (like the traditional mica capacitor). The number of plates, for a given dielectric, will alter with the capacitance value required.

The manufacturing process is shown schematically in Figure 4.11. The dielectric is a thin sheet of green (unfired) ceramic such as barium titanate. The properties of the barium titanate will be modified by reaction with a wide variety of compounds such as strontium or calcium titanate, calcium zirconate

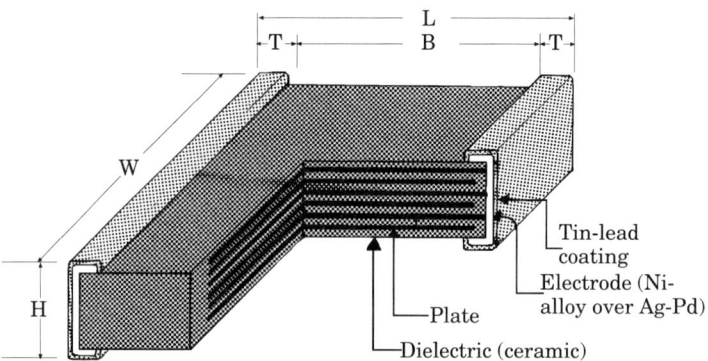

Dimensions (mm) (‡)

Size code	EIA-J type	L (approx.)	W	H(*) min.	H(*) max.	T(*) min.	T(*) max.	B (min.)
0805	732	2.0 ± 0.3	1.25 ± 0.20	0.5	1.25	0.25	0.75	0.4
1206	733	3.2 ± 0.3	1.6 ± 0.2	0.51	1.60	0.3	0.9	–
1210	734	3.2 ± 0.4	2.5 ± 0.3	0.51	1.90	0.3	1.0	–
1808	735	4.5 ± 0.5	3.2 ± 0.4	0.51	1.90	0.3	1.0	–
2220	736	5.7 ± 0.5	5.0 ± 0.5	0.51	1.90	0.3	1.0	–

(‡)Based on the current offer of several suppliers.
(*)These values vary with the capacitance. For reel packing H is usually either 0.65 or 1.0 mm with limited tolerance.

Fig. 4.10 Multilayer ceramic chip capacitor.

or niobium pentoxide to give different dielectric properties. The thickness of the ceramic sheet is 20 μm (0.0008 in.) or less. (Thinner dielectric sheets allow a proportionate increase in capacity within the same volume.) The sheet is screen printed with a conductive ink in an interleaved pattern. The conductive portion of the ink is a metal which will not melt at the firing temperature of the ceramic; platinum, melting point 1774°C (3225°F), or palladium, melting point 1552°C (2826°F), is normally used.

The screen printed sheets are stacked to form an interdigitated structure and an un-screened sheet placed on top so that top and bottom surfaces will be solid ceramic. The stacks are dried under pressure, cut to size, and sintered at a temperature of the order of 1300°C (2372°F) to form a solid block. The ends are then metallised to form terminations, each making contact with one set of electrodes. This is done by screening or dip coating with a silver-palladium paste. After firing, the capacitor has a totally enclosed electrode system, protected from the atmosphere. To eliminate the possibility

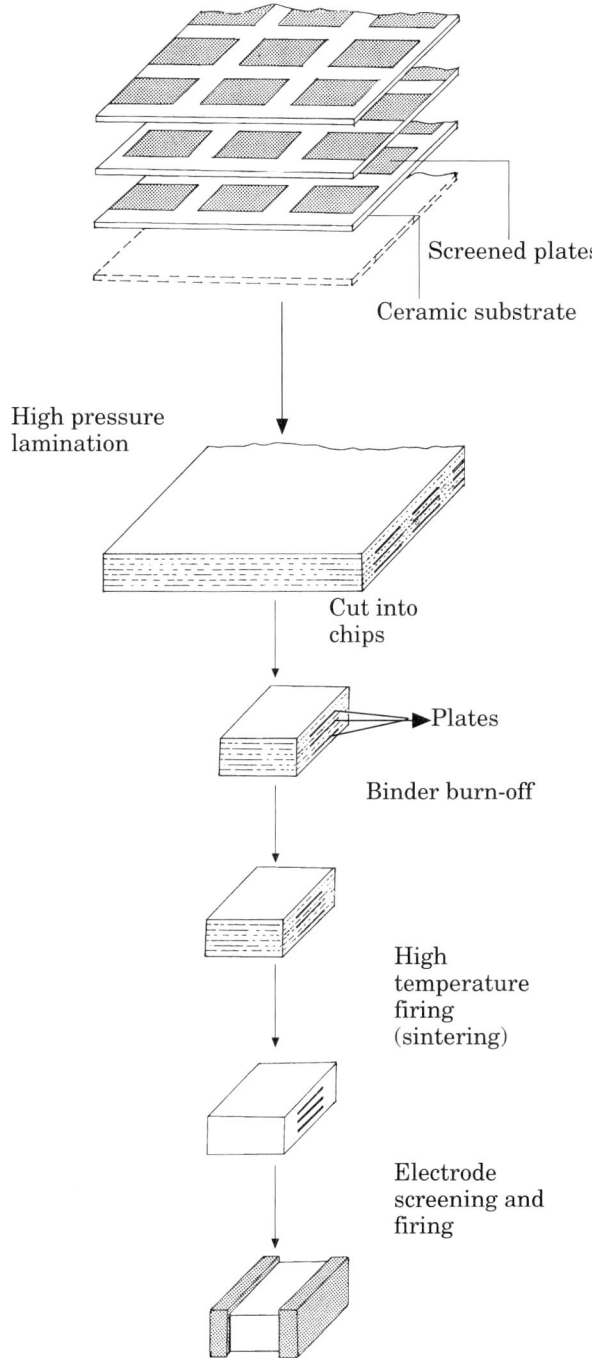

Fig. 4.11 Manufacture of ceramic chip capacitors.

of leaching of the terminations by molten solder during assembly, a nickel barrier coat may be applied followed by a solderable alloy.

There are several classifications of capacitor dependent on the ceramic used as dielectric. One such, described by Lea,[3] gives three classes. The first of these, designated NPO or COG, has a dielectric constant ranging from 30 to 150, giving a capacitance value of 1 pF to 10 nF with tolerances of 1 to 10%. The second class, X7R, has a dielectric constant from 500 to 2000, and a capacitance range of 100 pF to 200 nF with tolerances of 3 to 20%. The third class described by Lea, designated Z5U, has a dielectric constant greater than 4000, giving a range of capacitance values of 10 nF to 1.5 µF with tolerances of 50 to 100%. All three classes have a voltage rating from 50 V to 200 V. Operating temperature range is from -55 to +125°C (-67 to +257°F) for the first two classes and +10 to +85°C (+50 to +185°F) for the last. The designations COG, X7R and Z5U are those described by the American EIA and describe the temperature characteristics of the dielectric, e.g., in the case of Z5U, Z = a lower temperature range of +10°C, 5 = an upper temperature range of +85°C and U signifies an allowable capacitance change over the temperature range of +22/-56%. Other EIA designations which became quite popular in the late eighties/early nineties are Y5V operating over the range of -30°C to +85°C with an allowable change of +22/-82% and Y4T which operates from -30°C to +65°C with an allowable change of +22-33%.

Lea also shows that the NPO/COG dielectric has excellent temperature stability, less than 0.2% change in capacitance from -50 to +100°C (-58 to +212°F). The other types vary much more and too much reliance cannot be placed on temperature coefficients quoted in data sheets and specifications. While these are usually determined by measuring the values at 20°C and 80°C and quoting the average value over that range, the coefficient at a given temperature may be positive or negative, the greatest changes occurring with Z5U and Y5V materials.

Each manufacturer gives information on the degree of short time overheating his chip capacitors will withstand without permanent damage. Typically, these devices will resist solder wave immersion at 250°C (482°F) for up to 5 seconds, reflow soldering at 250°C for 8 seconds, and vapour phase soldering at 215°C (419°F) for no more than 40 seconds. The supplier's recommendations must be followed to avoid severe damage to the chip.

4.3.4.5 MICROCHIP CAPACITORS

In the eighties the Murata Manufacturing Company became the first to offer a significant range of very small ceramic multilayer capacitors. Three sizes were originally offered, the dimensions (L x W) in millimetres being 0.25 x 0.25, 0.50 x 0.50 and 0.90 x 0.90. In 1994, the range had increased to cover sizes from 0.25 x 0.25 to 2.29 x 2.29 (mm) and capacitances from 0.1 pF to 3000 pF.[4] The thickness of these capacitors varies with the capacitance value and is quoted as 0.18 mm maximum for the smallest chip size to 0.35 mm maximum for the largest.

The microchip capacitors have excellent frequency characteristics and are suitable for use at frequencies over 10 GHz. Because the electrodes are gold, these capacitors are assembled using semiconductor methods of die attachment and wire bonding rather than by soldering.

4.3.4.6 CHIP INDUCTORS

Murata also makes chip inductors, that is inductors manufactured by methods similar to those used for capacitors. These are not to be confused with surface mounted wire wound inductors, described later, which are sometimes erroneously referred to as *chip inductors* simply because they are suitable for surface mounting. Miniature wire wound inductors are typically about 3 mm cubed. True chip inductors made by thin film multilayer technology closely resemble chip capacitors and have dimensions (L x W x H) in millimetres of 3.2 x 1.6 x 0.5 with a tolerance of ±0.15 mm on L and W. These components are widely used in radio communication equipment, e.g., as noise suppressors at very high frequencies, up to around 400 MHz.

Taiyo Yuden supplies multilayer chip inductors which were developed by combining ferrite material technology with 'green-sheet' ceramic multilayer wiring technology. Conductors are applied by thick film techniques. The smallest of these chip inductors is (mm) 1.6 x 0.8 x 0.8 with tolerances of ±0.15 on all dimensions (EIA Code 0603).[5]

4.3.5 Other Flat Components

This term is used to describe other components which have a roughly parallelepiped shape. They are inaccurately called 'chip' components as they are produced one by one, as are traditional components. There is little or no standardisation of outlines.

4.3.5.1 METAL FILM SQUARE CHIP RESISTORS

The resistor shown in Figure 4.12 is supplied by Corning Glass Company as a *metal film square chip resistor*. The shape is a cross between that of a MELF and a chip, being rectangular with cylindrical end caps. It is manufactured like a MELF but the final lacquer coating is replaced by injection moulded plastic.

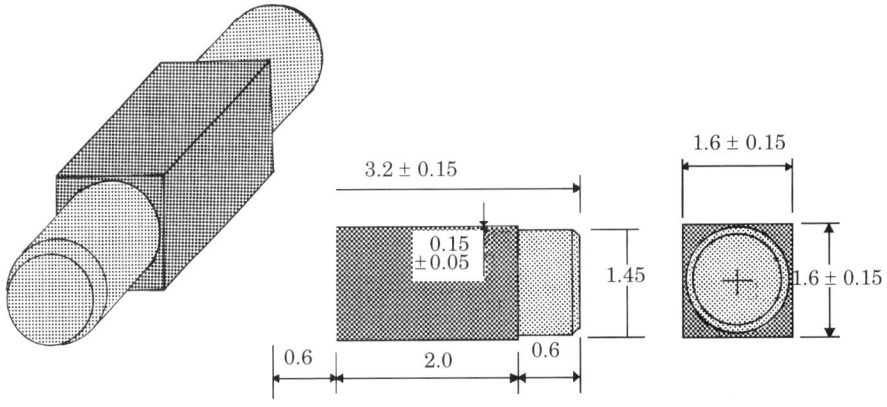

Fig. 4.12 Metal film square chip resistor.

The range of values is from 10 kΩ to 1 MΩ with a power rating of ¼ watt. The use of a metal film gives high stability and close tolerance control, 2% or 0.5% being standard. Zero ohm resistors (jumpers) are also available.

The footprint needed for these resistors on a printed board is virtually identical to that of the Mini-MELF, which has similar overall dimensions, and the components are very light, 1000 of them weighing about 25 grams (about 0.9 oz).

4.3.5.2 HIGH POWER RESISTORS

High power resistors are available, dissipating 1.5, 2.5, 3 or 5 watt according to the type. Resistance values range from about 10 Ω to 10 MΩ. Dimensions (and cost!) are much greater than those of MELF or chip resistors, Figure 4.13. However, the only other alternative to using conventional, wire-in-hole, components is to use a series of 5 to 10 (or more) low wattage MELFs or chips. The resistive element is either a film or, when low resistance and/or low thermal coefficients are needed, a wound wire.

These resistors will withstand contact with molten solder at 260°C (500°F) and are suitable for automatic assembly and wave soldering. To reduce stress on the solder joints caused by differential thermal expansion, different rates of heating or flexure of the substrate, 'J' type leads are used by Philips, Figure 4.13. Two types are shown in the figure. The standard pedestal mount gives a good bonding surface which improves adhesion during mounting. The recessed foot mount has two stand-offs on the bottom surface which enable solvents to penetrate under the device, to aid cleaning after soldering, and minimise transfer of heat to the laminate surface in use.

4.3.5.3 METALLISED POLYESTER CAPACITORS

Metallised polyester capacitors are also found in versions suitable for surface mounting. Sometimes called *chip-foil capacitors*, they are *not* chips. They are made in essentially the same way as conventional metallised film capacitors.

The body is rectangular injection moulded plastic, which must be heat resisting, with 'J' leads bent under the body. The leads are as large as possible to maximise the soldering area.

Body dimensions change according to the capacitance. Typically, capacitance values from 0.01 µF to 1.0 µF can be covered by six sizes, ranging from (L x W x H in mm) 4.9 x 4.5 x 2.5 to 9.4 x 8.0 x 6.0 (0.19 in. x 0.18 in. x 0.1 in. to 0.37 in. x 0.32 in. x 0.24 in.).

These capacitors give high dielectric strength, self-healing capability, and a low dissipation factor, in a small space at reasonable cost.

4.3.5.4 WET TYPE ALUMINIUM ELECTROLYTIC CAPACITORS

Surface mount versions of the 'wet' aluminium electrolytic capacitor differ from the conventional types in the form of the case. The construction is illustrated in Figure 4.14. The active element is anodically etched aluminium foil, interleaved with paper impregnated with electrolyte, rolled round a pin which protrudes on one side. The roll is inserted into an aluminium case,

Pedestal mount

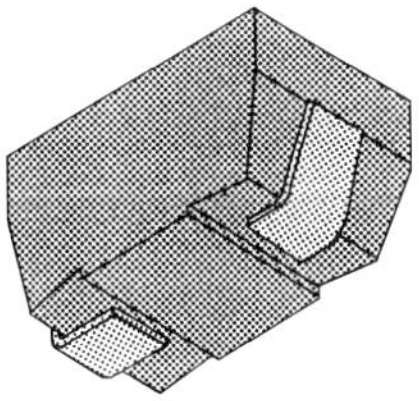

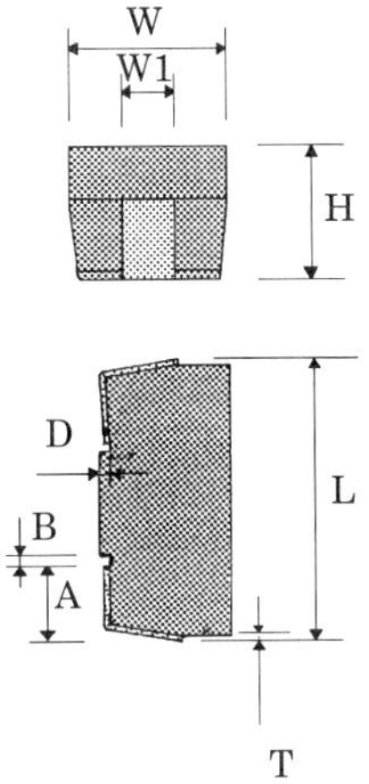

Recessed foot

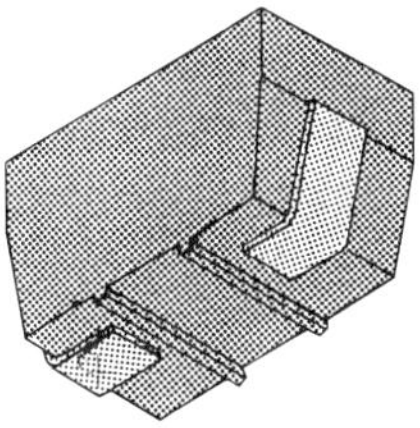

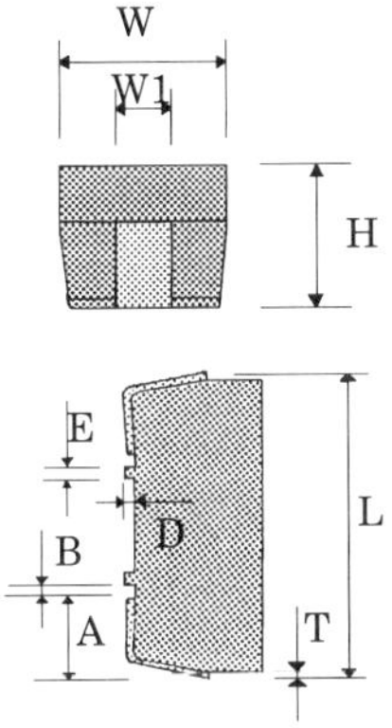

Nominal Dimensions (mm)

	Pedestal mount type 1.5 W	Foot mount type 1.5 W	Foot mount type 2.5 W
L	10.8	10.8	20.5
W	5.8	5.8	7.0
H	5.8	5.8	7.0
W1	1.5	1.5	1.5
T	0.15	0.15	0.15
A	1.6	1.6	2.35
B	0.5	0.5	0.6
D	0.45	0.5	0.75
E	–	0.75	1.55

Fig. 4.13 High power resistors.

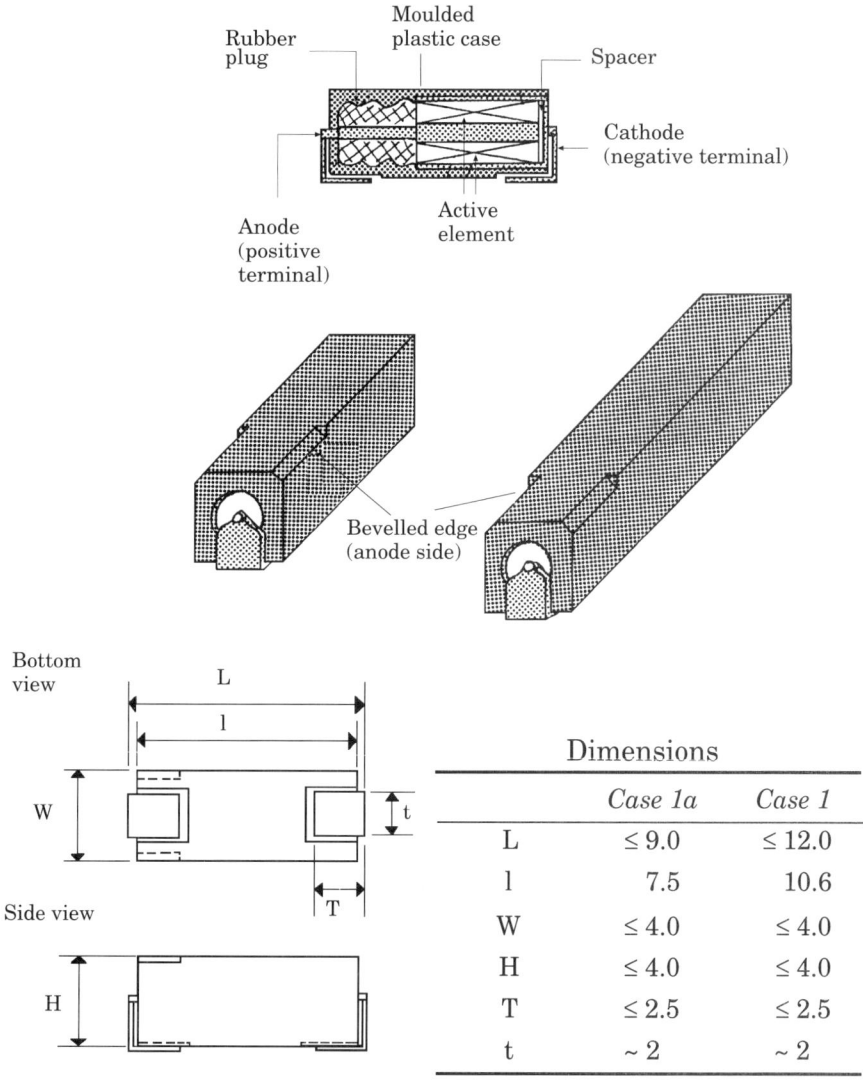

Fig. 4.14 Wet type aluminium electrolytic capacitors.

which acts as the cathode, and sealed with a rubber plug, leaving the pin protruding to act as the anode. The assembly is encapsulated in plastic and solder coated leads attached by soldering or welding to anode and cathode. It will be noted that the moulded case is bevelled at the anode end to identify the positive terminal. The leads are bent round under the body such that there is a stand-off gap of 0.2 to 0.5 mm between the body and the leads. The gap between body and leads assists cleaning and removal of fluxes and, in addition, enables the joints to cope more adequately with board flexure and differential rates of expansion between the component and substrate.

Values from 0.1 µF to 22 µF are obtainable, with voltage ratings of 6.3 V to 63 V. The components will resist immersion in molten solder at 260°C

(500°F) for 10 seconds without damage. The major advantages of these capacitors are high capacitance-to-volume ratios and relatively low cost.

4.3.5.5 SOLID ALUMINIUM ELECTROLYTIC CAPACITORS

The construction of this SMC is shown in Figure 4.15. After forming the anode from anodically etched foil, it is inserted into the cathode and all spaces filled with solid manganese dioxide, the electrolyte. The active element is potted in a temperature resistant, glass-filled plastic case and the anode and cathode soldered to identical solderable end caps. The anode is identified by a '+' to ensure correct polarity on mounting.

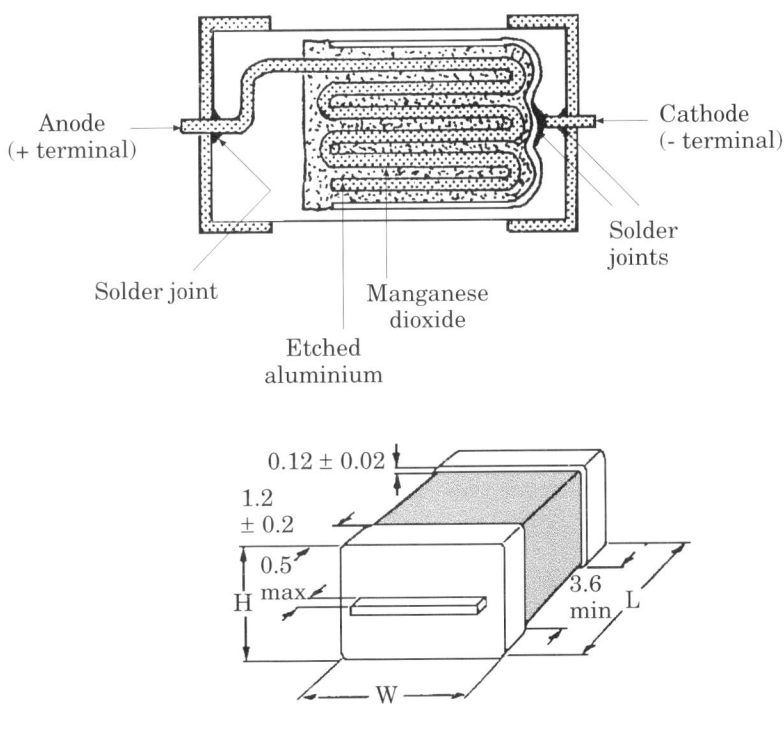

Case size	H_{max}	W_{max}	L_{max}	mass g
20	3.0	4.5	6.7	0.25
30	3.5	5.8	6.7	0.30
40	4.1	5.8	6.7	0.35
50	4.1	7.9	6.7	0.50
60	5.2	7.9	6.7	0.60

Fig. 4.15 Solid aluminium electrolytic capacitors.

Capacitance values from 0.1 µF to 68 µF are available with voltage ratings from 6.3 V to 40 V. Five different case sizes are tabulated in Figure 4.15. These SMCs have very high capacitance-to-volume ratios, excellent stability, long and reliable life, and a very low leakage current. The shape is very suitable for automatic assembly and they can be immersed in molten solder at 260°C for up to one minute without degradation. Wave soldering gives no problem.

Major uses for these capacitors include filtering, coupling and decoupling where a need for high reliability and/or other specific properties outweighs the cost.

4.3.5.6 TANTALUM ELECTROLYTIC CAPACITORS

Tantalum electrolytic capacitors are available in surface mounting form with values from 0.1 µF to 100 µF and voltage ratings from 4 to 50 volts. The technology employed is fundamentally that used for conventional components. A solid, high purity, tantalum wire — the anode — is embedded in a near rectangular pellet of sintered, anodically oxidised, tantalum particles. The anodised film, tantalum pentoxide, is the dielectric and the underlying tantalum is the cathode. The sintered pellet is impregnated with manganese dioxide as electrolyte. The cathode is electrically connected to the external can and the component embedded in heat resistant plastic. Electrodes are welded to the anode and cathode to project on opposite faces of the package.

Several types of case are available. A very popular type is shown in Figure 4.16. It is rectangular with flat surfaces, ideal for automatic pick-and-place assembly. The sturdy terminals are partially embedded in the plastic body and the 'split' anode termination is easily distinguished from the 'solid' cathode termination. Length and width vary from 3.8 to 7.3 mm and 1.9 to 4.3 mm respectively, and height from 1.6 to 2.8 mm (tolerances are generally about ±0.02 mm).

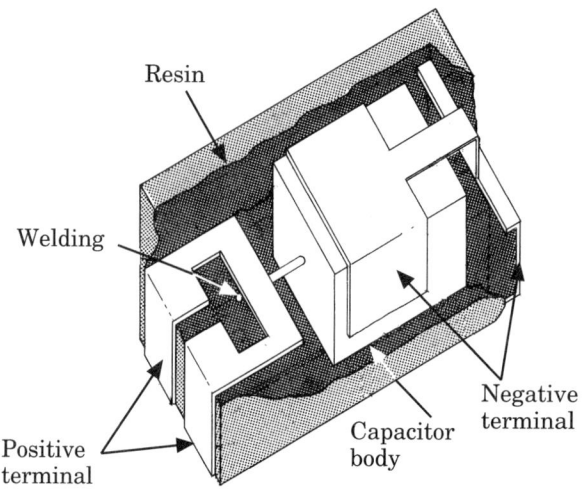

Fig. 4.16 Tantalum electrolytic capacitor.

Other outlines available differ in shape and in the method of indicating polarity. The anode may be identified by shape, by a '+', or by a coloured stripe at or near the anode termination. A typical tantalum 'chip' capacitor is illustrated in Figure 4.17. These are available in various 'chip' sizes (depending on the capacitance value) to be compatible with modern mounting machines.

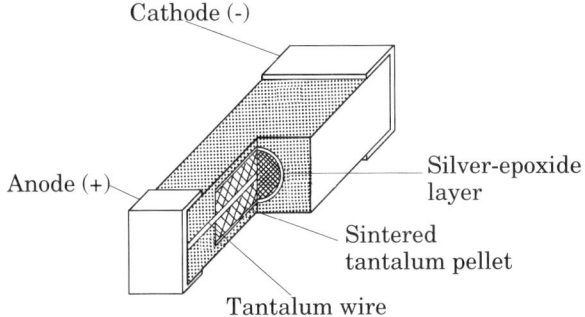

Fig. 4.17 Tantalum 'chip' capacitor.

The capacitance-to-volume ratio can be as high as 100,000 µF per cubic inch, almost as high as that achieved with wet aluminium capacitors. The devices are noted for high reliability, long life, low self-inductance and low electrical loss, as well as excellent resistance to mechanical shock, vibration and pressure. Where such criteria are critical, the higher cost of these devices is readily accepted.

4.3.5.7 INDUCTORS

True chip inductors were described earlier in this chapter. Several other types of inductor designed for surface mounting assembly are available. These are often called *chip inductors* or *chip coil* SMCs. These are not made by chip technology but are all constructed by winding a coil of wire on to a suitable core. Standardisation of outline and shape is virtually non-existent. Some typical constructions and outlines are shown in Figure 4.18.

Figure 4.18(a) shows a wire wound ferrite core embedded in plastic or resin. The terminals are wrapped round the ends and bent under the body. In Figure 4.18(b) the ends of the ceramic or ferrite core are rectangular and the terminations extend the full width of the base. The device is not encapsulated. Another common type, shown in Figure 4.18(c), has a rectangular core, axis horizontal, the winding being terminated on strip leads at either end. This type may or may not be encapsulated. On the encapsulated version the terminals are bent under the body as in the first type.

SM inductors are available with values which range from 0.01 µH (micro-Henry) to 2.2 mH (milli-Henry). Internal resistance is low, from 0.25 Ω to 13 Ω maximum. The rated current may be as low as 15 mA or as high as 500 mA, dependent on the type. In most cases the tolerance will be 453 or ±10%, although tighter tolerances can be obtained when necessary.

The majority of these devices will be guaranteed to resist molten solder temperatures attained in wave or reflow soldering. Where the assembler

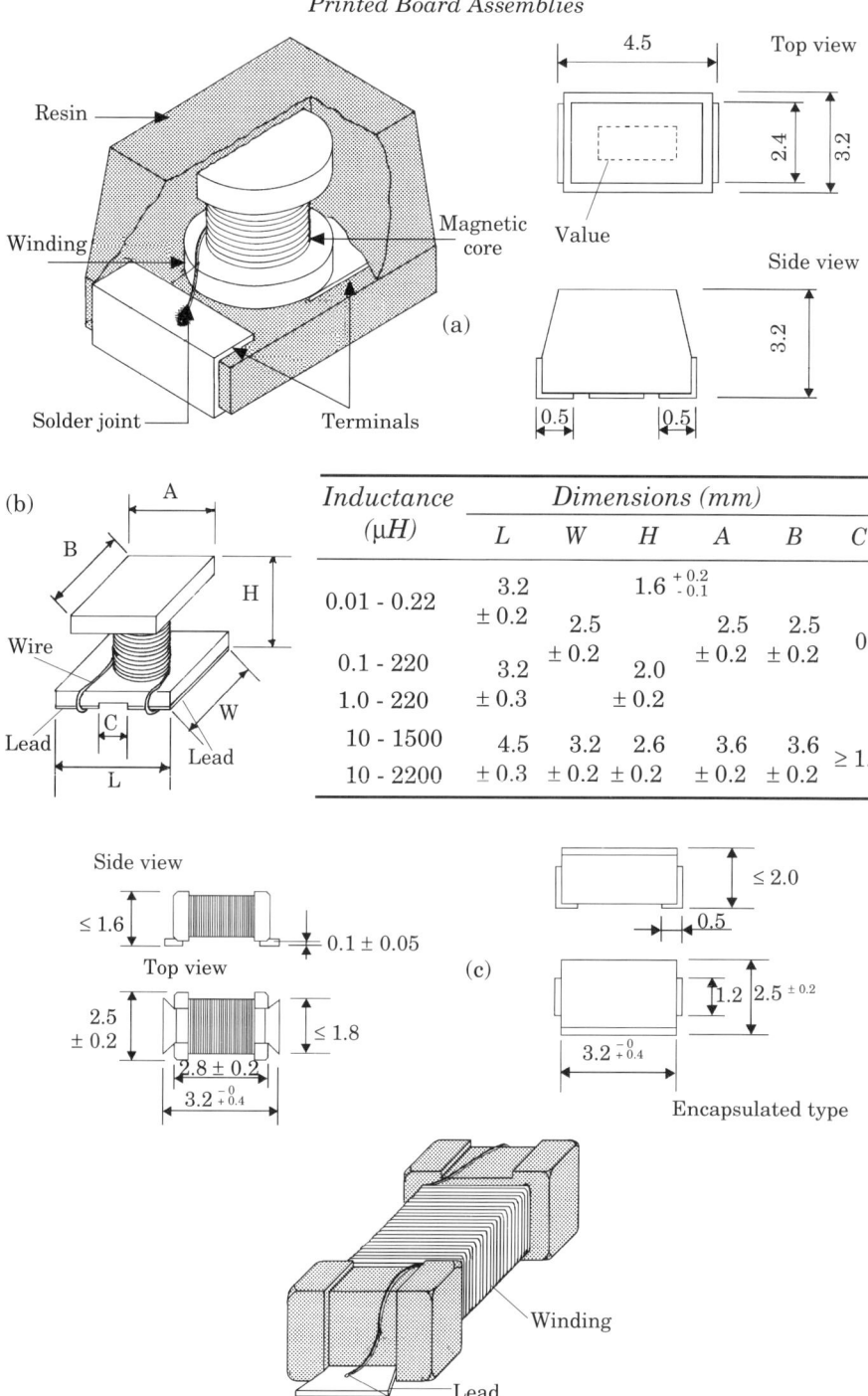

Inductance	Dimensions (mm)					
(μH)	L	W	H	A	B	C
0.01 - 0.22	3.2 ± 0.2	2.5 ± 0.2	1.6 +0.2 -0.1	2.5 ± 0.2	2.5 ± 0.2	0.7
0.1 - 220	3.2 ± 0.3		2.0 ± 0.2			
1.0 - 220						
10 - 1500	4.5 ± 0.3	3.2 ± 0.2	2.6 ± 0.2	3.6 ± 0.2	3.6 ± 0.2	≥ 1.0
10 - 2200						

Fig. 4.18 'Chip' or 'chip coil' inductors. (a) Encapsulated inductor; (b) unencapsulated inductor; (c) single-layer inductor.

prefers to make his own SM inductors, special magnetic cores can be purchased. These do not lose their properties at soldering temperatures, unlike normal ferrites.

Surface mounting inductors are used in radio frequency circuits in which space or volume is at a premium. Typical applications are found in car and portable radios, TV tuners, cameras and miniature TV sets.

4.3.5.8 OSCILLATORS AND FILTERS

The demand for increasingly compact FM (frequency modulated) radio sets stimulated development and production of miniature oscillators for use with appropriate ICs. The *ceramic discriminator*, offered by Murata, is a good example. Slightly larger in outline than a 0805 chip resistor, it gives good discrimination of frequencies around 10.7 MHz and does not require adjustment. Several types are available with slightly different centre frequencies. These devices can be reflow soldered but not wave soldered.

The same manufacturer also developed a flat ceramic filter, suitable for small radios. Shown in Figure 4.19, it consists of a piezoelectric ceramic element sandwiched between two sheets of plastic. It has three electrodes. The common electrode (ground) is in the centre of the bottom face and the other two (input and output) are on the two shorter sides, as on a chip resistor. Several types are available with centre frequencies around 10.7 MHz. This component can be assembled automatically and can be reflow soldered.

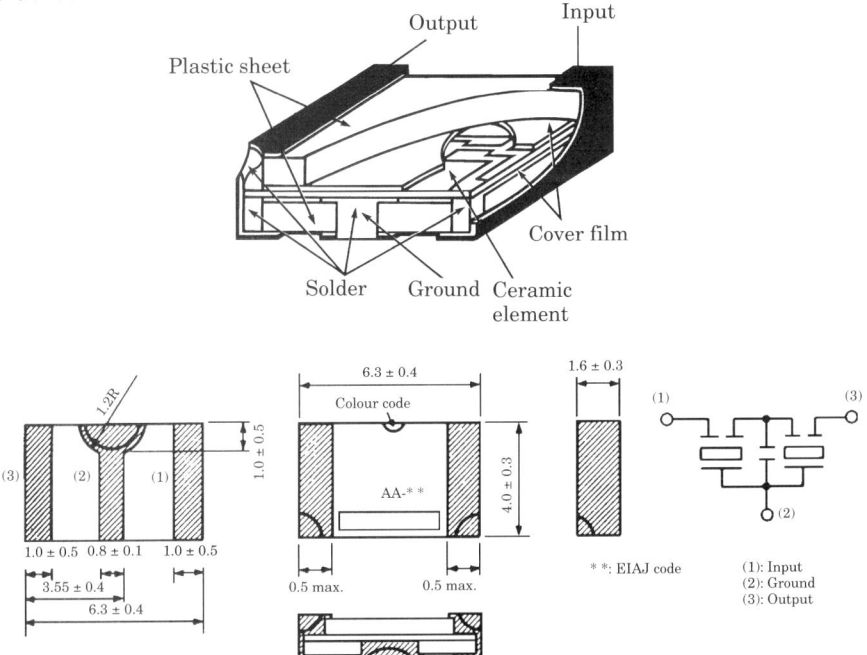

Fig. 4.19 Ceramic filter for FM radio. (Courtesy of Murata Manufacturing Company Ltd)

4.3.6 Miscellaneous Passive Components

There are a great number of devices which do not fit readily into one or another of the outlines previously discussed. These odd shaped components may give problems in automatic assembly as they may need special care and, possibly, special tooling.

The number, and variety, of such devices is enormous and increasing every day. This ignores those manufacturers who make their own SMCs by bending the leads of conventional components into SM forms! Standardisation of these components is almost non-existent, and is unlikely to improve significantly in the immediate future. Care must be taken, when using such components, to ensure that the device will not change in specification or go out of production during the design life of the assembly.

4.3.6.1 ALUMINIUM ELECTROLYTIC CAPACITORS

'Tombstoning' was, and still can be, a frequent problem in SMA, see Chapter 12, 12.10.4.4. It is, however, untrue that this is why Panasonic produced a vertical 'chip' or *vertical leadless* aluminium electrolytic capacitor! This component, Figure 4.20, has a square insulating base, bevelled on the anode side, with two terminal strips on the base, bent upwards on to the sides. The main body is cylindrical.

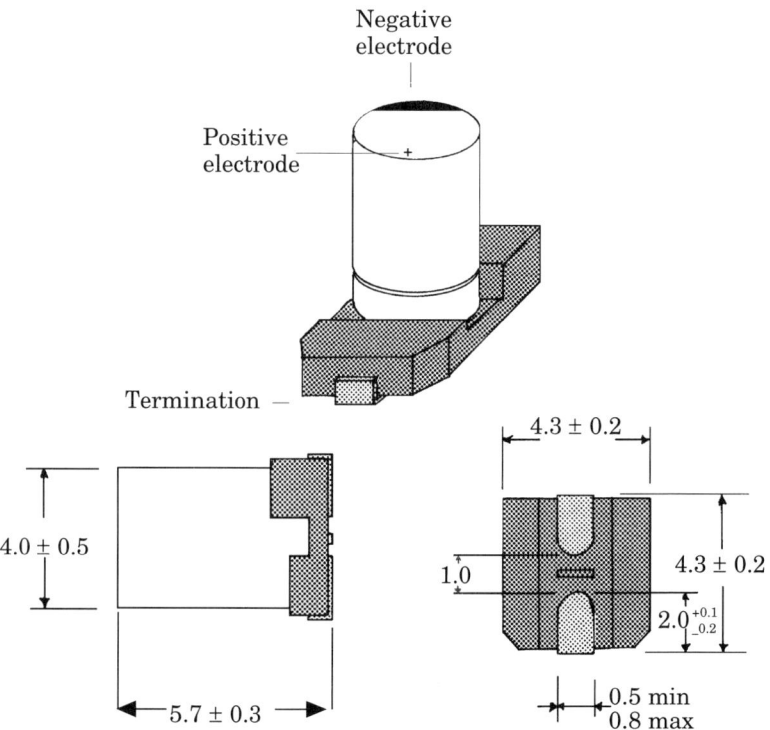

Fig. 4.20 The Panasonic vertical, leadless, aluminium electrolytic capacitor.

The component is available with capacitance values ranging from 0.1 µF to 33 µF and voltage ratings from 5 V to 63 V dc. The device has good capacitance-to-space ratios as a result of the unusual outline.

This package can be assembled with most automatic equipments but cannot be wave soldered (it must not be immersed in molten solder). It has good resistance to other methods of soldering and to cleaning.

4.3.6.2 POTENTIOMETERS

SM trimming resistors or 'pots' were developed at an early stage to support surface mounted assembly of circuits needing adjustment during manufacture or testing. These devices are not intended for use as tuning potentiometers, which must withstand many thousands of adjustments; they are designed to be used only a few times during the life of the equipment or module.

Few two-lead versions (variable resistor) are available. The three-lead device (the potentiometer) is obtainable in many versions, most of which will be similar to one or another of those shown in Figure 4.21. The two-lead type may also have three terminals, two being connected internally. The shape and size change with the supplier and type but are usually between (L x W x H in mm) 3.0 x 3.0 x 1.5 and 3.8 x 4.5 x 2.8.

The screwdriver slot, for adjustment, on the upper side may cause difficulty in automatic assembly by interfering with the operation of vacuum pick-ups. This can be resolved by asking the supplier to cover the slot with a thin plastic film which will be easily disrupted by the screwdriver on the first adjustment. This may also help to avoid damage during cleaning after assembly.

These trimmers are usually very sensitive to temperature. Wave soldering cannot be used, unless specifically allowed by the manufacturer. Reflow soldering is acceptable with gradual preheating to avoid thermal shock. Cleaning of the soldered trimmers is almost mandatory as any flux residues on the resistive pattern can cause problems. Some components are hermetically sealed, except around the adjustment slot which can be covered as mentioned in the previous paragraph. If, however, hermeticity is not good, the film or tape may create a problem by inhibiting circulation of the solvent.

If permitted by the supplier, ultrasonic cleaning will improve cleaning efficiency.

4.3.6.3 TRIMMER CAPACITORS

Trimming capacitors, Figure 4.22, consist of two parallel, asymmetrical plates and a ceramic or plastic film dielectric. One plate is held stationary while the other is attached to the rotor so that, when the rotor is turned with a screwdriver, the facing area of the two plates changes, altering the capacitance in proportion to the area.

The maximum capacitance available is normally from 1 to 15 pF and is from 5 to 10 times the minimum obtainable with the device.

Some types are hermetically sealed, the rotor being covered with heat resistant film as shown in Figure 4.22. This facilitates soldering, even by immersion, and cleaning. Non-hermetic types require careful handling and the supplier's instructions for soldering and cleaning must be followed.

(a)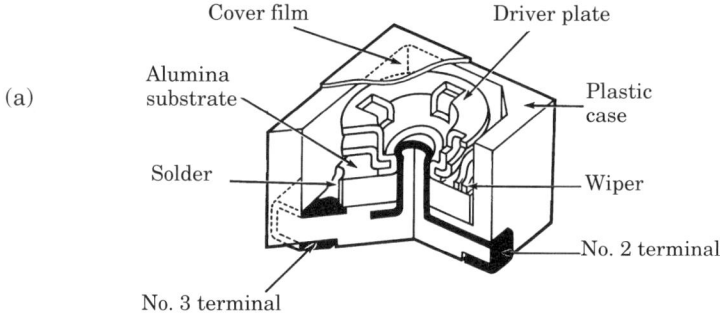

(b)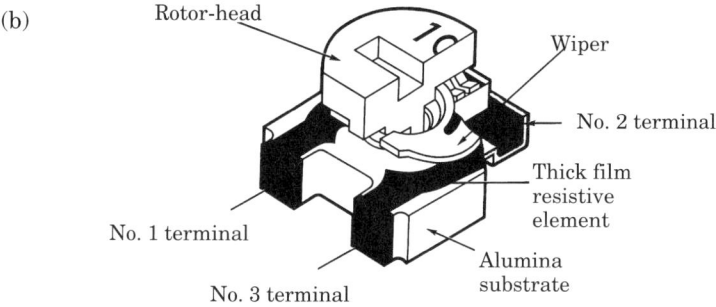

(c)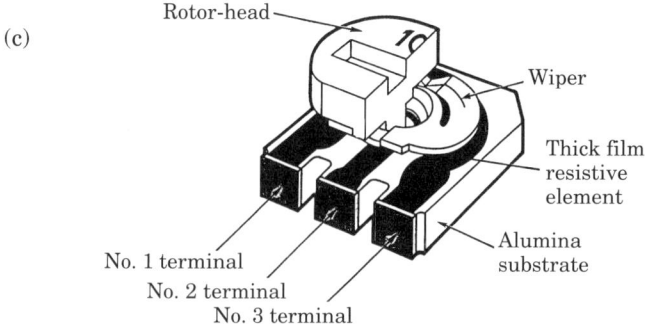

Fig. 4.21 Some trimming resistor constructions. (a) 3 terminal, in plastic case with cover film; (b) 3 terminal, open construction; (c) similar to (b) with alternative terminal arrangement. (Courtesy of Murata Manufacturing Company Ltd)

4.3.6.4 CONNECTORS

Connectors for surface mounting, with gull-wing leads, are obtainable from several sources. These may be used only if the stresses of insertion can

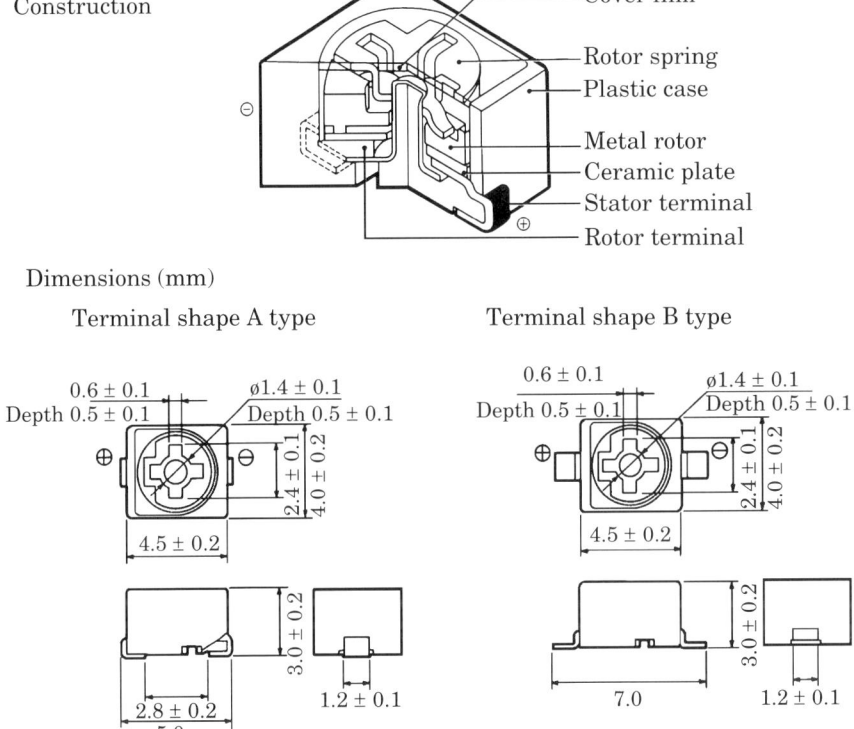

Fig. 4.22 A sealed type trimmer capacitor which can be immersed into flux and a solder bath. (Courtesy of Murata Manufacturing Company Ltd)

be controlled in some way. If this is not practicable, screw attached connectors are to be preferred, with the consequential need to provide mounting holes with a loss of board area.

SM connectors have smaller pitch between contacts, e.g., 1.25 mm or 0.05 in. or less, than conventional connectors. If the connector is screw attached, heat expansion or flexure of the board can apply abnormal stresses. The best types of connector have a 'floating' contact arrangement which allows deviation from its theoretical position. During reflow soldering the surface tension of the solder is enough to centralise the leads with reasonable accuracy.

A viable solution to mounting block connectors on a SMT board is to use a flexi-rigid board, Volume 2, Chapter 11. If the connectors are on the flexible portion, it is very unlikely that abnormal stress can be applied to the joints.

An alternative, seldom exploited, is to use an elastomeric connector of the type shown in Figure 4.23. The key feature of this connector is the insulating rubber block on to which thin conductors have been deposited, using carbon or silver powder inks or a vacuum deposited metal. This is held in a plastic frame and compressed into contact with the conductors on the two boards by means of two screws. The contact resistance produced is 0.02 Ω to 1 Ω per connection, dependent on type. This value can be tolerated by most circuits.

The same principle can be employed to connect a large IC package to its socket, as outlined later in this chapter.

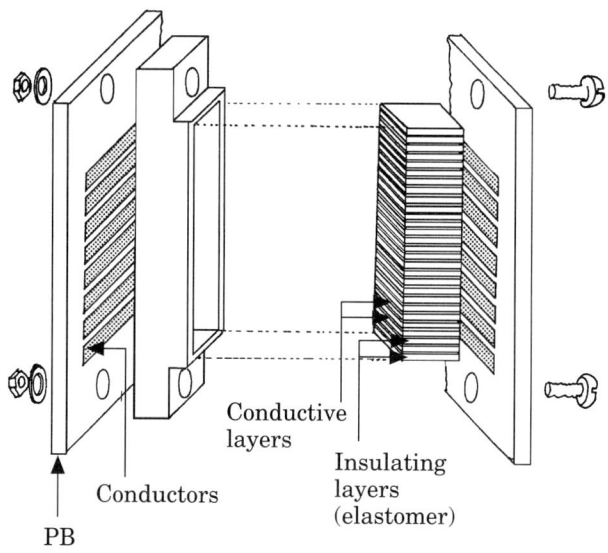

Fig. 4.23 An elastomeric connector (not drawn to scale).

Hopefully, continued development will produce more elegant solutions, but, at the time of writing, it is common to see mixed technology boards/assemblies where all components are SMCs except the connectors.

4.4 ACTIVE SURFACE MOUNTING COMPONENTS

The conventional pin-in-hole packages for ICs are 300 to 3000 times larger than the chip which is the actual active device. If, for example, a transistor chip has a volume of $0.6 \times 0.6 \times 0.2 = 0.07$ mm^3, the TO-18 case in which it may be contained will have a volume of some 100 mm^3. In the instance of a large IC, the chip volume may be $2.5 \times 2.5 \times 0.2 = 1.25$ mm^3 while the TO-116 in which it is packaged has a volume around 700 mm^3! These figures ignore the volume needed for the holes and the projections on the far side of the board.

The need for smaller packages for hybrid circuits was not met by the IC manufacturers. The best solution then was to mount bare chips on the substrate and cap or encapsulate these with the other components. In the late sixties, the SOT-23 package appeared. This had a volume of less than 5 mm^3, plus the space for the leads, and solved many of the hybrid manufacturers' problems. As die attach and wire bonding could be replaced by plain soldering, the SOT (*Small Outline Transistor*) was an immediate success.

As PCB technology began to provide higher packing densities, producing finer tracks and smaller hole sizes, other SMCs became available and the small outline packages for ICs started to be used on PCBs where space and/

or speed were important. This spurred the production of other packages to solve specific problems.

When dealing with SM active devices, and particularly high lead count ICs, it is essential to remember that all leads must be co-planar within very close limits, that is, the mounting feet must be in one plane or missed joints may result. In addition, the smaller pads on a surface mount board mean that there is more chance that a laterally skewed lead will miss its mounting pad. Leads must be co-planar and correctly aligned at source and great care has to be exercised in all subsequent handling. Solderability of the terminations must be very good. Because of the small joint sizes, inspection and touch-up or repair will be slow and fatiguing.

Traditional pin-in-hole packages can be converted for SMA by suitably shaping the leads. Much of the advantage of surface mounting will be lost due to the larger sizes and the following notes focus on packages developed especially for surface mounting.

4.4.1 Diodes

Zener, Schottky, signal and other diodes are available in cylindrical hermetically sealed, glass-to-metal structures, similar to the MELF and Mini-MELF outlines. The latter outline has been registered with JEDEC as the SOD-80. These are shown in Figure 4.24 and it will be noted that the tolerances are large, although individual suppliers are normally closer to the mean values. While the cases are glass, these packages are much more robust than glass cased diodes with leads.

While voltage rating can be quite high, total power dissipation is restricted by the MELF construction. A MELF diode can dissipate up to 1.3 watt but the SOD-80 is limited to 0.5 watt. As with other MELF types, higher ratings may be achieved by the use of alumina substrates or special PBs.

These packages may be asymmetrical but the orientation changes with the supplier and the type of device, so that a colour band indicating the cathode is applied as a reference mark. For automatic assembly, these components are supplied in a packing which keeps them correctly orientated. (Some machines can adjust orientation by checking the component electrically.)

High power diodes cannot be packaged in MELF and Mini-MELF cases because of the low power dissipation. The *Small Outline Diode* configuration, typified by the SOD-6 and SOD-15, illustrated in Figure 4.25, are more suitable. In these devices, the chip is mounted on a thick copper lead frame and electrically connected to the terminations by a copper alloy conductor. To improve heat transfer, the die is attached to the leads using high melting point solder.

The body of the component is formed with a high temperature resisting plastic, often an epoxide resin, in order to resist immersion in molten solder. The leads are formed under the body in a way that allows for thermal mismatch with minimum stress on the body. The leads are finished with a solderable coating such as tin-lead.

The SOD-6 is suitable for continuous dissipation of up to 3 watt; the SOD-15 is used for diodes dissipating 5 watt or more. While these cases are commonly used for Zener diodes and rectifiers, uni- or bi-directional transient suppressor diodes rated up to 100 A and up to 1,500 watt (for a very short time!) are also supplied in them.

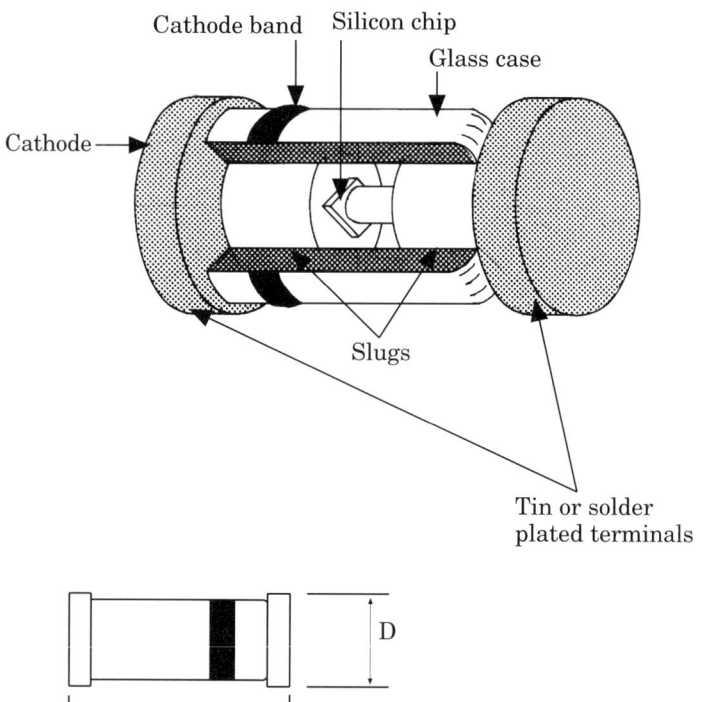

	Dimensions (mm)			
	'MELF'		'Mini-MELF' (SOD-80)	
	Min.	Max.	Min.	Max.
L	4.8	5.2	3.2	3.6
D	2.4	2.8	1.3	1.7
P	0.4	0.6	0.4	0.5

Fig. 4.24 Construction and outlines for cylindrical packages for diodes.

Tuning diodes for several ranges of frequency, from VHF (170 to 303 MHz) to satellite TV (950 to 1,750 MHz) are supplied in a MELF pack or in a case very like an SOT-60. The latter is similar to the SOD-6 but has gull-wing leads and is smaller, L x W x H in mm being 3.7 x 1.55 x 1.35 (the length L includes the projection of the leads by 0.5 mm at each end). These devices are also available in the same body with straight leads, total length about 14 mm, and may be obtained as single devices or matched pairs.

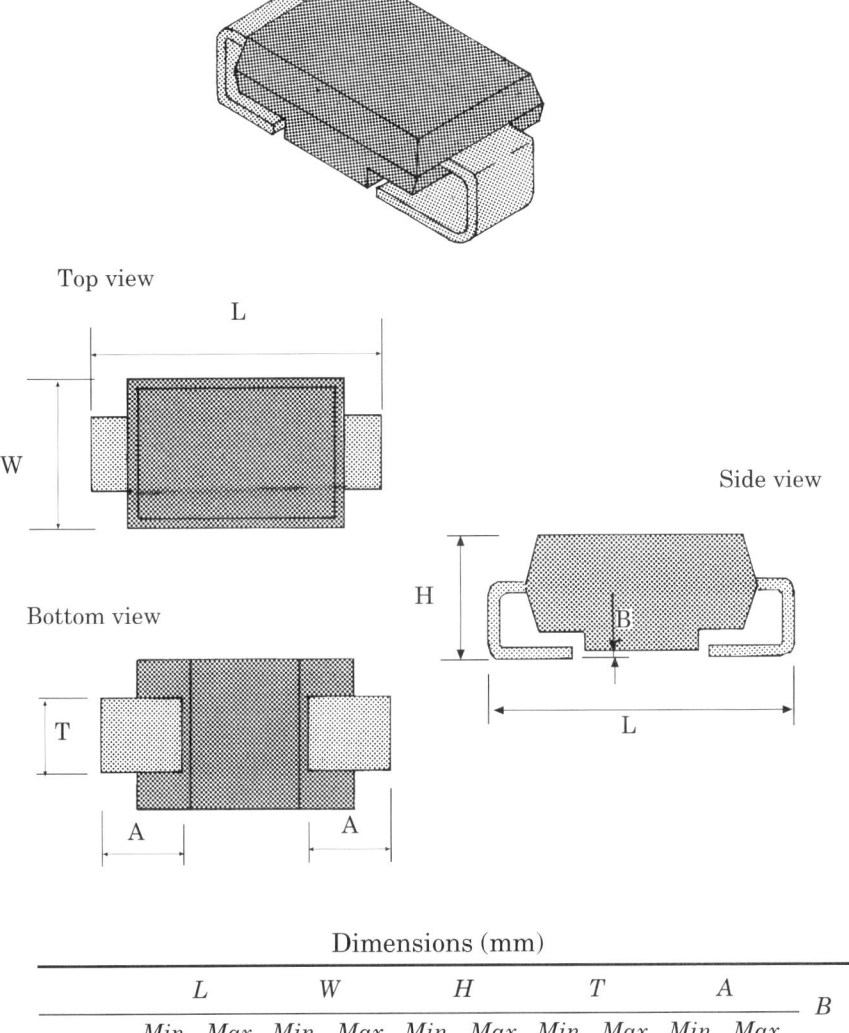

Fig. 4.25 Small outline diodes (SODs).

	L		W		H		T		A		B
	Min.	Max.	Min.	Max.	Min.	Max.	Min.	Max.	Min.	Max.	
SOD 6	6.0	6.4	3.8	4.2	2.5	3.1	2.8	3.2	0.9	1.3	0.1
SOD 15	7.6	8.0	4.8	5.2	2.5	3.1	2.8	3.2	1.0		0.1

Dimensions (mm)

4.4.2 SOT-23

The SOT-23, the first three-terminal case for SMDs, is still one of the most popular, being used for most types of transistor from small signal to SCRs and TRIACs. It is also used for diodes of several families, one electrode being a dummy. It owes its popularity to the fact that it is cheap, reliable, easy to handle with automatic equipment, compact and light (each weighs about 0.01 gram).

It is limited by the power which it can dissipate. In free air with a T_{jmax}, the maximum allowed temperature at the *p-n* junction in the chip, of 150°C (302°F), the maximum power which can be handled is 200 mW, not enough for some applications. If the case is assembled on a highly conductive substrate, e.g., alumina, instead of normal epoxide/glass, the power rating can be increased to 0.5 W. (Thermal aspects of assemblies are discussed in Chapter 11.)

Construction and dimensions are shown in Figure 4.26. The chip is mounted on the central terminal by eutectic bonding to maximise heat

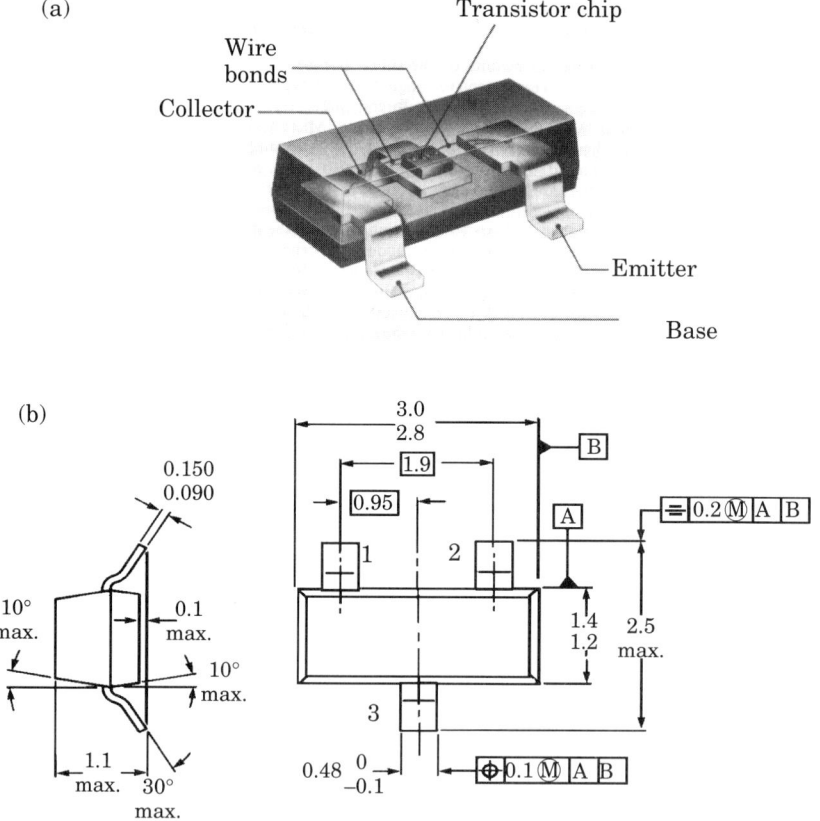

Fig. 4.26 The SOT-23 three terminal package for transistors. (a) See-through view of SOT-23 encapsulation (Courtesy of Philips); (b) dimensions of SOT-23 encapsulation.(Courtesy of Mullard)

transfer. Ball bonding and stitch bonding are used to connect the pads to the side leads. Encapsulation is carried out by injection moulding a flame retardant resin, e.g., glass-filled epoxide. The sturdy SOT-23 should withstand all soldering methods, including wave soldering, although the supplier's data sheet should be consulted for the maximum time of immersion in the molten solder. The package can be assembled by most automatic equipment, does not need high precision or special tooling, and allows easy testing and inspection, before and after assembly. Moreover, it is relatively cheap.

Special versions of standard components, having different pin assignment, can be obtained, ex stock or to order. Some suppliers prefer to conform to the TO-236 case, which has approximately the same outline, the main difference being the tolerance.

4.4.3 SOT-89

In the late seventies, the SOT-89 was introduced. The case is only slightly larger than the SOT-23 but can dissipate much more power. In free air with a T_{jmax} of 150°C the package can dissipate 0.5 watt on epoxide/glass, rising to 1 watt on a more conductive material such as alumina. It is a good alternative to the SOT-23 case for power components such as transistors, SCRs and TRIACs.

The device is illustrated in Figure 4.27. To improve dissipation of heat, the centre lead, on which the chip is mounted, is much larger and exposed on the base and the rear of the pack. This allows a much larger area for removal of heat generated in operation. The normal pin assignment is as shown in the figure but reverse pinning, in which the base and emitter are interchanged, is available to order. It will be noted that the leads are formed so that one side is in the same plane as the base of the package.

While this package, including lead projection, needs twice the space of the SOT-23, Figure 4.26, it is the smallest available for power devices. It can be assembled by the same automatic equipment and soldered, inspected and tested with similar ease.

4.4.4 SOT-143

The SOT-143 is derived from the SOT-23. It has exactly the same outline dimensions but has four leads instead of three, having been developed initially for packaging MOSFET and other four-lead devices. The fourth lead is wider than the others, 0.88 mm as opposed to 0.48 mm, and is the lead on which the device is mounted. It acts as the source of the device (these terms are explained in Chapter 3, 3.6).

Heat dissipation and other characteristics are as for the SOT-23, including methods of assembly and soldering. However, limits stated in data sheets for immersion in molten solder must not be exceeded.

4.4.5 SOT-223

Introduced in 1987, this is a four-lead plastic package for medium power semiconductors. Slightly larger than the SOT-89, the length being 6.3 mm, the SOT-223 can house chips up to 1.5 x 1.5 mm. It has four gull-wing leads, three on one side of the body, the fourth being on the opposite side and much wider than the others. The fourth lead acts as a heat conductor rather than a terminal and, in the case of transistors, is usually connected to the collector (the central lead on the other side).

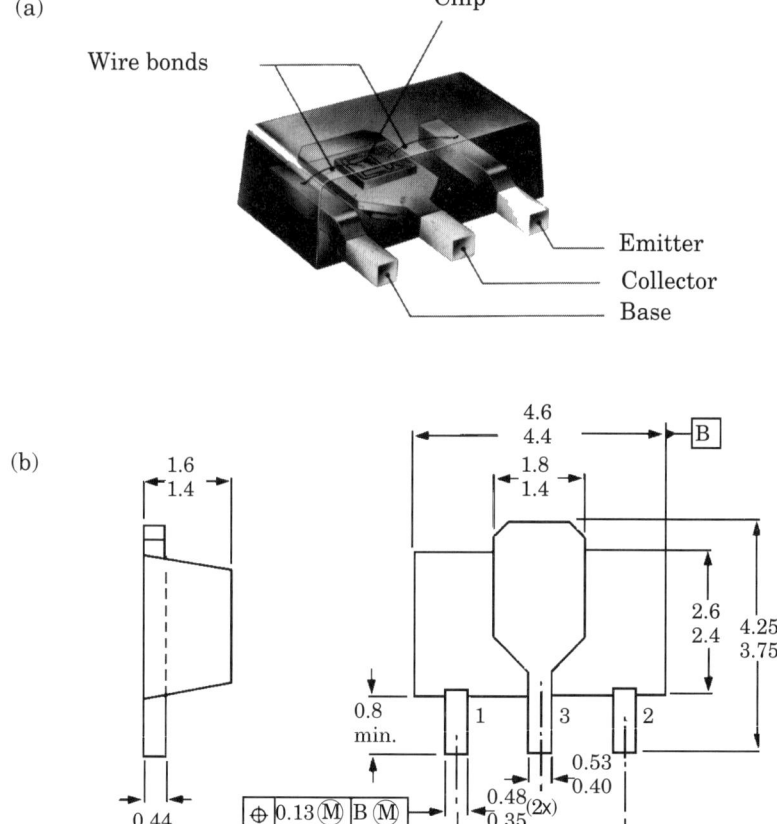

Fig. 4.27 The SOT-89 package. (a) See-through view of SOT-89 encapsulation (Courtesy of Philips); (b) dimensions of SOT-89 encapsulation. (Courtesy of Mullard)

The package is often used in RF circuits which can require a double connection to the emitter of the device. In order to meet this need, the device can be supplied with the larger lead connected to the emitter terminal.

4.4.6 Adapted Packages

TO-126 and TO-220 packages, see Chapter 3, can be adapted to surface mounting by using suitable leads formed to gull-wing shape. Both have been developed to be mounted with the case in contact with a heat sink. However, neither the SOT-82 package (with roughly the same dimensions as the TO-126) nor the SOT-186 (closely resembling the TO-220) is supplied with short leads for surface mounting and must be adapted before assembly.

The SOT-194 case, again similar to the TO-126, is supplied with leads formed for surface mounting. Very large chips (up to 3 x 3 mm) can be packaged in this case. Up to 8 A current can be handled, with a voltage of up to 500 V. In normal conditions the power dissipation may be 1 watt.

Similar conversions may be made whenever high power devices are involved. The adapted packages take up a lot of board space unless mounted vertically. Horizontal mounting is used only when the height above the board is restricted.

4.4.7 Small Outline Integrated Circuits — SOICs

The *Small Outline Package*, SOP, was developed for surface mounting and bears a strong likeness to the DIP, as can be seen from Figure 4.28. A major use of the package is for ICs and the term SOIC is now a universally accepted synonym of SOP, whatever the content.

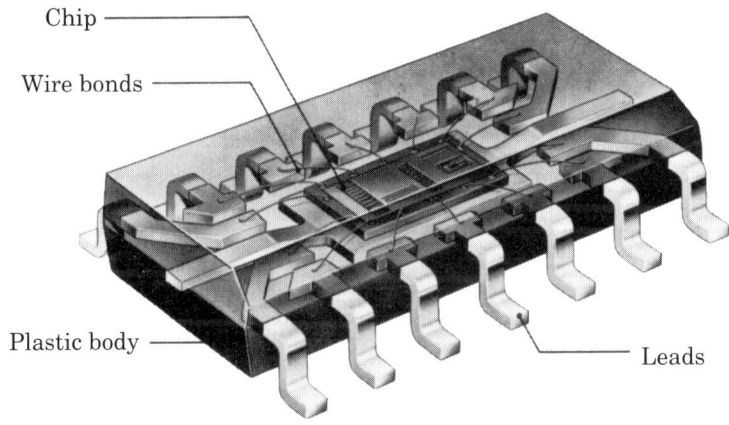

Fig. 4.28 See-through view of small outline integrated circuit (SOIC). (Courtesy of Philips)

SOICs are a family of components in packages having a dual-in-line structure. The number of pins and width and length of the body vary to suit the device packaged as do DIPs. The SOIC may have from 6 to 36 leads and is called accordingly an SO-6, SO-8, and so on up to the SO-36. Devices requiring 6 to 14 pins are encapsulated in a body approximately 4 mm wide and 1.45 mm thick, the length being a function of the number of leads, Table 4.1. The 16 pin package is available in two versions, a small one, the SO-16, having the same height and width as the 6 to 14 pin devices, and a wide body version, the SO-16L, which has a width of 7.6 mm (0.3 in.), i.e., that of the DIP. The SO-16L also is thicker and can accommodate larger chips.

20 to 36 pin SOICs have the same wider and thicker body of the SO-16L, the length increasing with the pin count. Dimensions of some of these packages are given in Table 4.1. These larger packages are sometimes called SOL or *Small Outline-Large Body*, a contradiction in terms. To avoid such contradictions, the terms '150 mil SOIC' and '300 mil SOIC' are quite commonly used.

Table 4.1

Selected Dimensions of some SOIC and VSO Packages

Package Name	Pin Number	Dimensions (maximum) in mm				
		Body Length	Body Width	Total Width	Lead Spacing (mm)	(in.)
SO-6	6	3.75	4.0	6.2	1.27	0.050
SO-8	8	5.00	4.0	6.2	1.27	0.050
SO-10	10	6.25	4.0	6.2	1.27	0.050
SO-14	14	8.75	4.0	6.2	1.27	0.050
SO-16	16	10.00	4.0	6.2	1.27	0.050
SO-16L	16	10.50	7.6	10.65	1.27	0.050
SO-20	20	13.00	7.6	10.65	1.27	0.050
SO-24	24	15.60	7.6	10.65	1.27	0.050
SO-28	28	18.10	7.6	10.65	1.27	0.050
VSO-40	40	16.00	7.6	12.80	0.762	0.030
VSO-52	52	19.00	9.1	11.60	0.762	0.030
VSO-56	56	22.00	11.1	15.80	0.762	0.030

The SOIC is popular because it is cheap, reliable and space-saving. While the SO-16L occupies twice the area of the SO-16, the ratio of total volume to chip volume is still good (and much better than that of the equivalent DIP). As all these packages have gull-wing leads on two sides only, the production and testing equipment used for the plastic bodied DIPs is easily adapted for use with SOICs. They retain the DIP structure, well suited for handling and testing, the major difference being in the lead pitch. This was reduced to 1.27 mm (0.050 in.), (a) to save space and (b) to take advantage of the lack of need for mounting holes which restricted increase in packing density. It is pertinent to note that lead pitches are shrinking all the time. However, a survey of contract assembly manufacturers carried out in 1993 showed that more than half were at a lead pitch of 0.50 mm (0.020 in.) and none were below 0.40 mm (0.016 in.).[6]

Identification of pins is simple and reliable. The body has a bevelled edge on the top. When the component is viewed with the bevelled edge vertical and on the left of the viewer, the pin count starts from number 1, the top left one, and proceeds in a counterclockwise direction. The large body also has a notch which serves the same function as in the DIP.

These packages have much stricter tolerances than have the plastic bodied DIPs and, further, manufacturers observe these quite closely. Availability of ICs in an SO case is extensive, every manufacturer wishing to cater for the expanding SM market. While conventional components are still widely used in all kinds of assemblies, SMT is taking the major share of new designs and new components are supplied in SO versions whenever applicable. Other components, e.g., resistor arrays, are supplied in SO outlines and can be mounted with the same equipments.

While the SOP or SOIC is a most popular case for medium complexity SMDs, it is not used above a pin count of 36. Above this figure, the space occupied on the PB becomes too large and the length of chip-to-pad connection becomes long enough to have an adverse effect on high frequency performance. Some SO packages with higher pin counts have been produced but have had little success unless pitch is reduced. The SOP-42 has leads on 1.27 mm pitch and the SSOIC (*shrink small outline IC*) packages have up to 56 pins on a 0.635 mm pitch or up to 64 pins on a 0.80 mm pitch.

The height of the typical SOP is of the order of 1.45 mm (0.057 in.) minimum and for many applications a thinner device package is essential, in particular the PCMCIA (Personal Computer Memory Card International Association) PC card Type 1.[7] The *thin small outline package* (TSOP) has a height of 1.2 mm (0.047 in.) and can be mounted on both sides of a double-sided printed board within the confines of the specified thickness (3.3 mm) for the PCMCIA Type 1 card. In addition, a TSOP with a lead pitch of 0.5 mm (0.020 in.) allows high packing density (but demands exacting control of design, placement and processing!). TSOPs are available in pin counts ranging from 16 to 76 and 0.3, 0.4, 0.5 and 0.65 mm lead pitches. They are unusual in that the leads protrude from the shorter sides of the plastic body. The *paper thin small outline package* (PTSOP) with a height of 0.5 mm doubles the potential capacity of the Type 1 card.

4.4.8 Very Small Outline Packages — VSOs

By reducing lead pitch from 1.27 mm (0.050 in.) to 0.762 mm (0.030 in.), it became easy to achieve a significant reduction in body length. By this means, a new family of packages was developed retaining the dual-in-line structure. The reduction in pitch allowed an increase in pin count while still giving a good ratio of total space to chip space. The new family was given the name of *Very Small Outline*, which refers to the fact that a large chip with many input/output pins is packaged in a very small outline. However, the name can be misleading as all VSOs are larger than the largest SO in one or more dimensions, as shown in Table 4.1.

Some component manufacturers include their VSO packages in the SO family. This generates confusion. The smaller termination pitch of the VSO demands more care in handling and has a considerable influence on assembly, soldering and testing techniques. There is little or no standardisation of the VSO packages and reference to the supplier's data sheet(s) is essential for design and production engineers.

The first VSOs introduced had 40 and 56 leads, the VSO-40 and VSO-56. These were followed by the VSO-52, shown in Figure 4.29. Note the identification of pin 1. The bodies of these packages differ in both width and length. Some VSOs were produced with a lead pitch of 0.8 mm (0.0315 in.) but had only limited success.

4.4.9 Flatpacks and Quadpacks — QFPs

The trend is to refer to these cases as *quadpacks*, *quad packs*, *quads* or *QFPs*. They have been called MFPs (*Mini Flat Packs*) and, incorrectly, FPPs (*Flat Plastic Packages*). These packages may also be called PQFP (*plastic*

52-pin DIL Plastic SMD Package — VSO-52
Weight approx. 0.9 g

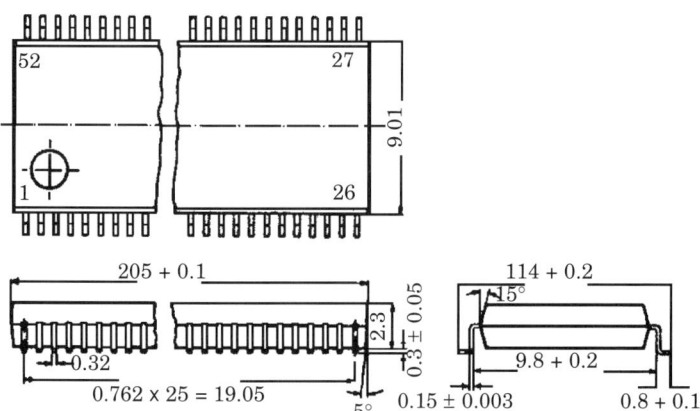

Fig. 4.29 A VSO-52 package. (Dimensions are in millimetres.)

quad flat pack — lead pitch 0.635 mm only), CQFP (*ceramic quad flat pack*), SQFP (*shrink quad flat pack*— lead pitch 0.5, 0.4 or 0.3 mm). The package may be rectangular or square. Because they have leads on four sides, they have also been included in the broad classification of PLCCs, dealt with later. The term 'QFP' or 'quad' will be used in this text to avoid possible confusion.

The case may be rectangular or square and is a natural extension of the SO package. It is suitable for a large number of terminals, from 32 to 440, 576 or more with lead pitch varying from 1 to 0.3 mm. An increase in pin count produces an increase in size which, for the same number of leads, will be greater in a structure having them on two sides only than in a structure which allows pins on all four sides. This is very apparent in Figure 4.30 in

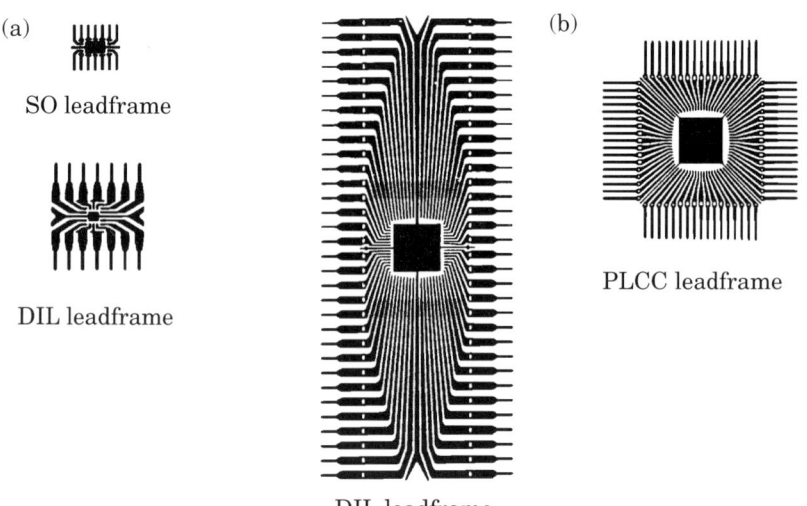

Fig. 4.30 (a) A 14 lead SOIC compared with a 14 lead DIP; (b) a 64 lead DIP compared with a 68 lead PLCC.

which a 68 pin PLCC (the same size as an equivalent QFP) is compared with a 64 pin DIP. The space saving resulting is at least 50% and can be more. The excessive length of the DIP leadframe conductors, with their consequential adverse effect on circuit speed, may also be noted.

The QFP solves many problems associated with mounting complex chips. Body size can vary to accept different sizes of chip, and the number of leads can be altered as required. Unfortunately, this makes standardisation very difficult as each manufacturer responds to his own needs in his own fashion. The only common features of these packages are:

(i) a square or rectangular body (one corner may be bevelled for orientation);
(ii) gull-wing leads projecting 1.0 to 2.0 mm from the body;
(iii) one of three alternative lead spacings and lead widths dependent on the number of leads;
(iv) body thickness in the range of 1.6 to 3.0 mm; it should be noted, however, that the *thin quad flatpack* (TQFP) has a maximum thickness of 1.2 mm;[7]
(v) when viewed from the top, the pin number starts from the marked corner and proceeds counter-clockwise.

Some of the more common structures available are given in Table 4.2.

4.4.10 Plastic Flat Packs — PFPs

The *Plastic Flat Pack* or *Flat Plastic Package* is frequently confused with the QFP. Both types have leads emerging from four sides and the differences between the packages may be unnoticeable. The main difference between them is that, while the QFP evolved from the SO package, the PFP or FPP was derived from the original metal flatpack, described in Chapter 3.

The metal flatpack is compact and highly reliable and the metal body and leads give very good heat dissipation. (If the normal nickel-iron alloy, *Alloy 42*, used for lead frames is replaced with a copper alloy, the dissipation of heat is exceptionally good.) However, the flatpack is very expensive and this promoted attempts to derive a cheaper plastic bodied version for the less demanding applications. The result is a package which is a cross between the traditional metal flatpack and the QFP.

Compared with the metal flatpack, the PFP has a larger body and smaller leads. The leads are normally longer than those of a QFP and are flat — not shaped, see Figure 4.31. The lead pitch is typically 0.8 mm or 0.65 mm (0.031 in. and 0.026 in. respectively) which may be compared with the original flatpack pitch of 1.27 mm (0.050 in.). The leads make the PFP difficult to mount. If they are long enough, they may be shaped but at the expense of using more board area. If they are short, they are very difficult to shape without damage to the plastic-to-lead seal. The net result is that PFPs are now available with pre-shaped gull-wing leads, although the length is longer than that of SOIC leads. As pin numbering of PFPs is now more often in line with a QFP case than with the flatpack standard, almost all differences between the PFP and the quadpack have disappeared and confusion is justified.

Table 4.2

Common Quad Flat Packages

Maximum Number of Leads	Typical Body Dimensions (mm)			Lead Pitch (mm)	Lead Width (mm)
	Length	Width	Thickness		
32	9	9	1.7	0.8	0.40
36	10	10	2.0	1.0	0.40
44	10	10	2.0	0.8	0.35
56	10	10	2.0	0.65	0.30
44	14	14	2.2	1.0	0.40
64	14	14	2.2	0.8	0.35
80	14	14	2.2	0.65	0.30
64	14	20	2.4	1.0	0.40
80	14	20	2.4	0.8	0.35
100	14	20	2.4	0.65	0.30
76	20	20	2.6	1.0	0.40
96	20	20	2.6	0.8	0.35
120	20	20	2.6	0.65	0.30
92	20	28	2.7	1.0	0.40
116	20	28	2.7	0.8	0.35
144	20	28	2.7	0.65	0.30
108	28	28	2.8	1.0	0.40
136	28	28	2.8	0.8	0.35
168	28	28	2.8	0.65	0.30
132	28	40	2.9	1.0	0.40
164	28	40	2.9	0.8	0.35
204	28	40	2.9	0.65	0.30
156	40	40	3.0	1.0	0.40
192	40	40	3.0	0.8	0.35
240	40	40	3.0	0.65	0.30

Ceramic flatpacks (CFP) are produced and range from 5.08 mm to 22.86 mm width with pin counts varying from 10 to 50.

4.4.11 Plastic Leaded Chip Carrier — PLCC

The PLCC was introduced by Texas Instruments to overcome the problems of thermal mismatch associated with leadless packages. Originally the acronym meant *Post-moulded Leaded Chip Carrier*. The interpretation of the acronym as *Plastic Leaded Chip Carrier* has gained universal acceptance

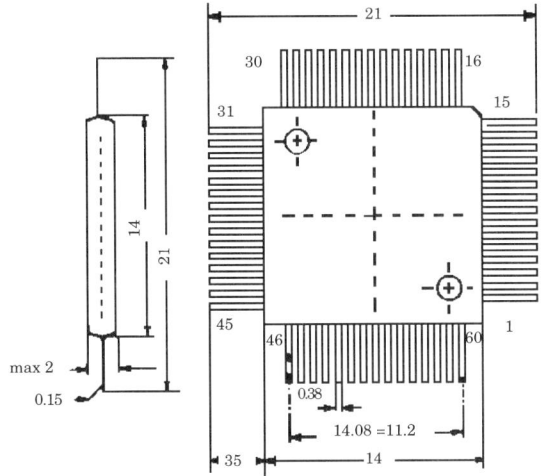

Fig. 4.31 A plastic flat pack, PFP. The pin numbering of a PFP is different from that of the original TO-84/86 metal flatpacks.

and gives a good indication of its structure, provided that one does not expect to find plastic leads!

The lead frame of a 68 lead PLCC is illustrated in Figure 4.30 and dimensions of the finished package are given in Figure 4.32. Key characteristics of the package are:

(i) a relatively thick, up to 4.8 mm (0.189 in.), square or rectangular plastic body;
(ii) J-shaped leads on four sides, rolled under the component body, which has nicks on the underside to retain the ends;
(iii) 1.27 mm (0.050 in.) pin spacing;
(iv) pin numbering starting from the centre of one side, on which the body has a bevel, a bevelled corner, a mark or all three, and proceeding counter-clockwise. The PLCC is viewed from the top and orientated so that the bevelled or marked side is at the top or the bevelled corner is at top left. With an odd number of pins on the upper side, the central pin is number 1. If the number is even, pin 1 is the first met starting from the centre of the row and going anti-clockwise.

Some popular JEDEC PLCC square outlines are shown in Figure 4.32(a). Rectangular packages are typified by the 22 pin version in Figure 4.32(b). The 'J' leads, while projecting less than 0.5 mm outside the body, are very successful in coping with thermal mismatch, so that these devices can be mounted on conventional PB materials without special precautions. An additional advantage over the leadless ceramic devices is that the use of copper alloy lead frames gives higher thermal conductivity and benefits in heat sinking ability into the solder lands.

The shape and position of the PLCC contacts give good contact when the device is mounted by insertion into a socket, the 1.27 mm lead pitch simplifies

FN Plastic Chip Carrier Package (28-terminal package used for illustration)

(a)

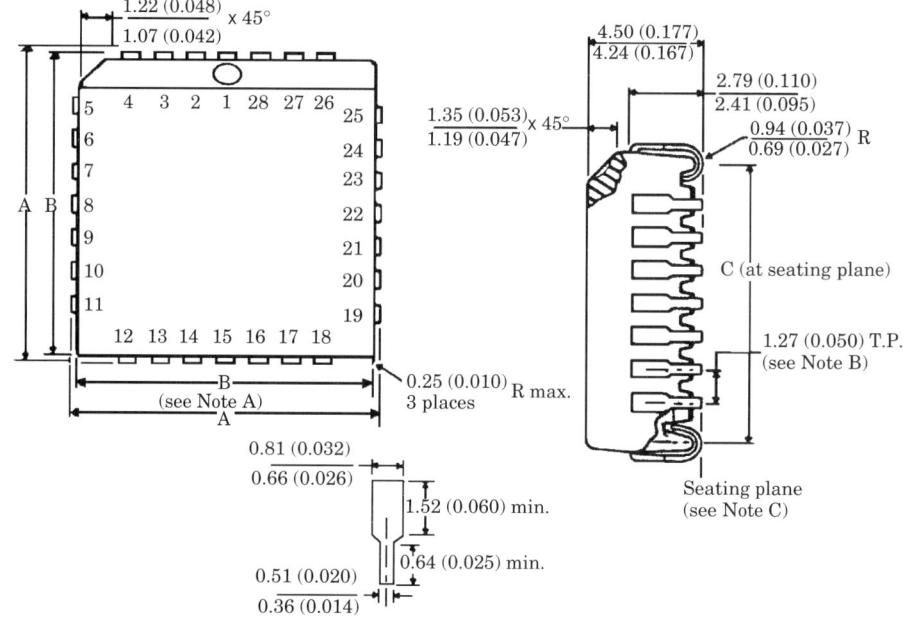

Lead detail

JEDEC Outline	No. of Terminals	A		B		C	
		Min.	Max.	Min.	Max.	Min.	Max.
MO-047AA	20	9.78	10.03	8.89	9.04	7.87	8.38
		(0.385)	(0.395)	(0.350)	(0.356)	(0.310)	(0.330)
MO-047AB	28	12.32	12.57	11.43	11.58	10.41	10.92
		(0.485)	(0.495)	(0.450)	(0.456)	(0.410)	(0.430)
MO-047AC	44	17.40	17.65	16.51	16.66	15.49	16.00
		(0.685)	(0.695)	(0.650)	(0.656)	(0.610)	(0.630)
MO-047AE	68	25.02	25.27	24.13	24.33	23.11	23.62
		(0.985)	(0.995)	(0.950)	(0.956)	(0.910)	(0.930)

All dimensions and notes for the specified JEDEC outline apply.
Notes: A. Centreline of centre pin each side is within 0.10 (0.004) of package centreline as determined by dimension B.
B. Location of each pin is within 0.127 (0.005) of the true position with respect to centre pin on each side.
C. The lead contact points are planar within 0.10 (0.004).
All linear dimensions are in millimetres and, in brackets, in inches.

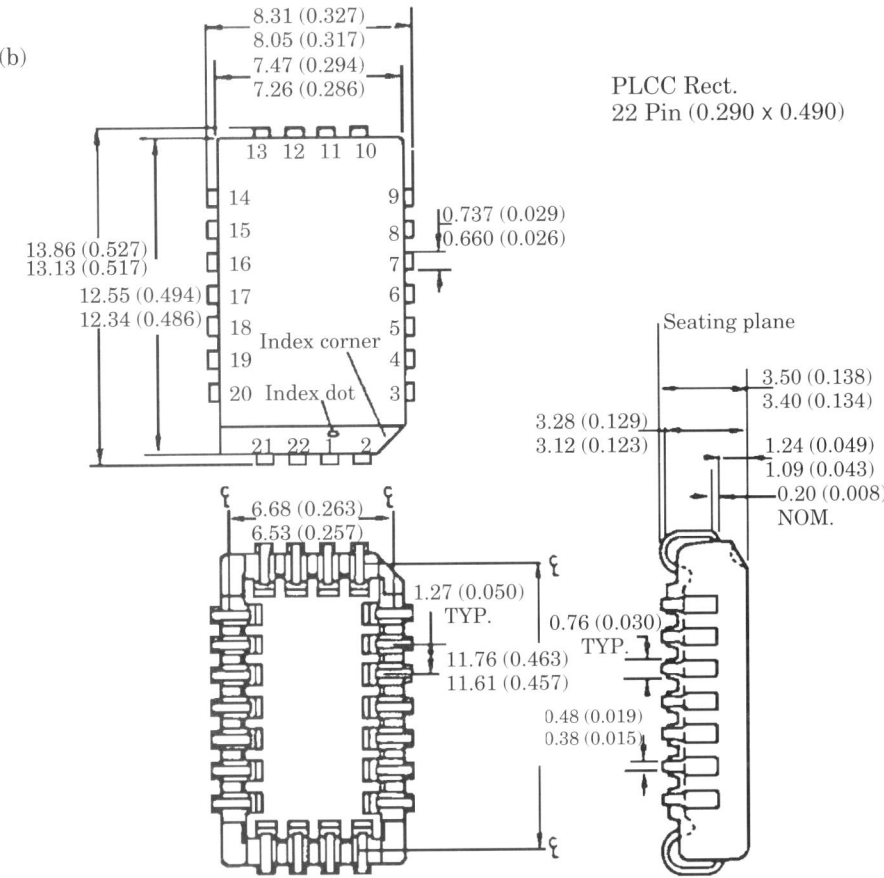

Fig. 4.32 (a) The PLCC square package. Pin counts from 18 to 164 are available; (b) the PLCC rectangular package, typified by a 22 pin version (0.290 x 0.490), refers to the nominal body dimensions in inches.

board design, and the package is space effective and cheap to produce. However, like all good things, there are some drawbacks.

When compared with SO and VSO packages, PLCCs are more difficult to inspect when directly mounted on the PB by soldering. The integrity of the joint is difficult to assess as it is mostly hidden under the board. Touching up and hand soldering is a tricky operation which demands a highly skilful operator. In common with other 'J' lead components, in-circuit testing will require special tools or the provision of test pads on the PB. With these exceptions the PLCC has been proven to be reliable, free from problems with thermal mismatch and board flexing, and continues to be one of the most popular device packages.

4.4.12 Small Outline J-leaded Package — SOJ

The SOJ is an attempt to combine the simplicity of the DIL structure with the advantages of the PLCC in modifying the SO or VSO package by incorporating J leads. A major rôle for the SOJ appears to be in packaging those ICs which have a large chip but relatively few connections. A typical example is its use as a package for 1 Mbit DRAMs, which require only 20 leads. The chip can be accommodated in an SOJ with body dimensions of (L x W x H in mm) 17.15 x 7.62 x 3.45 (0.675 x 0.3 x 0.136 in.). In this instance the leads, which in this style of package are still on a 1.27 mm pitch, are positioned as two groups of 5 leads at each end of the long sides (a 26 lead SOIC with the middle 3 leads missing on both sides).

The long-term success of this package is doubtful since it is competing with the simple, cheap and well proven PLCC.

4.4.13 Leadless Ceramic Chip Carrier — LCCC

This well known package appeared in the early days of SMT development. The structure consists of a small ceramic, usually 96% alumina, tile on which all internal conductors are deposited by screening and firing, in one or more layers. The top conductor is usually gold. An adequate soldering area is obtained by extending the conductors to the edge of the tile and, sometimes, to a limited extent on to the underside. The chip sits in a recess in the middle and connections are made to the conductors by normal wire bonding methods. Soldering to the external contact pads must be performed with consideration for the thin conductor layers, to minimise leaching of the solderable layer into the molten solder.

The cost, and reliability, of these packages depends not only on the metallisations used for conductors and terminals, but also upon the type of lid and seal used to enclose the chip. JEDEC recognises several types of LCCC, all with external contact spacing of 1.27 mm (0.050 in.). The JEDEC standards allow for packages on which the edge conductors are exposed or recessed into grooves on the body, and for capping varying with type from metal lids, flush or recessed, to ceramic lids. Dependent on construction, the lid may be on the top or bottom of the finished package.

Lea[3] points out that a three-layer construction with a flat gold plated lid sealed with a gold-tin solder preform is most reliable, and most expensive; a three-layer chip carrier with a glass sealed ceramic lid gives some cost saving; a single-layer chip carrier having a pre-glassed cavity and a cup-shaped ceramic lid gives a device which costs about half the price of the first. The least expensive solution, possible when there is not a need for a hermetically sealed device, is to encapsulate the device on its ceramic base with an epoxide resin.

Types with terminal counts ranging from 16 to 156 are available. Some of these are shown, with dimensions, in Figure 4.33. Figure 4.34 illustrates, quite dramatically, the difference in size between these device packages and a DIP.

The major problems arising with LCCCs stem from the different characteristics of alumina and common PB materials. Alumina is rigid and relatively inflexible. Any bending or flexing of the PB can impose considerable stress on the solder joints and break them or initiate cracks. During the life of the equipment such cracks may progress to cause open or intermittent connections. In addition, the coefficient of thermal expansion of alumina is much lower than that of PB materials. Abnormal stresses can be created on

(a)

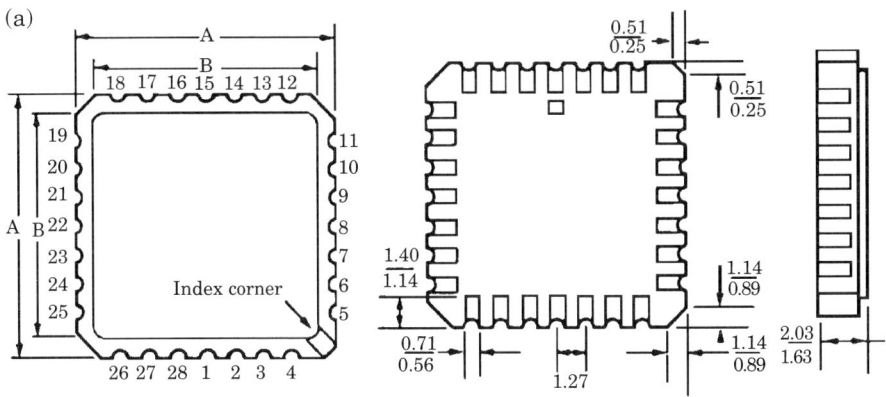

JEDEC Outline Designation	No. of Terminals	A Min.	A Max.	B Min.	B Max.
MS 004 CB	20	8.69	9.09	7.80	9.09
MS 004 CC	28	11.23	11.63	10.31	11.63
MS 004 CD	44	16.26	16.76	12.58	14.22

JEDEC Outline Designation	No. of Terminals	A Min.	A Max.	B Min.	B Max.
MS 004 CE	52	18.78	19.32	12.58	14.22
MS 004 CF	68	23.83	24.43	12.60	21.80
MS 004 CG	84	23.83	29.59	12.60	27.00

(b)

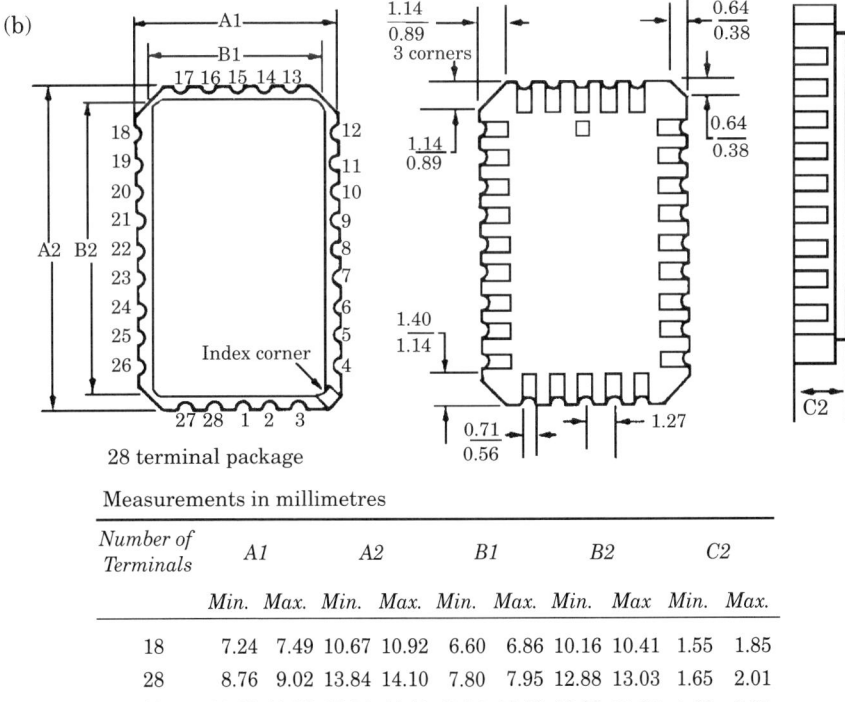

28 terminal package

Measurements in millimetres

Number of Terminals	A1		A2		B1		B2		C2	
	Min.	Max.	Min.	Max.	Min.	Max.	Min.	Max	Min.	Max.
18	7.24	7.49	10.67	10.92	6.60	6.86	10.16	10.41	1.55	1.85
28	8.76	9.02	13.84	14.10	7.80	7.95	12.88	13.03	1.65	2.01
32	11.30	11.56	13.84	14.10	10.34	13.03	12.88	13.03	1.65	2.01

Fig. 4.33 Leadless ceramic chip carriers, LCCCs. (a) FK-ceramic chip carrier packages; (b) FE/FG-ceramic chip carrier packages.

Fig. 4.34 LCCCs.

heating and cooling of the equipment and with large swings in environmental temperature. These problems can be minimised by using a mounting socket having spring contacts so that the LCCC can 'float' to some degree, without compromising the connection. An alternative solution is to use a *Leaded Ceramic Chip Carrier*, described in the next part of this chapter. Use may also be made of metal cored printed boards to reduce X and Y expansion. These can be very expensive and, unless there are other good reasons for using metal cored PBs, such as heat sinking or the need for restraint for other components, their use cannot be recommended.

4.4.14 Leaded Ceramic Chip Carrier — LDCC

The LDCC is derived from the leadless package by soldering or brazing J-type leads on to the sides of the substrate. The resultant device is more costly than the LCCC but retains the advantages and the leads provide the necessary resilience to minimise stressing. The outlines are often directly interchangeable with the equivalent LCCC case and there are instances where the LDCC provides a viable solution to mismatch problems, without redesign of the assembly or the use of sockets. In such cases it can be very cost effective.

4.4.15 Other Complex Packages

Under the name of *Cerquad*, one manufacturer (Thomson) introduced a ceramic package using the LDCC manufacturing method but having highly flexible J leads. To suit it for a large number of terminations, e.g., 148, the lead pitch is reduced to 0.635 mm (0.025 in.). It can accept quite large chips and needs limited area for mounting. The cost of the package is relatively low.

The packaging of chips requiring connections of 200 or more — such as gate arrays, standard cell devices and in general all complex ASICs, see Chapter 3 — creates severe problems for manufacturers and users of such components. On the assumption that the minimum lead pitch for economic mass production is of the order of 0.65 mm (0.026 in.), the dimensions of the package become too large compared with those of the chip.

The obvious solution is to have two rows of contacts on both sides of the chip. The result of this approach is the *Leadless Grid Array* (LGA) or *Pad*

Grid Array (PGA) package, see Chapter 3, Figure 3.45. This has two rows of staggered pads on each of the four sides. In this way it is practical to run a track between any pair of pads on the outer row, so that a PB can be designed to accept the package.

Other solutions have been proposed for high pin count packages. One example is an *Open Via Chip Carrier*, OVCC, developed by Thomson. This is similar to an LCCC case, with 1.27 mm lead pitch, with the addition of a second row of terminals on the top and bottom of the package. The ceramic substrate has a hole in it for each of these internal pads and connection may also be made by soldering a thin wire into it.

Continuous development is taking place in this field. The pin grid array, Chapter 3, has been developed far beyond what could be conceived in the eighties but problems with alignment of pins and the cost of highly complex packages may in the end restrict its further development. The *Ball Grid Array* (BGA), in which the pins of the PGA are replaced by solder balls, originated in the early eighties but in many cases gave severe quality problems although there is much interest in making BGA work reliably.[8] Improvements in technology and control of geometries and processing, coupled with the fact that solder balls are cheaper than Alloy 42 or copper pins, led to further development work and, by the late eighties/early nineties, several large semiconductor manufacturers were offering ICs packaged in ceramic BGAs and, later, plastic BGAs. While ball grid array is more expensive than quad flatpack, the BGA offers the ability to meet high I/O (input/output) needs with short interconnects, does away with problems due to distorted pins and is easier to assemble using standard pick-and-place equipment. Lau *et al.*[9] summarise the most frequently stated benefits of ball grid array packages as:

— better board assembly yield;
— better electrical performance;
— better package yield;
— cavity up or down options;
— easier to extend to multichip modules;
— faster design-to-production cycle time;
— higher interconnect density;
— higher I/Os for a given footprint;
— lower profile (smaller size);
— multilayer interconnect options;
— reduced co-planarity problems:
— reduced handling issues (no damaged leads);
— reduced paste printing problems (wider pitch);
— reduced placement problems (self-centring);
— shorter wire bonds.

At the same time those things which must be noted and understood are listed. Chief of these are:

— assembly inspection is more difficult;
— cracking has been reported during reflow;
— rework methods are more difficult;

— solder joint reliability is more critical;
— standardisation is in its very early stages;
— testability is not well established.

Other packages will doubtless be developed, aided by the increasing ability of printed board producers to provide boards with finer lines and spacings (FPT — *fine pitch technology*), greater dimensional accuracy and stability, and better impedance control. In the interim, independent forecasts all agree that by the turn of the century hundreds of millions of high I/O ICs will be packaged in BGAs.[10]

4.4.16   Tape Automated Bonding — TAB (See also Chapter 16)

In the early days of integrated circuits several different methods of providing the interconnection, IC to package, were considered. The major techniques developed were wire bonding, TAB, flip-chip and beam lead. Wire bonding has dominated chip interconnection. The technology for flip-chips has been perfected by IBM, and some others, and TAB has been used by a number of manufacturers who developed the method to a practicable technique. Beam leads were discarded as impractical technology for mass production of a wide variety of devices.

TAB was developed in the late 1960s, the first patents being awarded to General Electric. The major applications up to now have been in low cost/low density consumer products, such as calculators and watches, in which the planar profile of TAB is ideal. However, TAB can accommodate the high input/output (I/O) counts and higher operating speeds essential for VLSI chip packaging. Because of this, there is growing interest in TAB and many major component manufacturers are making increasing use of the method. This was very clear from the TAB activities of the top 10 IC manufacturers in 1989.[11] Three of these had been involved with TAB mounting of chips with an I/O count of from below 25 to 250. Another, Motorola, was already producing TAB packaged components with more than 250 interconnections. All 10 manufacturers were expending effort on the use of TAB in high I/O count package development.

From a manufacturing viewpoint, flip-chip bonding has a big advantage over wire bonding in that all connections to the chip can be made by one joining step. In the days of manual wire bonding this gave a significant labour/cost saving. Savings can still be made with flip-chips, except when surfaces to which the outer lead bonds are to be made are uneven, for example, multilayer ceramic packages. In flip-chip bonding for TAB the cost advantage remains for small chips but may diminish for large chips.

The 'bump' is the site at which the TAB lead frame is connected to the chip and is a critical part of the system. A typical bump structure is shown in Figure 4.35. As TAB mounting is essentially the mounting of a bare chip, the active area of the chip is protected with a passivation layer, usually silicon nitride, leaving only the contact pad exposed. Before growing the solderable bump a barrier layer is sputtered on to the bump site to prevent growth of intermetallics and assist adhesion. Quite often a further adhesion promotion layer is added before forming the bump by electroplating, typically with gold of 90 Knoop hardness. The bump height, profile and hardness are critical and

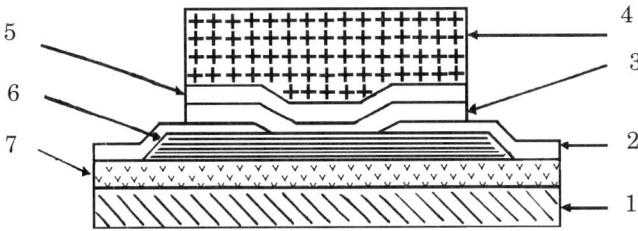

1. Silicon
2. Passivation layer (Sn)
3. Barrier layer (Ti/W)
4. Gold bump
5. Adhesion layer (Au)
6. Aluminium pad (Al +1% Si)
7. Silicon dioxide

Fig. 4.35 Structure of a flip-chip bump. In TAB the bare chip is exposed and, unless protected by a passivation layer, corrosion of the chip circuitry can result.

strict control of current density and deposition rate in electrodeposition is essential.

The chip is now ready for assembly to a very flat surface. It may be assembled to a rigid substrate, but only with close temperature control because the mismatch in coefficients of thermal expansion can be severe and may cause rupture of the solder bumps. In the IBM C4 flip-chip technique, the assembly is done in a temperature controlled environment. Use of the technique is viable because IBM mainframe computer modules are maintained at a constant temperature. (Stacked chip-to-substrate bumps, in which bumps deposited in the required pattern on a flexible substrate, e.g. polyimide, are interposed between the chip bump and the rigid substrate, can provide the necessary stress relief. However, by increasing the path length between the chip pad and the substrate conductor, they reduce a major advantage of flip-chips, the high electrical performance resulting from the short path length of the connection.)

Fortunately, in TAB the substrate is flexible. TAB is also known by different names, some of which are registered trade marks, for example, *Chip on Tape*, *Tape Pack* and *Mikropack*. The process of manufacture starts with an adhesive coated polyimide film which is punched along one or both edges with precision sprocket holes. In 1987 JEDEC established standards for such tapes using cinematic film formats, 8 mm, 35 mm and 70 mm. Windows are cut in the film to accept the component chips, with any other featured holes/slots needed. It should be noted that the individual tapes are normally produced in multiples on a wider reel of film.

Copper foil, slit to the active area width, is laminated to the polyimide film, photoimaged, etched and electroplated with tin or gold, as described for flexible circuits in Volume 2, Chapter 11. Custom-designed reel-to-reel equipment is used for the manufacturing operations, Figure 4.36. Finally, the original strip is slit into the finished tape widths.

The conductors on the strips are cantilevered over the cutout for the component so that, when the component is placed in its aperture, the chip bumps contact the projecting conductor. A pulse heated soldering or welding head presses the bumps on to the conductors and creates all joints

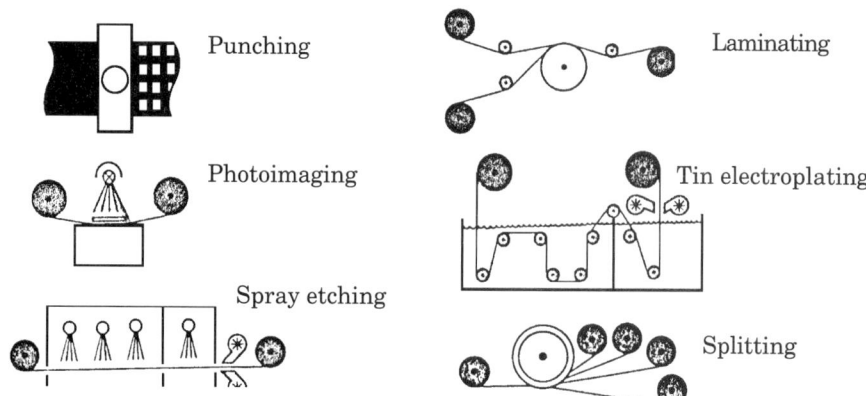

Fig. 4.36 Simplified sequence for reel-to-reel production of tapes for flip-chip mounting.

simultaneously. The joint produced is relatively strong when compared with a conventional wire bond.

After thorough testing, electrical and mechanical, the chip is further protected by 'glob topping' with a potting resin or by transfer moulding. (Many improvements have been made to methods and equipments for applying globules of resins for protection of the chips.) The finished height of the component is of the order of 0.6 mm.

The final result is a punched strip carrying many components, each with its own fan of conductors spreading out from the chip, Figure 4.37. It will be

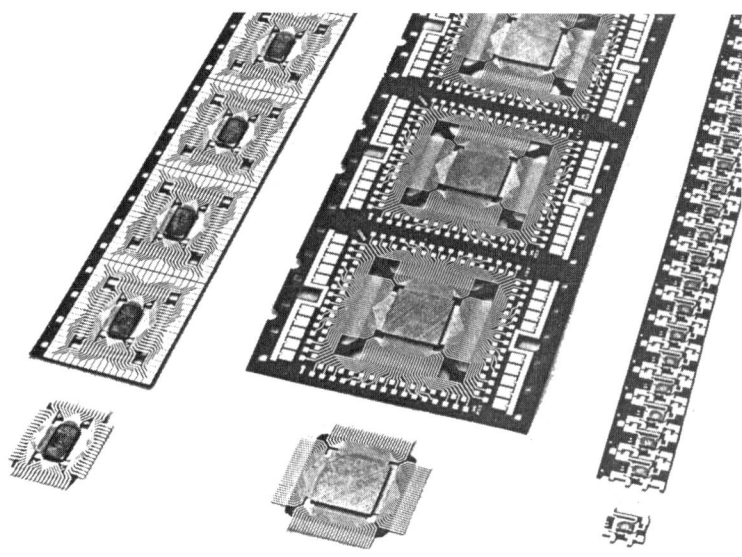

Fig. 4.37 Differing 'Mikropack' components; left, on 16 mm tape; centre, on 35 mm tape; right, on 8 mm tape. (Source: Siemens AG)

noted that there are two possible points at which the connection to the PB may be made. The inner connection point is as shown by the separate devices in Figure 4.37. This is the normal connection and requires that the components are removed from the tape by a special tool. This blanks the component from the tape, shapes the leads, and places it on the PCB where it is reflow attached, usually by pulsed tooling. The second area to which connection may be made is to the larger pads on the outer border of the tape. These pads are primarily intended to be used as test points but may be used as terminals in some assemblies.

TAB packages are cheap and very space efficient. TAB should be able to produce connections to chips at 25 μm pitch (0.001 in.), but most connections on high I/O chips are being made at 100 μm and 50 μm pitch. Even then the TAB package achieves a connection density 4 to 10 times greater than the LCCC. The methods of environmental protection may not as yet be accepted by military users although a two-part silicone with epoxide top covering has been widely accepted for general business environments.

Factors which still delay wider use of TAB are:

(i) no universal availability of bumped chips;
(ii) TAB is largely suited to a vertically integrated manufacturing operation;
(iii) infrastructure problems are more easily solved by OEMs with both silicon and system assembly capability.

The advantages of meeting performance needs of speed, small footprint, minimal connection length, testability, and ability to handle very high I/O counts make TAB an increasingly important solution to high performance packaging.

4.4.17 Chip-on-board (COB) and Multichip Modules (MCMs)

Both chip-on-board and multichip modules offer the potential for smaller and more powerful designs compared with traditional and normal SM technologies. In addition, COB, in common with TAB, is one of the cheapest packaging methods.

4.4.17.1 CHIP-ON-BOARD

Chip-on-board has been used for many years in watches and calculators and is a widely known and understood process. Advances made over the past few years mean that, while the technology looks fundamentally the same as it was in the late sixties and early seventies, most of the steps in production have been improved and automated. The cleaning systems (and cleanliness level of the process environment), die attachment methods and materials, wire bonding equipments and dispensers for adhesives and encapsulation materials have been greatly developed and refined. Repeatability of the various process steps is much improved and high quality, predictable results can be achieved.

The substrate is organic, for example, epoxide/glass or polyimide/glass laminate, and it can be of complex multilayer construction having as many layers as needed by the design. A suitable finish for the board conductors is a nickel deposit followed by a very thin deposit of gold, Volume 2, Chapter 6. Other finishes are discussed in Reference 6. After cleaning the substrate, the die is attached, e.g., by using an epoxide adhesive (which may be metal-loaded to give good thermal and electrical conductivity). The adhesive is then cured before proceeding to establish the electrical connections.

Typically, aluminium wire bonds are ultrasonically welded, see Chapter 16, to join the pads on the die to the appropriate conductors on the printed board. A mechanical pull test can then be carried out before electrically testing to ensure that the circuit is properly bonded and functioning correctly.

The die can now be encapsulated ('glob topped') to protect the chip and its joints from corrosion and other environmental damage. A number of resin types and curing methods may be used but all resins must be of high purity and the system must be fully cured before proceeding to the next stage, which may be attachment of packaged components as outlined in Chapter 7.

4.4.17.2 MULTICHIP MODULES

In the draft revision of IEC 194[1] released in May 1994, a multichip module is described simply as "A microcircuit module consisting primarily of closely spaced integrated circuit dice". In Chapter 3, the multichip module is referred to as "the pinnacle of hybrid technology". Although BPA (Technology & Management) Ltd, of Dorking, UK, found in 1991 that some 60% of MCM designs had 10 dice or fewer,[12] a multichip module can consist of many more bare IC chips attached to a high density interconnection substrate.

The major types of multichip module are:

MCM-C: a multichip module fabricated on ceramic substrates and multilayer ceramic packages built up using the thick film materials and methods generally associated with thick film hybrid circuit manufacture.

MCM-D: a multichip module in which, at a minimum, the signal lines are created by the deposition of thin film metals and dielectrics to form multilayer structures. The substrate is normally silicon or ceramic but other inorganic bases have been used with dielectrics such as polyimide resins or benzocyclobutene.

MCM-L: a multichip module fabricated on organic multilayer laminate structures.

Initially, the majority of multichip modules were produced by thick film techniques but the trend to high performance applications and increased speed requirements could favour thin film MCMs, as suggested by figures published by BPA (Technology & Management) in 1993.[13] Table 4.3 shows the number of MCM-D modules overtaking MCM-C by the turn of the century.

Multichip modules may be packaged in standard surface mount packages such as quad flatpacks and ball grid arrays and can give the power of a PC/AT in an outline size of 50 mm x 50 mm (2 in. x 2 in.) or less. Within such small footprints are circuits offering high density, high speed and improved reliability with decreased power consumption.

Table 4.3
Units of MCMs used Worldwide

	1990	1991	1992	1993	1994	1995	1996	2000
MCM-D modules	12 K	36.4 K	206 K	365 K	1.57 M	6.29 M	13.5 M	161 M
MCM-C modules	1.37 M	1.52 M	1.82 M	3.3 M	11.6 M	22.2 M	39.7 M	141 M
MCM-L modules	1 K	2 K	4 K	5 K	8 K	12 K	25 K	150 K

Source of values: BPA (Technology & Management) Ltd.

4.5 SUPPLY PACKAGING

Packaging of SMCs raises problems which do not arise with conventional components. The major of these is the small size. This often makes handling and identification of type, value and polarity very difficult. In addition, the great majority of surface mount assembly is done with automatic equipment at high speed and, to match this, the time taken to set the machine up must be as short as is possible.

Most suppliers of SMCs provide full details of packaging for delivery on the component data sheet. Where this is not the case, the supplier must be approached for full information before placing orders. In the extreme case, the type of packaging used may not be compatible with the production equipment available. In the following text the most significant styles of packaging for SMCs are described. The information is non-specific and precise details of type of package, dimensions and tolerances must be checked, case by case, for individual components and individual suppliers. The International Electrotechnical Commission (IEC) are continually, albeit slowly, creating standards, for example, for packaging on tape. Not every type is covered and in some instances the standards exist but are not used.

It is necessary to note that disposal of packaging materials, etc., should only be done as permitted by local bye-laws and anti-pollution legislation in the country concerned. If any doubt exists, the appropriate authorities and the supplier should be consulted.

4.5.1 Bulk

Bulk packing in cardboard boxes or transparent plastic bags is the cheapest method. While most applicable to sturdy, leadless, cylindrical components such as the MELFs and Mini-MELFs, it is also used for some non-polar leadless devices.

The assembly machines must be equipped with hopper feeds, specific to the shape and size of component. This separates one component at a time from the bulk and presents it properly to the mounting head. Polar components such as the SOD-60 may also be supplied in bulk, provided the assembly equipment is able to recognise the orientation and, when needed, alter it. Bulk supplied components can be loaded into tubes, rails or magazines by simple equipment.

Normal quantities per bag or box are in the order of 1,000 to 10,000 components. Smaller quantities can be obtained, as sub-division by weight is practicable.

4.5.2 Magazine

Several types of magazine or cartridge are available, often specific to particular outlines or assembly machines. Three types are shown in Figure 4.38. The linear type has a recess for each component which, although precise, holds only 30 to 50 pieces. The vertical or stack cartridge can hold more components, e.g., 200, but is suitable for only a few machines and can be used only for flat components.

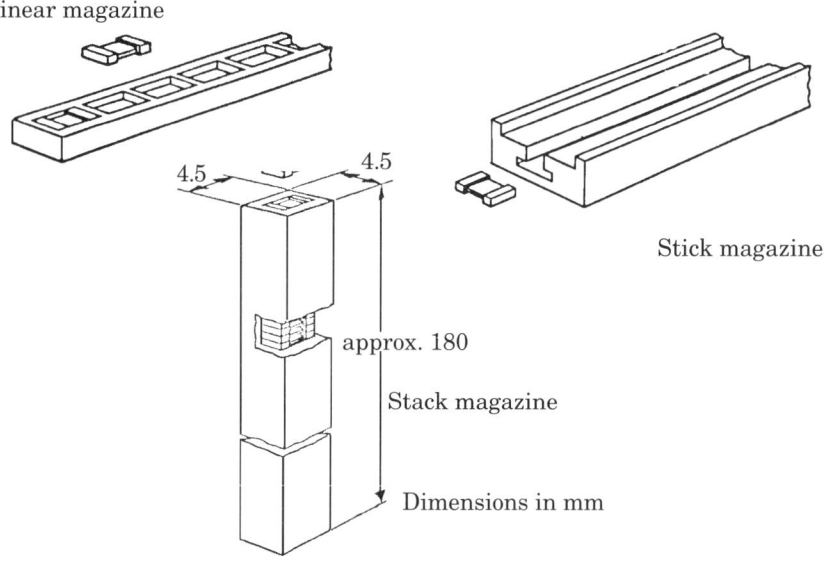

Fig. 4.38 SMC magazines or cartridges.

Magazine packages are constructed either of metal or plastic. The plastic carriers are disposable but less precise. Metal types are more sturdy and offer greater precision but are more expensive and, usually, returnable to the component supplier. Such magazines are frequently used in-house, being refilled from a bulk package.

4.5.3 Rail and Tube

The traditional method of packaging DIPs is applied also to SMCs. A simple stick magazine for flat components is shown in Figure 4.38, and the more expensive metal rail in Figure 4.39. Tubes are often plastic, which is cheap and disposable. When used for packaging static sensitive devices, the plastic will be conductive or a conductive coating will be applied.

Such packages can be employed whenever there is no likelihood of damage to leads by contact between adjacent components, and are often used for:

leadless devices — MELF, Mini-MELF, flat, LCCC, etc.;
J-leaded components — SOJ, PLCC, etc.;
devices with leads on one or two sides only — SOT, SOP, VSO, SOJ, etc.

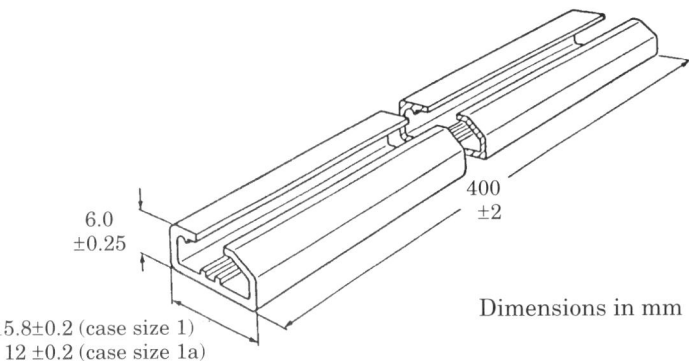

Fig. 4.39 A rail magazine for leaded components (leads on two sides only).

Because these packages hold only a limited number of devices, they are used either for large SMCs or as refillable magazines in-house.

4.5.4 Tray or Palette

Components which cannot be packaged in rails or tubes may be packaged in cardboard or plastic trays or palettes in which pockets have been vacuum formed, or drawn or pressed out. If needed, the trays may be made from conductive material or suitably coated. Each pocket holds one component and a tray may have many pockets, say up to 100, in a matrix of rows and columns. For example, 50 off 44 pin QFPs can be packed in a tray 200 x 104 mm (7.9 in. x 4.1 in.) and 20 off 64 pin QFPs in a tray 152 x 99 mm (6.0 in. x 3.9 in.).

The components may be sealed in their pockets by a thin plastic film to hold them in place in transit and protect them from spillage or environmental contamination. Such film will be removed immediately before assembly.

The trays are usually stacked one on top of another, several being delivered in a single box. Precision can be poor and the mounting equipment, if taking components directly from such a tray, must be capable of centring them accurately. On the other hand, metal trays can be very accurately aligned to provide precise feeding. Their cost normally restricts their use to within the assembly shop.

4.5.5 Adhesive Tape

This is a method employed for ICs by some manufacturers and extensively, as an alternative to trays, for other components which cannot be readily handled by other techniques, e.g., connectors. The components are stuck at defined intervals by a dot of glue or by adhesive tape to a punched paper or plastic tape, Figure 4.40. The tape is wound on to a reel. In use, smooth and accurate feeding is ensured by the transport holes on the tape.

Balancing the adhesion of component to tape is difficult. The adhesion must be strong enough to hold the component in place during handling and transport; it is equally essential that the bond strength is weak enough for the device to be detached easily from the tape when the mounting head picks

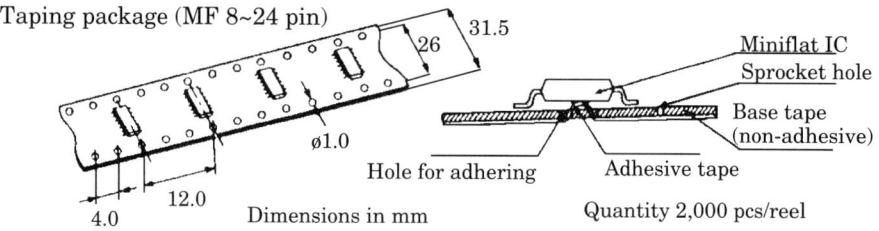

Fig. 4.40 A typical adhesive tape package. The adhesive tape contacts the body of the device package through a hole in the base tape. (Courtesy of Rohm)

it up. Adhesive residues from the tape on the underside of components can also cause problems.

4.5.6 Tape-on-reel

Next to bulk handling, tape-on-reel packaging is the most popular technique for small components. The carrier is a long tape, of cardboard, plastic, aluminium-coated plastic or pure aluminium, which is 8, 12, 16 or 24 mm wide, dependent on the component to be held. On one side there is a line of holes at 4 mm pitch so that the tape may be driven by an 8 mm movie film sprocket. Each 4, 8 or 12 mm, i.e., every 1, 2 or 3 sprocket holes, there is a 'pocket' with a component, held in place by a thin transparent film laminated to the carrier. The tape is wound on to a reel identical to that used for cine film.

Two sizes of reel are in general use, one of diameter 7 in. (178 mm) and the other of diameter 13 in. (330 mm). As both sizes have the same axle mounting they are essentially interchangeable. Each tape has an empty section at the beginning to act as leader in loading the automatic assembly machine and can carry many components, from a few hundred to several thousand, according to size. The perforation makes the tape asymmetrical and the reeling direction is standard, unreeling clockwise if viewed from the non-perforated side. This enables the manufacturer to assure that the polarity of devices is always in the same direction.

Two basic types of tape are used, cardboard and embossed. The cardboard tape is shown in Figure 4.41. The cardboard carrier tape is relatively thick and has slots at regular intervals. The bottom fixing tape is laminated to the carrier to complete the pocket for the component. The top fixing tape closes the pockets to retain the devices. It will be seen that the device cannot be thicker than the thickness of the cardboard which, in turn, is limited to a maximum of about 0.8 mm by the need to reel the tape on the core diameter of the reel (50 mm).

On the automatic placement machine, a sprocket mechanism feeds the component to the pick-up head in a fixed position by moving the tape forward one step, i.e., 1, 2 or 3 sprocket holes, at a time. The top film is peeled off just before the component reaches the pick-up head. In some cases a needle-like probe pierces the bottom film to push the component against the head.

The embossed tape is manufactured from aluminium or polystyrene, both of which are easily cold or hot drawn to form the pockets for the devices. While

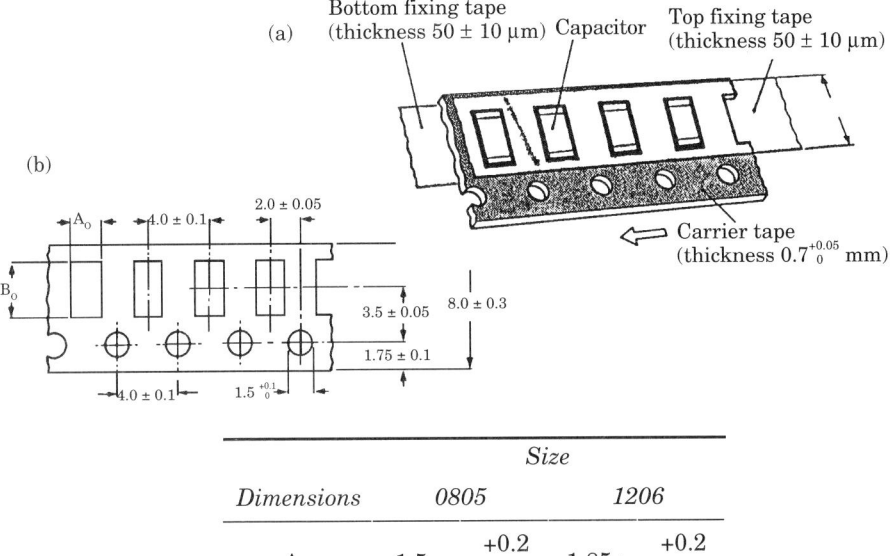

Fig. 4.41 Cardboard tape for tape-on-reel packaging. (a) Construction of tape for chip capacitors; (b) dimensions of carrier tape (mm). Cumulative pitch error 0.2 mm over 10 pitches.

the tape is very thin, it imposes no restriction on device thickness as the pockets can be formed to be as deep as is needed. Each cavity holds one device, as before, and a hole in the bottom of each formed pocket allows for a needle to push the device against the head. This tape is sometimes called *blister tape*, Figure 4.42.

The tape-on-reel method of packaging is simple, reliable and accepted by the great majority of automatic 'pick-and-place' machines. It is relatively expensive, especially if aluminium or metal coated tape is needed to eliminate static charges. Most problems with its use arise from the large numbers of parameters which must be specified and controlled, Figure 4.43, as tolerances can differ from one supplier to another and create difficulties in pick-and-place and assembly. Testing before assembly is not possible without removing the cover film (which cannot be replaced). It is, just, practicable to remove samples by cutting the tape but joining the cut ends takes time and care. A poorly made joint can seriously delay production.

Despite these drawbacks, tape-on-reel is a deservedly popular method of packaging devices for automatic assembly.

4.6 ADVANTAGES OF SMT

While design and assembly with conventional or traditional leaded components can draw on experience accumulated over some forty years,

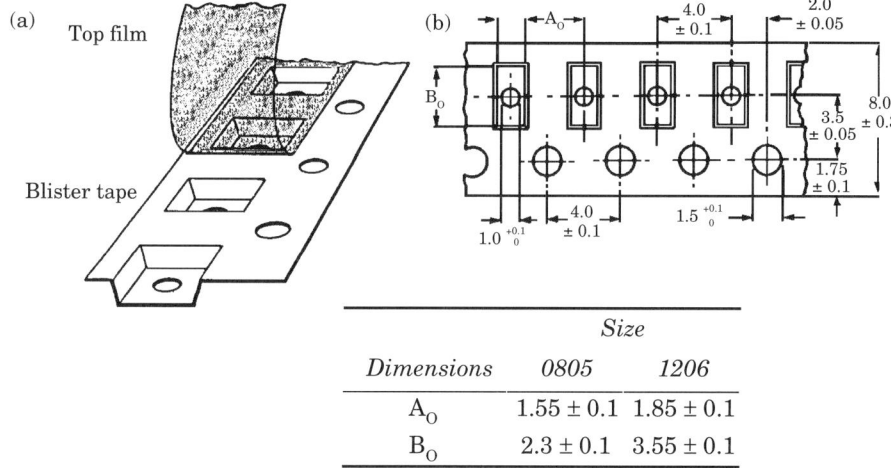

Fig. 4.42 Embossed or blister tape for tape-on-reel packaging. (a) Construction of tape; (b) dimensions of carrier tape (mm). Cumulative pitch error 0.2 mm over 10 pitches.

SMT is a much younger and faster changing technology. It is possible to 'jump in with both feet', but this is invariably the most expensive way to gain experience. If it is obvious, from customer pressure, market studies or the knowledge that business is being lost to a competitor who can make things smaller, cheaper or more reliable because of investment in the newer technology, action must be taken.

However, even when essential to improve competitiveness, the cost of a decision made without careful consideration of all potentially relevant factors can be measured in lost capital investment (in equipment and shop layouts), lost man hours at all levels, lost productivity and, most important in this competitive age, lost customers.

SMT demands a total review of all aspects of the business, design, manufacture, sales, plant layout, inventory management, training, and so on. Those aspects of SMT generally understood — smaller size, better high frequency performance, shorter path lengths — are the tip of the iceberg. All other pertinent factors must be keenly evaluated before making the final decision as to whether or not SMT will be advantageous to your business.

In the following text an attempt is made to compare SMT with traditional layout and assembly. Thin and thick film hybrid technology are not considered here.

4.6.1 Design Freedom

Some typical constraints in the design and layout of a conventional PB are:

(i) placing components on only one side of the board;
(ii) fixed grid for holes and allowances for hole diameters;
(iii) the number of tracks which can be routed between two insertion holes;

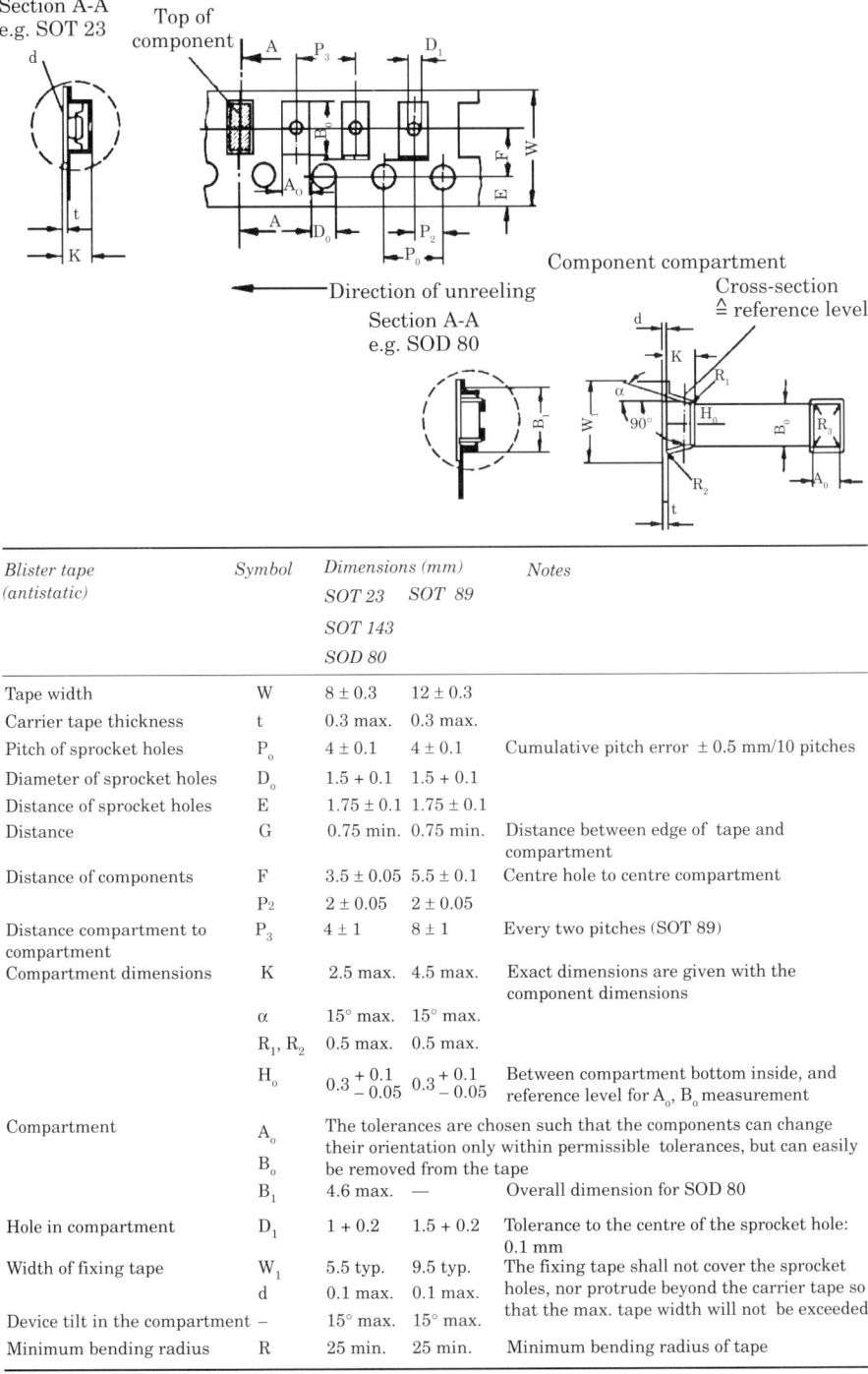

Blister tape (antistatic)	Symbol	Dimensions (mm) SOT 23 SOT 143 SOD 80	SOT 89	Notes
Tape width	W	8 ± 0.3	12 ± 0.3	
Carrier tape thickness	t	0.3 max.	0.3 max.	
Pitch of sprocket holes	P_0	4 ± 0.1	4 ± 0.1	Cumulative pitch error ± 0.5 mm/10 pitches
Diameter of sprocket holes	D_0	1.5 + 0.1	1.5 + 0.1	
Distance of sprocket holes	E	1.75 ± 0.1	1.75 ± 0.1	
Distance	G	0.75 min.	0.75 min.	Distance between edge of tape and compartment
Distance of components	F	3.5 ± 0.05	5.5 ± 0.1	Centre hole to centre compartment
	P_2	2 ± 0.05	2 ± 0.05	
Distance compartment to compartment	P_3	4 ± 1	8 ± 1	Every two pitches (SOT 89)
Compartment dimensions	K	2.5 max.	4.5 max.	Exact dimensions are given with the component dimensions
	α	15° max.	15° max.	
	R_1, R_2	0.5 max.	0.5 max.	
	H_0	$0.3 {+0.1 \atop -0.05}$	$0.3 {+0.1 \atop -0.05}$	Between compartment bottom inside, and reference level for A_0, B_0 measurement
Compartment	A_0			The tolerances are chosen such that the components can change their orientation only within permissible tolerances, but can easily be removed from the tape
	B_0			
	B_1	4.6 max.	—	Overall dimension for SOD 80
Hole in compartment	D_1	1 + 0.2	1.5 + 0.2	Tolerance to the centre of the sprocket hole: 0.1 mm
Width of fixing tape	W_1	5.5 typ.	9.5 typ.	The fixing tape shall not cover the sprocket holes, nor protrude beyond the carrier tape so that the max. tape width will not be exceeded
	d	0.1 max.	0.1 max.	
Device tilt in the compartment	–	15° max.	15° max.	
Minimum bending radius	R	25 min.	25 min.	Minimum bending radius of tape

Fig. 4.43 Tape packaging of discrete semiconductors. The table shows all the dimensions which must be specified and controlled to permit smooth production flow.

(iv) minimum allowable distance between components;
(v) possible need to break down the assembly into many boards to keep them within a reasonable size and, in consequence, have to provide space for connections between boards.

The use of SMT gives the designer more freedom. Some of the constraints listed are removed completely or are modified to permit easier resolution of the circuit. There may be a need for new CAD equipment and/or software, but the benefit can be enormous.

Figure 4.44, based on the paper 'SMT Squeezes Computer Power into Tighter Space' by D. W. Korf, *Electronic Packaging & Production*, January 1985, shows that, even then, the use of SMCs could reduce board area considerably. (It must be remembered also that a single multichip module occupying an area of some 2,000 mm^2 may contain a number of ASICs and VLSI devices which, individually packaged in quad flatpacks, will occupy a much greater area.) The area shown in the plots includes the space for holes or soldering pads, as well as the clearance needed for automatic insertion/placement and soldering. The surface mounting components are not only smaller than the conventional equivalent but also eliminate the insertion

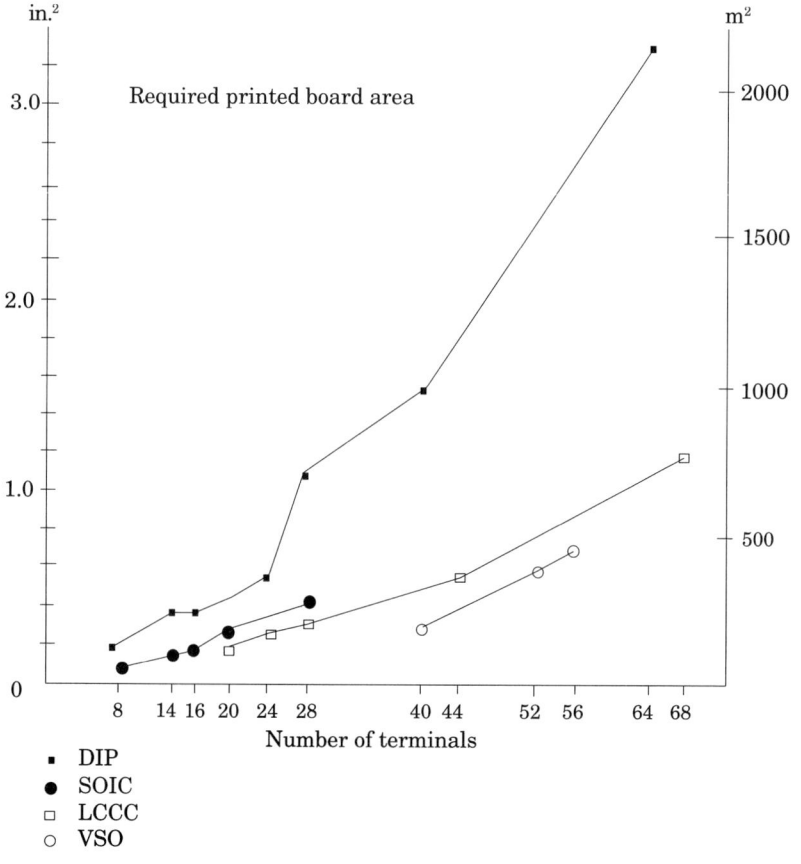

Fig. 4.44 Board area required by active components.

holes. It becomes possible, in many instances, to realise a circuit on a single-sided board instead of a double-sided PTH board. Frequently it is practicable to run tracks under the SM components, a trick which gives the equivalent of one and a half layers, Figure 4.45.

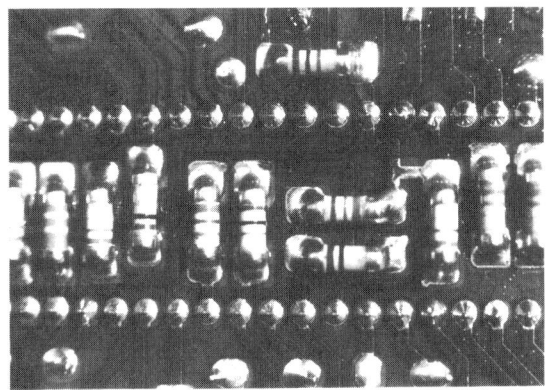

Fig. 4.45 A 'mixed technology' board. Note tracks under the SMCs.

Figure 4.45 shows another advantage of SMCs to the designer. This is a popular expedient to save space in consumer electronics. It relies on the resistance of the SM component to molten solder to realise the circuit on a single-sided board with the SMCs on the copper (soldering) side and pin-in-hole devices on the other. This gives great design flexibility with minimal effect on production methods.

It has been found practicable to replace multilayer boards with double-sided boards and, quite often, a number of boards can be replaced by one. In addition to the impact on cost, the restrictions imposed on the designer by elements such as connectors and edge contacts are minimised.

4.6.2 Size and Weight

Throughout this chapter the reduction in size and weight obtained by the use of SMCs has been emphasised. The illustration in Figure 4.46 is an indication of what can be achieved. To date, reductions of 30% to 90% have

Fig. 4.46 The reduction in size of SMCs is shown in this comparison of SOT, SOIC and VSO-40 packages with DILs.

been reported by various sources and, although no statistics have been noted, a fair assessment would be an average of about 50%. However, as the trend for components is ever smaller, the MELF being replaced by the Mini-MELF and Micro-MELF and the 1206 and 0805 giving way to the 0603 and 0402, the trend for assemblies will follow suit and further space savings will be achieved.

These savings have an impact also on such things as resistance to shock and vibration, in transport and in use. The smaller boards reduce the space requirement in assembly shops and stores and simplify handling. This means either less space is needed for the same production volume or more production output may be obtained in the same space. One manufacturer, Philips, estimated that, compared with manual insertion of conventional components, the reduction in shop area is about 40% for automatic insertion of leaded components and over 85% for SMT. Such savings are already important in areas where the cost of land is high, e.g., Hong Kong and Japan, and will be increasingly important in Western countries as costs escalate.

4.6.3 Reliability

There is an increasing amount of data being published, e.g., by Philips and by Texas Instruments, on the reliability of SM components. The reduction in the number of connections — in the component, on the boards, and between the boards — certainly means higher reliability.[14] However, doubts have been expressed in the past on the integrity of the new components, the hermeticity of the new packages,[15] and the condition of the solder joints, which in some cases can scarcely be seen, far less inspected.

Assembly, soldering and cleaning of SMAs are critical processes and, unless due precautions are taken at all stages, reliability may be adversely affected, see also Section 5 and Reference 3. In operation, factors such as adequate cooling of assemblies or modules, which may be very densely populated, may play a large part in dictating the life of components and assemblies. (Thermal design of assemblies is discussed in Chapter 11.)

4.6.4 Electrical Characteristics

The shorter connections between devices made possible by surface mount packages give significant improvements in the high frequency response of circuits. In addition, the smaller size of IC packages has a marked effect on their high speed switching performance. Apart from the reduction in resistance, both self-inductance and pin-to-pin capacitance are greatly reduced. These last two factors have a considerable impact on the switching wave form and on crosstalk between pins, i.e., the parasitic signals induced by one switching terminal in the others, and, therefore, on noise immunity. The lower they are the better and, if they are very small, simultaneous switching of several signals is possible.

Low capacitance means a lower load on driving gates and low self-inductance means less noise on reference signals, which permits long lines without signal deterioration. For both factors, SM packages exhibit lower values and a smaller spread than do the traditional DIPs, Table 4.4, and the resultant improvement in switching speed is of the order of 20%.

Table 4.4

DIP versus SO Package Comparison

Package	Number of Pins	Self-inductance (nH)		Pin Capacitance (pF)	
		Min.	Max.	Min.	Max.
DIP	14	3.2	10.2	0.38	1.13
SO	14	2.6	3.8	0.22	0.54
DIP	20	3.4	13.7	0.53	1.49
SO	20	2.5	4.3	0.45	0.85
PLCC	20	4.2	5.0	0.61	0.88

Source: Texas Instruments

4.6.5 Effect of SMT on Automation

With a few exceptions, conventional leaded components were developed for manual assembly. Automation of assembly was effectively 'mechanisation of manual assembly' as machines had to be developed to cope with established packages rather than machine and package being developed in parallel. With SMT the situation is very different. Packages were designed with automatic assembly in mind in parallel with relevant equipment, manual mounting being the exception rather than the rule. Real automation had arrived on the assembly floor.

The new machines are fast, silent and require less space for operation. Working environments are cleaner, less noisy and less crowded. In-parallel placement of devices became practicable, see Chapter 7, and productivity soared. A board, which took 100 minutes to assemble manually, took 40 minutes with traditional mechanisation and only 10 minutes if 80% of the components were SMCs. Amortisation of equipment, manpower costs, and repair or touch-up allowances are reduced by a similar amount. Because the new equipments have much better electronic/process control, error rates are lower and interfacing with CAD/CAM is much easier.

The availability of more versatile assembly machines, smaller and lighter components — giving faster machine set-up times — and easier board handling, means that shop areas for production and stores can be reduced and emphasis may be placed on process control rather than batch control, with reductions in manpower and test times.

JIT (*Just in Time*) or continuous flow manufacturing philosophies can be implemented more readily, as can information management and CIM. Therefore, although a switch of production, in part or whole, to SMT can dictate setting up an experimental shop or section to test and adjust the new concepts, the consequential benefits can be increasingly valuable to the company.

4.6.6 Cost

On average, components account for 50% of the total cost of an electronics assembly, the remainder being labour, direct and indirect, plant depreciation, design cost recovery and other overheads. Up to 1985, almost all SMCs were more expensive than the equivalent pin-in-hole device, the price premium ranging from 10% to 100%. Figure 4.47 shows this significant period. Between 1986 and 1988, the cost of SMCs became much more competitive with traditional components. Therefore, while in the earlier days the key factors in choosing an SMC rather than a conventional device were performance, size and weight, there is no reason now for an SMC to cost more than an equivalent traditional one. If the active part is the same, the SM version should cost less because it uses less material.

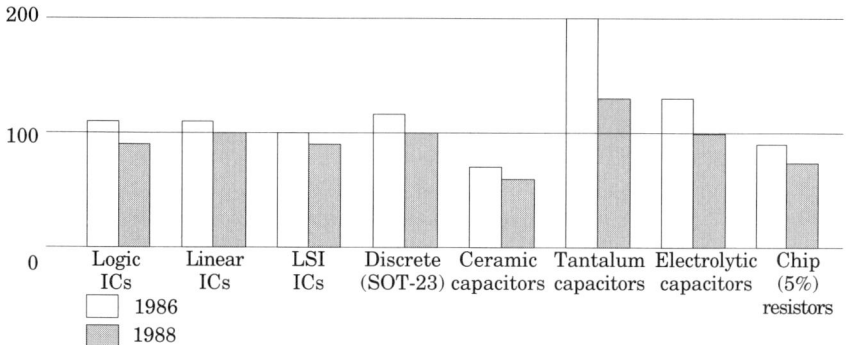

Fig. 4.47 Comparison of SMC cost with leaded components, taking the cost of the equivalent leaded component as 100%. (Source: Philips)

New devices must, of course, pay for development of device, package and setting up of manufacturing lines. All manufacturers, however, agree that in the longer term the SM component will not be able to demand premium prices unless there is some other unique feature which justifies the cost.

In the assembly shop, existing conventional equipment may be costing minimal amounts of depreciation and purchase of new equipment will have a considerable effect on gross margins as well as on cash flow (by additional capital spend). Costs of maintenance and machine down times may cost more or less, depending on the reliability of the original equipments, and it may be pertinent to delay until the older machines need replacing. (This assumes that competition is not making the change or addition necessary, whatever the apparent cost.)

4.7 DISADVANTAGES OF SMT

When asked "What is the main problem with SMT?", the manufacturing vice-president of a large electronics company replied "Headache!".

To the producers of traditional assemblies, surface mount technology is a new technology which has to be learned from basic principles. All personnel in sales, design, manufacturing, quality and service functions must be involved. The learning curve can take months or even years. The manufacturing process in particular is new. Very close collaboration between design and production engineers is essential and good team work is mandatory. SMT is not a simple extension of conventional technology — this cannot be over-emphasised — and every stage from the first circuit diagram to the final module assembly must be critically examined before proceeding.

It has already been said that investment costs will be high. Even so, many companies have found that the initial capital allowances were too low, in some cases costs escalated by hundreds of times not by factors of 2 or 3.

Incoming inspection/test of components can be extremely difficult and there is a strong and growing trend towards certification programmes being developed in conjunction with suppliers. The implementation of quality systems such as ISO 9000, see Volume 2, Section 4, is even more important, not just to components but also to all other material employed, from software to materials for bonding, soldering and cleaning.

There is much for the designer to study when first using surface mounting in a design.[16] Among the criteria which he must examine there is one which may be new to any designer who has not been involved with high density packaging or with exotic or, possibly, military circuitry. This is the problem of thermal dissipation. The combination of smaller packages and tight packing on substrates means that heat transfer and cooling become very important. The critical aspects of thermal design are dealt with in Chapter 11. Another result of the high component density is that in-circuit testing becomes more difficult. It has been said that in-circuit testing is the Achilles heel of SMT. The small pitch and design of SMA joints can make traditional bed-of-nails testing very difficult. Adding test pads can assist but, if overdone, will negate the space saving advantage of SMT.

Components for surface mounting are now available in many different types and shapes. Trying to establish who makes what can be difficult although there are now available several directories of SMCs, listing thousands of components and their suppliers. Remember also that, in many cases, the lack of real standardisation of outlines and parameters means that much checking and, possibly, testing may be needed before a component can be passed for use.

4.8 THE FUTURE OF SURFACE MOUNTING

Surface mounting technology is here to stay, at least until replaced by even more advantageous design/assembly methods. (FPT *(fine pitch technology)* may be viewed as an extension of SMT.)

SMT has replaced conventional design and assembly in many areas and has made designs possible which could not have been realised by using traditional methods and components. As IC technology advances, interconnection technology will respond and, for now, this means that the use of surface mounting components and methodology must grow.

The technology provides an eminently workable solution to the problem of mounting devices having high I/O lead counts by automatic assembly methods. So successful has this been that it is believed that through-hole mounted devices are a dying breed.[17] It is shown in the paper referenced that electronics production worldwide rose from a value of $220 billion in 1980 to $654 billion in 1990 (using 1991 exchange rates). On the same basis, it is predicted that the value will be $1403 billion in the year 2000. In 1993, surface mounting was already the predominant method of attaching component terminations to the printed board. Since it is predicted that mounting terminations on pads will grow at a rate of 14% per annum to the year 2000 while mounting terminations in PTH will grow at a rate of only 3% per annum in the same period, it is obvious that SMT, and hence SMCs, will fill the major share of the market.

In the same paper it is predicted that the use of technologies such as COB and TAB will grow, a view which is shared by all.

REFERENCES

1. Draft Revision of IEC 194: 'Terms and definitions for printed circuits and a proposal for a Decimal Classification Code (DCC) for the categorisation of the terms and definitions in IEC 194', Committee Drafts 52/607/CDV-I and 52/607/CDV-II, International Electrotechnical Commission, Geneva, Switzerland, October (1995).
2. Guidance Document CECC 00 802, 'CECC Standard Method for the Specification of Surface Mounting Components (SMDs) of Assessed Quality'.
3. Lea, C., 'A Scientific Guide to Surface Mount Technology', Electrochemical Publications Ltd, Asahi House, Church Road, Port Erin, Isle of Man, British Isles (1988).
4. Murata Product Information, Murata Manufacturing Company Limited, June (1994).
5. Taiyo Yuden Products 5, 'Chip Components', Taiyo Yuden Company Limited (1994).
6. Landis, D. and Notman, J., 'PWB Surface Finishes for COB', *Circuit World*, **Vol. 20**, No. 2, pp. 5-7, January (1994).
7. Ueltzen, K., 'Pushing the Packaging Envelope', *Circuits Assembly*, pp. 30-35, March (1992).
8. Becker, G., Consultant, Sweden, Private Communication, April (1994).
9. Lau, J., Miremadi, J., Gleason, J., Haven, R., Ottoboni, S. and Mimura, S., 'No Clean Mass Reflow of Large Plastic Ball Grid Array Packages', *Circuit World*, **Vol. 20**, No. 3, pp. 15-22, March (1994).
10. *Electronics Weekly*, April 27 (1994).
11. Reynolds, R., 'The reawakening interest in TAB', *Printed Circuit Fabrication*, October (1989).
12. Pearne, N., 'Multichip Modules — an Overview', *Electronic Production*, December (1991).
13. Clark, E., 'Transitioning to MCM Production', *Printed Circuit Fabrication*, **Vol. 16**, No. 2, pp. 68-73, February (1993).
14. See, for example, MIL-HDBK-217, Handbook: Reliability prediction for electronic equipment, US Department of Defense.
15. For a comprehensive discussion of this subject see 'Handbook of Microelectronics Packaging and Interconnection Technologies', edited by F. N. Sinnadurai, Electrochemical Publications Ltd, Asahi House, Port Erin, Isle of Man (1985).
16. Boswell, D., 'Surface Mount & Mixed Technology PCB Design Guidelines', Technical Reference Publications Ltd, Asahi House, Port Erin, Isle of Man (1990).
17. Tyler, R. L., 'PCB Investment Strategy into the 1990s', *Circuit World*, **Vol. 20**, No. 3, pp. 46-52, March (1994).

Section 3
Assembly Techniques

Chapter 5

MANUAL ASSEMBLY

GIOVANNI LEONIDA
Milan, Italy

W. MACLEOD ROSS
Welnorth Ltd, Bishop's Stortford, UK

5.1 GENERAL

Up to the mid-seventies manual assembly, in particular, manual insertion or placement, was the dominating technique and still accounts for some 10% of the total number of mounted components. The major uses of manual methods today are in:

(i) prototypes and small quantity production, including preproduction batches;
(ii) modification of assembled boards due to technical changes;
(iii) completion of automatically inserted boards with components which cannot be easily inserted by machine or which cannot be mass soldered;
(iv) replacement of damaged components;
(v) production in countries/areas where labour costs are still extremely low.

Item (iii) accounts for the largest volume as, while in theory any device may be inserted automatically, in practice there are several components which cannot be handled by normal insertion machines and require special tooling or equipment. If the production volumes are not very large and the number of such components per board are relatively few, manual assembly of these is economically justifiable.

An example of a board which needs manual completion is shown in Figure 5.1. This is the main board of a TV set. This board includes logic, power and high voltage circuits. Early radio and TV sets had a metal chassis or framework to which the tube and all other components were attached. The term 'chassis' was extended to the complete assembly and was still used after PCBs were introduced. The chassis of a TV set is now its main board (often the only one except for a small board on the rear of the tube). On such boards there will be several types of component, such as power transistors and resistors, transformers, large capacitors, remote control receiver, LED display and subassemblies, e.g., the tuner, which have odd shapes and are used in quantities of one or two per board.

Fig. 5.1 A television main board (chassis) showing the variety of odd shaped components which may have to be manually assembled.

It may be noted that, while manual placement is frequently employed for small batch quantities, mass soldering methods can sometimes be applied, both in conventional and surface mount technologies.

5.1.1 Manpower and Manual Assembly

As the name implies, manual assembly is labour intensive, not only in the primary assembly operations but also in supporting functions such as inspection. The first time error rate is normally greater than in automatic assembly, the incidence of missed components and wrongly placed or orientated components being much higher.

Operator training must be to higher levels, particularly where manual soldering is employed. Often, in the more affluent societies, it becomes increasingly difficult to find operators who are prepared to undertake fairly tedious tasks and have the intelligence and skill to be able to do these consistently well. When skill levels are low, random errors are more prevalent, even when using visual aids to assist placement and mechanical aids for lead forming and cropping. In the worst case every assembled board will have to be treated as a 'one off' and inspected accordingly.

Traditionally, female operators are inherently more suited to handling small items and seem to cope with repetitive work better than their male counterparts. In these sexist times, however, it is not always practicable to dictate the type of labour wanted! As far as is possible, the job must be made attractive, varied and interesting. Commonly used methods of creating operator interest and thereby more attention to the task are:

(i) to alternate assembly activity with inspection and/or rework;
(ii) involve the operator in workbench and component kit preparation;
(iii) have the operator complete the whole manufacturing cycle — assembly, soldering/joining, inspection/testing, rework, and so on — or as much of the cycle as possible;
(iv) in larger companies, establish small groups or cells of workers who have full responsibility for all manufacturing activities, direct and indirect, including job assignment, material administration, progressing work and scrap control.

The last suggestion, stemming from Volvo experience, may involve special agreements with unions and major changes in work practices. It can be highly successful if linked to productivity bonuses based not only on quantity and quality of group product, but also on factors such as material efficiency, response to company needs, and customer response.

5.2 LOW VOLUME ASSEMBLY

In its simplest form, manual assembly of conventional components requires only a few simple tools such as soldering irons, pliers, wire strippers and cutters, and cored solder wire. Manual assembly of surface mounting components will necessitate the addition of a variety of tweezers, adhesives and applicators, shaped bits for soldering irons and, at a minimum, a good binocular microscope. For dealing with 0603 and smaller chips, a vacuum operated pick-up head is essential.

To mount an axial leaded component, the leads are bent at 90° to the body at a pitch approximately that of the holes into which the leads are to be inserted. In many cases, the leads may be bent by hand, care being taken to provide an adequate radius. A variety of simple hand tools are available, or simple jigs can be made to ensure uniform bends and, when needed, provide stand-off. (Some typical lead forms are shown later in this chapter.) The component is pressed down gently to be flush with the surface of the board. To retain the component when the board is turned over, the leads may be bent over at about 45°, by hand or with a suitable tool. The board is then turned upside down, the leads soldered, and the excess lead cut off with flush or semi-flush end cutters to leave about 1 mm protruding above the solder meniscus.

While simple, much time is lost in turning the board, picking up and putting down the iron and the cutters. A simple board holder, as shown in Figure 5.2, speeds up the operation by allowing the operator to 'stuff' the board, i.e., to insert all or nearly all the components and bend the leads on the solder side, and then rotate the board to make all the solder joints and trim the leads. A good operator can achieve a rate for total assembly of about 20 seconds per component.

Fig. 5.2 Assembling components with a board holder to make bending of leads on the soldering side possible. The tool is part of a rotating support on which the assembler builds up a complete power supply, which includes the two heavy heat sinks on the back.

The method can be further improved, very easily if all components are about the same height above the board, by inserting all possible components at the same time, without bending the leads on the solder side. Pressure is applied to the components, in the simplest form by a synthetic rubber foam on a steel backing plate, so that, when the board is turned over, the components are retained in place, Figure 5.3. The assembly may then be completed by soldering the leads to the lands and trimming the leads. Alternatively, using special cutters, the leads may be end-cut at a standard height above the lands, or cut and bent or clinched, and the joints made by hand, wave or well soldering, see Chapters 14 and 15.

If wave soldering is employed it is advisable to remove excess lead length. Leads projecting by more than about 7.5 mm (0.3 in.) need a very high solder wave which is not easy to control. There is a danger that the leads will impact on the fluxer or on the solder wave nozzle(s) and the joints are poor because the smooth flow of solder is disturbed. Solder wells remove these difficulties but, while a wave solder machine can be used with advantage even on prototype boards, solder wells are more appropriate to mass production of conventional boards. In any case soldering with long leads gives results which are poor compared with those obtained by soldering properly trimmed leads. Trimming after soldering can shock the leads and disturb or crack the joints so that thorough inspection and rework is essential. There is no point in using well or wave soldering to speed up production when the end result is a need for up to 60% rework!

A further disadvantage of over-long leads is that the solder clinging to them makes manual removal of the excess even more time consuming and

Manual Assembly

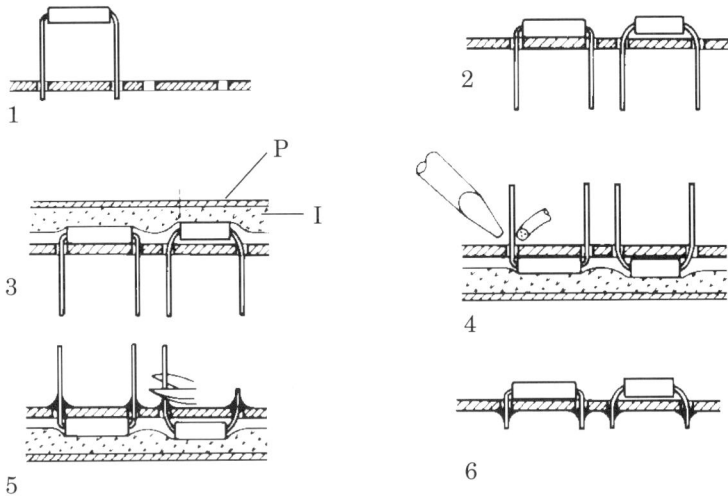

Fig. 5.3 Sequence for manual assembly of axial components. Component leads are bent manually and inserted in the printed board: when the board is completed, the components are held in place by means of a cushion, I, bonded to a steel plate, P. The operator then turns the board upside down and solders and trims the leads.

tiring for the operators. The use of air-operated cutters helps but, in larger scale production, an automatic cutter is needed. Some companies have developed their own equipment for this purpose but it is difficult to ensure accuracy of the height remaining, usually 1.0 to 1.5 mm (0.039 to 0.060 in.). One of the most successful proprietary machines is illustrated in Figure 5.4. Lead trimming systems are available from a number of sources and depend upon the use of tungsten carbide cutter rings with variable cutter speeds and very precise height adjustment. The assembled board is held on the machine by mechanical or vacuum fixturing and traversed over the blades which may be rotating at 7,000 rpm or more. If the blades are sharp and in good condition, the leads can be cut with little or no distortion or disturbance of the joint.

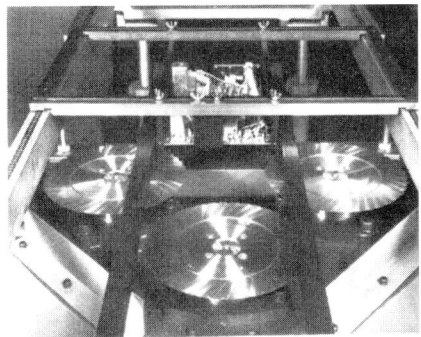

Fig. 5.4 An automatic lead trimmer. (The safety cover has been removed.) (Courtesy of Hollis)

Whatever method is employed for lead trimming after the joints are made, it is a potentially destructive operation which must be done with care. Hand tools must be checked regularly for clean cutting and operators trained to cut accurately without applying tensile or flexural stress to the joints. Automatic trimming requires fully tested and well maintained machines with regular checking of all factors which can influence their operation, for example, board carrier rigidity and planarity, lead lengths and diameters, relative speed of tooth and lead, and cutter sharpness.

Unless the trimming is perfectly done, joints can be cracked. Very small cracks can propagate in service and lead to serious degradation of the assembly's reliability. In some instances, the assembler assumes the worst and re-solders the board after trimming, as shown in Figure 5.5. The first solder wave is very high to cope with untrimmed leads; the second soldering machine is normal in wave height and repairs any damage caused by trimming. The first machine may be replaced by a wave unit filled with a petroleum wax to hold components in place while trimming. The wax is replaced by solder from the second soldering machine.

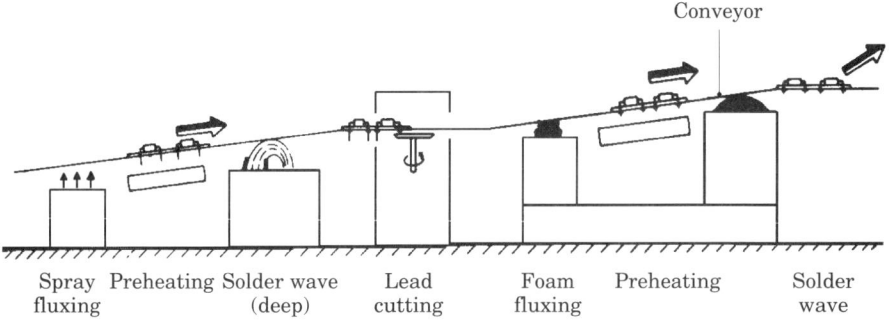

Fig. 5.5 A solder-trimming-soldering line. If the boards are still hot when they come to the second fluxing station, it may be better to use a second spray fluxing unit instead of the foam unit.

Surface mount assembly removes many of the problems associated with leads but is in many ways more demanding on operator skills when carried out manually. There is the problem of handling components which are very small. The PBs have much finer tolerances for component placement. Retaining the components in place, particularly cylindrical components, is difficult unless adhesive is being used. (Solder paste is helpful but will not hold a component against, say, careless handling of the board until the paste has been reflowed.) Making joints by hand soldering demands great skill and more attention to heat control,[1] requiring a variety of soldering bits or tools to cope with different SMCs. Solder masks can cause problems on conventional boards,[2] and can cause even more problems with SMCs, see also Section 5.

The primary steps for a surface mounted assembly depend on the type of assembly, i.e., single-sided, double-sided or mixed technology. They depend equally on the method of joining to be used. Manual soldering is seldom employed, the major application being in repair or rework; wave soldering and reflow of solder pastes by infra-red, hot plate or the vapour of boiling fluorocarbons are the preferred methods.

Before commencing assembly, the sequence of operation must be decided. Whereas in conventional assembly, after board cleaning and checking the solderability, etc., of the components, the first operation is to load the components, in surface mount assembly another operation must be performed first. (There is one possible exception. In single-sided mixed technology boards, the leaded components may be inserted first.) The first operation will be either to screen print the mounting lands with solder paste or to use syringe-type dispensers or screen printing to apply adhesive 'dots' between the mounting pads to hold the components until soldered.

Where the number of components to be attached and the total number of boards are small, it may be practicable to hand-position and manually solder the SMCs. In most cases it is of considerable help to glue the components to the board before attempting to solder. Not only does the cured adhesive hold the component firmly during the soldering operation, it allows inspection for correct placement, polarity, and so on, before the adhesive is cured. SMA adhesives are described in more detail in Chapter 7 and it is sufficient here to point out the need for proper selection and positioning of the adhesive. Factors such as solder mask type and thickness must be taken into account and it may be necessary to carry out trials with a number of different applicator nozzles to be certain that the glue droplet is correctly sized. It is essential to ensure that there is enough glue to bond the SMC in position and not enough to spread over the lands to which the device is to be soldered.

The ideal situation is one in which the adhesive is centrally positioned between the mounting lands, makes good contact with the component body when placed, and does not 'spew' over land and/or mounting face of the device when cured. The adhesive may be applied to the board base material, to solder mask over the base material or over bare copper, but never to solder mask over a tin-lead coated conductor. In the latter case there is a strong possibility that the component will move during the soldering stage.

If the components are to be 'reflow' soldered, it will be necessary to deposit a solder paste on the lands. This is best done immediately before the next stage of placing the SMCs. Solder paste is too easily contaminated by dusts, fibres, etc., to be left 'open' for long. Contamination in solder joints does not aid reliability.

Placing the SMCs can be done by hand using one of the wide range of tweezers available, although, as pointed out before, the smaller devices such as Micro-MELFS and 0402 or 0603 chips can be very difficult, if not impossible, for the average operator to handle. In addition, mishandling with tweezers can easily damage ceramic chips by cracking or nicking and the damage may not show until the assembly is tested. Several sources now supply vacuum pick-ups with a variety of heads and, at reasonable cost, X-Y tables — some with CNC facility — which assist materially in placement of the majority of devices without handling damage. Binocular microscope or CCTV attachments will help to ensure accurate alignment and correct polarity.

When the assembly is complete it must be cleaned (and tested). This is also true of mass assembly but in prototype work it is frequently found that machine cleaning is not available. With SMA, cleaning residues from beneath devices which are very close to the board surface is difficult. In addition, the growing concern about the effect of CFCs and chlorinated solvents on the atmosphere limits the solvents which may be used.[3] It has been shown that brush cleaning, that is the 'dip and brush' method used mostly for manual cleaning, is the least effective method of cleaning.[4] It is expensive in labour and materials and lacks the essential elements for thorough cleaning, namely virgin solvent and thorough rinsing. There are many other methods which can be adapted for use in a prototype or short run facility but each assembly shop will have to find the solution which fits best with its production environment. The subject of cleaning is dealt with in detail in Volume 2, Chapter 15.

It should be noted that, for minor rework and replacement of the odd component, a small SMA 'starter' kit is obtainable from a number of suppliers. If, however, the intention is to assemble prototype and preproduction batches, the initial investment will increase from some hundreds of dollars (US) to some thousands.

5.3 MASS PRODUCTION

In mass production using manual insertion, some tooling is needed to achieve reasonable rates of insertion and minimise operator fatigue and error. SMA is not dealt with here as pick-and-place equipment is essential to obtain adequate speed and accuracy, see Chapter 7.

5.3.1 Without Pre-cut and Formed Components

Trimming leads manually is time consuming, tiring, and when done after soldering can damage the joints. In the early 1960s a technique was developed which became very popular and still survives in some establishments today.

The board to be populated is positioned over a set of hardened steel templates which have been drilled to match the component holes in the printed board. The operator bends the component leads by hand and inserts them into the holes of the board and, in so doing, into the holes in the templates. When all relevant components have been inserted, they are kept in place by pressure from a steel plate faced with foam rubber. Operation of a pneumatic actuator forces the lower template to move with respect to the upper and cut all the projecting leads, Figure 5.6. If required, the upper template can then be moved to clinch the leads and hold the components in place.

The method is simple and effective. The plates are easy to produce, tolerances being such that they can be machined before hardening. All the leads are cut in one quick operation, reducing the time for mounting to 6-10 seconds/axial component. When the time for mass soldering and rework is added on, a total of 8 to 12 seconds to mount each axial leaded component is achieved, a reasonably cost-effective standard. An additional advantage is that no lead trimming needs to be done after soldering, removing any doubts about the reliability of the joints.

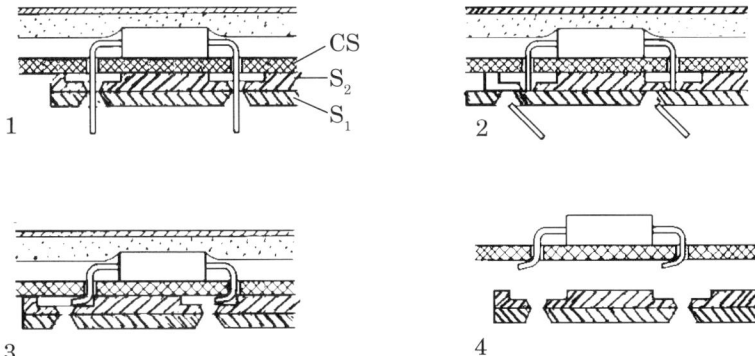

Fig. 5.6 Cutting and forming components after insertion. The two steel plates (S_1 and S_2) which are under the printed board CS are moved by a pneumatic piston to cut and bend the leads.

There are, however, some important disadvantages, among which are:

(i) the template hole pattern is specific to a given board and tooling cost can only be recovered by amortising over a large number of boards;
(ii) flexibility is low: a change in position of one hole means reworking the templates;
(iii) if the template is not well engineered, the components may be damaged;
(iv) the equipment works well only with flush mounted components, i.e., stand-offs must be inserted separately;
(v) manual bending of leads (for insertion) is very tiring and with this technique it is performed continually without a break for less tiring operations.

To overcome most of these disadvantages, the method is used in a different form with automatic positioning tables.

5.3.2 Automatic Positioning Tables

An automatic positioning table is a bench fitted with a mechanical or pneumatically operated system designed to minimise operator errors. The method is economic only if the bench is relatively inexpensive so that CNC tables are normally ruled out.

The bench is a desk size, complete and independent workplace. The operator sits in front of the board to be assembled which is fixed on a compound work-holder driven by an X-Y positioning device. The sequence of insertion and board movement is predetermined so that insertion takes place in the same position relative to the operator.

At the back of the table there are a number of small bins, each holding a given type of component. As each bin corresponds to one insertion position on the board, if a specific component has to be inserted in N positions, it must be present in N bins. The bins travel on a closed loop chain such that the operator can take only one component at a time from that bin which corresponds to the position into which the component is to be inserted. Under the workholder there are pneumatic cutters which trim the leads and clinch them to anchor the component in the board.

If the insertion sequence is A, B, C, etc., the operator takes component A from its bin, bends the leads at 90° to the body, and inserts it into the appropriate holes in the board. By pressing a control pedal, the cutters are actuated to cut and bend the leads. The table moves the board to the next position, the next bin (component B) appears, and the operator starts a new cycle.

By this method an axial component can be inserted in 5-6 seconds. Times as low as 4 seconds/component have been recorded. Unless batches are very small, such equipment will pay for itself in a year or less when time lost in assembly errors is taken into account.

5.3.3 Using Pre-cut and Formed Components

Manual bending of leads is easy if the leads are ductile. Accuracy of bending is difficult to achieve consistently and the components can be overstressed at the body to lead junction, both during bending and when the component is pressed down on to the PB.

To overcome these problems, components can be purchased with pre-cut and formed leads, at a slightly higher price. Most PCB assembly shops prefer to buy standard components and preform the leads in-house.

Assembly using preformed leads is simplified as no means of trimming or forming is needed. The workbench may be similar to the automatic positioning bench, minus the cutters, or can be reduced to a worktable with a jig for holding the board and as many bins as required to hold all the types of component in use on the assembly. The bins must be clearly identified and easy to access.

If different boards are to be assembled at the same work station, sets of bins can be mounted on a frame for each type of board so that they can readily be changed. The set of bins is normally prepared in advance in a separate area with the specified types and numbers of each type for a particular batch of boards.

Insertion time for preformed axial components is about 5-7 seconds each. (All insertion times quoted exclude the time taken to equip the workstation.) Special care is needed to handle the assembled boards prior to soldering.

5.4 LEAD PREFORMING

The main objectives of lead preforming are to:

(i) make insertion of the component in the holes in the board as easy as possible;
(ii) avoid dangerous loads on the component during bending and assembly;
(iii) make it possible to mount components flush, or with a specific board clearance, and so on;
(iv) reduce total assembly time;
(v) improve mass soldering performance;
(vi) make the assembly more orderly and uniform.

Lead forming prior to insertion is the most common approach to professional assembly when automatic insertion, see Chapter 6, is not in use.

5.4.1 Types of Preformed Shape — Axial Components

To decide on an appropriate shape for the lead, the following points must be considered:

(i) vertical or horizontal mounting;
(ii) flush mounting or mounting with clearance;
(iii) components retained by gravity (after insertion and prior to soldering) or locking to the board;
(iv) mounting hole pitch (insertion span).

Most of these factors are considered in the design of the PB and the board designer must be aware of the effect of his decisions on assembly costs and difficulties.

The choice between horizontal and vertical mounting is discussed in Chapter 8. The selection of flush or stand-off mounting can be made only after consideration of:

(i) the type of PB — conductors on one or both sides;
(ii) the type of component — heat dissipated, whether or not the body is conductive, other considerations;
(iii) soldering and cleaning processes;
(iv) service conditions — temperature, humidity, shock and/or vibration, etc.

The method to be used for retaining the components prior to soldering depends on the production process, i.e., how often the boards are handled and/or transported before soldering, how easily the components may be restrained by, e.g., pressure pads or vacuum formed covers, and how easy it must be to remove components.

Pitch is decided during design of the board but the board designer can sometimes give the production engineer a choice of pitch for a component before finalising the design.

Many different shapes are in use, some common ones for horizontal mounting being shown in Figure 5.7. The shapes shown are compared in Table 5.1 for difficulty of preforming, insertion, handling the assembled board prior to soldering, loading and soldering, and replacement of faulty components.

The most commonly used shapes for flush mounting on PBs are the U type and spread apart U, Figure 5.7, (a) and (c), the latter being employed where handling can cause problems. Where there is repeated handling, the 90° with lock-in, (d), is to be preferred. Of the stand-off types shown the double bend, (j), or the snap-in, (g), is usually preferred. The former works best when the lead and hole diameters are closely matched but can give problems when inserted as it can rest at two different heights, one of which is incorrect.

The general purpose snap-in, (g), is self-adjusting if inserted incorrectly. The spring-back of the leads ensures that the correct height is obtained. The

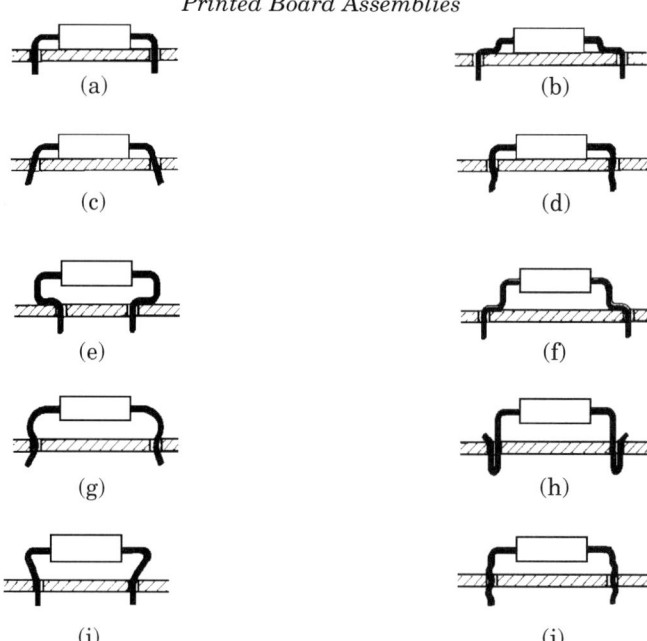

Fig. 5.7 Common preforming shapes for axial components: (a) U type (or at 90°); (b) at 90° with double bend; (c) spread apart U type; (d) at 90° with lock-in; (e) at 90° with internal stand-off; (f) at 90° with external stand-off; (g) snap-in; (h) snap-in, double lead; (i) Z type; and (j) double bend.

doubled lead snap-in, (h), can give problems in soldering because of the doubled lead but is invaluable when the hole diameter is 2.5 (or more) times the lead diameter.

Shapes (e) and (f), 90° with internal or external stand-off, give good control of height but can cause trouble if the board has conductors on both sides as shown in Figure 5.8.

At first glance the spread apart U, (c), and the snap-in, (g), are difficult to insert as the span between the tips of the leads is greater than the pitch of the insertion holes. However, if one lead is inserted first and then bent within elastic limits to allow the second lead to enter, no problem is encountered, Figure 5.9.

There are many more shapes to be found, as many assembly shops make their own preforming tools. In addition, suppliers of preforming equipment supply interchangeable tools which can be custom made at low cost.

A similar variety of shapes can be found for preforming for vertical mounting. Some styles are shown in Figure 5.10. Selection of the most appropriate shape should take into account the following factors:

(i) preforming is more difficult and less precise for vertical shapes than for horizontal ones;
(ii) vertically mounted components should never be flush as in Figure 5.10, (a) and (b). The shorter lead may be non-wettable close to the

Table 5.1
Comparison of the Most Common Preformed Shapes for Axial Components (Horizontally Mounted)

	Reference Fig. 5.7	Preforming Shape	Preforming	Insertion	Difficulty of Handling	Soldering	Repairing
Flush Mounting	a	U type (90° type)	3	4	9	7	3
	b	90° double bend	5	4	8	7	4
	c	spread apart U	4	6	5	5	5
	d	90° with lock-in	6	5	5	5	4
Stand-off Mounting	e	90° with internal s.o.	10	7	10	9	8
	f	90° with external s.o.	6	4	9	8	5
	g	snap-in	7	5	3	4	4
	h	snap-in, double lead	9	5	3	6	6
	i	Z type	7	6	9	7	6
	j	double bend	8	7	4	4	5

Key : 3 = easy; 10 = very difficult

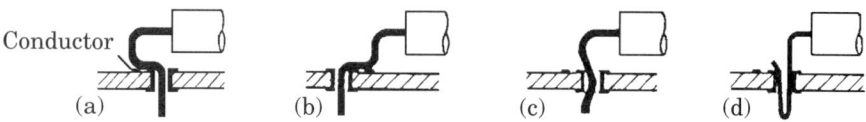

Fig. 5.8 On double-sided printed boards, preforming shapes such as (a) and (b) may cause a short between lead and conductors. This does not happen with other shapes such as (c) and (d).

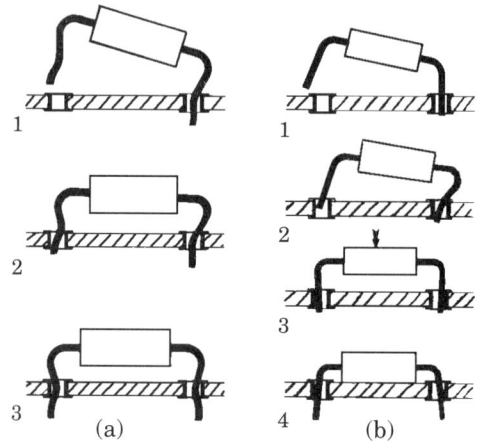

Fig. 5.9 Sequence of manual insertion of formed components: (a) snap-in shape; and (b) spread apart U shape. In both cases the two leads are inserted one after the other.

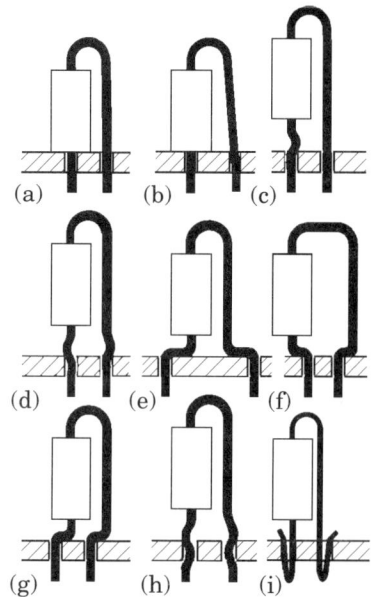

Fig. 5.10 Common preforming shapes for vertically mounted axial components. Shapes (a) and (b) should be avoided because soldering takes place too close to the component body and the body may plug the hole and cause blowholes. Shapes (d), (h) and (i) are preferred when boards must be handled a lot before soldering.

body, the body can be overheated or thermally shocked in soldering, and the forces on the lead during thermal cycling in service can strain or even rupture the inner joint;
(iii) handling boards with vertically mounted components is more difficult at all stages of assembly;
(iv) the permissible range of hole to lead diameters is narrower, which can add significantly to cost by making a larger range of hole diameters necessary.

Shapes (e) and (g), Figure 5.10, improve component stability by bringing the centre of gravity closer to the middle of the insertion span. Type (g) off-centres the body without any increase in insertion span, so reducing the board area (*footprint*) needed for the component, Figure 5.11.

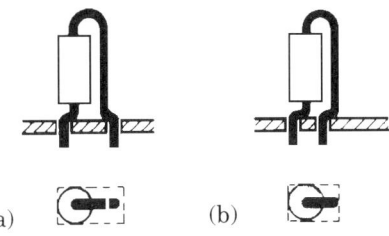

Fig. 5.11 Shape (b) takes less area of the board than (a) and provides good stability prior to soldering.

Lead clinching is the most common method of locking components into the board when inserting manually and is essential for automatic insertion, see Chapter 6. As will be seen, automatic insertion machines also carry out preforming as part of the insertion cycle. The most common shapes are the U and a U with an internal stand-off. Both must be clinched, internally or externally at 30° or more as shown in Figure 5.12, for the following reasons:

(i) the component can be extracted by the insertion tool on its return stroke;
(ii) on some machines, the inserted components can be misplaced by abrupt movement of the X-Y table holding the board during insertion;
(iii) vibrations generated during insertion of components can misplace those already mounted;
(iv) mounted components can be disturbed during subsequent handling, e.g., when manually inserting those components which cannot be inserted automatically.

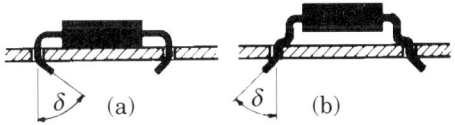

Fig. 5.12 Clinching of leads: (a) U shape with internal clinching; and (b) forming at 90° with stand-off and external clinching. On most automatic insertion machines the clinching angle is adjustable.

Internal clinching gives higher packing density, Figure 5.13, but locks the component to the board less efficiently than does external clinching, Figure 5.14. Opinion on the best clinching angle varies. A clinch angle of 90°, which brings the lead into contact with an elongated land, makes handling prior to soldering much easier. On the other hand, a 90° clinch is usually avoided as the need to elongate the pad takes up more space, soldering quality can be poor, and repair by changing a component is difficult and can damage the lands. For these and other reasons a normal round land is usually recommended with a clinch of about 45°. (If carried out on automatic insertion equipment, the clinch angle can normally be adjusted within fairly close limits.)

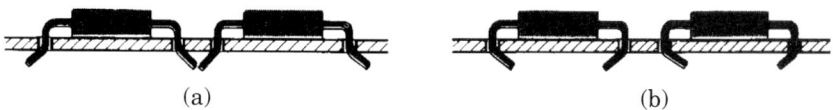

(a) (b)

Fig. 5.13 On high density boards, internal clinching (b) is preferred to external clinching (a) because it avoids the clinching interfering with previously inserted components and reduces the risk of producing shorts by bridging during soldering.

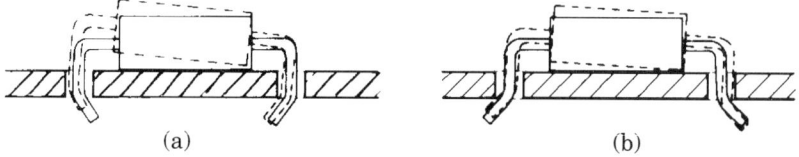

(a) (b)

Fig. 5.14 At the same clinching angle, a component with internal clinching (a) is locked to the hole less safely than a component with external clinching (b).

5.4.2 Manual Preforming — Axial Components

If the volume of components being assembled is large enough to justify the use of mass soldering, it will also justify the use of preforming equipment. Hand preforming is often used, however, for prototype and preproduction batches, for replacement of faulty components, and for very delicate components which might be damaged by preforming equipment.

Various types of cut and bend, cut and clinch, and lead forming pliers are available from most electronic/electrical tool suppliers. Many of these cut and form the leads in one operation. A very simple tool, illustrated in Figure 5.15, is a plastic section with slots for component bodies and leads. The leads are bent against the sides of the section and cut close to them with end cutters. While this tool is adequate for bending leads for insertion of flush mounted components, it will not, of course, cope with stand-offs or forming for retention in the board prior to soldering.

5.4.3 Preforming Equipment for Axial Components

Machines for preforming components are available with different modes of operation and degrees of automation varying from hand feeding and

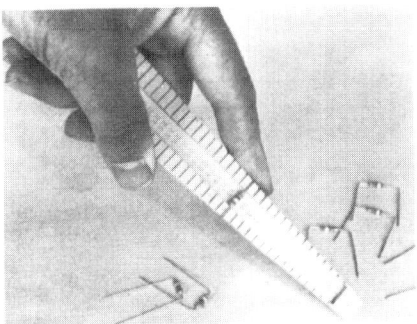

Fig. 5.15 Simple manual tool for forming axial component leads. (Courtesy of Rhopoint)

loading to magazine feed with fully automatic loading and operation. Most fall into one of the following classes:

(i) pneumatic reciprocating machines;
(ii) non-stop rotary machines (cone machines);
(iii) stopping rotary machines;
(iv) in-line (belt) machines.

The major methods of feeding components to the machines are manual, chute or gravity, and lead taping, see Chapter 6. Less frequently used for leaded components are magazine feed and body taping. The method of feeding has a major effect on the productivity of the equipment.

A typical *pneumatically operated machine* is shown in Figure 5.16. This type of equipment is either single stroke operated, allowing inspection of the work and, if necessary, loading components to ensure that markings will be visible when mounted, or automatically operated with the addition of a feeder.

Good tool design will preform leads without significant stress being applied to the component body or to the lead-to-component joint. To ensure this, the tool can be designed to clamp the lead, close to the body, against the fixed part of the die before cutting and bending take place. Even so, tensile stresses up to around 22 kg, 50 lbf, can result and, unless dies are properly tooled and maintained, compressive stresses may be generated.

Because leads must be clamped, they cannot be bent closer to the component body than 2-3 mm, about 0.1 in., so that the span between the bent leads is about 5 mm more than the body length. Some machines need only 1.5-2 mm for clamping. Below 1.3 mm, damage to the leads may result.

The forming rate of these machines depends upon the method of feeding used. With manual feeding, 600 to 900 components per hour can be achieved. Feeding by vibrating chute gives a rate of 2,000 to 3,000 per hour and with taped components as many as 10,000 can be preformed in the hour.

Non-stop rotary machines, Figure 5.17, are normally fed with taped components and have production rates of from 10,000 to 30,000 items per hour. The method is essentially restricted to simple shapes, e.g., U. The components are taken through one or more pairs of opposing conical rolls which revolve at high speed. The first set cuts the leads to length. Further sets then form the leads by forcing them through gaps of the required shape.

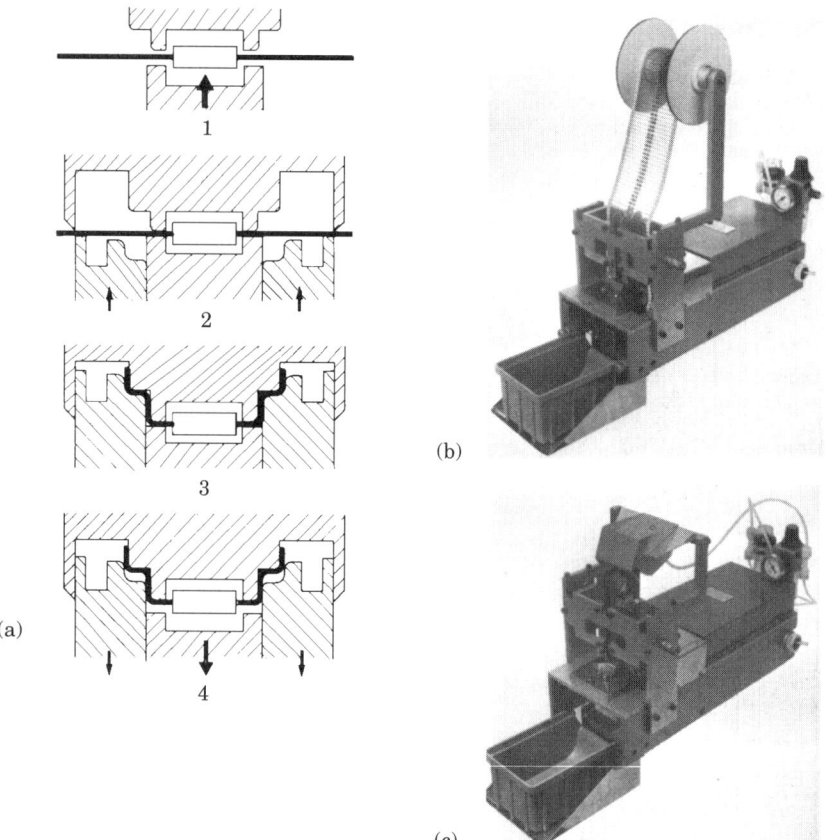

Fig. 5.16 Pneumatic reciprocating lead forming machine: (a) working cycle; (b) equipment tooled to process taped components; and (c) equipment tooled to process bulk components.

At the inlet to the forming rolls the gap is large. As the rolls rotate, the gap reduces to the minimum before opening again to receive the next component leads. In this way, smooth and gradual lead deformation is obtained, even at very high speeds.

While complexity and precision of form are limited, these machines have the advantages of low cost, high reliability (because of the mechanical simplicity), ease of adjustment and maintenance, and quick change of lead pitch.

Stopping rotary machines are usually fed by a vibrating chute which delivers one component at a time. The components are delivered to two rotating toothed discs which are parallel to each other and keyed on to the same shaft (which is parallel to the component axis). The chute leaves a component, or more correctly a lead, between each tooth on the rotating disc. As the shaft rotates, the components are delivered to one or more forming stations where a reciprocating mechanism cuts and forms the leads.

The discs must stop at each cutting or forming station and therefore rotate step by step. At each step, one component is discharged at the outlet and a

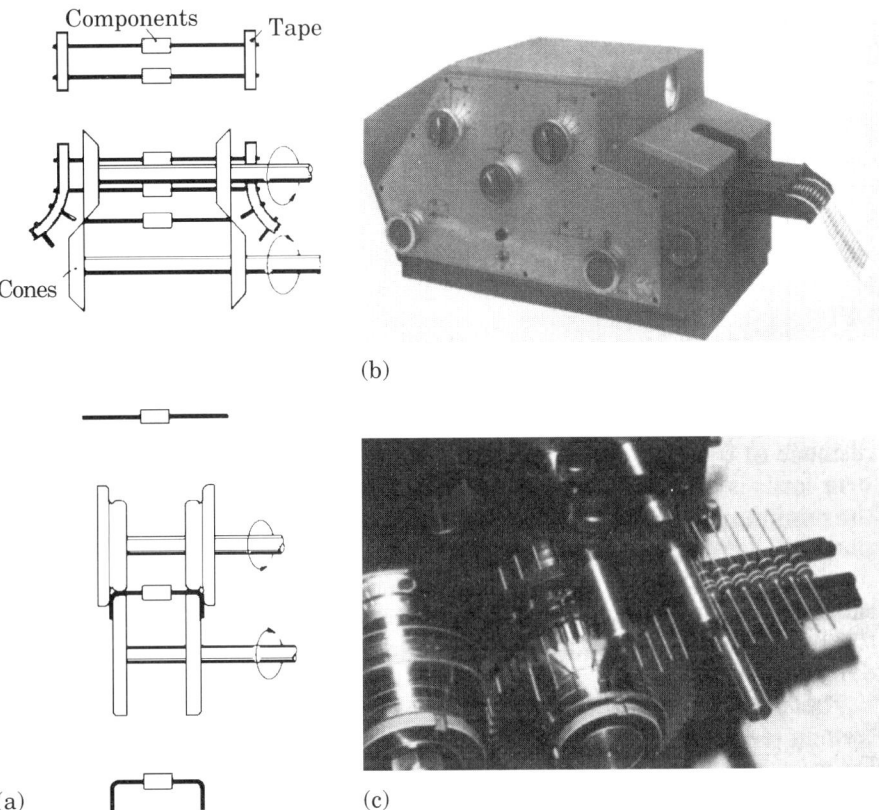

Fig. 5.17 Non-stop rotary lead forming machine: (a) working cycle (two pairs of cones cut the component leads, two further ones bend them); (b) and (c) overall and close-up view of a machine. (Courtesy of Loupot)

new one is picked up by the discs. Both chute width and distance between the discs can be altered to accept components of different body lengths. Cutting and bending stations can also be adjusted to give the desired insertion span and projection of lead on the solder side of the board. It is useful to be able to change the forming tools quickly to cope with different preforming shapes.

Production rate depends on complexity and precision of the form required but averages 10,000-12,000 pieces/hour if fed with taped components and 3,000-6,000 per hour if the chute is manually loaded.

In-line machines are slower than the rotary types but are more precise. Components are usually tape mounted and are held firmly during all operations so that they cannot be misplaced or rotate about their axis on the machine. Components are transported through the machine by two strong, shaped, rubber belts.

The cutting and bending stations are aligned on the sides of the belts which stop at each work station. Many more stations can be used with in-line machines than with rotary machines and the most complex shapes can be formed with great accuracy. To minimise generation of forming stresses on the component body, the machine can be arranged such that the same operation of cutting or bending can be carried out on the two leads at consecutive stations.

The production rate of such machines is of the order of 6,000 items/hour if fed with taped components. Feeding with single components is slow and difficult. The main disadvantages of the in-line machines, apart from the relatively slow speed, are cost and the lengthy setting up times needed.

Table 5.2 compares the most important characteristics of these machines. The table should be read in conjunction with the foregoing text. For example, while the last two types can be used for any shape by changing the forming tools, the time taken to make the change is much longer than that for changing the die of a pneumatic machine and the tools are a lot more expensive.

Table 5.2

Comparison of the Basic Types of Preforming Machine for Axial Components

Type of Machine	Capability	Number of Spans	Types of Shapes	Capital Investment	Precision
Pneumatic reciprocating	Low	Few*	Few*	Low	Good
Non-stop rotary	Very high	Very high	One or a few	Medium	Fair
Stopping rotary	High	Many	Few	Medium to high	Good
In-line	High	Many	Few	High	Very good

*A die is necessary for each shape and insertion span.

The pneumatic reciprocating machine needs a die for each shape and insertion span and may need special dies for very large or very small component bodies. The dies are, however, relatively cheap and can be changed in a few minutes since they need no adjustment.

Stopping rotary and in-line machines allows stepless adjustment of insertion span over a wide range. This adjustment takes time and frequent changes can reduce the production rate to an uneconomic level. The span of non-stop rotary machines can be adjusted in seconds.

5.4.4 Preforming Non-axial Leaded Components

The large variety of packages in current use has inhibited development of general purpose machines. Only the more common cases mentioned in the previous chapters are considered here.

TO-18 Package

This package has three leads placed on a circle of 2.54 mm (0.100 in.) in diameter. It can be mounted flush with the PB on a 1.27 mm (0.050 in.) spaced grid, Figure 5.18, or on a 2.54 mm grid by spreading the leads.

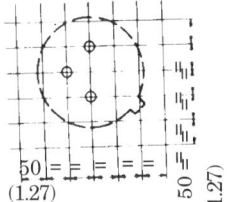 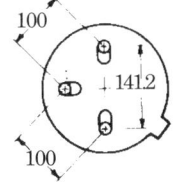 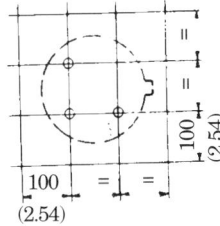

Fig. 5.18 Hole pattern for a TO-18 package: (a) insertion on a 1.27 mm (0.050 in.) spaced grid; and (b) insertion on a 2.54 mm (0.100 in.) grid by spreading the leads apart. Dimensions on the drawings are shown in thousandths of an inch and (in brackets) in millimetres.

As supplied, the leads are too long for flush mounting. It is possible to bend them after insertion and trim after soldering but better results are obtained by cutting to length before mounting. The leads must be straight for cutting. However, flush mounting of the devices usually packaged in TO-18s either has to be avoided because of high thermal shock to the devices on soldering or is not practicable, e.g., on PBs with conductors on both sides.

Several types of insulating spacer can be obtained to keep the case away from the board. Some of these are designed to eliminate stressing the fragile lead/body junction. The cost of spacers cannot be disregarded and the extra operation of fitting takes 3-7 seconds/item.

If lead straightening and cutting equipment is necessary, it is easy to add a forming station. This can provide preformed components with a stand-off and, if desired, a lock-in feature. Many shapes, similar to those of an axial component, can be chosen. One such is shown in Figure 5.19. If lock-in is needed, the shape must be very accurate, the forming tool being designed for a specific board thickness, hole diameter, and hole diameter tolerance.

TO-18 leads have a small diameter and are very ductile but are so close together that the forming mechanism can be quite complex. In forming, great care is needed to avoid stressing the fragile glass-to-metal seal. Two basic techniques are used. In the first, the leads are clamped firmly between the formed section and the body. The clamping device must operate before the shaping and cutting tools. The minimum thickness of the clamping jaws is about 1 mm (0.04 in.), and the first bend is about 1.5 mm (0.06 in.) from the body. This means that the component's body is mounted at a distance, dimension D of Figure 5.19, of approximately 1.8 mm (0.07 in.) from the board

 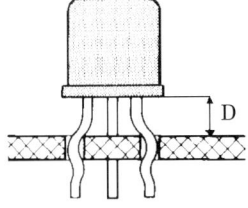

Fig. 5.19 Two TO-18 packages formed to provide a stand-off and inserted on a 2.54 mm grid. Their distance D from the PB is 1.7 mm (0.067 in.).

surface. The second approach forms one lead at a time, with the body floating on an elastic suspension. In this case the distance D can be reduced to of the order of 1.27 mm (0.05 in.). Most commercially available equipments give a D dimension within the range of 1.27 to 1.78 mm (0.05-0.07 in.).

The equipment is loaded by hand or by a vibrating hopper and positioned with reference to the orientation tab on the case. The leads are straightened and then cut and formed. Many types of equipment are available with several options and tooling for different forms. Although designed to avoid damage to the component, this equipment must be closely controlled and well maintained. It must be tested every time the tools have been dismantled for maintenance and every 1 or 2 million operations. The most important test on the formed package is measurement of body leakage, e.g., by a standard helium leakage test on a reasonable sample before and after forming.

The production capability of such equipment is in the range of 2,500-5,000 pieces per hour.

TO-5 and Similar Packages

Packages such as the TO-5, TO-9 and TO-39 are treated in the same way as the TO-18. The leads are normally placed on a 2.54 mm (0.100 in.) grid but, if necessary, can be spread apart to allow mounting on a 3.81 mm (0.150 in.) grid. Spacers and/or tooling, similar to those for the TO-18, are normally employed.

Plastic In-line Packages

Packages of the TO-92 and TO-98 type are difficult to preform so that the ends of the leads remain in line and, after insertion, their bodies remain vertical. Two of the possible lead forms are shown in Figure 5.20.

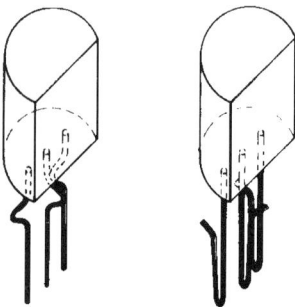

Fig. 5.20 Two typical forming shapes for TO-92 packages. The one on the right is preferred when the lead diameter is much smaller than the PB hole.

It is often preferred to mount these components by misaligning the central lead to allow insertion on a 1.27 mm (0.050 in.) grid. They may also be preformed to mount on the same hole pattern as the TO-18 package, either as an alternative or to replace TO-18 devices on old equipment.

TO-3 Packages

TO-3 and similar packages (TO-66, SOT-9, etc.) seldom need lead straightening and preforming. The leads of these packages are very sturdy and usually straight and the body is intended to be flush mounted, either on the PB or on a heatsink. Simple tooling is available for lead trimming to a precise length. The equipment can be fed by vibrating hopper as the TO-3 body is highly asymmetric.

Non-axial Leaded Capacitors

Disc capacitors are normally encapsulated by dipping into liquid resin which, after curing, usually covers part of the leads. To save board area, disc capacitors are mounted vertically and must be held vertical until the solder joints are made. Bending and soldering too close to the body must be avoided. This requires very good preforming with stand-off and locking to the PB. Typical shapes are shown in Figure 5.21.

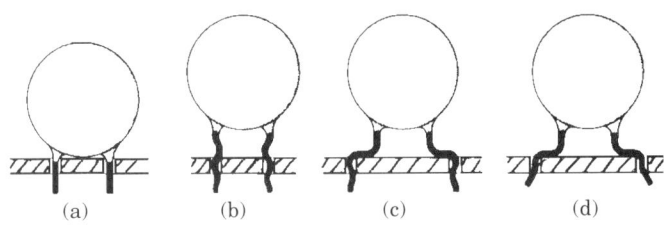

Fig. 5.21 Typical forming shapes for disc capacitors: (a) should be avoided; and (b) will cause problems during insertion.

The *dipped rectangular capacitor* and the *dipped drop capacitor* are more compact than the disc type but have exactly the same problems. Typical lead forms for these types are shown in Figure 5.22.

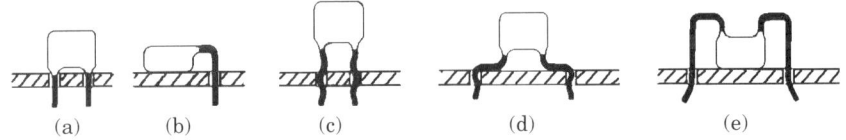

Fig. 5.22 Typical forming shapes for small silver-mica dipped capacitors: (a) should be avoided; and (b) is preferred when the board density is low.

Dual-in-line Packages

Normally, DIP leads do not need preforming or trimming. Straightening of misaligned pins may be needed on occasion and, if the PB is less than 1.6 mm (0.063 in.) thick, the protruding leads may have to be trimmed either before or after assembly.

DIPs are often supplied with their leads bent outwards at an angle of as much as 15°. This means the pin lines are more than 7.62 mm (0.300 in.) apart. For manual insertion, the centre lines of the rows of pins must be exactly 7.62 mm apart. A tool is needed to bend the pins inwards in excess and then outwards to the required limit. The degree of inward bending must be adjustable to cater for variation in lead spring-back between batches.

The need for this operation can be avoided if a tool is provisioned for manual insertion of DIPs. The tool is in essence reverse pliers (i.e., pliers which open when the handles are pressed together) which have a specially designed nose. This has a slot to accept the DIP body and two side legs which act on the pins, aligning their ends on two lines 7.62 mm apart, Figure 5.23.

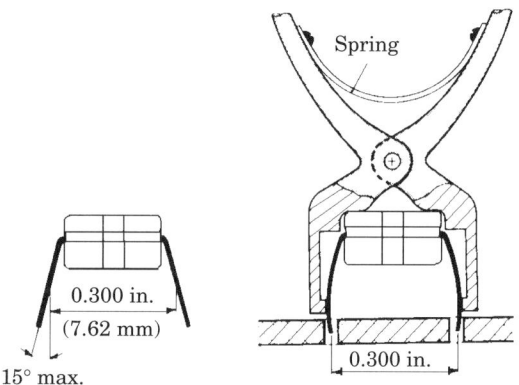

Fig. 5.23 Manual insertion of a DIP with a simple tool which reduces the distance between the ends of the pins. After the pin ends are inserted, the tool is removed and the component is pressed down. It remains in place because of the spring-back of its leads.

To use the pliers, the components are placed on a shaped steel section. The operator picks up a component by pressing the pliers on to it. As the component is grasped, the pins are bent to the correct position. The ends of the pins are inserted into their holes and the DIP is released by pressing the handle of the pliers. Finally, the DIP is pushed down on to the board, when the spring-back of the pins locks the component in its correct position by friction against the hole walls.

DIP insertion tools may be purchased from several sources or can be made by adapting standard circlip pliers. Using such a tool, the insertion time for a 14 or 16 lead DIP can be reduced from 20-25 seconds to about 10-14 seconds.

Leads which are misaligned outwards and/or sideways may be inserted using an insertion guide of the type shown in Figure 5.24. The guide is centred on the set of holes before pressing the DIP into the guide. If the pins are bent inwards, this tool will not work and may even exacerbate the situation by forcing the lead under the body of the component.

If leads are very badly distorted, it can be more economic to discard the component than to attempt to straighten them out by hand.

5.5 PRODUCTION AIDS

Mass production by manual insertion is still widely used, not only for bulky or odd shaped components, but also for assemblies where the cost of automatic equipment cannot be justified. Operator fatigue and errors must be minimised and, over the years, several methods have been developed to aid manual assembly. The methods now described are considered by many to be old fashioned and unpopular but, nevertheless, are likely to survive for a long time.

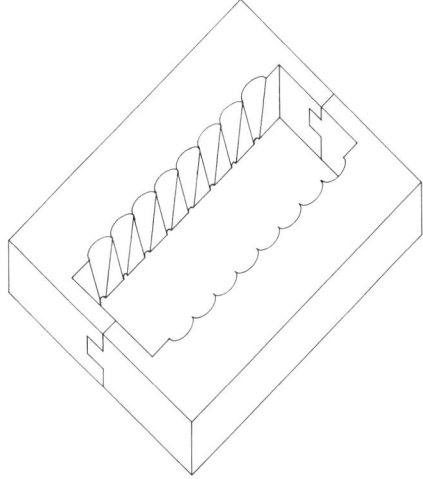

Fig. 5.24 Taper groove insertion guide for DIPs.

5.5.1 Layout Memorisation

If the board is simple and personnel turnover is low, the operator can be instructed to learn the board layout, that is, the positions into which all components are to be inserted.

Memorisation is made easier if (a) there are significant differences in visual attributes of the components such as colour, shape and dimensions; (b) components are arranged on the board in such a way that a sequence can be easily seen; and (c) the component side of the board has readily visible reference points, such as slots, cutouts, screen printed markings or large components already mounted.

The time H to learn a board layout depends largely on the type and complexity of the board and, of course, on the aptitude of the operator. In the absence of aiding factors, listed in the previous paragraph, H can be calculated roughly using the empirical formula:

$$H = 0.013C^2 + 0.3C$$

where H is in working hours and C is the number of components per board.

As the operator does some useful work while training, only some 30 to 50% of the total learning time is lost. However, if a board has 100 or more components, very few operators will be able to memorise the layout and work reasonably quickly without making too many errors.

When the board is laid out so that a sequence can be easily worked out, memorising the layout becomes much simpler. The small bins from which the operator takes the components can be put into the same sequence, Figure 5.25, and the calculated learning time can be reduced by 50 to 80%.

This method needs little investment but gives a relatively high insertion error rate. One misplaced component in 500 is a good result. The error rate increases sharply if components are similar in visual appearance or if the same operator is required to memorise the layouts of many types of board.

The method should not be relied upon if the board contains more than about 20 components.

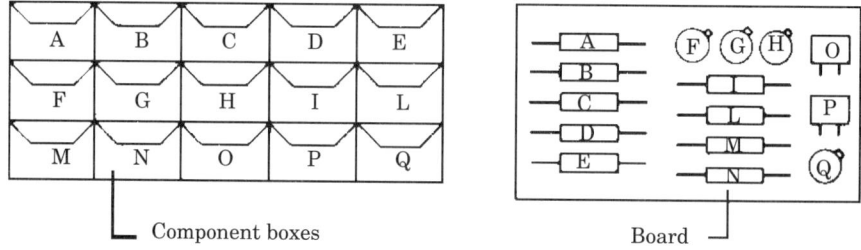

Fig. 5.25 Assembly with layout memorisation may be simplified by placing the component bins in the same order as the assembly sequence.

5.5.2 Stage Assembly

More complex boards can be assembled efficiently using the layout memorisation method if they are subdivided into two or more zones. Each zone is treated as a small board and, preferably, will have a minimal number of different types of component as each zone will be assembled by a different operator. The assembly line is set up as a series of work stations connected by a conveyor.

The first operator on the line takes an unpopulated (bare) board and inserts the components needed by the first zone. The board is moved on to the second operator who loads the second zone components and passes the board to the next stage, Figure 5.26. Normally, the last operator on the line inspects the assembled board before passing the boards to the soldering station.

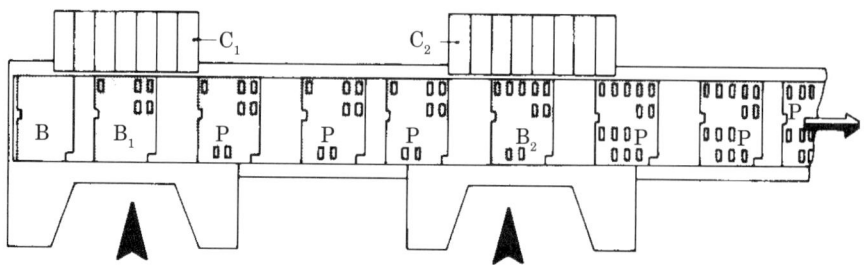

Fig. 5.26 Stage assembly on a push line. C_1 and C_2 — bins with components to be inserted; B — bare printed-boards; B_1 and B_2 — boards being assembled; P — part assembled buffer boards. Arrows indicate the working position of the assemblers.

To be effective, all stages must take the same time for assembly. Balancing the line must be done for each type of board to be assembled, keeping the number of different types of component at each stage to a minimum. Within reason, the higher the number of stations in the line, the shorter the individual stage time and the faster the production rate. A short stage time means operators need less training but their work will be more routine and boring which can cause discontent and high absenteeism. Moreover, it can be difficult to balance very short times, which *must* be the same for each station, and agree production rates with the operators.

The ideal stage time appears to be 2-4 minutes, although much shorter times, e.g., 20-30 seconds, are sometimes possible.

A common aid when many types of board have to be assembled is the use of cardboard, plastic or metal templates for each component type. The template is placed over the board and covers those areas in which the operator must not insert components. When the board type is changed, each operator changes the template set and can start assembling the new type immediately. Little or no time is lost in learning the new layout. It is possible to take this one stage further so that the template selects the corresponding component from a gravity fed magazine, Figure 5.27, or triggers an indicator on the appropriate bin.

Fig. 5.27 Stage assembly on a push line. Each workbench is equipped with 2 or 3 magazines of, e.g., DIP components (each magazine contains only one type). The correct component is selected and made accessible by placing a template over the board being assembled.

Stage assembly gives a fault rate of one wrong component in every 500-1,000 insertions. The use of templates and automatic selection of component type reduces errors to less than one in 3,000 insertions.

5.5.3 Sequence Assembly

Figure 5.28 illustrates an instance in which it has been easy to identify a logical sequence of component insertion. Used with a simplified automatic positioning table, described earlier in this chapter, a fixed board holder, and as many bins as needed arranged in the correct sequence on a closed loop hopper chain, insertion errors can be substantially reduced.

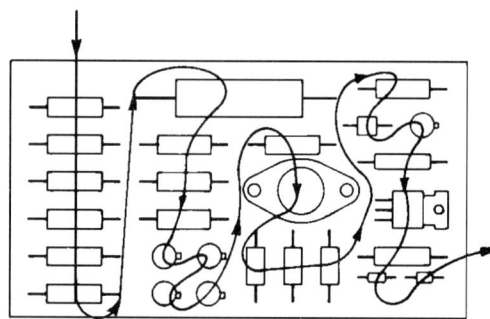

Fig. 5.28 An example of a logical insertion sequence on a board with several types of component.

Sequence assembly forces the operator to take one component at a time from the bins and insert it into the next vacant position on the board. The same motion is repeated for each component and insertion time is constant, i.e., a sequence in which the same type of component is present in several positions will not reduce the insertion time.

When it is not easy to plot out a sequence and layout memorisation is difficult, a Lazy Susan bench, Figure 5.29, can be used. This type of assembly bench is widely used by manufacturers of precision mechanical assemblies, such as watches and instruments, but is seldom employed in the electronics industry. The bench has a horizontal, closed loop, chain which carries the boards around the table. The chain is motor driven and moves one step each time the operator presses the control to bring a new board to the loading station in front of him. The component bins are mounted over the bench in easy reach of the operator. They may be fixed to a wheel which is manually controlled to bring the bin needed to the most convenient point.

The operator sits in front of the bench and starts by placing a bare board on the fixture in front of him. By operating the chain control, the loaded fixture is moved on and the next in line brought to the operator's position to receive the next board. When all fixtures have been loaded, the operator selects a component type and inserts it in all appropriate positions on the first PB. He presses the control to receive the next board and inserts the same component type in that. The operation is repeated for all boards on the bench. A second component type is now assembled and so on till assembly is complete.

The important advantages of the bench are that:

(i) the same component is inserted into the same position many times so that learning and memory retention required of the operator are minimal;
(ii) components may be loaded by keeping a handful of components in one hand and picking one and loading with the other. The lesser motion reduces fatigue and allows shorter insertion times to be achieved.

Manual Assembly

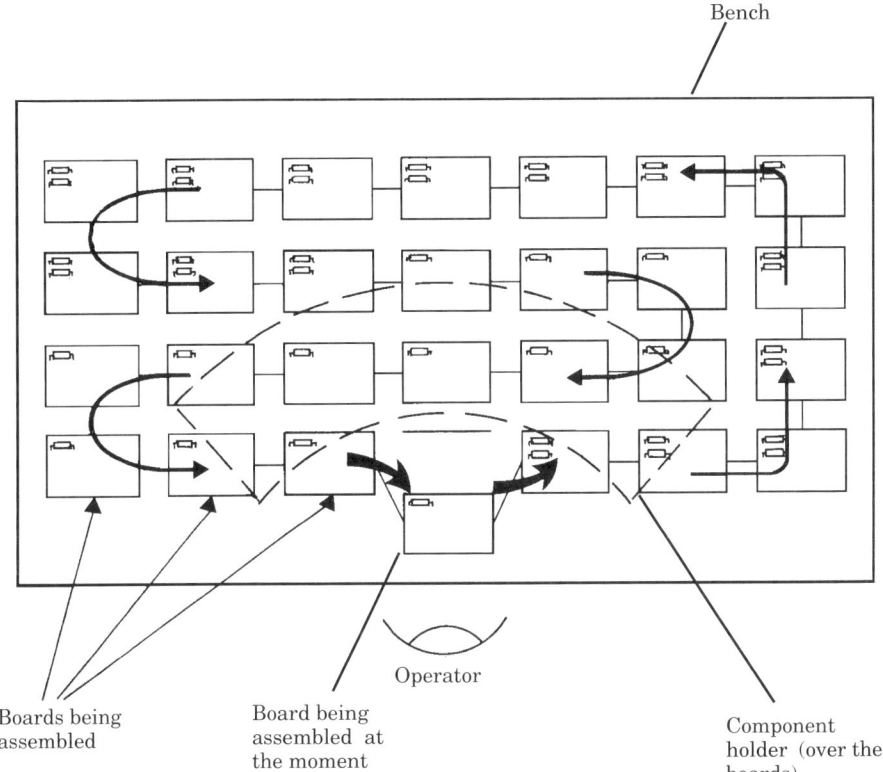

Fig. 5.29 The 'Lazy Susan' work station for sequential board assembly.

5.5.4 Component Sequencing

Insertion errors can be sharply reduced by allowing the operator access only to that type of component needed at that point in the assembly process. The simplest way to feed the components in the right sequence is the so-called *Streckfuss* bench, which has been used for a very long time in the assembly of mechanical watches.

The bench, Figure 5.30, looks like a normal table, on top of which there is a PB holder and bins for bulky or awkwardly shaped components. The majority of the components are beneath the bench top, in bins attached to a motor-driven chain. Only one bin at a time is accessible through a cutout in the bench top. Each time the operator presses the motor control, the chain moves one step so that the next bin, with the next component, becomes accessible.

In the simplest (and most foolproof!) method of working, each bin corresponds to one place in the insertion sequence so that if a component is used in 10 positions on a board it will be available from 10 bins. The bench is, therefore, limited to the assembly of boards having no more than about 50 components. This can be increased by using more than one bench and part assembling at each, or by enlarging the access hole and allowing the operator to pick up a handful of components from one bin to insert at all positions at

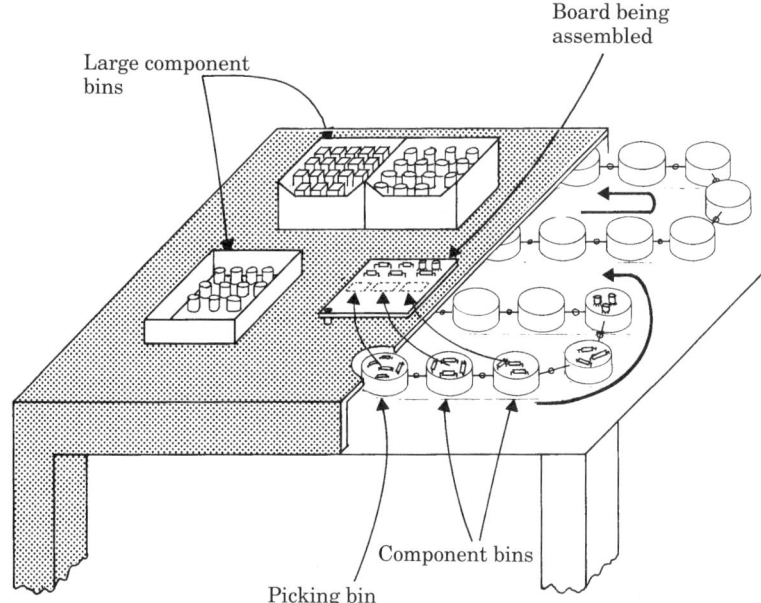

Fig. 5.30 The 'Streckfuss' bench — used for component sequencing for manual insertion.

which that component is required. The latter speeds up assembly but negates the advantage of component sequencing, since multiple insertion invariably leads to higher rates of insertion error.

5.6 VISUAL AIDS

A visual aid is something which assists the printed board assembler by indicating visually the location of component insertion holes. Most visual aids are simple in conception and construction and are often made by the PB assembler.

5.6.1 Screen Printing

In most cases all the identification markings needed to assist location of component positions can be printed directly on to the PB. (Very simple boards can be seen with the component identities shown in copper but increases in packing densities are enforcing the use of a non-conductive indicator.)

The markings may be simple symbols, such as a line between two holes, or they may be more complex, giving a component code (referred to the circuit diagram) and sometimes a descriptive outline as shown in Figure 5.31. Such information is useful not only in assembly (including preparation of the workbench) but also during test, service and repair.

It may be noted that screen printing of 'idents' is even more useful with surface mount boards. The components are often so small that it is impractical to put complete identifications on them and in a steadily increasing number

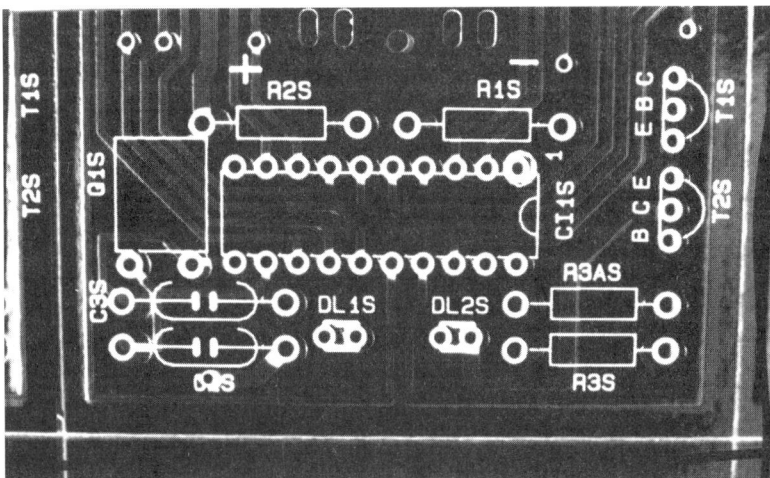

Fig. 5.31 Component identities and positions are easily seen on a properly screen printed board.

of cases marking of any sort is impossible, Figure 5.32. The only means of recognition in the field is to trace the circuit diagram all the way through, which may be difficult and time-consuming with a complex board, or to check the screen printed identity.

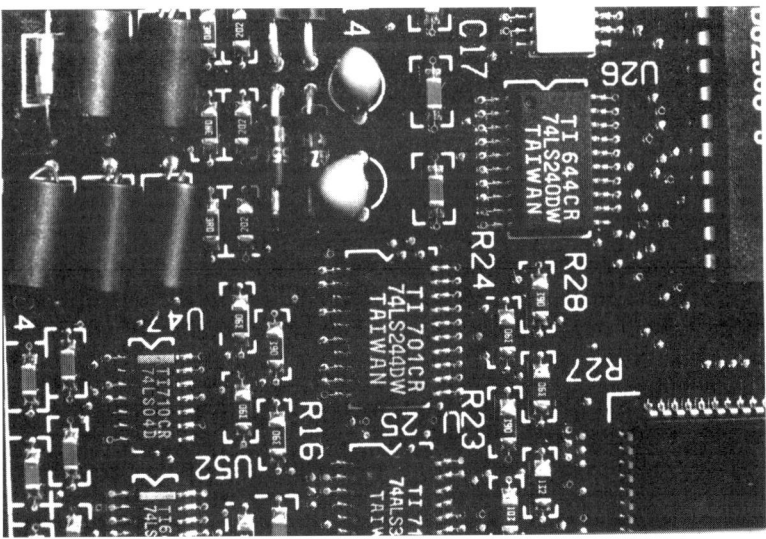

Fig. 5.32 Chip components without markings may be identified much more quickly if the board is suitably screen printed.

The accuracy required of screen printing is also increasing. Fortunately, stencil materials and screening equipment are being developed which can meet the challenge, see Volume 2, Chapter 3. Unfortunately, however, aside from prototype or small batch assembly, screen printing is not capable of producing a significant reduction in assembly errors economically and other means are needed to give operators quicker guidance to the assembly sequence.

5.6.2 Slide Projection

This technique employs a modified photographic film projector. It is normally fed with a 135 film (frame area 24 × 36 mm, about 1 × 1.5 inches). The projector is mounted over the workbench and projects the image at 1:1 size on to the board being assembled.

The film for a given board has a leader showing the board part number and description, followed by frames for each type of component to be assembled. Each frame is black with clear areas showing the position of each component of the given type to be assembled. When necessary, orientation is shown by making the transparency for that position asymmetrical, for example, in the form of an arrow. The film is made into a closed loop so that rewinding is not needed.

After checking part number, the operator goes to the first frame and inserts all components of that type. On completion, the next frame is projected and the component bin moved on to the correct type for assembly. Ideally, the bins are attached to a hopper chain which 'steps' in sequence with the film so that, on pressing the control to move the film, the correct bin is moved to the operator's pick-up position. A simpler method uses the film movement to trigger a lamp on the appropriate, fixed bin.

The films are easy to prepare. The images are made on a white or transparent sheet by drawing or by using self-adhesive preforms and photographed on high contrast film with a standard 35 mm camera. For prototype and small runs the film is used as developed; for mass production, contact copies on positive film can be made.

The major advantage of the method is simplicity. Limited skill is all that is needed to produce the film for a new board within a few hours of receipt of drawings and item lists. The main disadvantage is that the images can be seen only in subdued light. This may cause some errors which could have been avoided in normal lighting such as insertion of damaged components.

The lighting problem can be solved by using metal slides, produced by numerically controlled machining, mounted on standard heat resisting plastic frames, Figure 5.33. As heating has no effect on this type of slide, high power projectors may be used.

5.6.3 Lamp Displays

Lamp displays consist of a set of small lamps placed under the board to be assembled. They are lit in a preset sequence to show the specific component position. The sequence can be linked to the component bins such that either the bins move in step with the light sequence or a light shows on the correct bin to be used.

All components of the same type are inserted during one step of the sequence. To begin, the operator turns on all lamps A. These show where type A components are to be inserted. After inserting these, the control is activated to turn off lamps A and turn on the lamps for type B components.

Manual Assembly

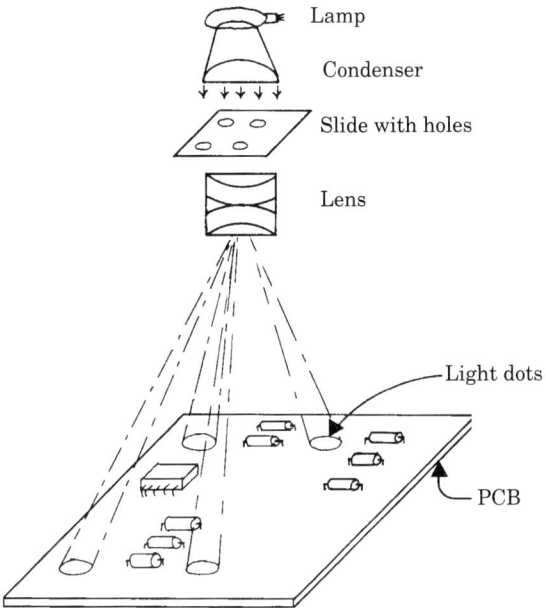

Fig. 5.33 Projection of light dots on to the PB gives an accurate guide to component position. Metal slides allow the use of high intensity lamps.

The operator inserts all the type B components and proceeds to component C and so on.

The equipment is easy to build for a given type of board. All lamps relating to one component type, which can include a lamp on the component bin or a relay to drive the bin movement mechanism, are connected in parallel to two wires. One wire is grounded and the other connected to one terminal of a multi-position switching selector which commutes each pulse it receives from the control.

The workstation becomes more difficult to design when it is desired to assemble more than one type of board because of the complexity of the lamp wiring. Two further disadvantages are: (a) limited application to boards with high packing density and many components, particularly if the components are small; and (b) the low reliability of miniature lamp bulbs.

5.6.4 LED Displays

The modern version of the lamp display uses an electronic controller instead of the selector and light emitting diodes instead of lamps. A suitable controller can be built using a microprocessor at very low cost.

Several types of high intensity LEDs are available in a variety of packages, small enough to assemble on a 2.54 mm pitch. Surface mounting LEDs, in an SOT-23 package, are also available. It can be practicable, therefore, to arrange a full plane of LEDs on an X-Y matrix for use with many types of board, the bench being programmed by software.

The colour of the diodes should be selected according to the light absorption of the PB laminate and its coatings. Double LEDs, which can provide red, yellow and green, are available in different packages.

5.6.5 Optical Fibre Displays

This is an improvement on the lamp display, having advantages in flexibility and reliability. It works on the same principle of back lighting the positions where the components are to be inserted, but the lamps are far removed from the board and only one lamp is needed for one component type. The light is taken by optical fibre light guides from the lamps to the back of the board being assembled, where they are fixed in the right position to a panel, Figure 5.34.

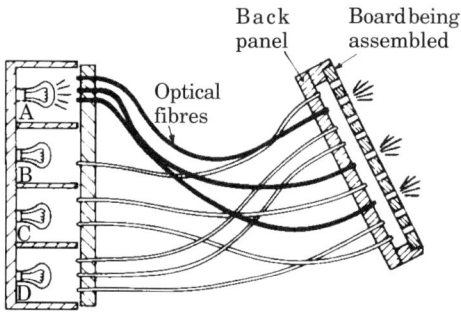

Fig. 5.34 Optical fibre visual aid: when lamp A is turned on, the assembler sees light in all the holes in which he has to insert component A.

The termination panel for the optical fibres is specific to a given type of board. It is easily made by enlarging, if necessary, the holes of a PB of the type to be assembled or by using the drilling program to drill an unprocessed panel. If a screen stencil for the identities of the components is available, this can speed the assembly of the light guides to the panel. Here again, the component bins can be fixed with no display, or stepping in phase with the insertion sequence, or lit by one supplementary optical fibre for each component type.

If it is desired to mount components with full length leads, the system can be modified by the addition of a dichroic mirror. This is employed to superimpose the light of the optical fibres on the image of the board being assembled, Figure 5.35.

There is no limitation on the size of lamps employed and high reliability lamps can be used. If a lamp does fail, it will be noticed immediately because no holes will be lit in that step of the insertion sequence.

The optical fibre display is a very reliable and versatile visual aid. It gives a minimal amount of error for a manual assembly technique, i.e., of the order of one error in 3,000-4,000 insertions.

5.6.6 Laser Scanning Systems

The latest, and possibly best yet, visual aid for assembly is the so-called soft laser system. This method depends upon a laser scanning system which

 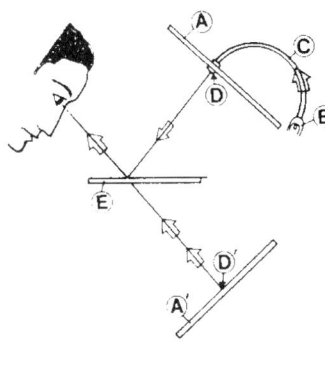

Fig. 5.35 Optical fibre visual aid with dichroic mirror. Optical fibre, C, ends on a panel, A, which is placed so that when the operator looks at the dichroic mirror, E, he sees the illuminated end, D, of the fibre superimposed on the hole D^1 of the board A^1 in which he has to insert the component. (Courtesy of Hollis)

uses electronics and mirrors to trace an unlimited range of clear and accurate component symbols on the printed circuit board being assembled. Symbols may be made up and held in memory for recall in the appropriate shapes, sizes and orientation required, while new symbols can be created readily as needed.

Some equipment manufacturers have been quick to combine laser scanning with assembly benches to produce microprocessor controlled semi-automatic units.

One such manufacturer has produced a series of printed board assembly (PBA) work stations of increasing versatility. The top model in the range offers laser scanning, ensuring accuracy and short assembly times on new or complex designs. The scanner is well out of the user's way and transmits from horizontal to the angled PB face with no perceptible distortion. The work stations have the ability to provide up to 76 different components, in the logical assembly order, from the computer-controlled, powered dispensing trays, with a further 40 larger components and ICs from surface dispensers.

A microprocessor generates the assembly instructions for each step, including the component type, the component dispensing location and, of course, the PB location. In addition, the display indicates the quantity of that type to be assembled and counts down to completion (with an audible signal on one). On-line programming is straightforward and is easily reviewed and modified. Off-line programming is carried out via an RS 232 port using a wide range of digitising pads.

Laser scanning systems have achieved rapid success in the USA, Europe, and in Australasia. The versatility of the symbols formed, the ability to operate from the top of the board in a wide variety of light conditions, and the ease with which the control of the laser can be combined with automatic, or semi-automatic, delivery or indication of the next component to be inserted (or mounted), make this an ideal system for small batch production with relatively low capital outlay. As the benches and component delivery systems can be adapted readily, the systems are equally suitable for use with PIH assembly and with SMA.

5.7 ASSEMBLY INSPECTION (see also Volume 2, Chapter 20)

In selecting any assembly method, the likely cost of inspection must be considered. If an assembly defect (wrong or missing component, wrong orientation, incomplete insertion etc.) is discovered before soldering the board, it can be rectified very quickly. If it is discovered after soldering, repair takes much longer and there is always a risk that the board may be damaged. If the fault is not found until the board reaches electrical test, diagnosis of the fault can take a time greatly in excess of that needed for the repair. As test equipment for assemblies is very expensive, it is usually loaded to near 100% of its capacity and the need to test repaired boards can lead to an overload on the test station. In addition to lost production and test time, a trivial assembly error can cause the failure of expensive components either on test or as soon as power is applied to the circuit.

Operators must be encouraged to be self-inspecting, to look continually for something 'which does not look quite right', and to query anything which looks suspicious with inspection and/or supervision. A decision has to be made whether or not to inspect the assemblies visually, at what stage, i.e., before or after soldering, and whether to inspect all boards or a sample only. An inspection plan must be prepared for each board or group of boards taking into consideration:

(i) cost of inspection;
(ii) estimated or known number of assembly errors;
(iii) repair/rework cost of assembly errors;
(iv) impact of assembly errors on finished product quality;
(v) expected changes in assembly techniques, personnel turnover and personnel training programmes.

The plan must be revised periodically as new data become available.

5.7.1 Visual Inspection

The most frequently used method of examination for assembly errors is visual inspection. Many standards and draft standards exist for the acceptability of electronic assemblies and are covered in Volume 2, Chapter 19. In addition, many of the major component, material and equipment suppliers can, and will, provide much useful advice and guidance.

Before the inspector handles the board in any way, he must be aware of any special provisions for handling, e.g., because of the presence of static sensitive devices on the board. None of the assembly aids mentioned previously is likely to be of much assistance to the inspector. Simple tools such as a template with cutouts for each component are more helpful, especially if each cutout can be labelled with the identity of the component which should appear in it, Figure 5.36. This allows immediate recognition of missing components and incompletely inserted components.

Fig. 5.36 Visual inspection of a complex board with the help of two templates (both shown turned up).

If the components are arranged so that a sequence can be singled out for assembly, this will also aid the inspector. Even a limited inspection, restricted to certain parts of the sequence, can give good results, provided that a sample of the batch is checked over the full sequence.

A common problem in visual inspection relates to the difficulty of reading component values, especially if these have been mounted such that the component marking faces the board. Consistent orientation of component markings is possible but expensive. If components are not preformed prior to assembly, orientation of the markings to allow easy identification after assembly increases the insertion time by about 0.5 second/component. If preformed components are used, orientation must be carried out before preforming. This can only be done using pneumatic reciprocating machines and reduces the forming rate to the order of 1,000 components per hour.

For a first evaluation of inspection cost, the following figures can be assumed for a board 254 × 254 mm (10 × 10 in.) with 150 components:

(i) overall inspection for missing components, damage, etc.:   40 s;
(ii) checking sequence accuracy on random components: 20-40 s;
(iii) checking every component — with a sequence: 150 s;
            — without a sequence: 300 s.

5.7.2 Equipment Assisted Inspection

Apart from magnifiers and the simple templates mentioned, there are a few instruments which are designed specifically for assembly inspection. One such is an optical system which compares the image of the board being inspected with the image of a reference board. The optics of the system give a stereoscopic view of the inspected board, differences between this and the reference board being very evident. With progress in digital image processing equipment, the process can be automated. One such system utilises a video camera input. The principle of operation is the subtraction of a picture of the board being checked from that of a known good board. Any difference or error is immediately highlighted. The system can be programmed to be discriminating in the degree or nature of the error. The electronic inspection process takes only 0.2 second to indicate pass or failure of *an assembly* which may be compared to about 0.1 second *per component* for the manual versions.

Inspection equipment of these and similar types varies considerably in cost, depending on the degree of automation, and application is limited by two facts:

(a) equivalent components can differ significantly from each other, for instance, two resistors supplied by different vendors;
(b) different components can look very similar to one another, such as two resistors which differ only in the colour of one band.

If such cases are frequent, much time may be lost in looking closely at indicated error points.

Whether or not such equipment should be provided for inspection depends upon the factors discussed for preparation of inspection plans. Only if the cost of inspection is less than the direct and consequential costs of no inspection can investment in hours and/or equipment be justified.

REFERENCES

1. Siegel, E. S., 'Where's The Heat? Thermal Process Control for SMT', *Printed Circuit Assembly*, November (1989).
2. Lea, C., 'The Effect of Solder Mask on PCB Solderability', *Circuit World*, **Vol. 15**, No. 1, pp. 12-21, October (1988).
3. Richards, B. P., Footner, P. K., Prichard, D. J. and Lea, C., 'An Assessment of Cleaning Options for Soldered Electronic Assemblies (Phase II)', *Circuit World*, **Vol. 19**, No. 3, pp. 4-17, March (1993).
4. Jones, M. D., 'Benchtop Cleaning', *Circuits Assembly*, September (1992).

Chapter 6

AUTOMATIC ASSEMBLY — CONVENTIONAL COMPONENTS

GIOVANNI LEONIDA
Milan, Italy

W. MACLEOD ROSS
Welnorth Ltd, Bishop's Stortford, UK

6.1 INTRODUCTION

The cost of machinery tends to increase at a relatively slow rate because of competition, greater productivity and better use of materials, all of which have a stabilising effect on prices. In contrast, except for some minor fluctuations, the cost of labour has been increasing in all industrialised countries since the Second World War. Labour costs are related to standards of living, taxation, social security burdens, desires for reduced working hours and other factors which have affected human reasoning since before Hammurabi, 1792-1750 BC, who protested bitterly about the effects of wage demands on inflation.

This explains the continued trend towards automation technologies because, with few exceptions, the major aim is to reduce the cost of a production hour. Other reasons for automation, such as scarcity of manpower, the difficulty of performing some operation manually and a desire for higher and more consistent levels of quality, have played a minor part in machine development. As soon as an equipment is successful and is produced in sufficient quantity, the price drops and its productivity tends to rise. The more skilful an operator, the higher the basic rate, sometimes with no corresponding increase in productivity.

6.1.1 Economics of Automation

It is easy to assess the point at which automation becomes interesting, if not mandatory, for an assembly company.

Assume that in a company the cost of direct labour employed in component insertion is equivalent to US $15 per hour. To this cost, all indirect costs must be added, from goods receiving to component preforming, with an overhead contribution for depreciation of buildings and equipment, heat, light and power, etc., which in a well run operation should total around 40% of the direct labour cost, i.e., about $6/hour. The total cost/hour is $21. In good conditions an experienced operator can insert an axial component in 4.5 seconds and a radial component in 5.0 seconds. This makes a total of 3600/4.5 = 800 and 3600/5 = 720 components/hour. At $21/hour, the cost of insertion of one component is:

axial: 21 x 100 ÷ 800 = 2.625 cents
radial: 21 x 100 ÷ 720 = 2.917 cents

and both figures are substantially independent of the number of components as the fixed costs are a minor fraction of the total. (Obviously, the breakdown of cost will vary from country to country and within a given country depending on the ratio of overhead to labour cost.)

If manual insertion is replaced with automatic insertion, the insertion rate rises to some 5,000 to 7,000 components/hour, i.e., 0.7 to 0.5 seconds per component. If it is assumed that a more skilled operator is employed, then the labour rate may be, say, $19/hour and the cost/insertion is 0.5 to 0.36 cents plus the cost of machine depreciation. Typical automatic insertion machines cost $200,000 to $300,000 and are amortised over 4 to 5 years. If 90% efficiency is obtained, in 220 days/year such a machine will insert from 8 to 33 million components/year, dependent on its speed and the number of shifts/day. Using this information and a value of $10/hour for indirect costs related to the actual time during which the machine is operating, the graph in Figure 6.1 can be derived.

It will be noted that the break even between manual and automatic insertion of axial and radial leaded components occurs at about 2 million to 5 million component insertions. However, it must be remembered that the costs postulated contain no allowances for the savings in rework and repair to be expected from automatic insertion.

6.1.2 The 'Mechanical Horse'

Automatic insertion of axial components started on a massive scale in the late 1960s and was followed a few years later by that of radial components. Equipments developed had to cope with components whose shapes and dimensions had been standardised for manual assembly. The equipment was forced to repeat, more or less exactly, the operations of manual assembly, often with rather inefficient results as if a horse were replaced with a mechanical replica instead of a car.

Automatic insertion of conventional components reached a peak in the early 1980s, just before the upsurge in surface mounting assembly. From 1985 its use started to decline in the USA, Europe and Japan, while it continued to grow in countries where the lower cost of labour and scarcity of skilled engineers/technicians still favoured manual insertion.

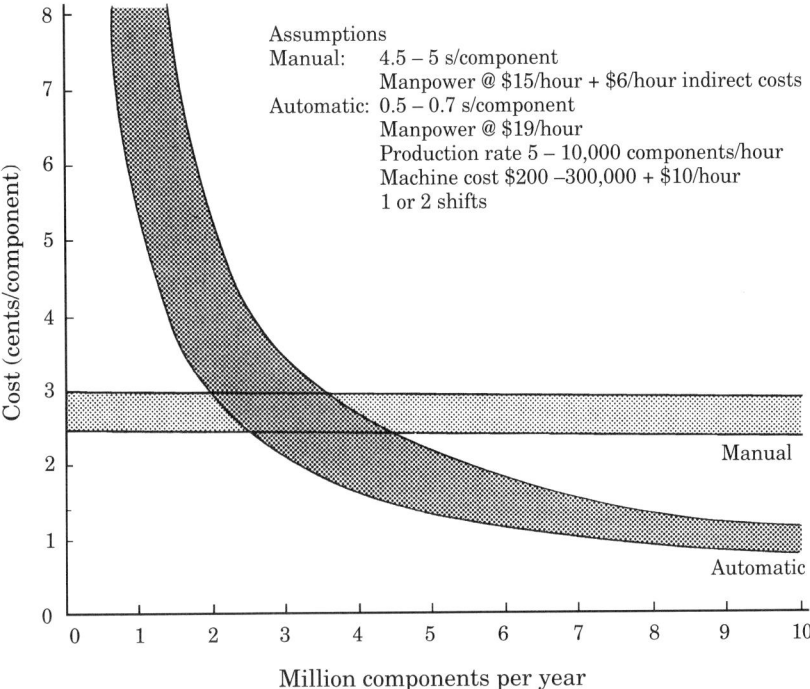

Fig. 6.1 Cost comparison — manual versus automatic insertion of axial and radial leaded components.

Except for automatic insertion of odd-shaped components, the rate of equipment innovation is slowing down considerably. Greater emphasis is now given to machine development for auxiliary functions such as board loading, unloading, stock buffering, automatic adjustment to board size, vision systems, interface with factory automation systems and so on.

The transition to SMT has accelerated the slow-down of sales of new conventional equipment directly and, indirectly, by releasing secondhand equipment on to the market. The availability of used machines at between 30 and 60% of the original price has a significant effect on the break-even point between manual and automatic insertion, so that more manufacturers in countries where labour costs are very low are abandoning manual insertion in favour of automatic insertion — to gain improved quality.

At the end of 1993, it was reported that, despite the fact that in Europe SMT machines were expected to account for more than 70% of capital equipment sales, there was still a strong demand for new conventional, 'pin-in-hole', insertion equipment.[1] Not only that, the demand for secondhand insertion equipment was such that the equipment was easier to sell than find. All major manufacturers of automatic insertion equipment, with one exception, were continuing to develop existing and new equipments to obtain higher insertion rates, greater reliability and cater for complex leaded components. This would not have been so in a highly competitive environment unless there was a continuing and expanding market for these machines.

For these and other reasons, we may be sure that the 'mechanical horse' still has a long life ahead and will co-exist with the newer technologies for a considerable time.

6.2 EARLY AUTOMATION

Even before the PB became the basis of electronic assemblies, there were attempts to mechanise assembly, for example, by automating the insertion of posts to which components were mounted. The same equipment was adapted for component insertion but there was little incentive to automate at the time. In the following text, only those developments which survived for a reasonable time and generated current technology are discussed.

6.2.1 Semi-automatic Equipment

The manual positioning machine, developed for inserting terminations, test points and similar items, is still in use to a limited extent. These machines were developed for a number of reasons, the most important of which are as follows:

(i) Before solid state technology became cost-effective (early 1960s), NC machines were very expensive and labour costs relatively low.
(ii) Mechanical technology was well advanced and mechanical pieceparts could be machined, formed or blanked with good precision.
(iii) The termination manufacturers also made the insertion machines, which were, therefore, fully compatible with their components, and rented them to the component user, thus promoting the use of their terminals.
(iv) Handling the terminals was easy because they had to be geometrically precise and could be supplied in rolls by leaving them attached to the metal strip used in the manufacturing process.

The simplest version is a C frame with an insertion head on top and a tool with a centring pin on the bottom. The operator places the PB under the head, centres and triggers insertion. This simple version has been replaced by most users with a pantograph type.

The pantograph positioning table is illustrated in Figure 6.2. It consists of a strong base plate on which is mounted a rigid support for the insertion head. The PB is positioned on a table, the board holder, which can move freely in the X and Y directions on roller bearings. The board holder is rigidly connected to a pantograph, the handle of which overhangs a template with countersunk holes which match the insertion points on the board.

The operator moves the handle over a hole in the template to move the relevant hole in the board under the head, centres and presses the 'fire' button. This locates a pin in the handle in the template hole to give precise centring and triggers the insertion cycle. Each cycle, the head blanks out a termination from the feeding strip, inserts it into the hole and moves the strip to present the next terminal ready for another cycle. The strip from which the elements have been parted is either cut into small lengths and drops away or is collected on a take-up spool.

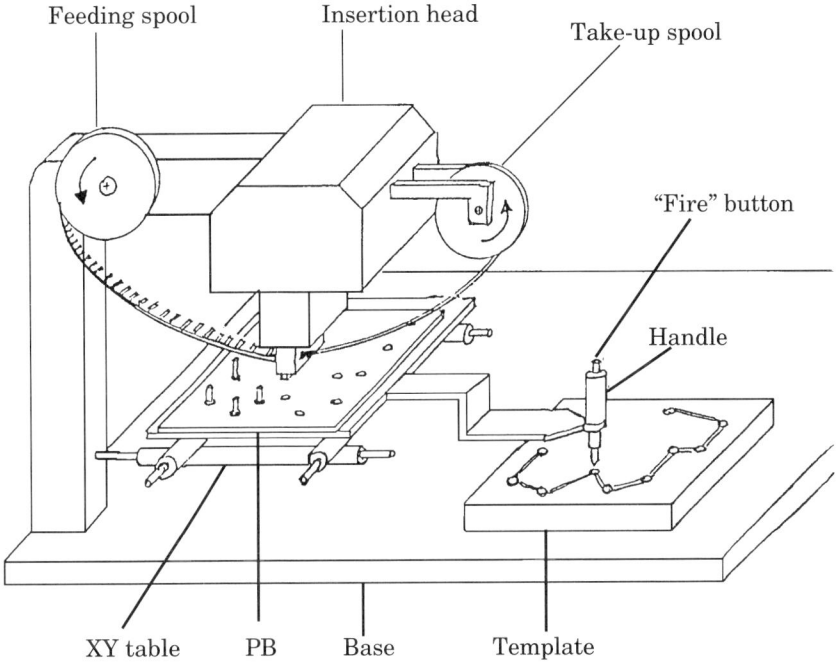

Fig. 6.2 Pantograph operated positioning machine for insertion of termination posts into PBs.

If the terminal shape is very simple and an overall plated finish is not needed, they can be blanked out immediately before insertion. In this case, the machine is fed with a plain, solid strip and the blanking tool is incorporated in the head. This obviously slows down the rate of insertion but reduces the prime cost of the terminal.

A popular variant inserts vertical posts made from copper wire which is cut, inserted and cropped on both sides to keep it in place. This can be used to make very inexpensive connections between boards.

The advantages of these manual positioning tables are low cost, low maintenance cost, low tooling cost (templates are very cheap to produce), good precision with low operator skill and the ability to insert components with controlled force. In some cases, this is one of the few methods available of inserting parts which are difficult to handle or make and insert in one operation. In addition to terminations, such machines may be employed, with some modification, for inserting relatively large numbers of identical components, especially those needing high insertion force and/or needing cropping and clinching or clenching to hold the component in place.

The average production rate of a simple machine is of the order of 600-800 parts per hour. The pantograph version can attain 1,500-2,000 parts per hour, which can be speeded up to some 2,500-3,000 insertions per hour, by fixing the insertion sequence and guiding the handle by machining a small groove between consecutive holes in the template.

6.2.2 The Transfer Line

This is another simple means of automating without involving complex systems, in which the total assembly operation is split into several elementary operations, each of which can be carried out by a relatively simple machine at a separate work station.

The board to be assembled is mounted on a workholder connected to a conveyor so that it moves along the workstations and stops at a precisely fixed point in front of each of them. There is nothing new in this idea. The basic principle of breaking down a task into several elementary operations has been well known for many years. The idea of splitting a complex process into a set of lesser tasks is usually attributed to the American engineer F. W. Taylor (1856-1915). However, the technique was practised long before then. The ancient city of Venice did just that in the Middle Ages. Venice kept only a few ships ready for battle. The others lay in harbour, unequipped and with a skeleton crew. In emergency, the ships slowly crossed the town via the Grand Canal, while the citizens threw supplies on board and the rest of the crew joined the ship. Within hours of the alarm, the ships, ready for battle, reached the open sea at the rate of one every two minutes.

The stations may be arranged in line or in a closed loop, according to needs, and each station assembles the same component in the same position and in the same manner. As there is a need for as many machines as there are stations, the line becomes very inflexible and expensive for other than very simple assemblies. Productivity may be high but one breakdown in the chain stops all production.

In the past, when sequential machines were not available or less reliable, transfer lines were used for large-scale production. These were made up from many stations, Figure 6.3, each dedicated to a single type of component. By equipping each station with an NC X-Y table, the same component could be inserted in many positions to give a relatively small and flexible line.

Fig. 6.3 A 20-head fully automatic transfer line for board assembly. Each head inserted only one type of component. The production rate was very high. (Courtesy of Universal)

Production rates up to 75,000 components per hour were achieved, but the technology was abandoned when the machines described later became available. However, the same approach is now used for assembly of odd-shaped components, either using dedicated machines or general purpose robots.

6.3 COMPONENT PACKAGING

In any automatic assembly operation there are three key steps:

(i) loading the board to be assembled in exact register with the machine;
(ii) feeding the part to be mounted to the assembly head in the correct position and with correct orientation;
(iii) picking up the part, placing it precisely on the assembly and fixing it by the most suitable method.

Step (ii) can be very expensive if the components are not supplied in a pack which simplifies handling. Component packing, of minor importance with fully manual assembly, is a critical contributor to successful automation.

6.3.1 Axial Component Taping

There are two basic taping methods in use for axial leaded components, body taping and lead taping. The first is used mostly for large components which are not usually inserted automatically and may be ignored. Lead taped components are arranged in a long strip, with their long axes parallel, and fixed in position by holding each lead by adhesive tape.

As shown in Figure 6.4, axial components can be machine taped. The taping unit has a spacing wheel, on which the precisely spaced slots hold the

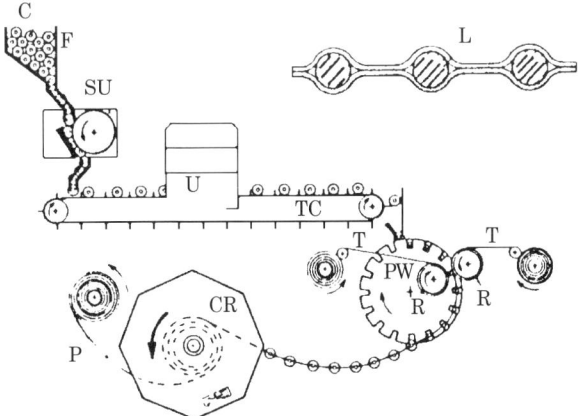

Fig. 6.4 Simplified schematic of a taping machine for axial components.
C — components; F — vibrating hopper; SU — lead straightening unit; U — orientation unit (for polar components); TC — conveying chain; PW — spacing wheel; T — pressure sensitive tape; R — pressure rolls; CR — collecting reel and P — paper strip. The detailed diagram shows how leads L are held in place by the tape.

components at a specified distance apart. On both sides of the wheel, pressure rolls apply tape strips to the component leads. The strip of components is then reeled together with a single-faced corrugated Kraft paper, which prevents contact between layers. Components can be fed to the taping machine by a standard vibrating hopper, as used for preforming equipments.

The taping machine is often mounted in line with other equipments, such as the unit used to apply the component marking in the last stage of manufacture. If set up as a separate unit, the taping machine must be fitted with a lead straightening device since the components may be received with distorted leads. Straightening is best done by passing the components between two facing steel or hard rubber rollers, with the surfaces arranged so that the straightening action starts near the lead/component junction and runs out gradually to the end of the lead.

As set up, taping units operate at a fixed component spacing which may be changed by changing the spacing wheel. The distance between the taping rolls, the taping span, is adjustable with good accuracy within a specific range. It is possible, therefore, to tape at a required span, although it is best to tape at a standard span, Figure 6.5.

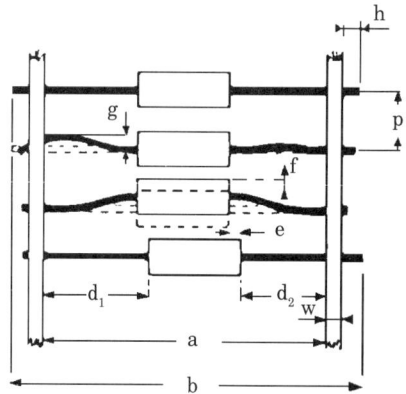

Tolerance:

a — constant within $1/16$ (1.59)
$d_1 - d_2$ — less than ± $1/16$ (1.59)
e — less than ± $1/32$ (0.8)
f — less than ± 0.050 (1.27)
g — less than ± $1/32$ (0.8)
h — less than ± $1/32$ (0.8)

	Class A	Class B	Class C
a (min.)	1 $15/16$ (49.2)	2 $5/16$ (58.7)	2 $5/16$ (58.7)
a (max.)	2 $5/16$ (58.7)	2 $11/16$ (68.3)	2 $11/16$ (68.3)
b (max.)	3 $3/4$ (95.3)	4 $1/4$ (107.9)	4 $1/4$ (107.9)
p	0.2 ± 0.015 (5.08 ± 0.38)	0.375 ± 0.015 (9.53 ± 0.38)	0.2 ± 0.015 (5.08 ± 0.38)
w	$1/4$ (6.35)	$1/4$ (6.35)	$1/4$ (6.35)

Fig. 6.5 Typical specification for taped components, based on EIA Standard RS 296. Dimensions are in inches and (in brackets) in millimetres.

Selection of spacing along the tape depends on body diameter. Up to 5.08 mm (0.2 in.), Class A or C spacing, 5.08 mm, is normally used. Class B spacing, 9.53 mm (0.375 in.), can be employed for body diameters up to 9.53 mm. Above this diameter, body taping is recommended. When a sequence of different component diameters is lead taped, 5.08 mm spacing can be used for a larger component diameter provided that the components on either side are small enough to leave sufficient clearance.

Joining tapes during a run is often necessary and is done by splicing. Tape joining by using staples is not permitted for obvious reasons. A strip of tape may be used but the strength of the join must equal that of the base tape and the thickness of the spliced section must be less than four times the thickness of one layer of the original tape.

The purchaser may stipulate that there must be no missing components in any section of the tape and will pay a premium to ensure this. If the tape is for feeding a preforming machine, up to 20-30% missing components have been accepted, although problems arise in checking quantities at Goods Receiving! If used for sequencing machines, normally the customer will demand that there is not more than one missing component in 100 or in 1,000, depending on the complexity of the sequences.

All polarised components should be orientated in the same direction when taped. The anode and cathode of diodes must be identified, for example, by tapes of different colours. The permitted number of inverted polarised components must be lower than that of missing components, as sequencing equipment can only identify an incomplete sequence. Most suppliers will guarantee, and deliver, fewer than one inverted component in 5,000 diodes. The rate is lower for electrolytic capacitors since the orientating units are less reliable.

To facilitate the use of taped components on many machines, a minimum leader of 305 mm (12 in.) must be provided before the first and after the last component. To prevent damage to components during shipping and handling, a few layers of corrugated cardboard are usually wrapped on the outside of each reel.

Reels are disposable and made from metal, chipboard or plastic. EIA Standard RS 296 dictates a core hole diameter in the range of 13.9-38.1 mm (0.547-1.5 in.) and an outside reel diameter of 76.2-355.4 mm (3-14 in.). These dimensions are not fully standardised.

Each reel will contain from 2,000 to 12,000 components, depending on their diameter and the reel dimensions. The actual quantity is marked on the reel together with such information as:

(i) customer's and/or vendor's part number;
(ii) manufacturer's name or trademark;
(iii) manufacturing date or batch number;
(iv) purchase order number.

Several other types of packaging for taped axial components have been proposed with little success. The only one which has survived to a limited extent is a long cardboard box in which the tape is folded in concertina fashion. This has the advantage of being easy to handle and store and does not require interleaving to prevent contact between consecutive layers. This

218 A Comprehensive Guide to the Design and Manufacture of Printed Board Assemblies

type of packing is sometimes called an *ammopack* as it is similar to the packing of machine gun ammunition.

6.3.2 Radial Component Taping

Radial components may have a body of any shape with leads projecting from one side only and roughly parallel to each other. The major types of these are described in Chapter 3. In the past, radial leaded components were much less numerous than axial leaded types, but the increase in integration has reduced the number of resistors and, in modern conventional circuits, discrete components are usually fairly evenly balanced between axial and radial devices.

Two or three lead devices can be readily taped for automatic assembly, provided that the three lead components have their leads in one plane, for example, double LEDs and cases such as the SOT-42, TO-92 or SOT-54 (Figure 6.6), commonly used for transistors, SCRs, Darlingtons and others. It should be noted that packages having their leads on a circle cannot be supplied on tape unless the leads are preformed to allow this. This is seldom done as it is preferable to package the device in a case which can be taped directly.

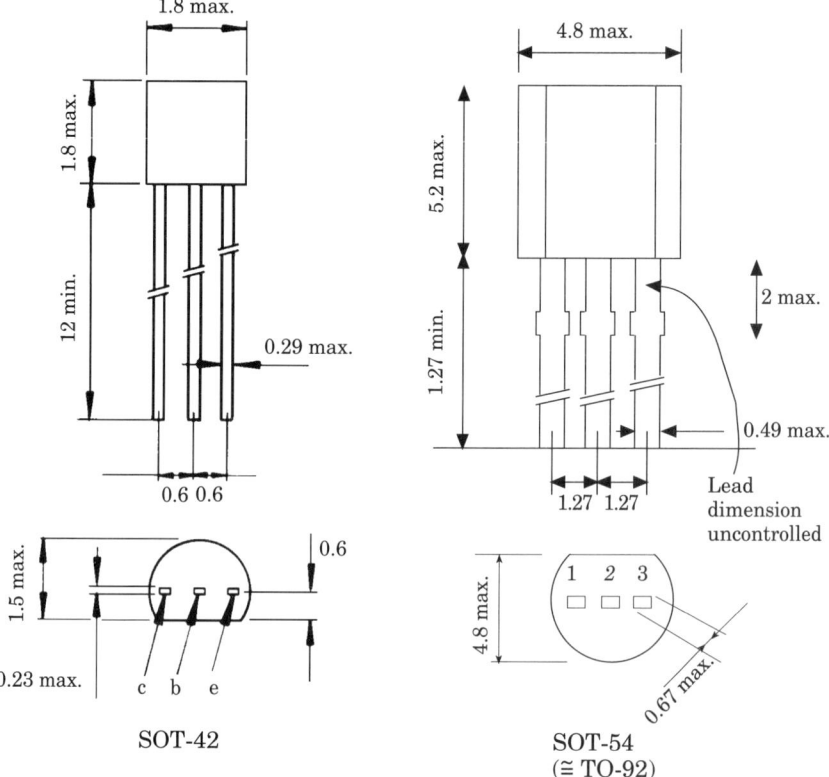

Fig. 6.6 Typical three-lead cases, suitable for tape packaging. Dimensions are in millimetres.

The tape carrier is a strip of paper, usually 18 mm or 3/4 in. (19.05 mm) wide, although this is not well standardised, perforated along the centre line with sprocket holes spaced 1.27 mm (0.05 in.) between centres. The component leads are placed on the strip with the component body midway between adjacent sprocket holes, Figure 6.7, and stuck down with an adhesive tape. The two leads, or outer leads, are an equal distance from the adjacent holes. The tape, therefore, can be fed by a sprocket, which engages the holes, and the component body and leads will be in a precise position with reference to the sprocket mechanism.

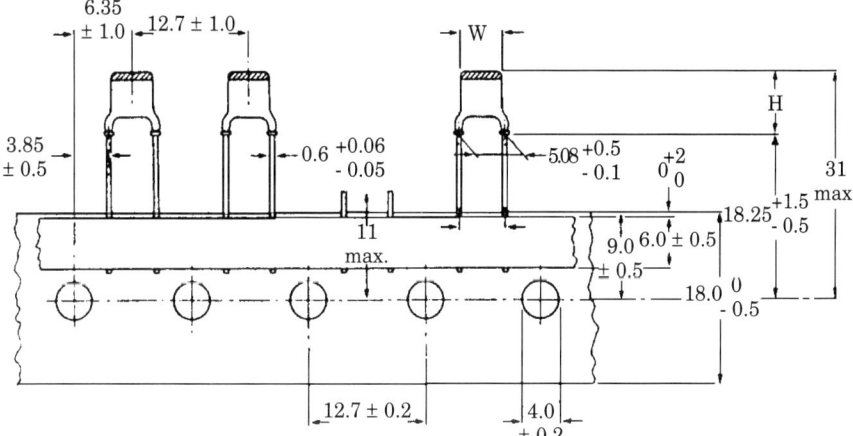

Fig. 6.7 Miniature ceramic capacitors secured to a carrier tape by a narrow adhesive tape. The component body is centred between adjacent sprocket holes. Dimensions are in millimetres.

The tape is then rolled on to a disposable reel or folded, concertina fashion, into a box, Figure 6.8. Depending on size, a reel may carry 100 to 5000 devices, while a box usually holds a slightly larger quantity. The package is labelled with quantity, part number and so on, as described for axial devices.

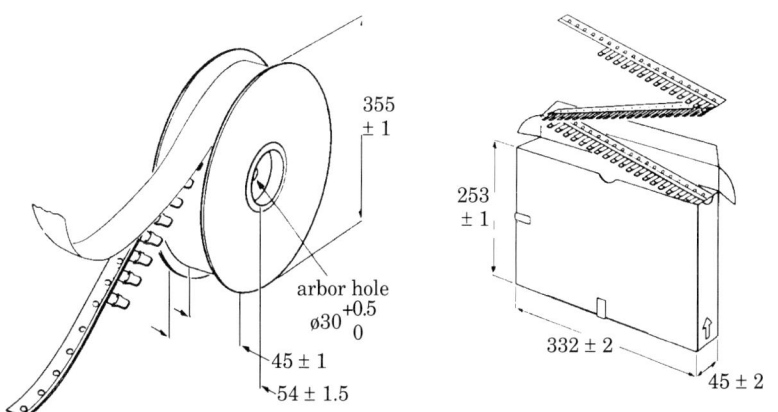

Fig. 6.8 Packing taped radial lead components. Dimensions are in millimetres.

220 A Comprehensive Guide to the Design and Manufacture of
Printed Board Assemblies

Several types of equipment can separate the components from the tape and shape the leads to form stand-offs before insertion. When the component is difficult to preform with automatic insertion machines, the component manufacturer may supply the preformed parts on tape, Figure 6.9. Axial leaded components to be mounted vertically can be treated as radial ones and be preformed by the component manufacturer before taping. While such components are usually sold at a premium price in comparison with those taped as axial devices, it is not easy to preform axial components for vertical mounting starting from normally taped supplies.

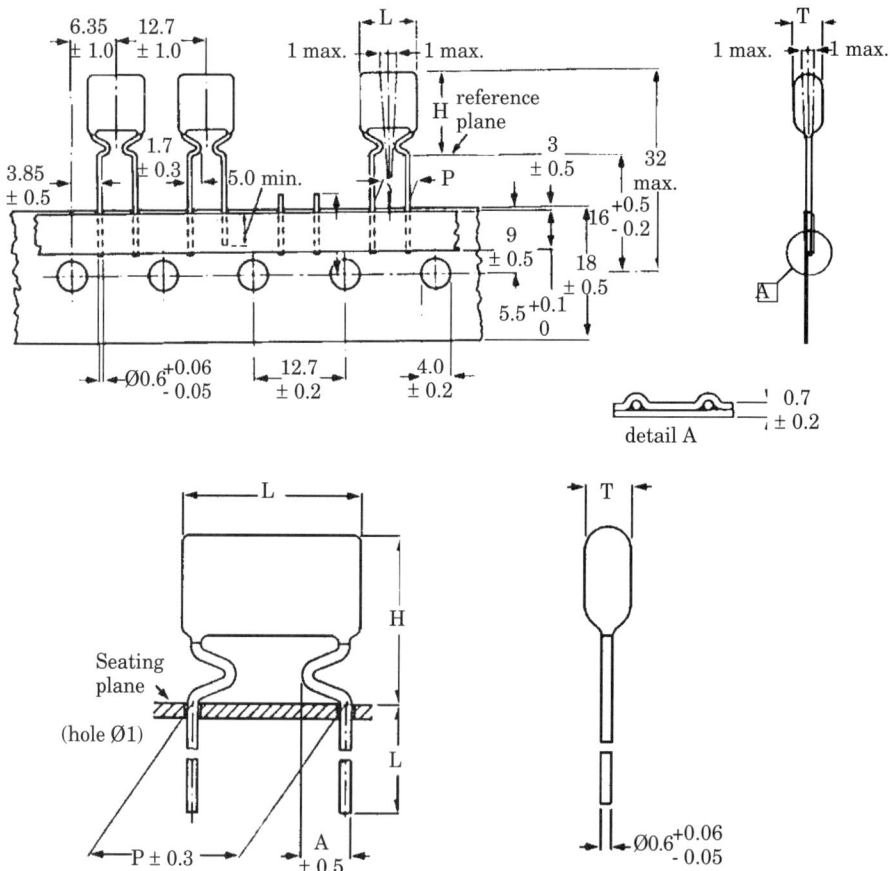

Fig. 6.9 Metallised film capacitors with leads preformed prior to tape packaging. All dimensions in millimetres.

To minimise problems with shorting, arising from distortion of the long lead of vertically mounted axial lead components, the long lead should be insulated. An insulating sleeve, as used with electrolytics, will help. Resistors can be obtained from suppliers with the long lead coated with the same resin as the body of the component.

6.3.3 DIP Components

Dual-in-line packages are relatively easy to handle because they have large bodies and short and robust leads. If placed on a rectangular inclined slide of appropriate width, they are easily moved by gravity. For this reason, the standard method of supply is in a plain tube or stick magazine, Figure 6.10, made from inexpensive plastic and closed at the ends by rubber or plastic plugs. If the component is sensitive to static electricity, an antistatic material or coating is used. Very often the stick may be opened at both ends. This is desirable as some insertion machines are unable to rotate the DIP for correct orientation and the magazine must be loaded from the correct side. The aluminium sticks are now seldom used for initial supply, because of cost, but are still employed as refillable magazines on some types of insertion machine.

Component	Size	Shape	Sectional view of package
IFT coil	5, 7, 10 square		
IC (DIP)	0.3 in., 0.6 in.	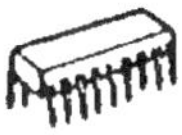	
IC (SIP)	5~9 P		
Trimmer potentiometer	8, 10 Type		
Connector	2 P~12 P		
Trimmer capacitor	Mould type		
Module resistor		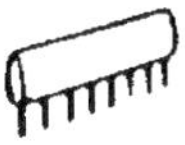	
IC socket	0.3 in., 0.6 in.		
Light touch switch	5, 8 Type		

Fig. 6.10 Typical stick carriers for a variety of components. The carrier is normally 508 to 610 mm (20 to 24 in.) long and shaped to hold the components without damaging the leads.

Two basic types of stick carrier are used for DIPs, depending on the width of the DIP body. The number of pins is not relevant. The small stick may hold components with 6, 8, 14, 16 or 20 leads, the number of components held changing with the body length. For the TO-116 package, the usual carriers hold 30 devices.

In the past, highly sensitive and expensive components were often shipped in boxes, with the leads embedded in a sheet of conductive plastic foam. Devices packed in this way must be mounted manually. Because of this and the reduction in device costs and more robust build, this form of packing is used only for prototypes or for very small quantities of custom-made devices.

Bulk packing of DIPs is not recommended as leads will be damaged and it is very difficult to load these components by a vibratory bowl.

The dual-in-line sockets used for DIP mounting may be treated in the same way as the DIPs, that is, supplied in the same types of tube and mounted by the same equipment.

6.3.4 Odd Components

Components with an irregular shape are difficult to handle and pack. There is no standard method of packing but the current most popular methods fall into one of four classes:

— *linear*, a shaped rail, tube, or stick carrier;
— *tray*, an embossed tray in which the components can be placed in an X-Y network;
— *tape*, a paper or plastic tape on which the components are placed at measured intervals.
— *bulk*, provided that a vibrating hopper or bowl is suitable for loading in correct orientation to a rail feeding the machine head.

Each of these methods may be used with automatic insertion machines, although particular machines may impose restrictions on the type of packing which can be employed.

6.4 MACHINE TYPES

While there is no accepted classification of automatic pin-in-hole (PIH) insertion machines, all equipments available on the market fall into particular basic types.

The operations of an automatic PIH inserter may be divided into the following major steps:

(i) selecting the right component from those available in the feeder for a specific insertion position;
(ii) picking up the component and delivering it to the insertion head;
(iii) moving the board (or the head) so that board and head are in the correct positions relative to each other;
(iv) inserting the component leads into the correct holes and fixing them.

Step (i) is the most critical and determines the size, cost and performance of an insertion equipment more than all the others.

6.4.1 Dedicated Inserters

An equipment dedicated to insertion of a single type of component can have a fairly simple structure, as it does not perform any selection. In general, it will be more economical to operate than a machine which can handle several types of device but has been derated to work with only one.

In essence, a dedicated automatic insertion machine is a numerically controlled X-Y table. This moves the board holder to position the hole(s) into which the component is to be inserted exactly under the head which performs the insertion. Such simple machines can give high productivity for limited investment, but, in most assembly shops, it seldom happens that one component is used on all boards in sufficient quantity to justify a dedicated insertion machine.

The most typical example is the vertically mounted square pin used for wire wrapping posts on large backplanes, as male connectors, test point terminals and so on. Conventional insertion machines for, say, radial components could not insert such pins, even if suitably packed square pins can be obtained. In this case a dedicated inserter can be less expensive, much faster and can perform additional operations.

One such machine is dedicated to the fully automatic insertion of square pins. The feed is a coil of plain or plated brass wire of section 1.0 x 1.0 or 1.2 x 1.2 mm with corner radii of 0.1 mm or less. The machine cuts off the appropriate length of wire and chamfers it as necessary before inserting it into the hole. The total length of the pin, 16 mm (0.63 in.) maximum, and its projection from the soldering side, 0.6 to 12 mm (0.024 in. to 0.47 in.), are prefixed by the manufacturer to customer requirements. The insertion positions are computer controlled, instructions being fed to the controller by a software program.

The machine cost reflects the simplicity of construction. Apart from the CNC (*computer numerically controlled*) electronics for the X-Y table, the insertion head and cutting/chamfering unit are straightforward. The feeder is elementary, advancing a predetermined length of wire, always in the same position. Material costs are minimised, as very little of the wire is wasted, and machine productivity is up to 6,000 pins/hour.

On the other hand, the machine is completely dedicated to pinning. Add-on units for loading and unloading boards are available, allowing the machine to be incorporated into a larger assembly line.

Another equipment starts with a reel of tinned annealed copper wire, 0.6 ± 0.02 mm diameter, and inserts jumper wires on to PBs. The basic structure is similar to that of the pinning machine with a head which cuts the wire to length, forms it to shape, inserts the jumper into the correct holes and clinches the ends on the soldering side of the PB to hold it firmly in place until soldered. The machine is fully programmed by the software for the co-ordinates of the jumper and for jumper length. Being limited to three possible lengths, the pitch of the jumper holes must be 10, 12.5 or 15 mm (0.39, 0.49 or 0.59 in.). While this simplifies the machine construction, it does impose tight restrictions on the board designer, see Chapter 8.

Insertion speed is up to 6,000 jumpers/hour and the head can rotate through 90° to insert jumpers in the X and Y axes. A detector for insertion errors is included. Being inexpensive when compared with a variable pitch inserter for axial components, this type of machine is of interest to shops assembling large numbers of single-sided PBs with several jumpers on each.

6.4.2 Off-line Sequencing

Insertion machines can be simplified if fed with the correct component in the right sequence. The technique is applicable to reasonably similar component outlines when presented to the machine in the same way and in the correct sequence.

Components can seldom be obtained from a supplier in this form and the assembler has to make his own sequence off-line. This means two equipments instead of one, more work in progress, more planning activity and more operators. To offset this, the machines are simpler than pick-and-place machines and the total cost — and downtime — is usually less than that of a single unit.

Off-line sequencing is still used for axial components, which can be untaped, put into the proper order, and retaped in that order for feeding the inserter.

6.4.2.1 AXIAL COMPONENT SEQUENCING

Sequenced components must be kept in a fixed position relative to each other so that they can be fed at high speed to insertion equipment. While appropriate carriers can be used, lead taping is the simplest solution. The equipment needed is, therefore, a sequencing unit and a taping unit.

In the simplest form a taping unit may be employed with the components placed in small bins as for manual assembly, see Chapter 5. A slightly more elegant solution is illustrated in Figure 6.11. If the sequence needed is

Fig. 6.11 Manual sequencing of axial components. The external ring, which may be freely rotated, carries 84 component bins. The operator picks up components in the right sequence and places them on the taping unit in the centre. (Courtesy of Universal)

A-B-B-C-A-C-D-D-C-E, that is, ten components of five types in the order given, the operator picks up component A and places it on the chain of the taping unit. Two B components are now placed on the chain and so on. Manual sequencing is used for small production runs, test runs, or, in some cases, to prepare a few sequences which have many types of component.

High volume sequencing requires fully automatic equipment, both to speed up the work and reduce the incidence of operator error. Such equipment will have several component stations, a transfer chain, a taping unit and an electronic control unit. The number of stations is usually modular from 20 to 100 or more, in modules of 20 heads. A typical equipment will have 40 or 60 stations. The number of components composing the sequence is unlimited but the maximum number of component types equals the number of stations. If a polarised device is to be placed with opposite orientations, it must be treated as two different components. One or more stations may be equipped with a special head to prepare jumper wires. Each component station has a reel holder and a head which is fed by the reel. When triggered by the sequence control unit, the head cuts the leads of one component, close to the tape, so that it drops on to the transfer chain. The chain is synchronised with the heads so that each component falls into a specific position on the chain. This moves step by step under the heads, collects the components as they fall, and takes them in the right order to the taping unit.

Considering the sequence A-B-B-C-A-C-D-D-C-E, if component A's reel is loaded on head one, B is on the second, C on the sixth, D on the seventh and E on the ninth, the numeric sequence is 1-2-2-6-1-6-7-7-6-9. This is entered to the electronic controller by punched tape, magnetic tape, disk or keyboard. As the control unit generates each pulse at the appropriate time, the desired sequence is obtained, Figure 6.12. The program entered will control the number of identical sequences to be produced. When a sensor is provided to detect missed components, the controller will stop the machine and display on the control panel the number of a head which has not worked correctly. The panel will also display the most likely reason for the missed operation (empty reel, jammed head, etc.) so that, when the operator has completed the sequence manually, he can clear the fault and restart.

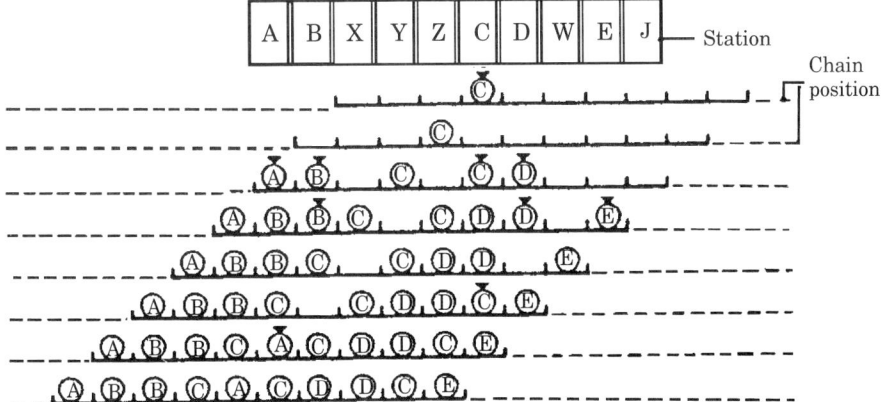

Fig. 6.12 Working cycle of an automatic sequencing machine for axial components. Arrows indicate when a component will drop on to the conveying chain.

Manual completion of a sequence creates the risk of reduced quality. If a sequencer stops only occasionally, say once every 1,000 components, the total rate of errors (components missing, wrong, supply reels poorly taped, damaged, etc.) may be as low as 50 ppm. If stops are more frequent, for instance once every 100 components, the total defect rate can be as high as 0.5-1%. Poor quality taping has a dramatic effect on productivity and, if component reels of adequate quality cannot be obtained, it is better to exclude these from the automatic sequence preparation. If there are many stops for missed components, there is an increasing probability that the operator will complete the sequence with a wrong component and such errors cannot be found by sample inspection.

The cutting heads are the major weak point of these machines. The transfer chain may also cause stoppages unless the carrier design is such that the leads cannot jump out because of vibration or other influences. Some machines use a positive placement system which is more reliable than the traditional units which allow the component to drop on to the conveyor by gravity.

Poor or badly adjusted taping units can lead to insertion defects, especially as the checks for sequence are carried out on most machines before the taping station.

If an axial component is assembled by automatic sequencing followed by insertion, the leads must be extra long because they are cut twice, first for sequencing and then for preforming/insertion. As the commonly used tape is 6.35 mm (0.25 in.) wide, each lead is reduced by 12.70 mm (0.5 in.). In addition, to minimise problems, the lead should be cut at least 0.25 mm (0.01 in.) from the tape and the maximum insertion span P_{max} for automatic insertion becomes much less than is possible using manual insertion, as shown in Figure 6.13.

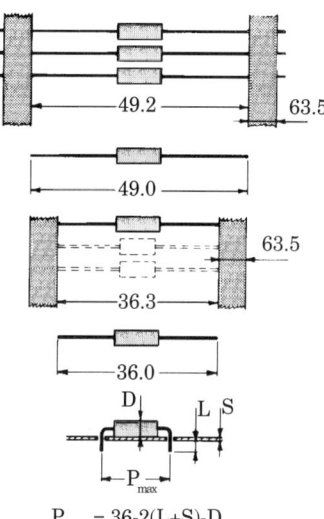

$P_{max} = 36-2(L+S)-D$

Fig. 6.13 Maximum insertion span (P_{max}) of a Class A taped component inserted automatically. Dimensions are in millimetres.

Components can be taped with leads spaced at 5.08 or 9.52 mm (0.2 or 0.375 in.). This means two different types of cutting head. These are easily interchangeable but time is lost in changing and spares for both types must be held.

When setting for new sequences, if all required types of component are present on the machine, it is necessary only to feed a new program to the control unit. If other types, or values, of component are needed, either reels can be changed or any unused heads can be loaded. Planning for sequencing is essential to reduce the number of reels to be changed and to minimise setting-up times. Setting-up times can be reduced by using auxiliary equipment for quick changing reels at the expense of higher investment on equipment, more off-line work and increased component inventories.

A 40 head sequencer can deal with 12,000 components/hour and one operator can run two machines. Allowing for setting-up time and machine downtime for maintenance, the productivity should not be less than 8,000 components/hour/machine. If the cost of work is US $14/hour x $1/2$ and amortisation and maintenance cost $11/hour, then the sequencing cost is about $0.2/100 components plus the cost of the tape.

A good quality tape can cost as much as 10 cents/100 components if taped at 5.08 mm. If the sequence includes one or more larger diameter devices which means using the wider spacing of 9.52 mm, it can pay to exclude these from the sequence and insert manually or in a sequence dedicated to the larger components. In the latter case, insertion will need two passes per board.

The machine now described is a good example of an off-line axial sequencer. Introduced many years ago, this machine was progressively refined as a result of the experience gained from many installations. The basic machine consists of a 20-station component dispensing unit, a taping unit, and a control unit which can be a satellite of a larger computer. In a configuration which includes 160 stations (8 modules), a production rate of 25,000 components/hour is achieved in a length of only 10.97 metres (36 feet).

This machine has quick change reels or can use ammopacks and the heads are quickly adjusted for different tape widths. Components are singled out by strong air-operated cutters and are positively placed on the transfer conveyor. An optical sensor may be fitted to check that the component has been fed to the conveyor and, if necessary, fire the head again until the component is dispensed.

As well as including features to reduce time lost on changing production runs, the machine can be fitted with an on-line component verifier. This fits over the conveyor, between the dispensing heads and the pitch wheel. The device checks components for parameters such as opens, shorts, resistance, capacitance, voltage and leakage and gives good results with resistors, capacitors, diodes and zeners. The unit is software programmed and can make multiple tests on each component. A component which is wrong or outside the set tolerances causes the sequencer to stop so that the fault can be remedied. Ignoring the time taken to replace faulty items found, the testing slows the sequencer by some 10%. This 'lost' time is justified by the gain in overall efficiency as the reject rate at in-circuit test may be dramatically high if there is no 'pre-filter'.

6.4.2.2 SEQUENCE VERIFIERS

Stand-alone testers are available to check the component sequence off-line after taping and reeling, either on a sample or on a 100% basis. The taped components are fed from their reel over a motor driven insulating pitch wheel on to a take-off reel. The insulating pitch wheel holds the components, one by one, in a fixed position where they are contacted by two test probes and checked for the required parameters. While the test position is such that the operator has easy access to it, replacement of faulty components is a tedious and time-consuming operation because the component is taped.

This equipment can test up to about 36,000 components/hour but the high testing rate is largely negated by the difficulty of replacing faulty components. On-line testing is preferable unless the average quality of the produced sequences is good and a sample test only is needed to avoid expensive rework due to human error on the sequencers, such as using a wrong dispenser reel or a wrong program step.

A sequence verifier is a useful tool in very large shops where several sequencing machines are installed. In smaller shops, it can be used for checking sequences and for normal goods receiving testing of non-sequenced components.

6.4.3 In-line Sequencing

Most types of insertion machine are equipped with a sequencing unit which prepares the correct sequence as it is required and forwards it to the insertion head. The required types of component are stacked in their packing in a feeder and the sequencing unit must be able to pick up each of them when needed on the PCB. For DIPs, in-line sequencing is almost obligatory, as there is no simple way of packing and handling a long sequence of such devices.

6.4.3.1 CHAIN SEQUENCING

For axial components, the sequencers already described may be used with the differences that (a) its speed is governed by the speed of the insertion head and (b) the sequenced components are fed directly to the insertion head in a fixed position instead of being taped again. This eliminates the cost, and time, of taping and allows a larger insertion span for the component, Figure 6.13.

The major disadvantages are that the machine is more complex, more expensive and less reliable. Stoppage of the sequencing unit stops the whole equipment, drastically reducing the theoretical throughput. This also occurs when component types have to be changed to cater for a new type of assembly.

Taped radial components can be handled in similar fashion but cannot be transported to the insertion head by 'V' notches. On removal from the tape, the component must be firmly held for transport. The chain depicted in Figure 6.14 is typical, being made up from a number of pliers or clips, each capable of holding all component types. In one cycle of the chain, each plier passes in front of each component dispenser and can be loaded with the correct one. The components are carried in sequence to a fixed point at which the insertion head picks them up. The length of the sequence is not limited by the number of elements on the chain, since there is a continuous flow from the feeder station to the insertion head.

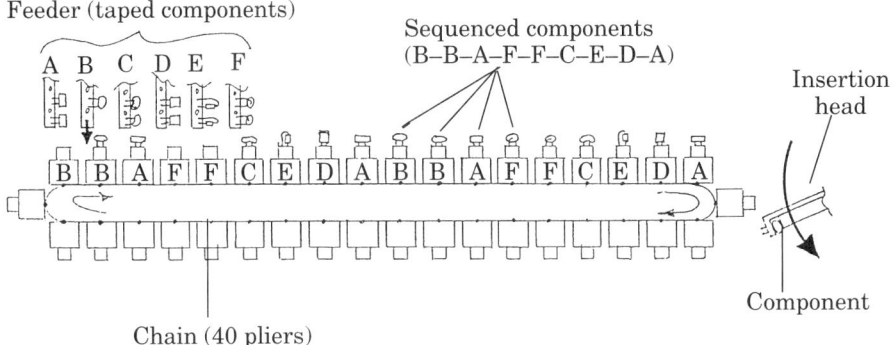

Fig. 6.14 Transport chain for taped radial and similar components. The components are gripped by pliers until picked up by the insertion head.

With this type of machine, programming is more simple. The sequence of insertion on the board is defined and the identity of the feeder for each position is stated. The equipment will then select the right component at the right time to be fed to the insertion head.

The weak point in these machines is the feeder or picking stage, both because the component taping may be poor or may differ from supplier to supplier and because there are many picking stations working in parallel.

This type of equipment is also known as a *multi-pick and single place machine*.

6.4.3.2 AIR DELIVERY

Air delivery, Figure 6.15, is an interesting variant of the chain transfer system. The component dispenser reels are placed along a tube in which the components are moved by air to a fixed point at which they are picked up by the insertion head. As a component is picked up, the next in sequence is parted from the tape and injected into the delivery tube. A blast of air moves it at high speed to the pick-up point. The component feeders can be placed on both sides of the pipe, reducing the overall length needed for a given number of stations and making reel changes simpler.

A complete feeder, with all components and cutting units connected to the delivery pipe, may be prepared off-line so that machine set-up for a new production run is reduced to disconnection of one tube and connection of the new one.

The method is cheap and rational in concept but has not as yet been developed fully by a major manufacturer of insertion equipment. The

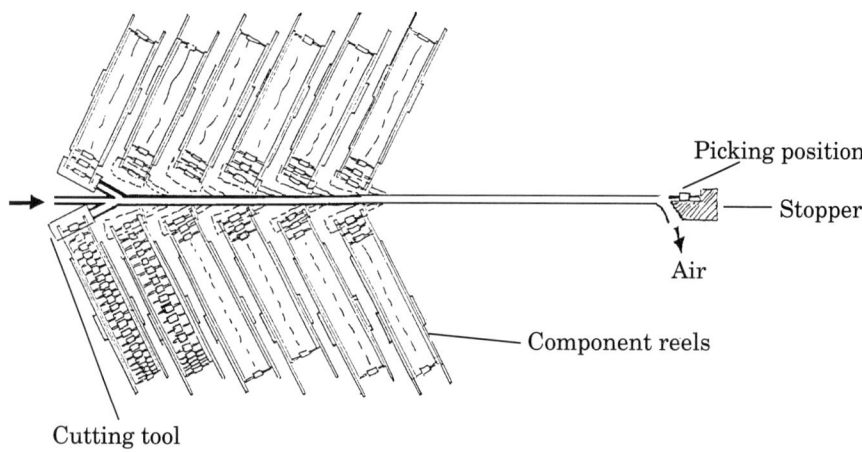

Fig. 6.15 Air delivery of components to the insertion head is cheap but does not yet work reliably on all body sizes.

primary reason appears to be the difficulty in refining the concept, so that it works reliably on all body sizes, from a jumper to a large capacitor. A contributing factor may be the emphasis being placed on development of assembly equipment for surface mounting.

6.4.3.3 SHUTTLE DELIVERY

Shuttle delivery is used on many high speed insertion machines for feeding the correct sequence of dual-in-line packages. The component feeder is stationary and consists of a bank of stick magazines connected to a mechanism which allows one DIP to drop when triggered by the controller.

The shuttle has a 'pocket' to receive the device and moves on a rail from which it can access all magazines, Figure 6.16. (If the inserter has to process two different body sizes, e.g., 7.62 and 15.24 mm wide, the shuttle can have two pockets, to be used one at a time, depending on the component.) On instruction from the computer control, the shuttle moves to the designated magazine, the release is triggered, and a component drops into the pocket. The shuttle moves to the insertion head and releases the component. The time required to move to a magazine, receive the component and deliver it to the pick-up point, depends largely on the distance between the relevant magazine and the head. The travel of the shuttle is very fast and the machine program can be written to minimise delays, but, on occasion, the machine speed can be adversely affected. To cope with this, at least one machine manufacturer supplies a double shuttle system.

6.4.3.4 GRAVITY CHUTE

This low cost and reliable method is widely used by at least one major manufacturer of DIP inserters. It relies on gravity, assisted by the machine vibrations, to transfer the components from the feeders to the pick-up point.

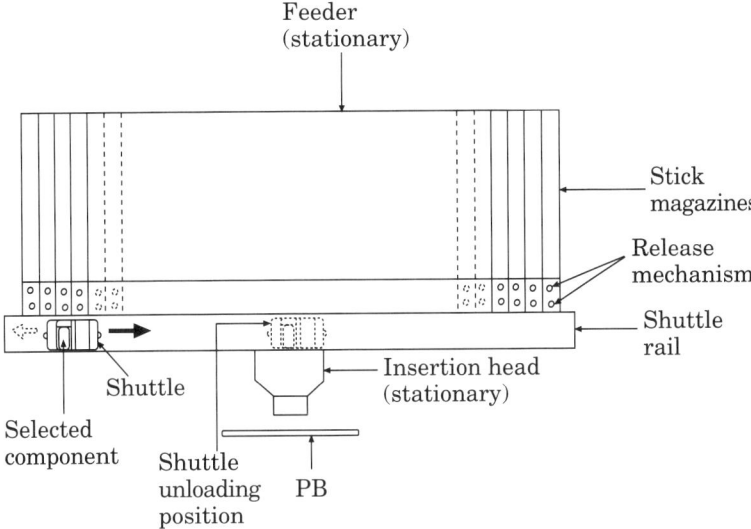

Fig. 6.16 The shuttle delivery system is widely used for sequencing dual-in-line and similar packages.

As shown in Figure 6.17, the stick magazines are positioned along a rail, inclined at 45° to the horizontal. On receipt of a pulse from the controller, a component is released on to the track. To be effective, the gravity feed rail

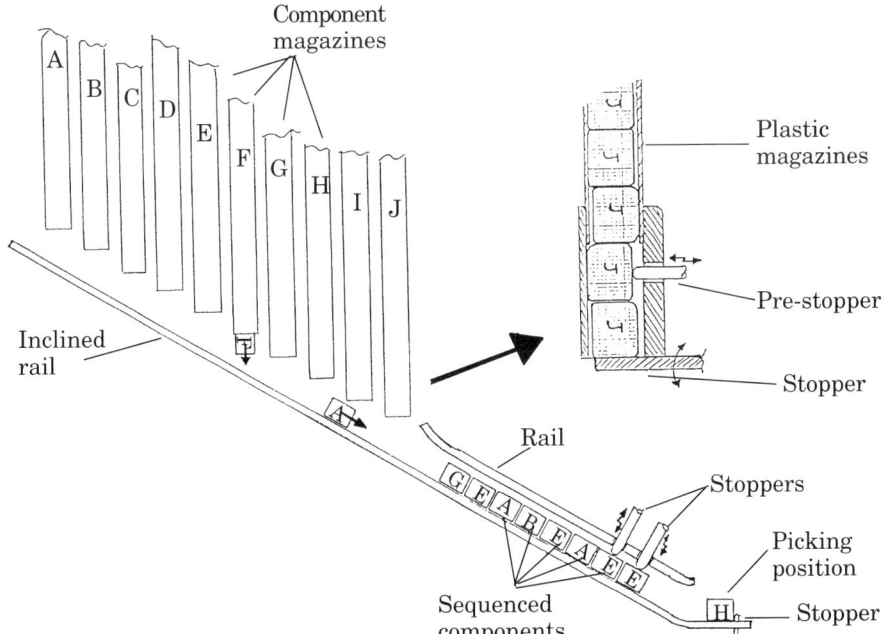

Fig. 6.17 Gravity feed for DIPs and similar components is simple and reliable.

must be reasonably short as components from magazines far from the pick-up point can take a long time to slide along the rail. This can be overcome to some degree by sorting components in advance of starting insertion, so that a ready-to-insert sequence is always available. It is also possible to have 2-4 rails supplying the head through a secondary sorting mechanism.

6.4.3.5 SORT AND PLACE

An alternative for DIPs, still popular on the more simple equipments, mounts the magazine sticks on a rail so that they can be moved over the head and drop a device on to the head feed.

This approach suffers from the penalties of slow speed of magazine movement, giving a low insertion rate, and the restricted number of components which can be held by the moving feeder. One large manufacturer of insertion equipment overcame such difficulties by using a compact feeder, which moves at high speed and holds up to 15 sticks per position, with automatic ejection of empty ones. As the components are actively picked up by the head instead of being dropped, this machine is more correctly classified as a *pick-and-place* machine.

6.4.3.6 PICK-AND-PLACE

This approach uses a moving feeder which moves in the desired sequence to present the correct component to the head, which picks it up and inserts it. It is used for virtually all types of component, including SMCs, see Chapter 7.

With a few exceptions, the component holders are mounted on a rotating drum (axis vertical) feeder which can rotate in both directions to reduce positioning time. This arrangement is cheap and simple for components packed in long sticks, e.g., DIPs, but non-standard packing may need to be adopted for discrete components.

Most insertion machines for odd-shaped components use this principle as it is usually easier to pick up the device than to feed it to the insertion head.

6.4.4 X-Y Tables and Rotary Tables

Most insertion machines for conventional assembly have a fixed insertion head so that the PB must move to allow the component to be placed in the correct holes. This is achieved by mounting the PB on a table which is moved in two orthogonal axes, termed the X and Y axes, by numerically controlled stepping motors driving 'endless' worm screws. The PB fixture holds the board in a precise position with respect to the table so that datum points can be set exactly.

X-Y or co-ordinate tables are well established and the units incorporated in insertion machines have no special requirements other than precision, speed (often greater than 20 metres/minute), smooth movement and reliability. Most tables give no problems and are capable of positional accuracy equivalent to, or better than, that of modern PB drills. An important factor, which must be known, is the value of one step, that is, the minimum increment between two positions which can be recognised by the machine. This will be quoted in metric or imperial units depending on whether the table has metric or imperial lead screws. Many modern control units can be switched from indexing and reading in metric values to imperial values, but in older machines it may be necessary to convert. Whenever possible the machine should be programmed using the original unit.

The total movement allowed in the X and Y directions governs the maximum size of board which can be inserted. Inserters are available which will handle boards up to about 600 x 600 mm, of which the insertable area will be about 510 x 510 mm, although most large machines can accept an insertable area no larger than 455 x 455 mm. The smaller machines may have a usable area of no more than 305 x 305 mm. (The insertable or usable area is always less than the board area, because of the need to provide reference holes for fixing the board to the table. The size of these holes and the non-insertable area resulting is specified by the machine supplier.)

Axial components quite often have to be inserted in two directions at 90° to one another. Suitable devices are available to rotate the board automatically as needed. Such tables are sometimes called X-Y-θ tables, where the Greek letter theta refers to the rotation angle. While rotation is fairly fast, about 1 second for 90°, it is normal to program to insert all components in one direction and then all those at 90°. A few tables can rotate by 180° and 270° (or in four steps of 90°) in order to insert polar components in a direction at 180° to that for which it was sequenced and possibly taped.

On machines with automatic loading and unloading of boards, the rotary table feature may be relatively expensive, although its cost should be recovered in a reasonable time. If the machine is loaded manually, the prime cost is less and recovery faster. However, the operator must always stop the machine to unload finished boards and load the next one. If the rotary table is large enough to hold two PBs and can rotate through 360° in four 90° increments, boards may be changed while the machine is still running. It is possible to insert the components in the X direction on both boards at the same time and those in the Y direction on one board at a time while the other, completed, board is being replaced with an unpopulated PB. This works because the two boards together are treated as one sequence which can become very long.

If polar components are on the board, some will be needed twice on the sequencer, as one board is reversed with respect to the other. Either, if acceptable to the designer, all polar components can be placed in the area which is inserted one board at a time, or, if the table will hold four boards, the same sequence can be followed for each board and the problem is resolved.

Very close co-operation between the PCB designer and the production engineer is always essential to avoid manufacturing costs arising from the position of a few components which could have been placed elsewhere.

On most machines, the turntable is a simple mechanism driven by a reliable but inexpensive motor. Alignment at 90°, or multiples, is assured by precision stops. A very few types are fitted with a numerically controlled motor which can rotate the table in small increments, down to 1 second (dividing the 360° into 21,600 steps). Such machines can insert components at any angle.

A rotating head can be used instead of a rotating table, a common solution for inserters of radial and DIP components.

6.4.5 Positioning Aids — Optical Verifiers

Automatic insertion gives good results only when tolerances are properly specified and controlled throughout the process, from production of a PB artwork to insertion of the components. In particular, insertion may become difficult when the radial clearance, hole to lead, is less than 0.5 mm (0.02 in.) for a number of reasons, such as:

  (i) off-centre leads due to imperfect straightening or distortion after forming;
 (ii) burred leads due to blunt or incorrectly sharpened cutters;
(iii) burrs or projections in the holes from the PB manufacturing process;
 (iv) misalignment of lead and hole because the board is warped;
  (v) off-centring of the holes with reference to their theoretical position in relation to the insertion head.

Of these, the last item is the most relevant. The insertion holes can be off-centred by a significant amount for a variety of reasons, in particular: (a) tolerance between the insertion holes and the reference holes; (b) tolerance on the diameter of the reference holes; (c) tolerance of the board fixture with reference to the X-Y table; and (d) positioning error of the X-Y table. (To ensure first-time insertion, the PCIF guide to PCB design and manufacture recommends that the minimum hole diameter should be not less than $W + L + V + M$, where W = maximum component lead diameter, L = maximum diameter of board locating holes - minimum diameter of locating pin, V = total variation of component hole to datum hole and M = total allowable variation of insertion machine.[2])

Optical verification systems are now available which measure the actual position of an insertion hole and, when necessary, correct the position of the table. The check is usually made on only one hole for discrete components, with checks on two holes being preferred for DIPs.

Check, feedback and correction of individual holes may slow the insertion operation by too much, while the larger part of the error is almost invariably repetitive and due to shrinkage of the PB laminate, especially if paper based material is specified. In such a case, if each lot of boards is made from the same batch of laminate, using the same processes, the chances are that the shrinkage will be reasonably constant for a given lot. It may then be adequate to make several checks in critical positions on a few boards, calculate the average off-centring, and correct all theoretical positions by programming in a standard offset. This usually recovers most of the error and allows satisfactory insertion.

In the more modern equipments, the averaging operation and correction are assisted by the electronic controls so that the operator has only to decide on the critical positions for checking (e.g., the centre, borders and some holes where clearance is at a minimum). In some cases, the optical verifier will check for board rotation or skew and take that into account in its calculation.

The optical verifier has other uses. On some machines it can be used for feeding in the insertion program by means of teach-in utility software. On other machines it may be used for automatic centring on the whole board by checking the true position of markers and correcting the insertion co-ordinates.

6.4.6 Cutting and Clinching

As for manual assembly, the final operation of the insertion cycle is to trim the leads on the soldering side and, as required, clinch them.

The cutting operation is performed by air operated cutters placed under the board holder. During the insertion cycle they rise to contact the PB. At this stage they also support the board against pressure from the insertion head. When the component has been placed, the cutters trim the leads. The same unit may also clinch the leads to keep the component in place during the insertion of the other components and subsequent handling for transport, manual completion and soldering.

Depending on the machine, the projection of the leads from the solder side of the PB may be preset at a value fixed typically between 0.9 mm and 1.5 mm (0.035 in.-0.06 in.), or adjusted for each component to a value dictated by the program for each insertion position, say from 1.27 mm to 2.54 mm (0.05 in. to 0.1 in.). The clinching angle can also either be set within a fixed range, usually between 30° and 40°, or programmed component by component to a value selected in the range 0° to 90°. The best equipments give a choice, preset or programmable, between inward and outward clinching, see Figures 5.12 and 5.13. The least expensive machines offer no choice and may also impose the constraint that all leads must be clinched at a fixed angle selected from a set of values such as 0° (lead flush with the conductor), 30°, 60° and 90° (no clinching).

If the insertion head can rotate to insert components at different angles, the cutting/clinching unit must rotate accordingly. If a turntable is used the cutting/clinching unit can be fixed, as it is always vertically under the head.

DIP leads only require to be trimmed when the board is thin and/or the maximum allowed projection is less than 2.54 mm. Most DIP insertion machines are provided with cutting and clinching units, which can normally be preset or programmed. Outward clinching is preferred for DIPs as it utilises the natural spring of the pins. Standard machine tooling provides for cutting, if needed, and clinching of all leads, though this makes the replacement of a faulty DIP much more difficult. Selective clinching, which clinches only two diagonally opposed leads or the four corner leads, is a good compromise available on some equipments when requested.

Clinching is a delicate operation which can impose serious stress on lead to body attachment. This can reduce component reliability, especially when the leads are too strong or the body is fragile. In particular, flush clinching should be avoided as far as is possible since, on automatic equipment, the insertion head will exert strong pressure on the body of the component to allow the leads to be clinched flush. While the equipment manufacturer may claim no damage will occur over a wide range of situations, the production engineer must check all types of component and ensure that the equipment is correctly adjusted and checked at appropriate intervals.

Another factor comes into play when the component leads to be cut and clinched are on a mixed technology surface mount board.[3] The presence of SMCs on the surface from which the component leads protrude makes post-soldering mass lead trimming impractical and, in most cases, it will still be necessary to cut and clinch the protruding leads. Advances in cut and clinch technology in the last few years enable the production engineer to specify individual lead lengths and clinch angles and program these into machines having smaller cutter footprints. However, thorough evaluation of the need and the available equipment is essential.

6.4.7 Insertion Verification On-line

The most interesting improvement in PIH insertion machines — in addition to higher production rates — is the increase in machine reliability. In the early seventies, an axial component inserter was considered to be working properly if the total reject rate was around 1-2%. This is measured as the ratio of insertion defects (components wrong, missing, misinserted, damaged, etc.) to the total number of insertions at the end of the assembly process. It is expressed in per cent or in parts per million (1% = 10,000 ppm). In the mid-eighties, the figure was 150-500 ppm and, in the nineties, 100 ppm or better is expected.

Not only are machines and components better, the insertion process can be controlled on a closed loop basis. Most inserters can be fitted with a sensor which detects a lead passing through the hole by contacting it on both sides of the PB. Either the end of the insertion head or the anvil of the clinching tool consists of two parts electrically insulated from each other so that the inserted lead closes two independent circuits.

In operation, as soon as the leads are pushed through the holes, the circuits are activated to monitor the insertion process. At the end of the cycle, if both circuits have not been closed because one or more leads failed to contact the anvil, an error signal is generated. Usually the machine will stop so that the operator is alerted to repair the board and restart the machine. Alternatively, the machine will carry on with the sequence and record the error for later corrective action.

Originally introduced for axial components, insertion verification was then extended to radial components and, later, to DIP inserters. Tooling is much more complex for DIPs as all leads are checked independently for proper insertion. This does, however, eliminate defects such as leads bent under the body, Figure 6.18, which are difficult to recognise on visual inspection.

The output from DIP inserters, in addition to the usual problems, suffers more from human error than that of other component inserters. ICs are generally packed in small quantities, e.g., 30 per stick, and many sticks have to be loaded (manually!) on to the inserter. The similarity of packs and component packages in outline and colours contributes to the probability of an error.

An optional on-line verifier is available for some machines which tests each component before insertion and checks the behaviour of each pin against a truth table stored in the memory of the control unit. If it tallies, insertion proceeds. If it does not tally, it may have failed, be the wrong type, or be misorientated. Before the machine rejects the component, it retests assuming that the device has been turned through 180°. If the component now appears good, a REVERSE alarm is displayed for the operator.

DIP verification takes time and, in order not to slow down the machine, it is carried out on the next device to be inserted. However, in the meantime the machine is selecting yet another component and delivering it to the head. The

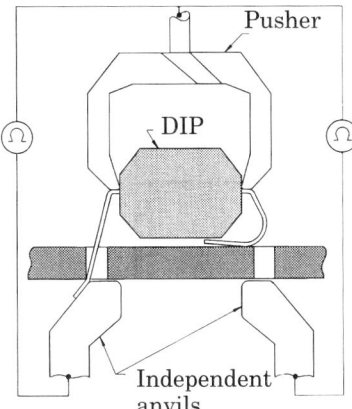

Fig. 6.18 A lead bent under a DIP is very difficult to detect visually but is easily found by an insertion verifier.

result is that most inserters cannot immediately carry out a new pick-and-place operation in the same position when a device is rejected. In such cases, the machine may be provided with an 'auto-repair' feature whereby the empty position on the board is memorised and, after the board has finished the insertion sequence, the machine returns automatically to that position to insert the missing item.

6.4.8 Control Unit

The control units of inserters are fairly fast and reliable minicomputers which store the program and supervise all machine operations. On older equipment the program is fed in by punched tape or by an ASCII terminal. (An ASCII terminal is any input/output device communicating by the American Standard Code for Information Interchange. It may be quite complex or as simple as a Teletype. The RS-232 interface is a communication protocol popular in the personal computer field.) More recent insertion machines employ a keyboard, a disk, cassette tape, a teach-in feature, or are fed directly from the main computer via a data link. While old machines had limitations on the number of program steps (300 to 1,000), new ones can store several programs and recall any of them in milliseconds. To avoid data loss on mains failure or when the machine is turned off, an auxiliary non-volatile memory or a disk unit interfacing with the central unit is needed.

Modern minicomputers have much more computing power than is needed to control an inserter. This can be maximised by using one minicomputer to run a number of machines or by timesharing the computer between its basic tasks and support activities such as:

— component testing (for sequencer-inserters);
— insertion verification;
— positional correction (optical verifiers);
— displaying information to the operator;
— program input or correction;

- communication with a supervising computer;
- selection of component reels to be changed for the next production run;
- working out optimum insertion sequences;
- keeping statistics on components/PBs processed, failures, etc.;
- monitoring maintenance cycles, downtimes and so on.

All this can be done with suitable software, usually supplied by the manufacturer, without significantly slowing the basic cycle of the machine.

An alternative keeps the local minicomputer as small as possible and dedicated to the basic job of driving the machine. This is the first level, the so-called PLC or *Programmable Local Controller*. All PLCs are connected on line with a fairly large host computer, the second level, which is shared between 2-8 PLCs and carries out the support activities mentioned above.

In turn the local host computer may be connected to a large process computer, the third level, which may carry out several jobs, driving automatically guided vehicles, storing and retrieving bills of material, computing material requirements, supervising material flow and overall quality data, and so on. This computer may in turn be connected by data links to receive other inputs, e.g., from R & D departments, or to transmit data such as production output. (Similar computer structures are also popular on machines for assembly of SMCs.)

There is an abundance of good and reliable minicomputers available, enabling the intending user to consider building up his own system. However, it is best to ensure that the control unit has been proven in this specific application. While high computing power for this unit is not required, the environmental conditions, which may involve vibration, dust, metal chips and so on, are demanding.

6.5 AXIAL COMPONENT INSERTION MACHINES

These machines are fully dedicated to axial components and cannot process any other device except jumper wires. The great majority of machines available fall into one or another of the classifications shown in Figure 6.19. From the work performed, they are divided into:

- pure insertion machines, which need to be fed with off-line sequenced components;
- sequencer-inserters, which sequence the components on-line immediately before inserting them.

From the standpoint of restrictions imposed on the design and layout of the PCB, the pitch classification is important, dividing the machines into:

- fixed pitch, inserting components with a fixed or hardware preset pitch;
- VCD, *variable centre distance*, allowing insertion on a pitch which can be programmed by software, component by component, or, at least, be selected from a number of predetermined pitch values.

Finally, the inserters can be classified by number and type of insertion heads as:

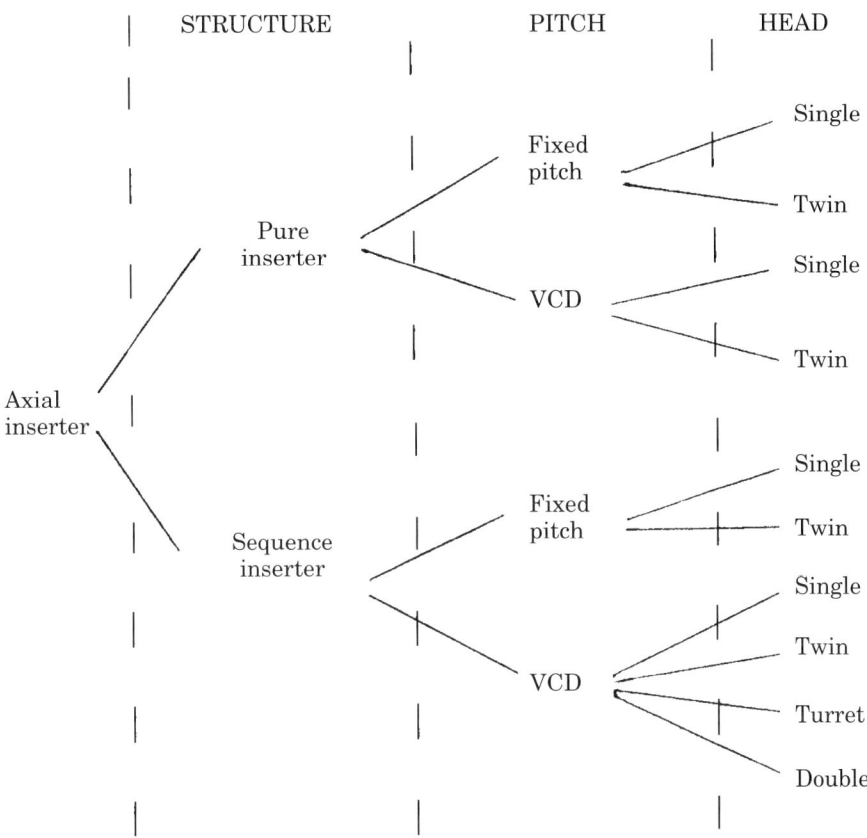

Fig. 6.19 Classification of the main types of axial component insertion equipment.

— single head;
— double or twin head;
— turret.

It will be noted that no attempt is made to classify the equipments as 'fully automatic' or 'semi-automatic' since, as this is a mature technology, all machines are fully automatic and differ only in the degree of mechanisation of ancillary operations.

While all of the above classes are available, the trend has been towards single head, VCD sequencer-inserters. The main reasons for this may be seen in the following sections.

6.5.1 Inserters versus Sequencer-inserters

All earlier axial component insertion machines were based on off-line sequencing, mainly because there was a large difference in productivity of the two steps, one inserter needing up to two sequencers, and the reliability of both machines was low, in particular, the sequencer. Also, in the early seventies, the stress placed on manpower saving and production flow organisation was not as great as it became some ten years later, and the off-line approach offered a good payback on investments. Now, with the increased emphasis in the nineties on reducing labour costs, it has many disadvantages, such as:

(i) greater need for direct labour and slower (e.g., 0.06 seconds/component against 0.04 for a similar machine level) because of the need to handle the sequences;
(ii) more indirect labour needed as two machines mean (a) two programs to prepare and load, (b) two production schedules to be provided and (c) the sequences require administration;
(iii) higher stock levels are needed, because of sequences held ready for insertion and the components 'locked' in these which are not available for alternative production runs;
(iv) lower production flexibility as any change of schedule has to be carried out on two machines;
(v) longer production lead times due to the need to allow two operations instead of one;
(vi) slightly higher maintenance and repair costs, due to the need for a taping unit and extra lead cutting;
(vii) cost of tape;
(viii) additional floor area required, for machines and stocks of sequences;
(ix) more difficult to implement Just-in-Time and continuous flow manufacture as the sequences peg inventory and introduce an additional level in the bill of material.

The major advantages of off-line sequencing appear to be:

(i) the lower machine complexity is more suitable for 'entry level' automation;
(ii) the production-to-cost ratio is higher when component quality is poor, mechanically or electrically — when one machine is stopped, the other may be kept running, and components out of tolerance can be removed from the sequence;
(iii) the machine cost for the same production throughput is lower, as a taping unit costs less than a feed conveyor to take components to the head;
(iv) more flexibility in start-up investment and replacement since there is a wider choice of new and used machines at reasonable prices, and sequencers and inserters from different manufacturers are often compatible;
(v) better cost-to-performance ratio with low volume production runs mixed with a few high volume jobs;
(vi) more flexibility in factory layout — machines are smaller and sequencing may be centrally placed to deliver to different assembly shops.

For these reasons, sequencer-inserters are preferred in larger companies and when the cost of labour is high. The use of off-line sequencing is more suited to smaller shops and in newly industrialised countries where labour costs are relatively low. While both types will survive for a long time, see 6.1.2, the increasing trend to SMA is reducing interest in automatic assembly of axial components.

6.5.2 Fixed versus Variable Centre Distance

Axial components are becoming smaller. As example, excluding leads, the old standard 0.5 watt carbon resistor was about 2.5 mm diameter and 6.5 mm long. A modern, equally rated, metal film resistor is of the order of 1.9 mm diameter and 3.5 mm long. However, as not all components have been reduced by the same amount, the number of body lengths has increased. In the past, the PCB designer could use a standard insertion pitch of, say, 12.7 mm (0.5 in.). To do this now would result in much wasted space on the PB. It is worth noting that cost trends today are such that the component often costs less than the cost of the area it occupies on the board!

As a result, design and production engineers started laying out boards with two or more insertion pitches, to be processed in two or more passes on different insertion machines, or on the same machine by changing the insertion tool. At the same time, equipment manufacturers began producing variable centre distance, VCD, machines, capable of inserting components at different pitches in one pass.

Two different types of VCD inserter are available:

(i) insertion pitch variable over a wide range, e.g., from 5 mm (0.2 in.) to 30 mm (1.2 in.), in steps of 0.025 mm (0.001 in.), the value being selected by the insertion program;
(ii) insertion pitch selected by the software from a set of predefined values such as 5, 7.5, 10, 12.5 and 15 mm or, if imperial units are in use, 0.2, 0.3, 0.4, 0.5 and 0.6 inch.

Both types of machine are substantially equivalent. The PCB is normally laid out on a standard grid, see Chapter 8, and the number of insertion spans is usually restricted to 3 or 4 unless there is good reason to adopt a larger number.

A VCD machine is necessarily more complex (and expensive) than a fixed centre machine (FCD), since insertion and clinching tools must be adjustable. In addition, too frequent adjustment of pitch slows the machine down and accelerates wear on the tools. To minimise these effects, it is better to program to insert all components having the same pitch before changing pitch to insert the next set. This will mean more table movement but wear on a well maintained X-Y table is minimal and, in general, time will be saved and replacement costs reduced.

The sole advantages of a fixed centre distance machine are lower prime cost and simpler maintenance. Selection of type, however, is not straightforward as it also affects product development cost and time, and the cost of boards and assemblies.

6.5.3 Insertion Programs

Manual programming has been largely replaced by automatic generation of the program from CAD software, see Chapter 10, or by using a dedicated software package which operates on the CAD data base. Alternatively, the program may be developed directly from the control unit of the insertion machine on the basis of simple information. If such utilities are not available, the program can be hand written in the following way.

1. Identify, from a board layout drawing or master drawing, the components to be inserted on the axial inserter. This is usually done with a coloured pen, a different colour being used for each pitch.
2. Decide how many boards will be inserted at the same time, by checking PB dimensions against board holder and insertable area. If the machine is fitted with a rotating table, the insertion sequence may also depend on the degree of rotation available. For ease, it is assumed here that components are mounted in one direction only, have different pitches, and that two boards can be accommodated on the workholder.
3. Identify a datum point on the board. Where the X and Y axes cross is the *board zero point*, Figure 6.20. Selection is arbitrary, but should be such

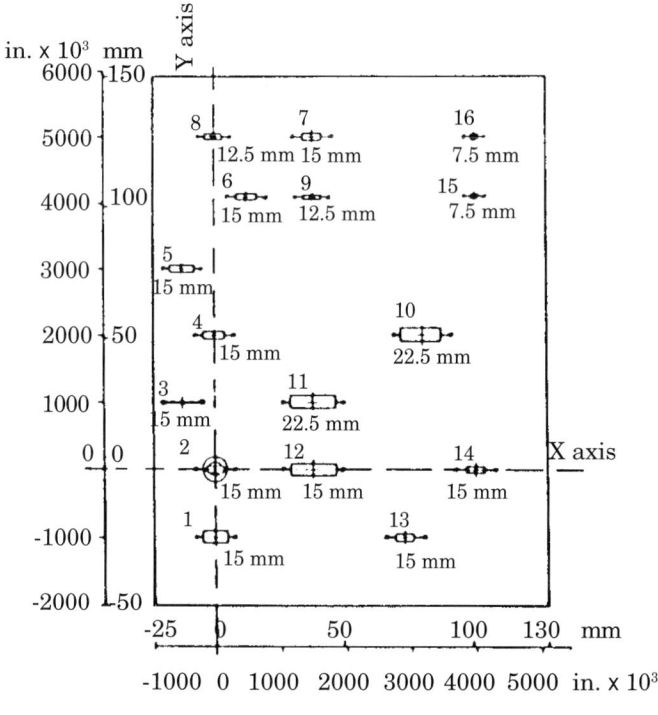

Fig. 6.20 Board adopted as an example for the insertion program described in the text. The sequence number and the insertion span (in thousandths of an inch) are shown near each component. Board zero is the centre of the second component.

that the X and Y co-ordinates of each hole will be as simple as is possible.
4 Specify the board holder, Figure 6.21, by defining:

— its outline, usually the same for all boards, and its reference pins to the X-Y table; as a consequence, the machine zero point (i.e., the zero point of the X-Y table) will be marked on the board holder;
— the slots in which the board, or boards, will be placed;
— the position of reference pins which match the reference holes of the PB;
— the position of the zero point of the board, or boards, with reference to that of the machine;
— the position of the point where the machine must stop at the end of insertion so that assembled boards may be replaced by bare PBs.

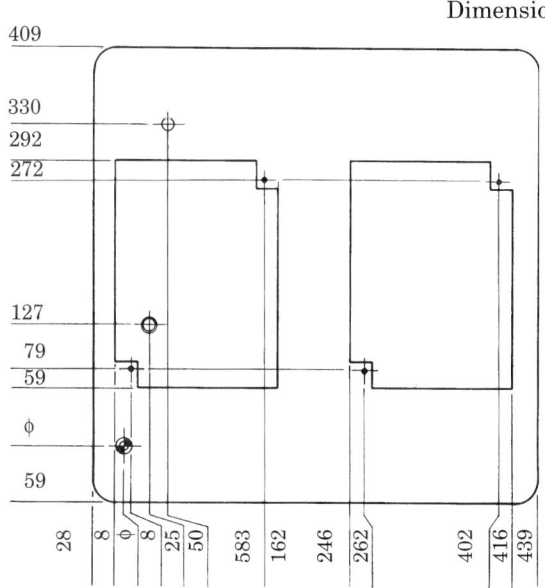

⦵ Machine zero point
⊕ Board zero point
φ Loading position
+ Reference pins

Fig. 6.21 Layout of a board holder to accommodate two PBs on an axial insertion machine.

This work should not take long, but it involves preparation of a drawing of the board holder for the machine shop, machining the tool, and storing it close to the insertion machine for a quick change of production run. Standardisation of board size and reference holes is essential to avoid proliferation of board holders.

5 Define the insertion sequence according to two simple criteria. These are, in order of priority, (a) the minimum number of pitch changes and (b) the minimum length of path to reach all components, Figure 6.22. There will be cases where exceptions must be allowed.

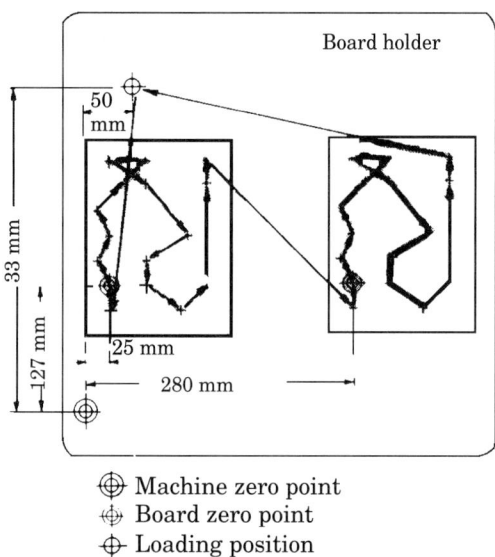

⊕ Machine zero point
⊕ Board zero point
⊕ Loading position

Fig. 6.22 Example of board holder for the sample board of Fig. 6.20. Two boards are placed on this holder and inserted without stopping the machine. The insertion pattern is indicated by arrows.

6 List the insertion sequence in order, writing down for each component to be inserted:

— step number (1, 2, 3, ...);
— type (for sequencer-inserter only);
— X co-ordinate and Y co-ordinate with respect to the board zero; the co-ordinate must be expressed according to the machine user's manual, in mm or in imperial units; conversions from one unit to the other must be done with care to avoid cumulative errors; if possible, it is best to avoid conversions by designing boards with a grid in imperial or metric units, dependent on the insertion machine;
— insertion span;
— body diameter.

The X and Y co-ordinates give the position of the central point between the centres of the two insertion holes. Writing all these down, without mistakes, can be tiring and the work can only be simplified if the board zero point overlaps the zero point of the master drawing.

The body diameter of each component must be known to instruct the machine in how far it can press the component down on the board without causing damage. The insertion span is influenced by the lead diameter, as the formers of the insertion head have a 'V' shaped groove to retain the lead. On some machines this is taken into account by correcting the span manually, e.g., by writing 590 mils instead of 600 mils (the 'mil' is the American term for 0.001 in. and is often used as a convenient unit in programming machines in imperial units), using a table supplied by the machine manufacturer. Other machines will carry out the correction automatically when programmed with the lead diameter.

7   Complete the program by adding auxiliary instructions, such as:

— sequence start;
— position of board zero, first board, with reference to the machine zero;
— offset of board zero, second board, with reference to the first;
— position of the board loading point, that is, where to stop after a complete sequence has been inserted on the two boards;
— number of sequences to be inserted;
— stop on insertion error;
— sequence end;
— repeat the sequence if there is one missing component; this instruction is used on old machines as each sequence is separated from the next by an empty position in the component tape;
— any others specific to the machine type.

8   As soon as the board holder is available, store the program on a non-volatile memory such as a floppy disk, load it into the control unit, test it, correct it as necessary and file it in the program library. It is good practice to file the drawings of the board holder, the working paper(s) of step 6 and a copy of the program away from the production shop, in a location easily accessed by design and production engineering departments.

6.5.4 Insertion Cycle

In order to execute a program, it must be loaded into the working file of the control unit, by inserting the appropriate disk or by recall from another memory. The board(s) and the correct sequence of components are then loaded and the machine started.

Each time the X-Y table moves to present a new insertion position under the head, the inserter performs the following cycle, Figure 6.23:

(i) *Feeding*: a pair of sprockets feed the next component of the sequence to the cutting position.
(ii) *Cutting*: both leads of the component are cut by pairs of shear bars to free it from the tape: the component is temporarily held by a jig called an *inside former*. On a sequencer-inserter this step is not needed and the component is taken directly to the head by the transfer chain of the sequencing unit.
(iii) *Forming*: two jigs called *outside formers* move downwards and bend the leads into a staple-like shape. During this operation the component

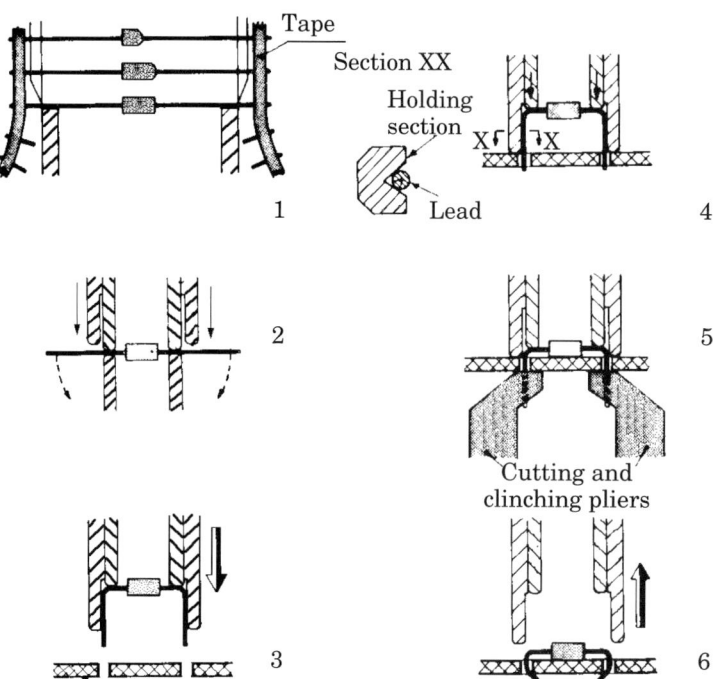

Fig. 6.23 Simplified working cycle of an automatic insertion machine for axial components. The complete cycle may last less than 0.2 of a second.

is held firmly by the inside formers and the *pusher* or *driver*. Bends are made on the outside of the clamped section to avoid any stress on the component body. The inside formers are shaped so that the leads are bent at a suitable radius.

(iv) *Positioning*: the inside formers move away and the driver and the outside formers move downwards with the component. This is held by the friction of its leads in the 'V' shaped grooves of the outside formers, pressure against them being assured by the lead spring-back.

(v) *Insertion*: the outside formers stop very close to the PB, while the driver goes on moving and pushes the component down. When the leads are inserted fully in the holes, the driver stops.

(vi) *Cutting*: at the start of insertion the clinching unit, under the PB, moves up to stop very close to the underside of the PB. The component leads pass through holes in the anvils of the clinching unit and, when the driver stops, the leads are cut by two pairs of blades.

(vii) *Clinching*: one of each pair of the cutting blades has an extra travel and pushes on the leads to clinch them at the desired (adjustable or preset) angle.

(viii) *Sensing*: if an insertion sensing device is fitted, the sensing circuit checks if the driver is electrically shorted to the clinching unit by the lead. If there is an error, the control unit stops the machine.

A production rate of up to 15,000 components/hour is not uncommon and insertion rates of 32,000 components/hour on a two head machine were quoted by one manufacturer in early 1992.[4] The cycle must, therefore, be performed at very high speed, e.g., in less than 0.2 of a second. (It should be noted that, while machine production rates under ideal conditions are quoted throughout this chapter and the next, for estimating purposes it is usually safe to assume an actual production rate of between 70 and 80% of the rated speed.) The speed is increased by partially overlapping the positioning of the X-Y table with the insertion cycle (steps (i) to (iii) and part of (iv) but no more). The table must be stopped before the formers move too far down, or the machine would not be able to work on boards where there are previously inserted components. Similarly, it cannot start to move to the next position before the cycle is complete as the final steps of clinching and verification require a steady component.

6.5.5 Typical Axial Inserters

As a specific axial insertion equipment may have an insertion rate (components per hour) of less than 3,000 or more than 16,000, it is difficult to define or depict a 'typical' axial inserter. In general, the major insertion equipment manufacturers, among whom are companies such as Amistar, Dynapert, Panasonic ('Panasert' machines) and Universal Instruments, supply high volume axial inserters. Some other well-known companies, such as Fuji, are now concentrating on the surface mount market but have sold many excellent machines for which full spares and equipment servicing will be available for many years, either through agents or specialist suppliers of secondhand machines. Yet other companies offer excellent small modular or semi-modular equipments which are ideal for start-up or small batch situations. The information given here covers only two basic types of inserter. It is offered as a starting-point to an engineer looking to fill a need. The prospective purchaser of an equipment to fill specific insertion requirements must make a proper and thorough study of *all* potentially suitable machines and suppliers before committing himself (or his company's money!) to a particular machine or insertion line.

Starting with a basic machine which may be enhanced at a later stage by adding on options to improve productivity or capacity, an axial VCD inserter is available with a production rate of up to 16,000 insertions per hour. This can be uprated to 32,000 insertions per hour by adding a second insertion head and a 'productivity enhancement' option. With a maximum insertion area of 18 x 18 in. or 457 x 457 mm, the basic machine is attractive even to the smaller user. Options available include a rotary table, optical positional verification system, detector for faulty boards, insertion sensing circuit, and automatic loading and unloading equipment.

One of the best known machines, having been in production for a number of years with continuous improvement, is a high speed inserter rated up to 15,000 insertions/hour. This machine also inserts over an area of 18 x 18 in. (457 x 457 mm) and is fitted with a high speed rotary table indexed from 0° to 360° in four 90° increments. It is a software programmed VCD machine with a comprehensive specification and an electronic controller which can provide a pattern library, production rates and management information, as well as interfacing with higher rank computers.

A nice touch is that this machine, like others in the range, reduces noise levels by using high impact resistant plastic shielding over the work area. If the cover is moved or lifted while the machine is operating, the machine stops. However, it does not lose memory or synchronisation and may be restarted at the same point in the cycle and sequence at which it stopped.

A wide range of options are offered, including an optical position verifier and correction unit (board error correction system), which can be extended to detect faulty boards in a multiple panel or 'break-out' assembly. (An insertion verification system is a standard fitting.)

A 60-80 head sequencer is offered for use with this machine, converting it to a 15,000 component/hour sequencer-inserter. A modified transfer chain takes the untaped components directly to the insertion head. For those looking for very high production rates, up to 30,000 insertions/hour can be made on a dual head VCD machine which uses the same head and offers much the same features.

6.6 RADIAL COMPONENT INSERTION MACHINES

The major suppliers of automatic inserters for radial leaded components include Amistar, Panasonic, TDK Corporation (Avisert) and Universal Instruments. Radial component inserters appeared some years after those for axial leaded components. With a few exceptions the sequencer-inserter approach was followed, partly because it is difficult to retape a radial component. Obviously, the radial inserter has many similarities to the axial inserter but there are differences.

The radial inserter also relies on an X-Y table to position the board under the insertion tool. However, the insertion head is more complex than that of an axial inserter because of the different shapes and dimensions of the component bodies and leads. As the simplest way to hold such components is by the leads, this is the normal method of holding the device for insertion. A further difference, which has been mentioned in the discussion of chain sequencing, is that the components cannot be held on the transfer chain by V grooves. They are retained during transfer to the pick-up point by clips or pliers.

The differences in structure of radial inserters is most readily shown by reference to some typical machines.

6.6.1 Machine A

This equipment is interesting as it features a 6-position turret unit which takes the components from the transport chain and moves them to the insertion head. During this transfer, the component may be inspected, tested, rotated and so on.

The standard cutting unit leaves a projection on the solder side of 1.5 ± 0.3 mm and spreads the leads outwards. (Optional clinching units give 45° bends or flush leads.) The machine will also insert transistors packaged with in-line leads, e.g., the TO-92, and an option is offered for clinching the middle lead. The insertion verification circuit can check all three leads. As well as all the normal options, the equipment offers:

— a square pin insertion unit to replace the normal head; it is fed by a reel of square wire and cuts the terminal, inserts it, and checks for its presence;
— a 'between IC' insertion head which inserts components such as disc capacitors between closely mounted ICs;
— a high density insertion head to increase the allowed component density (for insertion) by around 20%;
— an axial part insertion unit which enables the machine to insert small axial lead components.

With an insertion rate of up to 8,000 components/hour, high versatility and reliability, at a reasonable price, this machine has been deservedly very successful and is continuously updated in performance and options.

6.6.2 Machine B

This machine has a linear feeder which moves to left or right to present the correct component to the head mounted in the middle of the machine. However, the feeder movement is much less than that needed for axial components. This is because the radial component tapes are driven through a rail which is shaped so that the tape is gradually turned through 90°, reaching the head in a vertical position, component up, and taking only a small amount of space in front of it.

The insertion head picks up the component by the body with a chuck, instead of clamping the leads. With cylindrical bodies this gives much better grip and allows the head to insert the components more closely to the PB. Another version is similar in construction but holds the component by clamping the leads to allow insertion in high packing density boards. A third version has a head specially designed to insert ceramic capacitors between closely mounted DIPs.

Other models, based on this machine, are offered. These have the ability to accept larger boards. The insertion rate of this series of machines is of the order of 6,000 components/hour.

6.6.3 Machine C

The last example is a radial inserter with a production rate under optimum conditions of 7,000 insertions/hour. The horizontal sequencing unit is behind the machine, at right angles to the insertion unit/board transport. Each component station (up to 4 modules of 20 stations may be present) has a sprocket feed and a cutting unit to free the component from the tape and attach it to a clip on the chain conveyor.

The head has an end resembling the pressure foot of a sewing machine. This clamps the component by its leads to centre them to the holes while a tampon-like pusher or driver pushes the body down. The end of the head is made up of three parts; the central part is fixed and the two lateral ones perform the clamping action. This works at a fixed span, usually 5 mm or 0.2 in. As it is very small, insertion at a spacing of 2.54 mm between two holes of two different components is acceptable and high component densities can be achieved. The head may also rotate through 90° in either direction in order to insert components in the X or Y axis during the same pass. The lead cutting and clinching unit, offered in fixed or adjustable versions, also rotates in accordance with the head.

The replacement for this machine offers a production rate of up to 9,000 insertions per hour, an increase in insertion density and enhanced versatility.

6.7 DUAL INSERTERS

For many years after the introduction of automatic pin-in-hole assembly, all inserters were dedicated to one type of component — terminals, square posts, axial components or radial components. In particular, the last two required totally incompatible equipment although the machines could be similar in construction and use some common items, such as the X-Y table.

As almost all assemblies contain both axial and radial discrete components, all assembly shops, including the smaller ones, were forced to purchase at least two automatic insertion machines. It might have been possible to mount all axial components vertically but there are good reasons for avoiding this method of mounting, quite apart from the higher costs of taping, see Chapter 8.

Among the types of insertion equipment mentioned up to now, the only one which can insert both axial and radial devices routinely is described in 6.6.1. This remains basically a radial component inserter with a limited capability for inserting axial devices.

Inserters designed specifically to cope with both axial and radial devices were introduced in the early eighties, but do not seem to have gained wide acceptance. A typical dual sequencer-inserter is a pass-through machine with feeders at the back and a central bridge holding two heads. Each head is dedicated to one type of component. For example, on a typical machine the head on the right inserts axial components and that on the left inserts radial types. The feeder is therefore divided into two sections, each feeding one head only with the appropriate devices.

Because the chances of having the correct holes presented to both heads at the same time are minuscule, the heads must work sequentially and the production rate depends on the head with the longer cycle time.

This type of machine is of obvious interest to small and medium size assembly shops as it can insert a large variety of components in one pass.

Those seeking inserters with this degree of flexibility will have to study available equipment very carefully. Potentially suitable machines may be obtained from Panasonic, Siemens or TDK among others.

6.8 DIP INSERTION MACHINES

In the early days of ICs, metal packages such as the TO-99 and TO-100 were used extensively and are still superior to the standard DIP in properties such as heat distribution and high frequency behaviour. Unfortunately, they are extremely difficult to insert automatically and the DIP has been almost universally employed. However, the DIP has disadvantages and insertion machines dedicated to its insertion are expensive, slow and difficult to use.

The TO-116 was designed to be simple, cheap and easy to handle and insert automatically. The last statement is incorrect because, while having the pins on two lines makes insertion less of a problem than for other packages, the lack of strict standardisation of body dimensions, Figure 6.24, demands complex insertion machines.

Ceramic packages are made from the more expensive white ceramics which give better repeatability of body dimensions. This is negated somewhat

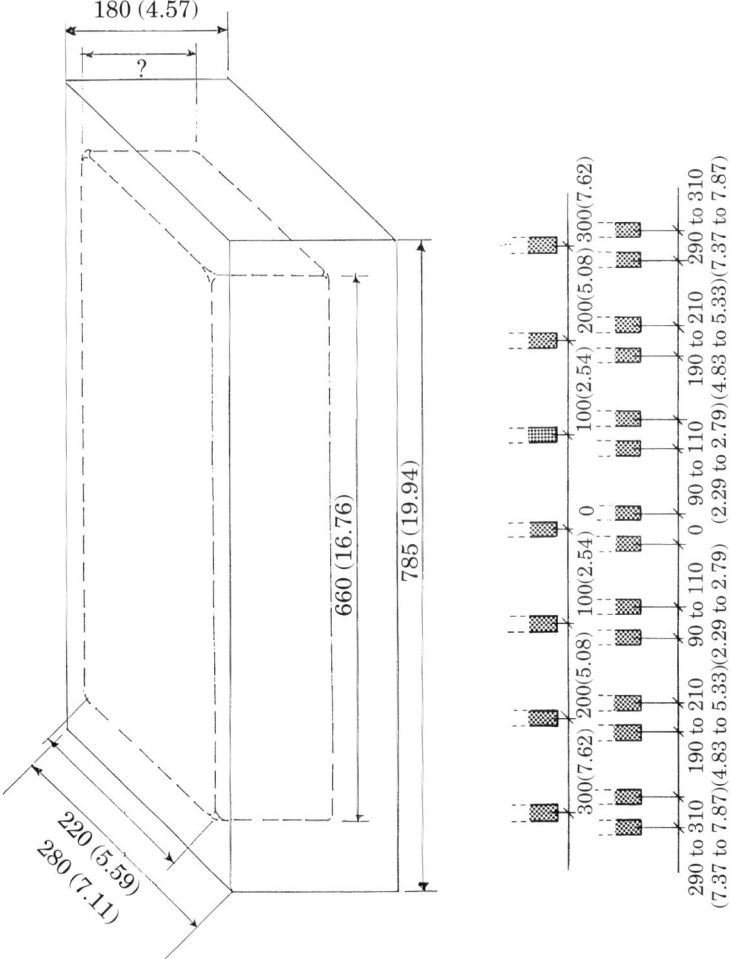

Fig. 6.24 Spatial representation of the 'standard' 14-lead TO-116 package. Dimensions are in mils (1 mil = 0.001 in.) with the mm conversion in brackets. The wide variation in body size and lead pitch is clearly shown. (The '?' indicates that no lower limit is specified).

by the proliferation of sub-classes. Most ICs are supplied in plastic DIPs. Each supplier has his own moulding dies within the maximum dimensions of the standard and a comparison of the dimensions of DIPs from different suppliers shows wide variation, one to the other. Some manufacturers have, for the same basic TO-116, two or three dies with significant dimensional differences!

As it is not usually practicable to assemble boards with devices from a single supplier, a given insertion machine must be able to handle and insert both ceramic and plastic DIPs from different sources. In these circumstances, the only dimensions which can be assumed to be strictly controlled are those relevant to the pins and their positions. All other dimensions must be assumed to vary within the allowed limits.

The insertion tool must refer to the pins and not to the component body. Before 1970, most DIP inserters referred to the end of DIP bodies, i.e., a surface perpendicular to the lines of the pins. They required such frequent adjustment that in many cases their productivity did not justify the investment. A modern machine first combs and straightens the pins and then uses them, at their attachment to the body where they are stronger, as a reference for insertion. Even then, problems arise due to poor pin geometry and the difference in elastic modulus of the pins, different materials having different mechanical properties.

As damage to PBs and components may be frequent, the machine cycle must be performed fairly gently and with several checks, although this slows the machine down. As a result, DIP inserters operate at a much lower speed than those for axial or radial components.

6.8.1 Component Feeding

Off-line sequencing of DIPs is difficult, so all equipments are of the sequencer-inserter type. Any of these can handle the 'small' TO-116 package, i.e., that with 7.62 mm spacing between the lines of pins, regardless of the number of pins (6, 8, 14, 16, 18 and 20). Many of the inserters can also handle the larger DIP, that having 15.24 mm spacing between the rows and 12 to 40 pins, and some can handle all three currently available types (the third having 10.6 mm between the pin rows).

Most machines can also insert DIL sockets, some types being able, in one pass, to insert: (a) DIPs in the PB holes; (b) DIL sockets in the PB holes; and (c) the DIPs into the sockets.

The feeder is necessarily complex. The machine must be able to handle the large number of variations of spacing, number of pins, and combinations of sockets and DIPs. In addition, the packing of DIPs is not good. Stick and rail magazines have to be opened on the correct side, need careful handling, are easily reversed — leading to mounting the wrong way — and hold only a small number of devices.

If the inserter works at a theoretical speed of 4,500 insertions/hour with an overall efficiency of 80%, it will require loading with 3,600 components/hour. Feeding it with stick or tube carriers holding, on average, 30 DIPs each means that the operator must load $3,600 \div 30 = 120$ tubes/hour, an average rate of one every thirty seconds. The result is that an operator can attend only one machine and the risk of stoppage, when two positions run out of components at the same time, is quite high. This may be overcome by installing equipment which can hold more than one carrier in each feed position, so that there are perhaps 8 sticks at each point. Naturally, this adds to complexity and cost.

To exacerbate the situation, the cost and performance conscious electronic designer usually takes full advantage of the enormous variety of IC types available and a board with 100 ICs of 50-90 different types is not a rarity. This has led the development of DIP inserters towards greater flexibility, accepting all DIPs and sockets, with feeders having a large number of positions and holding many devices in each position, rather than towards an increase in insertion speed. For this reason, the drum feeders, which were very popular in the past, are largely replaced by gravity chutes on the more simple machines and by shuttle delivery on the largest machines.

6.8.2 Insertion Cycle

Besides the normal table positioning, the insertion cycle of an automatic DIP sequencer-inserter can be summarised as follows:

(i) *Selection*: a component is selected from the appropriate feeder position and delivered by gravity or shuttle, etc., to the insertion head.
(ii) *Escape and form*: movement of a retaining pin allows the component to drop into the insertion tool where its leads are straightened and formed by an inside former. The component is retained by a closed gate.
(iii) *Gate*: the gate opens and the component falls down a chute and straddles a transfer mandrel.
(iv) *Transfer*: a pusher forces the component along the mandrel to the pick-up point, where it is clamped on the outside by fingers which work as pliers. The mandrel then retracts.
(v) *Insertion*: the fingers drive the component down towards the PB. When the component leads enter the PB holes, the fingers open and a driver/pusher presses the device down to seat it.
(vi) *Cutting and clinching*: as the driver seats the component, the clinching unit, under the PB and vertically in line with the insertion head, moves up. While the component is still held down by the pusher, the excess leads (if any) are cut and are clinched as needed.

Outward clinching is normal, although inward clinching is used and is common for sockets. Several machines will limit clinching to two or four leads, for more easy removal of the package in service.

As with other sequencer-inserters, optional stages may include in-line component verification, optical hole position recognition and correction, and insertion verification independently on all pins. Each of these slows the insertion rate to a noticeable degree.

Visual recognition of hole position may operate on either one or two holes. In the latter case, skew of the hole lines may be detected although no machine can correct this properly, by rotating head or table through a few degrees, since neither head or table can rotate by only the small amount necessary.

6.8.3 Typical DIP Sequencer-inserters

6.8.3.1 EXAMPLE A

The first example has three gravity chute feeders, two of which take up to 32 types (each) of 0.300 in. (7.62 mm) DIPs or sockets. The third feeder is dedicated to other dual-in-line components and may hold 12 types of 0.600 in. (15.24 mm) or, on request, 0.400 in. (10.16 mm) DIPs/sockets or special stations for DIPs with 2, 4 or 6 pins.

A unique feature is a turret holding six different tools which can be selected by program control to insert six different types of component. A set of 16 different tools is on offer, although only six may be mounted on the turret at any given time.

The machine is rated at a maximum of 3,200 insertions/hour. While this is low compared with other equipments, the machine is very flexible and quite reasonably priced, making a good 'entry level' or general purpose machine for shops handling many types of board.

Most of the usual options are available and an ancillary magazine replenisher is worthy of mention. This is a random access unit having 32 magazines, each of which can hold up to 14 commercial DIP tubes. Each magazine has a buffer which holds the content of one tube, allowing the unit to operate while refilling. The unit can hold up to 12,000 components of 32 different types and can supply them to the inserter, in a sequence defined by the machine control unit, at a rate of 3,600 devices/hour. Up to three of these units may be attached at one time to the machine.

6.8.3.2 EXAMPLE B

This is a high performance DIP inserter. It is a conventional machine with a fixed linear feeder and shuttle delivery. The feeder is organised in 7 blocks, each holding either 15, 13 or 11 types of DIP, depending on the spacing between pin rows. Each of the standard blocks may be replaced by an automatic stick feeder which holds 10 sticks of the same component instead of one. When a stick is empty, the machine starts to pick from the next one.

The machine is rated at 4,500 insertions/hour and is equipped with two shuttles which work alternately to keep the insertion speed independent of the position of the component in the feeder.

6.8.3.3 EXAMPLE C

The third example is a two head machine which can insert all types of DIP. With appropriate tooling, capacitors, transistors and all devices packaged in 0.300 in. DIPs (2 or 4 pin) can also be inserted. The insertion rate is between 3,420 and 4,800 components/hour.

The structure of the machine is conventional with a linear feeder, a two pocket shuttle being used to feed both heads. The two heads work independently on the same board and, due to the design of the X-Y table, while one head is inserting the other is receiving the component from the shuttle. If one head is shut down for any reason, the maximum insertion speed drops by less than 10% to 4,500 insertions/hour.

The feeder holds 70 0.300 in. DIP magazines or a proportionately lower quantity of the larger DIPs. For large production runs, the standard magazines may be replaced with cartridges which hold 8 sticks and advance automatically when a stick is empty. Cartridge replacement is quick and easy.

Insertion of sockets requires the fitting of optional tooling and dedicated magazines, which are compatible with the auto-load cartridges.

6.8.3.4 EXAMPLE D

This machine is significantly different in structure from those just described. It is fitted with a moving linear feeder and a bridge which goes over both feeder and the board to be assembled. Two insertion heads are mounted on the bridge and work alternately, collecting the component from the pick-up point and inserting it on the board. One head is dedicated to 0.300 in. spacing DIPs and the other to 0.600 in. DIPs. Components are loaded into special cartridges, each of which can hold up to 15 magazines: when one is empty, it is ejected automatically and the next full one is brought to the ready position.

The rated productivity is 4,500 insertions/hour, regardless of the number of component stations, since the time to position the feeder is overlapped by the insertion time. The rate of insertion is influenced by the DIP size as, while a 0.300 in. DIP is picked and inserted in about 0.8 second, a 0.600 in. package has a cycle time of about 1 second.

6.8.3.5 EQUIPMENT MANUFACTURERS

Among the major manufacturers of DIP insertion equipment are Amistar, Dynapert, Panasonic, Siemens and Universal Instruments. While Panasonic announced in 1993 that the insertion side of the company's business was reducing in favour of surface mount equipment, it was also stated that the company was continuing development of insertion equipment.[1]

6.9 ODD COMPONENT INSERTION MACHINES

What is an 'odd' component? For the majority of technicians involved with electronic assembly, it simply means 'any component which cannot be inserted automatically by standard insertion machines for axial, radial or DIP components'.

Using this definition, a wiring harness, a connector, a transformer, a heat sink, a small 'daughter' board and so on are all odd components as well as a stand-off mounted axial resistor. The variety is so large that any single classification is not possible in simple terms.

Any engineer will agree that, however complex a component may be in outline, number of leads and so on, equipment can be developed to assemble it automatically. The important criterion is whether or not it is economically justifiable when compared with manual insertion, taking all relevant cost factors into account. Quite often the automatic assembly of particular components looks more like an engineer's masterpiece or plaything rather than an economically motivated investment, as it purely replaces manpower with capital and involves additional risk for the company because:

(i) the component packaging may change;
(ii) the technology involving that component may become obsolete;
(iii) the company may discontinue the family of products using that component;

and the conversion of equipment is usually much more costly than the retraining of operators. To illustrate (i), in one of many cases, Giovanni Leonida was running a shop assembling hundreds of thousands of TO-18 transistors per day. He developed equipment to form and insert these automatically. Then most of the boards were converted to ICs, the price of which had dropped sharply. The TO-18 case was virtually abandoned as the remaining transistors became available in plastic cases, like the TO-92, which were cheaper and could be mounted by standard radial inserters. Another 'white elephant' on the floor!

This does not mean that the production engineer should stop thinking of ways to mechanise assembly operations. The cost of labour is increasing, in real terms, in almost every country, while the cost of automatic equipment is decreasing. It does mean that premature automation can result in financial loss. Better equipment may be available, more cheaply, one year later and the resale value of older machines is dramatically low.

The decision to produce an automatic equipment to insert odd components or, more usually to adapt an existing machine, should be taken only after investigating the economics of automating and:

(i) projecting the cost of manpower and the cost/performance of the machine over the next 3-5 years;
(ii) talking, together with design engineers, to current and potential suppliers of components to determine what will be available in components and packages in the near future;
(iii) involving marketing and planning departments, to ascertain that the product family and/or technology will sell for as long as is needed to recover the projected investment.

6.9.1 Dedicated Insertion Machines for Odd Components

A machine can be dedicated to any component, no matter how odd the outline or pin configuration. Any part of the machine, except standard parts such as the X-Y table or the electronic control unit, can be tailored to the particular device/package.

In addition to the in-house custom-modified machines, several types of dedicated insertion machines are available to mount whole families of component, rather than a single device. Frequently, the component manufacturer will develop a suitable machine, not only to promote the sale of his product (e.g., connectors) but to help to amortise his own development costs.

Radial insertion machines have also improved in flexibility and many components, previously 'odd', are now offered in packages suitable for automatic insertion without functional changes.

One machine developed to handle odd components will insert only 5 types of different device having the same shape and size. It is relatively slow, taking 3 seconds per component, and limited in scope but is simple and low in cost. It is compact in size, allowing easy connection in series so that a line of 10 such machines can be created to insert up to 50 different components at a theoretical rate of 12,000 insertions/hour. The same manufacturer now offers a radial inserter which also inserts a range of odd shaped components with lead pitches up to 7.5 mm (0.295 in.).

Another manufacturer produced a very versatile machine, claimed to insert components such as IFT coils, DIPs, trimmer potentiometers and capacitors, SIL resistor arrays, switches, stand-off resistors, large axial components and tab terminals, provided that they are packed in properly shaped, Figure 6.10, 508 mm (20 in.) long magazines. This machine is also relatively slow, rated at 2.4 seconds per insertion, but is very flexible and can be used on a stand-alone basis or as part of a larger line.

Major manufacturers of such equipment are TDK and Panasonic. However, while it is not possible to generalise, the major suppliers of components and equipment, especially those in Europe, the USA and Japan, are usually open to proposals to modify their product in order to facilitate automatic insertion.

6.9.2 General Purpose Robots

The idea of totally automated assembly dates back to the early eighties when it became a reality for a few highly engineered boards. The need for new types of insertion machine was linked to three factors:

(i) while the range of traditional insertion machines was still expanding, it was at a much slower rate than in the past;
(ii) several packages which were difficult to handle, e.g., the TO-3, were gradually replaced by others, such as the TO-220, which are more manageable but are odd shaped;
(iii) component manufacturers were developing new components for surface mounting and expending less effort on making conventional devices easier to handle and insert.

As a result, adaptation of general purpose handling/assembly equipments or *robots* was often the only way to improve further the degree of automation. Attempts to adapt robots to electronics assemblies are not new. In the mid-seventies DIP inserters were rather rough machines, accepting only a small number of component types and unable to control insertion force such that components and PBs were being broken. In addition, discarded components were being mixed up and automatic loading and unloading of boards was not available. Under some circumstances, a suitably equipped general purpose assembly robot performed better, technically and economically.

Experience based on old, heavy robots designed for mechanical assembly was seldom encouraging. Today, robots designed for electronic assembly are smaller and better equipped to handle the small loads and forces encountered. Moreover, as a result of the improvements in computer technology, they are easier to program and are able to communicate, with each other and with a host computer, when several are used to build a large assembly line.

Figure 6.25 shows a good example of a machine developed specifically for insertion of electronic components. It features a four-axis movement which

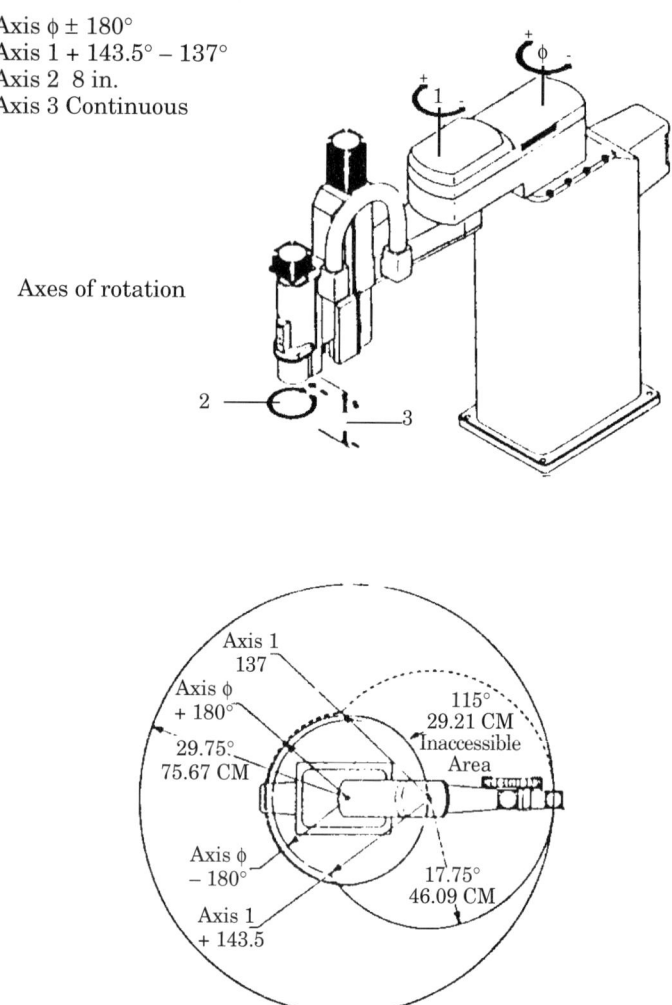

Fig. 6.25 The MicroSmooth 440 robot for assembly of electronic components. The diagrams show the general outline of the machine, the four axes of rotation, and the access area. (Courtesy of Intelledex, Inc.)

gives it access to a 755.7 mm radius circular area around it, except for two smaller inner circles. The maximum operating area, which must include both the board(s) being assembled and the feeder outlets, is nevertheless quite large.

For a robot, the insertion cycle time is reasonable at 2.3 seconds, equivalent to 1,565 components/hour. Accuracy is very good and the repeatability is impressive. The payload of the operating arm, which must include the end-effectors and the tool change devices, is 10 pounds (about 4.5 kg).

A major advantage of robots is that the tool change is programmed by simple software and imposes almost no limit on the shape and size of components. The tool used to handle the component is itself considered a component, to be picked up by the arm when needed.

The machine illustrated has an advanced end-effector system equipped with sensors which enable the controller to make adjustments for small variations in part dimensions, while an optical recognition system supports fine adjustment of position. A load cell can be fitted to detect mechanically poor components by measurement of insertion force. *Continuous microstepping* or fractional stepping motors give smooth and precise movement. Diagonal movements are also smooth and are always made on the shortest path.

Two of the four axes are used for X-Y positioning, the third (Z) axis being used to insert the component. The fourth axis gives program controlled rotation around the Z axis and is used to rotate the component or to turn screws, an operation not performed by other types of insertion equipment.

A *Robot BASIC* was developed by the manufacturer by adding over 150 specific, user friendly commands to *Microsoft* BASIC, a language well known to most users of small business, personal and home computers. Programs can be prepared off-line, on a simple home computer, to minimise robot downtime when changing board type. As this robot was conceived as a peripheral to a computer, extensive communications capability is built in. Five sequential RS-232-C ports are provided, two being dedicated (one to the host computer and one to 'daisy chain' other robots), leaving three for general purposes. In addition, four 8-bit parallel ports are available for high speed communication with peripheral devices.

Another machine from the same manufacturer has 6-axis movement with fully sealed joints. The electronics are located inside the operating arm. For these reasons, it is particularly suited to clean room use in the microelectronics industry, where robots have proved to be better than humans for an additional reason — they are less prone to cause particulate contamination!

Another well-known machine is capable of inserting axial, radial and DIP components in a single pass and has an insertion rate of 2,000 components per hour for axial components, 800 for radial and 1565 for DIPs. The machine can handle batches from 1 to 1,000 or more with great flexibility.

Specialists in design and manufacture of robots for electronic assembly include Blakell Ambotech and Intelledex Incorporated.

6.10 ASSEMBLY INSPECTION

The reliability of current automatic insertion machines is relatively low, although it is claimed to be getting better. With the proper components and good machine maintenance, the error rate due to missing or badly inserted components is between 0.2 and 1.0%, Table 6.1, which is far from acceptable when the total number of rejections is added up. If no assembly inspection is carried out, inspection will not take place until the test and inspection stage, by which time the figures will be masked by handling damage occurring after assembly and repair will be very expensive.

Inspecting prior to soldering, at least on a sample basis, is good practice. Such inspection may take place after each automatic insertion operation or

Table 6.1

Error Rates of Automatic Insertion Machines*

Component Type	Without Manual Recovery**		With Manual Recovery***	
	Minimum (%)	Maximum (%)	Minimum (ppm)	Maximum (ppm)
Axial (1)	0.5	1.0	150	500
Radial (2)	0.8	1.2	100	300
DIP (3)	0.2	0.5	50	200

* Data provided by Giovanni Leonida. Based on interviews, these represent practical figures achievable under normal operating conditions. They relate closely to the tolerances of components supplied, including PBs, and to machine maintenance.
** The operator simply restarts the machine when it stops on error.
*** The operator repairs the fault when machine stops on error (or later for sequenced axial components).
(1) Data related to commercial VCD inserters.
(2) Data related to commercial fixed pitch sequencer-inserters.
(3) Data related to commercial sequencer-inserters for 0.300 in. (7.62) DIPs.

at the end of the assembly process, which may for example include the following steps:

(i) automatic insertion of axial components;
(ii) automatic insertion of DIPs;
(iii) automatic insertion of radial components;
(iv) automatic insertion of odd components;
(v) placement of surface mounted components, see Chapter 7, on the soldering side of the board;
(vi) manual completion, see Chapter 5.

If steps (i) to (v) are performed on the same line, no intermediate checks, other than those carried out by the equipment, are possible or convenient, since one position-checking equipment at the end of the line will be adequate. If the steps are carried out on stand-alone machines, the operator can visually check a board from time to time and provide immediate feedback. Alternatively, a complete inspection may be carried out off-line, visually or using automatic equipment.

In all cases, a complete inspection, 100% or on a sample, is usually located after all assembly operations but before soldering, as this gives the maximum benefit/cost ratio. The data ensuing will be related to a specific equipment by the position on the board and, when needed, handling damage can be determined by comparing the results of the final inspection with the checks on the output of each machine.

6.10.1 Visual Inspection

Visual inspection of manually assembled boards, see Chapter 5, is meaningful since the assembly error rate is quite high. The value of visual inspection for automatically assembled boards is questionable, because the inspector's error rate is the same or worse than the machine error rate.

Visual inspection may pay for itself, if restricted to looking for simple, easily seen faults, such as damaged boards or components, skipped operations, bad clinching and the like. No general rule can be applied and a costing analysis must be done in each individual case.

6.10.2 Inspection Equipment

Fully automatic inspection equipment is not widely used in electronics assembly shops despite improvements in price and performance. A very basic reason for this is the difficulty of recognising components because of the lack of strict standardisation of shape, dimensions, colour, markings and so on. In addition, as far as PIH components are concerned, the board and hole tolerances are such that the component position is loosely determined. As a result, the automatic equipment often needs the involvement of an operator and a relatively large amount of labour.

Image recognition, Figure 6.26, is a tough job and is still expensive if performed by traditional sequential computing. New, highly parallel computer architecture, such as neural computing, could eliminate the restrictions and provide fast, cheap and reliable image recognition electronics.

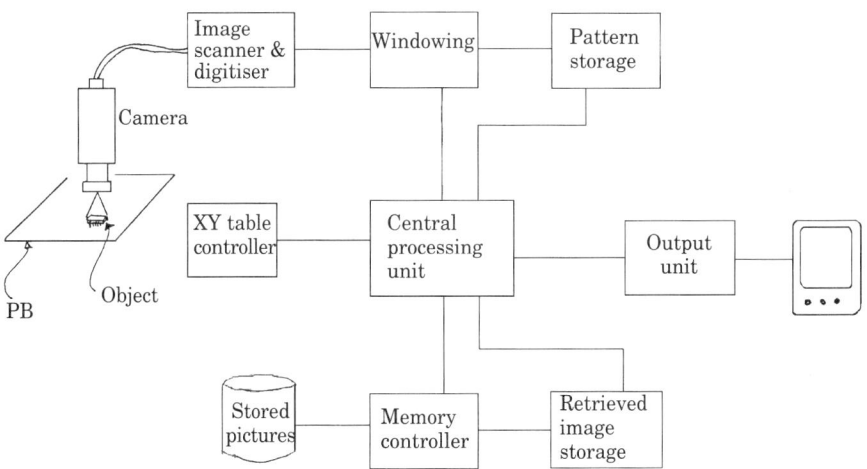

Fig. 6.26 Image recognition is a complex procedure which is not helped by inadequate standardisation of components.

Best results are obtained with surface mounted components as their outlines are more standardised and their position on the PB is more precise. In some assembly shops, automatic inspection equipment was introduced for SMDs and extended to PIH components on mixed technology boards.

The majority of machines available utilise a video camera to capture a picture of the board being examined. The resulting image is then compared with a reference picture in the memory of the image recognition unit. The board must be accurately positioned when using this technique in order that components are not erroneously flagged up as out of position or missing. This can be done either by using simple jig fixtures or by using an X-Y table, with position error correction when necessary, as for insertion machines.

Colour is relevant, even when monochrome cameras and monitors are used, as these machines resolve colours into 64 levels or shades of grey. A threshold setting allows the user to determine the amount of grey difference between the reference picture and the board being inspected at which an error will be indicated. This is necessary to take account of variations in lighting conditions or texture of the components. For the best performance, the light level should be uniform and free from casual shadows.

Various methods of error indication are offered. The machine can be programmed to indicate an error of, or greater than, the level set by flashing on a monitor screen, by stopping and indicating the position for repair or by downloading data to a repair station. Audible alarm signals can be set to attract the operator's attention.

Automatic visual inspection machines are now available from about US $6500 upwards, dependent on the degree of automation, image processing capability, and flexibility. Even the more basic machines detect missing components easily and very quickly. Detection of the wrong type or value of component is much more difficult owing to the poor standardisation of markings and the limited number of shapes which can be identified by the machine without operator assistance.

The more powerful machines are capable not only of grey level mapping but also of morphological analysis, identifying randomly orientated parts, optical character recognition, reading bar codes (e.g., for board identity) and so on. One manufacturer, Intelledex Inc., demonstrates the inspection of the solder side of an assembled board, to detect the presence of component leads as well as their clinching angle!

6.11 BOARD HANDLING

When component insertion has been largely automated, board handling becomes one of the most labour intensive operations. It is always a major cause of damage and has a considerable influence on quality and productivity. For these reasons, the advantages of automation in an assembly shop cannot be fully realised until diligent attention is given to materiel flow and information management. Information management is facilitated in the electronics assembly shop by the fact that most, if not all, modern electronic controllers either supply such information directly or may be linked to a host computer and, through this, to a larger process computer. This subject is discussed more fully earlier in this chapter (6.4.8).

When several sequencer-inserters from the same manufacturer are connected into a complete assembly line, the machine manufacturer is capable of supplying all necessary ancillary equipments and ensuring that all machines are compatible with one another.

Automatic Assembly — Conventional Components 263

If stand-alone equipments, possibly from different manufacturers, are used, the task of ensuring compatibility falls on the production/assembly engineers. An obvious first priority is to select a suitable magazine for board handling and to check that it is acceptable, as far as is practicable, throughout the full manufacturing process.

If, for example, all or the majority of the boards have at least two parallel sides, a simple rack magazine or cassette can be constructed on a rigid metal base of suitable size. If the verticals are made adjustable for board width, by putting a series of slots or holes in the base to which the verticals are fixed, boards of different sizes can easily be accommodated in the 'standard' rack. The spacing or pitch between the grooves or runners into which the boards slide can be standardised at, say, 10 mm (0.4 in.), allowing about 25-30 boards of normal component population in a comfortable height of magazine. If there are larger components on the board, they can be placed in alternate runners, halving the magazine capacity. It should be noted that, if plastic runners are employed to ease insertion of boards into the rack, they may cause problems with static electricity. They can be coated with conductive paint but the use of aluminium or another suitable metal runner is to be preferred. An earth contact must be provided wherever necessary.

To load the first stage of assembly, a loading unit as used in PB manufacture may be employed, Figure 6.27. Once components have been placed, the boards can no longer be stacked and must be transported to the next position by transfer conveyor or stacked in rack magazines. All large suppliers of insertion equipment offer complete sets of equipment for materiel handling. A complete range will include, for instance, the following units, Figure 6.28:

(a) bare PB loading unit;
(b) inserter-to-inserter connecting conveyor;
(c) magazine-to-machine loading unit;
(d) machine-to-magazine unloading unit;
(e) machine-to-machine buffer stock unit;
(f) tilt-transfer unit;
(g) transfer-to-stock unit.

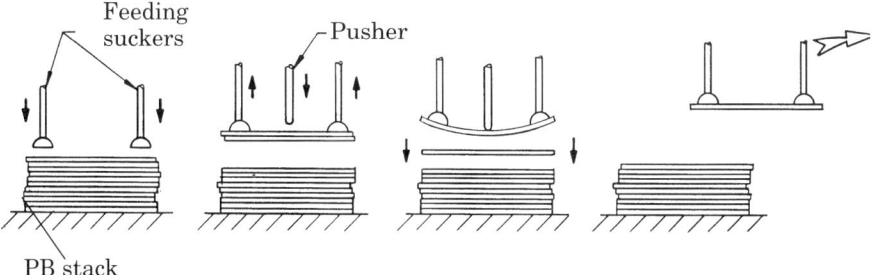

Fig. 6.27 A PB board feeder may be used to load the first stage of assembly. The action of the pusher, shown in exaggerated form, is needed to break any partial vacuum between the top board and the next.

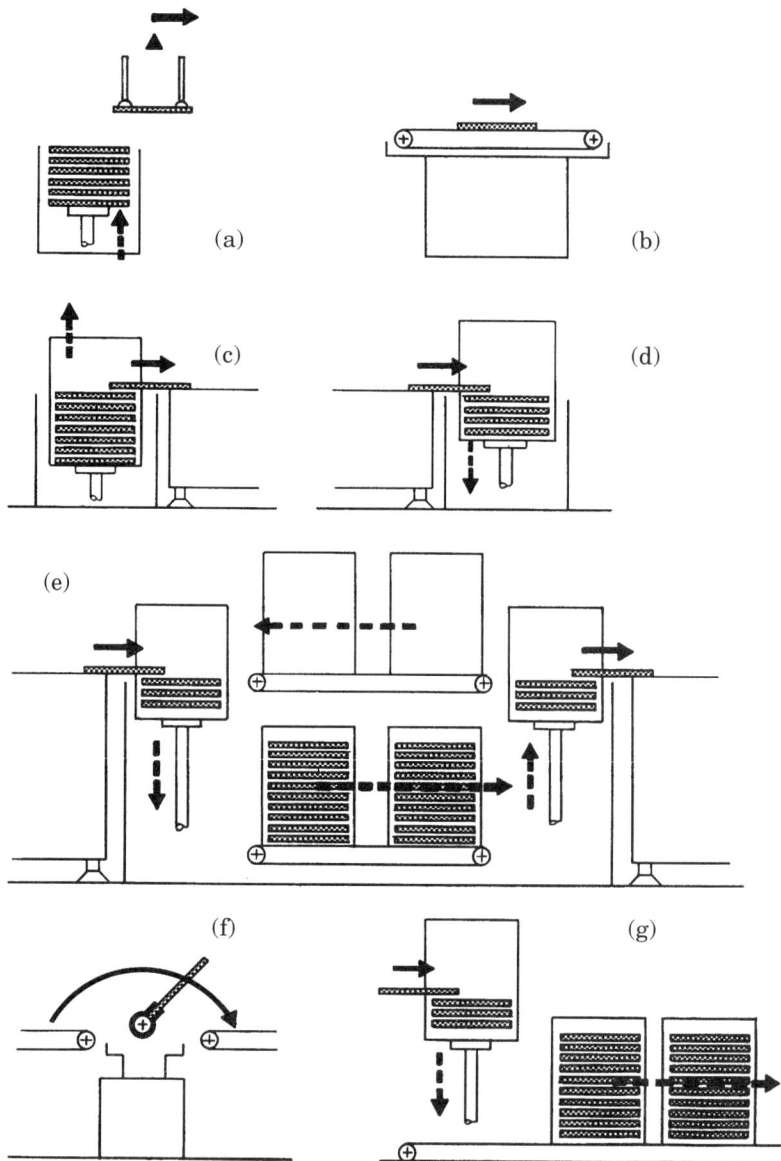

Fig. 6.28 Commonly used handling equipment may either be constructed in-house or purchased from equipment manufacturers. (a) Bare board loading unit (pusher is not shown); (b) transfer conveyor; (c) rack/magazine-to-machine loading unit; (d) machine-to-rack loading unit; (e) buffer stock unit; (f) tilt-transfer unit; (g) machine-to-stock unit.

Other units such as magazine-to-conveyor, conveyor-to-magazine and conveyor-to-conveyor are essentially the same as (c), (d) and (e) respectively.

A large buffer between two machines or two assembly operations can be provided by using a tall hopper chain equipped with lengths of pressed steel channel section. All such equipment is intended for buffering and not for long-term storage. If the need to store assembled boards for long periods cannot be avoided, it may be done economically by stacking the cassette magazines manually.

When no tilting or buffering is necessary, a conveyor is usually the most economical handling device. When practicable, the conveyor can also be used to transfer the completed assemblies to the soldering machines and load them directly.

Many solutions to material handling problems are offered by equipment suppliers and the assembly engineer can select those which are most appropriate to his shop's needs. With such equipment, apart from cost and fitness for purpose, the major points to look for are reliability (the consequences of handling equipment failures can be traumatic!), smoothness of movement and, when necessary, good electrical bonding to avoid static build-up.

REFERENCES

1 Buckley, D., 'Insertion in the Nineties', *Electronic Production*, pp. 9-12, November (1993).
2 'Fundamentals of Printed Circuit Design and Manufacture', The Printed Circuit Interconnection Federation (PCIF), London, United Kingdom (1993).
3 Gustin, J., 'SMT Presents Challenges for Cut/Clinch Machines', *Circuits Assembly*, p. 108, February (1992).
4 Daniels, R., 'High Volume Placement and Insertion Equipment', *Circuits Assembly*, pp. 50-67, February (1992).

Chapter 7

AUTOMATIC ASSEMBLY — SURFACE MOUNT COMPONENTS

GIOVANNI LEONIDA
Milan, Italy

W. MACLEOD ROSS
Welnorth Ltd, Bishop's Stortford, UK

7.1 TYPES OF SURFACE MOUNT ASSEMBLY (SMA)

Surface mounting assembly is completely different from traditional pin-in-hole (PIH) assembly. Not only are the components usually much smaller, making manual handling extremely difficult, but leads are either non-existent or shaped for planar mounting. Other notable differences are:

(i) components must be attached to the PB with some kind of adhesive because they are simply placed on the surface of the board and not held by leads in holes;
(ii) components may be placed on only one side or on both sides of the mounting substrate;
(iii) assembled boards may have to undergo two or more soldering operations;
(iv) in some cases, e.g., where the component body cannot withstand the temperature of molten solder, wave soldering may not be possible and alternative joining methods may be preferable and/or more economical;
(v) mixprint boards (i.e., boards with a mixture of PIH and SM components) are possible and often convenient in design;
(vi) cleaning of soldered boards can be very difficult because clearance between the SMCs and the board surface may be very small, while the high packing density of SMAs demands a low degree of contamination;
(vii) the small size of components and their footprint on the PBs means that mounting machines must be capable of very high precision in handling and in placement.

Taken individually, these differences do not seem dramatic. When combined, however, as they usually are, they mean that the manufacture of surface mount assemblies is a whole new discipline which must be learned from the beginning — starting for instance from the basic types of assemblies, Figure 4.1, and continuing with the new materials and equipments required.

Apart from SMA, the most common abbreviations used in the following text are: SM = surface mounting, shortened by common (ungrammatical) use to 'surface mount'; SMC = surface mount component; SMD = surface mount device (= surface mount component); SMT = surface mount technology.

7.1.1 Single-sided SMA (Figure 7.1)

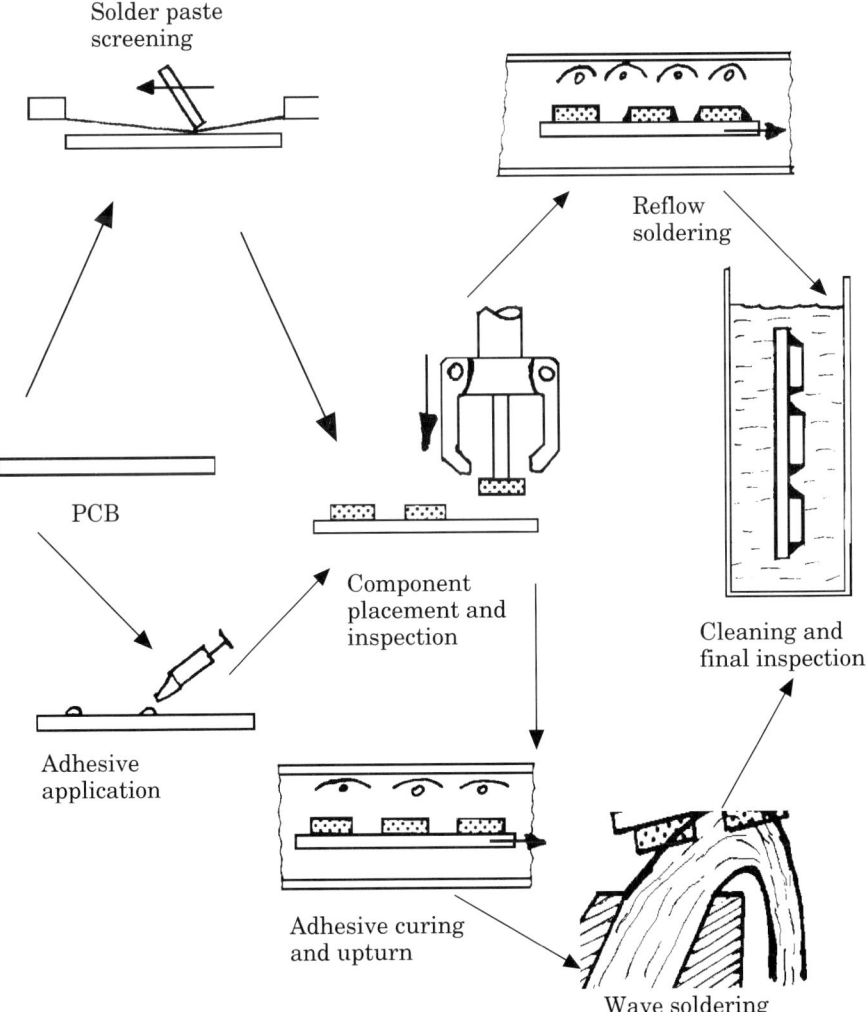

Fig. 7.1 Assembly of single-sided surface mounting assemblies.

The simplest type of SMA has components on one side only, all components being SMCs. The board may be single-sided with only one layer of conductors or it may be a complex multilayer circuit.

If it is established that all SMCs can withstand the temperature of molten solder for sufficient time, wave soldering can be used and the production cycle will take the following form:

(i) prepare the PB by printing identification marks, cleaning, etc.;
(ii) apply adhesive by a suitable method, e.g., pin transfer, syringe or screen printing;
(iii) place the components on the board using 'pick-and-place' equipment; (manual placement, e.g., with hand-held tweezers, is seldom used, even for very small quantities, because of the accuracy needed);
(iv) inspect the board, visually or automatically, to see if all components are properly placed on glue dots of the correct size;
(v) cure the adhesive by the stipulated method;
(vi) invert the board;
(vii) solder the joints by passing the boards over suitable wave soldering equipment (the shape of the wave is highly important and, for SMAs, a double wave is recommended, see Chapter 15);
(viii) clean the boards by a method which does not damage the components, idents or boards;
(ix) inspect the assembly.

If only a few components cannot resist molten solder, they may be added and hand soldered at (or just before) the final inspection stage. Alternatively, such devices can be placed automatically and soldered, individually or collectively, by reflow, local heating, laser, etc. It may be necessary in such instances to protect their footprints during the wave soldering operation, to avoid uneven build-up of solder.

If there are more than just a few such components, it is likely to be more economic to use a cycle based on reflow soldering as follows:

(i) prepare the PB;
(ii) apply solder paste to the soldering pads (footprints) by screen printing, metal stencil, syringe dispensing or pin transfer;
(iii) place components on the board;
(iv) if required, dry the solder paste;
(v) inspect the boards to ensure proper placement of the components;
(vi) fuse ('reflow') the solder paste, using radiant heat, condensation soldering, hot-plate reflow or other suitable method — see Section 5 of this book;
(vii) clean the boards;
(viii) inspect the assembly.

Although the footprints on the PB are normally coated with tin-lead during manufacture of the bare boards, the coating is not thick enough to supply sufficient solder to make a sound joint between pads and component terminations. Application of solder paste or cream is a convenient method of providing a reservoir of solder. In addition, as the pastes are usually tacky as applied, or become adhesive after drying, there may be no need for a separate adhesive to hold the components in place until the solder joints are made — provided the boards are handled with reasonable care! Solder pastes and creams are described later.

7.1.2 Double-sided SMA (Figure 7.2)

This is the most typical surface mount assembly, which takes maximum benefit from possible reductions in size and weight. As no insertion holes are needed, via holes on double-sided or multilayer boards can be very small, leaving most of the space available for conductors.

There are five alternative manufacturing sequences, dependent on the soldering method:

(i)   simultaneous reflow soldering of both sides;
(ii) wave soldering on both sides (in separate passes);
(iii) reflow soldering of both sides in separate passes;
(iv) wave soldering on side 1 followed by reflow soldering of side 2;
(v)   reflow soldering of side 1 followed by wave soldering of side 2.

Method (i) is seldom used and *never* for large boards. Placing components on the second side may cause those on the first to fall off (unless glued in place); on heating in the oven, the solder paste on the underside may lose adhesion; it is very difficult to control the temperature on both sides of the board with sufficient accuracy.

Method (ii) cannot be used unless all components can withstand immersion in molten solder. Method (iii) can be used if all heavy components are placed on the second side to be reflowed; during reflow of the second side, the components on side 1 are held in place by the surface tension of the molten solder.

Methods (iv) and (v) are direct alternatives which imply that only heat resistant components will be placed on the side to be wave soldered. Method (v) is preferred as (a) it is easier to apply solder paste to an unpopulated board and adhesive to a partly assembled board than *vice versa* and (b) during wave soldering the opposite side heats up less than during reflow soldering.

The most likely manufacturing cycle is therefore as shown in Figure 7.2, involving solder cream application, component placement, drying (if required) and pre-soldering inspection, reflow soldering, and perhaps a solder joint inspection. The semi-assembled boards, usually without intermediate cleaning, then go through the second sequence of turning over, application of adhesive, component placement, adhesive curing, wave soldering and final inspection.

7.1.3 Mixprint — SMDs One Side Only (Figure 7.3)

This type of board is very popular in consumer electronics as it exploits the advantages of SMCs and the traditional assembly cycle. In essence:

(i) the traditional distinction of component side from solder side is retained;

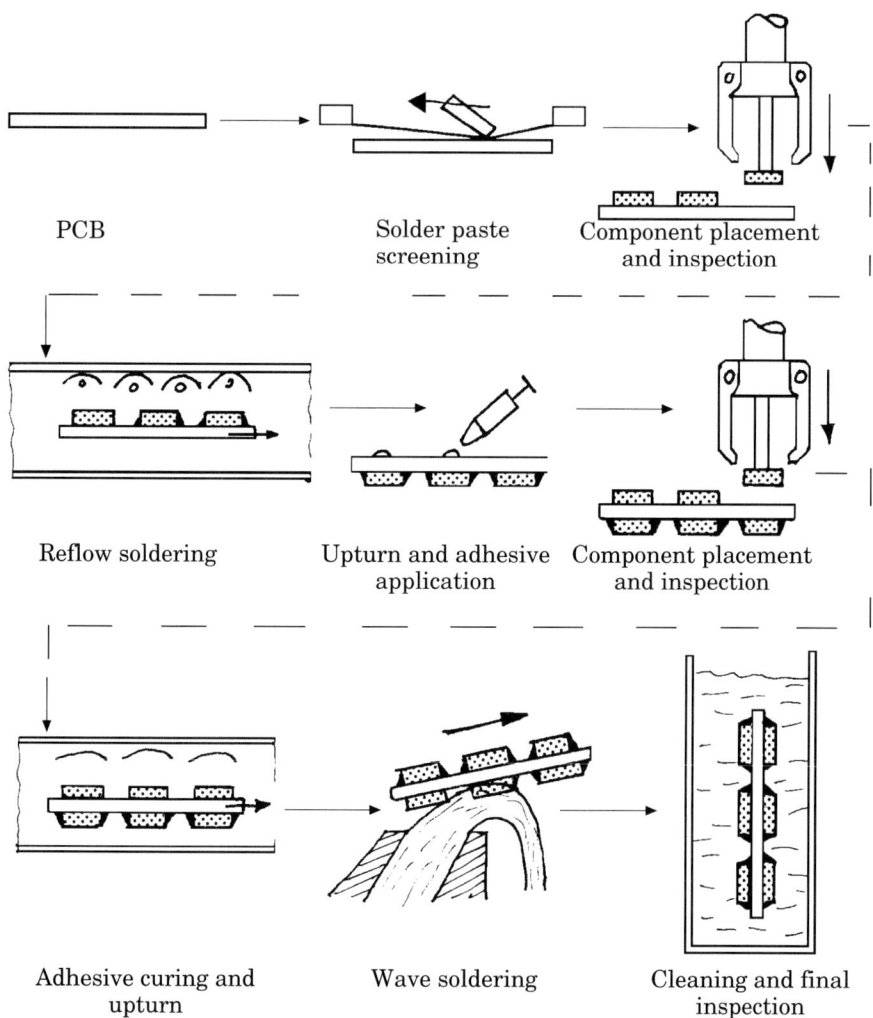

Fig. 7.2 Outline process cycle for assembly of double-sided SMAs.

(ii) PIH components are inserted as usual on the component side and secured by clinching their leads;
(iii) all SMDs are mounted on the soldering side and must, therefore, withstand immersion in molten solder;
(iv) both types of components are soldered in a single wave soldering operation.

The resulting cycle is effectively an extension of the traditional one for PIH components. Initially, all PIH components are automatically inserted in a suitable technological sequence (see Chapter 6) such as axial, then radial, followed by DIPs and, finally, odd components. The board is then turned upside down and the soldering side assembled with the surface mounting

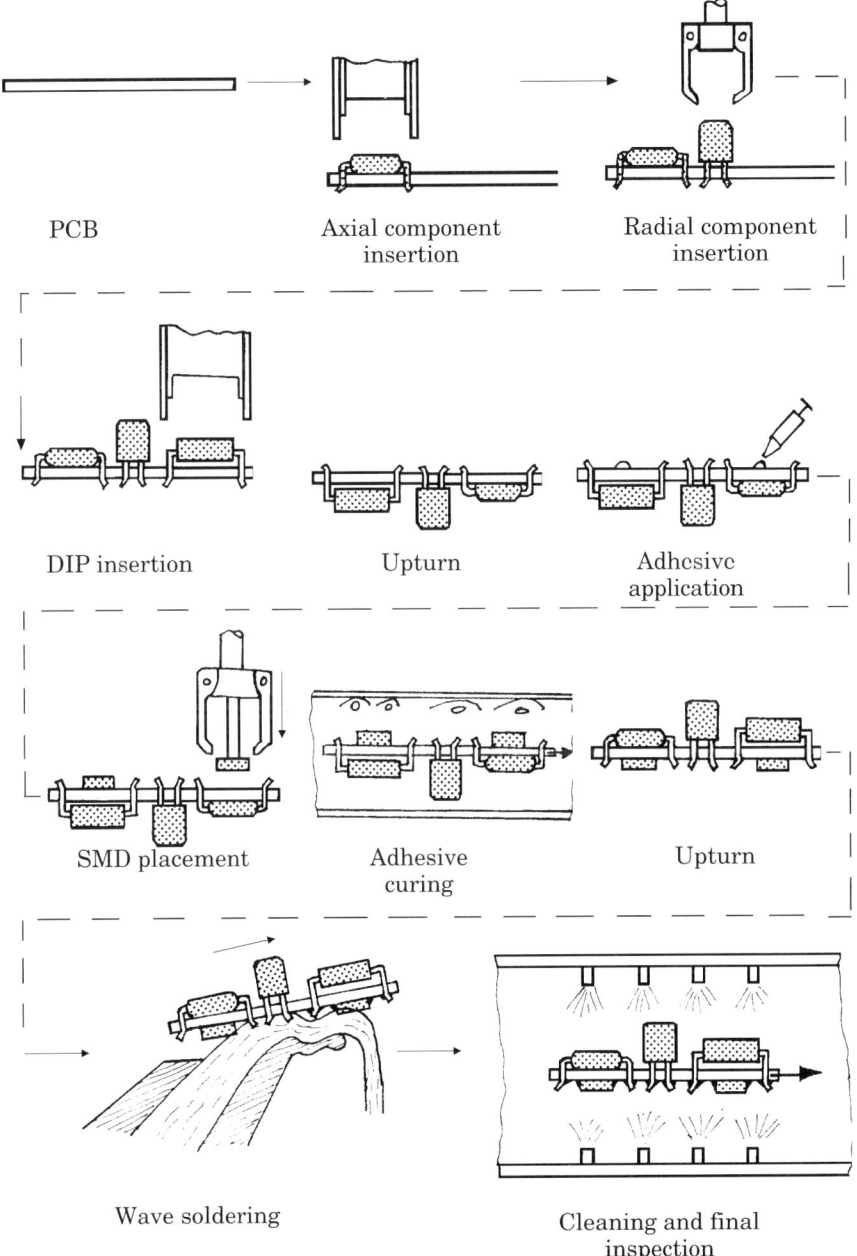

Fig. 7.3 Process cycle for a mixed print board with SMDs on one side only.

components — adhesive application, SMD placement, adhesive curing and pre-soldering inspection.

The board is turned over again and manually completed (if necessary), wave soldered, cleaned and inspected. As both PIH and SM components are passing through the same wave, the design of the board layout needs additional care and the shape of the wave(s) and the soldering parameters require ultra-careful adjustment to achieve good results.

This type of mixprint is used extensively in consumer electronics because it employs all existing equipment and adds just one more step. SMCs can be applied by parallel mounting which gives very high productivity with a limited investment. The limited increase in complexity of the manufacturing cycle arising from placing SMCs on the solder side of the board gives remarkable advantages, such as:

(i) reduction of total PCB area, often resulting in one board replacing two or more, giving savings also in terms of connectors and wiring harnesses;
(ii) reduction in board complexity — a single layer board can replace a double layer PTH board;
(iii) increased design freedom with components being placed on both sides, SMCs taking less space, jumpers being mounted on the solder side, and so on;
(iv) reduction in labour and space;
(v) increased quality and reliability.

Reasons such as these explain why the massive move to SMT started when SMCs were much more expensive than the equivalent traditional leaded components.

7.1.4 Full Mixprint

A full mixprint board has PIH components on one side and SM components on both sides. This allows the designer to obtain maximum advantage from surface mounting technology even when not all components can be supplied in surface mounting versions.

The first step in the typical manufacturing cycle for these assemblies, Figure 7.4, is to mount all SMCs required on the 'component' side of the board, usually by reflow soldering. PIH components are then inserted and clinched, SMCs placed on the 'soldering' side and the board wave soldered. When, as sometimes happens, the only PIH components used are inserted manually, this operation is carried out after adhesive curing and just before wave soldering.

7.2 SURFACE MOUNTING ADHESIVES

The adhesives with which one is concerned here are those which are used to keep the components in position from the moment they are placed on the board to the completion of the soldering operation. These materials are non-conductive and must not be confused with conductive adhesives used in some assemblies to replace soldering (see Chapter 16).

Selection of the adhesive must be consistent with the application method and equipment, the components and the substrate, and the manufacturing process.

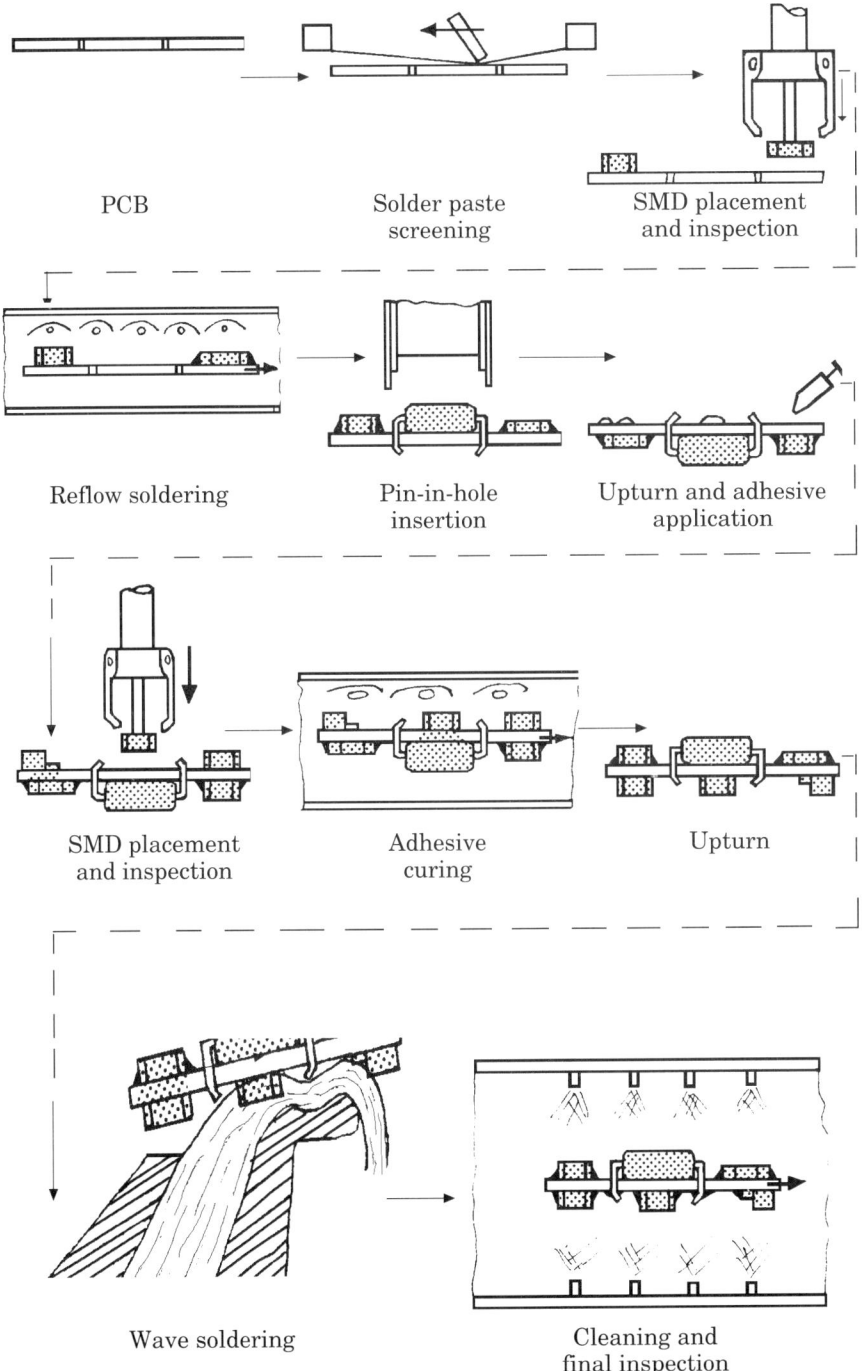

Fig. 7.4 A typical process cycle for a full mixprint SMA.

7.2.1 Requirements for Adhesives

There are many adhesives which are potentially suitable for bonding components to the substrate and their characteristics must be carefully evaluated.

Electrical

The adhesive must be non-conductive, both as applied and during the life of the equipment. It must not release ionic contaminants during curing or as a result of environmental conditions.

Chemical

As supplied, the adhesive must have a very long shelf life at normal room temperatures. Before curing it should be highly soluble in a non-hazardous solvent to allow easy cleaning of application equipment and tools. If applied to the PB off-line, the adhesive must not cure or change properties significantly, even after many hours' exposure to air. (Certain materials may have to be stored in conditions free from UV [ultra-violet] radiation.)

The adhesive should be readily cured when required, for example, by heat or by UV radiation. It shall be non-corrosive at any stage and, when cured, must not react with other materials it can contact, such as flux, solder, cleaning agents and so on. To fulfil its primary function, it must resist wave soldering temperatures for as long as is necessary (e.g., 260°C, 500°F, for at least 10 seconds) without degradation or outgassing.

Rheology

The adhesive must have flow properties suitable for the application method and must be thixotropic, retaining these properties as long as possible when heated (for curing). It must be capable of being dispensed in small quantities to form dots which fill gaps fairly easily but do not slump, spread, or flow over adjacent surfaces, in particular over the soldering lands. Many thixotropic materials become more fluid and less tacky on heating. If this is the case, the adhesive must cure rapidly so that it does not drain away from the component surface and fail to bond.

Bond Strength

The adhesive must develop a reasonable bond strength when cured, say 3 to 5 grams per dot, to all possible component and substrate materials, including solder resists. Too high a bond strength when cured can make it impossible to remove components for repair without damaging the substrate.[1]

Colour

The adhesive must have a distinctive colour so that its dots may be easily seen on all types of board material and solder resist.

Health and Safety (see also Volume 2, Chapter 18).

Ideally, the adhesive and its components should be non-toxic, non-flammable, and non-hazardous at every stage of use. If this is not practicable, methods of handling, storage and use must be carefully documented and all operators thoroughly trained so that hazards are minimised, if not eliminated.

Automatic Assembly — Surface Mount Components

No adhesive can match all requirements perfectly although several families of material can come close to the ideal. Table 7.1 gives an outline of the advantages and disadvantages of several types which have been used in electronics applications. Of these, only two systems have found widespread use.

Table 7.1

Advantages and Disadvantages of Various Adhesive Types

Adhesive	Advantages	Disadvantages
Epoxide (one- and two-part systems)	• Excellent moisture resistance	• Complex application system (two-part)
	• Excellent solvent resistance	• Higher cure temperature (one-part)
	• Good void filling characteristics	• Limited shelf-life
	• High temperature use	• Longer cure time (one-part)
	• Proven history in electronics	• Refrigerated storage (one-part)
	• UV cure systems available	• Single application method (two-part)
Cyanoacrylate	• Long shelf-life	• Bad void filling
	• One-part system	• Fair moisture resistance
	• Room temperature storage	• Hazardous application
	• Very fast bonding	• Single application method (syringe)
Acrylic	• Moderate cure time	• Application system may be complex
	• Good moisture resistance	
	• Good solvent resistance	
	• UV cure systems available	
Anaerobic	• Good solvent resistance	• Chemical activity
	• High temperature resistance	• Incompleteness of cure
	• One-part system	• Low bond strength
	• Room temperature storage	
	• Simple, inexpensive cure	
	• Unlimited shelf-life	
	• UV cure systems available	

Source: Signetics bulletins

These systems are a thermosetting type, cured by heating, and a UV sensitive type, cured by ultra-violet radiation. A major difference between these systems is the rheological behaviour on curing. While this depends on the precise chemistry of the product and on any fillers employed to influence the rheology, the general form of the viscosity versus cure time curve is dictated by the type of cure. If the material is heat cured, the viscosity will initially decrease with rise in temperature. This can lead to slumping of the glue dot, unless this is well controlled by the additives used to procure a thixotropic state. On the other hand, the UV curing systems will exhibit a steady increase in viscosity with increase in cure time.

The pot life (usable life) of the adhesive is also affected by the curing mechanism. A heat curing adhesive, usually an epoxide system, will cure, albeit very slowly, at room temperature and refrigeration in storage will be necessary. In most cases, a usable life on the line of one working day (or an 8-hour shift) can be achieved without raising the curing time/temperature beyond reasonable limits.

On the other hand, UV cured systems, usually acrylates or epoxides with photo-initiators, have a very long shelf life provided that UV radiation at the appropriate wavelength is excluded. Such systems are easily cured in about 30 seconds by conveyorised UV tunnels at wavelengths around 350 nanometres. It is essential, of course, that sufficient adhesive protrudes from under the component to start polymerisation. Once initiated, curing can proceed by chain reaction, aided by the exothermic nature of the mechanism, to achieve cure in areas shielded by the component body from the radiation source.

7.2.2 Application Methods

Several alternative methods are available for in-line and off-line application of adhesive, depending on the type of adhesive, the component(s), placement equipment and method of cure.

7.2.2.1 PIN TRANSFER

Pin transfer is one of the most widely used methods and is extremely versatile. The adhesive is contained in a reservoir and maintained in a flat, level condition by a squeegee which moves over the surface before the pin enters the adhesive to pick up a drop for transfer to the PB, Figure 7.5. This ensures that the pin is immersed to a constant depth. The dimensions of the drop depend on the diameter and shape of the pin, the depth of immersion in the reservoir, and the rheology of the adhesive.

The pin is moved over the board and lowered very close to it so that a dot of glue is transferred. The pin is then lifted and returned to the reservoir to start a new cycle. If necessary, the pin can be cleaned by a brush or other suitable means before return to the reservoir.

The critical controls are those on the adhesive and on the pin. The amount of glue taken up by a given pin depends on the physical properties of the adhesive. These must be kept reasonably constant, bearing in mind that the properties will change with temperature and exposure to air and dust. In addition, heat curing materials in particular will be advancing in cure and changing in viscosity and the reservoir must be cleaned out and the contents discarded at least at the end of each working shift.

To ensure a suitable size and shape of dot, the pin must stop at a closely controlled distance from the PB and not come into contact with it. This would be impossible to achieve with warped boards but for the fact that it is possible

Automatic Assembly — Surface Mount Components 277

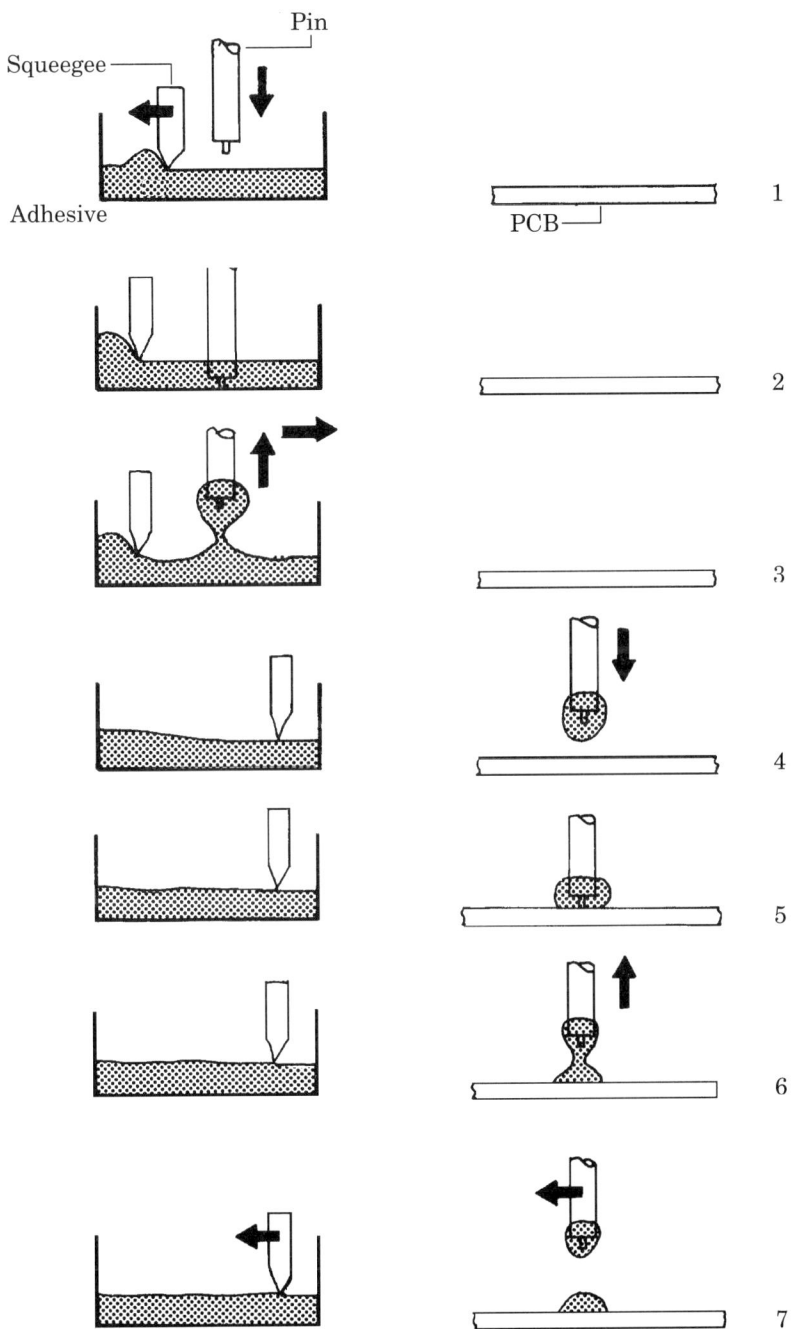

Fig. 7.5 Application of adhesive by pin transfer. Single or multiple pins may be used, depending on the size of board.

to have a small projection on the end of the pin, Figure 7.5, which assures the correct distance without changing the shape/mass of the drop significantly. The pin, of course, must be spring loaded to avoid excessive pressure on the PB.

Provided these elements are controlled, the method is simple and works quite well. It is useful for boards pre-assembled with PIH components, as in mixed assembly or mixprint boards, as projecting leads do not interfere with the pins, Figure 7.6, and for large components — packaged in cases like the SO, SOJ or PLCC — two or more dots of adhesive can be applied. In such instances, a heat setting adhesive may cure more reliably.

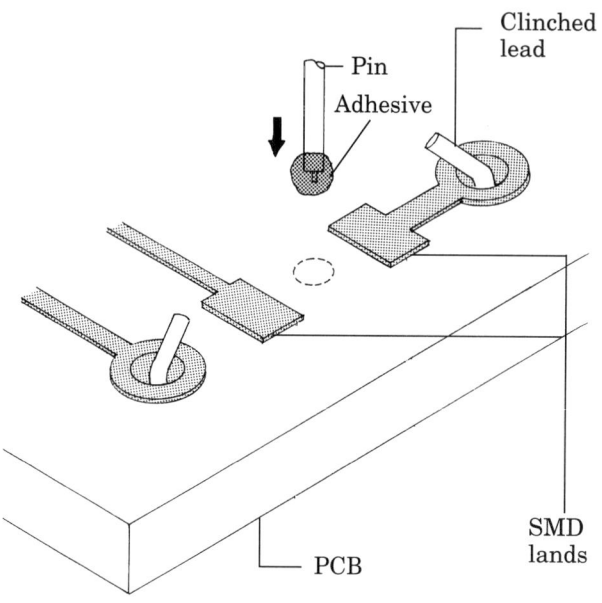

Fig. 7.6 Pin transfer of adhesive does not interfere with projecting leads on mixed assembly (mixprint) boards.

Relatively simple CNC equipment can be programmed to apply the glue dots in the desired pattern, either on- or off-line. On-line application of adhesive is to be preferred as this will minimise the risk of contamination and/or premature curing of the adhesive.

Multiple or *parallel* pin transfer is an extension of this method, in which many pins operate in parallel so that a dot of adhesive is applied to all positions at the same time, Figure 7.7. Spring-loaded pins are fitted into a plate in all those positions where a dot of glue is needed on the board. The plate moves over the adhesive reservoir and then, in a fixed position, over the PB. Up to a few hundred dots may be applied simultaneously by relatively simple equipment, which can be made up as a station of a large assembly line.

Automatic Assembly — Surface Mount Components

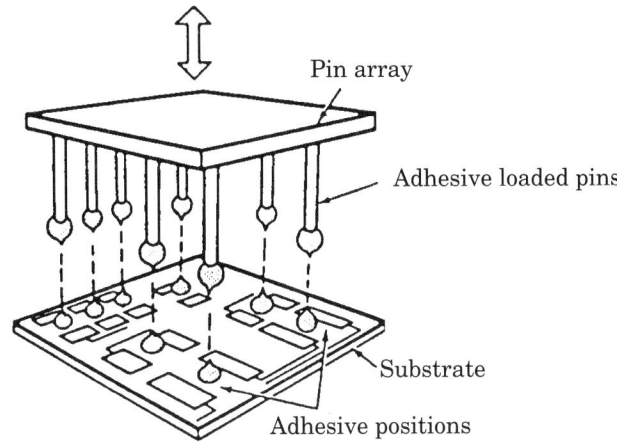

Fig. 7.7 Pin transfer of adhesive by pin array directly on to the substrate.

Obviously, a plate specific to each type of board is needed and the method is most suited to mass production of large batches of a few part numbers.

Another variation of pin transfer, *reverse pin transfer*, still relies on a pin to pick up the adhesive, which is then placed on the bottom of the component and not on the board, Figure 7.8. The operation is always performed on-line in a fixed position, as soon as the component has been picked up by the pick-and-place equipment, and can be activated by the same software

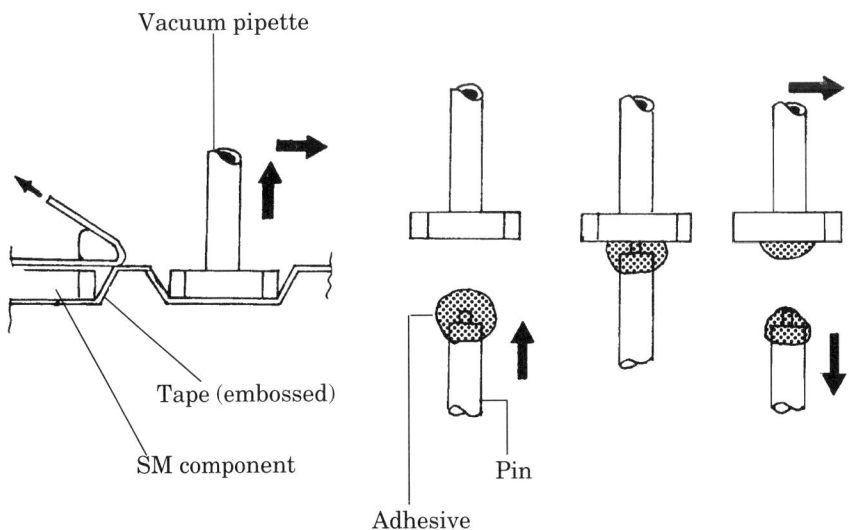

Fig. 7.8 In reverse pin transfer the adhesive is applied to the component, not to the board. This operation is always performed on-line by the mounting equipment.

program as the mounting machine. It does not, therefore, need specific tooling and is very suitable for small batches.

Reverse pin transfer has disadvantages, however. The operating mechanism is necessarily more complex and the method is not really suitable for cylindrical components. In addition, the amount of adhesive which can be applied is limited and may be inadequate where reasonable gap filling is needed. (This may be overcome to some extent by provision of dummy tracks on the board — which dictates recognition of the need at the design stage.)

7.2.2.2 SYRINGE

Dispensing the adhesive from a syringe, Figure 7.9, is usually confined to small boards and an assembly rate of less than 5,000 components per hour. However, high speed automatic dispensers are available which will produce between 12,000 and 18,000 (or more) glue dots per hour.[2] The syringe can be part of the programmed mounting machine, essential for high speed dispensing, or it can be hand-held for prototype and very small batch work.

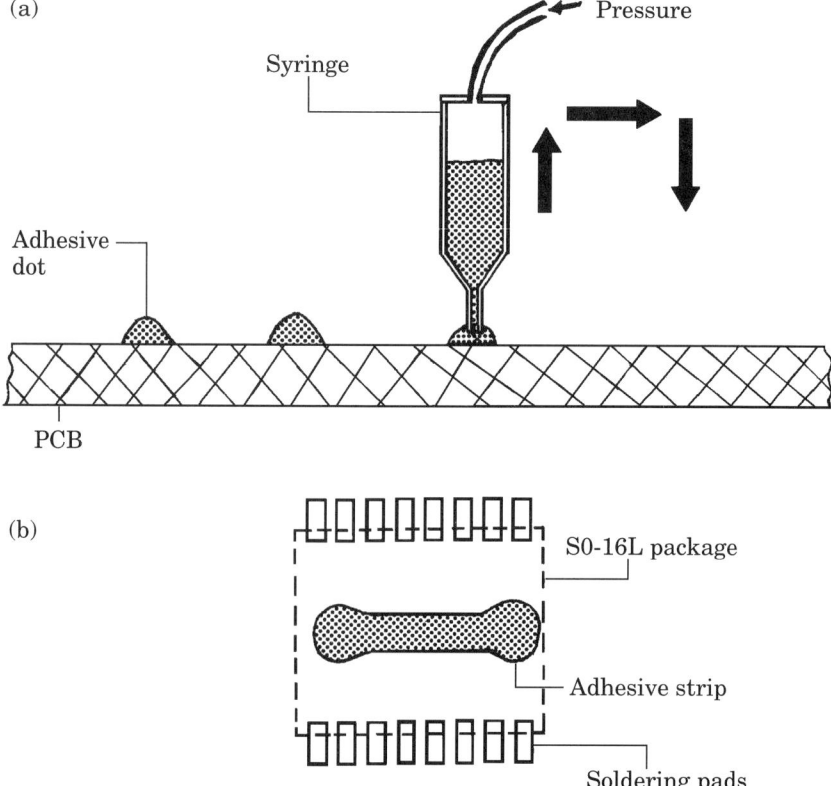

Fig. 7.9 A syringe may be used to dispense a wide variety of adhesives. (a) Dot size is controlled by the nozzle size, the pressure pulse strength and duration, and the properties of the adhesive; (b) on larger packages, either several dots or a strip of glue may be applied.

In either case, the syringe is operated by pulses of air pressure, each pulse dispensing a drop of adhesive.

The method allows for a large variety of glue types, including cyanoacrylates but excluding anaerobics. If necessary, the syringe can be operated by inert gas to avoid contact of the adhesive with air. By adjusting the pressure pulse, it is possible to control the amount of glue to give different dots for different sizes of component.

For the largest devices, a strip of adhesive, Figure 7.9(b), may be applied by moving the syringe during the pressure cycle.

By using several syringes controlled by independent CNC mechanisms, it is possible to have a limited degree of parallelism. However, the process needs to be tightly controlled and the syringes require considerable maintenance.[2]

7.2.2.3 SCREEN PRINTING

Screen printing is described in detail in Volume 2, Chapter 3 of this book. The method is intrinsically parallel, all dots being deposited in one operation. Tooling is straightforward and can be prepared quickly with simple equipment.

For adhesive application, it does have two important drawbacks. The first is that screen printing can only be used on bare PBs, before any components are assembled. The second is that control of adhesive thickness is rather limited. The latter problem may be reduced by replacing the screen with a metal stencil with holes where the adhesive is to be applied. Although tooling is more complex, this gives more control over thickness and the metal stencil is easier to maintain than a screen stencil.

A comparison of screen printing with other adhesive application methods is given in Table 7.2. Although it has some advantages, it remains the least used method. On the other hand, it is the most common technique for applying solder pastes and creams.

7.2.3 Adhesive Dot Criteria

The applied dot of adhesive must have the right shape and dimensions for the component concerned. In general terms, the dot should be roughly hemispherical with a base diameter of from less than 1 mm up to 2.5 mm (0.04 in. to 0.1 in.) and a height of about 0.3 mm to 0.8 mm (0.01 in. to 0.03 in.) giving a volume of about 0.1 mm^3 to 2 mm^3. The smaller figure is suitable for chip resistors or capacitors of up to 1206 case size, the higher values being preferred for larger and cylindrical components.

Sometimes the terminations of a flat component are quite thick and, together with soldering pads of, say, 100 µm thickness, can raise the body of the component to a height above the board surface which needs a dot of adhesive too high for the chosen method of application. In such cases, if a conductor is not present, a dummy track may be placed between the soldering pads to lift the dot, as shown in Figure 7.10.

It was pointed out earlier that devices in large packages, such as ICs in SOJ, PLCC or QFP cases, might require two or more dots or a strip of glue. A further consideration arises with such packages. To assist cleaning of the assembly, the leads of such packages are shaped to provide a clearance of

Table 7.2
Adhesive Application Methods — Advantages and Disadvantages

Method	Advantages	Disadvantages
Pin transfer	• Compact system	• Dependent on adhesive uniformity
	• Control of adhesive quantity	• Flat surface only
	• Easy cleaning	• Open system
	• Little maintenance required	
	• Simple process	
	• Simultaneous dot placement	
Screen printing	• Easy cleaning	• Flat surface only — with no obstructions
	• Simple process	
	• Simultaneous dot placement	• Open system
	• Small thickness	• Screen maintenace required
	• Uniform dot size	
Pressure syringe	• Accommodates irregular surfaces	• Difficult cleaning
	• Closed system	• Fewer dots simultaneously
	• Control of adhesive quantity	• Large size of system
	• Handles most adhesives	• Lot of maintenance required
	• Uniformity	

Source: Signetics bulletins

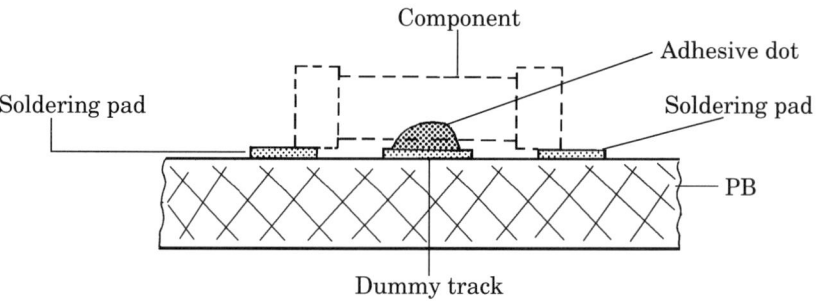

Fig. 7.10 To avoid using excessive amounts of glue, which may slump and contaminate soldering areas, a real or dummy track can be sited under the component.

0.25 mm to 1 mm (0.01 in. to 0.04 in.) between the body and the seating plane. When added to the thickness of the soldering pads, this again means that the dot height may be too low. The substrate under such components is usually densely populated by conductors and a solder resist may be used to fill the gaps between them or to bridge them, as in Figure 7.11.

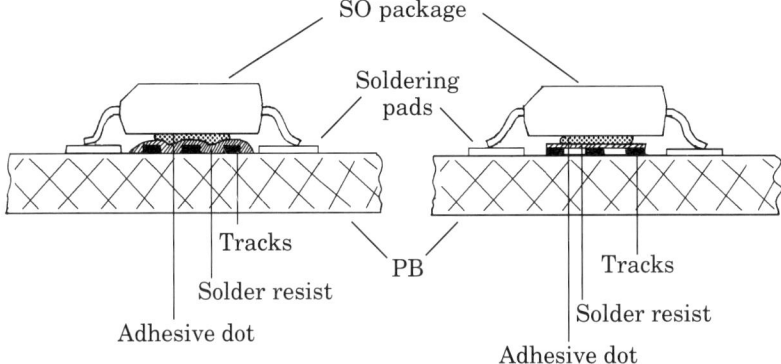

Fig. 7.11 Solder resist may be used to fill the gaps between tracks under large packages to reduce the need for extra adhesive.

Special care must be taken to avoid contamination of soldering areas with adhesive, including mounting holes in mixed technology boards.

7.2.4 Storage of Boards

After adhesive application, boards should not be stored at all, other than for a very short time. Even if the adhesive is stable with regard to curing, surfaces can be contaminated by gases, moisture, dusts, etc., with detrimental effects on the ability of the glue to wet the component body and develop the expected bond strength. To minimise problems, on-line application of the adhesive is recommended, preferably immediately prior to placing the components.

After device placement, the adhesive must be cured as soon as is possible, ideally in a tunnel oven, in-line. If boards have to be stored prior to curing, they must be stored horizontally as uncured adhesives are liable to creep. (Creep is a slow deformation under stress or pressure which occurs frequently with thixotropic fluids under their own weight.)

7.2.5 Adhesive Curing

Adhesives may be cured in a variety of ways, depending on their type. The great majority of adhesives employed for component security are cured by heat, or by UV radiation, or by a combination of these.

Experiments have been performed with two-component systems in which the hardener is coated on to the PB and the base resin is applied (by pin transfer or syringe) immediately before the component is placed. The hardener diffuses rapidly through the resin, producing a cure at room temperature without added heat or radiation. The adhesive in the reservoir has a long life and the hardener system can be very active, producing rapid curing. However, application of the hardener to the board is an additional operation which must be carried out off-line and thorough removal of the excess is necessary during cleaning.

Heat curing is commonly employed with several types of adhesive. Conveyor belt ovens can be used, being loaded directly from the pick-and-place station. The essential requirements are close control of temperature profiles and a smooth transport mechanism. There has to be a compromise between the pot life of the adhesive and the cure temperature and time. An adhesive with a very long pot life may need to be cured at a temperature/time combination which exceeds that which the components can withstand.

UV curing is also popular. The equipment may be quite simply set up, with lamps of the recommended wavelength angled on both sides of a conveyor belt to give maximum coverage. Obviously, the radiation will not reach the adhesive under a flat component. One solution to this is to apply two dots of adhesive, one on either side of the body, as in Figure 7.12. The problem does not arise with cylindrical components as, if the dot is well aligned, the adhesive will squeeze out evenly on both sides.

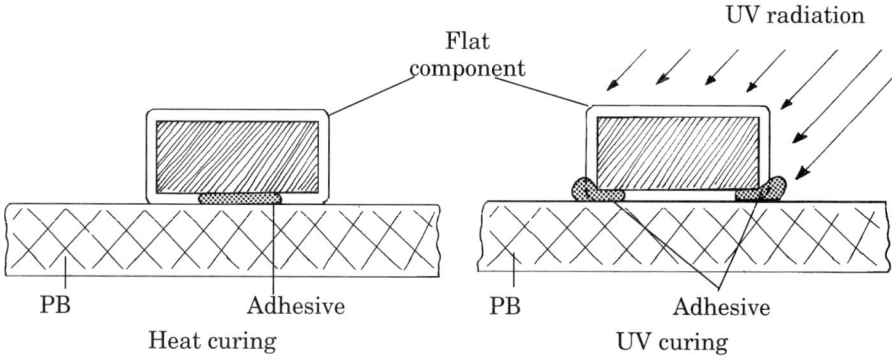

Fig. 7.12 Heat cured adhesives can be hidden by the component body but UV radiation needs to 'see' the adhesive to initiate cure.

A combination of UV radiation and heat radiation can solve many problems. For example, a typical acrylic SMD adhesive may be cured in several ways, such as:

(i) Circuit with MELF components only — high intensity UV for 30-45 seconds, or low intensity UV and heat for 3 minutes at 120°C, or heat alone, such as 5 minutes at 120°C or 3 minutes at 150°C or 30-50 seconds in an infra-red (IR) tunnel oven (but see Figure 7.45)

(ii) Circuit with MELFs and flat chips — high intensity UV for 30-45 seconds followed by heat for 5 minutes at 120°C or 3 minutes at 150°C or 30-50 seconds in an IR tunnel oven.

(iii) Circuit with flat components only — heat alone as recommended by the supplier.

7.3 SOLDER PASTES (see also Chapter 13, 13.6)

Solder paste or cream is used to supply the additional amount of solder needed to make properly formed joints when soldering is carried out by reflow techniques. It consists of small solder particles dispersed or suspended in a flux binder. The flux binder rheology and the shape and size of the solder particles are highly important and must be tailored to the method of application.

Solder pastes may be applied by several methods, although only two are used in common practice — syringe and screen printing. Syringe application is used for prototype and very small volumes as well as for cases where screen printing cannot be used because components or leads protrude from the board surface. Screen printing is the principal method.

The screen material is either stretched wire mesh impregnated with a photosensitive emulsion or a solid thin metal with etched or machined apertures. Solid screens for non-critical work can be cut from almost any flexible plastic material which will resist the solvents used in the creams. Mesh screens are described in Volume 2, Chapter 3, and the only point which will be made here is that, if the use of a mesh screen for minimal aperture dimensions cannot be avoided, the size of the aperture between the wires in the mesh will dictate the smallest accurate print size. If the aperture in the stencil material covers only two or three apertures in the wire, its location in relation to the mesh will be critical. This is not always possible to arrange and, when it is possible, it is always difficult and expensive. Typically, for a 300 mesh screen, the smallest recommended aperture width for reasonable control without special alignment precautions is 0.5 mm (0.02 in.).

Solid metal screens are more costly but last much longer and are preferred for large production runs. They deliver more accurate volumes of solder paste for very small apertures and the dimensions of the apertures are not restricted as they are by mesh screens. Solid screens are normally used as stencils in an 'in-contact' mode. There is no spring-back, the whole frame being raised away from the PB. Stencil screening requires a different type of paste, a different material for the squeegee, and more precise equipment.

Selection of a solder paste can only be made after exhaustive trials. A solder paste or cream is a complex material.[3,4] The shape and size of the solder particles have a major effect on the oxide content of the paste, the amount of solder balling, and the definition achievable by the paste on screen printing. The binder contains not only the flux but also additives to retain the solder particles in suspension, to control the rheological properties of the paste, and to support and retain components placed on it, without inhibiting the cleaning action of the flux during soldering. Most solder pastes also contain solvents so that their behaviour will change continuously when exposed to the atmosphere.

Recent developments, brought about by the need to phase out CFCs and by the cost of alternative cleaning materials, have produced pastes formulated to leave very low amounts of non-corrosive and non-conductive residues on the printed board assembly. Further development has produced rosin-free pastes, the flux, binder and solvents in which are almost completely removed by heat and a controlled reactive atmosphere in the IR furnace used to reflow the paste. As the atmosphere also reduces oxidation, solder wetting is promoted and, it is claimed, exceptionally good soldering without solder balls is obtained.[5,6,7] Further discussion of controlled atmosphere soldering and its advantages will be found in Chapter 15.

The result of this is that pastes and creams cannot be readily standardised, nor can one manufacturer's material be substituted easily for another's. Having selected a supplier, a composition and a method of application and use, no change can be made without a long and tedious series of tests to establish the new operating parameters.

Off-line application of solder paste is to be preferred, since inspection of the printed board is desirable. The layer of paste on the footprints should have a thickness of 100 to 300 µm (0.004 in. to 0.012 in.). If thinner than 100 µm, defective joints can result; if thicker than 300 µm, excessive slumping and/or bridging between adjacent pads may be experienced. It must be remembered that the pads to be printed are often very small and closely spaced. If a pad has not been properly printed with paste, because of, for example, a clogged screen, it can be very difficult to see the defective joint on inspection of the reflowed board. If the defective print is seen at the 'as-printed' stage, it is very easy to wash off the board(s) and rectify the screen before reprinting.

After the boards have been printed and checked, the components should be mounted as quickly as possible. Paste exposed to the atmosphere gradually loses its tackiness and the particles of solder may oxidise further. Reflow of the solder should normally take place within a few hours of printing and component placement, unless a minimum period of air drying is stipulated.

Solvents included in solder creams and pastes are usually of high boiling point. This is necessary to avoid too fast evaporation leading to printing difficulties and excessive thickness variations as printed. However, entrapped solvents can cause spitting and lead to joint disturbance and voids. This may be avoided by drying the paste. On-line drying in equipment positioned immediately before the reflow station may be adequate. A better result overall may be obtained by oven drying, typical guidelines given by manufacturers ranging from 1-2 hours at 50°C to 5-20 minutes at 90°C and even 10 seconds at 170°C. The drying method must be discussed fully with the manufacturer. Only the manufacturer will know the precise balance of solvents and how the evaporation of these is affected by the shape, size, and content of solder particles in the paste. There is no standard drying time. Too short a time or too low a temperature will leave solvent behind to cause voids. Too long a time or too high a temperature will cause excessive oxidation of the solder which will cause voids as a result of reaction products between the flux and the oxides.

7.4 CONDUCTIVE ADHESIVES

The use of conductive adhesives to replace solders is growing in popularity. This joining method considerably simplifies the SMA cycle by doing away with the need for fluxing, soldering and cleaning (to remove flux residues, etc.).

Conductive adhesives have been employed for a long time in electronics and are available from several sources. Their major uses are described in Chapter 16. In surface mount assembly, the SMCs are placed on to dots of uncured conductive resin in exactly the same manner as components for reflow attachment are placed on to dots of solder paste. The populated board is heated to cure the adhesive after which assembly is complete.

The most popular adhesives which have been employed are based on silver-loaded epoxide resins. Other conductive fillers, notably silver/palladium and copper, have been used. Gold is extremely expensive in this application (most conductive adhesives contain approximately 20% by volume conductive filler in the cured material).

The conductive resins can be applied to the PBs by pin transfer, syringe, and screen or stencil printing just like mounting adhesives and solder pastes. Curing temperatures are similar to those of heat cured mounting adhesives. The final properties of these materials depend upon the precise resin formulation but are very compatible with epoxide resin based laminates. T_g is of the order of 80°C to 100°C, high enough for normal service and low enough for easy repair. (Certain types of conductive adhesive are based on polyimide resins which have higher T_gs but have not as yet found much use.) Conductivity depends upon many factors, including the shape, particle size, nature and amount of the conductive material. For pure silver powder giving a metallic content of about 20%, the volume resistivity of the cured adhesive is in the range of 1 to 10 micro-ohm.m, compared with a value of 0.17 micro-ohm.m for 60:40 tin-lead solder.

This relatively high value of volume resistivity may be a major factor in deciding whether or not to use conductive adhesives for mechanical and electrical joining of components to tracks. The high resistivity will have an adverse effect on the high frequency performance of the circuit. Another major disadvantage is that it is very difficult to inspect the joints visually and there is no real alternative method of inspection available.

7.5   COMPONENT MOUNTING

Assembly automation has been a major reason for the success of surface mounted technology, having had an enormous impact on production costs and quality. In turn, SMCs have been developed with automatic placement on the board in mind. Their small size and shape, the frequent absence of identification marks on the body, and the need for very precise placement confine manual assembly to a marginal area, essentially prototypes. Even small production volumes are usually assembled automatically on simple software programmed machines or on highly versatile robots.

In the descriptions of SMA machinery which follow, the aim is to focus more on key equipment modules and types than on specific machines, the evolution of which is still too rapid to be able to be fully up-to-date in a book of this nature. In writing this chapter, over 70 different models of equipment offered by over 30 manufacturers have been checked. (Companies checked literally covered most of the alphabet of equipment manufacturers — Amistar, Arcotronics, Automelec, Blakell, Citizen Watch, CKD, Contact Systems, Dima, Dynapert, ECG, Eurosoft Robotique, Fabrilec, Fuji, Ismeca, Kyushu Matsushita, Mamiya Denshi, Matsushita, Mydata, Nissei Sangyo, Nitto Kogyo, Okano Electric, Peter Jordan, Panasonic Factory Automation, Philips, Quad Europe, Siemens, Takachiho, TDK, Tenryu Technics, Universal Instruments, Yamagata Casio, Yamaha Motors, Zevatron.) However, new suppliers and new models appear almost daily. Fortunately, it is fairly easy for the earnest reader to update on available machines as most suppliers participate in the major exhibitions for electronic production equipment.

7.5.1 Evolution of Pick-and-place Equipment from Manual Assembly

In Chapter 5 it was made clear that even manual assembly of SMCs requires some support. The simplest tools are tweezers, a syringe and an illuminated magnifier. The next tooling step is usually designed to aid component selection and reduce errors, by arranging a set of containers, one for each type and value of component, so that they repeat the assembly sequence. Pick-up of SMCs can be made more easy by using a vacuum pipette with a plastic or rubber tip and a switch on the handle to turn the vacuum on and off.

Unfortunately, this still requires an operator with a very steady hand, even at the end of a long and tiring day, and it is very slow. Manual assembly of SMCs is at least three times slower than manual insertion of PIH components and is very difficult with some ICs, which have many closely spaced leads on two or four sides. More equipment assistance is necessary.

The simplest way to steady the hand is to provide a pantograph type unit which has the vacuum pipette on one side of the arm and a handle with switching on the other, Figure 7.13. This may then be combined with a rotary carousel, holding a number of small dishes for components, a PB holder, and a mechanism which rotates the tip of the pipette when the operator rotates the handle.

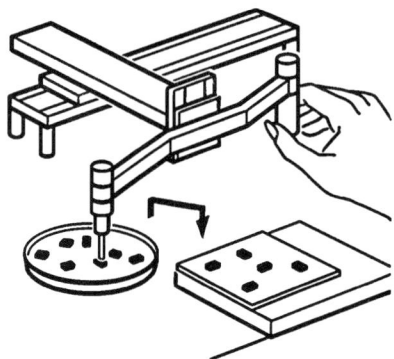

Fig. 7.13 A simple pantograph-style unit for assisting manual placement of SMCs.

This approach culminates in the type of equipment seen in Figure 7.14. The machine has a pantograph mechanism, an accurately located board holder, a template, a component holder, and a pick-and-place head with a vacuum tip, which can be moved up and down by a switch on the control unit. By moving the probe head assembly, the operator positions the tip over the selected component and picks it up by operating the switch with the other hand (the vacuum can be actuated by a foot pedal). The probe is moved to that

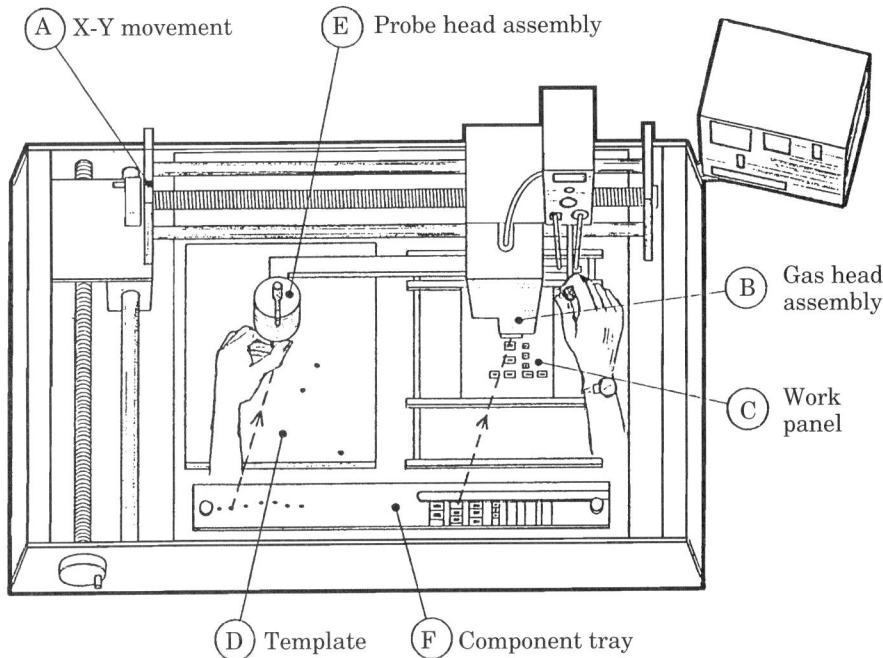

Fig. 7.14 A pantograph operated pick-and-place machine which may be used both to mount SMCs and to remove them for repair. (Courtesy of Computer Recognition Systems Ltd, Packman Division)

position on the template corresponding to the component position on the board and the component placed by actuating the switch.

The pantograph again leads the way to automation. The particular equipment shown in Figure 7.14 has a wide range of accessories applicable to surface mount, from a stereo microscope to a set of nozzles for connection to a hot gas supply to reflow solder the component after it has been placed. The equipment may therefore be used both to place and solder components and to de-solder and remove defective components, Figure 7.15. It may therefore be a useful adjunct to fully automatic equipment in any assembly shop.

7.5.2 Automatic Mounting

There are two basic families of automatic mounting machine, the *pick-and-place* family and the *simultaneous* family. The former picks up one component at a time and places it on the board; the latter operates on many components in parallel and is suitable for very large production volumes. The effect of mounting method on the relative placement cost per component is illustrated in Figure 7.16.

Selection of mounting equipment is a critical step which can only be taken after careful consideration of the functions which it can perform. A simple point, which has been overlooked on occasion, is the size of board which is to be assembled. Some machines and modules are derived from earlier units

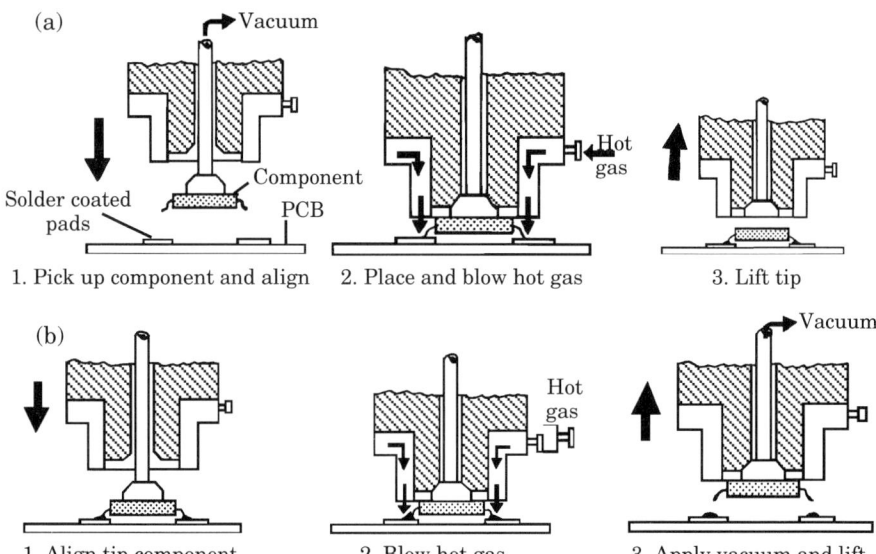

Fig. 7.15 Hot gas may be used efficiently for soldering and desoldering if jets and shrouds are well designed. (a) Reflow attachment using hot gas; (b) desoldering components using hot gas.

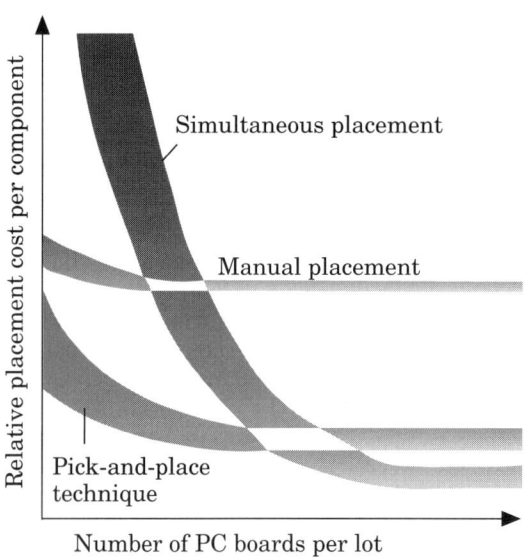

Fig. 7.16 The effect of mounting method on placement cost per component. The actual values and crossover points depend on many factors such as type of component, number of components per board, methods of attachment, and so on.

used in hybrid IC manufacture and can accept only small boards; others have been developed specifically for SMA on PBs and will assemble boards of 460 x 460 mm (18 in. x 18 in.).

7.5.3 An Automatic Mounting Machine

Equipment for automatic surface mounting assembly is very complex and includes many functions and modules dedicated to specific tasks. Such a machine is shown in Figure 7.17, a simplified diagram of the *Siemens MS-72* (first developed in 1982 and still selling as a low cost, high quality entry level machine in the early 1990s).

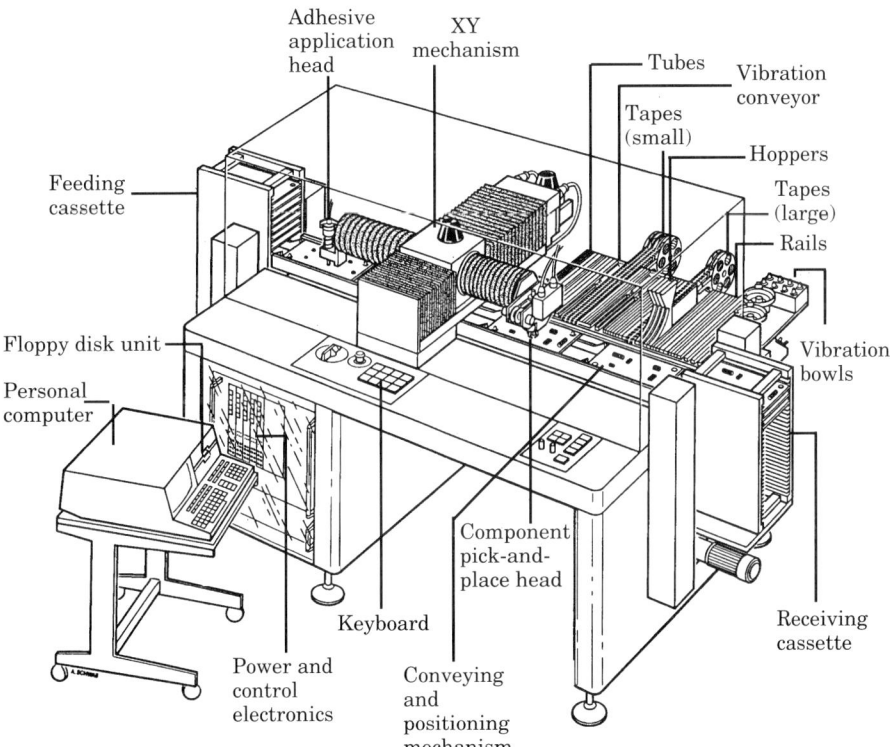

Fig. 7.17 A typical pick-and-place machine for surface mount assembly (based on the Siemens MS-72 equipment).

Preferably, such equipment should operate unattended, which includes being able to load the PB from a conveyor or cassette, position it accurately on the equipment and unload it, when assembled, on to a conveyor or cassette. When needed, it must also be able to apply adhesive by the preferred method before placing the components.

The machine must be capable of picking up several types of component, packed in a variety of ways (bulk, stick magazine, reeled tapes of different sizes, etc.), and selecting the right part number for each mounting position.

The operating head, i.e., the element which picks up and places the component, must be able to centre it with respect to its own axis and orientate it as required. Centring and rotation should occur while the head is moving to the placement point, in order to save time.

On most machines the head (and the adhesive applicator, if fitted) is the only moving part, PB and component holder(s) being fixed although in some cases they too may move in order to reduce the run of the head. In addition, if a component has not been picked up or has been lost between picking and placing, the machine should recover the error or at least signal the operator. The same should also happen if there is a fault on the board(s).

This requires well designed machinery and clever electronics. Modern machines have a lot of electronics, very often including a support minicomputer or personal computer. This helps in writing and loading assembly programs, in controlling the machine, fault finding, compilation of statistics, and so on.

7.5.4 PB Loading and Positioning

On some automatic mounting machines boards are loaded/unloaded manually, being precisely located in the placement area with the aid of pins which fit reference holes in the board, Figure 7.18. The mounting plate may be drilled and connected to a vacuum pump to assist in fixing the board, even if warped or twisted. If the underside of the board has been assembled, the board holder may have to be relieved by machining and becomes a tool specific to that board, unless specially designed to be flexible.

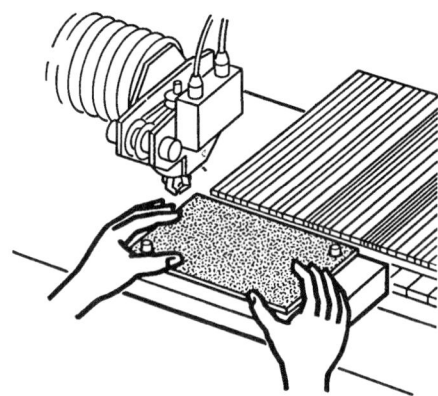

Fig. 7.18 Manual board positioning requires the full-time attention of an operator.

Manual changeover of boards enforces full time operator attendance and slows down the production rate of the machine. The usual techniques to increase productivity can be applied if the boards are small enough (multiple PBs, tools holding two or more boards) but these still need operator attention.

In certain cases of single-sided SMA, notably on ceramic substrates, automatic loading and unloading can be simplified by positioning always in the same way and at the same points with reference to the border of the substrate. However, several methods for loading the PBs and positioning them precisely on the work area are available. These fall into three basic classes, shown in Figure 7.19:

(i) cassette to work area to cassette;
(ii) cassette to work area to conveyor (or *vice versa*);
(iii) conveyor to work area to conveyor.

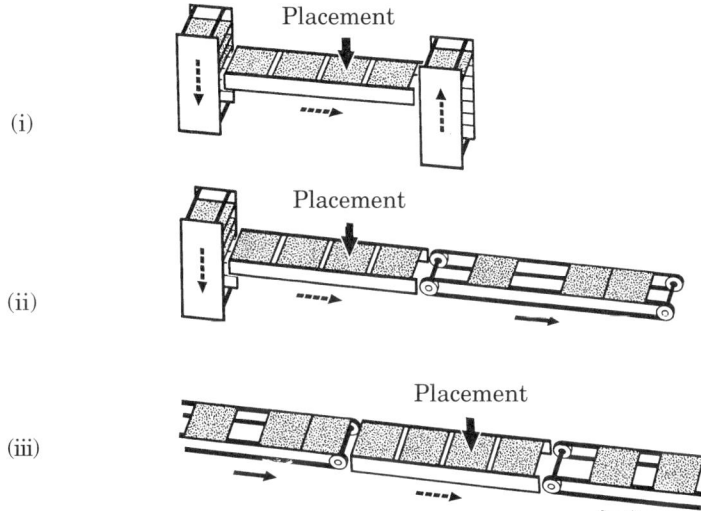

Fig. 7.19 Feeding the work area (placement area) automatically may be done (i) cassette to work area to cassette; (ii) cassette to work area to conveyor; (iii) conveyor to work area to conveyor.

The last method is most suited to very high production volume in large batches. As discussed under 'Board Handling' in Chapter 6, if cassettes or magazines are used for the boards, they should be standardised throughout the assembly shop. Automatically driven vehicles can be used to take the cassettes from one machine to the next, saving indirect labour and eliminating conveyors.

With automatic board changeovers, a key area is that of the reference fixing holes. Many equipments require one slot and one hole on the same side of the board, with an area free from components around each of them, Figure 7.20. The exact positions and dimensions must be checked in the manual for the specific machine. The feeding mechanism and the carrier will impose restrictions on boards pre-assembled on the underside. These will be similar for most equipment, Figure 7.21, but must be checked.

7.5.5 Component Feeding

The large variety of supply packs for SMDs, see Chapter 4, is a matter for concern with automatic equipment, as most types of machine will accept only a few types of pack, which can restrict the choice of component suppliers.

Re-packing to suit the machine is possible but is time consuming and the necessary equipment is seldom available.

Machines capable of accepting several types of pack, Figure 7.17, have a considerable advantage over those which are limited to one or two alternatives

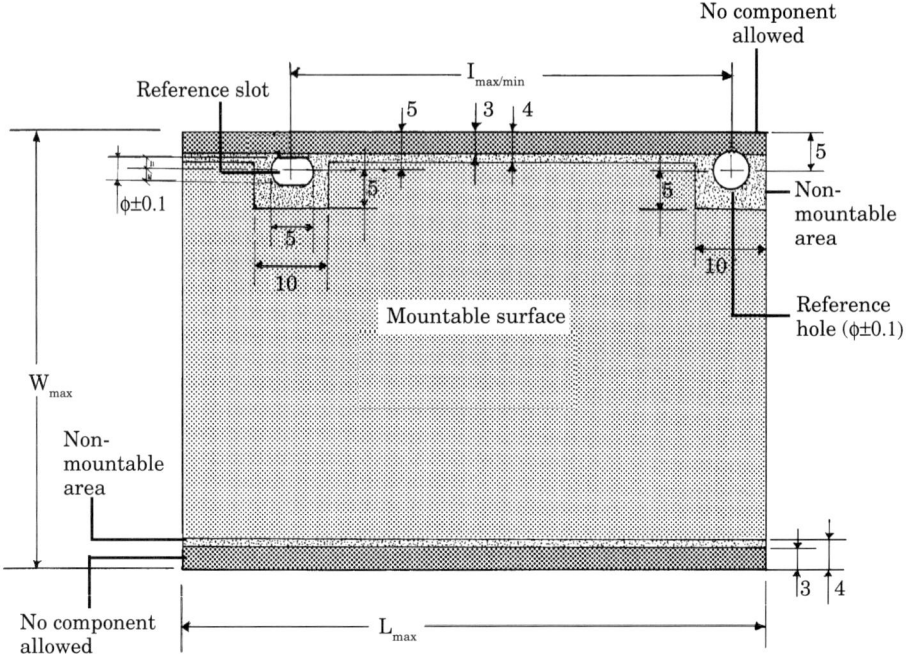

Fig. 7.20 Typical layout of reference holes for automatic positioning of boards for component mounting. (Dimensions are in millimetres.)

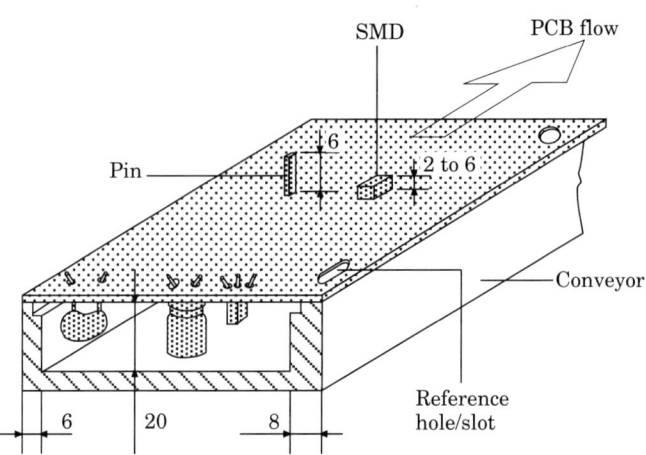

Fig. 7.21 The feed mechanism and carrier impose restrictions on the height of components pre-assembled on the underside of the boards. Typical dimensions are shown in millimetres.

(such as only 8 mm and 12 mm wide tapes) as the same component can be packed in different ways by different manufacturers.

The ideal machine should be capable of using *all* the feeding mechanisms mentioned in the following text.

7.5.5.1 RAILS AND TUBES

On automatic machines, feeding from rails and tubes, generically called stick magazines, is used only for components needed in small quantity, e.g., one per board, or for the largest packages that cannot be obtained on tape. Bulk packed components are sometimes pre-loaded into magazines. However, if they are used in large quantity, other methods are available.

The sticks are slightly inclined towards the pick-up end, so that, when a device is removed, the remainder slip down as a result of the machine vibrations. This works very effectively on machines which move the component holders. On other machines, especially with large components in plastic sticks, trouble may be experienced, and the sticks may have to be placed on a vibrator which holds several of them, depending on their size.

7.5.5.2 VIBRATORY BOWL

Loose components, received in bulk packs, may be fed to the picking head by a rotary vibratory bowl, Figure 7.22. Such a unit can be adapted to several types of component, e.g., flat, MELF, mini-MELF, etc., by adjusting a few small baffles. Special versions, for packages like the SOT-23, SOT-89, etc., are also supplied.

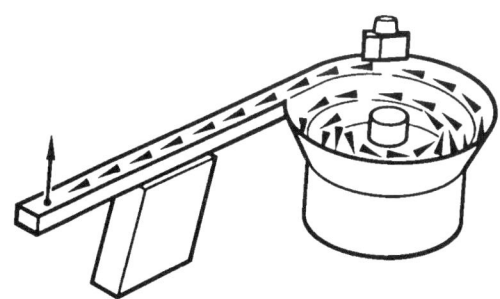

Fig. 7.22 A vibratory bowl feeder for components packed in bulk.

A single bowl holds about 10,000 devices and can be refilled very quickly. Unfortunately, it occupies a lot of space, about the same as 9 small sticks, and limits the number of component types the machine can have on board at a given time. For this reason, all bowls are usually arranged together and staggered to save space.

The other drawback of the vibrating bowl is the rather high cost, which can, however, often be recouped quite quickly because of the lower cost of bulk packed devices.

7.5.5.3 VIBRATING CONVEYOR

This unit, Figure 7.23, does a similar job to the vibrating bowl, but is more compact, about one-third of the width. The capacity is correspondingly smaller and it requires more frequent refilling. It is designed primarily to handle chip components, although special versions can be obtained to feed other devices such as SOT-23s.

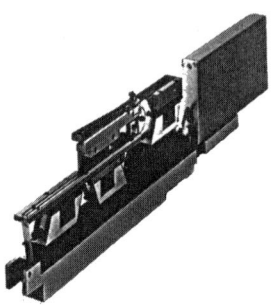
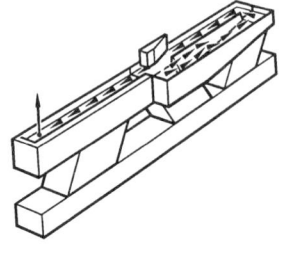

Fig. 7.23 The vibrating conveyor is more compact than a vibrating bowl and is cheaper.

7.5.5.4 HOPPER

The cheapest and least space consuming method of feeding bulk packed cylindrical components, particularly MELFs and mini-MELFs, is a simple vibrating hopper, several of which can be set side by side on one vibrating unit, Figure 7.24. The capacity of such units is quite high.

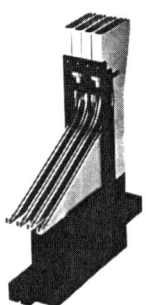
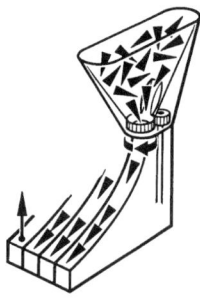

Fig. 7.24 The cheapest method of feeding bulk packed cylindrical components is a simple vibrating hopper. Several hoppers can be mounted side by side on a vibration unit.

Non-vibrating hoppers are obtainable. Despite a low cost, these are not recommended as there is a tendency for the components to stick, aggravated by the fact that the same devices from different sources can have significantly different dimensions.

7.5.5.5 TAPE ON REEL

This type of packing, Figure 7.25, is the most suitable for the majority of mounting machines as it was developed specifically to feed them. While the tape perforations are always the same and components are set at a fixed pitch (one hole, two holes, etc.), a different width of tape will need a different loading module, unless an adjustable type is used. All feeders can accept the standard 178 mm (7 in.) diameter reel, and some also accept the large 330 mm (13 in.) one. The same module will often deal with both cardboard tapes and embossed tapes.

 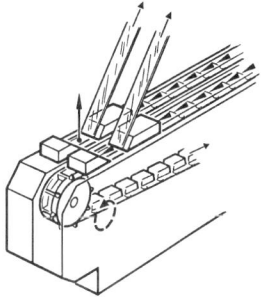

Fig. 7.25 Tape-on-reel dispensers are very reliable and are accepted by the majority of automatic mounting machines.

Tape feeders are very reliable, compact and relatively cheap. On the other hand, taped components are more expensive than those packaged in bulk because good quality tape is not cheap.

Problems which may arise are related to lack of smoothness in removal of the cover tape (any jerk can disturb the component waiting to be picked up) or, more usually, to the different tolerances on tapes coming from different suppliers. Reference to Chapter 4, Figure 4.43, shows the number of parameters which must be defined and that opportunities for divergence are many.

7.5.5.6 OTHERS

Other delivery packs are supplied, such as palettes, or trays, and adhesive tapes. These may be fed to suitably equipped automatic placement machines, although relevant feeders must be discussed on a case by case basis with the machine supplier.

7.5.6 Application Heads

The application head (or *pick-up head* or simply *head*) of an automatic mounting machine is the key module of the machine. It must perform a variety of tasks on the components — pick up, hold, centre, test, rotate, place and so on — at high speed and with great reliability, Figure 7.26. The head must be compact (to place components close to each other) and light (in order to minimise the load on the X-Y mechanism).

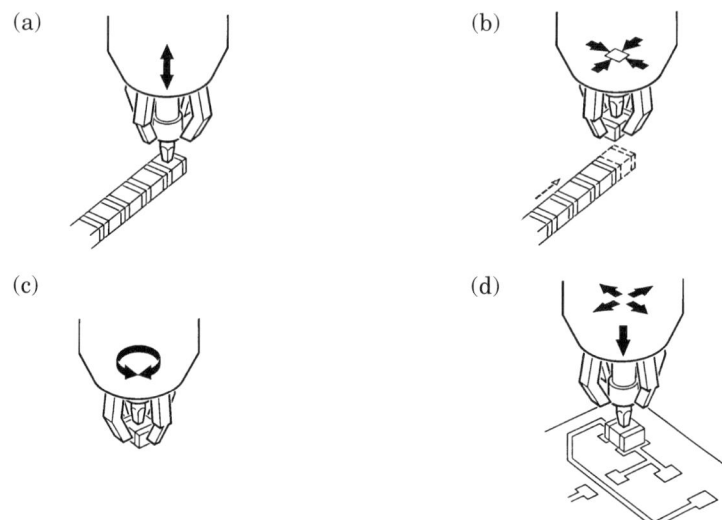

Fig. 7.26 Some of the tasks performed by the head of a mounting machine: (a) pick-up; (b) hold; (c) rotate; (d) place. Such tasks must be carried out with speed, accuracy and high reliability.

The variety of shapes and sizes of SMCs is such that the picking up and holding arrangements must be very versatile in order to be able to assemble a full board. In addition, the head must accept components with imperfect shapes, such as those devices which have uneven surfaces because they have been potted or resin dipped.

Several heads are supplied with automatically interchangeable tips, although changing tips will reduce machine productivity and it is normal to program to minimise the number of changes. Each manufacturer has his own solution to the problem and claims it is better than all the others. Most, however, closely resemble each other.

7.5.6.1 PICKING UP

To enable the head to pick up a large variety of devices, flat and cylindrical, it is equipped with a vacuum nozzle, the tip of which may be changed, automatically or manually, to cope better with the type of component or to develop a different suction, e.g., from 5 to 50 grams. The picking position is controlled on three axes to avoid undue stress on the component body. (Conventionally, the X and Y axes refer to the work plane and the Z axis refers to vertical movements. Rotation — often referred to as the fourth axis — if present, will be named theta, θ, the eighth letter of the Greek alphabet.) On some machines, the bottoming force can be programmed and the descending speed of the head reduced in the last part of travel.

When the tip is lowered until it comes very close to a component, this will be sucked against the nozzle. A vacuum or flow detector informs the control unit that the component has been picked up. If the picking cycle is void, the machine may be programmed to repeat it two or three times before stopping and signalling the operator (because a device may be missing temporarily in a bowl feeder or a tape may have an empty position). Alternatively, the machine may print out a message and continue with the next step in the sequence.

The flow detector may be activated from time to time during the movement of the head, to check that the component is still in place, and again after placement to ensure that the component has been properly released.

There are pick-up methods other than a vacuum nozzle but these are used only when strictly necessary, for instance, on components with a slot on the top, like trimmers, which cannot be closed with a cover. It is difficult to design a mechanical device which will work with many different shapes and sizes of bodies and, at the same time, will be very small to allow for close placement of components on the boards.

7.5.6.2 CENTRING

In theory, the picked-up component will be perfectly centred with reference to the axis of the nozzle. In practice, because of the tolerances of components and feeding devices, including tapes, it is always offset and rotated to some degree. These errors must be recovered before the component is placed on the board.

A commonly employed technique uses two pairs of alignment jaws, Figure 7.26, which close, one pair at a time, and gently press on the sides of the component to centre and align it. For devices which do not have four flat sides, only one of the two pairs is activated. In this case the feeder must be accurate enough to ensure that the position in the other direction is within the limits stated in the design of the PCB. If the component has a very odd shape, centring may be impossible or undesirable and both pairs of jaws will be inactivated by the software.

Another technique, which works well with most regular shapes of component, uses a vacuum chuck to centre and align the component. A number of different sizes of chuck copes with a wide range of components. An alternative method employs an optical recognition system to measure the offset and rotation of the picked-up component with respect to its ideal position. The control electronics uses the data to modify the co-ordinates (X, Y and theta) to place the component correctly on the board.

This last device is more expensive than the jaws and does not allow for testing of the components (covered later). On the other hand it has no moving parts and can place components very close together since the pick-up nozzle can be very slim.

7.5.6.3 ROTATION

Not all polar components are mounted in the same direction on the printed board. This would impose an excessive restriction on the designer of the PCB and almost certainly reduce the density of the board. If the devices are supplied on tapes, pick-up orientation is fixed by the supplier and it is undesirable to have different directions of polarity on the same tape. In addition, SMCs are often mounted in two different directions at 90° to each other, like PIH ones, see Chapter 8.

This leads up to the point that the picking head should be able to rotate the device by 90, 180 or 270 degrees before placing it on the board, Figure 7.27. Some machines can be programmed in eight steps of 45°, while the most flexible can rotate to any angle between 0° and 360° in steps of one degree. Rotation is, of course, specific to each individual component and is built in to the assembly program. To save cycle time, device rotation is performed while the head (or the board) is moving and the four jaws, if employed, are closed to avoid displacement of the centred component during that movement.

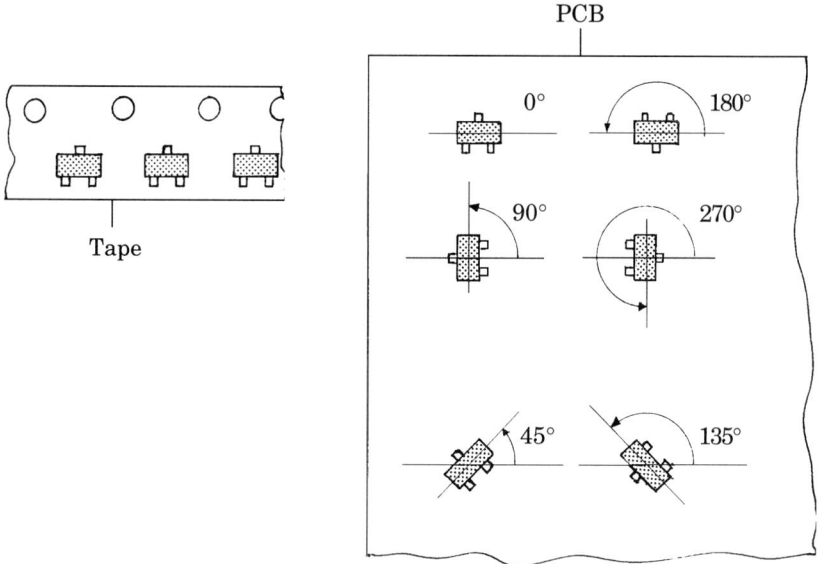

Fig. 7.27 The designer must have the freedom to mount components in different directions on the PCB. This means that the mounting head must be able to rotate, at a minimum, in steps of 90° and, preferably, in steps of 45° or better.

On some parallel mounting machines, described later, both centring and rotation (except at 180°) are performed mechanically before picking up so that the head is simpler than described in the preceding paragraphs.

7.5.6.4 TESTING

In order to test components, two opposing centring jaws may be electrically insulated from the head and connected to an instrument, Figure 7.28, which tests the device and instructs the machine electronics to accept the component, to discard it because it is out of tolerance, or to turn it through 180° because the polarity is wrong. The measurement must be simple and give a 'go' or 'no-go' verdict, since time is limited (typically 50 milliseconds) and the test conditions are not the best possible. Even so, it is easy to check polarity of diodes, output voltage of Zener diodes, and measure a wide range of resistance and capacitance with a precision of about +10%.

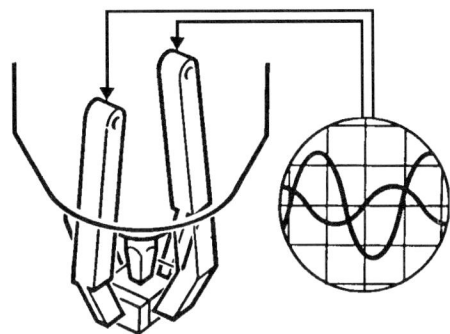

Fig. 7.28 A variety of components can be tested quickly during placement if an opposing pair of jaws is used as the test clamp.

Testing is usually performed on machines with rotation capability, as the most common error is a polar component turned by 180°. Providing the machine can rotate the component, it may be possible to economise on some components, such as diodes, by purchasing in bulk and feeding them randomly. Even if all components are bought on tapes, from suppliers who consistently deliver high quality, testing during placement can prevent large numbers of boards being wrongly assembled because of human errors, e.g., wrong labelling of reels, loading the wrong part on a feeder, or faulty programming.

This type of testing can give much higher final assembly quality for limited cost and significantly reduce the workload on in-circuit testing. (The latter operation may even be omitted, provided that ICs have been 100% tested before assembly.)

7.5.6.5 PLACEMENT CONTROL

It is important for the machine to be able to recognise that a component has been placed on the board. It is possible to have heads fitted with optical recognition equipment but this is seldom done because of the cost. Simpler indirect methods are normally used. These methods are usually based on measurement of the pressure or flow in the vacuum nozzle.

However, while indirect methods will determine that the component has been carried to the correct point on the board before release, they cannot show that the component has not been moved from that point, by jarring etc., before the soldering station is reached, and a visual or optical check is desirable.

7.5.6.6 TEACH-IN OPTIONS

If the program for placing the components is not generated directly by the CAD software used for the board design/layout, it must be hand written. This tedious and time-consuming job requires a skilled person, and mistakes are quite frequent. The most difficult part of the work, writing tens or hundreds of co-ordinates, can be replaced by a teach-in option applied to the mounting head.

The vacuum nozzle is replaced by a TV camera attachment, a precision reticle being clamped in the centring jaws. The operator uses the manual control, usually a joy stick, to move the head over the PB, the pattern of which appears enlarged on a monitor screen, with the reticle superimposed to give the exact position of the head, Figure 7.29. When the position of the component is perfectly aligned, a command to the programmer includes the co-ordinates in the program.

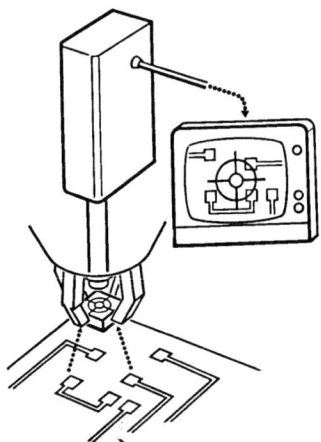

Fig. 7.29 A teach-in option allows accurate 'on machine' programming of component placement.

When the placement program is running, the pick-up nozzle of the machine will move to the 'taught-in' XY values, assuming that they are the centre of the component. If the component is asymmetric, the operator/programmer must define a centre.

If the board is not available at the time of programming, an artwork master can be used. This must, of course, be placed in the same position as the PB on the mounting machine and it will not be practicable to recover errors in placement of the mounting holes. (An advantage of on-machine programming by teach-in using a PB is that positional errors in reference holes or machine fixture will be recovered.)

On some equipments, this sort of option is replaced by a precision light beam (in the nozzle position) which projects a small spot of light in the XY position of the centre of the head. This is less expensive but is also less precise as an enlargement of at least 10 x is needed for fine centring.

7.5.6.7 BAD CIRCUIT DETECTOR

This option is particularly useful when two or more circuits are being processed on one panel for assembly, and repair or rework of a faulty circuit or wiring pattern is impractical or uneconomic. When intermediate testing/inspection shows some of the circuit boards to be faulty, e.g., broken tracks, the boards concerned are marked with a dot of ink or paint in a fixed position. A small camera attached to the placement head, Figure 7.30, can see the indicating dot and trigger the software to block mounting of components on the faulty board(s).

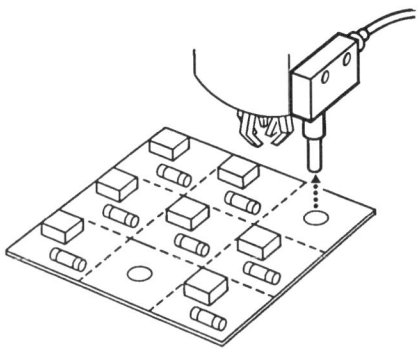

Fig. 7.30 A small camera attached to the placement head can be used with suitable software to prevent mounting components on faulty circuit boards.

7.5.6.8 OPTICAL RECOGNITION EQUIPMENT

It was said earlier (under 'Placement Control') that it was seldom economic to use optical recognition equipment. However, for the most complex boards, it can be economic to use a high resolution camera and pattern recognition software. It is practicable with such equipment to recognise the footprint of a component on the PB and to centre it directly with its soldering lands, recovering the whole chain of design/manufacturing tolerances in one step.

Additional uses can be found for optical recognition equipment, in particular in inspection. For example, it can be very useful with devices having closely spaced leads to inspect the leads optically and discard the component if they are bent. In addition, the equipment can be employed to check for missing components at the end of the placement cycle.

7.5.6.9 TURRET HEADS

If all or some of the options mentioned are added, the head may become too complex. It may be preferable to use a turret head, designed only to pick up, hold and place the component, and to assign the ancillary functions to fixed stations. A simple solution, Figure 7.31, involves four stations, devoted to picking, testing, orientation and centring, and placement. Defective or wrong components may be rejected between the testing and orientation stages by applying pressure instead of vacuum to the nozzle.

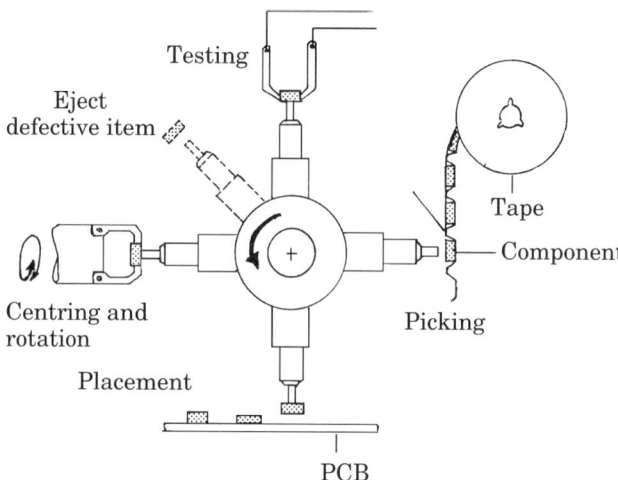

Fig. 7.31 A turret head reduces complexity by acting only as a pick-and-place head, all ancillary functions being at fixed stations, to each of which in turn the head presents the component.

This construction has the advantage of simplicity and permits effective testing and centring stations to be used (as they are not constrained by available space or by being limited by the head). It does, however, affect the overall structure of the machine because the position of the head remains fixed and both components and PB have to be moved. Some versions use turrets with a large number of vacuum nozzles, so that there can be a reasonable number of feeders in addition to the stations for centring, testing, and other auxiliary functions, Figure 7.32. It should be noted, however, that, while turret head machines can execute several operations in parallel on different components, they are still sequential placers. Only one tip at a time will be in the correct position to place a component on the board.

7.5.6.10 ADHESIVE APPLICATORS

Some heads may be equipped with a syringe so that they can dispense adhesive just before placing the component. This technique does slow the machine. A better method equips another head with a syringe which places the glue in the same position as the component being placed, but on another board which is at a fixed distance along the PCB conveyor, Figure 7.33. This second head is rigidly connected to the main head and, although primarily intended for glue application, can be modified to mount components in parallel to it.

7.5.7 Control

Automatic placement equipment may be controlled either by software or hardware. The use of hardware gives higher operating speeds but requires specific tooling for each board part number and is usually confined to large

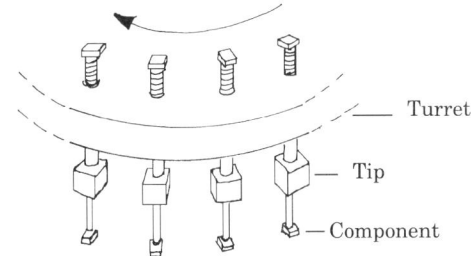

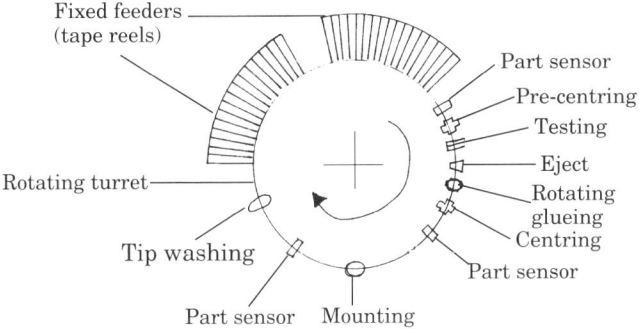

Fig. 7.32 The carousel head is a turret which revolves in a horizontal plane and can carry many pick-up tips.

equipments suitable for mass production in large batches. Software control is the usual solution for small and medium productivity machines and is also used for the large volume equipments when high flexibility is needed.

The assembly program is read step by step by the machine controller, which translates it into commands to the motors which move the head, table, and/or component feeders, apply vacuum, test and align components and so on. Auxiliary functions, such as teach-in routines, bad circuit detection, etc., also rely on the machine software.

Small and medium size machines are normally programmed on-line and no production is possible while the program is being fed in. Large machines are programmed off-line, either on another computer or by using a multitasking capability on the machine itself. Several programs may be stored in the computer memory and recalled very quickly. They can, of course, be copied and saved in a program library, e.g., on disk.

Machine precision depends on the type of control (open or closed loop, resolution, etc.) as much as on the mechanical construction. While XY positioning accuracy is often quoted as ±0.02 mm, placement precision, i.e., the maximum deviation between the component lead and the mounting pad, including X/Y and rotational offset, is usually about ±0.15 mm with a rotating head (but can be ±0.06 mm with optical alignment). Repeatability is usually much better than the precision. Rotational accuracy is generally around ±3°, although machines with ±1° are available.

These values refer to the machine controls and mechanical structure. The final position on the board depends also on other factors, such as board

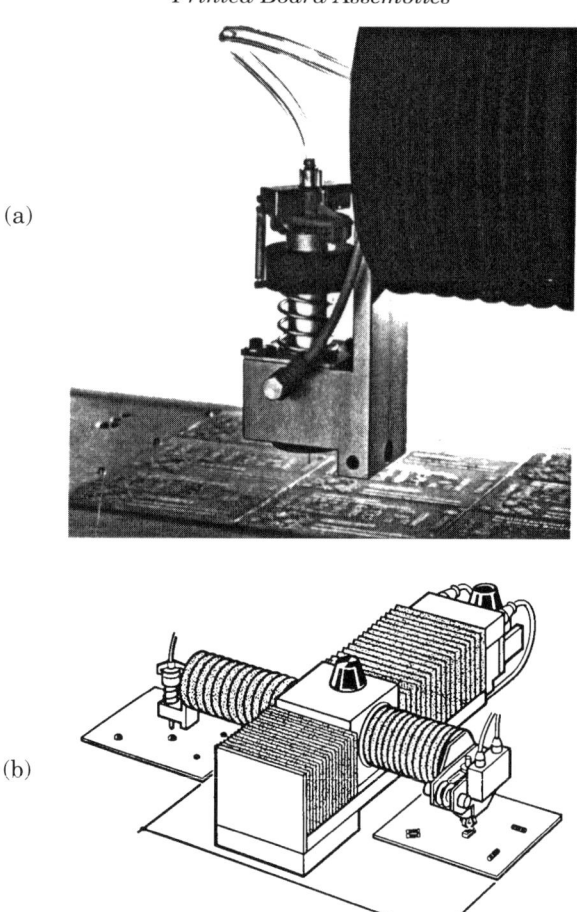

Fig. 7.33 (a) A typical placement head. (b) The first head (on the left) places adhesive on one board while the second head places a component in the same position on a board glued earlier.

positioning and component movement due to the adhesive. If the glue dot is uneven or off-centre (or the board warped), the adhesive may move the component after the head has released it. Some types of glue tend to do this more than others. The basic reason is that the shear/stress diagram of a non-Newtonian fluid is non-linear and may be time dependent. Solder pastes may exhibit similar behaviour although, when reflow soldered, the surface tension of the molten solder will usually centre the component (if the design is good). The result is that, even with a good, well maintained machine, the final tolerance to be expected on XY position cannot be much better than 0.15 to 0.2 mm.

The pick-up tip on the head can have a significant influence on the accuracy of placement. If the end is made of a soft material or is equipped with a rubber sealing ring, it makes the vacuum more effective. It seals better on all types of component, so that the same tip can be used for all sorts of devices, including cylindrical ones and those with an uneven top. On the other hand, the elasticity which improves sealing also reduces placement accuracy. After the action of the centring jaws, the component will return partially towards its original position. Soft tips should, therefore, be used only on machines fitted with optical recognition equipment.

7.5.8 Software Packages

A major advantage of machines available today is their ability to cope with a factory automation system through a standard interface, e.g., RS-232, Figure 7.34. In addition to control software, and, indeed, often integrated with it, software packages can be provided to perform tasks such as:

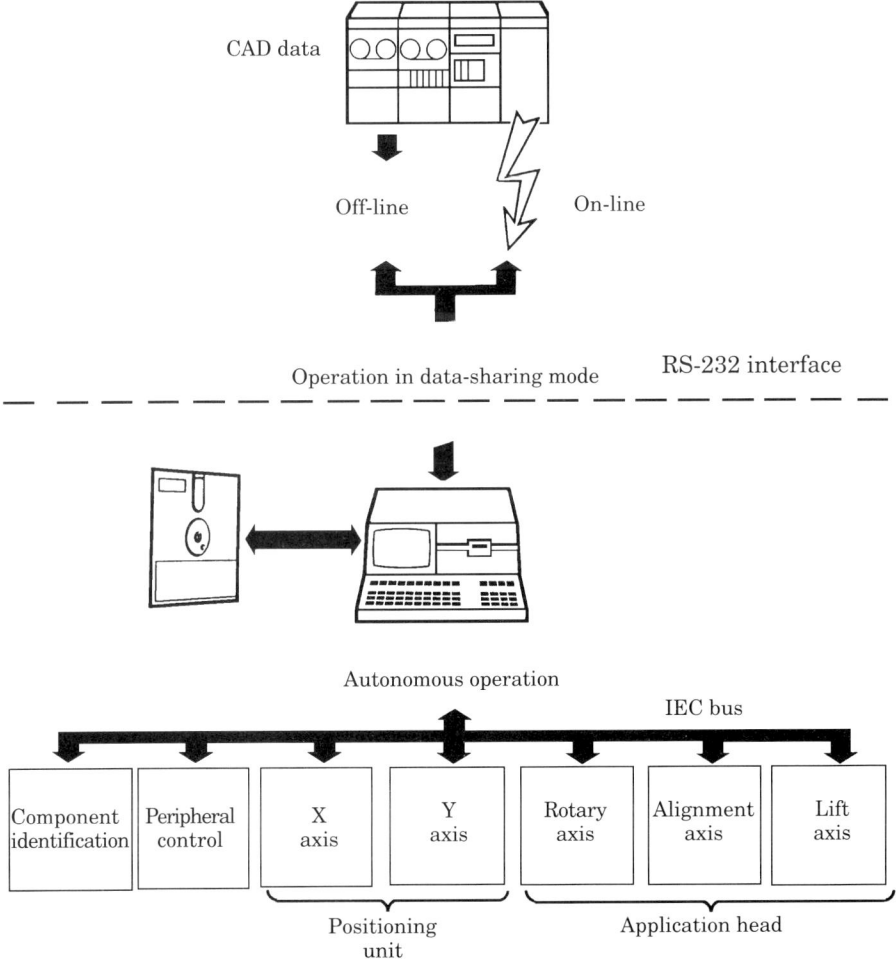

Fig. 7.34 The control system for a modern automatic placement machine can interface with other computers to become part of a factory automation and data system.

(i) on-line or off-line interfacing with a CAD/CAM system;
(ii) production data and statistics;
(iii) automatic setting up — if all necessary components are available — on change or production;
(iv) analysis and derivation of the optimum placement sequences;
(v) optimisation of the number of components to be changed in the feeders;
(vi) fault diagnosis on the machine, ancillary equipments and components;
(vii) planned maintenance programmes;

The only limit on the number of tasks is that imposed by financial returns.

7.6 MOUNTING EQUIPMENT

Equipment for mounting SMCs ranges from simple desk-top machines or aids to complete multimillion dollar assembly lines. It can be classified in a number of ways, according to different parameters. The following six classes are commonly defined:

(i) table-top;
(ii) pick-and-place (or *one-by-one* or *sequential*);
(iii) multiple arm pick-and-place;
(iv) in-line;
(v) parallel (or *multi*);
(vi) parallel-sequential.

While this classification is rather arbitrary, as it refers to several parameters at the same time, it is reasonably clear and widely accepted.

7.6.1 Table-top

There are a number of table-top machines on the market, varying in layout and ability to mount automatically, semi-automatically or manually depending on the complexity of mechanics and electronics. The machine shown in Figure 7.35 is primarily intended to be used in an engineering laboratory for A/B model production but is, like several other types, employed in sub-contract assembly shops for very small runs.

Such machines consist of an XY positioning mechanism controlled by a small computer, a simple vacuum pick-up head, and banks of component feeders of various types. The same machine can usually apply adhesive dots before mounting the components, if the pick-up head is replaced with a syringe and air pressure system.

Board size may be restricted, for example, to about 385 mm x 255 mm (15 in. x 10 in.) or less and the boards will have to be loaded and unloaded manually. The total number of component types available to the head at one time does not normally exceed 20. This is not often a drawback for the type of work for which such machines are used. Some more complex features, such as centring, interchangeable tips, rotating heads, etc., are frequently offered at reasonable cost.

Automatic Assembly — Surface Mount Components 309

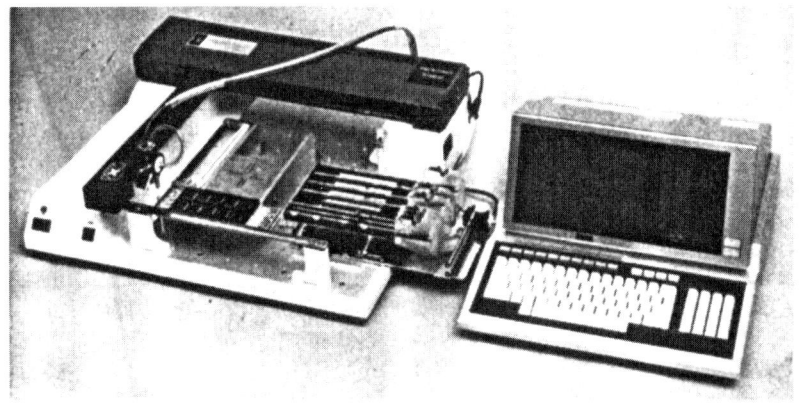

Fig. 7.35 The Mamiya Denshi Superhand Trainer ECM83. (Courtesy of Hedinair Ltd, Essex, UK)

Placement time per component varies with the type and model. It may be as fast as 2 seconds or less or it can be 5 seconds or more. While these machines are essentially pick-and-place machines, the compact size, low price and restricted performance justify a separate classification.

7.6.2 Pick-and-place

The pick-and-place machine derives its name from the mounting cycle which they execute in sequence — pick up, centre, orientate, test, place — on one component at a time. This is the largest classification and covers equipments which perform in very different ways, Table 7.3.

Table 7.3
Performance of Sequential Mounting Machines (typical values)

	Low End	*High End*
Minimum board size (mm)	20 x 20	20 x 20
Maximum board size (mm)	100 x 150	460 x 510
Number of component types	6 to 20	20 to 180
Size of tapes (mm)	8	8, 12, 16, 24, 32, 44
Other types of feeders	no	several
Speed (seconds/component)	1 to 4	0.2 to 1.0
Production (components/hour)	800 to 3600	3600 to 16,000
Precision (± mm)	0.2 to 0.25	0.1 to 0.2
Repeatability (± mm)	0.1 to 0.2	0.05 to 0.15
Possible rotations	1 to 8	8 to 360
Rotation precision (degree)	3	1
Automatic board loading/unloading	no	yes

The basic ways in which a pick-and-place cycle can be performed are illustrated in Figure 7.36: (a) the head moves in X, Y and Z axes and rotates; (b) the head moves in X and Z directions and rotates, the table moves in the X and Y, and the feeders move in the Y; (c) the table is fixed, the head moves in the X, Y and Z and rotates, and the component feeders move in the Y direction so that the picking position is constant (or simply to hold a larger number of types without slowing pick-up time); (d) the head moves in the Y and Z and rotates while the table moves in the X.

Many machines adopt the first option although it requires a rather complex head. The second option requires the feeder to move or else a device to be fitted to feed the component directly to the head. Option (c) makes the machine more complex but increases the speed of operation and allows more space for the feeder without increasing the maximum travel of the head. The final option simplifies the drive of the head so that its run in the Y axis can be quite long, allowing it to access many different component types.

The actual implementation of these basic options results in other variations, some of which are shown in Figure 7.37. This shows (a) a turret head, (b) a carousel head (with multiple pick-up tips), (c) two (or more) independent heads, each with its own component feeders, and, finally, (d) two (or more) heads working in sequence.

A turret head has the advantage of performing different tasks in different stations which work in parallel, so increasing the overall speed of the cycle. It does mean that components must be held on a drum feeder, which restricts the component feeding method to tape-on-reel only. The carousel is also a turret head, although it rotates in a horizontal plane. It may have a large number of vacuum tips performing several different tasks in parallel and can be fed in many ways, as the feeder remains steady, any tip passing over any individual feeding position.

The productivity of a sequential machine can be increased by using two or more independent heads, working in sequence, that is, one head picks up a component while the other head places one and *vice versa*. The throughput will increase but cannot double as the electronics will sometimes have to slow one head in order to avoid interference with the operation of the other. Having independent feeders for the heads minimises loss of productivity.

Alternatively, the two heads may move in only one direction (horizontally) and work in sequence on the same board. This requires the support of a reciprocating sliding table to move the board under the head which is to place a component while the other head picks up a new component. The component feeder must also move in the same axis as the PB. Two heads working in tandem, that is, in the same position on two different boards at a fixed distance from each other, should actually be classed as a parallel machine, since the heads perform the same action at the same time. Since the second head is often used solely to apply glue dots, this arrangement is frequently included in the pick-and-place classification.

7.6.2.1 EXAMPLES OF SEQUENTIAL PICK-AND-PLACE MACHINES

Among the many representatives of this classification, the following examples are either truly representative of a particular type or have some unusual and interesting feature. As mentioned earlier in this chapter, over

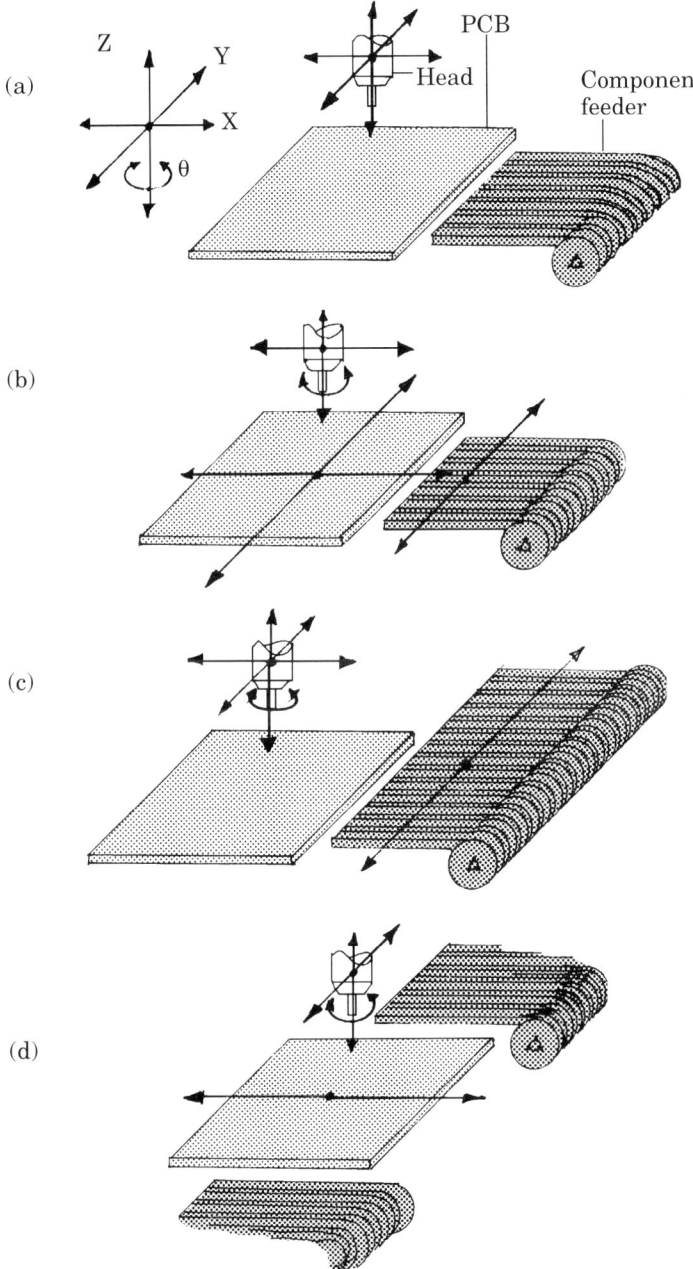

Fig. 7.36 The different ways in which 4 axes motion can be realised on a pick-and-place machine: (a) the head does everything; (b) the table has X-Y movement and the component feeders move in the Y direction, the head in X, Z and θ; (c) the table is fixed and the component feeders move; (d) the table moves in the X direction and the head in Y, Z and θ.

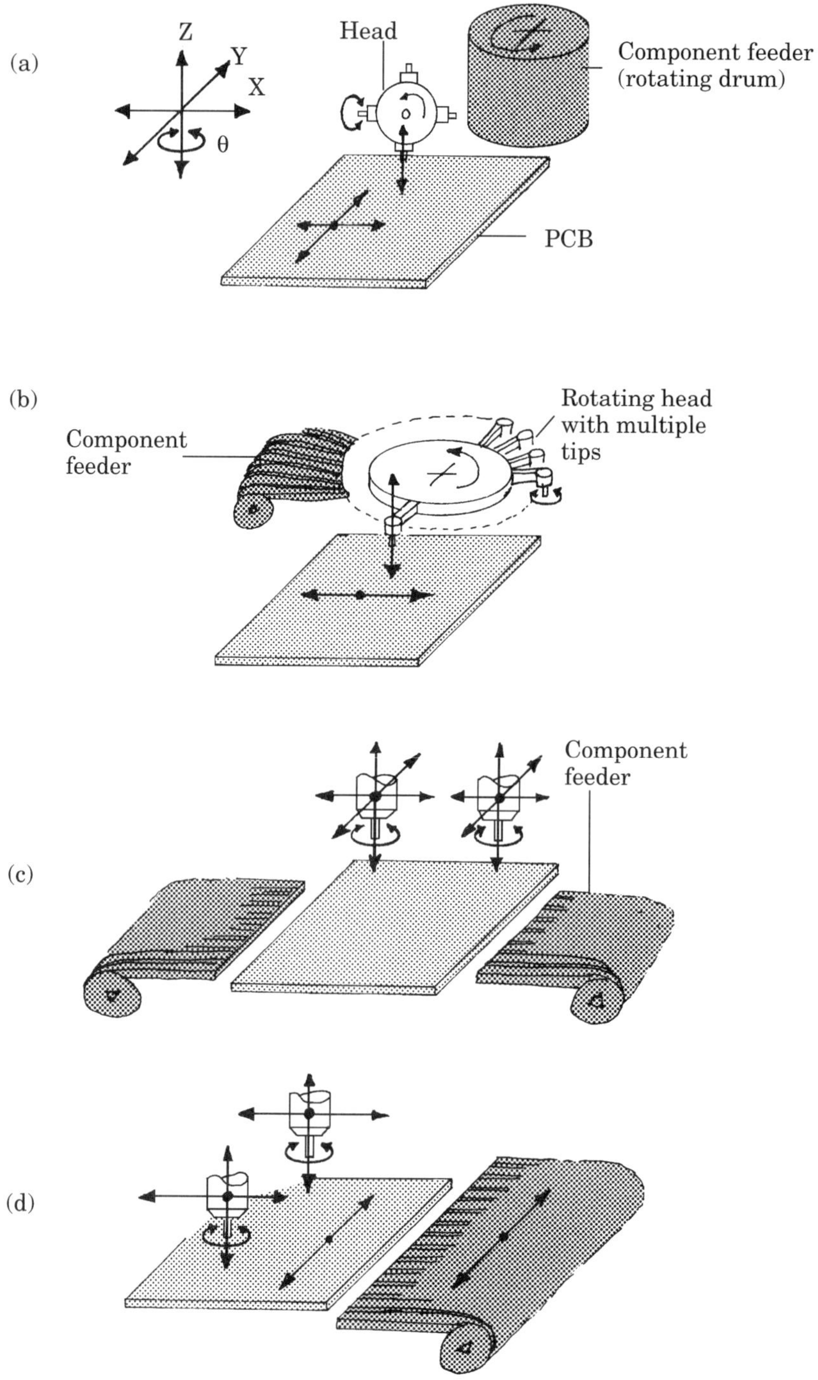

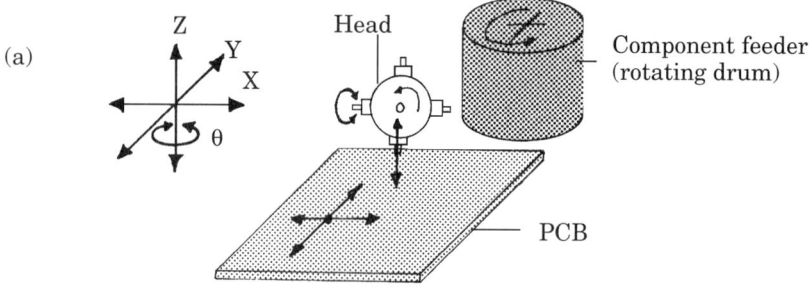

Fig. 7.37 Variations on pick-and-place mechanics: (a) turret head; (b) carousel head; (c) two independent heads; (d) two heads working in sequence.

70 different machines were examined before this chapter was written. More than 30 manufacturers were involved. If a particular machine is not mentioned, it does *not* mean that it was not worthy of consideration or mention. In addition, it must be realised that, in the natural course of development, updating and so on, any particular machine illustrated or described may no longer be available in identical form (or, indeed, from the original manufacturer). Any engineer, manager or buyer looking for the solution to a specific business need must survey all available machines, taking performance and price into consideration. The following descriptions are intended to do no more than outline the flexibility and productivity of existing, available equipments. In addition, they may give some guidance on the overall utility of the various types.

Example 1 — Figure 7.38

This equipment, on which the 'typical' equipment of Figure 7.17 is based, was an excellent compromise between speed, accuracy and flexibility. It had a maximum placement area of 270 x 240 mm when fitted with automatic loading and unloading and up to 350 x 310 mm in manual changeover mode.(Its replacement takes a PB up to 460 x 460 mm.)[8]

Despite a relatively high operating speed (4,000 components/hour — remember to use figures around 75-85% of this for estimating — with one head), the placement accuracy was very good (±0.08 mm absolute and ± 0.02 mm repeatability). The placement head had a centring device with four fingers. The traversing mechanism on one variant has a second arm which holds a dispensing syringe at a fixed distance from the component placement head to dispense glue dots in tandem. Alternatively, this arm may be fitted with a second placement head and, as mounting will then be in parallel, the production rate *will* double (8,000 components/hour in theory).

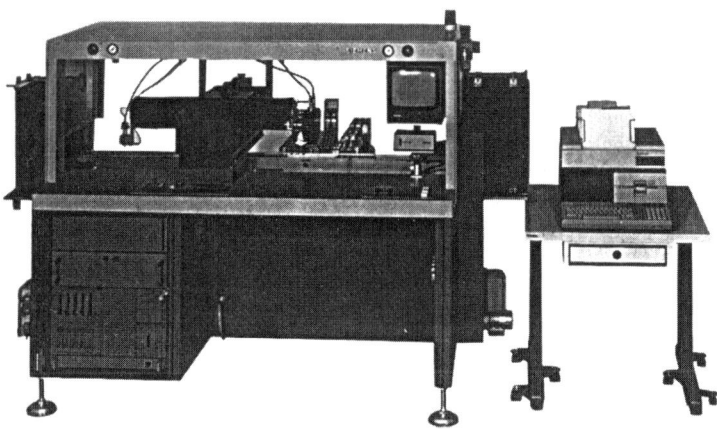

Fig. 7.38 A popular pick-and-place machine with good flexibility — the Siemens Automatic Placer MS-72.

The feeder capacity is 72 x 8 mm tape feeders for one model, and 72 x 8 mm tape feeders plus, e.g., 54 x 4.5 mm stick magazines for another. The wide range of options offered (nozzles, bad circuit detector, component testing, teach-in, automatic loading, and so on) and the variety of components which can be handled (flat, cylindrical, SOT, SO, PLCC, etc.) make this a flexible general purpose machine, ideal for entry level use or supplementing dedicated machines. (The MS-128 has a capacity of 128 8 mm tapes.)

It is of interest that the manufacturer produces SMCs and, in parallel, has developed and produced a range of machines for mounting them. The range includes machines specially designed to have high feeder capacity and a 12 position turret machine capable of placing 7,200 components/ hour in the single head version.

This manufacturer also produced a versatile insertion machine capable of inserting up to 4,000 axial, radial or DIP components/hour. Such a machine could be the ideal complement to the surface mount range for dealing with mixprint boards.

Example 2

Example 2 features a four position turret head and a drum feeder holding up to 64 tape reels (8 mm only) on two levels. Boards of size 457 mm x 407 mm (18 in. x 16 in.) can be accepted for populating at a rate of about 10,000 devices/hour (8,000 with testing) with a precision of ±0.20 mm (0.04 in.). The centring station can deal with rotation in steps of 1° over 360°.

The machine may be used as a stand-alone, operator assisted unit or as part of a modular system connected by a board transfer unit. A typical configuration includes the transfer system, an adhesive dispensing module and a placement unit. A more complex set-up can include several machines connected in-line to make up a fully automatic assembly line.

Software interfacing is easy, up to 7 or 8 machines being connected to a local processor, which provides co-ordination of the working cells, mass storage, line interface and several utility programs. In turn, up to 8 local processors may be connected to a host computer, a mainframe unit which communicates with other host computers, the material transport system, and other equipments.

The manufacturer concerned produces a number of different machines having maximum placement rates ranging from 5,100 components/hour to 25,700 components/hour on a board size up to 508 mm x 457 mm (20 in. x 18 in.). All machines have full in-line capability.

Example 3

This example is a typical sequential machine. Although it has two heads, these work alternately on the same board, which is moved by an XY table. While one head places, the other is picking up. Each head has its own feeder mounted on a carriage, which moves along one axis to deliver the selected component to a fixed picking position.

The two feeders can hold up to 120 types of component in tape (up to 32 mm), bulk or stick, and are placed on opposing sides of the board holder. The machine can accept reasonably large boards, 457 mm x 356 mm (about 18 in. x 14 in.), and operates at about 6,000 components/hour with an accuracy of ±0.2 mm. A glue dispensing station can be fitted, which works in tandem and does not reduce the placement speed.

Example 4

Example 4 is a pick-and-place machine which features a carousel head. The carousel disc, Figure 7.32, holds 120 vacuum tips equally spaced around the periphery.

The PB, up to 330 mm x 250 mm (about 13 in. x 10 in.), is on an XY table, Z axis movement being provided by the individual tips. Each tip picks up a component and goes through many stations, Figure 7.32, before reaching the placement position. While one head performs one step, the next head is performing the preceding one.

This arrangement gives high speed, 12,000 components/hour, i.e., 0.3 seconds each, and assures good accuracy, ±0.20 mm. To minimise lost time, the feeder is mounted on two castors for quick machine set-up and batch changeover.

Options include glue dispensers, PCB loading/unloading, testing, and a vision camera to check for skew and missing components.

7.6.3 Multiple Arm Pick-and-place

A machine of this type basically consists of a board loading, conveying and positioning system and several pick-and-place arms, each with a mounting head. It is an *in-line machine* or *progressive placer*, since the board travels in the X direction, passing under each head in turn. The heads move in the Y direction (and are capable of Z and rotational movement). The board is completed when it has passed under all heads.

For very high production volumes, each head mounts only one component, which is presented at a fixed position by an appropriate feeder. Alternatively, each head may mount a few components, delivered by a feeder having limited travel along the X axis.

If each head is dedicated to a component and a fixed position, both picking and placing can be controlled by hardware (mechanical stops) which will increase the operating speed and reduce the cost. Flexibility will be heavily penalised. If a more flexible line is needed, it may be established, at much higher cost, by putting several machines in-line, connected by a single board feeding and positioning system.

Usually this class does not include machines of fixed configuration. These machines consist of a basic structure which is made up into as many units as needed to meet the customer's specification. It is, in fact, more commonly used in the assembly of hybrid integrated circuits than for the assembly of SMCs to printed boards.

An Example

The sample selected uses two heads — each carrying 10 suction pins — to mount up to 20 different types of flat (size 2.0 mm x 1.25 mm up to 7.6 mm square) and cylindrical (diameter 1.25 or 2.3 mm) SMCs at the rate of 11,250 components/hour, equal to 20 components per 6.4 second cycle. If desired, 16 such units can be connected in line to reach a throughput of 320 components in the same 6.4 second cycle, that is, 180,000 components/ hour or 50 every second!

The same machine can be engineered to mount odd shaped components supplied on tapes or in sticks. The two heads are each fitted with 5 suction pins for mounting packages such as SOP, QFP, PLCC, LCCC, and many other odd shaped devices. A single machine places up to 10 components in a 5.4 second cycle or about 6,700 components/hour. Several machines can be connected in-line to increase the production rate.

7.6.4 Parallel Equipment

If the mounting equipment places an entire set of components (say from 2 to 320) on a single board at the same time, it is said to be *parallel* or *simultaneous* or *multi*. Over a wide range of components, the board can be completed in one pass under the head. For this reason, the speed of such machines is measured in seconds/board rather than in components/hour. Generally available machines have cycle times in the range of 5 to 20 seconds and can place from 60 to 200 components per cycle.

A multiple head, having a vacuum tip for each component to be mounted in the cycle, picks up all components at one time and places them in the correct position on the board. The head, or at least that part carrying the vacuum pins or tips, must therefore be specific to the board being assembled.

The major problem with parallel mounting is that the feeding mechanism must bring all components under the head, in the correct position and orientation with reference to each other and to the PB. This is because the head must be simple in order to accommodate the required number of vacuum tips and can only pick up and place.

One possible method is to use vertical stack magazines — with a device which pushes the component to the top of the magazine — arranged in the same relative position and orientation as the components on the board. While this does work, it is complex and applicable only to assemblies of low component density. A more viable alternative requires that mounting be carried out in two steps. First, all the components are loaded in a template which has slots to keep them in the correct position and orientation. Second, they are transferred to the board, on which adhesive dots have been placed, usually by a multiple pin transfer unit, Figure 7.39.

Parallel machines have very high production rates and acceptable accuracy for most assemblies, but they often require expensive tooling for each board part number and setting up can be difficult and tedious. A further drawback is that they can mount only a limited range of components, such as chips, MELFs and the like. For such reasons, the use of parallel equipments is restricted to mass production and they are popular for consumer electronics, especially in the Far East where large batch quantities are common.

Production of parallel machines is essentially confined to Japan and, although amounts produced are relatively small, this is one of the fastest growing classes of placement equipment. Two examples of parallel equipment are described, one for each of the feeding methods mentioned.

Example 1

This machine is unique in that, instead of the placement head taking the components to the printed board, the board is brought to the components. This solution allows 200 components to be placed in a 5 second cycle, that is, 40 components per second or 144,000 in one hour. This means more than one million devices in an 8-hour shift!

Automatic Assembly — Surface Mount Components

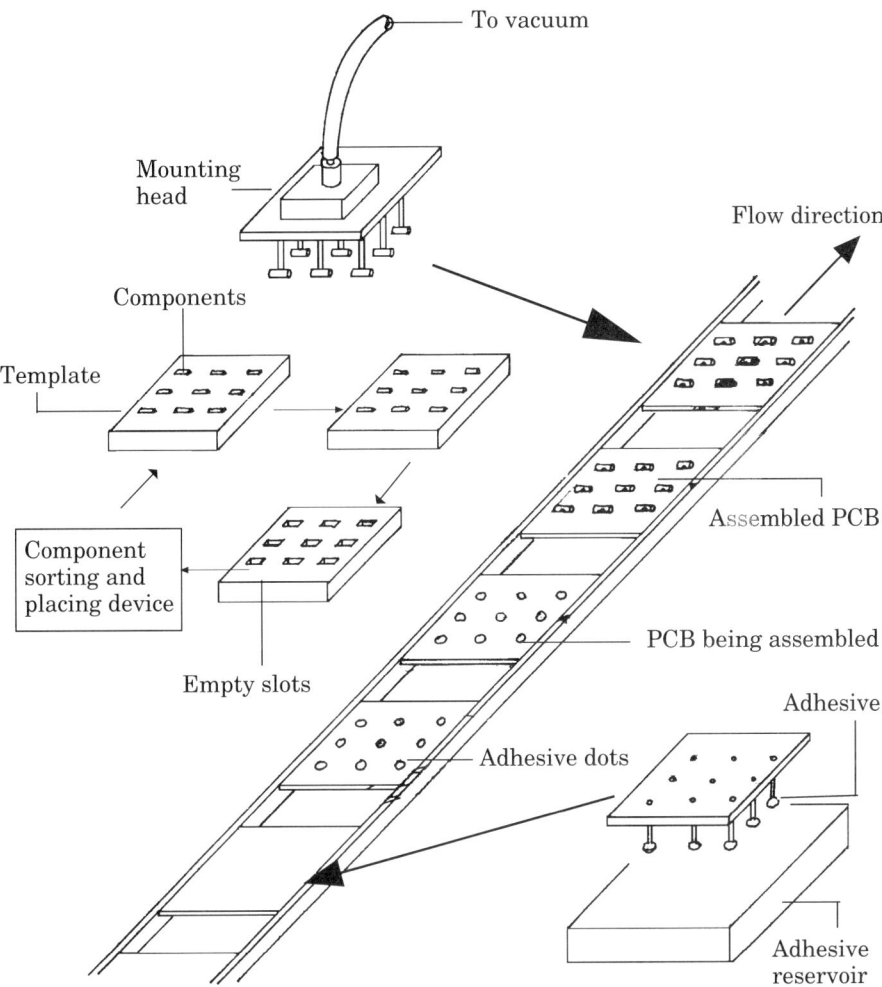

Fig. 7.39 Feeding parallel mounting equipment is complex and often requires expensive tooling. In the method illustrated, a multiple pin transfer system applies glue dots to the PB. The components are loaded into a template having properly positioned and shaped slots and are picked up from the template by the placement head.

The components, basically chips and SOTs, are loaded in stack magazines. The magazines are arranged vertically in an array which mirrors exactly the component position on the board. (Multiple pick-up from the same magazine is not possible as each position must have its own stack to permit parallel placement.) The PB, with previously dispensed glue dots, is positioned over the stack array and pressed against it, while a set of pins pushes the components in the stacks upwards. The uppermost component in each magazine will stick to the PB in its final position so that the board is loaded with its components in one pass.

The magazines may be arranged on a grid of 2.54 mm x 2.54 mm (0.100 in. x 0.100 in.) and the board must be designed so that the centre of each component is on the same grid. This restriction may reduce board density significantly. The equipment handles boards up to only 210 mm x 210 mm (about 8.3 in. x 8.3 in.) but accepts up to 200 different components and 200 mounting positions, each of which can be in any one of four directions at 90° to each other.

Example 2

This example is made by a Japanese company, formerly specialising in equipment for component manufacture, which produces a range of placement machines. The model chosen is a parallel placer working on the template method, Figure 7.39. The equipment is very popular for cylindrical components, performing best with mini-MELFs and the like. The cycle time is 14 to 20 seconds per board, with a maximum of 320 devices in 20 seconds. This equates to 180 boards/hour or up to 57,600 components/hour.

The maximum board size is 330 mm x 250 mm (about 13 in. x 9.8 in.) and the mounting accuracy is ±0.15 mm in both X and Y axes. Rotational accuracy is determined by the direction of the slots in the template and skew accuracy is within ±5°.

The heart of the machine is the drum feeder, Figure 7.40. This may have up to 80 types of bulk-supplied cylindrical (or chip) component held in the non-vibrating hoppers. A small pipe projects into the hopper, Figure 7.41, which moves up and down relative to the pipe such that correctly orientated components fall into the pipe and keep it full. (A special version of the component feeding module includes on each feeder tube a device to stop the component, check its polarity and sort it into the right direction. While this module is more expensive, being more complex, it allows the use of bulk-packed polar devices, such as diodes, with a significant saving in cost.)

The components drop by gravity through the pipe until stopped by a separator. When triggered, a component drops through a sensor and is retained by a perforated disc. To fill a template, the disc rotates and one component from each loaded pipe falls through a perforation into a small flexible tube which takes it to its position in the slot of the template. In this way 60 or 80 components may be loaded on to the template in one operation. The loading step may be repeated up to three times in sequence (a total of four feeding steps) and the disc may rotate by a different angle each time. This allows the same hopper to feed, through four different tubes, up to four different positions on the template. The cycle time for component feeding, which dictates the machine cycle time, is 14 seconds. If 2, 3 or 4 devices have to be supplied by the same hopper, the overall time increases to 16, 18 or 20 seconds respectively.

The template is specific to each board type and must be carefully and precisely machined. As the component falls with the long axis vertical, it is fed to one side of the slot so that it can drop into its mounting position with long axis horizontal, Figure 7.42. This movement is assisted by vacuum applied to a small hole in the bottom of the slot at the opposite end to that to which the distribution pipe is connected.

On this machine, while the filled template is passing to the position at which the multiple head will pick up all components, it passes under a line

Automatic Assembly — Surface Mount Components

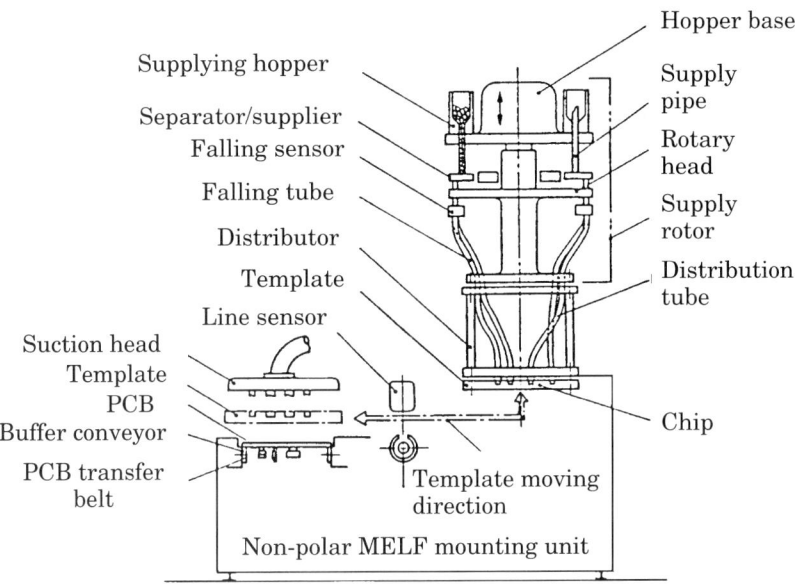

Fig. 7.40 The component feeder mechanism for the STM 3 template-operating parallel placement machine. (Courtesy of Nitto Kogyo, Japan)

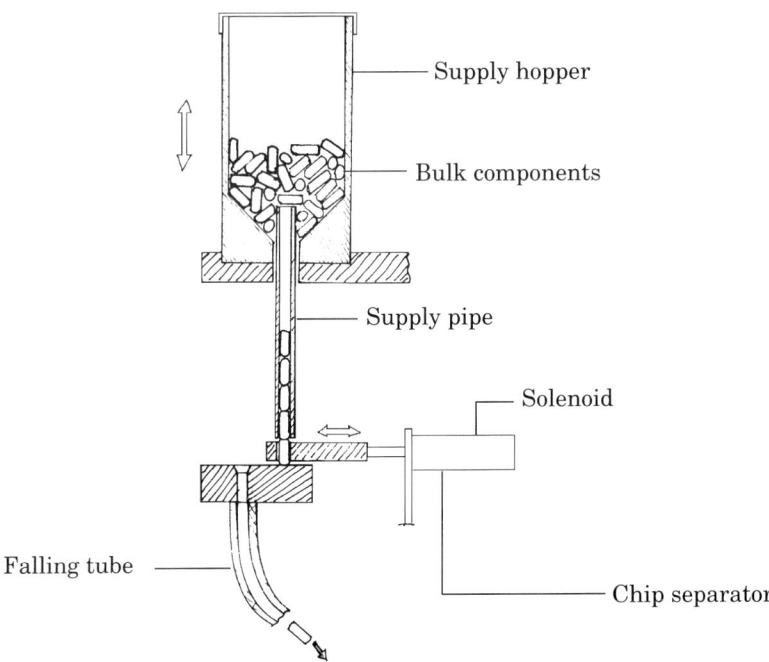

Fig. 7.41 Operation of a supply hopper for a template-operating parallel placement machine. The hopper, filled with components, moves up and down on the supply pipe so that the components drop into the pipe. The separator is solenoid operated and, when triggered by the controller, passes a component into the falling tube for delivery to the template.

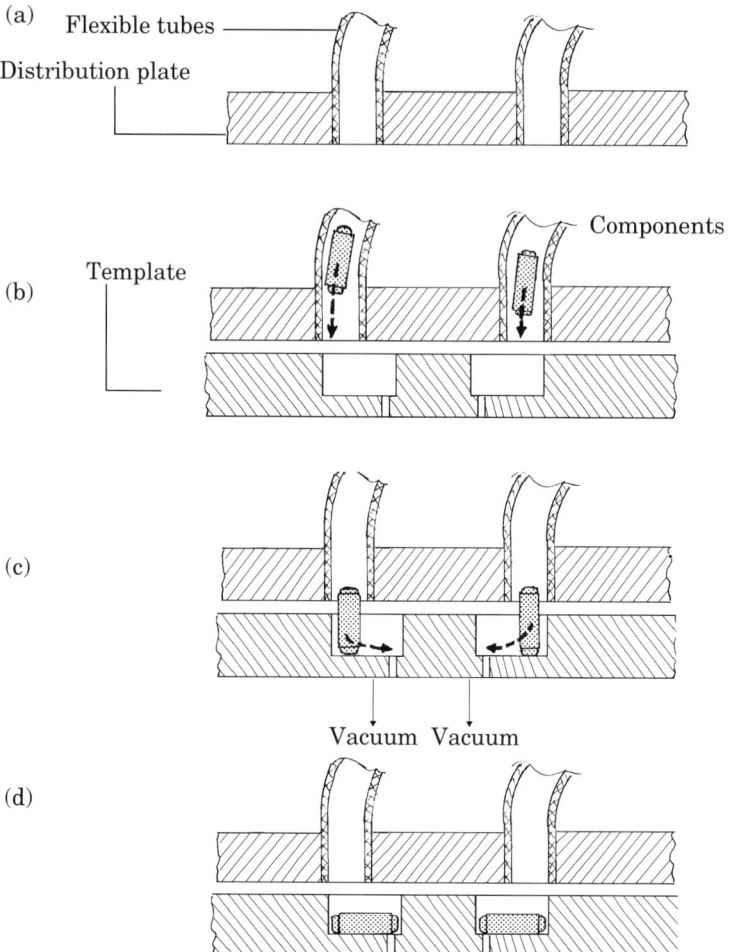

Fig. 7.42 The distribution plate and template for the STM 3 parallel placement machine must be designed to allow the component to assume its correct position for mounting — long axis horizontal.

sensor aligned with a lamp, Figure 7.43. Any empty slot is detected because light passes through the small vacuum hole. The electronics will recognise which device is missing and flash its position on a suitable monitor. The same check can be used after pick-up of the components to ensure that the multiple head has not left any components in the template.

Originally designed for cylindrical devices, the machine can be supplied with a module for chip components, using small tubes with rectangular section. (A basically similar machine was designed from the outset for chips.)

These are typical hardware controlled machines, particularly suited to the placement of passive components and diodes in the mass production of large or multiple boards. When the board type changes, it is necessary to change the glue application head, the placement head, the templates, the distributor with its pipes, and the control program. Even the most efficient shops will

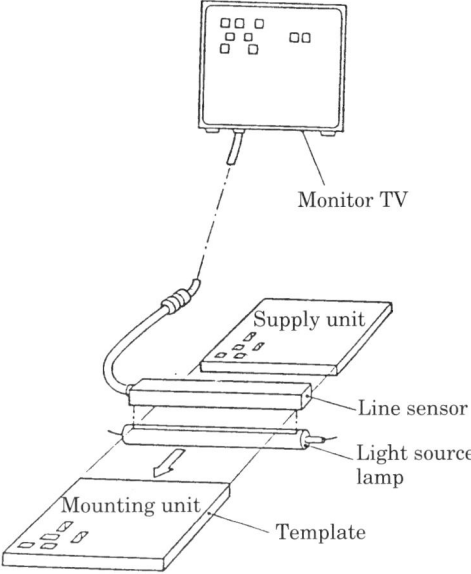

Fig. 7.43 Templates passing from the loading point to the multiple pick-up head on a parallel equipment may be checked automatically for missing components using optical inspection with an appropriate sensor.

need downtime of 15 to 20 minutes if the boards are of the same width, and much longer if the width of the conveyor has to be changed. As a result, there is great benefit to be gained from a standard board width.

7.6.5 Parallel-sequential Equipment

A machine of this type picks up many components at a time and places them on the PB individually, in sequence. In this way, the pick-up time is reduced. For example, if the head has 24 vacuum pipettes or tips, the time is divided by 24. One drawback is that, although software controlled, these machines are less easily set up than pick-and-place machines because the components to be picked up must be presented to the head in a precise position, one beside the other. This means that, when the board type changes, the whole component feeder has to be rearranged. However, the operation can be carried out off-line by changing the complete feeder, which is mounted on castors.

The PB is mounted on an XY table, and the machine controller decides on its movements and the sequence in which each tip has to be fired to place its component. It may happen, especially with properly designed multiple boards, that, while the board is positioned for the component held by one tip, other tips are in the right positions and can be fired at the same time. In this case, the mounting becomes partially parallel with a consequential increase in speed.

Parallel-sequential machines generally have heads with 10 to 32 pipettes, which can give theoretical production rates of 10,000 to 40,000 components per hour. A large variety of devices can be mounted by such equipments — provided the devices are packed on tapes. The productivity achievable suits parallel-sequential machines to medium-large production shops and many units may be combined to build a line with a very high production rate.

7.7 ANCILLARY EQUIPMENT

In addition to the options offered for the mounting units, such as vision cameras, test and inspection equipment, and so on, manufacturers of assembly machines will have ancillary equipments available, most of which have been mentioned earlier in this chapter. (Many manufacturers and/or their agents also offer equipment for subsequent process steps — soldering, cleaning, final test, etc., — which are beyond the scope of this chapter.) The major items offered are:

(i) automatic loading/positioning/unloading units from cassette to cassette and from conveyor to conveyor;
(ii) a tilting unit to turn the PCBs over as required;
(iii) a turntable to turn the PCBs through 180°;
(iv) adhesive application units and solder paste dispensing equipments of various types;
(v) a range of curing ovens of different sizes and types;
(vi) robots or other equipments for assembly of odd-shaped components, electrical and mechanical, or for performing other operations;
(vii) data processing equipment and software.

7.7.1 Curing Ovens

While any item to be purchased must be assessed fully before investing, particular care has to be taken when selecting the curing oven. Each adhesive needs its own curing cycle, which must be followed as closely as is possible. The temperature profile along the oven, or during the curing cycle, must be closely controlled and monitored by the oven controller, to avoid undercure of the adhesive or overheating of boards and/or components. Temperature uniformity is also important and must be checked (and monitored in use).

As mentioned earlier in this chapter, adhesives of several types may be employed, some being cured by UV, some by heat, and some by a combination of both. For this reason, the most commonly used oven is an in-line tunnel oven which can combine both UV and IR sections, Figure 7.44. This will cope very well with most curing options.

It is essential to check the operating characteristics very thoroughly, and, in particular, rate of temperature rise, maximum temperatures achieved and for how long, and the rate of cure of the adhesive under specific conditions. A typical experiment for checking temperature rise and bond strength is illustrated in Figure 7.45. This is specific to the oven, the type of lamps used, the power supplied and the board thickness and, to some degree, the configuration and surface coatings present. The positions of the thermocouples are indicated in the sketch at the top left of the figure. It can be seen that, under the conditions of the experiment, the adhesive was

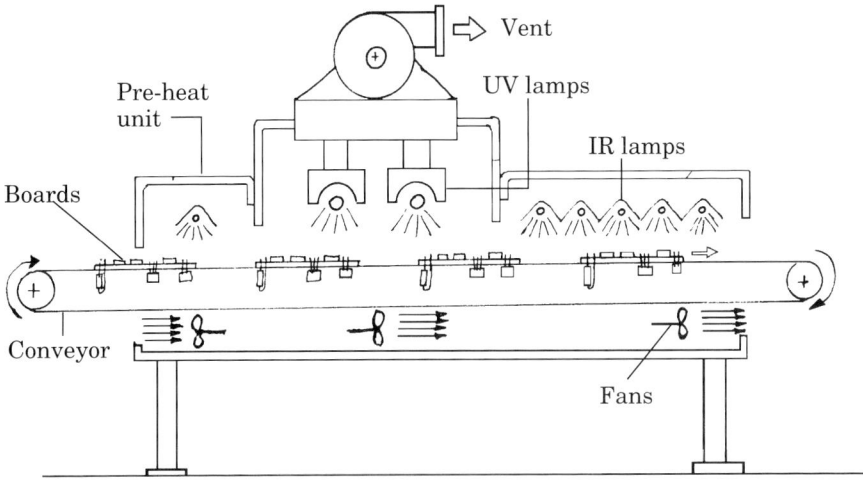

Fig. 7.44 Diagram of a typical in-line oven for curing adhesives. A cooling section may be added at the end of the tunnel or the assemblies may pass straight to the wave soldering stage.

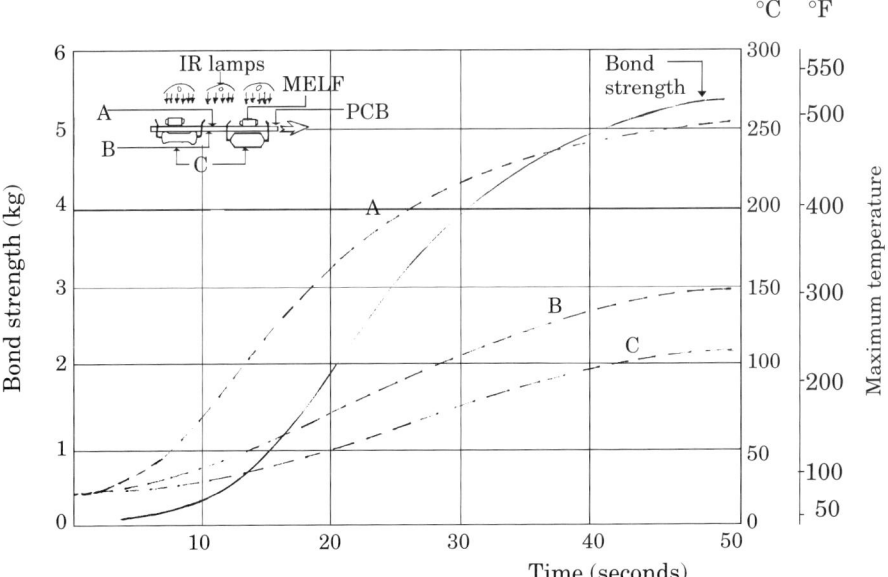

Fig. 7.45 The state of cure of a heat-cured adhesive is dependent on time and temperature. The higher the final temperature achieved, the faster the adhesive will cure. Too high a temperature for too long will damage components and assemblies. Temperatures reached can be measured by mounting thermocouples at the points indicated by the arrows.

reaching full cure after 45-50 seconds. However, the surface of the board facing the lamps had reached 250°C and the component temperatures were between 110° and 150°C. Such information can dictate, for example, the stage at which certain devices are mounted to avoid damage or it may be necessary to modify the curing cycle, which is especially easy when UV options are available.

As well as heating rates and uniformity, ovens must also be examined for proper ventilation/extraction and, in the case of tunnel ovens, for smooth movement of conveyors and ability to transport partly-assembled boards.

7.7.2 General Purpose Robots

In Chapter 6, general purpose robots developed for electronics assembly are described. The operating arm of such machines is provided with an extended range of software controlled movements and may be equipped with a camera and software for image processing and object recognition, interfaced with that on the major assembly equipment.

In addition to being able to mount PIH and SM components of any shape, with accuracy of ±0.05 mm or better, they can perform operations which dedicated placement machines cannot, e.g., screwing. Moreover, the number of component feeders is virtually unlimited, as the picking position can be programmed to be any point within the area which can be covered by the arm.

One disadvantage is their relatively high cost. These machines are essentially sequential pick-and-place units and their cycle time is rather high, typically 2 to 4 seconds, compared with their cost. Because they are general-purpose equipments, they are fairly complex and cannot compete in price/performance with dedicated machines in simple operations.

Robots are employed by specialist prototype and small batch production assembly shops and prove to be cost effective in this environment, particularly when overall costs — including quality costs — are considered. In most other assembly shops, while on many boards there will be several components (heat sinks, transformers, large capacitors and the like) which cannot be mounted by dedicated placement machines, the boards will be completed manually rather than invest in a robot. The situation is gradually changing, largely because the cost of labour is always increasing while the cost of robots is steadily decreasing in real terms. In addition, the error rate of robots is much smaller than that of a manual assembly line.

7.8 INTEGRATED ASSEMBLY LINES

Placement machines and ancillary equipments are quite often put together in one line to establish an assembly line which can perform the complete assembly of a board. The line will usually start with unpopulated PBs, or with mixed technology PCBs to which the PIH components have been assembled, and proceed through stations for adhesive application, component mounting and adhesive curing. (The same line will often carry on to provide soldering, cleaning and testing stations — discussed in later chapters.)

Almost all suppliers of placement equipment will assist in setting up a line to suit the needs of the assembler, with different machines to mount different components or to increase throughput. Such a line is shown in Figure 7.46.

Automatic Assembly — Surface Mount Components 325

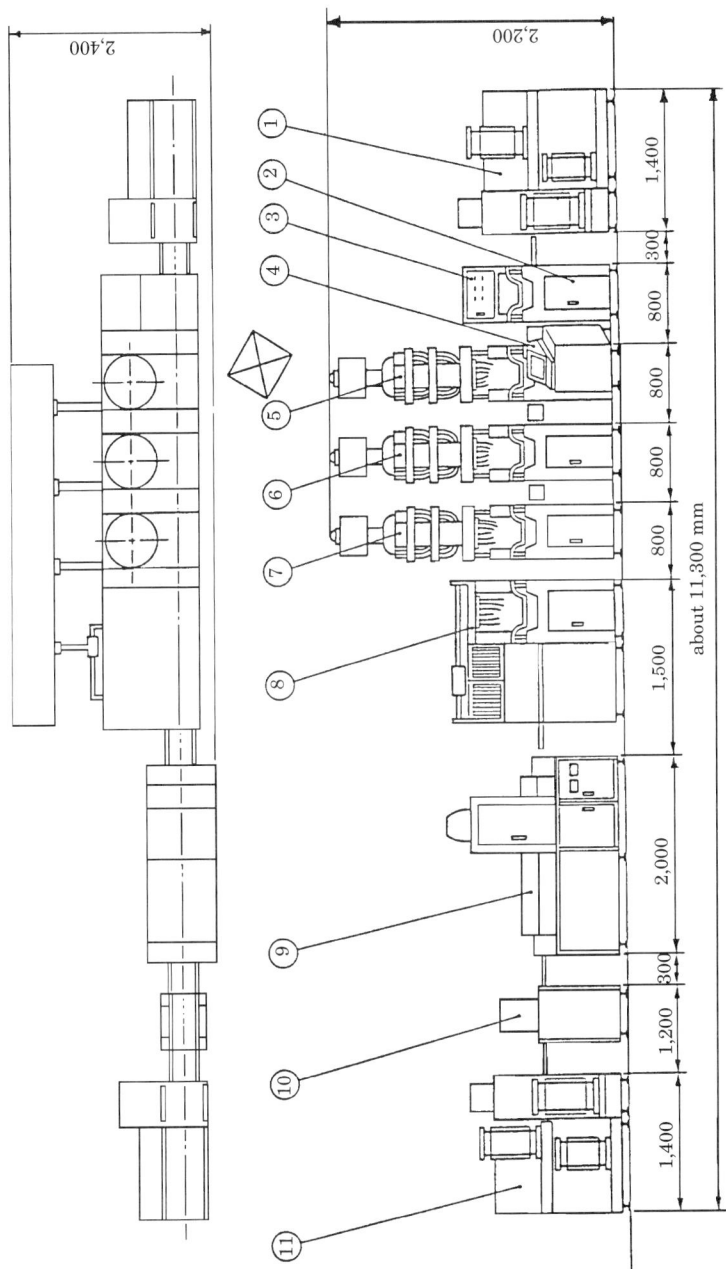

1 PCB loading unit 2 Adhesive transfer unit 3 Control box 4 Inspection unit 5 Non-polar MELF mounting unit 6 Non-polar MELF mounting unit 7 Polar MELF mounting unit 8 Square chip (taped) mounting unit 9 Curing oven 10 PCB stocker 11 PCB loading unit

Fig. 7.46 An integrated line for surface mount assembly based on the Nitto Kogyo STM 3 parallel placement machine.

Obviously, this is a line for mass production. Other set-ups can be based on sequential pick-and-place machines to give greater flexibility and, if sufficiently modular in structure, can usually be rebuilt to cope with changes in products or volumes by adding new stations or replacing some of them.

An integrated line can achieve substantial savings of indirect work in mass production but is usually a second step in SMT. To select the correct machines, install them and achieve the desired level of productivity demands much experience and a deep knowledge of SMT throughout the company, coupled with suitable design of the PB assemblies (PBAs).

7.9 EQUIPMENT SELECTION

As always, selection of equipments must take into account all relevant factors. The main ones which dictate the type of equipment needed are:

(i) projected production rate (boards/year and SMCs/board);
(ii) intended production organisation (batch size, number of changeovers per day, number of shifts per day, etc.);
(iii) types of boards envisaged (sizes, degree of standardisation, full SM or mixprint, number of different SMDs per board, etc.);
(iv) likely types of component (flat, cylindrical, odd, etc.) and supply packing;
(v) joining method(s) to be used (conductive adhesives, soldering — wave, reflow, etc.);
(vi) effect of PCB design rules employed on placement accuracy (XY and skew).

Having derived a short list of potentially suitable machines, a closer comparison can then be made based on:

(i) placement reliability (if known) based on the number of missing or wrongly placed components per million;
(ii) theoretical and actual production rate;
(iii) flexibility with respect to changing board type (same size and different size, same components and different components, etc.);
(iv) adaptability to potential changes in production needs — organic laminates or ceramic substrates, integration into a line?;
(v) options available now and later, can modules be added and/or changed?;
(vi) cost, service and other purchase conditions, including the experience and references of the supplier.

From all the figures obtained, it should then be possible to summarise the cost of purchase and operation of one make of machine against the others considered, and to reach a reasoned conclusion as to the best purchase for the company's requirement. Typically, headings should include:

(a) Capital investment — including depreciation, interest expense, etc.
(b) Installation and commissioning — number of days and cost/day.
(c) Operator training costs — days/year, cost/day.

(d) Operation costs — operator requirement/machine, operator requirement for manual assembly, operator pay, floor space occupied, floor space cost, power costs, air costs, costs of ancillaries.
(e) Maintenance and cost of spare parts.
(f) Actual capacity — machine capacity based on a typical benchmark board with allowances for in-line unreliability, service/ maintenance, set-up time from one job to the next, actual manual assembly capacity, from which the actual production rate can be derived.

Note: A typical board could conform to 176 components of 78 types, made up as 10 x 0603s of 3 types; 50 x 0805s of 13 types; 80 x 1206s of 32 types; 10 x SOIC-16s of 10 types; 10 x SOIC-28s of 8 types; 5 x PLCC-44s of 4 types; 5 x PLCC-68s of 4 types; 1 x QFP-64; and 5 trimpots of 3 types.

This choice can, of course, be amended to be absolutely typical of the intending purchaser's production. As shown, it is fairly typical of a small sub-contract assembly shop.

(g) Repair work — difficult to determine for a given machine without discussion with actual users, but some figure may usually be derived for the number of boards needing repair because of machine deficiency and the cost of such repairs.

From these figures, the total annual operating costs can be calculated and, from that and the production rate, a figure may be obtained for the cost per mounted component. This may seem a great deal of effort to be expended before deciding which machine to purchase. It must, however, be stressed that the variety and number of competing placement machines and systems are very large and the cost of a wrong purchase can be very high. Quite often it will turn out that the machine which is least expensive in initial cost is one of the most expensive in true cost per mounted component. There are no short cuts to assessing what that cost will be.

REFERENCES

1 Harria, N., 'SM Adhesives — Just Strong Enough', *Electronic Production*, pp. 13-16, October (1991).
2 Kirchner, G., 'Problems in Processing SMT Adhesives', *Circuits Assembly*, pp. 54-61, February (1991).
3 Lea, C., 'A Scientific Guide to Surface Mount Technology', Electrochemical Publications Ltd, Port Erin, Isle of Man, British Isles (1988).
4 Park, C. H., 'Understanding Solder Paste Fundamentals', *Circuits Assembly*, pp. 28-36, August (1992).
5 Bandyopadhyay, N. and Marczi, M., Technical Paper on Controlled Atmosphere Soldering, SMT Conference, San Jose (1989).
6 Data Sheets on BX 32 Solder Cream, Multicore Solders.
7 Trovato, R. A., 'Inerting the Soldering Environment', *Circuits Assembly*, pp. 48-52, April (1991).
8 'Buyers Guide, SMD Placement Machines' *Electronic Production*, pp. 15-18, November. (1994).

Section 4
Design and Layout of Printed Boards

Chapter 8

GENERAL PRINCIPLES OF DESIGN AND LAYOUT (OF PRINTED BOARD ASSEMBLIES)

GIOVANNI LEONIDA
Milan, Italy

W. MACLEOD ROSS
Welnorth Ltd, Bishop's Stortford, UK

8.1 GENERAL

The essential elements of a printed board (PB) are:

the base: a thin board or panel of insulating material (the base material), rigid or flexible, which supports all conductors and components.

the conductors: a set of very thin metallic strips, normally high purity copper with or without a solderable finish, having appropriate shapes, firmly attached to the base material and, where necessary, drilled or punched together with it.

The base provides mechanical support to all copper areas, all components attached to the copper, and any other component or part which is fastened to the base material. The base affects the electrical properties of the completed circuit by virtue of its dielectric properties, which must be known and controlled within reasonable limits. The conductors have two functions: first, to provide electrical connections between components and, second, to provide solderable (or weldable) attachment points for the components.

The completed board, therefore, provides mechanical support and all necessary connections to the components. In more recent years other functions have been assigned to the PB — for example, the dissipation of heat generated by the components.

When the board acts only as support and wiring harness, it is a *Printed Wiring Board* or PWB. However, it is possible to realise certain components (such as small inductors, small value capacitors, and resistors) and the PWB is then more correctly a PCB. If doubt exists about the nature of the printed

General Principles of Design and Layout (of Printed Board Assemblies) 329

board, it is better to use the term PB (printed board), which covers both meanings.[1] Many other terms are also used, such as printed circuit, card or board.

The term 'printed' arose because the conductive areas are usually generated by means of a printing process (screen printing, photoengraving, etc.) such as used to print inscriptions, drawings, ornamental designs and so on.

8.2 BASIC DEFINITIONS

If designers, makers and users of printed circuits are to understand one another and be able to communicate in an unequivocal manner, the terms employed must mean the same thing to all parties involved. The international document giving terms and definitions for printed circuits is IEC Publication 194, the first draft of which was discussed by IEC Technical Committee No. 52, Printed Circuits, in 1971. Since then it has been revised, reaching its third edition in 1988, and is in the process of revision and amalgamation with IPC-T-50E.[2] The terms given here are based on those of the IEC (where they exist) but may not correspond exactly. Several of the terms are illustrated in Figure 8.1.

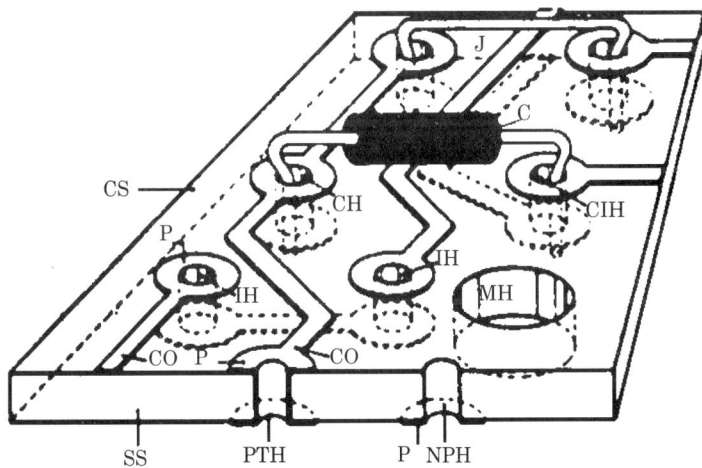

Fig. 8.1 Basic elements of a printed circuit board. C — component; MH — mounting hole; CH — component hole; CIH — component and interconnecting hole; IH — interconnecting hole; PTH — plated-through hole; NPH — non-plated-through hole; CS — component side; SS — soldering side; P — pad; CO — conductor; J — jumper.

IEC member countries have equivalents to Publication 194 and it is advisable for those who design, specify or buy printed boards or assemblies to refer to an official version. For example, the subject is covered in the United States by IPC-T-50E[2] and in the United Kingdom by BS 4727, Glossary of Electrotechnical, Power, Telecommunication, Electronics, Lighting and Colour Terms, Part 1, Group 11, Printed circuits.

Additive Process
A process for obtaining conductive patterns by the selective deposition of conductive material on a clad or unclad base material.

Annotation
  Text or legend pertinent to a board design; text appears off the board area and consists of lettering and symbols while legend appears on the board.

Artwork Master
  An accurately scaled image of the conductive pattern of a PB which is used to produce the 1:1 production master. The scale is chosen to provide the necessary degree of accuracy. (With modern 'front end technology' or CAT (*computer aided tooling*), the production master is usually photoplotted 1:1 using digital data.)
  Synonyms: original artwork, master artwork.

Base Copper
  The original, thin copper foil present on one or both sides of a copper clad laminate. During PB manufacture part of this base copper will be removed by etching. Conductors on the simplest PBs consist of base copper only.

Base Material
  The insulating material, rigid or flexible, on which the conductive pattern may be formed. This also supports all components after assembly.

Branched Conductor
  A conductor which connects electrically two or more leads on a printed board assembly. Some branched conductors, notably ground, supply and reset signal, connect many leads.

Component Hole
  A hole in a PB through which a component lead passes in order to be soldered or connected mechanically to the printed circuit and electrically to the conductive pattern.
  Synonym: mounting hole.

Conductive Pattern
  The configuration formed by the conductors of a PB or their image.

Conductor
  A thin conductive area on a surface or internal layer of a PB. Usually each conductor is made up of lands or pads, to which the component leads are attached, and conductive strips or areas which form an electrical connection between lands. A PB has at least one layer of conductors.
  Synonyms: path, trace.

Conductor Spacing
  The average or minimum (as specified) distance between the adjacent edges of conductors on the same layer of a printed board.

Copper Clad (usually written *copper-clad*)
  A material, usually supplied in large sheets, consisting of a base material to one or both sides of which a thin copper foil is bonded.
  Synonym: laminate.

Electroless Copper
A layer of copper plated on to an insulating or conductive surface of a PB by chemical reduction, that is, without the use of applied electrical current.

Electrolytic Copper
A layer of copper plated on to conductive areas of a PB using an applied electric current.

Etching
Removal of metal from the surface of a PB by chemical dissolution. The process is normally carried out selectively by masking areas of metal which are to be left on the PB.

Grid
An orthogonal network of two sets of parallel equidistant lines for positioning connections on a printed board. The intersection of any two lines of the grid is a possible connection point, or point for the centre of a mounting hole. Conductors may or may not follow the grid lines. (Grid spacings are defined in IEC Publication 97 or the national equivalents, e.g., the United Kingdom's BS EN 60097.[3])
Synonym: design grid.

Footprint (see also *land pattern*)
A set of properly sized and placed pads of a PB on which a surface mounted component can be placed and soldered. Alternatively, the footprint is the board area occupied by an SMC *and* its mounting pads.

Jumper
A connection added to a PB after it has been manufactured. The connection is made with a wire or, in the case of SMT boards, by using a zero ohm resistor.
Synonyms: jump wire, jump staple, link.

Land
That portion of a conductor commonly, but not exclusively, used for the connection and/or attachment of components. In many cases a hole is formed through the land and the board (essential for PIH boards): in other cases, for mounting flat packs or surface mount devices, for example, no mounting hole is required in the land. Lands can be of many shapes (circular, elliptical, triangular, rectangular, etc.) and sometimes have no clear distinction from the conductors to which they belong.
Synonyms: pad, terminal area.

Land Pattern
A combination of lands that is used to mount, interconnect and test a particular component.
Synonym: footprint.

Legend
Lettering or symbols on the printed board, e.g., part numbers, component locations.
Synonyms: notation, annotation.

Locating Hole; Locating Notch; Locating Slot
A hole, notch or slot in the panel or printed board to enable it to be positioned accurately during manufacture and/or assembly.
Synonyms: fabrication hole (or notch or slot), indexing hole, location hole, manufacturing hole, outrigger hole, tooling hole.

Mixprint
A printed board assembly which includes both pin-in-hole (PIH) and surface mounting components, see Chapter 4.
Synonyms: mixed print, mixed board, mixed technology board.

Panel
This is the workpiece that passes through the production process. For example, a sheet of copper-clad laminate, cut from a full size sheet, having a regular outline and the required dimensions for PB manufacture. It may contain one or more PBs which are extracted from the panel, by blanking or routing, after all other manufacturing steps are complete.
Synonyms: production panel, manufacturing panel.

Pattern
The configuration of all conductive and/or non-conductive areas on a PB. Letters and inscriptions may also be included. Pattern also denotes the circuit configuration on related tools, drawings and masters.
Synonym: image.

Physical Layer
A conductive board layer or artwork image representing a complete conductive layer.

PIH Assembly
Pin-in-hole — a printed board assembly made up of components with leads which pass through holes in the board and lands.
Synonyms: traditional assembly, conventional assembly.

Populated PB
A printed board on to which all passive and active components have been assembled.
Synonyms: printed board assembly (PBA), card, assembled board.

Printed Board (PB)
The preferred term for a completely processed (but *not* populated) printed circuit or printed wiring configuration. The term includes single-sided, double-sided and multilayer boards fabricated with rigid, flexible and rigid-flex base materials.

Production Master
A photographic film or plate with a 1:1 scale image of the pattern, which is used to transfer that image on to the panels. It can include more than one PB.

PTH — Plated-through Hole

The holes on a PB can be metallised or 'plated-through' on the hole walls either to provide a larger area for soldering (to resist vibration better) or to provide a 'via' hole, or for both purposes.
Synonym: THP (through-hole-plated).

Resist

A resist is an agent, e.g., a coating, which prevents a specific action on a panel or a part thereof. The most common resists, described more fully in later chapters of this book, are:

Etch Resist: an organic ink, lacquer, photoresist, self-adhesive plastic tape, metal deposit or other material which will prevent specific areas of the metal on a panel from being attacked by an etchant.

Plating Resist: an ink, lacquer, photoresist, etc., resistant to plating solutions, which prevents specific areas of a panel from being plated.

Solder Resist: an ink, lacquer, photoresist or metal coating which is not wetted by molten solder. It is applied to specific areas of a PB to stop them from being solder coated (usually when mass soldering).

Semi-additive Process

A process for obtaining conductive patterns by a combination of electroless metal deposition with etching and/or electroplating. A semi-additive process is used in conjunction with a metal-clad base material.

Side

A finished printed circuit board has two main surfaces or sides, no matter how many layers of circuitry are buried within the board. In PIH boards, the sides are called *component side* — the side on which all or most of the components are mounted, and *solder side* — the side on which all or most of the solder joints are made.
These terms are not appropriate for SMT boards and the following terms should be used:

Primary Side: that, external, side of the printed board which is defined as layer 1 on the master drawing. It is frequently the side which is most complex or contains the greatest number of components.

Secondary Side: that, external, side of the printed board which is opposite the primary side of the board. It is the same as the solder side in PIH technology.

Step-and-repeat

This is a technique of multiple exposure of a single image to produce a multiple image, Figure 8.2, in accordance with a precise scheme of positional translations or rotations (or both) within a plane. It is used to

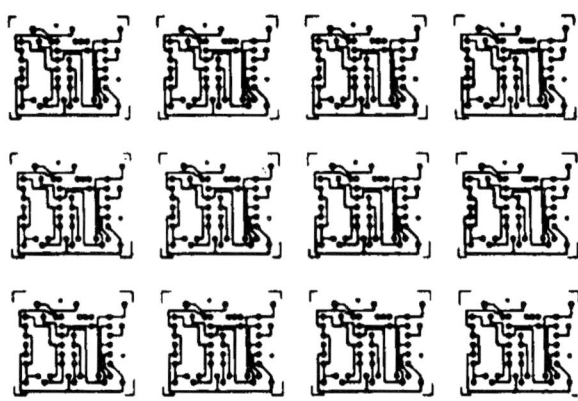

Fig. 8.2 Multiple image of the master artwork of a small PB, produced by step-and-repeat. The production master included 35 of these images, each 43 × 50 mm (about 1.7 × 2.0 in.).

produce multiple image production masters so that large production panels can be used.

Subtractive Process
A process for obtaining conductive patterns by selective removal of unwanted areas of conductive foil from a metal-clad base material.

Through Connection
A through connection is an electrical connection between two or more conductors on different conductor layers. It is commonly achieved by plating the internal wall of a hole drilled through the conductors and the insulation separating them, that is, by means of a plated-through hole. Other methods of providing through connections include wire links, e.g., 'C' links, and various types of eyelet.

Via Holes
A via hole is a plated-through hole that is used as an interlayer connection but in which there is no intention to insert a component lead or other reinforcing material.

8.3 CLASSIFICATION OF PBs

PBs can be classified by reference to several different and overlapping criteria, often with ambiguous results.

Traditionally, they are divided into classes according to use and referred to as consumer, professional and high reliability boards. This attempts to make a clear division between materials and methods of manufacture employed for different end uses.

Consumer PBs are found in radio and television sets, cheap measuring equipments, and so on. In general, the base material will be less expensive and have lower resistance to environmental factors. Larger tolerances can be allowed for manufacture and the need for very good and consistent electrical properties will not be as great.

Professional boards are more demanding, needing better material properties, tighter tolerances in fabrication, and electrical properties orders higher than those of consumer PBs. High reliability boards, abbreviated to 'hi rel', possess very closely controlled base material, tightly controlled manufacture — with 'SPC' a must — and the best of electrical properties.

These vague distinctions may well have meant something in the sixties and seventies and, indeed, are still used today to classify standards and specifications (for example, the build standards shown in Figure 8.3) but the terms are too loose to be of real use for describing PBs. In addition, the distinctions between 'consumer' and 'professional' markets have disappeared. Many 'professional' products, such as personal computers, terminals, etc., are less demanding in the standards required of the PBs than are 'consumer' products like camcorders and picture cameras, especially in terms of packing density, complexity and reliability.

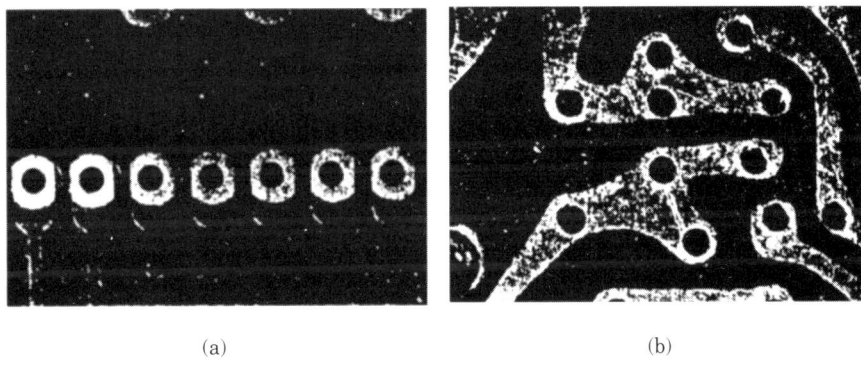

Fig. 8.3 Examples of single-sided PBs: (a) professional type; and (b) consumer type. Note that the professional board shows properly centred holes, better defined conductors and large pads partially covered with solder resist for improving solder joints.

The barrier between professional and consumer electronics is disappearing — professional products such as compact disks, digital audio tape and laser disks were derived from consumer products. The advent of SMT and the expansion of automatic assembly means that even the lowliest consumer board must be manufactured to strict dimensional and flatness tolerances.

A more understandable classification relates to the number of layers of conductors and to the presence (or absence) of plated-through holes. While far from exhaustive (a complete board description requiring definition of 10-20 parameters), this method of classifying boards has the advantage of being simple and can be related readily to board specifications. The major distinguishing constructions are as follows:

Single-sided PB: The single-sided PB has conductors on one side only. This is, of course, the soldering side, Figure 8.4. Single-sided boards were for many years extremely simple PBs but, spurred on by the advent of SMT and fine

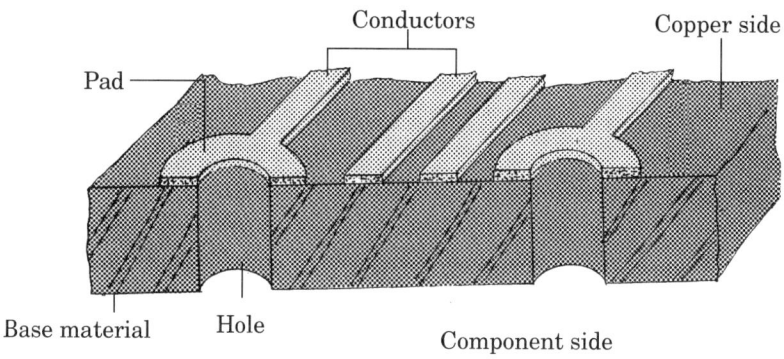

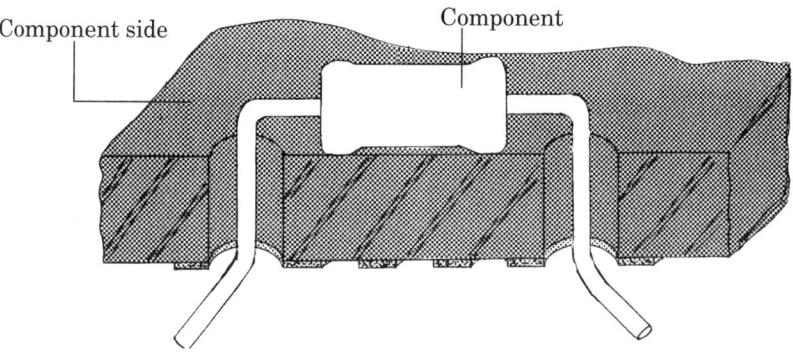

Fig. 8.4 A single-sided board has only one layer of conductors. The copper side is the soldering side of the board.

line/fine pitch technology, they now include very complex types. Single-sided (and double-sided) boards are dealt with in Volume 2, Chapter 8.

Double-sided PB: These have conductors on both sides of the board. The boards may or may not employ plated-through holes to give electrical connections between one side and the other. Figure 8.5 shows some alternatives to PTH which can be used to connect the two sides, but nowadays the great majority of double-sided boards have plated-through holes, Figure 8.6, and Volume 2, Chapter 9.

Multilayer PB: Multilayer boards have at least three layers of conductors, and in common practice the minimum is usually four (to achieve a balanced construction). A multilayer is used when the density of connections needed is too high for two layers or when other reasons intrude, such as a demand for accurate control of line impedances or for earth screening. Multilayers can be simple or very complex. They can be divided into several sub-classes, for example based on the method of layer interconnection, the major techniques for which are clearance holes, Figure 8.7, in which the connection is made by solder, PTH, Figure 8.8, the best known method, now enhanced by the use of

General Principles of Design and Layout (of Printed Board Assemblies) 337

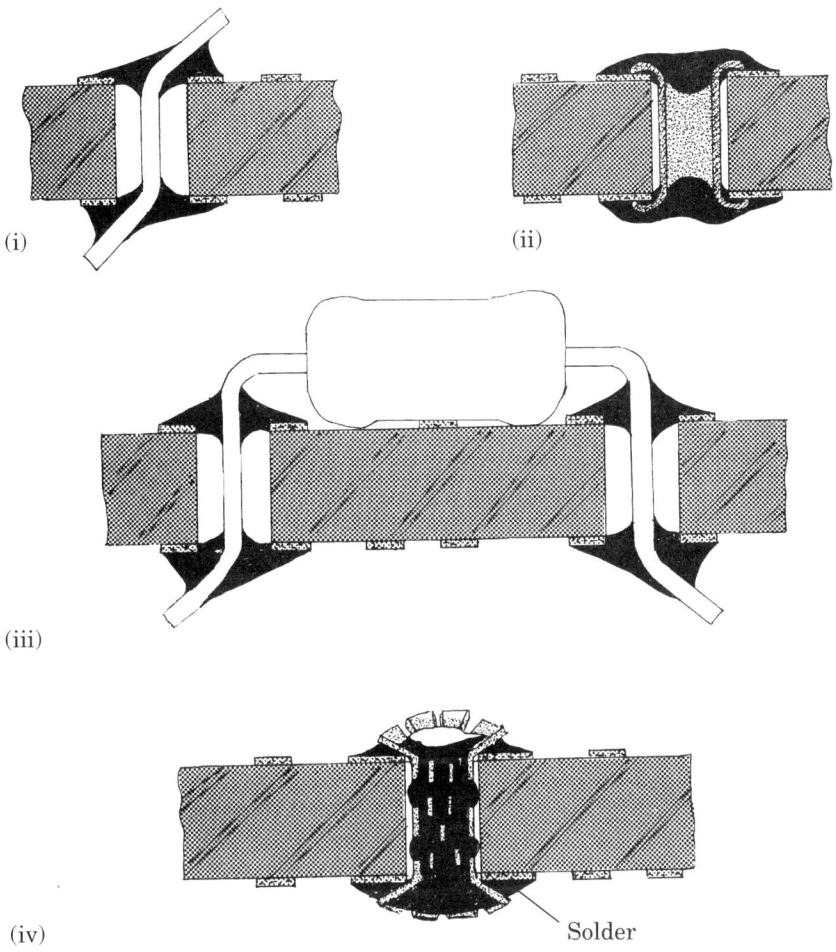

Fig. 8.5 If the holes are not plated-through, electrical connection between the two sides may be made by: (i) inserting a wire into the hole and soldering it on both sides; (ii) inserting an eyelet into the hole and soldering it on both sides; (iii) (non-preferred) inserting a component lead into the hole and soldering it on both sides; (iv) inserting a split funnel eyelet or a braided wire into the hole and soldering manually or automatically on one side only. Capillary flow of the solder produces the joint on the other side.

buried via holes, Figure 8.9, blind via holes, and sequential lamination, Figure 8.10. The last method to be mentioned here is the plated pillar method, Figure 8.11. This type of board is built up layer by layer by:

(i) plating the conductors;
(ii) applying a dielectric layer;
(iii) 'growing' the pillars,

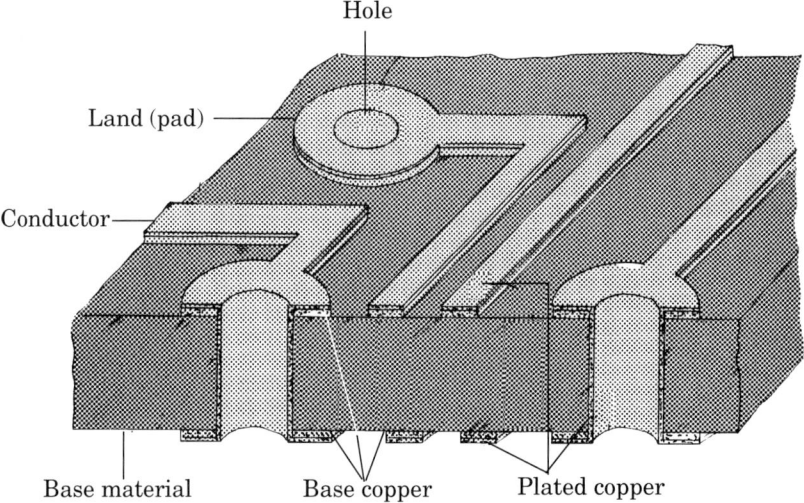

Fig. 8.6 In the majority of double-sided PBs the two conductor layers are interconnected by metallising the hole walls to form a plated-through hole.

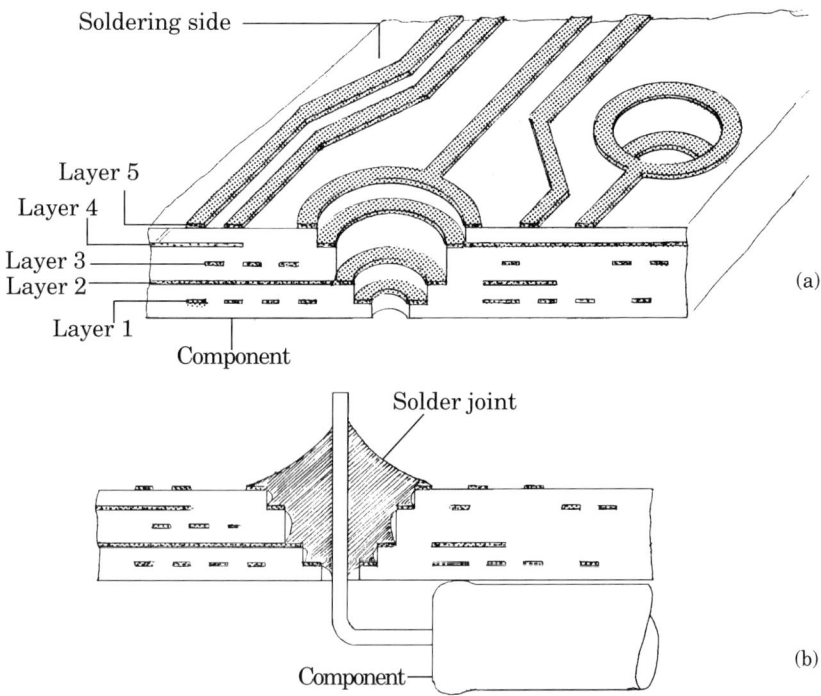

Fig. 8.7 The clearance hole method of interconnecting conductor layers in a multilayer board. The hole diameter is increased from layer to layer so that, where necessary, each pad on a layer is accessible through the holes in the layers above (a). When a lead is soldered into the board, the layers are connected by solder (b).

General Principles of Design and Layout (of Printed Board Assemblies) 339

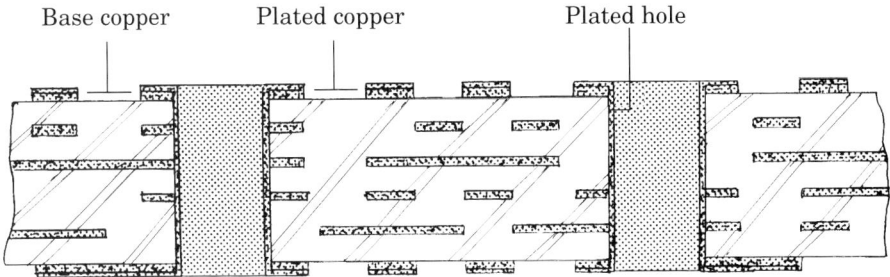

Fig. 8.8 The PTH method of interconnecting the layers in a multilayer board is deservedly the most popular.

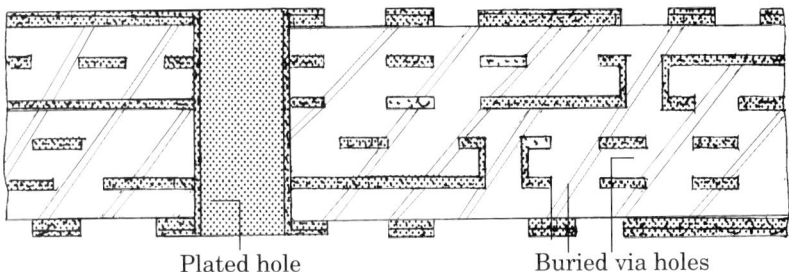

Fig. 8.9 The use of buried via holes allows a higher density of conductors than possible with the normal board seen in Fig. 8.8.

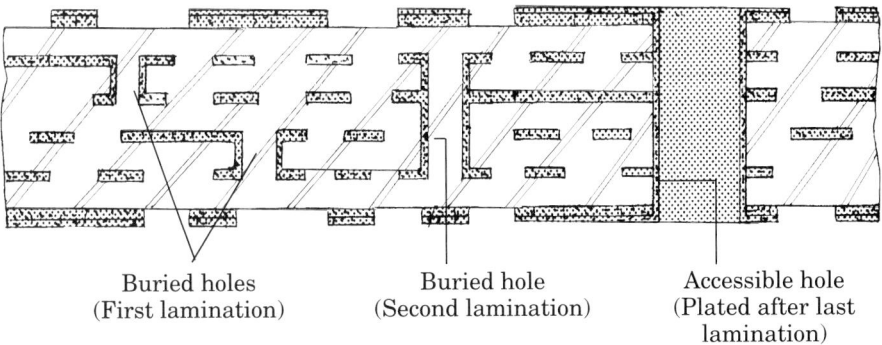

Fig. 8.10 By sequential lamination, the multilayer board can incorporate buried via holes which connect 2, 4, 6 or more conductor layers.

and repeating the sequence as many times as necessary. The final result is a thin multilayer with high density since the conductors can be very precise and the diameter of the plated pillars can be smaller than that of buried via holes.

The fabrication of multilayers is described in Volume 2, Chapter 10.

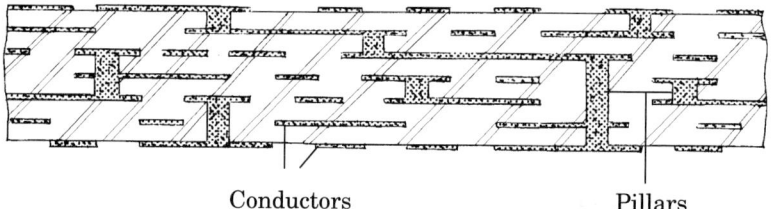

Fig. 8.11 The plated pillar multilayer allows very high packing density. Large pillars may be included for use as thermal conductors from soldering pads to a metal substrate or core.

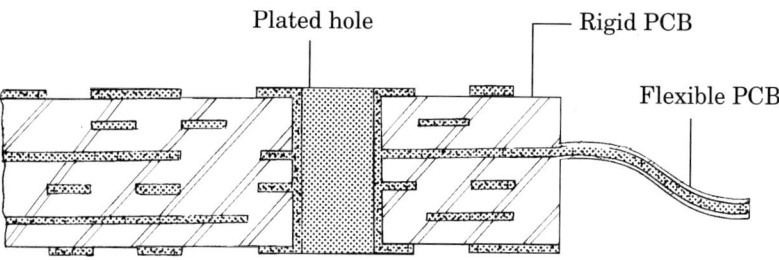

Fig. 8.12 A flex-rigid multilayer board. The flexible circuit may have one or more layers of conductors and is bonded into the rigid assembly to form a connection to other parts of the circuit assembly.

Flexible PB: Flexible PBs may be single-sided, double-sided (PTH or non-PTH) or multilayer. They may be fully flexible or the flexibility can be limited to one or a few portions of the board — the remainder of the board being rigid. In the latter case the PB is called a flex-rigid (the term 'rigid-flex' now seems to be preferred[1]). A flex-rigid multilayer is shown in Figure 8.12 and flexible circuits are described in Volume 2, Chapter 11.

Other types of PB: Other types of PB are described in Volume 2, Section 2. Each type has its advantages and disadvantages, mechanical and electrical, which may recommend it to or deny it from a particular circuit realisation. Each aspect must be considered carefully before making a final decision.

8.4 INITIAL ASSESSMENT

The first step in the design of electronic equipment is the development of an electrical diagram specifying all necessary components and the relevant connections. It is not the function of this book to deal with basic electronics design, well catered for by a multitude of text books. This assessment starts at the next task, a complex one which may take several trial-and-error steps to resolve — the design of the printed circuit assembly.

Based on the experience of the designer, the available equipment and assembly technology, the first decision will be to select the assembly method, see Section 3.

General Principles of Design and Layout (of Printed Board Assemblies) 341

The assembly technology will dictate the selection of packages. However, as most devices are available in 1 or 2 versions for PIH assembly and up to 3 versions for surface mounting, there may still be a degree of uncertainty to be resolved either by experience or by carrying out parallel evaluations.

The area occupied by each component must be calculated. This consists of the area determined from consideration of the component outline, those factors which affect either the placement or insertion of the component, and the area needed to ensure that there is adequate insulation between one component and the next, Figure 8.13. The factors which must be considered may be summarised as follows:

(a) the orthographic projection of the component on the board;
(b) a safeguard area around the component for automatic insertion or placement, which depends not only on the component outline but also on the equipment used, see Section 3;
(c) the area needed for the solder lands or footprints, where this exceeds the area defined by (b);
(d) the area needed to ensure insulation from other components, where this exceeds that calculated from consideration of (b) and (c).

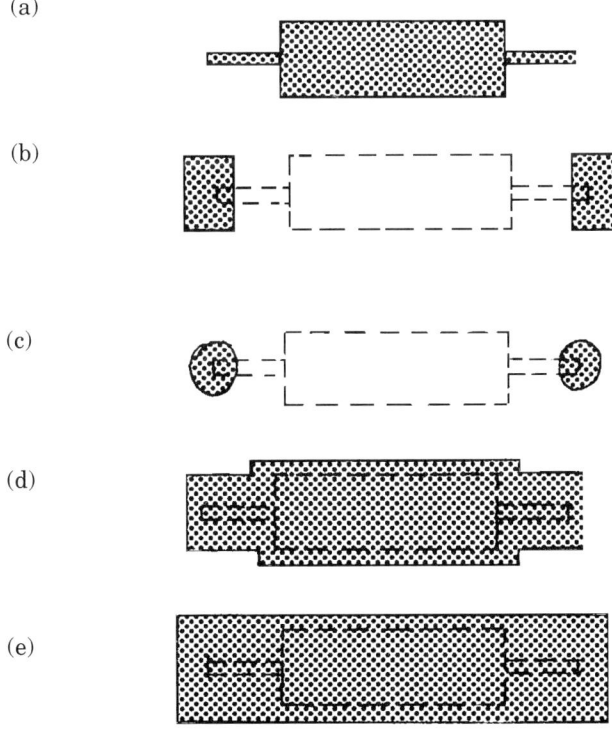

Fig. 8.13 In establishing the area on a PB occupied by a component, attention must be given to (a) its orthographic projection, (b) a safeguard area for lead insertion, (c) soldering lands, (d) insulation from other devices. The result is presented as a rectangle (e) which includes all other areas.

Whenever possible, the component dimensions should be established by checking the manufacturer's catalogue both for sizes and tolerances rather than simple measurement of an existing component. Alternative manufacturers should be identified at this stage since the 'same' component may well be offered in different sizes and tolerances. In many cases it will pay to examine any published standards, as these may allow larger dimensions and tolerances than those offered by some suppliers.

The same care must be given to the specification of insertion pitch of all axial components (and of some of the radial ones). The minimum pitch is determined by the component structure, see Section 2, while the maximum pitch is usually limited by the insertion machine. The number of different pitches to be allowed is determined by the available components and the versatility of the insertion machine(s) employed.

The actual dimensions of components and their minimum and maximum pitch must be recorded for later use in final board design or, alternatively, the information can be saved in the CAD system for recall when needed. At this stage — definition of the total area needed for the assembly — approximation is still acceptable. Components may be considered as simple geometrical figures. All axial components and DIPs will be rectangles, TO-18 and TO-92 transistors will be circles, and so on. To speed the process, similar components may be grouped together as an error of 10-20% is still acceptable.

8.5 TENTATIVE DIMENSIONING

The minimum area of PB required by an assembly is worked out by adding together the areas needed by all components, as previously calculated. The total must then be increased by:

(i) a non-insertable (or non-placement) area around the board — on two sides at least — the size of which depends on the insertion/mounting equipment and handling devices used;
(ii) a non-insertable area around the reference holes to be used during assembly and testing, as well as around all other holes, slots, etc. used as tooling or reference holes when assembling the board into the final equipment;
(iii) areas needed for clamping for soldering, testing, and so on, or which must be clear of components to avoid interfering with any part of the main assembly;
(iv) areas for finger connectors, test points, part number identification, bar code marking, and so on;
(v) areas for running tracks for component connections; this is usually assumed as a percentage of the component area based on previous board designs.

Frequently the uncertainty about these elements, particularly the last one, is such that no reliance can be placed on the result. If no similar board is available on which a calculation may be based, an alternative method must be used.

One such method starts with the orthographic projection of the components and assumes the total PB area needed is a multiple of this area. The

General Principles of Design and Layout (of Printed Board Assemblies) 343

multiplying factor depends upon many elements, such as the type of PB, the assembly equipment, the test equipment, cost objectives, operator skills, etc., and can only be worked out with real precision for a given assembly regime. Table 8.1 gives some figures which have been found sensible in the past. If these are used as a starting point, they may be refined by adapting them to the actual product family and manufacturing facility.

Table 8.1

PB-to-Component Area Ratio*

Type of Assembly	Type of Printed Board		
	Single-sided	PTH Double-sided	Multilayer
Discrete components (ICs less than 5% of the area)	2-3	1.5-2	n.a.(**)
Mixed (ICs from 35% to 50%)	2.5-4	2-3	1.3-2
IC (discrete components less than 20% of the area)	4-6	2-3	1.3-1.8
Mixprint (split side)	2-5	2-3	1.3-1.8
Mixprint (full)	n.a.(**)	2-3.5	1-1.5
Single-sided SMA	2-4	2-3	1.5-2.5
Double-sided SMA	n.a.(**)	1.5-4	0.8-2(***)

    *   Square millimetres of PB needed to assemble a given number of components whose orthographic projections amount to 1 square millimetre. Components may be assembled on one or both sides of the PB.
  ** This type of PB is not applicable for this type of assembly.
*** The PB area may be less than the component area, since boards are assembled on both sides.

While the table may be misleading if it is not adapted to the actual situation, it does show how much the type of printed board influences the density of the assembly. The fact that the type of PB also has a great influence on cost is shown in Figure 8.14.

8.5.1 Vertical Mounting

In most cases it will be assumed that axial leaded components will be mounted with the long axis horizontal. It is possible to mount them vertically but most designers will avoid this because it results in:

(i) lower reliability of the assembly, especially when vibrations are present;
(ii) for manual assembly, more difficult preforming of leads;
(iii) for automatic assembly, more difficult and expensive taping; a radial component insertion machine will be essential;
(iv) more insertion defects and more awkward board handling.

Considerable area can be saved by mounting a component vertically on a 5 mm pitch instead of horizontally on a 10 or 15 mm pitch. If the board has

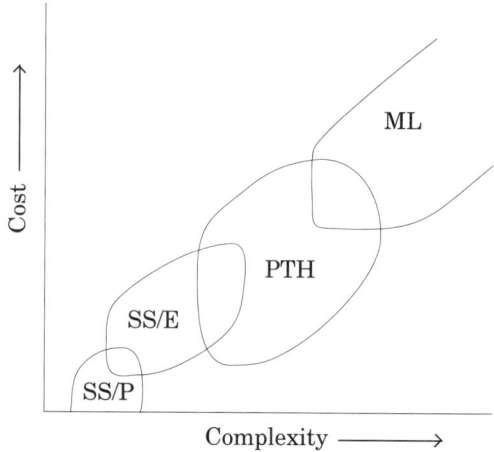

Fig. 8.14 The cost of a PB depends upon its complexity and on its technology. (SS/P = single-sided, paper base laminate; SS/E = single-sided, epoxide glass laminate; PTH = double-sided, plated-through hole epoxide glass laminate; ML = multilayer).

plenty of vertical clearance, due to either the shape of the final equipment or the presence of large devices, the saving achievable by vertical mounting can greatly reduce the PB area and avoid the need to split the assembly into two or more boards.

In such cases, vertical mounting of axial components can be an acceptable practice. The increase in cost is partially balanced by the reduction in PB area, and any reduction in reliability resulting will be less than that involved in a board-to-board connection.

Unless the density of axial components limits that of other devices, it is good practice to start the design by mounting them horizontally, and amend the design later only where the change to vertical mounting produces unquestionable benefits. In general, vertical mounting is seldom encountered on logic boards, where most components are ICs, or on SM boards, which benefit from small outline components. It is more usually found on boards which include at least one power section with large components.

8.6 WORKING OUT THE VOLUME

When the total PB area needed for the assembly has been assessed, the packaging engineer can make a rough calculation of the total volume required. As it is often the case that, in the final equipment, the volume available for electronics is limited, the calculation must be made with care. The most important thing to establish is the maximum volume which the board can occupy in the worst conditions, taking into account safety clearances.

For axial components, horizontally mounted, the height from the board can be assumed to be the diameter of the largest component plus the clearance between its body and the board (if any is required). If the component is vertically mounted, the height is usually not less than 5 mm (0.2 in.) greater than the body length. The forming shape must be checked to determine the exact figure.

General Principles of Design and Layout (of Printed Board Assemblies) 345

The way in which the PCB is fixed to the rest of the equipment is very important. Supports and fixtures also have tolerances and the board may be warped or twisted. If the board is a rectangle, fixed in position along two parallel sides of length W, the maximum deflection along the other two sides can be as much as 3 or 4% of W, according to some manufacturing specifications. This must be allowed for in the estimation of maximum height above the mounting plane. Some benefit may be obtained by locating larger components as close as possible to the PB mounting fixtures, where warp and twist deviation from the theoretical plane will be at a minimum.

8.7 MULTIPLE BOARD ASSEMBLY

Although the packaging engineer may have total freedom to decide the shape and size of the final equipment, in most cases he will have to cope with restrictions in shape and volume. If he has enough freedom, the next step in the design process is to decide whether to realise the circuit on one board or to split the electronics into two or more boards. This is a very important decision, with many points to be considered.

8.7.1 Advantages of a Single Board Solution

The single board assembly, used for many years in consumer electronics, became very popular in the world of computers when the increase in integration made possible the production of a single board personal or small business computer. Since then it has exercised an undoubted fascination for all designers despite severe limitations such as the total size of the board, which cannot exceed the size accepted by automatic assembly and mass soldering equipment. (This is usually between 18 x 18 in. (457.2 x 457.2 mm) and 20 x 20 in. (508 x 508 mm) — see Chapters 6 and 7.)

Where applicable, the single board approach has several advantages:

(i) less overall board area, as none is lost for connections to other boards;
(ii) lower PB cost, unless the designer is forced towards a complex multilayer board to replace, say, two or three much less complex PBs;
(iii) lower assembly cost (connections can be expensive);
(iv) higher reliability because there are fewer connections;
(v) less assembly work;
(vi) lower materials management costs as there are fewer part numbers to handle.

The disadvantages of the single board assembly are best seen by looking at the advantages of the multiple board assembly.

8.7.2 Advantages of a Multiple Board Solution

Unless there are predefined company or customer standards, the decision to split the assembly into two or more boards, and thereby increase the *immediate* cost, may be quite difficult, even though such a solution may have several advantages, for example:

(i) the PCB design may be simpler and take less effort and time;
(ii) PB cost may be less, especially for very simple boards when careful size selection can reduce laminate scrap or for very complex multilayers where the number of boards scrapped tends to increase with board size (less true today with the increase in process control and better inspection aids);
(iii) test and repair of smaller assemblies is usually simpler and demands less expensive tooling (conversely, it is sometimes impossible to test small assemblies until they are assembled to the next stage);
(iv) equipment maintenance is simpler (often a field repair can be carried out by replacing boards until the failed board is found; this may then be returned to base or to the manufacturer for repair or replacement);
(v) engineering changes can be simpler and less expensive;
(vi) certain equipment functions can be restricted to a specific board which is used on more than one product;
(vii) stand-alone boards previously designed for other products may be incorporated into the new design;
(viii) different boards may incorporate different technologies and different design standards, enabling the best use to be made of past, present and future technologies;
(ix) PBs may be made of different base materials, each suited to the specific needs of a particular part of the electronic circuitry;
(x) better use may be made of the permissible volume for the design;
(xi) when there is a shortage of a given component, boards which do not use it can still be built and tested;
(xii) a basic unit can be made, which can be expanded or specialised, either at the end of the assembly line or in the field, by the addition of one or more boards;
(xiii) it may be possible to make the new product compatible with add-ons and/or peripheral equipment for a previous model.

To add to the list of decisions to be made, it may be more important to achieve earlier marketing of a new or revised model by using existing sections of circuitry than to optimise the cost and performance of the design. This lengthy list of consideration points testifies to the potential difficulty of reaching a decision which basically involves *the total cost of the equipment, including all related indirect costs, during its whole life cycle*, from the beginning of the design work to the end of service.

If the decision is taken to use two or more boards to realise the circuit, the next step in the process relates to how many will be involved and how these will be connected one to the other.

8.7.3 Point-to-point Wiring

Figure 8.15 shows the typical structure of a large electronics unit, introduced since the first generation of solid state computers. It consists of many small or medium size boards assembled vertically in a metal frame (a *housing* or *rack*) like books side by side on a bookshelf. Like books, the boards

General Principles of Design and Layout (of Printed Board Assemblies) 347

are always inserted and removed from the same side of the rack, the front side. To maintain each board in a precise position, the top and bottom edges slide in metal or plastic guides fixed to the frame and normal to the front.

The rear or back of the assembly is made up from rows of as many connector blocks as needed and each board has one or more sets of edge contacts (edge connectors) which mate with the relevant connector block(s) at the back of the rack. Insulated jump wires are fixed to the connectors by soldering, plugging, or by terminals and screws, etc., to give all the necessary connections between the boards.

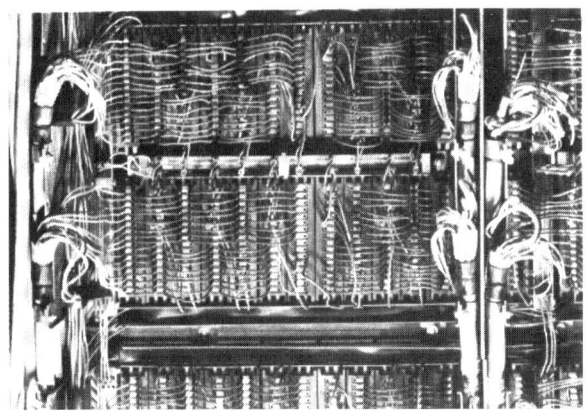

Fig. 8.15 This rack of boards was the central unit of a second generation computer. The boards were connected by point-to-point wiring.

This type of structure was widely used for equipments such as CNC machine control racks, process controllers, exchange equipments and so on, but, while very flexible (as each unit can be wired differently from the others), there were disadvantages. Point-to-point wiring is time consuming. Errors are easy to make and difficult to find.

An improved method, the *wire wrapped backpanel* or *backplane*, was used for many years in low volume complex equipment (large telephone exchange equipments, large computers, and so on). This is still used for prototyping, dedicated test equipment, and the like, as it provides flexibility at low cost with high reliability.

The edge connector blocks are mounted on the backplane, a board as large as the back side of the assembly (or a major portion of it), and secured mechanically with no electrical connection except, possibly, to specific conductors such as power supplies and earth. The connector block leads are very long and project like square posts on the opposite side of the backpanel. The posts are interconnected as required by wire wrapping, each post accommodating up to 5 or so connections, dependent on its length. (Wire wrapping is a technique of mechanical connection in which a bare solid wire is wrapped around a square section post, the corners of which are reasonably sharp. The tension on the wire during wrapping is controlled to produce enough pressure on the post corners to deform both wire and post to give a gas-tight joint. Wire wrapping can be performed manually or automatically. For further details see Volume 2, Chapter 14.)

Coupling the contacts on the boards to the connector blocks may be a problem if the boards are warped or the guides and the backplane are not perfectly aligned. The golden rule that 'any time there are two connectors, at least one should be floating' cannot always be applied.

8.7.4 Soldered Backplanes

For mass production, wire wrapped backpanels evolved into soldered backpanels. These are normal double-sided or multilayer PBs on which connectors are mounted and soldered by wave soldering. While the concept is simple, the number of connections to be made on a complex equipment often dictates the use of an expensive multilayer backplane.

Soldering must be performed carefully, often with the support of specific tooling, to ensure that all connectors are normal to the PB and in their proper position within close tolerances. Soldering of the connector to the panel leads to a rigid assembly which may give problems during board insertion or during the life of the assembly due to joints which remain under stress.

Selection of connector type is important, the housing, shape and load of the spring contacts, etc.[4] playing an important rôle in controlling ease of mating. The springs of the connectors exert a positive retention force on the boards, which is often sufficient to keep them in place. Correct extraction of the boards from the rack needs a special tool, a *card puller*, to overcome the retention force without damage to the boards.

A more professional solution is to mount two triggers on the front edge of each board, to engage the frame of the rack. These have a dual purpose; they lock the inserted board into place and operate as a card puller during extraction. The two triggers may be connected to each other by a strip of plastic fixed to the board edge. This assists in reducing warp of the board, can hold visible information on the status of the board and, when many boards are mounted one beside the other, it closes the front of the rack to assist in keeping out dust and simplifying cooling.

Cooling is a problem with this type of assembly, a lot of heat being generated in a relatively small volume. The simplest way to operate is to test the rack in the worst conditions and check the most critical points with a multichannel thermograph or with labels that change colour at given temperatures. If control of airflow is not enough to give the required conditions, the assembly must be redesigned, paying special attention to thermal aspects, see Chapter 11.

Assemblies with racks and soldered backplanes are intended for boards which will undergo a limited number (10-20) of insertions and extractions during the life of the equipment. The spring contacts in connector blocks are usually gold plated to give a reliable and low contact resistance with the edge contacts on the board. However, the gold on the spring contacts is very thin and is gradually worn away by the abrasive edges of the boards, Figure 8.16. Although it is possible to polish the board edges, a reliable solution requires the use of different connectors.

8.7.5 Book Connection

This is an alternative to the rack and backplane method, extensively used in the early years of electronics, Figure 8.17. It allows access to every component on every board without having to disconnect the board or use

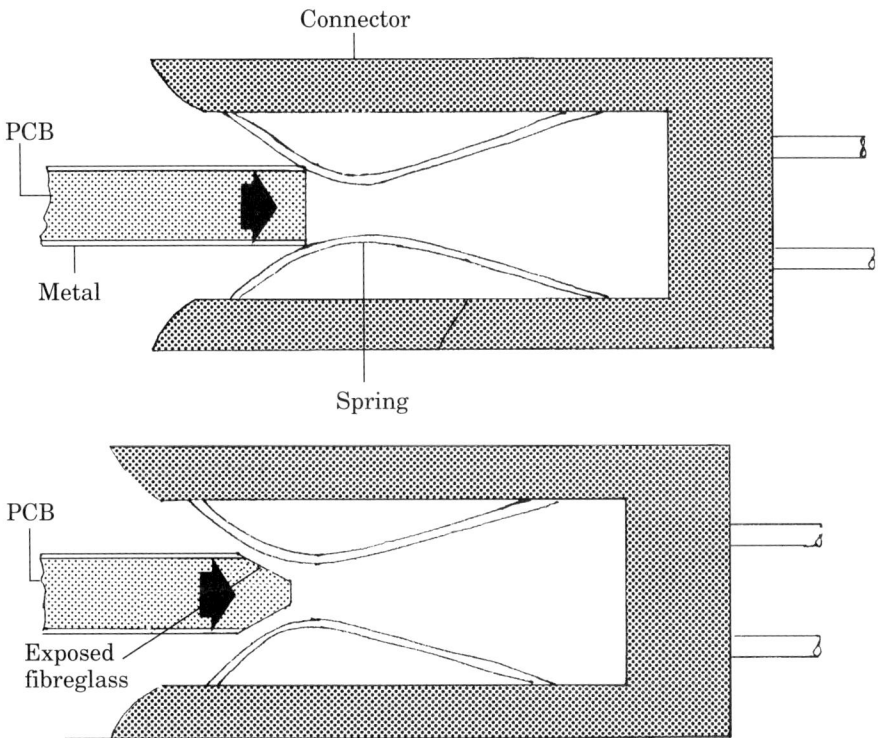

Fig. 8.16 Repeated insertion of a board into a connector block will wear away the plating on the springs. Top: the springs slide against sharp metal edges. Bottom: the springs slide against abrasive fibreglass laminate.

special tools. The boards are mounted on a simple frame, having connector blocks on one side such that the boards are connected to each other like the pages of a book. Board-to-board wiring may be performed by jump wires, wire wraps, or by connectors which mate with the fingers of the edge connectors.

This structure was virtually abandoned in the late sixties but was 'rediscovered' in the late seventies with the advent of ribbon cables, multi-wire connections and the rebirth of flex-rigid circuits.

A popular book type connection between two boards can be found in several pocket calculators, connecting their two parts which can then be closed like a book. The best type uses flat cable and avoids any stressing of joints by running a length of the cable parallel to the hinge. Cheaper types make direct board-to-board connection across the hinge, with ribbon cable or

350 A Comprehensive Guide to the Design and Manufacture of Printed Board Assemblies

Fig. 8.17 Book connection of four large boards (an arithmetic unit built in 1960 on double-sided non-plated-through hole PBs).

flat cable, and are much less reliable. The cheapest type of assembly relies on a simple array of jump wires soldered to connect two boards. This sort of construction should be avoided since, when the book is opened, the solder joints are stressed and repeated bending of the wire will work-harden the wire until it breaks.

8.7.6 Motherboard

As the name suggests, a motherboard assembly relies on a 'mother' board, which holds all the basic functions of the equipment, and on one or more 'daughter' boards, which will add other functions such as additional memory, co-processors, modems, graphics capability, and so on.

From the standpoint of production of the basic unit, it is simply a single board assembly, with all the relevant advantages. For instance, when the board has been tested, the whole equipment has been tested except for its peripheral units. The presence of spare connectors does not pose any problem, and may even simplify testing as they can be readily accessed by probes.

From the marketing viewpoint, this approach gives great flexibility, as it allows the production of many models or variants, simply by plugging in different boards at the end of the manufacturing line. Field upgrading is also straightforward and can be carried out by the dealer or even by the customer himself.

If the interface with the motherboard is simple, the equipment user may develop his own add-on boards to perform specific tasks and independent manufacturers may offer a large variety of add-ons to extend the use of the product far beyond the original field of application. (The enormous success of the first generation of IBM personal computers was almost certainly due in no small measure to the flexibility built in to the 'motherboard and add-ons' structure.)

In the home computer field, the same approach is the basis of all 'cartridges' sold as add-ons to the basic unit which include either a real expansion (such as a memory, modem, etc.) or simply hard-wired software.

8.7.7 Soldered Add-ons

The basic principle that 'if an assembly is split into two or more boards, they should be disconnectable', may be ignored if one board is quite small compared with the other and can be tested along with it. In such cases the smaller board can be simply plugged into the larger one and soldered as a normal component. This structure can be convenient in many situations, for example:

(i) modification of an existing board without upsetting the whole layout and test fixtures;
(ii) temporary design changes, pending the next major revision of the large board;
(iii) alternative circuits, to be assembled on the same board in order to produce several part numbers starting from the same basic board (often incorrectly termed the 'motherboard');
(iv) restrictions on the area of the main board, and the need to find space for additional components;
(v) the need to mount a component at 90° to its normal position.

It is good practice to support the add-on board, unless very small, on two vertical edges by two guides which are mechanically fastened to the main board. Several types of male-male connector, with two pin rows at 90° to each other, are available for such connection. The designer must remember that these are designed for electrical connection and that their mechanical resistance to shock, vibration, etc., should be carefully checked (and trusted to only a limited extent).

Another possible method of fastening the add-on board to the main board relies on solder joints. The small board fits into a slot on the main board and the boards have corresponding solderable pads which may be soldered together by hand or wave soldering. This is a cost-effective solution but one which must be used with considerable care. The solder joints will be stressed by the different shrinkage of the laminates (shrinkage in the Z direction is higher than shrinkage in the X or Y directions) and by deflection of the boards. In addition, handling, unless very careful, often causes fractures in the solder joints.

8.7.8 Sandwich

Cases may be encountered where the area required by the components is larger than the space available in the complete unit while there are few vertical restrictions. In such cases a 'sandwich' structure, i.e., two boards held parallel to one another, may solve the problem. The most reliable joining method will have a mechanical fastening, e.g., a normal bolt and spacer technique, and a separate electrical connection which imposes no stress on either board, even when mechanical alignment is not perfect.

Elastomeric connectors may also be used and can provide a very high density of vias at reasonable cost. In this instance the fastening of the two boards also provides the electrical connections.

One method to be avoided relies on soldering jumpers to connect the boards, both mechanically and electrically. In addition to the difficulty of connecting and disconnecting the two boards, which requires repeated soldering and desoldering on the same pads, some of the solder joints will always be stressed by board warpage, externally induced stress, and so on. Moreover, the amount of labour needed is so great that it will exceed the cost of a connector.

There are some small professional assemblies which employ the jumper technique in order to cope with severe space restrictions. Reliability is assured by careful assembly, a thick glass reinforced laminate, a plated-through hole PB and a large number of connections between the two boards.

8.7.9 Others

The types of assembly described represent only a few of those possible, although they are the most popular. Situations may still arise in which it is not practicable to use one of these, and boards will have to be connected by the older method of wiring harnesses.

A low cost (but unreliable) solution is to strip the end of an insulated wire and solder it into a hole on the PB, like a component lead. If connections are made where required on the PB, board design becomes very simple.

A more reliable alternative, which requires more design work, makes use of connectors and flat cables or ribbon cables. Several fast termination methods are available for connecting cable to connector in one operation, with considerable savings in labour.

8.8 DESIGN OF BOARDS

Board design is the key to success in electronics assembly. Quite often electrical designers underestimate the time and effort required to do a good job. This can cause delays in production start-up and much hidden cost during the life of the product.

The design of boards influences so many areas that it can be extremely difficult to take them all into consideration and strike the right balance between conflicting objectives.

8.8.1 Partition of the Assembly

Assuming that the end assembly has to be split into two or more boards and that a connection method has been defined on the basis of experience and the final requirements of the finished product, the designer must plan how many boards there should be and what will be on each board. In order to be able to do this, he must evaluate at least the items which follow.

8.8.1.1 NUMBER OF CONNECTIONS

Connections are undesirable in electronics assemblies since they reduce reliability and add a cost which is not strictly necessary to the functioning of the circuit. For these reasons the first instinct is to reduce them as much as possible, by locating on each board a defined section of the whole circuit, having the smallest number of connections to the remainder.

On the other hand, when the overall structure of the assembly has been defined — say, rack and soldered backpanel — the maximum number of vias per board is dictated by the type of connector selected. Saving a few via holes does not normally produce significant cost reductions. (If a gold-plated finger is associated with the via, the cost of the gold will be saved, but the actual saving is small.)

In addition, the actual point of partition depends largely on the ability to test, since functional testing of the board is only easy when a whole function is included. Different considerations will apply to in-circuit testing when a large number of connections may simplify the program and the fixture.

8.8.1.2 ASSEMBLY TECHNOLOGY

Taking maximum advantage of the technologies available to the company means using those which are most cost effective and best known. The assembly would therefore be partitioned, as far as practicable, into boards which utilise those technologies. The appearance of simple single-sided boards in the same rack as complex multilayer boards should not be uncommon and may prove very cost effective for the company.

The expansion of automatic assembly frequently leads to boards which can be manufactured on a fully automatic line, with the exception of a few components which are not available in the right package or are too awkward to handle. If such components can be confined to one board, all other boards benefit from a higher degree of automation with simpler material handling and management.

The expected volumes of the product will play a rôle in the selection process. If the volumes are small, the board should be capable of being processed on existing equipment; if they are very large, it may be the opportune time to acquire new equipment and/or technology.

(It is always advisable to acquire the know-how of a new technology well in advance of an actual requirement, and to test it on low volume products, where failure or delay will not have a traumatic impact on the economics of the company — the chance seldom arises, however, as most companies set a low value on assembly technology planning/ development.)

8.8.1.3 MAKING A LAYOUT

The more complete a function the board performs, the easier it is to design its layout, see Chapters 9 and 10. On the other hand, there may be many reasons for setting a section of a circuit apart, e.g., heat dissipation, electromagnetic shielding, the need for a different assembly technology, accessibility for operation and/or maintenance, the existence of a previous design, shared components, product variants achieved by changing a single board instead of many, and so on.

Making a layout was a very important factor in the past as it was performed manually and had a significant influence on development time and cost. Extensive use of CAD, see Chapter 10, and the developments in hardware and software for CAD have reduced layout relevance to particular situations, such as care in dimensioning for current carrying capacity, careful thermal design, and so on.

8.8.1.4 TESTABILITY AND REPARABILTY

Depending on the product complexity, incoming component quality and the manufacturing process, a certain number of assembled boards will not work. These will have to be screened by appropriate testing and reworked. The board design must take into account the level at which tests may or should be performed, and must try to make such testing as simple as possible.

Testing and rework/repair are closely connected. If the test shows a failure, at what level is rework/repair viable and at what point should the board be scrapped? Who should do the testing? What equipment is to be used and what diagnostic support is to be provided? What is a repair and what is rework? What can be replaced on the board and how many times can it be repeated? Will a given repair influence reliability?

The board designer cannot answer all such questions on his own, and he is not in control of all company operations. Even if he is supported by full history sheets and experience with similar boards, he must involve quality engineers in these aspects of his design.

8.8.1.5 MAINTAINABILTY

With a few exceptions — which, paradoxically, include the simplest toys and the most complex missile control systems — all electronics equipment is built to work for years and must be maintained during its lifetime.

Field service can take many forms, from a complete local repair to returning the complete equipment to the assembly point in the manufacturer's factory. In most cases a flexible approach is taken: simple repairs are undertaken at the customer's premises and more complex ones are dealt with at a specialised repair shop or returned to the original factory.

Multiple board assembly generally makes field servicing easier, provided that each individual board performs a specific function and may be easily replaced with a board known to be good, thereby allowing the faulty board to be traced. In such cases, easy replacement is defined as involving no major dismantling and minimal soldering/ desoldering. When found, the faulty board may be repaired either by skilled personnel or with the support of automatic test equipment.

Repair personnel are seldom very highly skilled in a particular product. They have to cope with many different products, many of which may well have been delivered many years before so that the design history is lost in the mists of time! It can be very useful, for the designer's development and for the company's reputation, for the designer to discuss the aspects of equipment maintenance and repair with the Field Service Department.

8.8.1.6 ENVIRONMENT

Boards in an equipment assembly may encounter a wide range of different conditions, both expected — such as temperature, humidity, shocks, vibration, dust, electromagnetic interference, stresses, etc. — and unexpected, for example, spilt liquids, insertion of foreign objects (even gold chains!) in cooling slots, and others that cannot be foreseen because of their highly individual nature.

The designer of the electronic assembly must be aware of what the final product will be and try to protect the most sensitive parts of it, as far as possible, from anything he can reasonably foresee. For instance, any high voltage circuit will be protected to prevent contact from outside, although accessibility for maintenance may be reduced. Careful location of components on the boards and of the boards in the product can do much to minimise the likelihood of damage by whatever agent.

Checks on the environmental situation must be carried out early in the design programme, before too many 'final' decisions on layout are made. The situations considered must cover all likely eventualities, from shipping to the customer, the operating environment, and conditions which may arise through routine or non-routine maintenance.

8.8.1.7 LOGISTICS

A designer may wonder why he should have to think of the logistics of the assembly shop. In electronics manufacture, material flow and logistics are highly relevant because of the following factors:

(i) components and materials are the largest single cost factor; a large inventory involves high carrying costs and can tie up a significant amount of the company's financial resources;
(ii) the life cycle of most electronics products is short and technical changes are frequent so that the risk of inventory obsolescence is high;
(iii) unlike many other industries, the cost of many electronics components is decreasing rapidly, and early purchase may incr*ease* the cost of the product;
(iv) component availability depends on a highly variable demand and procurement lead times may change rapidly;
(v) shortages of some components are quite frequent and, to avoid too much lost time, production should be taken as far as possible with incomplete boards;
(vi) incoming inspection cannot prevent some faulty components from reaching the assembly line and being discovered at a very late stage; clearing such situations must be carried out effectively and in a short time;
(vii) the market is highly competitive, making sales forecasting difficult; the ability to cope with unforeseen market demand is a major factor in success;
(viii) product innovation is a major competitive factor and, when a new design is ready, there is no time to 'fill up the pipeline' (it must be ready to flow);

(ix) production, even in small batches, is now often highly automated and mechanised materials handling is becoming more and more important.

With these factors to bear in mind, the electronics equipment designer must consider the logistics implications of his work. He must design to: minimise the number of component part numbers; ensure whenever possible adequate availability of all part numbers for the expected life of his product; minimise the number of packaging styles with a view to maximising the use of assembly equipments; enable the assembly to accept variants along the production line or to be customised at the end; be able to assemble and, even if only to a limited degree, test boards when a component is missing; maximise the ability to handle boards automatically.

8.8.1.8 COSTS

Cost is not, and never can be, an item on its own. It is the primary basis for any evaluation. It would, in fact, be impossible to balance so many factors (design time and cost, component type and cost, manufacturability, dependability — the sum of maintainability, maintenance support, including field service, and reliability — logistics, etc.) without a universal unit of measure. This can only be 'dollars', 'Eurodollars', or any other preferred local unit of currency.

In many cases it is difficult to derive even an approximate cost for many factors. (How do we measure the value of coming on the market two months earlier? Does it not depend on the product being designed and on what our competitors are doing? Do we know when competitors will reach the market with a similar product while we are designing ours? How can we . . . ?)

No matter how difficult it may be, during the design phase maximum effort must be put into evaluating the total product cost, from design to the end of field service, including the cost advantage/disadvantage of earlier or later marketing. The difficulty is obvious when it is remembered that it is far from easy to calculate the exact cost of a product *even after it has been produced, sold, and serviced for years.*

8.8.2 Gross and Net Area

Having assessed, at least on a tentative basis, the size, shape and content of a board, it becomes possible to start the actual design. First, it is best to locate on the board all components which must be in specific places because of the overall assembly design, such as connectors or contact tabs/fingers, fixtures, fasteners and other anchoring devices, and control devices (switches, potentiometers, test points, and so on) which must be accessible from the exterior.

The second step is to consider mounting holes, slots, cutouts, etc., as it is probable that a 'safety area' will have to be provided around many of them. Inside the safety area no component is allowed to be present and conductors must be routed so that they do not intrude.

In this way, the PB gross area is reduced to a net area, Figure 8.18, on which all other components are located after considering geometric constraints such as:

General Principles of Design and Layout (of Printed Board Assemblies) 357

(i) *Locating holes (for assembly)*: these are usually 0.125 in. (3.175 mm) or 3.20 mm diameter and are placed along the longest edge of the board, as far apart as possible. The centre of each hole must be at least 1.5 times the hole diameter from the edge and, in any case, not less than 2 mm (0.08 in.). Around each hole a safety area is usually

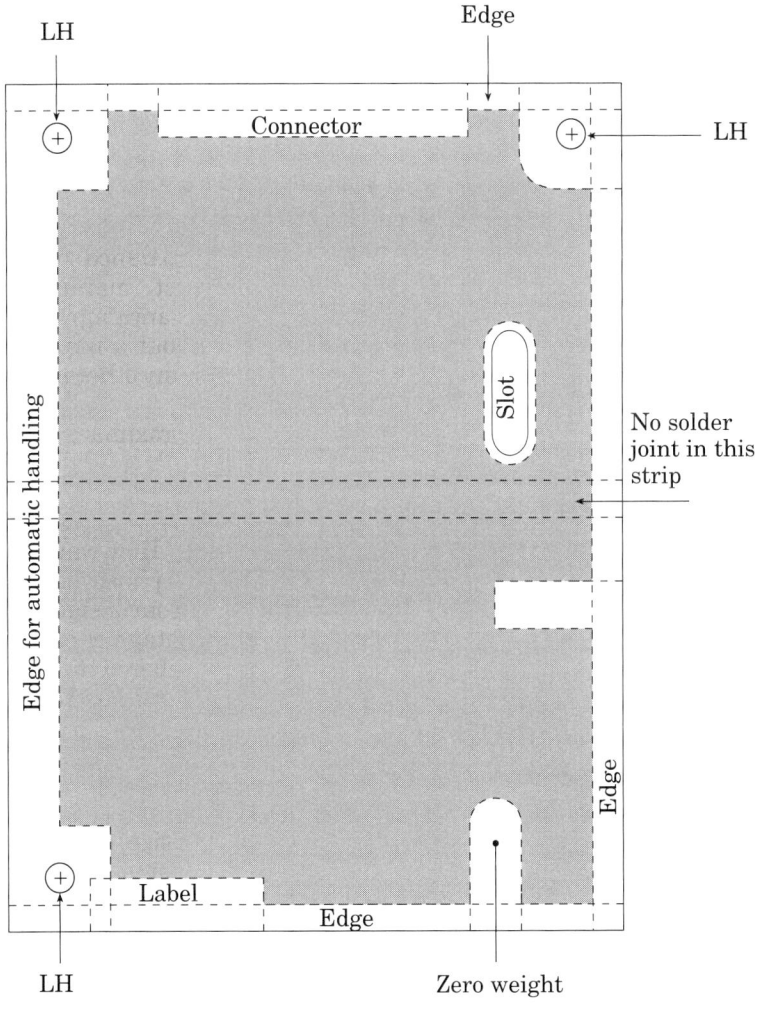

Fig. 8.18 The 'net' area (shaded) of a PB is obtained by considering all restrictions and all components which must be in a fixed position. On very large boards it can be useful to leave a strip, near the middle of the board, free from solder joints to permit extra support when wave soldering.

provided, of circular form with a radius which equals the hole radius plus 2 mm or more.

(ii) *Mounting holes*: if the board is fixed to the rest of the equipment by, e.g., metal washers and screws, a safety area is essential around the fixing holes. Conductors can only be placed in the area if insulating washers/fixings are used.

(iii) *Slots*: depending on their function, a safety area may not be needed around slots.

(iv) *Supports and bearings*: usually, these generate constraints on one side only of the board. Care must be taken to avoid traps being generated between conductors where dust, dirt, and/or moisture may accumulate. (Permanent solder masks or other insulating coatings help to prevent this.)

(v) *Edges*: depending on production methods, tooling, etc., a strip free from component bodies and leads may have to be provided along all edges. Such a strip is normally 2 to 5 mm wide (0.08 to 0.2 inch). If conductors are present, they should be kept at least 1.5 mm (0.06 in.) from the edge or, if the conductor width is less than 1.5 mm, at least 2.5 mm (0.1 in.). (The main reason for this is to avoid mechanical damage — see also (xi).)

(vi) *Identification markings*: whatever the form of the board identification mark, if it has to be read by a fixed scanner it must be in a fixed and precise location with reference to the indexing holes.

(vii) *Heavy components*: these must be placed close to the points at which the board is supported and where they are less affected by vibrations. Heavy components often need some form of mechanical fastening and, when the type to be used has been decided upon, the relevant area must be taken into account. (These fastenings may sometimes be used to improve board rigidity.)

(viii) *Hot components*: these must be located to take maximum advantage of the airflow — convection or forced — through the assembly and may need to be positioned with reference to any heat sinking provided for the PBA. Ideally, they should be positioned as close as possible to the point at which heat leaves the assembly, to minimise heat transfer to other devices.

(ix) *Cooling slots*: especially if the board will be horizontally supported in the final equipment, it may be necessary to put holes or slots in the board to allow more effective cooling of parts which cannot be placed at the edges.

(x) *EMI shielding*: if some parts of the circuit require to be shielded to avoid either active or passive electromagnetic interference, this may influence the board design. (A distance of a few additional centimetres between emitter and receiver can eliminate an expensive shield.)

(xi) *Handling*: there is usually a need for larger clearance from components on two parallel edges to allow for manual or automatic handling of the boards. The use of automatic insertion or placement machines may increase the need and may impose two indexing holes on the same side, at a fixed position with respect to each other. On mixprint boards there are also restrictions on the maximum height of components inserted in specific areas.

General Principles of Design and Layout (of Printed Board Assemblies) 359

(xii) *Soldering*: soldering techniques and equipments can impose many restrictions on board layout. Considering only the effects on net area at this stage, in wave soldering the maximum size of slots, edge clearances and handling clearances may all be affected. On very large boards, it can be useful to leave a strip free from solder joints across the board to allow additional support during soldering to minimise warping.

(xiii) *Test fixturing*: in-circuit testing of boards may demand simplification of the construction of test fixtures.

The result of these factors, and any others specific to the design, is to reduce further the total area available and to increase the restrictions on the location of the remaining components, which are in fact the large majority. Fortunately, not all boards have all these restrictions — otherwise design would take an extremely long time, even when assisted with CAD.

Having considered definition of the net area of the board and the location of a few specific devices, it remains to study the rules for placing all other components.

8.8.3 PIH Assembly

The rules which apply to the design of a board are determined by:

(i) assembly technology and machinery available;
(ii) joining technology and equipment;
(iii) testing technology and equipment;
(iv) reparability and maintainability;
(v) aesthetics.

The last item may look odd in this context but we have all heard the statement 'If it looks good, it is good!' and it is a fact that an ugly, messy design has usually been achieved hastily or carelessly, and will be difficult to produce at reasonable cost with reasonable quality.

The basic rules for locating PIH components are the same for manual and automatic design:

(i) Each component lead must terminate in a hole and be soldered to the PB, even if it is not connected to the circuit thereby; if there is good reason not to solder it, the hole must not be without land or plating.
(ii) One lead only shall terminate in any given hole, regardless of its diameter.
(iii) All components are connected to each other by their connection to the PB; direct lead-to-lead soldering shall not be permitted.
(iv) Holes for an axial component shall be aligned with its axis and spaced at the distance needed for correct bending of its leads.
(v) Each component must have its footprint area on the PB. This shall not interfere with the footprint of any other component.
(vi) The board must have an identifiable grid, preferably selected from IEC Publication 97,[3] and each hole must be on an intersection or *node* of this grid.

(vii) Axial components must be parallel to each other. When necessary, their axes can be at 90° to each other.
(viii) Ideally, axial components should all have the same insertion span or pitch. If this is not possible, the minimum number of discrete spans shall be chosen. All spans must be a multiple of the basic grid chosen.
(ix) Axial orientated components (diodes, electrolytic capacitors, etc.) must all be orientated in the same direction, at least according to type.
(x) The emitters of transistors must be oriented in the same direction; the same criterion shall be applied to all non axial components of the same type.
(xi) Components with more than two leads require their own hole layout; their insertion must be possible without stress (or with a fully controlled stress) on the leads.
(xii) The clearance between components must be such as to avoid short circuits or any unacceptable reduction of insulation, independently from the position they occupy in their holes.
(xiii) The clearance between any two adjacent components shall be such that it is acceptable for the majority of automatic insertion machines.
(xiv) The relative position of components should allow acceptable soldering conditions, see Chapter 9 and Section 5.

The first items in the foregoing list are obvious and part of the basic knowledge of any person who has been involved with electronics assembly work, Figure 8.19. As most boards are now designed by CAD, such rules are built into the software and any exceptions require specific intervention by the designer. Care must be taken, however, when carrying out urgent manual design changes to avoid infringing the rules.

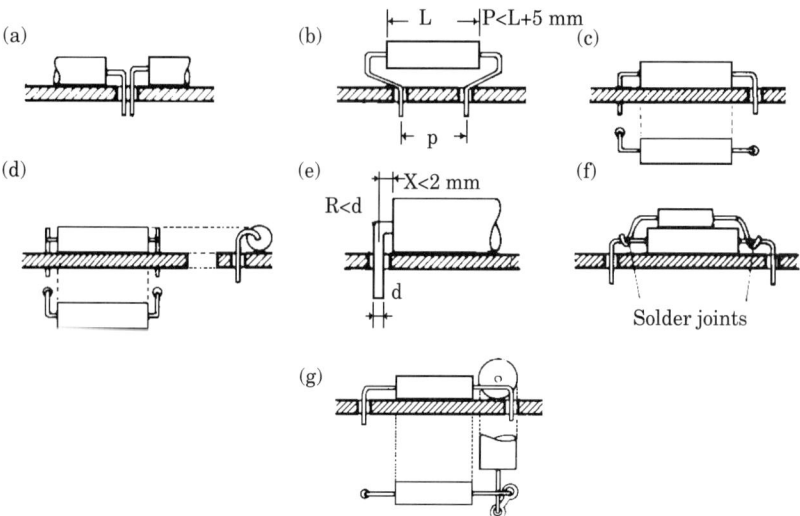

Fig. 8.19 Typical situations to be avoided in artwork design. Case (g) (superimposed leads) is not allowed even if the two leads are electrically connected on the electrical diagram.

General Principles of Design and Layout (of Printed Board Assemblies) 361

The rules for arranging components, Figure 8.20, are also quite obvious in their intention, since they originate from the specifications of automatic insertion machines, see Chapter 6, most of which can insert on a fixed pitch or on a limited number of pitches, either in one direction or in two directions at right angles to each other. If polar components are not orientated in the same direction, it may be necessary to feed the inserter with two tapes of the same component, reducing the number of feeders available and making inspection more difficult (and therefore more prone to error).

Adherence to these rules will simplify production and reduce the chances of error. Exceptions to the rules may limit the manufacture of the boards to

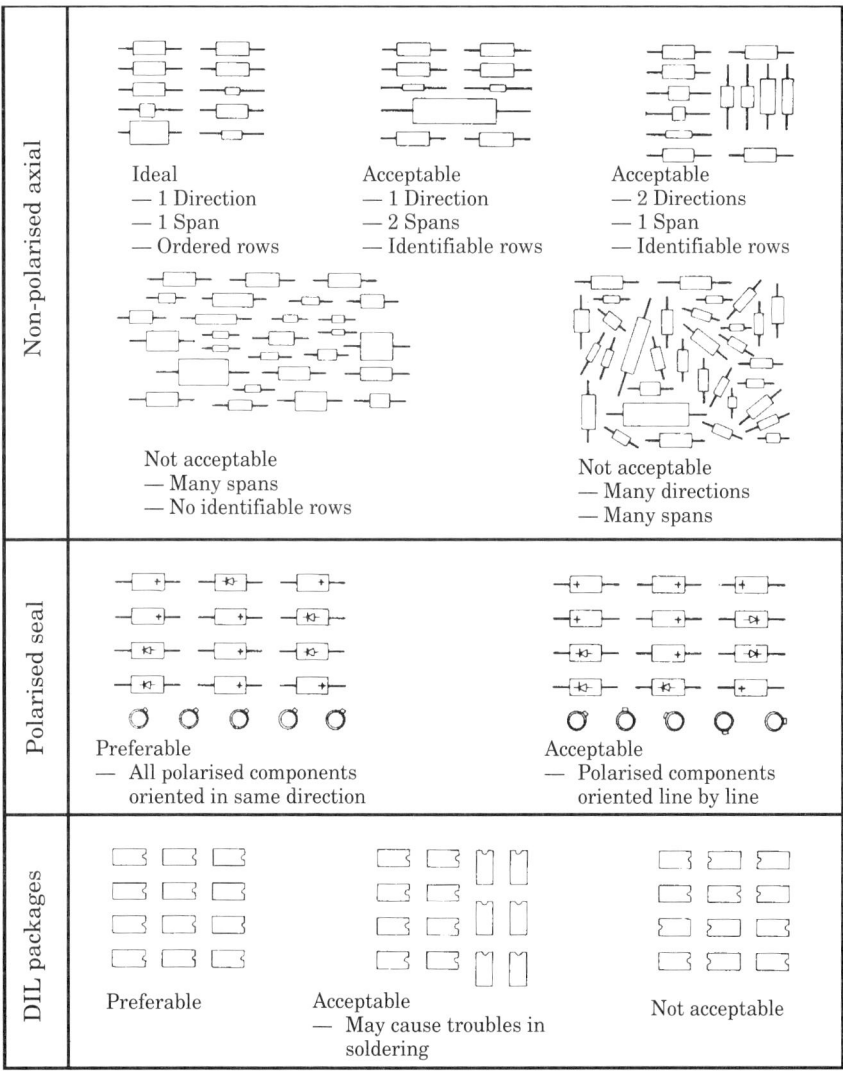

Fig. 8.20 Basic rules for arranging components on a board.

specific machines and reduce the attainable productivity. When dealing with double-sided and multilayer layouts, conductors are usually run at 90° to each other, see Chapter 9, and many potential track positions can be lost when components have different orientations.

The distance between two adjacent axial components is limited by the size of the tool on the inserter which drives the component down and holds it in position until it is fully inserted and clinched. Even in the worst conditions of tolerances on hole positions and component positions, there must be no interference with any previously inserted component, Figure 8.21. While it may be possible to increase the component density slightly by specifying the insertion sequence during the design of the board, the disadvantages of so doing are many — additional design work, restrictions on the manufacturing shop, additional documentation, more problems with design changes, and so on.

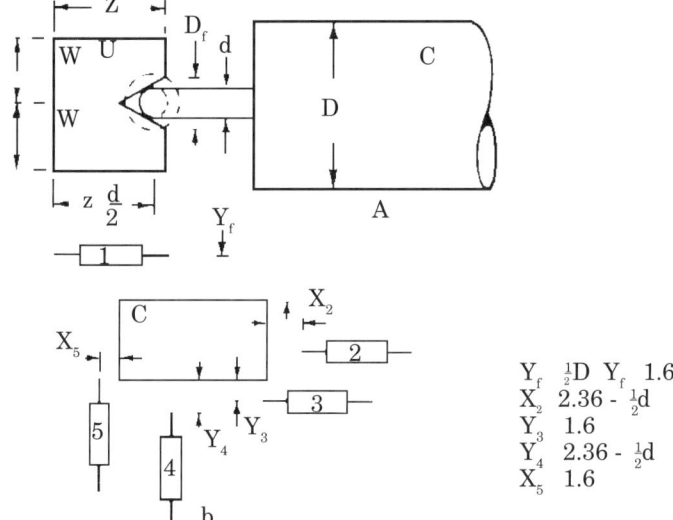

Fig. 8.21 Restriction of packing density for the automatic insertion of axial components (such as C) in the presence of a previously inserted component A. Values are based on an insertion tool U whose dimensions are Z = 2.36 mm and W = 1.60 mm (Z=0.093 in. and W = 0.063 in.). D_f is the hole diameter.

The designer must, therefore, set the design rules independently and take the view that components may assume several different positions on the PBA in relation to others inserted previously. For two axial components, as many as seven different positions must be considered, for each one of which the rule for minimum clearance has to be established. Since this depends on the specific machine, it is impossible to give universal values. A quantitative indication may give an excessive value as it must be based on the worst condition experienced with the most popular machines, Figure 8.22. The relevant values do not take account of the tolerance on insertion hole position — this must be added. If, for example, any hole is specified to be within 0.05 mm (about 0.002 in.) of its theoretical position, the minimum clearance between two components must be increased by twice that value, i.e., 0.1 mm (0.004 in.).

General Principles of Design and Layout (of Printed Board Assemblies)

Case	Figure	Manual insertion	Automatic insertion — A inserted first	Automatic insertion — B inserted first
1	(diagram with D_f, d, L, D_p, D, P)	$P \geq L + 5$ $P \geq 7L$	$P \geq L + 4$ $P \leq 25.4$	
2	(A above B, offset Y)	$Y \geq \tfrac{1}{2}(D_A + D_B)$	$Y \geq \tfrac{1}{2}(D_A + D_B)$ $Y \geq 1.6 + \tfrac{1}{2}d_a$	$Y \geq \tfrac{1}{2}(D_A + D_B)$ $Y \geq 1.6 + \tfrac{1}{2}d_a$
3	(A above B, offset Y)	Same as 2	$Y \geq \tfrac{1}{2}(D_A + D_B)$ $Y \geq 1.6 + \tfrac{1}{2}d_a$	$Y \geq \tfrac{1}{2}(D_A + D_B)$ $Y \geq 1.6 + \tfrac{1}{2}d_a$
4	(A above B, X and Y)	$Y \geq l_m + \tfrac{1}{2}(d_A + d_a)$ ($l_m = 2 \div 5$)	if $X \geq 2.36 - \tfrac{1}{2}d_A$ $Y \geq 1.6 + \tfrac{1}{2}d_a$ Otherwise see case 3	if $X \geq 2.36 - \tfrac{1}{2}d_A$ $Y \geq 1.6 + \tfrac{1}{2}d_a$ Otherwise see case 3
5	(A — B horizontal, X between)	Conditions depend upon minimum distance between lands of the PCB	$X \geq 2.36 - \tfrac{1}{2}d_A + \tfrac{1}{2}D_{fA}$	$X \geq 2.36 - \tfrac{1}{2}d_A + \tfrac{1}{2}D_{fB}$
6	(A horizontal, B vertical, X)	Same as 5	$X \geq 1.6 + \tfrac{1}{2}d_{fA}$	$X \geq 1.6 + \tfrac{1}{2}d_{fB}$
7	(A horizontal with X, X; B vertical)	$Y \geq l_m + \tfrac{1}{2}(d_A + d_a)$ ($l_m = 2 \div 5$) a) if $X > 1.6$ $Y \geq 2.36 - \tfrac{1}{2}d_B + \tfrac{1}{2}D_{fA}$ b) if $X > 1.6$ See case 8		a) if $X' \geq 1 + \tfrac{1}{2}d_{fB}$ and $Y' > 1.6$ $Y = l_m + \tfrac{1}{2}(d_A + d_a)$ b) if $X' \geq 1 + \tfrac{1}{2}d_{fB}$ and $Y' > 1.6$ $Y = l_m + \tfrac{1}{2}(d_A + d_a) + 1.6 - Y'$ c) if $X' < 1 + \tfrac{1}{2}D_{fB}$ $Y = 1.6 + \tfrac{1}{2}D_{fB}$
8	(A horizontal, B vertical, Y)	$Y \geq \tfrac{1}{2}D_A + d_B + \tfrac{1}{2}D_{pB}$	$Y \geq \tfrac{1}{2}D_A + \tfrac{1}{2}D_{fA} + 2.36 - \tfrac{1}{2}d_B$	$Y \geq \tfrac{1}{2}D_A + \tfrac{1}{2}D_{fb}$

Fig. 8.22 Restriction of packing density for axial components to be inserted manually or automatically. The first diagram explains symbols (D_f = hole diameter; D_p = pad diameter). The two components are identified as A and B. Dimensions are in millimetres. To obtain dimensions in mils (thousandths of an inch), 1 should be replaced by 39, 1.6 by 63, 2 by 79, 2.36 by 93, 4 by 157, 5 by 197, 25.4 by 1000. Factors remain unchanged. The insertion tool is assumed to have the dimensions specified in Fig. 8.21. If the head has different dimensions, values 1.6 and 2.36 mm should be replaced accordingly.

Some attention must be given to the possibility of shorts being generated by an inserted component leaning over before soldering. A general rule of thumb is that, at the maximum allowed inclination, the component remains within 15° of its theoretical position. If hole diameter and lead diameter are mismatched, this can rise to 20°.

Usually each organisation has a standard for minimum spacing between any two leads not electrically connected (not to be confused with conductor spacing). Figures of 1.5 to 2 mm are common (0.06-0.08 inch). Where voltages are in excess of 30 volts or where a short can cause failure of very expensive components, there are good reasons for increasing the spacing.

Other needs for larger clearances may arise from, for example:

(i) particular quality or approval requirements;
(ii) components which dissipate considerable heat (like many Zener diodes);
(iii) avoidance of electrical noise or crosstalk;
(iv) expectation of abnormal mechanical shock or vibration during transportation or the life of the equipment;
(v) high tolerances on hole positions and diameters.

Minimum spacing must be related to routine assembly practice. Consider a TO-18 transistor metallic package which is to be assembled on a board having 1 mm (0.04 in.) diameter holes with a 2 mm clearance between the base of the TO-18 and the board surface. With all tolerances unfavourable, Figure 8.23, the deviation from the vertical could be 20° (with no lead deformation). The device would then occupy an area of 10 mm diameter, or three times the theoretical area. This, however, is not likely to arise in practice since the clearance of 2 mm cannot be obtained without either using an insulating spacer or preforming the leads (which also holds the device vertical).

An added safety factor is that metal cases, such as the TO-18, are disappearing from designs, plastic packages being less expensive and easier to insert automatically. Even vertically mounted axial components are now less likely to be a hazard since it is virtually standard practice to insulate the longer lead and surrounding components may well be in plastic packages.

8.8.4 SMC Assembly

With the advent of surface mounting technology, the rules for the design of boards have changed significantly. In addition to paying attention to the differences in component footprints, on full SMC boards there may be a need to:

(i) avoid plated holes under the device body (or coat them with solder mask) to avoid contact of solder with the device body as the stand-off can be very small, Figure 8.24;
(ii) provide enough room for test points or pads for in-circuit testing;
(iii) provide sufficient area around large components for fan-out of all lead connections;
(iv) reduce packing density to avoid thermal problem, see Chapter 11;
(v) layout components to eliminate shadowing, etc., during wave soldering.

General Principles of Design and Layout (of Printed Board Assemblies) 365

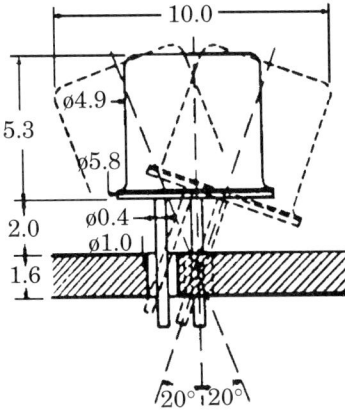

Fig. 8.23 Maximum deviation from the vertical of a TO-18 package, mounted at a distance of 2 mm from the PB, if the hole diameter is 1 mm. Dimensions are in millimetres.

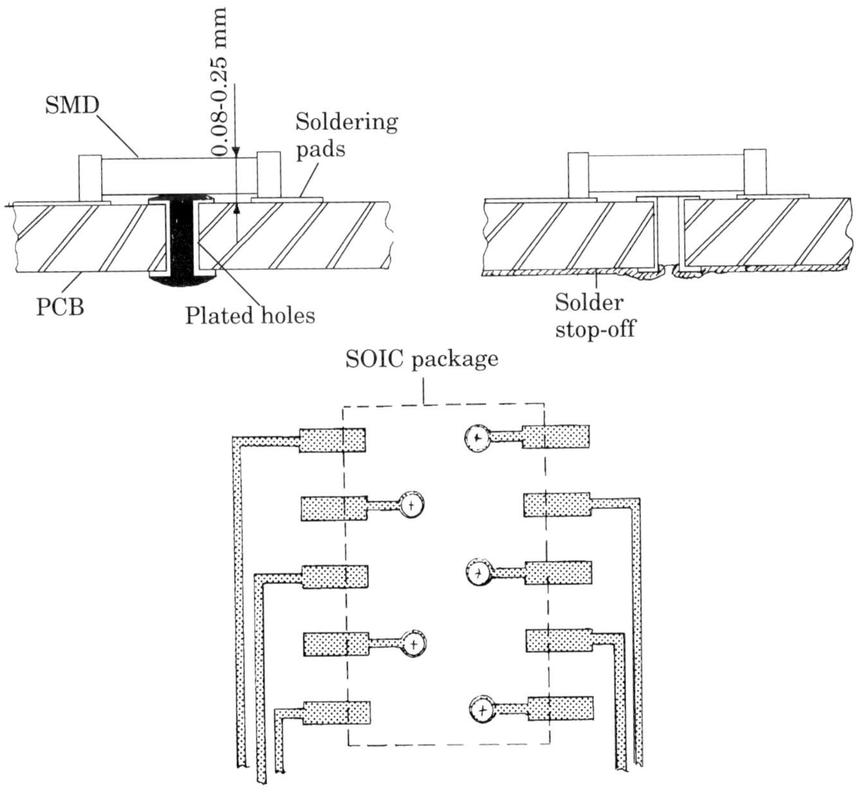

Fig. 8.24 On PBs with PTH vias, holes under the body of SMDs should be avoided or coated with a solder stop-off or tented. Therefore fan-out, as shown for an SOIC package at the bottom, should be avoided.

Other restrictions derive from the way in which the soldering pads must be designed, see Chapter 9. On the bonus side, placement machines for SMCs are more modern in conception than PIH inserters and provide some additional freedom. For instance, some high volume placers impose no restriction on component orientation. In addition, most SMC placers do not place strict requirements on the minimum distance between components as the component is usually held by a vacuum pipette operating on top of the body.

Obviously it is impossible to state hard and fast general rules, since each machine has different specifications which must be carefully checked. It is, however, possible to give some general guidelines, which apply to most machines, Figure 8.25.

Many SMCs will withstand the temperature of molten solder as long as is needed to wave solder the board, and the designer must try to select such types to permit the use of such proven technology. On a two-sided SMA, the designer may decide to confine to one side of the board all components which cannot be wave soldered, and then to specify a different soldering method for that side. This does, of course, restrict design freedom on that board.

Wave soldering of SMAs has generated considerable discussion and the creation of additional design rules. Improvements in end metallisation and developments in soldering techniques have meant that soldering SMAs is much more controllable and not as demanding as it used to be, provided that the soldering parameters are strictly controlled.

Chip components should preferably be placed so that movement of the board relative to the solder wave is normal to the long axis of the chip. If the bodies are small, as is the case with chip resistors and capacitors, they can also be placed in two directions at 90° to each other, Figure 8.26. However, the density of components is more important than their orientation, as excessive crowding makes skipping and shadowing much more prevalent.

MELF components pose similar problems and better results are obtained when the solder flow is normal to their long axis, unless, as before, the board is very crowded. In many situations, a significant improvement can be obtained by soldering at about 45° to the long axis, Figure 8.27, although a dedicated carrier will be required to take the board over the solder wave.

With discrete SMCs in SOT-23, SOT-89 and SOT-143 packages, the solder should flow in the same direction as the long axis of the SMC, Figure 8.28. With SOICs, wave soldering gives best results if they are placed to allow the solder to flow easily between them; a staggered layout is also acceptable, provided the density is fairly low, Figure 8.29.

It must always be accepted that the relative position of two SM packages must be such that the large component does not shadow the smaller one. The distance allowed for wave soldering therefore changes according to the position, Figure 8.30.

In general, wave soldering of a PIH board takes place on a flat surface, interrupted here and there by the projections of leads, whose diameter is small compared with the size of the solder joint. On the other hand, wave soldering of SM boards takes place on a surface having large projections. These disturb the flow of the solder and may also entrap large bubbles of flux, which cannot escape through the holes of the PB. Regardless of how solderable the pads and component terminations are, solder joints cannot be

General Principles of Design and Layout (of Printed Board Assemblies) 367

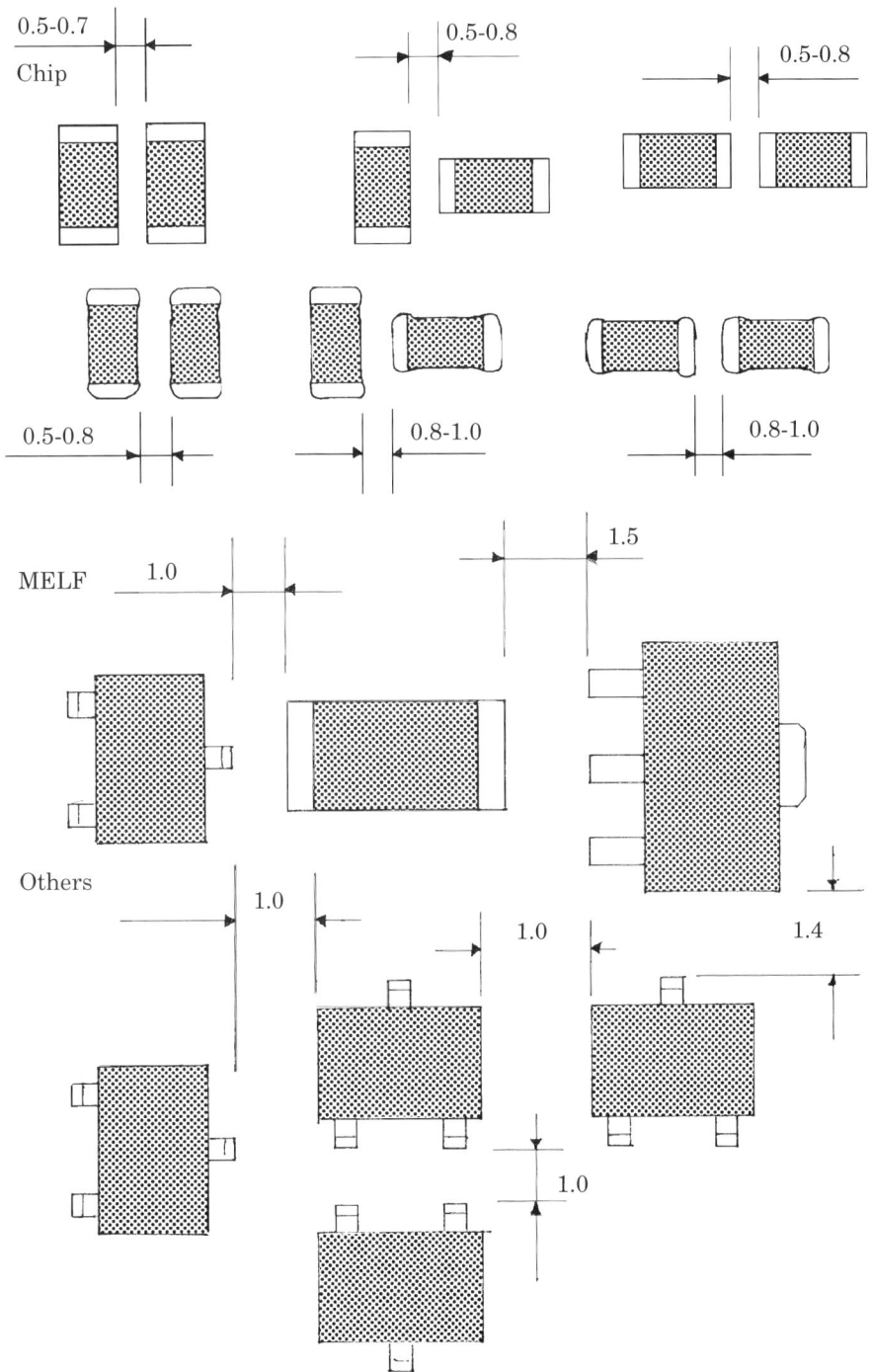

Fig. 8.25 Minimum distance (in mm) between SM components achieved by most automatic placing equipments.

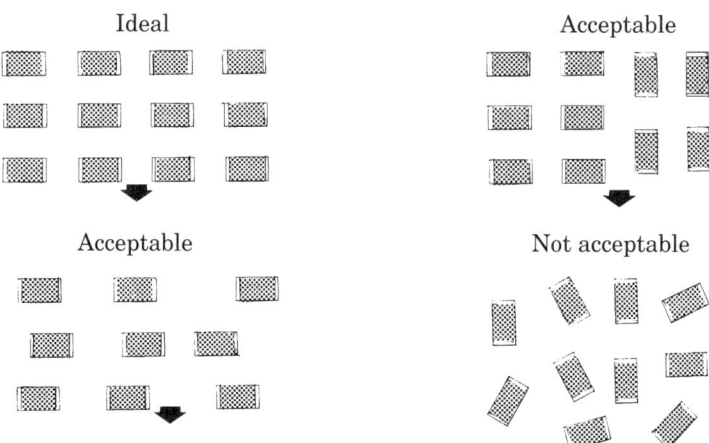

Fig. 8.26 Relative position of chip SMDs for wave soldering. The arrows show the direction of solder flow.

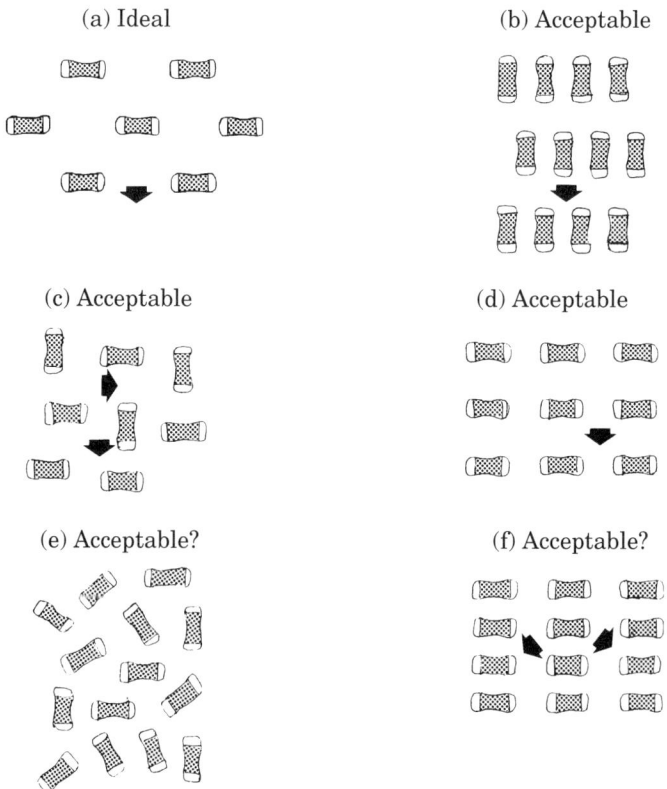

Fig. 8.27 Relative position of MELF SMDs for wave soldering. Arrows show the direction of solder flow [it does not matter in which direction the solder flows in (e)].

General Principles of Design and Layout (of Printed Board Assemblies)

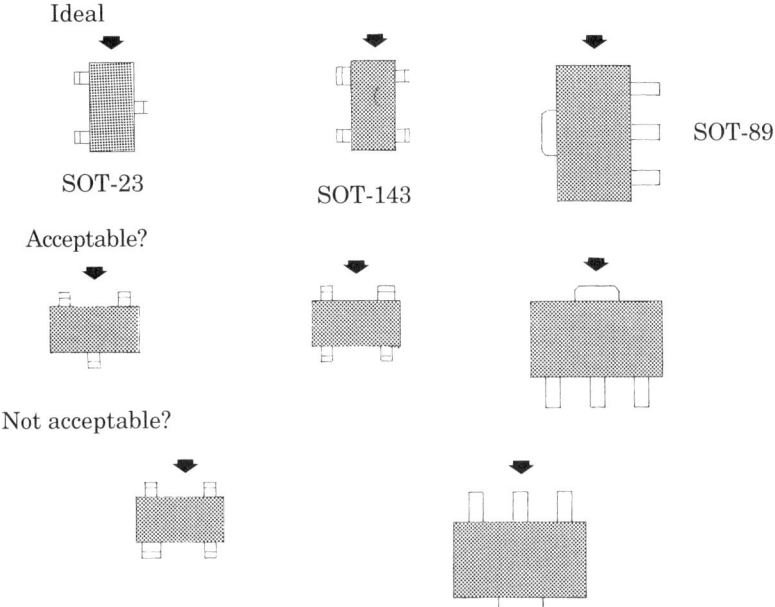

Fig. 8.28 Optimum direction of solder flow during wave soldering of SOT packages.

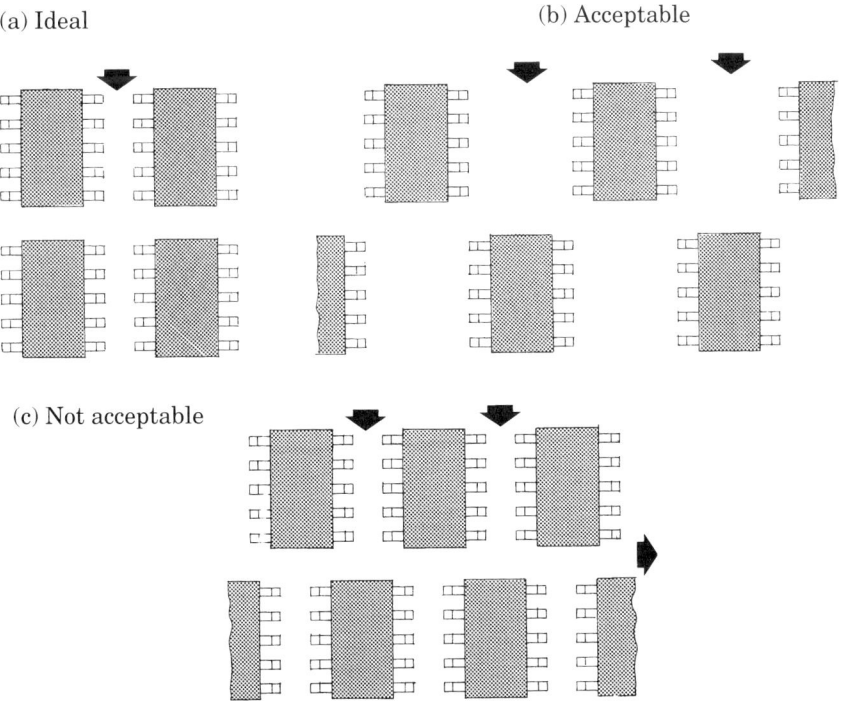

Fig. 8.29 Relative positions of SOIC packages for wave soldering. Arrows show the direction of solder flow.

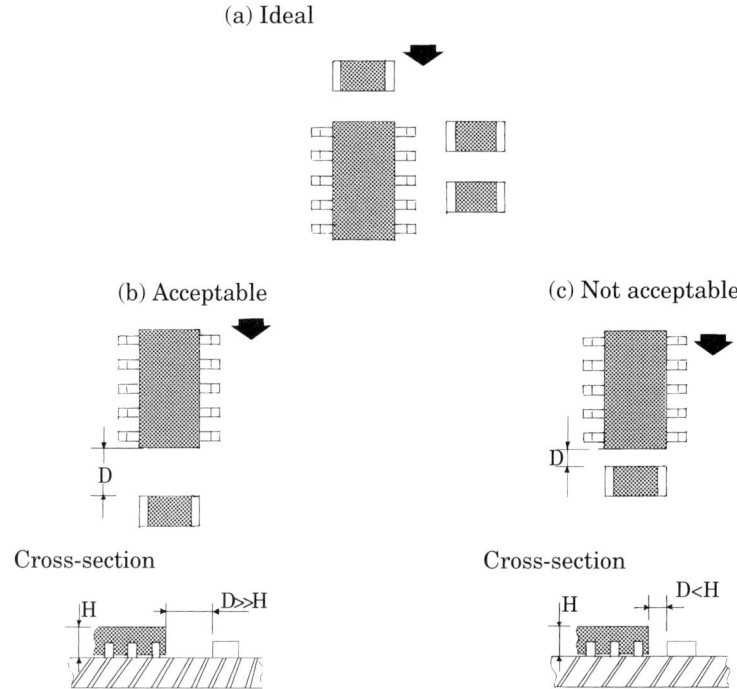

Fig. 8.30 Relative position of a chip and a larger package for wave soldering. Arrows show the direction of solder flow.

formed properly if the solder flow is too turbulent to allow the solder undisturbed access to the potential joint. The solder must remain in contact with the terminations for a finite time.

8.8.5 Mixprint Assemblies

This type of assembly includes both SM and PIH components and is, therefore, subject to the restrictions of both types of assembly, in mounting and in soldering. On the other hand, the lower density of SMCs and the presence of many relatively large holes can give benefits in soldering.

For component mounting the minimum distance between two adjacent components will be determined by the type of machine in use. Since the normal manufacturing cycle places SMCs after the insertion of PIH components and before the addition of odd components, the restrictions imposed are not particularly tight.

In laying out a mixprint board, it must be remembered that solder bridging can cause more than the usual number of problems, being related to the height and shape of the SMCs and the length and direction of clinching of PIH leads, Figure 8.31. Clinched leads are more likely to cause bridge formation and extra distance must be allowed.

General Principles of Design and Layout (of Printed Board Assemblies) 371

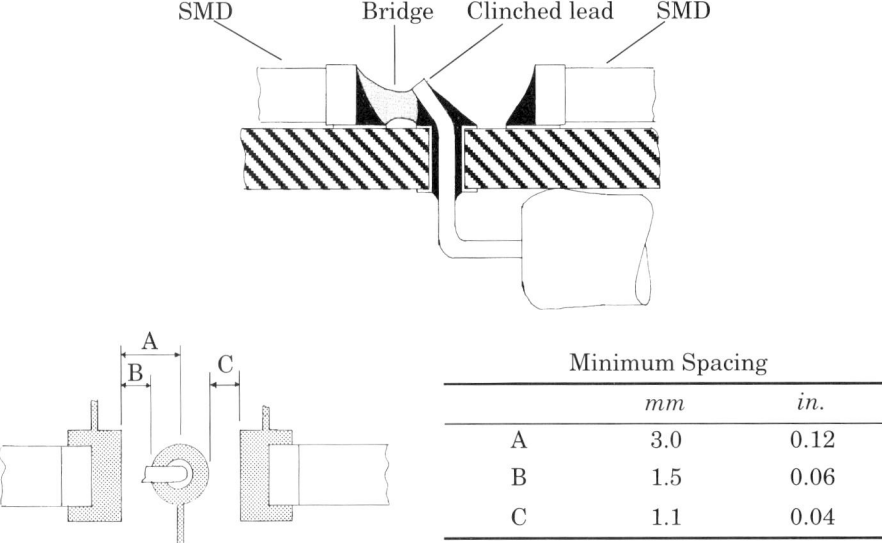

Fig. 8.31 To avoid solder bridges forming during wave soldering of mixprint boards (top) adequate clearance must be given (bottom).

8.8.6 Heavy Components

Bulky or heavy components need particular attention. Usually some method of anchoring them to the board will be essential, Figure 8.32. It should be practicable to calculate the maximum acceleration the board must withstand from the specification for the completed equipment. For most equipments this will be in the range of 10 g to 50 g, where g is the acceleration due to gravity. From these data the force acting on the component leads may be determined.

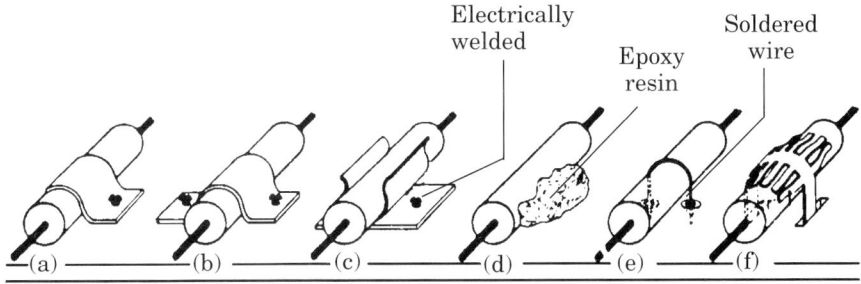

Fig. 8.32 Some common methods of fastening heavy components to the board. Solution (d) is the most economical, but component replacement is difficult.

Consider a component of mass 70 grams (0.154 lb). If the acceleration experienced is 20 g, the force F acting on the leads is:
F = 0.07 x 20 x 9.8 Newton = 1.4 x 9.8 Newton = 1.4 kgf = 3.1 lbf

When generated by cyclic vibration, even such apparently low loads can cause fatigue fracture of quite strong leads. Therefore, if any component weighs 30 grams (just over 1 ounce) or more, consideration should be given to fastening it to the board.

It has been found that loads generated by shock or vibration can easily lift the pads of single-sided PBs. If packing density permits, failure may be prevented by increasing the pad area for large components. Another expedient sometimes used in the past was to provide dummy or spare pads so that the component might be mounted on them if one of the original ones delaminated, Figure 8.33. (A larger pad area seems to make much more sense.) Better practice for single-sided boards is to use one of the many proprietary eyelets to give the component better mechanical anchorage. Where high reliability boards are needed, the holes in single-sided boards can be through plated. The metallised hole wall gives excellent anchorage and the absence of a land on the component side is of little consequence.

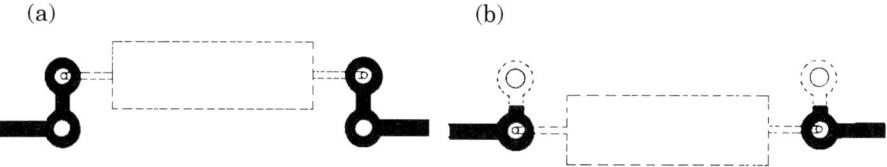

Fig. 8.33 Use of dummy pads to help in repairing boards. The heavy component is assembled as in (a); if a pad delaminates, it can be mounted again as in (b).

8.8.7 Other Points

There are many other points to note in terms of the requirements imposed on the board designer. To illustrate the sort of thing which can arise when trying to define the component layout, the following examples are given.

As different suppliers can supply equivalent components of very different sizes, it may be necessary to provide two or more different insertion spans for the same component by introducing dummy pads and holes. On occasion the designer may have to allow for the use of a single component or for two in series, a single component or two in parallel, a single component or three in parallel and so on, Figure 8.34. Such requirements may have little impact on the layout design, other than to decrease the number of possible alternatives for conductor routing and, perhaps more importantly, to reduce the packing density.

It is often desirable to substitute four independent diodes for a potted full wave, solid state rectifier (because of procurement problems, price changes, ease of field repair, etc.). Figure 8.34(d) shows how the board layout can be arranged so that a cylindrical rectifier, having four leads on the vertices of a square of side 5.08 mm (0.2 in.), can be substituted by four single diodes. By introducing four supplementary holes (and pads) the four diodes can be inserted with a predetermined span (which may be larger than 5.08 mm).

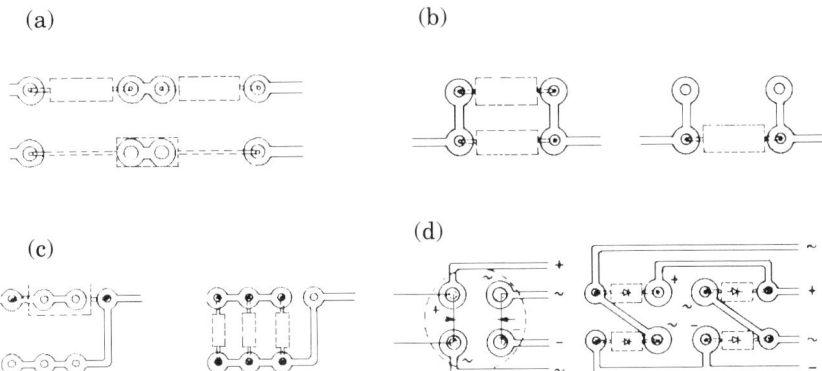

Fig. 8.34 Design of artworks for alternative assemblies: (a) one component or two in series; (b) one component or two in parallel; (c) one component or three in parallel; and (d) a full wave rectifier or four spare diodes all orientated in the same direction.

8.9 CONDUCTOR DIMENSIONING

In most cases, conductor width is dictated by:

(i) the component packing density;
(ii) the minimum spacing between conductors and/or components;
(iii) the number of rejects during PB manufacture (usually from consideration of established designs);
(iv) geometrical constraints due to component outlines or fan-out;
(v) other non-electrical factors, see Chapter 9.

Current-carrying capacity is often disregarded because the conductor dimensions are normally much larger than needed to carry the currents involved. However, there are cases where the conductor width has to be calculated (such as inductors or ground or supply conductors).

The resistance of a conductor, considered as a copper bar of rectangular cross-section, depends on the specific resistivity, r, of copper. This may be assumed to be :

$$r = 1.76 \times 10^{-6} \text{ ohm-cm} = 0.693 \times 10^{-6} \text{ ohm-inch (at 25°C, i.e., 77°F)}$$

This value is about 2% greater than the resistivity of pure annealed electrolytic copper and will be used in all subsequent calculations. Alternatively, the minimum conductivity values (and maximum resistance values) allowed for copper foils for laminates may be used. These are stated together with other properties in IEC 249-3A — and in its intended replacement, IEC 1249-5-1,[5] a positive vote for which was registered in June 1994. Obviously, if conductors are made from other metals, it will be necessary to know the resistivity of the metal used. The contribution of surface coatings, e.g., gold, gold over nickel, or tin-lead, is ignored, unless frequencies involved are very high.

The resistance R of a conductor of length L, width W, and thickness d is:

$$R = r \frac{L}{W d}$$

The resistance (at 25°C) of a copper conductor can now be written as:

$$R = 17.6 \times 10^{-3} \frac{L}{W d}$$

where R is in ohms, L and W are in mm, and d is in microns (μm)

$$R = 0.693 \times 10^{-3} \frac{L}{W d}$$

where R is in ohms, L and W are in inches, and d is in mils (1 mil = 0.001 inch)

For the most common values of copper foil thickness, which are 1 oz/ft² (305 g/m²) and 2 oz/ft² (610 g/m²) or 35 and 70 μm (0.0014 in. and 0.0028 in.) respectively, the expression may be simplified to:

or
1 oz/ft² copper R = 0.50 × 10⁻³ (L/W) ohms
2 oz/ft² copper R = 0.25 × 10⁻³ (L/W) ohms

where (L/W) is dimensionless. In the sheet resistance concept, the sheet resistance R_s of the copper foil is:

1 oz/ft² copper R_s = 0.50 × 10⁻³ ohms per square
2 oz/ft² copper R_s = 0.25 × 10⁻³ ohms per square

where the number of 'squares' equals the length/width (L/W) ratio of a conductor.

For example, a conductor 500 mm (19.69 in.) long and 0.3 mm (0.0118 in.) wide in 1 oz/ft² copper has a resistance of

$$R = 0.5 \times 10^{-3} (500/0.3) = 0.83 \text{ ohms } [= 0.5 \times 10^{-3} (19.69/0.0118)]$$

When determining the cross-section of a conductor, account must be taken of differences between nominal and actual width, conductor defects such as pits and nicks which reduce the effective width, uneven thickness of copper and so on. To cater for the effects of such defects, the theoretical width is normally increased by up to 30% if the width is less than 0.5 mm, 20% if it lies between 0.5 and 1.0 mm, and 10% if it is more than 1.0 mm.

If a conductor has different widths, e.g., necking down to pass between pads, it is possible to calculate the resistance of all portions having the same width and then add them together as for resistors in series.

When a length of the conductor has a trapezoidal shape, an average width may be determined. If W_1 is the minimum width and W_2 the maximum width, as a first approximation the arithmetic average can be used for W:

$$W = (W_1 + W_2)/2$$

If the ratio $W_2:W_1$ is less than 1.5, the error is negligible. The larger the ratio, the greater the error and a more accurate value will have to be determined by using the logarithmic average between W_1 and W_2. The average width is given by:

$$1/W = \frac{1}{W_2 - W_1} \ln(W_2/W_1)$$

where $W_2 > W_1$ and ln is the natural logarithm.

As an example, consider a conductor for which $W_1 = 0.5$ mm and $W_2 = 1.3$ mm. The arithmetic average is $(1.3 + 0.5)/2 = 0.9$ mm, and $1/W = 1.11$ mm^{-1}. The logarithmic average is:

$$1/W = \frac{1}{1.3 - 0.5} \ln(1.3/0.5) = 1.25 \ln 2.6 = 1.19 \text{ mm}^{-1}$$

from which $W = 0.84$ mm

It will be found that the arithmetical average gives larger values of W (and hence smaller resistance values) than does the logarithmic average — in the example, 0.9 mm against 0.84 mm, some 7% higher.

Resistance of a conductor can be determined quickly using a nomograph of the type shown in Figure 8.35. By aligning conductor width, left line, with foil thickness, middle line, the resistance of a unit length of conductor is given by the intersection on the right hand line. If the values of any two parameters are known, the third can be found. The conductor width W read from the nomograph is theoretical and must be adjusted by 10-30% as mentioned earlier.

All values are true at 25°C (77°F). When current flows in a conductor, the Joule effect causes an increase in temperature. Unfortunately, the resistance of copper is not constant with change in temperature but increases as the temperature of the copper rises. This means that all values determined at 25°C must be adjusted to allow for possible temperature rises in the conductors due to the Joule effect and/or heating caused by components.

If a conductor has a resistance R_1 at temperature T_1, at temperature T_2 its resistance R_2 will be:

$$R_2 = [1 + \alpha(T_2 - T_1)] R_1$$

where α is the coefficient of thermal resistivity of copper. This also changes with temperature. Over the temperature range normal for electronics equipments, an average value can be used:

$$\alpha = 0.004°C^{-1} = 0.0022°F^{-1}$$

A more precise value may be obtained by expressing the coefficient as a function of temperature by the empirical equation:

$$\alpha = \frac{0.0045}{1 + (T/180)} \quad \text{where T is in °C,} \quad \text{or}$$

$$\alpha = \frac{0.0045}{1 + (T - 32)/324} \quad \text{where T is in } °F$$

Once T_1 and T_2 are known, the two relevant values α_1 and α_2 can be derived and an arithmetical average can be used. A precise calculation will need detailed experimental data on copper resistivity, obtainable from reference data books.

In the majority of cases, the degree of accuracy needed is satisfied by the use of the following empirical equation:

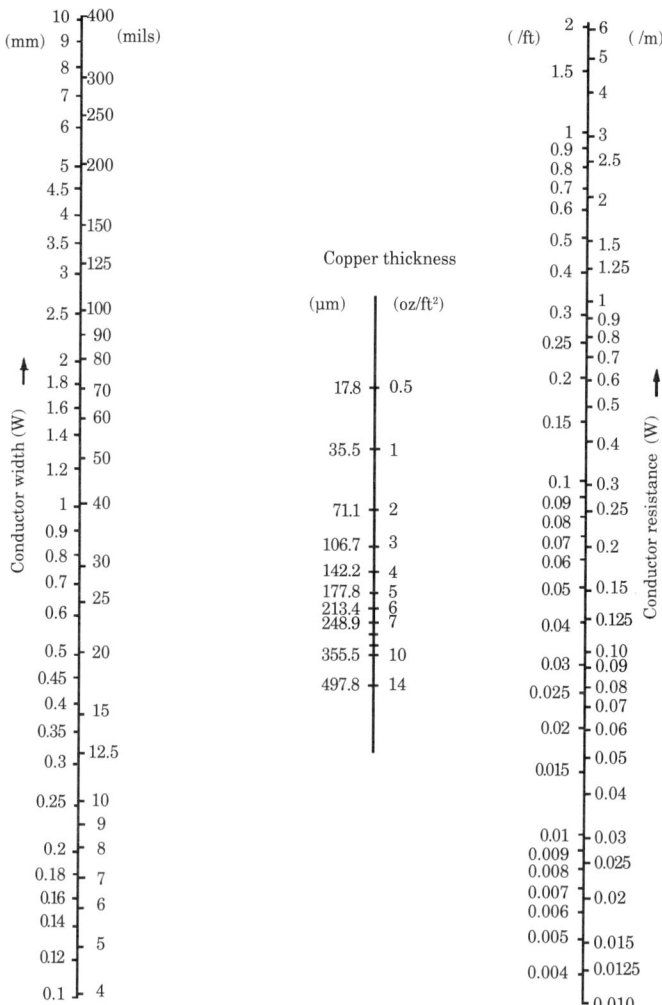

Fig. 8.35 Nomograph for determining the resistance of a PB conductor as a function of its width and copper thickness. Scales are logarithmic.

$$R_2 = R_1 \frac{235 + T_2}{235 + T_1} \quad (T_1 \text{ and } T_2 \text{ in } °C)$$

As an example, consider a conductor which has 0.5 ohm resistance at 25°C (= T_1). If in operation its temperature rises to 65°C (= T_2), i.e., from 77 to about 150°F, the resistance at 65°C can be calculated using the three methods as follows:

1. $R_2 = [1 + 0.004 \,(65 - 25)] \, 0.5 = 1.16 \times 0.5 = 0.58$ ohms

an increase of 16%.

$$\alpha_1 = \frac{0.0045}{1 + 25/180} = 0.00395; \alpha_2 = \frac{0.0045}{1 + 65/180} = 0.00331° C^{-1};$$

$$\alpha = (\alpha_1 + \alpha_2)/2 = 0.00363° C^{-1}$$

2.

By using the same equation as in the first method:

$R_2 = R_1(1 + 0.00363 \times 40) = 0.5 \times 1.145 = 0.573$ ohms

which means an increase of 14.6%.

3. $R_2 = R_1 \dfrac{235 + 65}{235 + 25} = 0.5 \times 1.154 = 0.577$ ohms

an increase of 15.4%.

This shows that, if temperature changes are small, the three methods give similar results, the differences being less than the error permitted for copper resistivity. (It should be noted that PB copper [base copper plus electroplated copper] can have a resistivity very different from that of annealed copper.)

In calculating changes in resistance, the temperature of the conductor when the equipment is working must be known. The atmospheric temperature inside the equipment is normally known as it is a basic factor for design. Often a very close estimate can be made from comparison with similar equipment.

All conductors which carry current will be at a temperature higher than the ambient by a temperature rise ΔT which depends upon the current carried (I), conductor width (W), and thickness. An empirical formula has been derived for a 1 oz/ft² copper foil PB. Assume a 40°C maximum ΔT and determine the maximum current permitted in the conductors from the equations which follow:

$$I = 3 \sqrt[3]{W^2} \quad (\text{where I is in amperes and W in mm})$$

$$I = 26 \sqrt[3]{W^2} \quad (\text{where I is in amperes and W in inches})$$

Figure 8.36 shows temperature rise above ambient as a function of current and conductor width for a number of copper thicknesses. Diagrams such as these can be faster and easier to use and, depending on scale, can be very precise. Figures for other thicknesses of copper can be derived by considering that temperature rise for a given current and conductor width is inversely proportional to the copper thickness. For instance, if a 2 oz/ft² conductor has a rise of 40°C, a 4 oz/ft²conductor will have a temperature rise under the same conditions of about 20°C.

Figure 8.36 can be used to find the current carrying capacity of a conductor or to determine the conductor width and thickness needed to carry a given

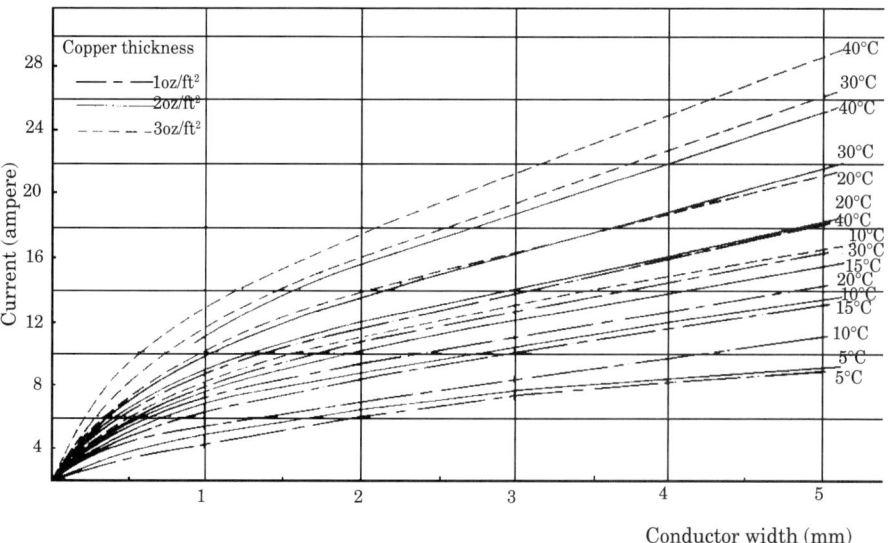

Fig. 8.36 Temperature rise (over ambient temperature) of a PB conductor as a function of conductor width, current (dc) and copper thickness.

current. In both cases the maximum theoretical temperature rise which may be allowed has to be established. Usual values are about 40°C (72°F) for equipment with forced air cooling and 10 to 15°C (18-27°F) for equipment without forced cooling.

To guard against local hot spots due to nicks, pits, etc., in conductors and to allow for tolerances in conductor width and thickness, the values determined from Figure 8.36 should be amended. If current carrying capacity is determined, the value obtained should be decreased by 30 to 50%; if conductor width/thickness for a given current is calculated, the current value should be increased by 50 to 100%.

The following example illustrates the use of Figure 8.36. Given the following data, the task is to determine the conductor width which will give a voltage drop of less than 1 volt along the conductor:

General Principles of Design and Layout (of Printed Board Assemblies) 379

Air temperature	= 25°C
Maximum allowed ΔT	= 30°C
Copper thickness	= 35.5 microns (1 oz/ft²)
Current	= 4 amps
Conductor length	= 400 mm (0.4 metres)

The maximum resistance R of the path is easy to determine by applying Ohm's Law:

$$R = \frac{1 \text{ volt}}{4 \text{ amperes}} = 0.25 \text{ ohms (at } 25°C)$$

Resistance increases when temperature increases, and the worst condition will be at 25 + 30 = 55°C. From 25 to 55°C the resistance increases from R_1 to R_2. The empirical formula for R_2 gives:

$$R_2 = R_1 \frac{235 + 55}{235 + 25} = 1.12 R_1$$

If the resistance is 0.25 ohm at 25°C, it will reach 0.25 x 1.12 = 0.28 ohm at 55°C. To stay within the limits, the resistance at 25°C must not exceed 0.25/1.12 = 0.223 Ω, which means a specific resistance of the conductor of not more than 0.223/0.4 = 0.558 ohm/metre.

Using the nomograph, Figure 8.35, width W = 0.9 mm. By increasing the figure by 20% for safety, a width of 1.08 mm is obtained, which can be rounded up to 1.1 mm.

Checking thermal limits, in Figure 8.36 for a 1.1 mm width, a 30°C ΔT and 1 oz/ft², the current carrying capacity is about 6.4 amps. As the actual current is to be 4 amps the derating is 37.5%, which can be considered adequate.

Another way to approach the problem is to determine the minimum width for the permitted temperature rise, and then check the nomograph for resistance. First let the current increase by 70% from 4 to 6.8 amps. For a 1 oz/ft² thickness, a 6.8 amp current, and a 30°C temperature rise, a minimum conductor width of 1.1 mm is found. Checking the nomograph for a specific resistance of 0.558 ohm/metre, a conductor width of 0.9 mm is given. The proposed value of 1.1 mm exceeds this by 22%, a reasonable safety factor.

The diagram in Figure 8.36 can be replaced by a numerical approach, using the empirical formula:

$$W = K\, I^{1.86}$$

Table 8.2
Values for Coefficient K for Computing Conductor Width

Copper Thickness		W in microns (μm) Temperature Rise (°C)				W in mils (0.001 in.) Temperature Rise (°C)			
oz/ft^2	g/m^2	10	20	30	40	10	20	30	40
1	305	91.4	50.0	36.8	30.5	3.6	1.97	1.45	1.20
2	610	50.0	28.0	20.3	16.5	1.97	1.10	0.80	0.65
3	915	33.0	20.3	13.4	11.4	1.30	0.80	0.53	0.45

where W = conductor width (in μm or in mils)
K = a factor which depends on copper thickness and the allowed ΔT (Table 8.2)
I = current-carrying capacity of the conductor with the allowed ΔT.

If it is desired to find the current I which can flow through a conductor of a predetermined width W for a given maximum ΔT, the formula becomes:

$$I = (W / K)^{0.538}$$

Table 8.3 gives some values for the exponential formulae. The following examples should make the calculation involved quite clear:

1. Copper 1 oz/ft², ΔT = 30°C, I = 5 amp, W = ? mm;
   W = 36.8 x $5^{1.86}$ = 36.8 x 19.95 = 734 microns or 0.73 mm

2. Copper 2 oz/ft², ΔT = 10°C, I = 15 amp, W = ? inches;
   W = 1.97 x $15^{1.86}$ = 1.97 x 154.0 = 303.4 mil or 0.30 inch

3. W = 1.6 mm, 1 oz/ft² copper, and ΔT = 40°C, current-carrying capacity of conductor = ?;
   I = $(1600/30.5)^{0.538}$ = 8.4 amps

4. W = 0.15 inch, 3 oz/ft² copper and ΔT = 40°C, current-carrying capacity of conductor = ?;
   I = $(150/0.45)^{0.538}$ = $333.3^{0.538}$ = 22.8 amps

To the values thus obtained a derating factor has to be applied by:

(i) increasing the given value for current by 50-100% before computing the conductor width;
(ii) decreasing the value of current-carrying capacity by 30-50%.

Normally, maximum derating will be assumed for small conductors (width less than 0.8 mm, about 0.30 in.) and minimum derating will be assumed for large conductors.

Table 8.3
Values of Function $Y = X^{1.86}$

X	Y	X	Y	X	Y	X	Y	X	Y
0.100	0.014	1.00	1.000	10.0	72.44	19.0	239.05	28.0	491.72
0.150	0.029	1.50	2.126	10.5	79.33	19.5	250.88	28.5	508.17
0.200	0.050	2.00	3.630	11.0	86.49	20.0	262.98	29.0	524.88
0.250	0.076	2.50	5.498	11.5	93.95	20.5	275.34	29.5	541.84
0.300	0.107	3.00	7.717	12.0	101.69	21.0	287.96	30.0	559.04
0.350	0.142	3.50	10.279	12.5	109.71	21.5	300.84	30.5	576.50
0.400	0.182	4.00	13.177	13.0	118.01	22.0	313.98	31.0	594.20
0.450	0.226	4.50	16.405	13.5	126.60	22.5	327.39	31.5	612.15
0.500	0.275	5.00	19.957	14.0	135.46	23.0	341.05	32.0	630.35
0.550	0.329	5.50	23.827	14.5	144.59	23.5	354.97	32.5	648.79
0.600	0.387	6.00	28.013	15.0	154.00	24.0	369.14	33.0	667.48
0.650	0.449	6.50	32.510	15.5	163.69	24.5	383.57	33.5	686.41
0.700	0.515	7.00	37.315	16.0	173.65	25.0	398.26	34.0	705.59
0.750	0.586	7.50	42.424	16.5	183.87	25.5	413.20	34.5	725.01
0.800	0.660	8.00	47.835	17.0	194.37	26.0	428.40	35.0	744.67
0.850	0.739	8.50	53.545	17.5	205.14	26.5	443.85	35.5	764.58
0.900	0.822	9.00	59.551	18.0	216.18	27.0	459.55	36.0	784.73
0.950	0.909	9.50	65.852	18.5	227.48	27.5	475.51	36.5	805.13

Values of Function $Y = X^{0.538}$

X	Y	X	Y	X	Y	X	Y	X	Y
0.010	0.084	0.10	0.290	1.0	1.00	10.0	3.45	100.0	11.91
0.015	0.104	0.15	0.360	1.5	1.24	15.0	4.29	150.0	14.82
0.020	0.122	0.20	0.421	2.0	1.45	20.0	5.01	200.0	17.30
0.025	0.137	0.25	0.474	2.5	1.64	25.0	5.65	250.0	19.50
0.030	0.152	0.30	0.523	3.0	1.81	30.0	6.23	300.0	21.51
0.035	0.165	0.35	0.568	3.5	1.96	35.0	6.77	350.0	23.37
0.040	0.177	0.40	0.611	4.0	2.11	40.0	7.28	400.0	25.11
0.045	0.189	0.45	0.651	4.5	2.25	45.0	7.75	450.0	26.76
0.050	0.200	0.50	0.689	5.0	2.38	50.0	8.20	500.0	28.32
0.055	0.210	0.55	0.725	5.5	2.50	55.0	8.64	550.0	29.81
0.060	0.220	0.60	0.760	6.0	2.62	60.0	9.05	600.0	31.24
0.065	0.230	0.65	0.793	6.5	2.74	65.0	9.45	650.0	32.61
0.070	0.239	0.70	0.825	7.0	2.85	70.0	9.83	700.0	33.94
0.075	0.248	0.75	0.857	7.5	2.96	75.0	10.20	750.0	35.22
0.080	0.257	0.80	0.887	8.0	3.06	80.0	10.56	800.0	36.46
0.085	0.265	0.85	0.916	8.5	3.16	85.0	10.92	850.0	37.67
0.090	0.274	0.90	0.945	9.0	3.26	90.0	11.26	900.0	38.85
0.095	0.282	0.95	0.973	9.5	3.36	95.0	11.59	950.0	40.00

8.10 FURTHER READING

The only real way in which a designer can keep up with technology is by reading, attending seminars and conferences and, most importantly perhaps, playing an active part in an appropriate association.

It is often said that published information cannot help, because it will be out of date by the time it is published. While this may contain an element of truth for a 'state of the art' technology, many poor designs, in terms both of performance and cost of manufacture, would never have seen the light of day if only someone on the design team had referred to published material!

International documents are typified by IEC Publication 326: Printed Boards, Part 3: Design and Use of Printed Boards. Unfortunately, the difficulties and time delays associated with circulating proposals for inclusion in or amendment of such documents and obtaining agreements between many member countries mean that these documents will always lag behind national and 'commercial' documents. For example, the full revision of IEC 326-3 published in 1991[6] gave fundamental information on PIH printed boards, their design, characteristics and applications, but omitted any information on surface mounted boards. Work to improve the situation and status of IEC publications is ongoing and information for designers of boards and circuits is receiving much attention, see Volume 2, Chapter 19.

At a national level progress can be faster and, for example, in the United States and in the United Kingdom progress has been made towards producing more comprehensive information for the board designer. In the UK., while IEC 326-3 is issued as BS 6221: Part 3 (in order to comply with international agreements), BS 6221: Part 22 covers printed wiring substrate materials — with specific reference to their use for SMA boards, and Part 23 deals with design for surface mounted assemblies.

The United States Department of Defense has produced many definitive guidance documents. MIL-STD-275 was the military standard for printed wiring for electronic equipment but was replaced in July 1995 by IPC-D-275. Probably the most famous, and most comprehensive, standards and/or guides on the subject are not military but are produced by the Institute for Interconnecting and Packaging Electronic Circuits, the IPC. Thought of by many solely as an American Institute, the IPC has over 1800 members worldwide so that most of the documents issued incorporate comment and input from companies in other countries. Most of the basic design guidance needed will be found in the Design Guide Manual, IPC-D-330, which is supplied in two volumes, one dealing with design and fabrication of boards and the second with printed wiring assemblies.

Several component and/or equipment manufacturers provide information on design for use of their products. In particular, in SMT, the names of Philips and Siemens spring to mind as both have produced excellent guidance documents under the generic title of SMD Technology.

There are, of course, several excellent journals and magazines. Rather than name specific titles, which may possibly do injustice to those not mentioned, it is recommended that a good start may be made by obtaining those books and publications referenced in relevant chapters of this book!

It must be remembered that there are specifications which have a real impact on the design of an assembly, possibly in addition to those directly

General Principles of Design and Layout (of Printed Board Assemblies) 383

requested by a customer. One of the many examples could be IEC Publication 65 (which corresponds to BS 415). This deals with safety requirements for mains operated electronic and related apparatus used in the home (radio and TV receivers, stereo amplifiers, tape recorders, video recorders, and so on), including requirements for equipment provided with protection against splashing water. All such documents which may dictate to the designer must be sought out, easily done through the relevant trade association, and studied for possible impact.

REFERENCES

1 Revision of IEC 194: Terms and definitions for printed circuits and a proposal for a Decimal Classification Code (DCC) for the categorization of the terms and definitions in IEC 194. Committee Drafts 52/607/CDV-I and 52/607CDV-II, International Electrotechnical Commission, Geneva, Switzerland, October (1995).
2 IPC-T-50E, 'Terms and Definitions for Interconnecting and Packaging Electronic Circuits', Institute for Interconnecting and Packaging Electronic Circuits, Lincolnwood, IL 60646, USA.
3 BS EN 60097: 1993 ($\equiv$ IEC 97 : 1991) Specification for grid systems for printed circuits, British Standards Institution, London W4 4AL, United Kingdom.
4 Clayton, P. A., 'Handbook of Electronic Connectors', Electrochemical Publications Ltd, Asahi House, Church Road, Port Erin, Isle of Man, British Isles (1982).
5 IEC 1249-5-1: Base materials for printed circuits. Part 4: Conductive foil and film. Specification No. 1: Copper foil for use in the manufacture of copper-clad materials.
6 IEC 326-3: 1991, second edition, 'Printed boards. Part 3: Design and use of printed boards', International Electrotechnical Commission (IEC).

Chapter 9

LAYOUT OF PRINTED CIRCUIT BOARDS

GIOVANNI LEONIDA
Milan, Italy

W. MACLEOD ROSS
Welnorth Ltd, Bishop's Stortford, UK

9.1 INTRODUCTION

Layout means the actual design of a PCB. It starts from the electrical diagram (*circuit diagram* or *schematic*) and ends with an *artwork*, or series of artworks, which includes the location of each component and the routing of every conductor on the printed board (PB).

Manual design of artwork is rare today. CAD software is available even for personal computers and costs are easily recovered by the time saved on a few boards. The art of manual design — it is more an 'art' than a 'science' — virtually died out in the late seventies. Manual design of PCBs is now largely restricted to very simple boards and to old-fashioned amateurs.

However, since those with some knowledge of manual layout should be better equipped to understand and evaluate the performance of CAD software and hardware, see Chapter 10, the first part of this chapter is devoted to that art.

9.1.1 Manual Layout

Once the basic principles have been established, manual layout requires only time and patience as the designer proceeds by trial and error, aided by experience. The best way to illustrate the process is to take a simple circuit as an example. One such circuit is illustrated in Figure 9.1 and should be realisable as a single-sided board.

The circuit includes two transistors, taken to be in TO-18 packages, see Chapter 3, and seven components, axial and of the same dimensions — which must be known. A highly experienced board designer may go straight from the circuit diagram to the final artwork, but in most cases he will make one or more draft artworks to assist in minimising board area and producing a cleaner design.

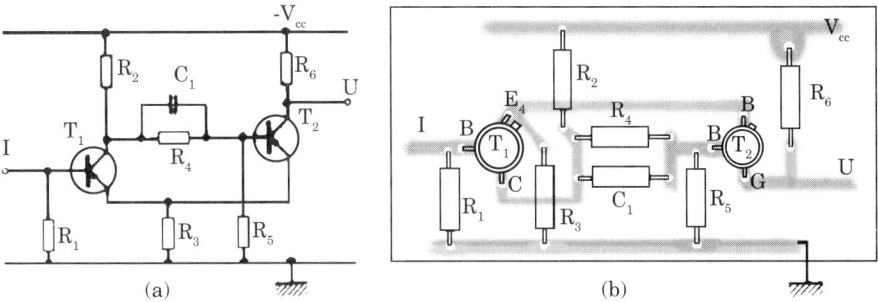

Fig. 9.1 A simple circuit (a) and one realisation as a single-sided board (b). (The shaded conductors are on the other side of the board.)

All that is needed to start with is a sheet of graph paper with a suitable grid and a set of drawing instruments. For trial drafts the grid need not be a standard grid but the final artwork demands this. The board is usually designed as seen from the component side, and conductors will be seen through the laminate, which is imagined to be transparent. This means that the pinning of active components will be a mirror image of the electrical diagram. The circuit diagram may be designed with the pin-out of the component seen from the bottom, while the board is assumed to be designed as seen from the component side, i.e., the component is viewed from the top. In Figure 9.1, when the base of a transistor is on the left hand, the emitter is on the top of the electrical diagram and on the bottom in the sketch of the board.

The simplest way to sketch a layout is to redraw, if necessary, the electrical diagram to eliminate crossovers, and then to locate the components in the same position as shown in the new diagram, with as few changes as possible, Figure 9.1, in order to:

(i) assign the required area to each component;
(ii) avoid crossing conductors (if this is not possible, it will be necessary to add a dummy component, i.e., a jump wire or link);
(iii) allow sufficient spacing between conductors.

If the circuit is intended to operate at low voltage (< 35 volts) and only small currents (a few milliamps) are involved, any specific design of conductors can be avoided by taking their width to be about 0.5 mm (0.02 in.) and a copper thickness of at least 1 oz/ft^2 (35 μm). Furthermore, since such a board is very straightforward, production can be simplified by using a conductor width and minimum spacing of 1 mm (0.04 in.). In this instance also, a hole diameter of 1 mm in a pad diameter of 2.54 mm (0.1 in.) may be used.

After making a freehand sketch to determine the basic conductor routing, the circuit can be drawn at enlarged scale (2:1 or more) on grid paper to check minimum clearances between tracks and between components. The result is a board which will work, even if the layout can almost certainly be improved (by reducing waste board area, for example).

From this starting point, the art of layout is to rearrange the components to reduce the PB area and produce an ordered and clean assembly, Figure 9.2. The effectiveness of the later designs may be judged by:

(i) space saving (increase in percentage of board area occupied by components);
(ii) ease of assembly (components orientated in one or two directions, mounted on the same pitch, and so on);
(iii) number of components, including added jump wires;
(iv) total length of conductors;
(v) number of critical placement and/or soldering positions (points where the clearance between components and/or copper areas is at the minimum stated value).

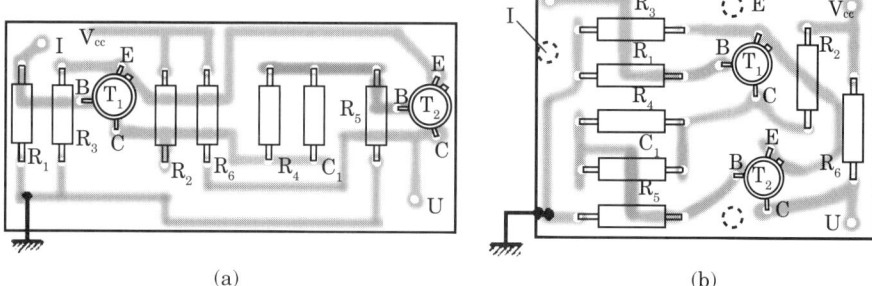

Fig. 9.2 Rearrangement of the initial layout in Fig. 9.1(b) to produce smaller and neater assemblies.

Improvements can be made only by trial and error and achieving them may take a considerable time, even for a simple circuit like the example shown. Pareto's so-called law applies — 80% of the result is achieved by the first 20% of the effort applied. (Vilfredo Pareto (1848-1923), an Italian engineer of noble family, left his job as general manager of Italian Railways to study economics and sociology. Despite his important theoretical contributions to international trade, economy and circulation of capital goods, he was banished by the 'official' scientific community because of his social theories. By studying the distribution of wealth and income in Italy, in 1897 he stated that about 80% of the national income was in the hands of 20% of the population. Since he found approximately the same situation in all countries, he started to believe that the empirical 20-80 relationship (the so called *Pareto Law*) could have a much wider application: in any group of items, the significant ones represent a small portion only, and the large majority are of minor significance. The application of Pareto's axiom to Inventory Control (by H. Dickie in 1951) led to the so-called 'ABC analysis', which was later extended to Quality and other fields. The Pareto axiom (not law!) implies that any time something is developed by cumulative effort (either learning how to play the guitar or designing the layout of a PCB!), 80% of the result is acheived thanks to the first 20% of the effort. By extension, whenever there are many factors which may influence a process, it is possible to acheive good results by listing all of them and selecting the most important 20% to be kept under strict control.)

By using conductors of constant width and circular pads all of the same size, a more professional look is achieved and the artwork will be simpler to lay out.

9.1.2 Amateur's Layout

Amateurs are little concerned with automatic insertion equipment and may accept waste of area, unless space for the assembly is limited. Often, they like to make their own PBs using the simplest possible equipment and materials.

Because of this, they may prefer to make a draft artwork in the so-called 'English style': instead of tracing all conductors, they draw only the clearance around them, leaving the conductors as large as possible, Figure 9.3. This also minimises the amount of copper which has to be etched away from the copper-clad laminate. Frequently such layouts ignore factors such as minimum clearance between components (they may be held in the correct position when soldered) or the package may be changed where there are problems (e.g., the metallic TO-18 may be replaced by the plastic-bodied TO-92)).

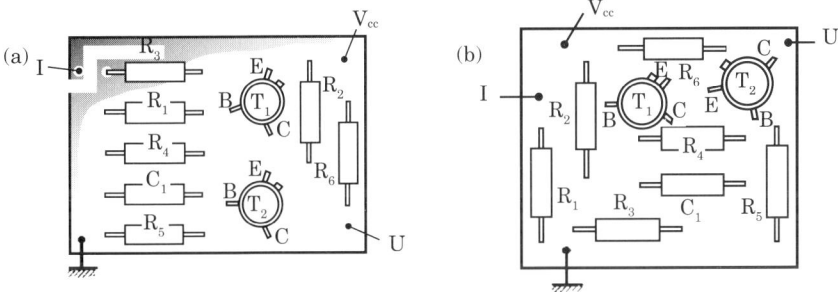

Fig. 9.3 Examples of draft artworks implemented 'English' style: (a) with the components in the same position as Fig. 9.2 (b); and (b) with component allocation modified to simplify the pattern and reduce the board area.

Moreover, the amateur is not concerned about possible contact between two electrically connected leads of bodies such as, in Figure 9.3(b), the metallic body of the T1, which is connected to its collector, and the left lead of resistor R4.

9.1.3 Manual Drawing of Artworks

Although simple, the example in Figure 9.2(b) will be used to illustrate the production of an artwork. As the layout is drawn looking at the component side, the transparent sheet showing component locations and conductor routing is viewed from the back. While the layout design only had to be precise in critical areas, the artwork must be drawn with much greater accuracy. The final result depends on this accuracy, and on the scale of the artwork (which is usually the same as the layout scale).

To achieve the required precision, work is carried out on a rigid support, using transparent, high dimensional stability films specifically designed for the purpose.[1,2] Although not mandatory, it helps considerably if the work is carried out on a co-ordinatograph. This is a light table over which a pen, for pen plots, or a special tool for cut and strip artworks, is supported by a precision XY movement. (The plotting pen or other tool may be linked to a digital display and/or a digitiser.) In any position, the operator can inhibit the movement in either the X or Y axis when he wants to draw a line in the other.

Before starting, the artwork scale must be decided. For most artworks this will be 2:1 or 4:1, but for very precise work it may be 50:1 or more. The scale is determined by the accuracy needed for the master pattern, the printed board dimensions and the photographic equipment available. Any photoreduction process introduces errors, the limits of which must be known exactly to ensure the desired precision.

Assume that a 3:1 scale has been selected and that the circuit has a 0.1 in. (2.54 mm) grid, see 9.3.2. A sheet of transparent, high dimensional stability film is placed on the co-ordinatograph table and fixed in position. The tool selected is a pin point knife, which can cut sharp uniform marks on the film. First, two orthogonal axes (X and Y axes) are marked, superimposed on the centres of at least two holes. Now a set of lines parallel to the X axis will be drawn, spaced exactly 0.1 x 3 = 0.3 in. (7.62 mm) apart. A similar set of lines is drawn on the Y axis so that a 0.3 in. base grid results. If there are sizeable areas in which no holes are needed, some of the grid lines may be omitted; in this case, at least one in every five grid lines should be dimensioned (on the edge of the sheet).

After the grid has been marked out, pads can be located. Special opaque self-adhesive labels or preforms may be used. These can be placed accurately by centring them on the grid before pressing down. To aid centring, most preforms have a small hole in the centre which allows the grid crosspoint to be seen.

The transparent film with all pads in position is the 'dot master' or dotted artwork, Figure 9.4(b). It can be removed from the co-ordinatograph to make

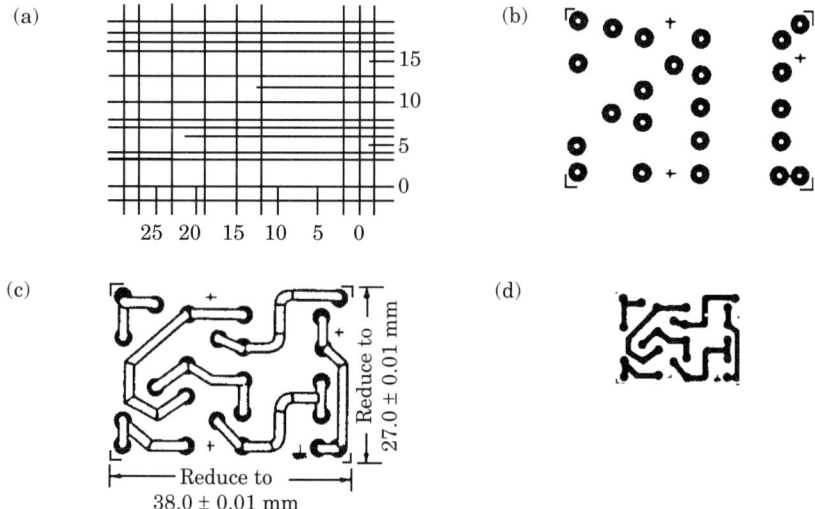

Fig. 9.4 Manual implementation of the master artwork of Fig. 9.2(b) (solder side view): (a) grid drawn on the co-ordinatograph; (b) dotted artwork or pad master; (c) complete artwork; and (d) master pattern obtained by photographic reduction of the artwork. Actual dimensions of the reduced image are shown in (c).

one or more contact prints which are very useful if the layout has to be reworked later. For double-sided boards, one contact copy is essential (using this method).

Back on the co-ordinatograph, opaque self-adhesive tapes of the correct width (three times the width required on the 1:1 master) are used to lay down the conductors. The tape is run from pad to pad and overlapped on to the pad to ensure continuous surfaces. Grid lines are very useful as guides for straight lengths and running the tape for a straight conductor is easy, provided it is placed and pressed directly on to its final position. Attempts to 'slip' the tape to adjust its position result in pick-up of dirt and dust by adhesive transferred to the backing film. Such contamination will ruin the artwork, rendering it unusable.

Bends can create problems. With large radii, it may be possible to stretch the outer edge of the tape to form the curve, but the chance of reversion towards a straight line will always exist. It is better (and essential with small radii) to use precut bends or to change direction by using straight tapes to form angles of 90° or, better, 45°.

It is most important that the tape adheres perfectly to the transparent film. The adhesive side of the tape must not be touched, nor, as far as practicable, the transparent film. It will pay to wipe the film surface on to which the tape is to be placed with a suitable solvent immediately before placement.

Where conductor spacing is critical, a transparent scale will be used. If a mistake is made, the length of tape must be removed, the surface carefully cleaned, and a new piece of tape put down in the correct position. Pushing the tape sideways into the correct position will be worse than useless. Not only will the stressed adhesive (which has a 'memory') try to take the tape back to its former position, but dust and dirt will be picked up by the adhesive residues left on the surface and will cause blurring and widening of tracks.

In the absence of a co-ordinatograph, a light table or a universal drafting device can be used. In this case, the transparent film can be placed on, and fixed to, a high-precision plastic film having the desired grid or, alternatively, a special, transparent, high dimensional stability plastic film with an integral grid. The latter are very useful for manual layout. Commonly used grid spacings are offered in a number of different scales. The grid colour, usually blue or violet, is easily seen by the naked eye but is readily filtered out during photoreduction.

Many useful aids are obtainable from the major suppliers of drafting preforms, such as strips of pads for DIP packages, rings of pads for TO-100, TO-99 or similar packages, bends of different radii and widths, and so on. Some of these are shown in Figure 9.5. The different shapes of pad available should be noted. Some of these assist in component type recognition. Others are intended to permit some degree of inaccuracy in placement or in drilling — for example, the elongated pad shown in Figure 9.5(b) can be used with a conductor to make misregistered drilling less likely to separate the conductor from the pad, Figure 9.6.

(In metric units it seems odd to have patterns with 2.54 mm diameter pads or 0.635 mm wide conductors, but it must be remembered that most drafting materials are available in imperial (or 'British') units. The figures mentioned correspond to 0.1 in. and 0.025 in., which are still internationally accepted

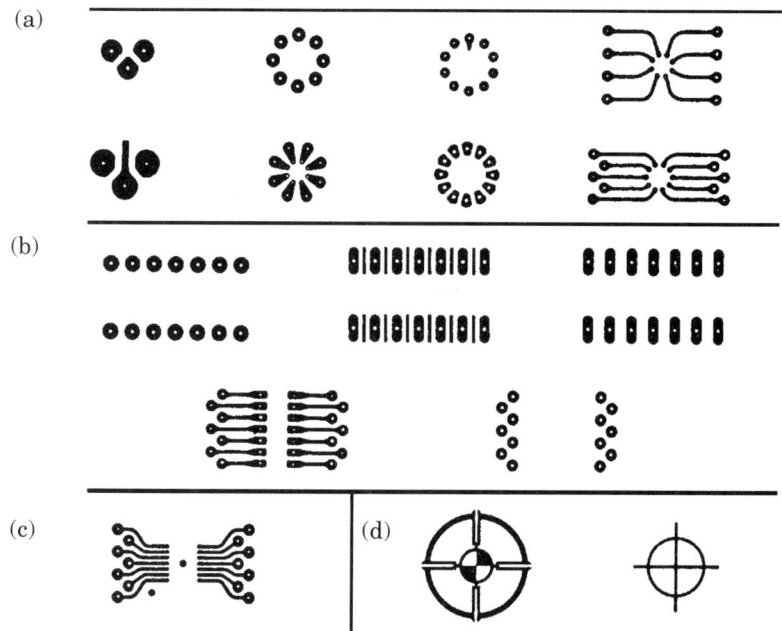

Fig. 9.5 Typical multi-pad pressure-sensitive labels for master artwork drawing: (a) circular TO packages; (b) standard and staggered dual-in-line packages; (c) flat packages; and (d) targets for photographic reduction. These patterns are not to a common scale. (Courtesy of Bishop Graphics)

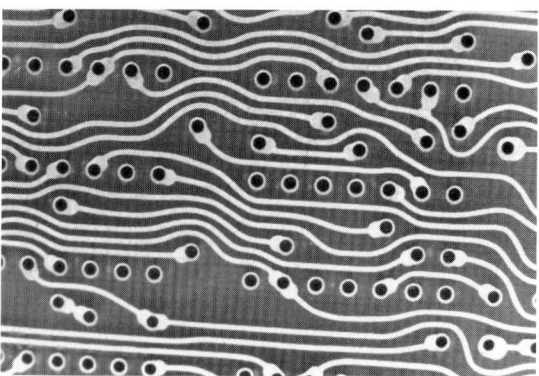

Fig. 9.6 The elongated pads seen here help to prevent off-centre drilling cutting the connection of conductor to pad.

grid sizes, although not now recommended — even by the British Standard![3] Common sense has prevailed and there is increasing use of true metric grids by all countries which support IEC 97.)[4]

Once all the pads and conductors have been placed, any notation — letters, symbols, etc. — can be added by using transfer letters, numbers and symbols of the appropriate sizes.

The artwork must now be checked. Checking to ensure consistency with the schematic is usually carried out by two people, one reading the circuit diagram while the other checks that the correct connection has been made on the artwork. To avoid excessive handling or marking of the artwork, it is best to make contact copies. While checks can be made on these copies, Figure 9.7, it is better if blueprints or dyeline copies can then be made as positives, that is, showing pads and tracks as dark traces on a clear background. It is then easy to mark the copy with the component or signal names, tick off checked connections, sketch out component outlines and pin numbers, and so on.

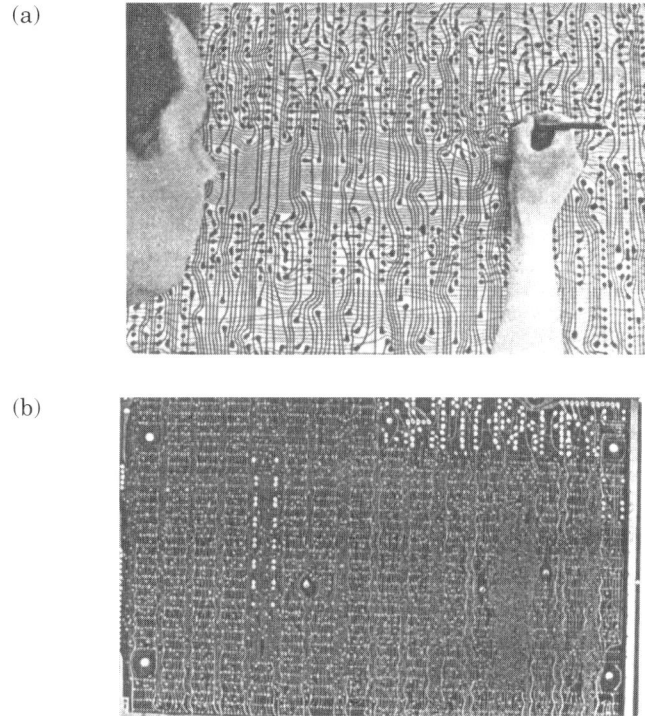

Fig. 9.7 Examples of manually drawn artwork: (a) checking a copy of the final artwork on a transparent film; and (b) the solder side of the manufactured board (the actual size is 245 x 390 mm, about 10 x 15 in.).

When checking and any reworking are complete, the artwork is photoreduced to working size. Photoreduction is a highly professional job for which the appropriate equipment is essential, Figure 9.8. It should feature the best quality lenses, two-directional tilting of the photographic film or plate support, a very fine focusing screen and so on. If the right equipment is not available in-house, the photoreduction should be sub-contracted to one of the many specialist companies.

The master artwork is placed on a horizontal (or vertical) light table, paying particular attention to obtaining absolute flatness and freedom from trapped air bubbles. The choice of photographic material depends on several considerations, one of the most important being dimensional stability of the base. Size changes, as a function of temperature only, are of the order of 0.005% per 10°F (5.56°C) for glass and 0.015% per 10°F for polyester film

Fig. 9.8 Photoreducing a master artwork.

base. Glass is almost unaffected by change in relative humidity, while plastic films show size changes of the same magnitude as those due to temperature. An important point to remember is that, whereas a change in temperature produces a change in dimension within seconds, a change in humidity is much slower in producing an effect. Emulsion and plastic film bases take up to 8 hours to equilibrate when humidity changes, but 90% of the change will take place within the first two hours.[5] The movement of plastic films under the influence of temperature and humidity changes can be calculated,[6] and it has been shown that maintenance of the 30 in. (762 mm) diagonal of an 18 in. x 24 in. (457.2 x 609.6 mm) within ±0.001 in. (0.025 mm) requires temperature control within ±2.38°F (1.32°C) and humidity control within ±2.78% RH.

Glass, obviously, has advantages over plastic films in that the temperature and humidity controls required are much less exacting. On the other hand, glass does create problems in handling and shipping and is normally only used when very accurate photoreductions are needed.

Double-sided PBs require a master artwork for each side. If the layout is drawn using a component-side view, it will be necessary to draw:

(i) the solder side — by looking at the draft artwork from the back;
(ii) the component side — by looking at the draft artwork from the front.

It is common practice to create the component side first, making a copy of the pad master before laying down the conductors, as described earlier. The solder side is obtained by taping the copy pad master on its back. This ensures good registration back to front, and also makes certain that no pads are omitted, even when there are no connections on the solder side. (Note: the component side is side 1 and the solder side is side 2.) There is nothing to prevent side 2 being drawn up first — except convention and long practice.

By using special tapes, it is possible to create both sides of the artwork on the same piece of transparent film. For each side, a different colour will be used, pads common to both sides being put down in black on one side only. By employing different colour filters, only black and one of the other colours at a time will be transferred to the photographic film or tape.

Some companies have developed a standard grid, on which all crisscrosses are occupied by pads, no matter which specific PB is involved. With this system, PBs differ only in the conductor traces. This eliminates the need for work with the co-ordinatograph to place the pads. Once the pad pattern has been established, as many copies can be made as are required for all PB drafting.

Such a degree of standardisation is highly desirable — but rarely possible. A compromise which reduces the amount of work needed to realise a circuit may be reached. For example, if a large set of similar circuits using DIP packages has to be designed, all major component positions may be fixed at the beginning so that all their mounting holes will be in the same positions. It is possible to create a common pad artwork for all component insertion points. From this common master, contact copies for all specific artworks can be made. Any extra pads needed will be added using preforms before taping the conductors.

The artwork for a solder resist coating can be produced using one of the following methods:

(i) drawing, by using a co-ordinatograph as previously described;
(ii) if solder resist has to cover the entire pattern (except pads), taking a contact copy of the pads and enlarging all the pads with suitable preforms; when necessary, unwanted pads can be scraped off the copy.

Care must be taken to observe any differences in number or shape of pads on side 1 and side 2 to ensure that the solder mask is derived from the pad copy for the side on to which it will be placed. For instance, if solder resist is to go on side 2 only, then the mask must be derived from the side 2 pattern.

The final result of the artwork design process will be a master pad image, a film for each layer of conductors, and, when specified, a film or films for the solder resist pattern(s) and a notation film of the inscriptions to be screened on the component side of the board. From these 1:1 master films the production films will be prepared, usually by contact printing.

9.1.4 Master Drawing

Design of the PB is not complete until a master drawing has been prepared (although it can be difficult to convince some customers of this!). A printed board is a mechanical device as well as the base for an electronic circuit and has to be dimensioned. Borders and fixing holes are not always included in the artwork pattern and the relation between these and the conductor pattern must be specified to obviate problems of fit on assembly. In addition, there may be slots and/or cutouts which must be dimensioned and related to the board pattern. All these points must be set out on a master drawing.

Many electronics designers overlook the importance of a proper drawing, frequently making prototypes and sometimes making production boards using inadequate drawings or even no drawing at all! This is not recommended.

At a minimum, the master drawing of a PB, Figure 9.9, should include:

(i) nominal position, diameter and tolerance on position and diameter of all holes, including locating holes;

(ii) all dimensions and tolerances of the outline;
(iii) positions, dimensions and tolerances of other significant elements such as slots;
(iv) all data needed for identifying the base material to be used (type, thickness, copper thickness and type, relevant specifications and standards, approved suppliers, etc.);
(v) all data on metallic and non-metallic coatings required (gold on edge board contacts, solderable finishes, solderability protective coatings, solder resist, colour and type of ink for coding/notation, etc.);
(vi) references to all films needed to manufacture the PB, such as conductor layer films, solder resist films, notation films, non-plated or differently plated pattern areas, and so on;
(vii) all references needed to identify specific tools (punching dies, blanking dies, drilling tapes, etc.) required to manufacture the PB;
(viii) references to all documents needed to inspect/test the PB.

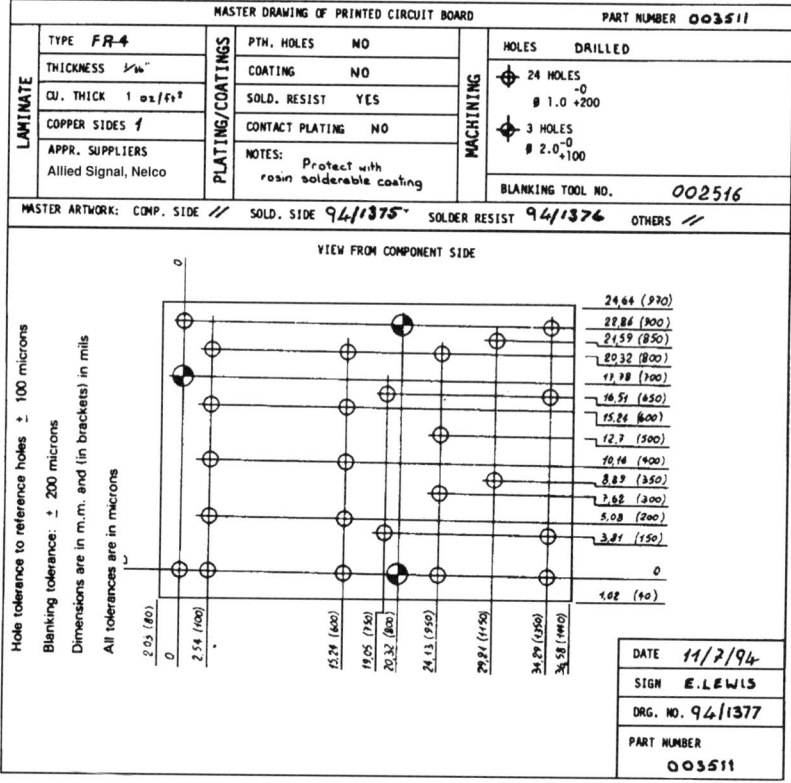

Fig. 9.9 The master drawing for the PB of Fig. 9.2(b).

In addition, the drawing can state any special requirements peculiar to the board.

The master drawing is usually prepared from a component-side view. Whichever side is shown should be clearly stated on the drawing. For example, hole drilling or punching of single-sided boards is normally performed with the copper side up so that the operator's view of the board during these operations is the opposite of the mechanical drawing. For this reason, some companies produce mechanical drawings of single-sided boards viewed from the copper (solder) side of the board. This lack of uniformity provides a good reason for giving a clear indication of the viewing direction on the drawing.

9.1.5 Component Map

Having focused on the preparation of phototools and mechanical drawings for PB production, it is now necessary to look at the demands of assembly. Both for assembly and for repair, a map of components must be produced. This drawing does not need to have the same accuracy as the PB documents since it simply states into which holes components must be inserted or, for SMDs, on to which pads components must be placed. (When required for, say, automatic insertion, accurate positions can be derived from the master drawing.)

A typical component map, Figure 9.10, shows the profile of components, with clear indication of polarity when relevant, as seen from the assembly side. A simple code is used to identify each component, e.g., R1, R2, R3, etc. for resistors, C1, C2, C3, etc., for capacitors, and so on. Since this code refers to position on the board, identical components will have the same alpha prefix but a different number. The drawing may show a component list, which brings together all identical devices and gives the company's part number for each device. The table may be arranged as follows:

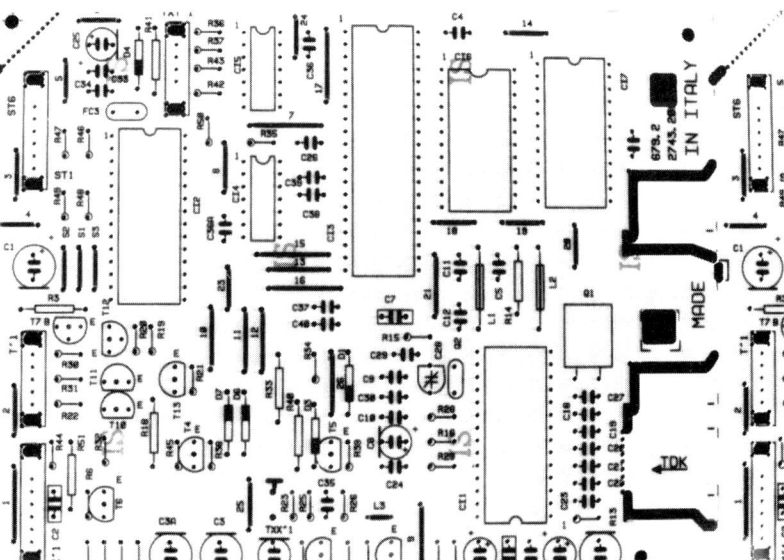

Fig. 9.10 The 'component map' shows the position and orientation of all components on an assembled board by means of simple outlines and identities. The map may be screen printed on reasonably simple boards to act as an assembly/inspection aid.

Position	Part Number	Description
R1, R7, R8, R12	27-385-479	Carb. res. 10% — ¼ W — 1.74 K
R2, R3, R15	27-385-200	M. film res. 2% — ¼ W — 3.92M
C1, C5, C9	27-764-369	Film cap. 150 p — 25 V
IC12	28-300-792	Op. amp. Philips LM 124

If the map lists all the components, including the PB and any mechanical items, it may replace the bill of materials for that board. The component map may also carry other information, such as:

(i) reference holes for automatic insertion;
(ii) direction of flow on automatic conveyors and on soldering machines;
(iii) reference to special tools needed to assemble particular devices, and so on.

The component map drawing must make reference to the PB part number, which is normally printed on the board, etched in copper or, rarely, added on a label. If the board is quite simple, the component map may be reduced to a 1:1 scale and applied to the PB by screen printing to act as an aid to assembly and inspection.

9.2 AUTOMATED PRODUCTION OF FILMS

The quality of a manually produced master film depends very much on the skill and patience of the operator. Taping takes a long time and a taped artwork is easily contaminated or damaged. Storage of taped masters is often troublesome as the adhesive 'memory' can cause tracks and pads to creep from their correct position. Any adhesive residue exposed as a result of such movement will pick up dust and dirt to blur or enlarge the track outline.

In addition, regardless of how accurately they have been placed, pads are distorted during sticking down (unless dry-film transfer pads are used) and the sides of tracks will never be perfectly parallel lines. While this can be tolerated on low density circuits, the increasing density of PBs and finer line widths (even before the advent of SMT) had already forced designers to adopt automatic methods of artwork production.

This was an important step forward in circuit layout since it allowed an increase in circuit complexity and contributed significantly to the cost reduction of multilayers. Although now largely replaced by CAD and CAE, which are dealt with in Chapter 10, the technique does still survive. Automatic film production can be broken down into three major steps — digitising, editing and photoplotting.

9.2.1 Digitising

The (manual) design having been finished, the relevant information is 'digitised'. The requirements for use of a stable film base, a grid, and for placement of pads on grid intersections still apply. If the tracks can be run in both directions (X and Y) in a fixed position with reference to the grid lines, digitising will be more efficient. (The artwork does not need to be absolutely accurate as minor digitising inaccuracies are corrected on editing.)

The artwork can be digitised using a *reading table*, Figure 9.11, a relatively simple equipment (which, with some additional items, can also be used as a co-ordinatograph). The operator moves a cursor over the draft artwork, centres a reticle on a specific element — the centre of a pad, the end of a straight length of conductor, etc. — and pushes a button. The X-Y co-ordinates of the point are determined and fed to the support computer memory. An auxiliary keyboard, usually small enough to be mounted on the cursor itself, is used to input the nature of such points as well as additional information such as conductor width, land diameter and so on. This device may be fitted with a display which shows a menu for the operator's selection. Special patterns, such as the land pattern for a TO-116 package, may be digitised in one shot. Other such 'macro' instructions are available. For example, it is possible to instruct the computer that a list of X-Y positions to be digitised in sequence refers to a single track made up of a series of straight segments.

Fig. 9.11 A manually drawn master artwork is digitised to produce a 1:1 scale artwork by computer.

The computer will arrange all the input information into a suitable format and store it in a non-volatile memory, usually a floppy disk. Additional information, such as the type and part number of the component to be assembled in each position, may be fed in at this stage or added while editing.

9.2.2 Editing

While some digitising tables are equipped with a multicolour plotter which draws a diagram of all connections for off-line editing, this operation is usually carried out on-line. A high resolution raster scan monitor is required. This shows an image of the board layout with the actual size of tracks and pads. A zoom function is normally provided to facilitate checking the critical points. The operator can edit the artwork by adding or deleting connections, changing the size of elements of the pattern, re-routing conductors, and so on. (The editing function is also used for implementing design changes — it is not necessary to redigitise the board.)

Most editing software also provides additional facilities, such as ground plane generation, the addition of special symbols including the company's logo or factory code, and the ability to enter information specific to the design, e.g., approved suppliers of a particular component.

Even when due care is taken in pad/track placement, the error rate on X-Y positions can be quite high. If the board is designed to a grid, the computer will automatically correct the digitised co-ordinates to the theoretical values and recover all placement tolerances. For example, if all holes/lands are on a 0.050 in. grid, any co-ordinate between X1 + 0.0024 in. and X1 - 0.024 in. will be assumed by the computer to be X1 and will be registered accordingly. The same correction will take place for the Y co-ordinate, Figure 9.12.

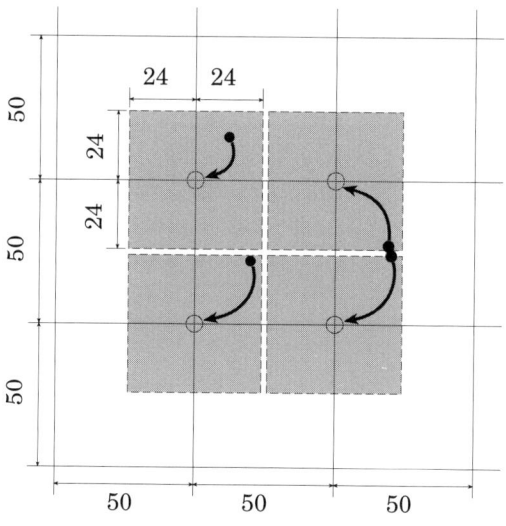

Fig. 9.12 While digitising an artwork laid out on a 0.050 in. grid on a digitiser having a resolution of 0.001 in., any point within 0.024 in. of the true grid will be moved automatically to that position.

Digitising and editing are simple, straightforward operations which give excellent results when a grid is used. However, preparation is still tedious and time-consuming. Checking the digitised artwork is also a tedious but essential task as the operator may miss a hole or even a length of track. On complex boards, it is not unknown for such a mistake not to be discovered until the boards have actually been produced. In this case, not only is the cost of producing the boards lost, but the design process has to return to the editing stage and valuable time is lost as well.

It is for such reasons that CAD software for laying out boards has been so successful. The software generates all, or the great majority of, the data needed to design and produce the master artwork.

9.2.3 Photoplotting

Once all the data have been fed into the computer and stored on disk, they can be transferred to a photoplotter — by data line, telephone line, or by transfer of the disk — in order to draw the master films directly on a 1:1 scale.

A photoplotter is a CNC (Computer Numerically Controlled) machine consisting of a flat table on which a sensitive film or plate is placed and firmly held by vacuum. A carriage moves over the table, in the X and/or Y directions, as instructed by the electronic controller. The carriage holds a 'pen' which projects a beam of light on to the film via a precise aperture or via a diaphragm mounted on a rotating turret, Figure 9.13, which carries a large number (e.g., 24 to 64) of these apertures.

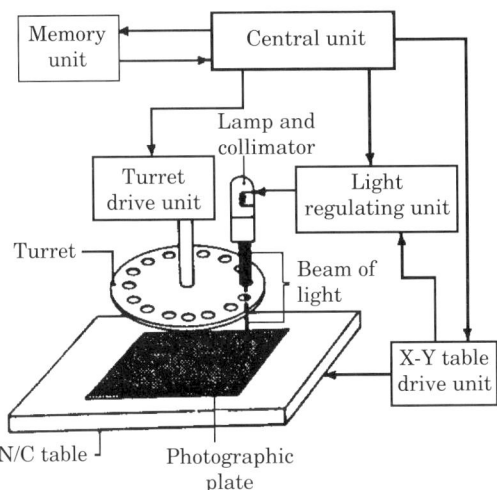

Fig. 9.13 A simplified diagram of a numerically controlled machine for generating 1:1 master artwork by optical pen.

Elements like pads, holes, connector strips, company logo, gauge marks and so on are projected as *fixed images* or *flash images*, since the carriage remains stationary over the relevant X-Y co-ordinate for the (short) exposure time. On the other hand, elements such as conductors, circles, boxes, ground planes and similar items are drawn by projecting the beam of light while the X-Y carriage is moving. Constant width of these elements is assured by electronic controls, which select the aperture and modulate the light intensity according to the table speed and movement.

Since the diaphragms are interchangeable, the same equipment can cope with any kind of pattern and a large variety of symbols. The accuracy of a photoplotter is usually very high, with typical values ranging from 0.01 mm to 0.02 mm (0.0004 in. to 0.0008 in.) and repeatability of the order of 0.013 mm (0.0005 in.). Plotting speeds are not very high at 2.5-7.6 metres/minute (100-300 inches). Drum photoplotters, similar to those used in CAD equipment, can achieve much higher speeds, over 25 metres/minute (i.e., over 1000 inches/minute) with accuracy of about 0.064 mm (0.0025 in.) and repeatability around 0.05 mm (0.002 in.). The reduced accuracy is still acceptable for many artworks, even on a 1:1 scale.

Early photoplotters had to be routed in the X or in the Y direction, as a synchronised X-Y movement was not possible. Tracks running in other directions assumed a stepped profile, which became very evident when there were several parallel tracks, Figure 9.14. As a result, tracks were longer than necessary and their density was lower.[7]

Fig. 9.14 Pattern of a PB, the film of which was produced on a photoplotter which could move only in the X or Y direction.

This problem, now no longer prevalent thanks to improvements in equipment, should not be confused with the edge roughness typical of raster photoplotters. A raster photoplotter exposes the image with a flying laser spot switched rapidly on and off to expose the film in raster fashion (similar to the working of a television screen). The big advantage of the raster plotter is speed — it can create a 455 mm x 610 mm (18 in. x 24 in.) image in less than 5 minutes. The edge roughness all but disappears once the image has been transferred on to dry film, developed and etched.

Modern photoplotters have additional features. For instance, when a dot master pattern is needed for two or more layers, contact copies should be made emulsion to emulsion to avoid parallax error. Since this provides a mirror image of the original, most photoplotters now have a 'mirror' command.

In addition to the circuit layout films, it is possible to photoplot films for solder mask, legends, component layouts, master drawings, drilling drawings, and a hard copy of the bill of materials, parts list and so on. In addition, drilling co-ordinates may be fed to the controller of the drilling machine, either by direct link or via a floppy disk or cassette.

Digitising and photoplotting artworks reduces development times and increases accuracy of the final films — by a factor of at least ten when a grid is used. When producing PTH double-sided and multilayer boards the improved accuracy of each film, increased consistency of the layers, and the excellent alignment of drilling positions to pad centres allow the adoption of smaller pads and smaller pad-to-hole diameter ratios. In turn, this permits smaller clearances between pads and conductors. The good quality of conductor edges, added to the steady improvements in photoresists, allows the use of finer lines.

Photoplotting was a key factor in making possible a gradual, steady increase in the density of tracks and components on PCBs, and allowed the adoption of smaller and smaller grid pitches. This was essential to cater for the increasing space reduction made possible by the development of surface mounting components.

9.3 GENERAL PRINCIPLES

A printed board is 'custom-designed', in other words, with very few exceptions, it is specific to a particular product design. It may, therefore, seem impossible to apply any form of standardisation. So many conflicting requirements are imposed that each PB will be unique.

Fortunately, this is not strictly true. Each time a board is designed there will be similarities to others. In the easiest case, it will be similar to an existing board, having more or less the same types and amounts of components, the same number of holes and conductors, similar input and output arrangements, and so on. Much time and money can be saved if the design of the PB and its artwork can be referred to a standard or to a set of design rules, either developed within the company or published nationally or internationally. (Some of the better known design manuals are mentioned at the end of Chapter 8.)

By insisting that designers use an external and widely accepted standard, a company can gain advantages such as:

(i) gaining customer confidence by relating the design to nationally and/or internationally accepted criteria;
(ii) saving the cost of developing and proving its own standards;
(iii) being reasonably certain that most design problems will find an 'off the shelf' solution;
(iv) being able to estimate the cost of the PB, hopefully with the help of a PB manufacturer, even before the design is started;
(v) being able to estimate, with some accuracy, the cost of rejects, rework, and scrap during manufacture and assembly of the board;
(vi) being able to estimate the reliability of the assembly.

Despite the availability of published standards developed by those who have the skill and experience to cover the majority of situations, many companies (or at least their designers) claim that these are not applicable to their specific assemblies or products and develop, at no little cost in time and testing, their own set of design rules. Such situations are seldom in fact unique, and most problems can be resolved by making minimal changes to existing standards. To illustrate this, the following text discusses several points which still arise frequently, and indicates where possible an appropriate solution.

9.3.1 Why Use a Grid?

Discussion about whether or not to use a grid and what the grid size should be has been going on for many years. Experience has demonstrated that using a grid has many advantages and the majority of PBs are now designed to a grid, with the possible exception of very small boards and occasional prototypes.

Even in manual preparation of artworks, the adoption of a grid permits the use of a simple light table and grid paper instead of a co-ordinatograph. The grid is only used for pads since conductors will run where there is room for them. This makes it easier to produce the master drawing and to state the drilling co-ordinates, insertion co-ordinates and so on, since all dimensions will be multiples of the grid selected.

When a manual layout is digitised to produce films by photoplotting, the grid allows automatic correction of the pad position and recovery of the original tolerances. It is possible to use the same approach for conductors, provided they can be run in a given number of fixed positions with reference to the pad positions.

While it may restrict design flexibility slightly, there is no simple way in which the sinuous manual conductor pattern can be digitised: conductors must be made up of straight segments, running mainly in two orthogonal directions and, where necessary, at a defined angle to the two main directions. (The same criteria apply to CAD boards, see Chapter 10.)

9.3.2 Grid Selection

When alternative grids are available, the compulsory starting point is to select the simplest one which will cope with the PBA. If a grid has to be developed specially for the circuit, the amount of work and interaction with other departments (PB manufacture, automatic assembly, soldering, testing etc.) is enormous.

The first basic decision to be made is whether to use imperial (inch) or metric units (Table 9.1 gives rounded-off conversions between mils (1 mil = 0.001 in.) and millimetres). After many years of argument (and procrastination!), the IEC member nations made the metric grid the preferred grid in 1991 (IEC Publication 97),[4] retaining the 'metricated' inch grid as a non-preferred alternative. It can be argued that, despite the general trend towards the use of metric units (fuelled by the expansion of Japanese industry), printed circuit design is simpler in imperial units because:

(i) many major components still have all dimensions in inch units, often in simple figures: for example, the pins of a TO-116 package are spaced 100 mils (0.100 in.) in one direction and 300 mils (0.300 in.) in the other; (the 'metric' equivalents, 2.54 mm and 7.62 mm, are not conducive to easy calculation, although a *true* metric grid is);
(ii) photoplotters, PB drilling machines and PB routers are available in both metric and imperial units;
(iii) equipment for component insertion and placement is usually designed in imperial units, the metric version being a variant (although quite common);
(iv) past designs made by the company are probably in imperial units and to change may be expensive; interchangeability can create major problems;
(v) in most equipments, the interface between the electronic part and the mechanical part is usually quite simple and the use of two different measurement systems does not pose a major problem.

For such reasons (and because of in-built inertia) many PB designers continue to use imperial units to state dimensions in 'thou' or mils. Since the dimensions of many components are multiples of mil, and since most NC equipment used in the manufacture and assembly of boards has a maximum resolution of one mil, only integer figures are used, making calculations quite straightforward.

On this basis the minimum grid pitch could be one mil = 0.001 in., which is seldom, if ever, used. (Publication 97 states that the smallest grid spacing

Table 9.1

Mils to mm Conversion (rounded-up)

mils	+0	+1	+2	+3	+4	+5	+6	+7	+8	+9
0	0.000	0.025	0.051	0.076	0.102	0.127	0.152	0.178	0.203	0.229
10.00	0.254	0.279	0.305	0.330	0.356	0.381	0.406	0.432	0.457	0.483
20.00	0.508	0.533	0.559	0.584	0.610	0.635	0.660	0.686	0.711	0.737
30.00	0.762	0.787	0.813	0.838	0.864	0.889	0.914	0.940	0.965	0.991
40.00	1.016	1.041	1.067	1.092	1.118	1.143	1.168	1.194	1.219	1.245
50.00	1.270	1.295	1.321	1.346	1.372	1.397	1.422	1.448	1.473	1.499
60.00	1.524	1.549	1.575	1.600	1.626	1.651	1.676	1.702	1.727	1.753
70.00	1.778	1.803	1.829	1.854	1.880	1.905	1.930	1.956	1.981	2.007
80.00	2.032	2.057	2.083	2.108	2.134	2.159	2.184	2.210	2.235	2.261
90.00	2.286	2.311	2.337	2.362	2.388	2.413	2.438	2.464	2.489	2.515

Millimetres to Mils Conversion (rounded-up)

mm	+0.00	+0.01	+0.02	+0.03	+0.04	+0.05	+0.06	+0.07	+0.08	+0.09
0.00	0.00	0.39	0.79	1.18	1.57	1.97	2.36	2.76	3.15	3.54
0.10	3.94	4.33	4.72	5.12	5.51	5.91	6.30	6.69	7.09	7.48
0.20	7.87	8.27	8.66	9.06	9.45	9.84	10.24	10.63	11.02	11.42
0.30	11.81	12.20	12.60	12.99	13.39	13.78	14.17	14.57	14.96	15.35
0.40	15.75	16.14	16.54	16.93	17.32	12.72	18.11	18.50	18.90	19.29
0.50	19.69	20.08	20.47	20.87	21.26	21.65	22.05	22.44	22.83	23.23
0.60	23.62	24.02	24.41	24.80	25.20	25.59	25.98	26.38	26.77	27.17
0.70	27.56	27.95	28.35	28.74	29.13	29.53	29.92	30.31	30.71	31.10
0.80	31.50	31.89	32.28	32.68	33.07	33.46	33.86	34.25	34.65	35.04
0.90	35.43	35.83	36.22	36.61	37.01	37.40	37.80	38.19	38.58	38.98
1.00	39.37	39.76	40.16	40.55	40.94	41.34	41.73	42.13	42.52	42.91
1.10	43.31	43.70	44.09	44.49	44.88	45.28	45.67	46.06	46.46	46.85
1.20	47.24	47.64	48.03	48.43	48.82	49.21	49.61	50.00	50.39	50.79
1.30	51.18	51.57	51.97	52.36	52.76	53.15	53.54	53.94	54.33	54.72
1.40	55.12	55.51	55.91	56.30	56.69	57.09	57.48	57.87	58.27	58.66
1.50	59.06	59.45	59.84	60.24	60.63	61.02	61.42	61.81	62.20	62.60
1.60	62.99	63.39	63.78	64.17	64.57	64.96	65.35	65.75	66.14	66.54
1.70	66.93	67.32	67.72	68.11	68.50	68.90	69.29	69.69	70.08	70.47
1.80	70.87	71.26	71.65	72.05	72.44	72.83	73.23	73.62	74.02	74.41
1.90	74.80	75.20	75.59	75.98	76.38	76.77	77.17	77.56	77.95	78.35
2.00	78.74	79.13	79.53	79.92	80.31	80.71	81.10	81.50	81.89	82.28
2.10	82.68	83.07	83.46	83.86	84.25	84.65	85.04	84.43	85.83	86.22
2.20	86.61	87.01	87.40	87.80	88.19	88.58	88.98	89.37	89.76	90.16
2.30	90.55	90.94	91.34	91.73	92.13	92.52	92.91	93.31	93.70	94.09
2.40	94.49	94.88	95.28	95.67	96.06	96.46	96.85	97.24	97.64	98.03
2.50	98.43	98.82	99.21	99.61	100.00					

to be used on the non-preferred imperial system shall be 0.635 mm, i.e., 0.025 in. or 25 mils. It is interesting to note that the smaller *preferred* metric grid system has a nominal spacing of only 0.05 mm (1.9685 mils), which should provide a real incentive to use it for today's tightly packed designs.)

The maximum pitch normally required is 100 mils (the non-preferred 2.54 mm). The preferred pitch will be a multiple of 0.5 mm, say 2.5 mm, which can create problems. If, for example, a component manufacturer designs a component with its terminations on 2.50 mm spacing, such a component may easily be confused with the old standard of 2.54 mm spacing, and tolerances on components and PBs could soon find terminations and holes or land patterns not matching because of tolerance run-out. One method of avoiding this confusion will be to replace the component with terminations on 2.54 mm centres, with one having terminations on 2.00 mm centres. Alternatively, components to the new, true, metric standard can be marked as belonging to the preferred metric grid system.

Until there is a good supply of all, or most, components with terminations on true metric pitch, designers will have to carefully consider the implications of selecting a multiple of the preferred metric grid.

Many PIH components require two or more holes at 100 mil pitch and the size of holes, lands and tracks should be such that at least one track can run between two holes spaced at 100 mils. This seldom creates a problem, as most tracks are overdimensioned as regards the current they must carry, see Chapter 8. Their size is decided on the basis of the manufacturing cost of the board and the reliability of the assembly. Where a large amount of power is involved, e.g., in the power supply section, large tracks may sometimes be narrowed locally, although this is not very good practice and may, in some cases, be totally unacceptable.

In SMT boards, component terminations may be spaced at 25 mils and it is often necessary to run tracks between adjacent holes or lands to achieve a reasonable density of components and connections.

Within the imposed restraints of area and volume, length of connections and so on, the decision on grid pitch must balance the cost of increasing the board complexity by decrease in grid spacing against the cost of handling, assembly, testing, maintenance, and any other factors affected by the size and complexity of the board.

9.3.3 Grid Sizes in Common Use

Pin-in-hole components dictate the use of a fairly large grid and, for many years, a grid of 50 mils (1.27 mm) was found suitable for single-sided, double-sided, and simple multilayer boards. A grid of this size permits the use of fairly large holes and lands, while tracks as wide as 16 mils (0.4 mm) can be run between adjacent holes 100 mils (2.54 mm) apart, Figure 9.15. This grid size gives good results, except that packing density is much too low. On the other hand, even when retaining 16 mil track widths, large minimum clearances (e.g., 17 mils) and acceptable pad diameters (say 50 mils), one track may be run between two holes on 100 mil centres and four tracks may run in 100 mils, Figure 9.16.

The tracks run parallel to the grid lines but will not lie on them and, if 1 mil is taken as the minimum unit, it is not possible to develop a uniform spacing for the tracks. Variable spacing has to be adopted — 34 mils once and 33 mils twice. This does not create any significant problems, especially when the tracks connect tangentially to the pads. Such attachment, applicable to any grid, has other advantages:

(i) it assures a reliable track-to-land connection, even when the hole is off centre;
(ii) it increases conductor density;
(iii) identification of possible shorts during solder joint inspection is made easier, as the shape of the short will differ from the shape of the desired connection.

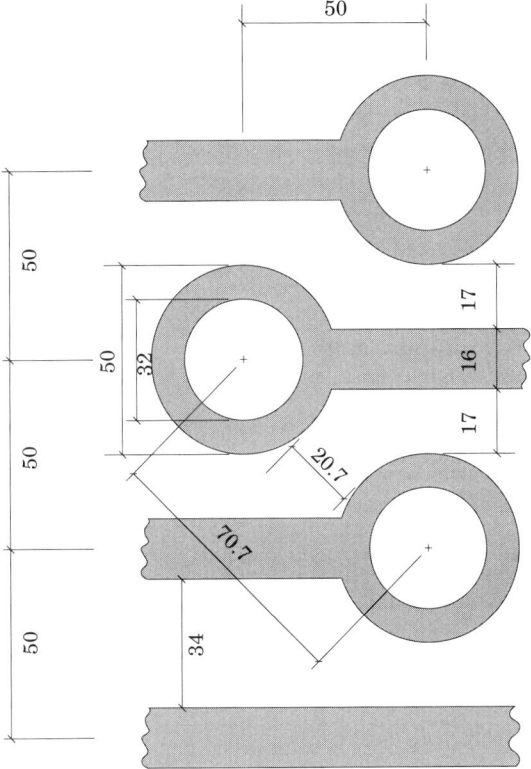

Fig. 9.15 On a basic 50 mil (1.27 mm) grid the holes are spaced on 100 mil (2.54 mm) centres or staggered at 50 mils in the X and Y directions. (All dimensions are in mils.)

The need for increased packing density led to the use of a 25 mil (0.635 mm) grid, which was at that time a preferred size in IEC 97. Tracks are typically 12 mils (0.305 mm) wide and usually only run along grid lines. Clearance between tracks is of the order of 13 mils (0.33 mm). The minimum track-to-pad clearance is greater than 13 mils, since any hole will interrupt two tracks, Figure 9.17, and this cannot be avoided by, for example, bringing the track to the pad tangentially.

Holes may be staggered at 50 mils in both directions or at 25 mils in one direction and 75 in the other, unless very small pads are used in which case the spacing may be 50 mils in one direction and 25 mils in the other.

The main problem with this grid arises from the large area occupied by the relatively large holes, since, in many cases, the minimum hole diameter to

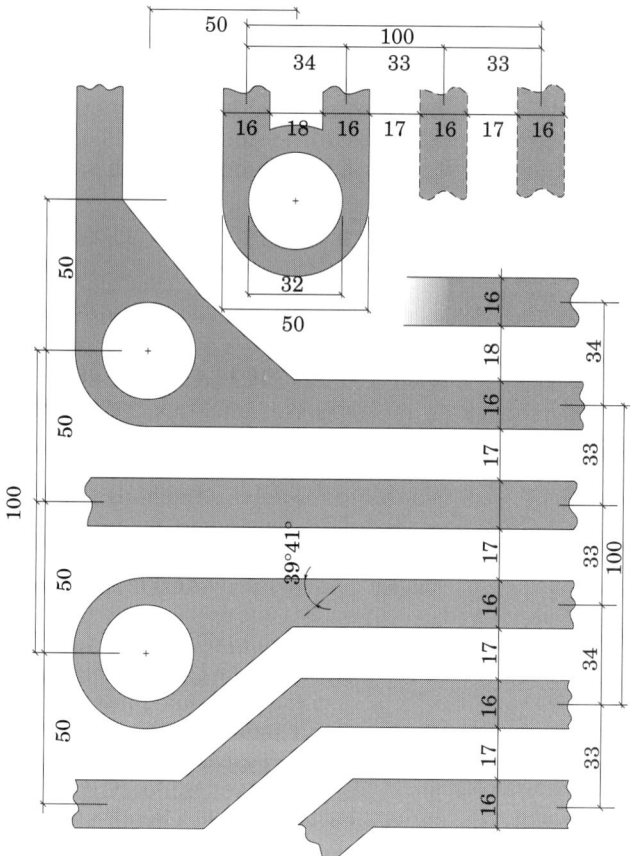

Fig. 9.16 An alternative 50 mil grid layout with 4 conductors in 100 mils. Note the alternative track-to-land layout. (Dimensions are in mils.)

allow component insertion cannot be reduced below 30 mils, as plated. However, a significant gain can be made by reducing the size of via holes.

When small via holes can be employed with relatively fine tracks, it is possible to achieve reasonably high packing density on all types of board, including mixprints. Typically, hole sizes around 16 mils (0.41 mm) diameter in a 26 mil (0.66 mm) pad are used with track widths of the order of 8 mil (0.20 mm).

This grid size affords good flexibility, as a via hole in the middle of a track need not interrupt the adjacent tracks, as shown in Figure 9.18, and via holes on adjacent tracks can be spaced at 25 mils in both directions.

It also provides an easy solution for fan-out of most SMCs on multilayer boards: drilling an array of holes just beside the mounting pads in order to shunt them out on the appropriate layer, Figure 9.19. If required, the fan-out may be achieved on one layer and, when necessary, on only one side of the component, as one conductor can be placed between adjacent holes.

The grid can be used for SMDs with 25 mil lead spacing (as for tracks), fan-out to other layers being achieved by two lines of staggered via holes.

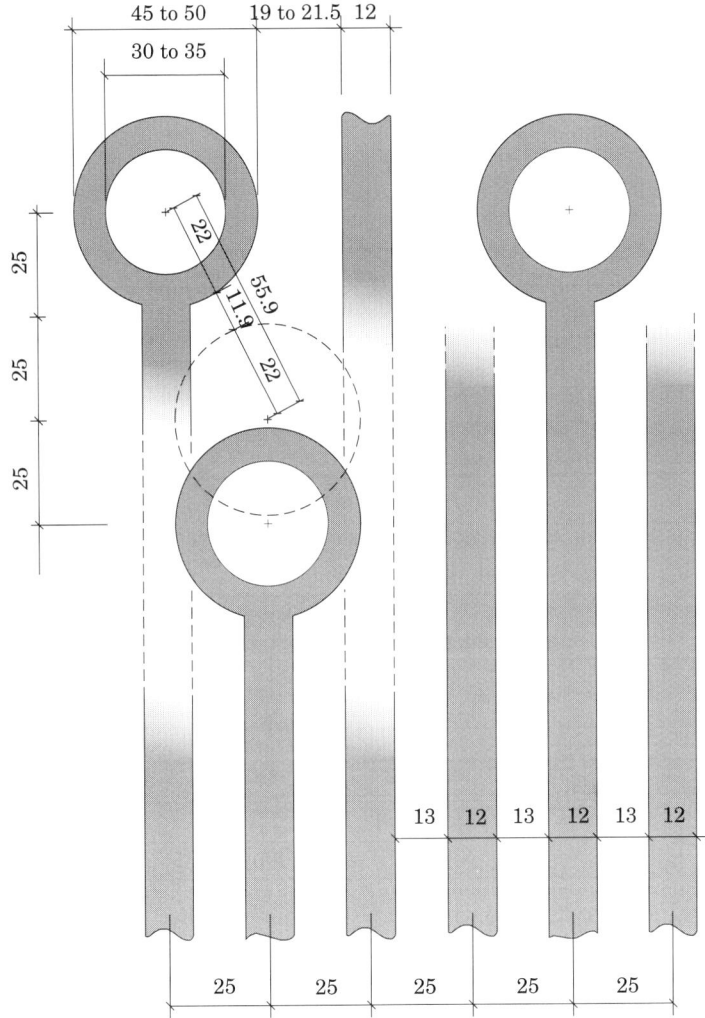

Fig. 9.17 Design rule for a 50 mil grid with insertion holes and via holes of the same size. Each hole interrupts two tracks. Holes must be 100 mils apart or staggered by 50 mils in each direction. Alternatively, they may be staggered by 75 mils in one direction and 25 mils in the other or, if lands are very small, 50 mils and 25 mils.
(Dimensions are in mils.)

While such a grid may be used for mixprint boards, the inability to run more than one track between adjacent insertion holes (on 100 mil pitch) is a serious drawback. There are solutions. One of these uses a track width of 4 mils (0.10 mm) which, even with a minimum gap width of 12 mils (0.30 mm) and fairly large holes and pads, allows two conductors to run in the same space.

While this is basically a variant of the 25 mil grid, close examination will show that all the holes are on an intersection of the 25 mil grid lines but

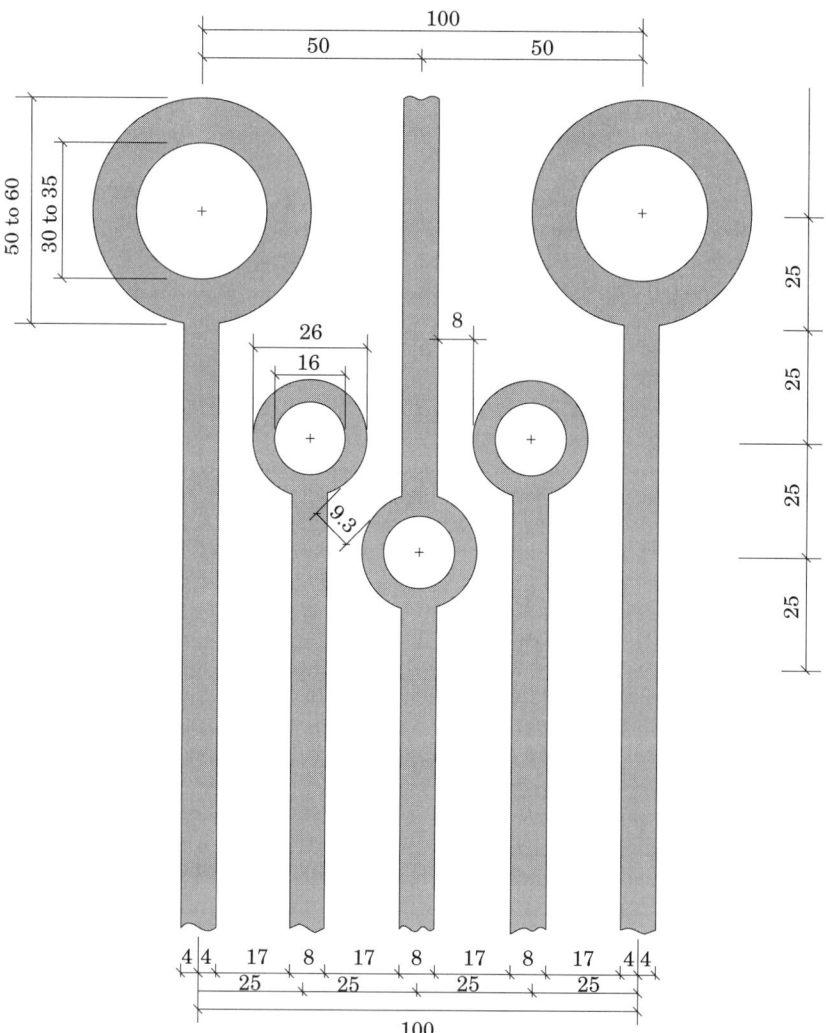

Fig. 9.18 A layout on a 25 mil (0.635 mm) grid, using smaller size holes for vias. When these are staggered at 25 mil in both directions (X and Y), they do not interrupt adjacent tracks. (Dimensions are in mils.)

conductors may run both along and off the grid lines (e.g., between two insertion holes the tracks will run at a distance of 42 mils from the hole centre).

This means that in some parts of the board the 25 mil grid is abandoned and a finer grid, which in this case can only be 1 mil, is adopted. While this gives great flexibility (for example, 7 tracks and 6 gaps can be achieved in a 100 mil span), any computer used to lay out the board will have to be much more powerful, since the number of intersections on a 1 mil grid is 625 times greater than the number on a 25 mil grid. (If a variable grid, such as 25 mils

Fig. 9.19 Fan-out of an SO package by shunting on the internal layers requires a line of PTHs on 50 mil (1.27 mm) pitch. (Note that 9 of the holes have 'stolen' solder from the joints, even though they are quite well separated.)

overall and 1 mil only where needed, can be handled by the software, the amount of memory and computing time will be more commensurate with the return in increased board density.)

On surface mounted boards, including mixprints on which SMCs are in the great majority, the diameter of via holes can be reduced to 16 mils, or even less, with pads of 25-30 mils diameter. This is an area in which drilling and metallising and electroplating development is critical for producing good through-plated holes and in which holes of 8 mils (0.2 mm) and finer are being produced. While this is stretching the limits of technology, it also allows greater board densities. In turn, this means that smaller grid sizes will be used.

9.3.4 A Bad Example

The board described here was designed and produced in the late seventies by a large computer manufacturer of good reputation. It offers a magnificent collection of examples of 'what you should *not* do'.

The board is laid out on a conventional 50 mil grid and fairly large square pads have been employed, Figure 9.20. Square pads were used to increase land area — quite a popular trick in the late seventies/early eighties, although the advantages were very marginal. In the case shown, the large size imposed a minimum hole-to-hole distance of 100 mils in both axes and limited board density.

The board is digital, so the track width of 16 mils is unnecessarily large, especially when compared with the gap width of 10 mils. This limits track density.

The track layout, seen to advantage (!) in Figure 9.21, is very random, no attempt having been made to minimise the number of critical clearances. This may cause rejects during the manufacture of the boards and will almost certainly lead to shorts by bridging during wave soldering. Many of these could have been eliminated by simple editing of the layout to move, where possible, tracks which were close to a pad to the next grid position, Figure 9.21.

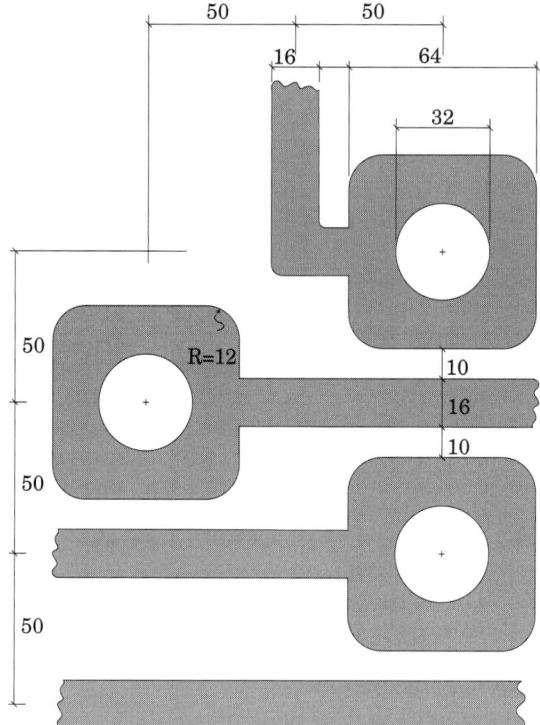

Fig. 9.20 A sample of bad design; large square pads limit board density.

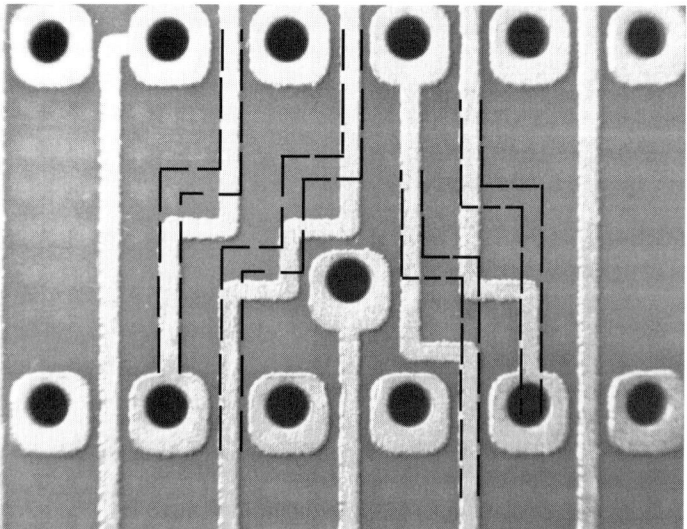

Fig. 9.21 Tracks are routed in positions likely to cause solder bridges, although room is available for a better design — as shown by the dotted outline. (This board was scrapped because of poor tin-lead plating and was acquired by Giovanni Leonida from the PB manufacturer after promising not to reveal the board designer's name.)

Some of the worst examples of poor clearance could easily have been avoided if the track had been laid at an angle to the X and Y axes, Figure 9.22. Angled conductors were used elsewhere on the board and it is therefore all the more surprising that those illustrated, and others, were missed or ignored. The small clearances between such large pads and the conductors running close to them will certainly cause bridging on soldering, creating much additional work in inspection, test and rework. Moreover, the inspector will have great difficulty in distinguishing shorts from the required connection, since the lands are frequently connected to the tracks in the position where the short is likely to occur, Figure 9.23.

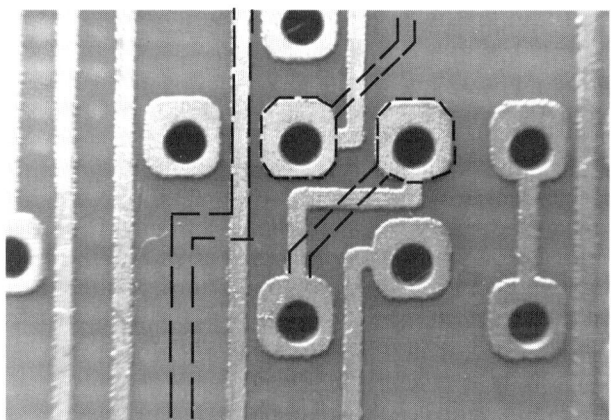

Fig. 9.22 Lines at an angle to the X and Y directions would have produced a marked improvement in this area of the layout.

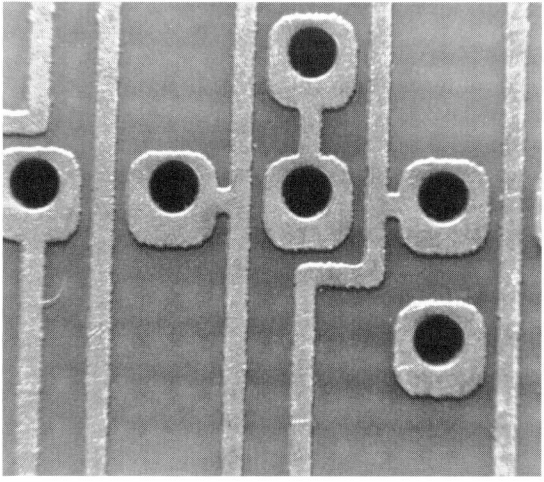

Fig. 9.23 Where lands have short, straight attachment to the conductor, as seen in the middle of the figure, it will be difficult to differentiate between them and shorts on solder joint inspection.

The connector zone, Figure 9.24, shows further unforced design errors. The 100 mil pitch connector requires a mechanical fixture, and the fixing hole takes up space in the most crowded area of the board. Tracks abandoned during design were not removed, one of them being visible in Figure 9.24. On the component side, Figure 9.25, large tracks are used for ground and power supply. Because of its position, the fastening hole imposes a restriction on both of these, although there is plenty of spare space in the area.

Fig. 9.24 A fixing hole near the connector causes problems in routing conductors, one of which has been abandoned.

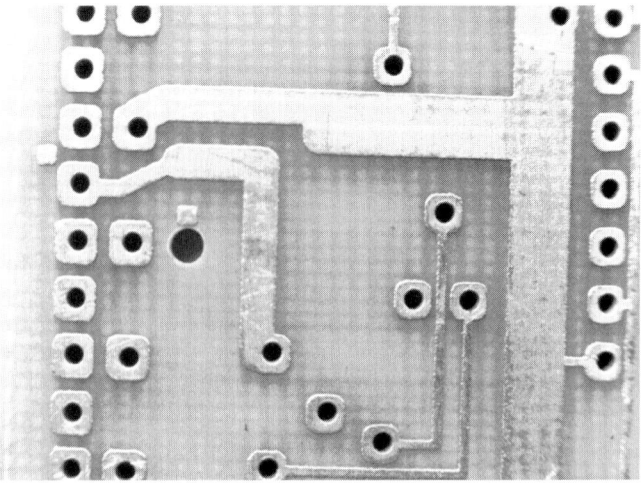

Fig. 9.25 The opposite face of the board seen in Fig. 9.24. The poor location of the mounting hole has forced slimming down of both ground and power supply tracks.

Again on the component side, holes are drilled directly in the large ground and supply conductors, Figure 9.26. The conductors will act as heatsinks during wave soldering and make these joints cooler than the others.

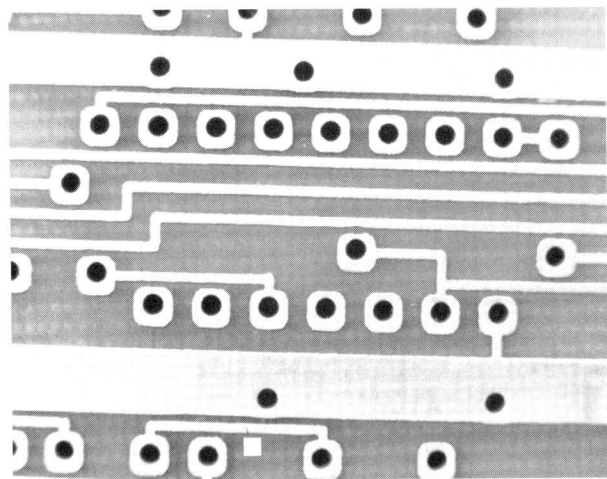

Fig. 9.26 These large tracks (on the component side) will act as a heat sink during soldering. The relevant joints will be cooler than all the others and the integrity of the joint may be reduced.

9.3.5 A Good Example

In the manufacture of PBs, drilling is one of the most expensive steps, as it is frequently the least automated operation. In addition, the number of holes needed for component insertion and the via holes to connect to internal layers cannot be significantly reduced, even by very careful design.

One large computer manufacturer developed the solution illustrated in Figure 9.27. All grid intersections on a 100 mil (2.54 mm) grid were drilled, regardless of whether or not a hole was needed in that position. The drilling pattern was, therefore, the same for all designs and could be produced at high

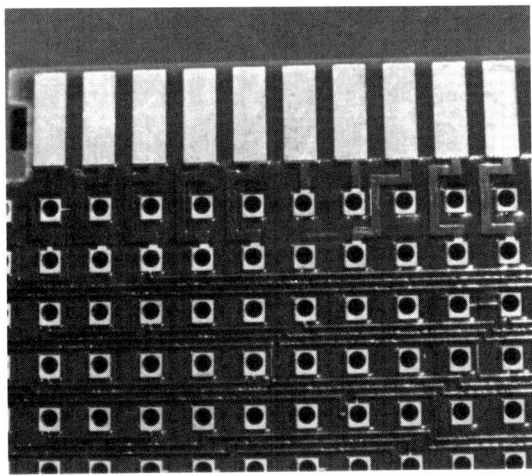

Fig. 9.27 A four-layer PB designed on a 100 mil (2.54 mm) grid. All grid intersections are drilled and plated. Up to three conductors run between adjacent lands.

speed on CNC machines with many drilling heads. All holes were plated through and, in this instance, square pads ensured adequate soldering area despite the relatively large hole.

By using a conductor width of 6 mils (0.15 mm) and a gap width of 8 mils (0.20 mm) three tracks could run between adjacent rows of lands, at a fixed position in relation to the lands but not along grid lines, Figure 9.28.

Although simple, this solution affords a high level of standardisation and allows assembly of any kind of PIH component, with high packing density. Similar 'matrix' types of board are still widely used today.

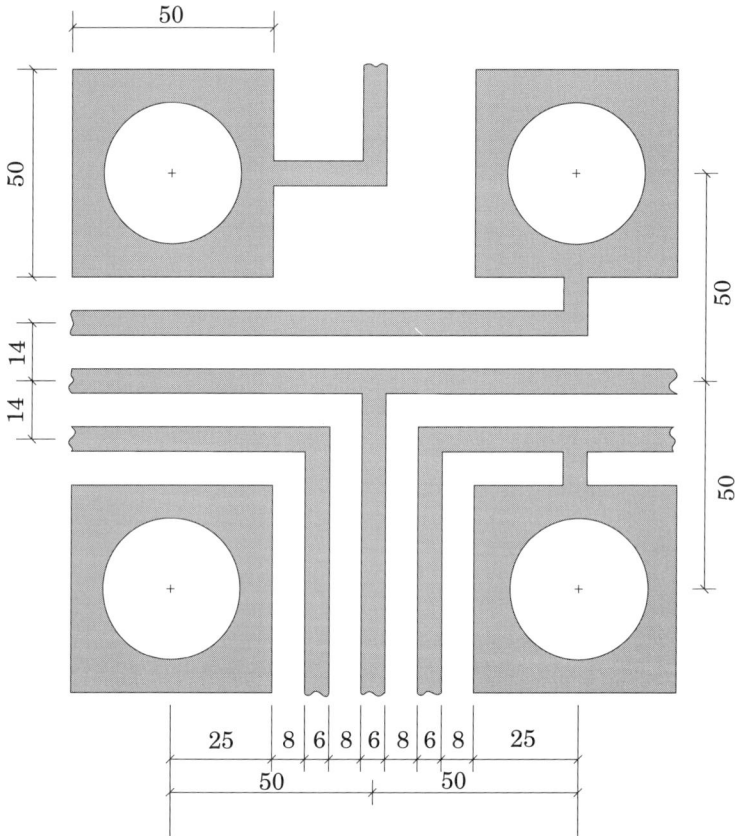

Fig. 9.28 The design rule of the PB shown in Fig. 9.27. (Dimensions are in mils.)

9.4 PIH (PIN-IN-HOLE) COMPONENTS

The selection of a grid cannot be divorced from the type of components to be assembled. Ideally, each component type has to have a specific hole and pad size. It is possible, once a grid has been selected, to introduce some holes and/or pads larger or smaller than the standard, obviously with some effect on the packing density. However, if the exceptions are great in number, and the basic grid has to be modified too much, it may be found that another grid will fit better overall with the prevailing type of component on that assembly.

The wise designer will start by classifying all the components to be mounted, listing insertion span, hole diameter and land diameter for each. From this listing he should be able to group the components into families and avoid an excessive number of varieties. This is highly desirable because:

(i) if a fixed centre distance inserter, see Chapter 6, is used, all components should be inserted on one pitch or on a strictly limited number of pitches;
(ii) if a variable centre distance inserter is used, the pitch can change only step by step and it is always preferable to limit the number of pitches;
(iii) in drilling the PB holes, each hole diameter involves a tool change and, therefore, additional time and cost;
(iv) if the PB holes are punched rather than drilled, additional diameters add to the complexity of the tooling;
(v) the larger the number of different pad sizes, the more difficult the layout design (manual or automatic).

For these reasons, it is always best to start with the grid which most closely fits the majority of the components and to change locally to another when necessary, according to the following criteria.

9.4.1 Axial Components

As outlined in Chapters 5 and 6, when preforming an axial component, either off-line manually or on-line by automatic insertion equipment, the leads must be firmly clamped and bent at the correct radius to avoid damage. This means that the minimum insertion span (P) is dictated by the maximum length of the component body (L) plus, on both sides, the clamping length (C) and the bending radius (R) with respect to the axis of the lead. The total is, therefore:

$$P = L + 2(C + R) = L + 2C + 2R$$

If we assume that a suitable bend radius is 1.5 times the lead diameter (D), the formula becomes:

$$P = L + 2C + 3D$$

which is easier to use, since it refers to elements which are specified on the component data sheets (L and D).

The remaining element (C) depends on the equipment used for preforming or automatic insertion. In most cases its value is:

nominal minimum: $C = 1.5$ mm (59 mils)

absolute minimum: $C_{min} = 0.76$ mm (30 mils)

By assuming one of these values, the absolute minimum being avoided as far as is possible, the minimum insertion span can easily be calculated. The next step is to check the result against the layout grid and to round it up to the next value compatible with it, Figure 9.29, (or to any other higher value of the insertion spans accepted by the insertion equipment), by increasing the value of the clamping distance (C).

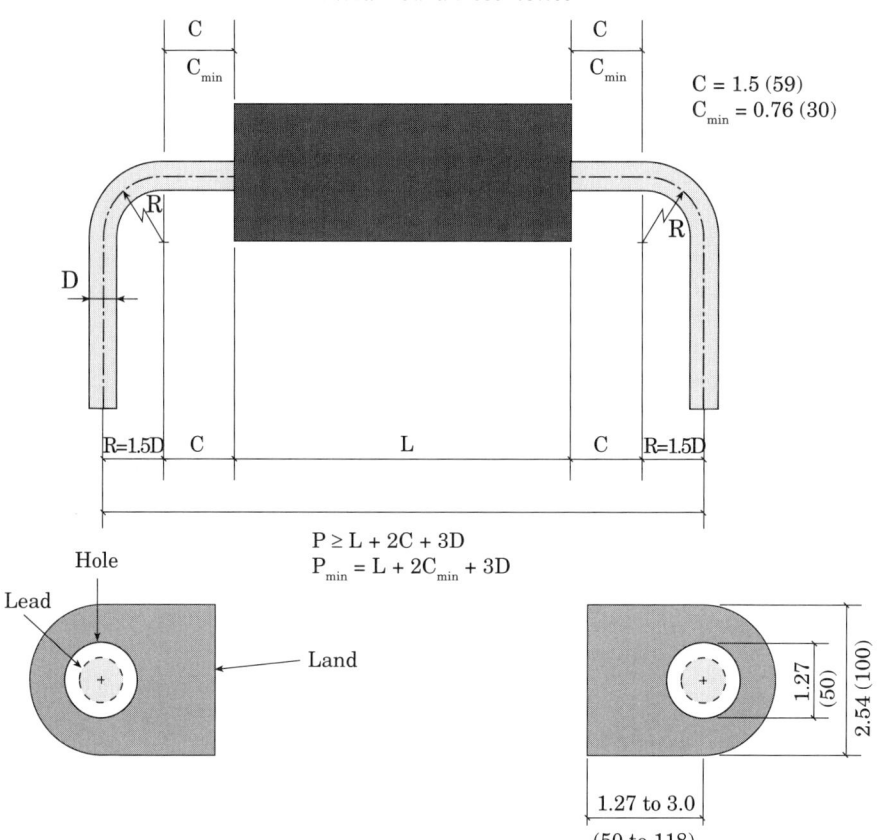

Examples (metal film resistors)				
Power	0.5W	0.5W	0.25W	0.25W
L_{max}	6.5 (256)	7.0 (275.6)	3.5 (137.8)	3.8 (149.6)
D	0.6 (23.6)	0.6 (23.6)	0.5 (19.7)	0.5 (19.7)
$P \geq$	11.3 (444.9)	11.8 (464.6)	8.0 (315)	8.3 (326.8)
$P_{min} \geq$	9.82 (386.6)	10.32 (406.3)	6.52 (256.7)	6.82 (268.5)
Grid = 100 mils $P \geq$	(500)		(400)	(400)
$P_{min} \geq$	(400)	(500)	(300)	(300)
Grid = 50 mils $P \geq$	(450)	(500)	(350)	(350)
$P_{min} \geq$	(400)	(450)	(300)	(300)
Grid = 25 mils $P \geq$	(450)	(475)	(325)	(350)
$P_{min} \geq$	(400)	(425)	(275)	(275)

Fig. 9.29 Derivation of the insertion pattern of an axial component. The lands are assumed to be on a single-sided board with inward clinching of the leads.

An alternative method of calculation, suggested by the IPC,[8] refers to the internal bending radius of the lead and gives a very similar result.

The optimum size of the hole and pad depends upon the type of board and is always a compromise between four conflicting elements:

(i) the need to standardise hole and pad diameters;
(ii) the requirements for component insertion (manual or automatic);
(iii) the requirements for wave soldering, including matching lead and hole sizes;
(iv) the density of components and connections on the board.

The compromise must, of course, be cost-effective in the broadest sense, although this can be difficult to achieve as the first three elements may pose different demands, Table 9.2. The requirements for wave soldering are usually the least critical, but they cannot be overlooked.

Table 9.2

Hole-to-lead Ratio*

	Single-sided PBs			Plated-through-hole PBs		
	Min.	Pref.	Max.	Min.	Pref.	Max.
Manual insertion	0.1	0.4	0.5	0.2	0.4	0.6
(straight lead)	(4)	(16)	(20)	(8)	(16)	(24)
Manual insertion	0.2	0.5	0.8	0.3	0.6	0.8
(preformed)	(8)	(20)	(32)	(12)	(24)	(32)
Automatic insertion	0.3	0.4	0.7	0.3	0.5	0.7
(clinched)	(12)	(16)	(28)	(12)	(20)	(28)
Wave soldering **	0.2	0.35	0.8	0.1	0.15	0.5
	(8)	(14)	(32)	(4)	(6)	(20)

* Difference between the hole diameter and the lead diameter, as a compromise between board density and insertion/handling cost. (Dimensions are in mm and, in brackets, in mils.)
** Acceptable range for good results in wave soldering.

Pad diameter depends on both the grid adopted and the hole diameter. On single-sided boards, a fairly large pad is desirable to ensure sufficient adhesion of the copper to the base laminate. Common practice dictates that the minimum copper annulus width left after allowing for all manufacturing tolerances, such as off-centred holes, should be about 0.25 mm (10 mils). (Some well-known specifications allow a minimum width of metal at any point round the hole of 0.15 mm (6 mils), but we are concerned here with common sense and not with the minimum limits in specifications.) As an example, if we assume a hole diameter of 1.0 mm and a total pad-to-land misregistration of 0.10 mm, the minimum pad diameter to be used should be 1.6 mm. However, if conductor density is not too critical, higher values are to be preferred as greater adhesion to the base laminate will result, with increased resistance to shock and vibration, and board manufacture will be less critical.

IEC Publication 326-3: 1991, second edition, see Chapter 8, recommends that the minimum land dimensions should be derived from:

$D - d = 1.0$ mm minimum for plain holes

$D - d = 0.5$ mm minimum for plated-through holes

where D = diameter of land and d = diameter of hole. At the same time it is recommended that the ratio $D:d$ for these lands should be:

2.5 to 3.0 for plain holes in phenolic paper boards;

2.5 to 3.0 for plain holes in glass epoxide boards;

1.5 to 2.0 for plated-through-holes.

This means that, on single-sided boards, a 1 mm diameter hole should have a land of diameter 2.0 mm minimum, which is acceptable if the base material is, say, FR-4. A diameter of 2.5 mm (or 2.54 mm = 100 mils) would be preferable.

Since most axial components have lead diameters in the range of 0.5 to 0.8 mm, a 1.1 mm diameter hole with a 2.54 mm pad can accept a wide range of components provided they are inserted automatically or are preformed. Larger components may require larger holes and lands.

On double-sided PTH boards, adhesion is greatly enhanced. This is reflected in the minimum width of metal permissible at any point around the hole, which is commonly 0.08 mm in specifications and may be assumed to be 0.1 mm in practice. (Obviously, this is not applicable to landless PTHs!) Taking the figure as 0.1 mm, a 1 mm hole in a PTH board will need a pad with a minimum diameter of about 1.3 mm, often rounded down to 1.2 mm = 50 mils. Here again, larger pads are desirable and the 1.5 mm derived from the IEC recommendations is to be preferred whenever possible.

If TO-116 packages predominate on the board, it may be possible to use only those discrete components having a lead diameter of about 0.5 mm (20 mils). These can be inserted, preferably by an inserter fitted with a positioning aid, in the same 30-32 mil (0.762-0.813 mm) holes used for the ICs. The resulting reduction in drilling cost and the increase in conductor density may well justify the closer manufacturing tolerances and the use of a more expensive insertion machine.

One point to be remembered in the design of the lands is that, if the leads are flush clinched, the lands must be enlarged along the direction of clinching as much as is necessary to keep the entire lead within the land. If clinching is at 45° or 60°, the enlargement can be smaller, provided the orthographic projection of the lead is kept within the land.

9.4.2 Other Discrete Components

Small radial components, such as many types of capacitor, TO-92 transistors and so on, are subject to the same considerations as the axial components since their lead diameters are quite small. If they are to be automatically inserted, the pitch is fixed by the component manufacturer.

With taped radial components, two further points must be considered. First, tolerance on the taping pitch is fairly large, e.g., 0.3 mm on a 10 mm pitch. Second, it may be necessary to use components taped on a pitch which differs slightly from the distance between centres on the PB. This happens, for example, when one dimension is a round figure in metric units, say 10 mm (394 mils), and the other is a round figure in imperial units, 400 mils (10.16 mm.)

Both problems can be resolved, at the expense of packing density, by using a fairly large diameter hole. For instance, a diameter of 1.0-1.1 mm (39.4-43.3 mils) may be used for leads up to 0.6 mm (23.6 mils), 1.3-1.5 mm (51.2-59.1 mils) for leads up to 0.8 mm (31.5 mils) and so on. Even with such large holes, components can be kept in position either by using a stand-off or by preforming the leads. However, lead clinching will be mandatory, both to avoid losing or misplacing components prior to soldering and to ensure reasonable solder joints.

9.4.3 Integrated Circuits

For years the TO-116 package (DIP) has been one of the most popular, despite one notable problem. The tolerances on the dimensions are so large that, in theory, it is impossible to design a PB on which it can be properly assembled!

To illustrate the point, the end portion of the leads can vary in thickness from 0.20 to 0.38 mm (8-15 mils) and in width from 0.38 to 0.58 mm (15-23 mils). This means that the diagonal, which dictates the hole diameter, can vary from 0.43 to 0.70 mm (17-27.46 mils), Figure 9.30. The smallest pins will fit easily into a 0.762 mm (30 mil) diameter hole while the largest just fit a 0.889 mm (35 mil) diameter hole. However, the body is meant to be spaced from the PB by the enlarged portion of each lead, which can vary from only 0.762 mm (30 mils) to 1.78 mm (70 mils). If the width is 0.762 mm and the thickness 0.20 mm, the diagonal of the portion which is meant *not* to enter the hole is only 0.789 mm, considerably less than the 0.889 mm hole needed to cater for the top size on the lower portion of the lead!

So far the tolerance on position and diameter of the hole has not been considered, nor has allowance been made for the fact that the TO-116 leads are specified to be within ±0.254 mm (±10 mils) of their theoretical position (which means adding 0.508 mm (20 mils) to the hole diameter). Taking all these factors into account and making some allowance for automatic insertion, a hole diameter of 1.65-2.03 mm (65-80 mils) would be needed. This is much too great for:

(a) correct seating of the package;
(b) good clinching to the PB;
(c) a correct hole-to-lead ratio for wave soldering;
(d) acceptable component density on the board.

Fortunately, most component manufacturers have adopted an informal 'standard within a standard' and provide a much tighter tolerance around a mean value (close to that given in the JEDEC recommendation). In consequence, the TO-116 package may be mounted easily, although it is

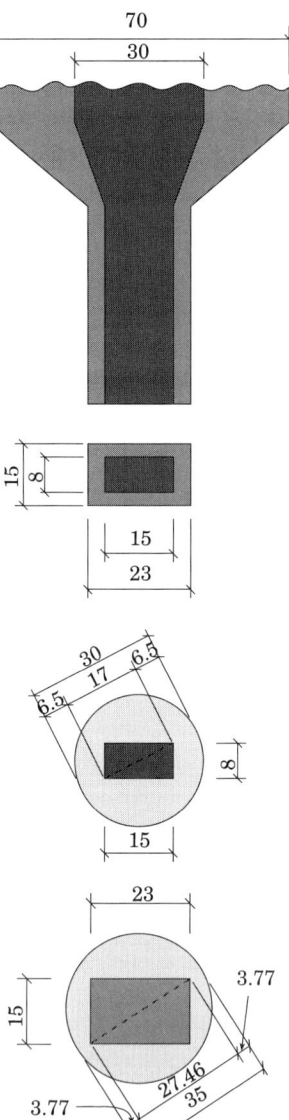

Fig. 9.30 The tolerances of the pins on TO-116 packages are so large that the optimum hole diameter may vary from 30 to 35 mils (0.768-0.889 mm), without taking other tolerances into account. (Dimensions are in mils.)

advisable to check any new supplier's devices carefully in order to avoid unwelcome surprises. (The large tolerances were connected to the original usage of cheap ceramic packages. The production technology of the alumina or plastic packages now used inherently gives better tolerances.)

In practice, therefore, on single-sided PBs, DIPs can be assembled using a hole diameter of about 1 mm (about 40 mils), with a spread of 0.762 to

1.27 mm (30 to 50 mils). Land diameter should be kept as large as possible, with elongated pads used to give extra adhesion to the base laminate — this being a potential weak point when the DIP fails and has to be replaced.

On double-sided PTH boards, hole diameters (plated) commonly adopted in practice are 0.762 mm (30 mils), 0.8 mm (31.5 mils), 0.81 mm (32 mils) and 0.85 mm (33.5 mils). Diameters as large as 0.9 mm (35.4 mils), 1.0 mm (39.4 mils) or even 1.02 mm (40 mils) are not unknown, especially when the devices are supplied by several vendors and are automatically inserted.

Pad size depends on the design guidelines adopted and whether two or three tracks are being run between adjacent lands, Figures 9.31 and 9.32, combined with the PB manufacturing tolerances. Pad diameters usually chosen are 1.27 mm (50 mils), 1.4 mm (55 mils) and 1.52 mm (60 mils), although diameters as small as 1.14 mm (45 mils) and as large as 2.0 mm (78.7 mils) are sometimes seen, as are square pads.

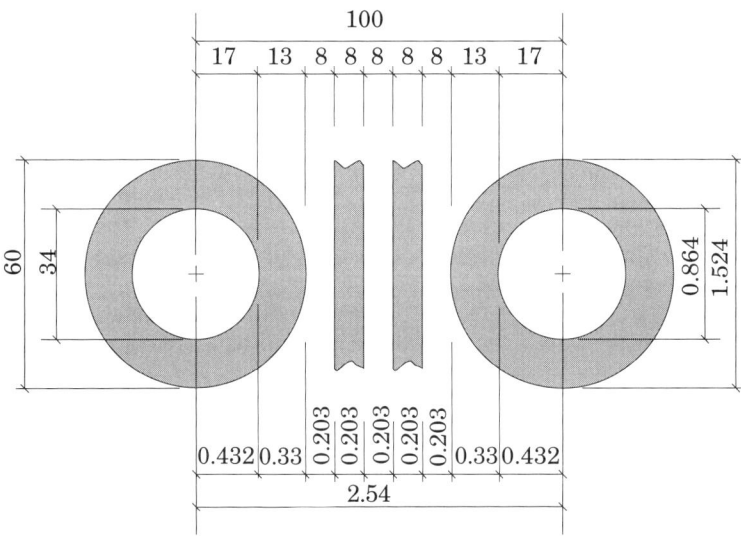

Fig. 9.31 A typical solution for running two tracks between adjacent leads of a DIP. Dimensions are in mils (top and left) and in mm (bottom and right).

On multilayer boards a fairly large pad-to-hole ratio (e.g., 1.52 mm to 0.762 or 0.813 mm, that is, 60 mils to 30 or 32 mils) is often preferred, the consequent reduction in packing density being largely offset by a lower cost/unit area for the PB.

9.5 SURFACE MOUNTED COMPONENTS

Designing a PB for surface mounted components is much more demanding than designing a traditional board and many failures experienced in the early days of SMT were directly related to poor design. In Chapter 4 it is shown that SMT requires a much greater degree of co-operation between the board designer and other departments in the company, since design, assembly, cleaning, testing, servicing and so on are closely related.

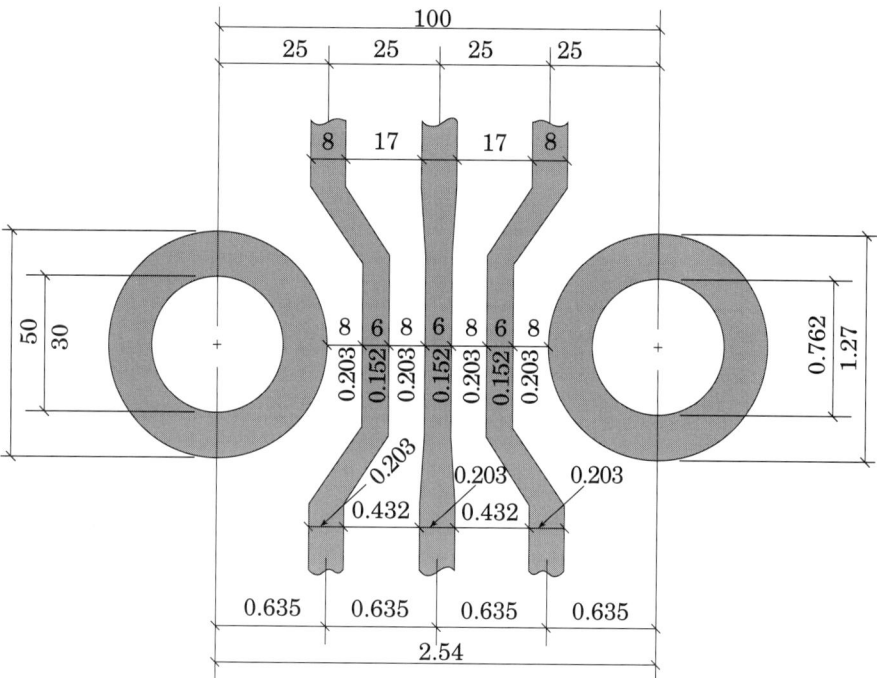

Fig. 9.32 One method of running three tracks between adjacent leads of a DIP. The PB is designed on a 25 mil grid and, locally, on a 1 mil grid. Dimensions are in mils (top and left) and in mm (bottom and right).

In addition to the general requirements discussed in Chapter 8, the design of SMT boards involves definition of the size and relative positions of the copper lands on to which the component must be soldered (or otherwise electrically bonded). In turn, these factors depend on the techniques and equipments used in assembly, soldering and so on.

Although PIH components offer an enormous variety of shapes and sizes, the presence of leads is a unifying factor. It is possible to design a mounting pattern, i.e., lead span, hole diameter and pad diameter, which will accept a very large number of component types and give good results throughout the entire manufacturing process.

SMCs offer a similar variety of shapes and sizes and a rather poor degree of standardisation of body shape, size, material, resistance to soldering, cleaning, etc. The situation is even worse for the shape, size and material of the leads or terminations. Over the years there have been many complaints about the lack of standardisation[9,10] although, to be fair, the growth of SM component types has been so rapid that it has been impossible for standards to catch up — far less keep up!

Because of the poor standardisation, it is difficult to achieve the highest packing density which should be attainable with SMT. Any design which has to be based on a 'worst case' situation demands large distances between components. One recommended approach[11] involves statistical analysis of all requirements to derive an optimum solution, which — because of the method used — may not be unique.

This approach, proven to work in practice, has some important implications, the foremost being:

(i) the board assembler must accept that a certain number of things just will not work, even when all tolerances are respected;
(ii) the PB designer must not rely totally on published data, which refer to other boards and manufacturing processes, and must support his decisions by carrying out a statistical analysis on the specific process for each board;
(iii) a good statistical process control system must be established to provide a constant flow of feedback data to improve the design.

Published recommendations, such as those which follow and those given in the guides referred to at the end of Chapter 8, must be used only as guidelines from which to develop one's own procedure for the design of surface mounted PBAs.

9.5.1 Is a Grid Essential?

The grid was developed as a guide for the placement of mounting holes. By definition, SMCs do not use mounting holes and the need for a grid is not immediately obvious. Pads for SMCs may be placed on a PB without using a grid or they may be placed so that the component centre lies on the intersection of two grid lines. In the latter instance, the packing density may be somewhat reduced, but quite significant advantages are obtained, for example:

(i) the board design, whether carried out manually or by CAD, is simpler;
(ii) manual programming of placement and insertion machines is easier;
(iii) test fixtures will be simpler to design and build;
(iv) visual inspection of the board is more straightforward and more effective;
(v) the assembly looks, and usually is, more orderly.

However, if the variety of components is very great, it may be better to abandon the grid or to use a really basic grid of, say, 1 mil (or 0.025 or, better, 0.05 mm). This is equivalent to a gridless design, as 1 mil is, commonly, the resolution of the placement equipment.

9.5.2 Minimum Distance

SMD placers do not make very rigorous demands as regards the minimum distance to be allowed between closely mounted components, since the device is usually held on its top by a vacuum pipette. A stricter requirement is imposed by the need to avoid solder bridges forming between the terminal of one device and the terminal, or mounting pad, of another.

Experience recommends that, even when the component is misplaced in the X or Y direction or rotated at a certain angle, a minimum gap of 1 mm (about 40 mils) for wave soldering and of about 0.5 mm (about 20 mils) for reflow soldering should remain between any two solderable areas that should not be connected, Figure 9.33. Most placement machines have an accuracy of ± 0.15 mm or better and an angular accuracy better than 3°, so no serious problem arises either with two termination components or, with a little extra attention, with devices in SOT-23, SOT-143 and similar packages.

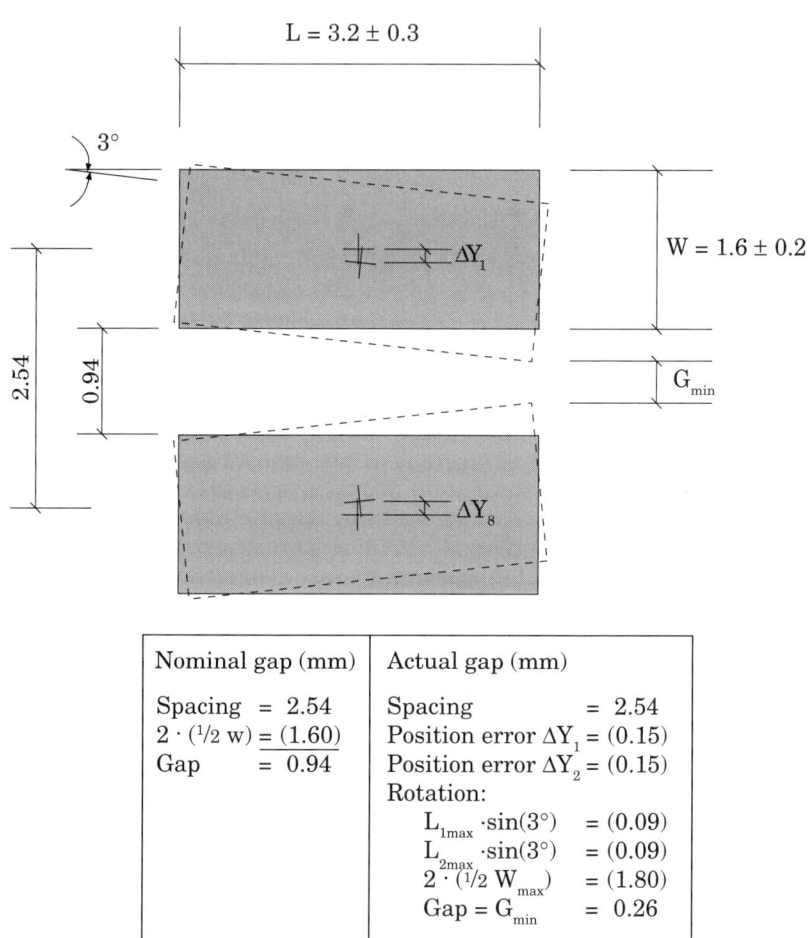

Fig. 9.33 The nominal and actual gap between two 1206 case components calculated by assuming a placing tolerance of ± 0.15 mm and 3° maximum rotation. The tolerance in body width has a significant impact. The effect of rotation is calculated assuming the maximum body length.

Most of the problems arise on large packages, such as the SO, VSO, SOJ, PLCC, etc., which have many leads spaced on a 1.27 mm (50 mil) or 0.635 mm (25 mil) pitch. Here the tolerances on lead positions are fairly large, while the small distance between leads (and perhaps the need to run a track between two lands) imposes narrow lands. In addition, even a small rotation of the packages can have a marked effect on the offset of some terminals to the lands.

9.5.3 Dimensioning of Lands

Providing the correct size for soldering lands is the most difficult part of the design of a surface mount board. Not only are there many different package and terminal shapes and sizes to be considered, but a high density of components and connections is essential if maximum benefit is to be obtained from SMT. Small lands reduce the area occupied by the components and make track routing between adjacent lands easier, but small lands mean that component placement must be more precise.

Even in the late eighties, there was very little advice available on design of pads or lands, and what *was* obtainable often contained errors. As a result, SM boards with poor solder joints due to faulty design of the soldering lands are still to be seen. Results largely depend on the soldering method used to establish the joint between the terminal and the land, wave soldering and reflow soldering having different requirements.[12] For instance, if one assumes, according to a widely accepted rule, that a reasonably low rate of rejection in solder joint inspection is achieved when the *minimum* overlap between the solderable terminal and the land is about 0.1 mm (about 4 mils) for wave soldering and 0.2 mm (about 8 mils) for reflow soldering, the size of the lands can be quite different. Regrettably, the designer cannot adopt a universal design which holds good for all processes.

9.5.3.1 WAVE SOLDERING

As the wave of molten solder provides as much alloy as is needed, the footprint is less critical and need only give a land of reasonable size and shape to establish a good solder fillet between the land and the component termination. Although a poor design can give a satisfactory result, Figure 9.34, it is always better to follow some simple guidelines:

Fig. 9.34 This small mixprint board with Mini-MELF, Cerachip, SOD-80 and SOT-143 components shows very poor design. It has ill-defined lands, large and irregular conductors, multiple lands, solder mask applied only to some very large copper areas. Nevertheless, because the combination of small board, cylindrical components and wave soldering is the least demanding for PCB design, no soldering defects appeared after wave soldering.

(i) avoid multiple lands, i.e., lands to which more than one termination has to be soldered, Figure 9.35;
(ii) avoid using large copper areas as solder lands, even when the soldering area is defined by a solder resist, Figure 9.35;
(iii) avoid the 'shadow effect', Figure 9.36, by using a properly shaped solder wave and, if necessary, extending the lands to which the device is soldered;
(iv) define a soldering direction for the board, place the components accordingly, see Chapter 8, and specify the direction on the drawings — a suitable mark on each PB is recommended;
(v) define suitable dimensions for the pads of each component and adhere to them throughout;
(vi) avoid the use of 'degassing holes' unless they have proved useful in the specific design situation.

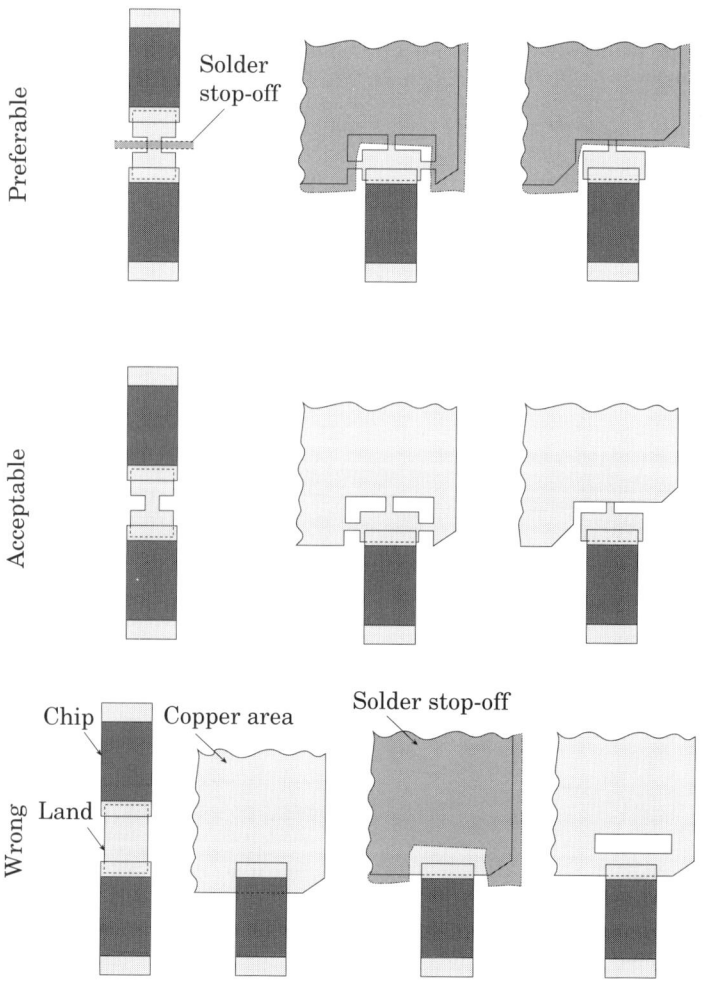

Fig. 9.35 PB layout for wave soldering chip components.

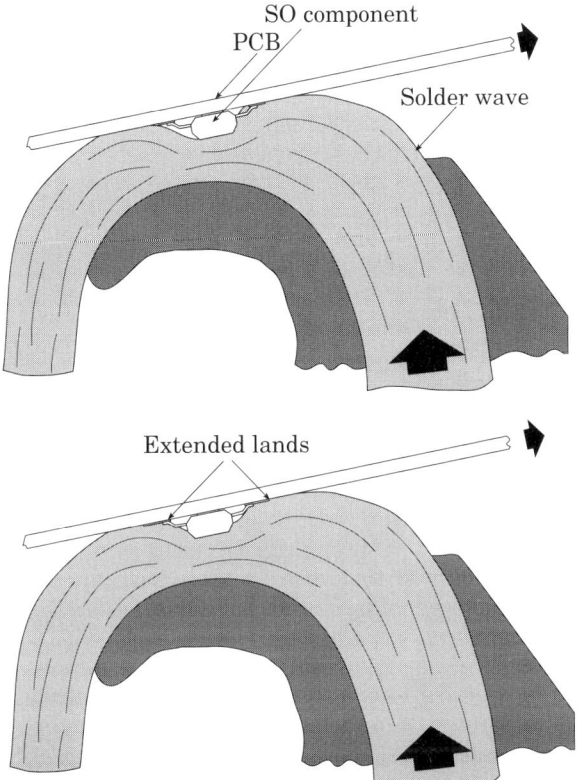

Fig. 9.36 The body of the component may prevent the solder wave from reaching the downstream leads (top). This is the 'shadow effect'. Extending the lands may be a cure since, once the solder contacts part of the land, the joint will be formed.

Degassing holes originated in the early days of wave soldering SM boards. Several soldering defects then prevalent were ascribed to entrapment of volatile flux residues between the PB and the wave. This could be avoided by letting the gases escape through holes. Consequently, all lands for SMCs had a 0.5-0.76 mm (20-30 mils) diameter hole drilled in them. (In emergency, the holes could be used, either in the factory or in a field situation, to replace a surface mounted component with a traditional PIH component of equivalent characteristics.) In some companies this habit became so entrenched that holes in SMC lands were also provided on mixprint boards, where their degassing function was absolutely unnecessary.

Now that soldering parameters are better understood and controlled, degassing holes should be avoided. The cost of drilling them must be added to the board cost and their benefit is highly questionable.

The use of a solder mask is not mandatory but in most cases it is worth adding to the board. In addition to separating adjacent lands, stopping solder build-up on large conductors and minimising bridging, it can prevent dirt and moisture lying on the surfaces and causing current leakage and corrosion of conductors.

9.5.3.2 REFLOW SOLDERING

In reflow soldering, enough solder alloy is placed on the land and on the component termination for a joint to be formed when both are heated, in contact with one another, to the melting point of the alloy. The solder is normally added to the land in the form of a solder paste or cream; alternatively, the land may be electroplated with a relatively thick coating of tin-lead which may be fused to form the solder alloy.

In both cases, the land acts as a reservoir for all or most of the solder needed to make the joint. The land must be large enough to hold the appropriate amount of solder and best results will be obtained when the pad size is matched to the type and size of the component.

A land size smaller than suggested, which can be desirable for increasing board density and achieving a degree of standardisation, may prove acceptable, but in this case the whole process becomes more critical. On the other hand, an excessive solder land area can be decidedly detrimental, since the surface tension of the molten solder will spread the solder over the whole solderable area and leave too small a fillet at the attachment of the component termination.

In general, since the amount of solder applied is limited, any situation that draws solder from the joint must be avoided. Each land should be independent from other lands, and connecting tracks, particularly to large copper areas, must be as thin as possible, Figure 9.35. Defining the land by means of a solder resist only may be inadequate, since molten solder may flow under the resist.

Plated-through holes can create serious problems by drawing considerable amounts of solder away from the lands. They should, therefore, be placed away from the land and connected to it by a narrow conductor, preferably coated with a solder stop-off or at least cut by a line of ink, Figure 9.35. How far from the land they should be placed depends on the soldering materials and processes. A general rule calls for a minimum of five times the conductor width (which is usually 0.1, 0.15 or 0.2 mm — 4, 6 or 8 mils). It must, however, be remembered that it is not advisable to use large conductors and try to limit the solder flow by ink alone.

Note: If the component map or 'ident' is screened on the board, it is possible to screen a line of ink between the lands and the holes without increasing the cost of the PB.

When a solder cream/paste is used, the size and shape of the lands must be related to the application method. When the paste is screened, it must remain inside the periphery of the land, and allowances must be made for the tolerances of the board and screening process, and for bleeding of the paste.

Solder resist is useful on boards with narrow conductors and high component density, since it protects the conductors, prevents shorts between conductors due to slivers and/or solder balls and allows better definition of the solderable areas. The type and the application method of solder mask can exert considerable influence on the board assembly since it must not contaminate the solderable areas and must permit the component terminations to sit down snugly on the lands, Figure 9.37.

9.5.4 Footprints

The SMC footprints — whether for wave soldering or reflow soldering — which are specified in the literature can be taken only as a guide. The suggestions which follow must be adapted to each specific process.

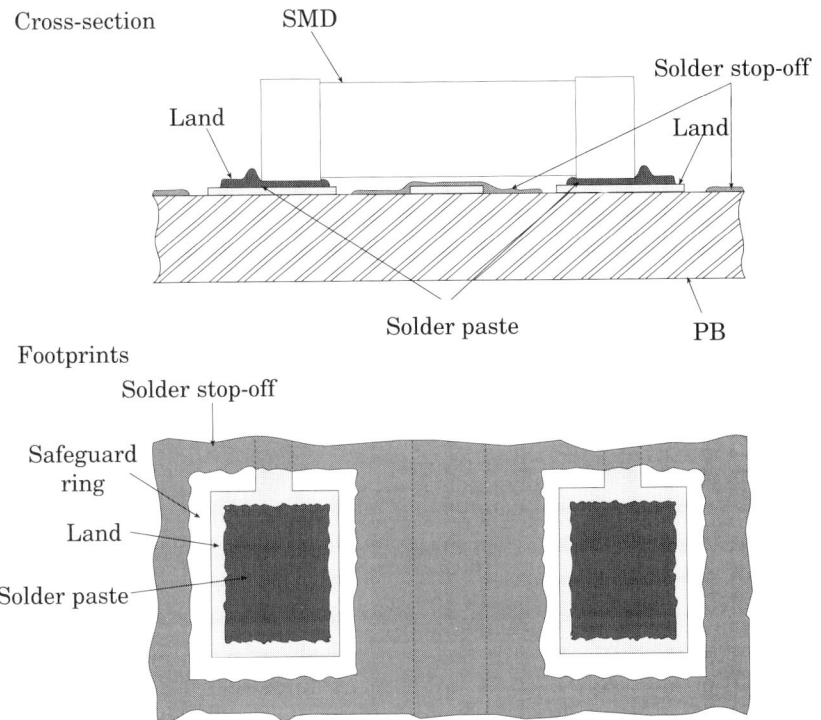

Fig. 9.37 For reflow soldering, the design of the footprint must include the pattern for the solder resist and solder paste. The height (or thickness) of the solder mask must allow the component to sit firmly on the lands.

9.5.4.1 CHIP SMD FOOTPRINTS

Footprints for chip components (resistors and capacitors) often include a track, genuine or dummy, running under the component body, Figure 9.38. This is primarily to improve the effectiveness of the adhesive. In reflow soldering this track can be omitted (if it is a dummy!), since the solder paste should be able to hold a chip component in place until the joint is made.

9.5.4.2 MELF COMPONENT FOOTPRINTS

The footprint for MELF, Mini-MELF and similar components is less critical than for chip SMDs because the round shape of the termination allows a wide variation in solder quantity while still giving good and easily inspectable solder fillets. On the other hand, the circular terminals require very accurate placement to ensure that they contact the lands, unless they are fairly large, Figure 9.39.

No track under the body is required to assist the adhesive as, thanks to the body shape, even a very large dot of adhesive should remain clear of the lands. On the other hand the adhesive contacts a smaller area and it is strongly recommended that, for maximum bond strength, it be applied directly to the laminate and not, for example, to solder resist.

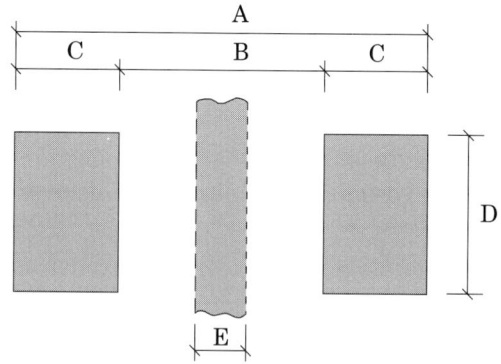

Case style	Case size	Wave soldering (if applicable)				
		A	B	C	D	E
0805	2.0 x 1.25	3.2-4.0	0.8-1.2	1.2-1.4	1.2-1.5	0.4
1206	3.2 x 1.6	4.3-5.0	1.5-2.0	1.4-1.6	1.4-1.9	0.5
1210	3.2 x 2.5	4.3-5.8	1.5-2.0	1.4-1.9	1.4-2.8	0.5
1808	4.5 x 2.0	6.5-7.8	2.5-2.8	2.0-2.5	1.4-2.3	0.5-0.8
2220	5.7 x 5.0	7.8-10.0	3.6-4.0	2.1-3.0	2.0-5.5	0.5-1.0

Case style	Case size	Reflow soldering			
		A	B	C	D
0805	2.0 x 1.25	2.5-3.8	0.8-1.0	0.85-1.4	1.25-1.4
1206	3.2 x 1.6	3.8-5.0	1.8-2.0	1.0-1.6	1.6-1.8
1210	3.2 x 2.5	3.8-5.0	1.8-2.0	1.0-1.6	2.5-2.7
1808	4.5 x 2.0	5.4-6.7	2.8-3.5	1.3-1.6	2.0-2.2
2220	5.7 x 5.0	6.8-7.9	4.0-4.5	1.4-1.7	4.5-5.4

Fig. 9.38 Suggested footprints for chip resistors and chip capacitors. When a range is given, the lower figure should not be reduced but, where component density permits, the upper figure may be increased. (Dimensions are in mm.)

9.5.4.3 OTHER DISCRETE SMC FOOTPRINTS

Components with small terminals, such as SOT-23 and similar outlines, require accurate dimensioning of the mounting lands, which, for wave soldering, must also be orientated in the proper direction, Figure 8.28.

Here again, size and shape may change according to the soldering method, Figures 9.40, 9.41 and 9.42. The dimensions may be enlarged when packing density permits. Size and shape must be defined before designing the PB, since it can be very difficult to reconcile values at a later stage. Sizing is less critical for wave soldering than for reflow soldering.

Layout of Printed Circuit Boards

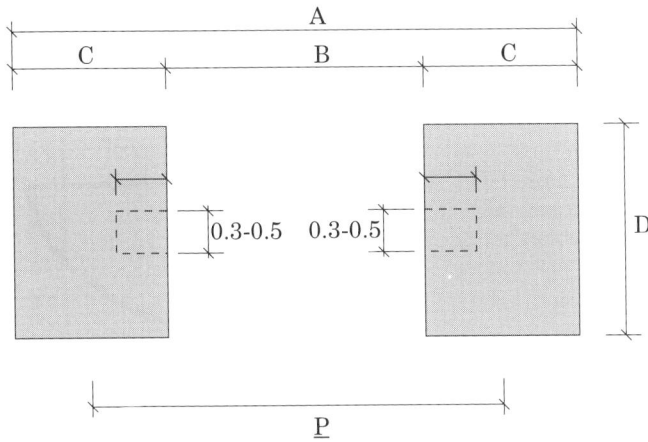

Size (L × Ø)	Wave soldering					Reflow soldering				
	A	B	C	D	P	A	B	C	D	P
MELF 5.9 × 2.2	6.6	3.8	1.4	3.0	5.08 typical	7.0	3.8	1.6	2.0-3.0	5.08 typical
Mini-MELF 3.5 × 1.45	5.0	2.5	1.25	2.0	3.81 typical	5.2	2.4	1.4	1.4-1.6	3.81 typical
SOD-80 3.4 × 1.6	4.8-5.5	2.0-2.5	1.3-1.6	2.0-2.2	–	5.2	2.4	1.4	1.4-1.6	–

Fig. 9.39 Footprints for common cylindrical components. Larger land sizes are acceptable both for wave soldering and, when the solder paste is thick enough, for reflow soldering. The notch shown (dotted outline) on the inside of the lands is optional. It is very useful for reducing skewing during reflow. (Dimensions are in mm.)

9.5.4.4 SM INTEGRATED CIRCUIT FOOTPRINTS

Integrated circuits, and other components in similar packages, require a specific footprint, Figures 9.43 and 9.44, and precise placement.

In reflow soldering, when it is necessary to reduce the width of lands in order to run tracks between them, the reduction in land area may be compensated for by increasing the length of the land. However, it must be remembered that, during the soldering process, the solder paste loses its tackiness and when the solder particles are completely melted the component may move because of the surface tension of the molten solder. This phenomenon (*floating* or *swimming*) is beneficial when the component self-centres on the footprint, but normal production equipment cannot control this. Indeed, the equipment can cause many problems at this stage if it transmits vibrations or shocks to the board.

In wave soldering SO and VSO components, the best results are usually obtained when the component axis is parallel to the soldering direction. In some cases a number of solder bridges may occur between those leads which

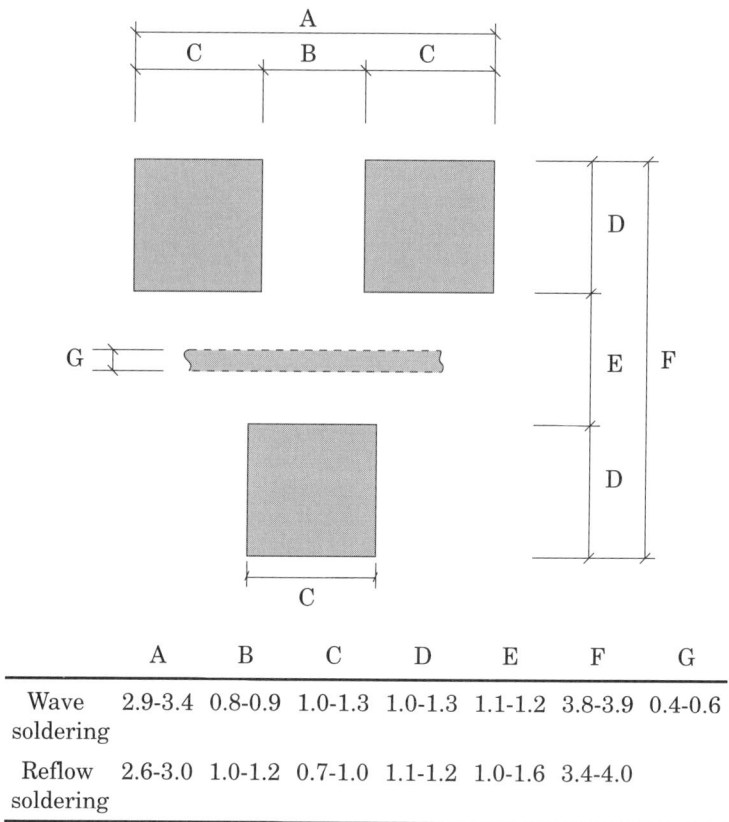

Fig. 9.40 Suggested footprints for SOT-23 packages. (Dimensions are in mm.)

are downstream with respect to the solder flow. One good remedy for this is the provision of a *solder thief*. This is a solderable area, Figure 9.45, which locally modifies the surface tension of the solder wave.[13, 14]

9.6 MULTIPLE BOARDS

The base laminate accounts for a significant part of the cost of PBs. Even 'standard' FR-2 or FR-4 material may contribute 50% of the cost of a simple single-sided board. Materials are manufactured in sheets of a specific size and designers should attempt to minimise laminate scrap. Obviously this can only be achieved by close liaison with the PB and assembly production engineers, as both will require 'scrap' allowances in their production and testing processes. In many cases, the production panel size will be dictated by, for example, the optimum size for a given insertion or placement machine.

A large PB will to some degree reduce the cost associated with its manufacture, as well as costs of handling, assembly, soldering and testing. Once an optimum size, or selection of sizes, has been decided, the PB designer must do his best to comply with the decision. If the boards required are small,

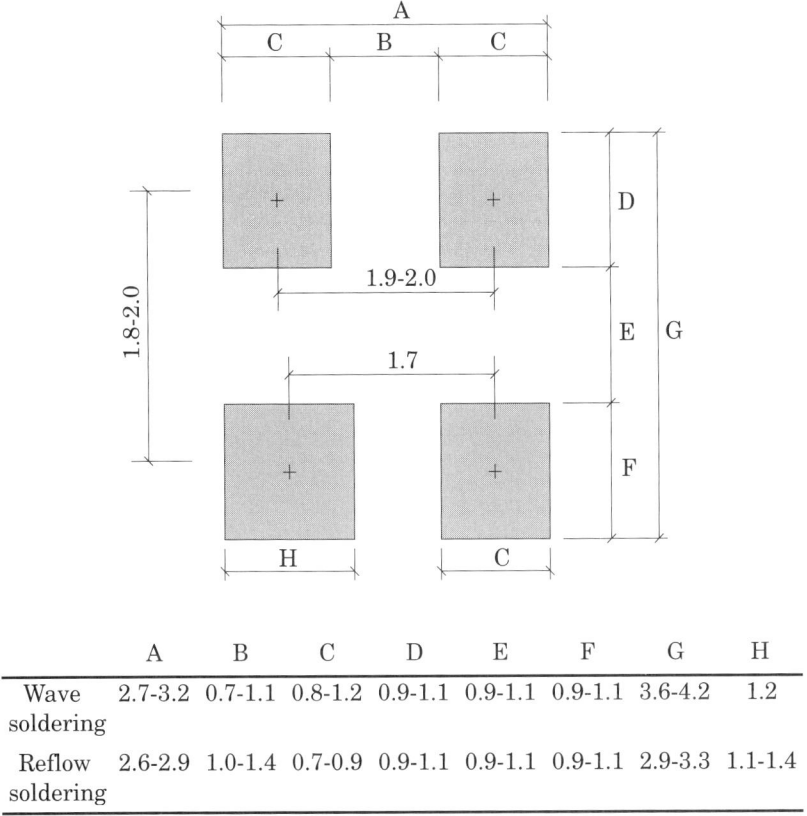

Fig. 9.41 Suggested footprints for the SOT-143 package. (Dimensions are in mm.)

	A	B	C	D	E	F	G	H
Wave soldering	2.7-3.2	0.7-1.1	0.8-1.2	0.9-1.1	0.9-1.1	0.9-1.1	3.6-4.2	1.2
Reflow soldering	2.6-2.9	1.0-1.4	0.7-0.9	0.9-1.1	0.9-1.1	0.9-1.1	2.9-3.3	1.1-1.4

they should be designed so that a given number can fit into one of the standard sizes. The board pattern can be duplicated as many times as necessary on a step-and-repeat machine (or photoplotted directly) and the resulting multiple panel will then be treated as a single board throughout the whole process, through to testing or just before it.

On completion of the process, the multiple board can be separated into individual boards by routing, sawing, blanking, or scoring and breaking out, see Volume 2, Chapter 2. More frequently, with automated insertion and placing equipments for assembly, the multiple board will be part profiled or scored so that assembly may be carried out on the multiple boards. After final functional testing, the individual assemblies are carefully removed from the panel.

9.7 CONCLUSION

In this and the preceding chapter, much of the emphasis has been on PIH or PTH technology. It has been noted in Chapter 6 that the technology of printed circuit boards is advancing steadily in the nineties and the demand

Reflow soldering

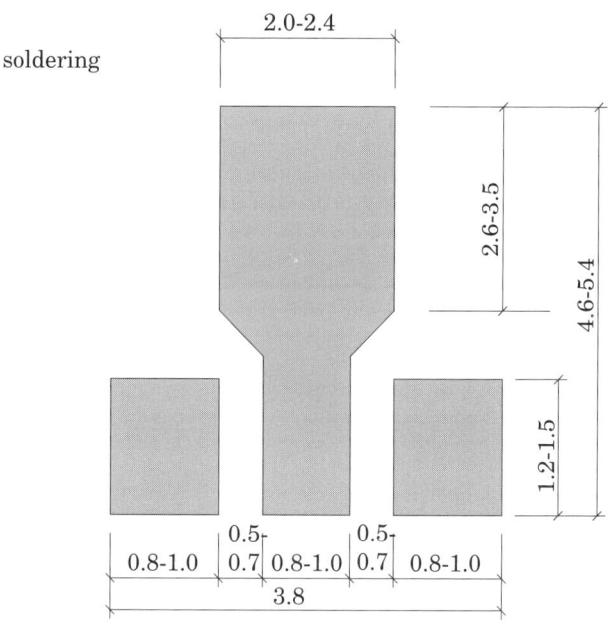

Wave soldering

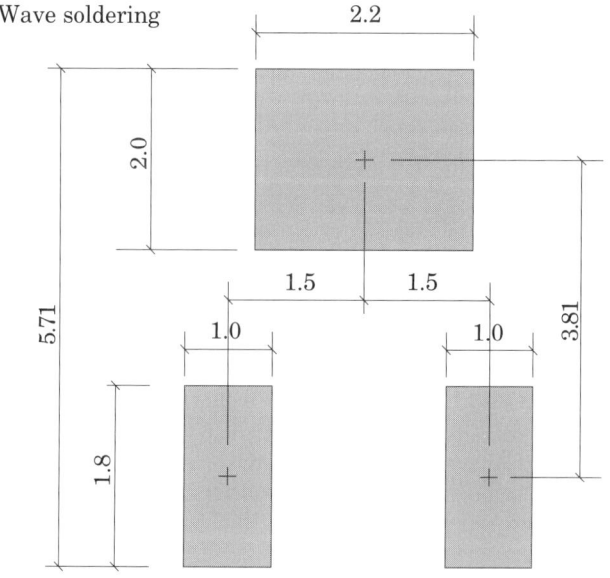

Fig. 9.42 Suitable footprints for SOT-89 packages. (Dimensions are in mm.)

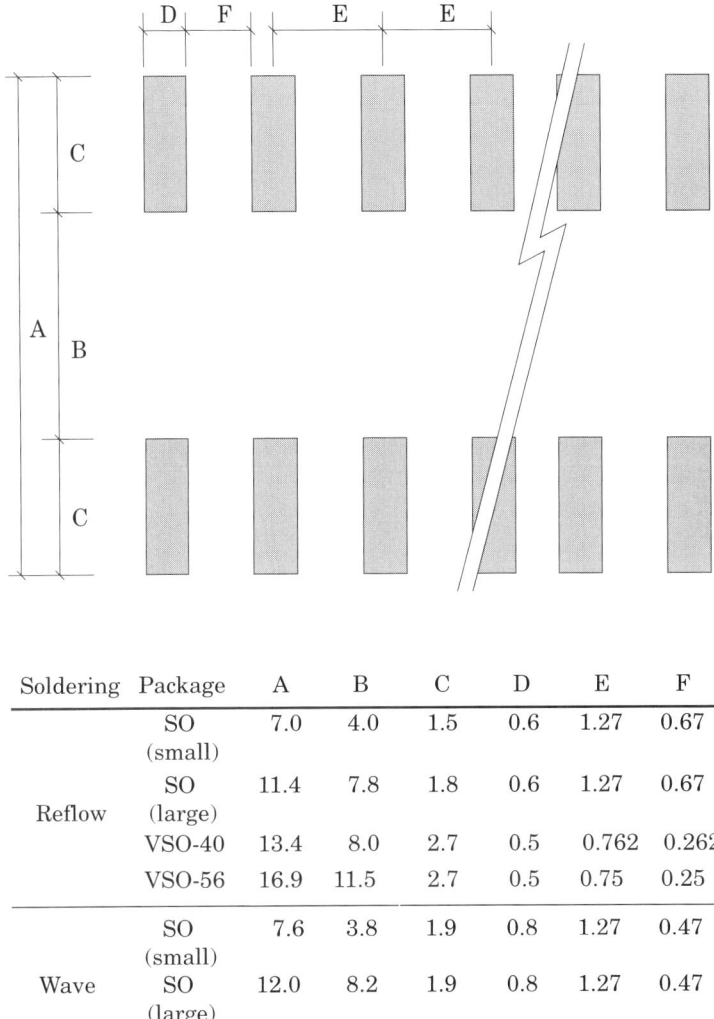

Fig. 9.43 Possible footprints for selected SO and VSO packages. In many cases, land dimensions can be larger. (Dimensions are in mm.)

Soldering	Package	A	B	C	D	E	F
Reflow	SO (small)	7.0	4.0	1.5	0.6	1.27	0.67
	SO (large)	11.4	7.8	1.8	0.6	1.27	0.67
	VSO-40	13.4	8.0	2.7	0.5	0.762	0.262
	VSO-56	16.9	11.5	2.7	0.5	0.75	0.25
Wave	SO (small)	7.6	3.8	1.9	0.8	1.27	0.47
	SO (large)	12.0	8.2	1.9	0.8	1.27	0.47
	VSO-40	14.2	8.0	3.1	0.5	0.762	0.262

for insertion equipment is still strong. The attention given to conventional PIH technology is not misplaced.

PIH technology will continue to give good results with products where size and weight are not of paramount importance, with analogue circuits such as power supplies and any high-current, high-voltage applications, and with a multitude of digital designs which run at relatively low clock speeds, say below 12 MHz. It is important to remember that PIH assemblies can be designed manually or with minimal assistance from CAD.

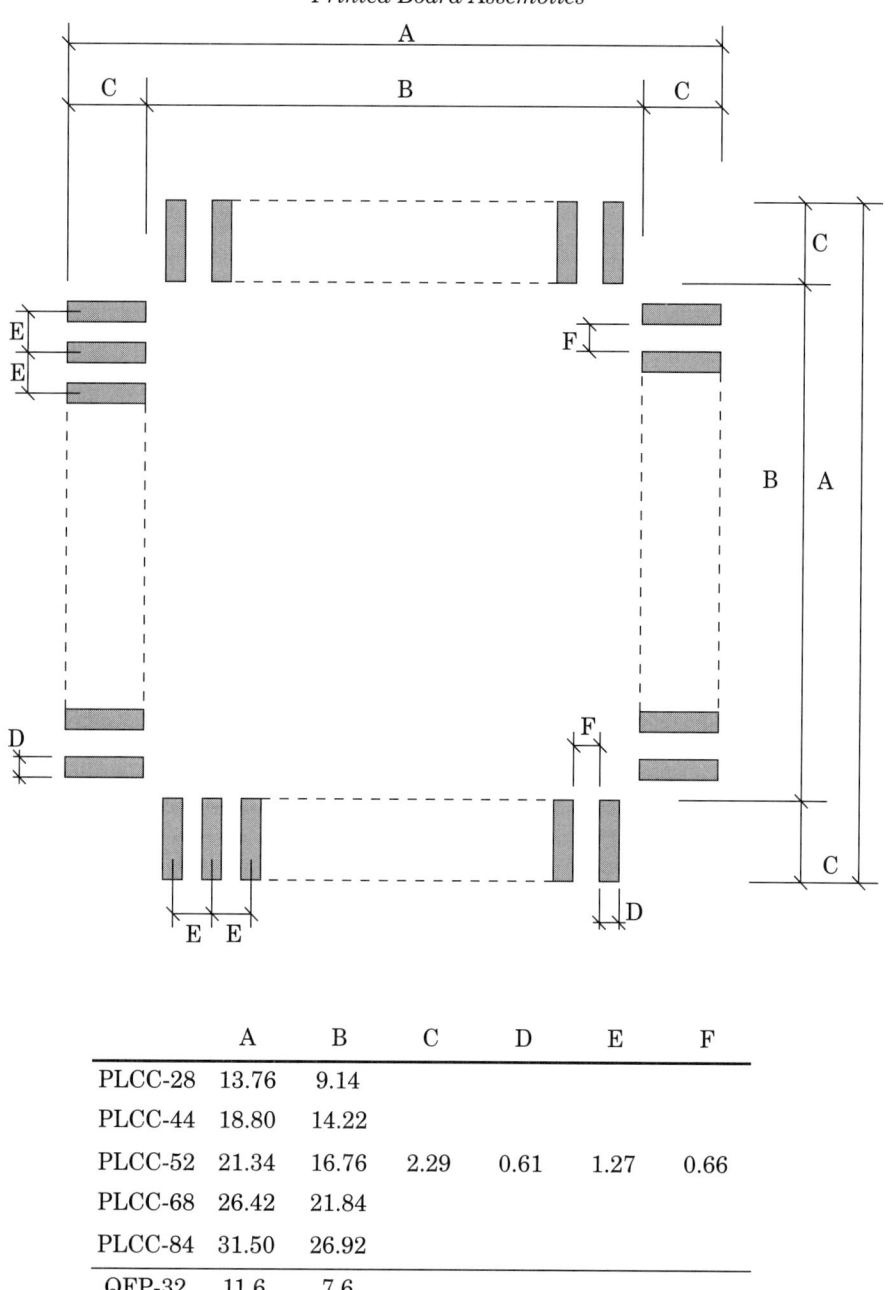

	A	B	C	D	E	F
PLCC-28	13.76	9.14				
PLCC-44	18.80	14.22				
PLCC-52	21.34	16.76	2.29	0.61	1.27	0.66
PLCC-68	26.42	21.84				
PLCC-84	31.50	26.92				
QFP-32	11.6	7.6	2.0	0.50	0.80	0.30
QFP-44	14.0	10.0				

Fig. 9.44 Suggested footprints for selected PLCC and QFP packages. Larger land dimensions are possible, provided placement is very accurate. (Dimensions are in mm.)

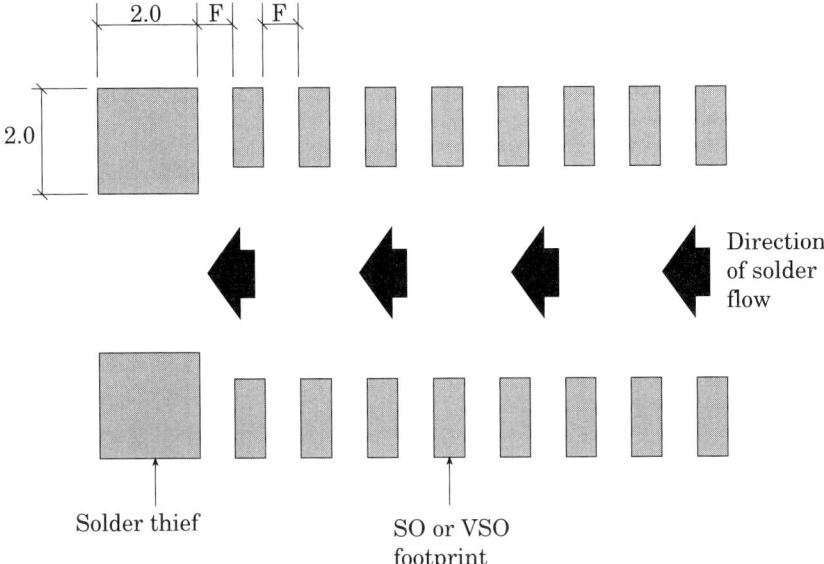

Fig. 9.45 On wave soldering SO and VSO packages, solder bridges may occur between the last pins to leave the wave. If space permits, provision of solderable areas downstream of the last lands may cure the problem by modifying the solder meniscus locally. (Dimensions are in mm.)

As complexity and circuit density increase, the need for more powerful assistance becomes apparent. Just as the progress from DIP through SO and LCC packages to TAB, COB (chip-on-board) and multichip modules results in a reduction in space requirement, it also provides increasing amounts of densely packed electronic circuitry and manual design becomes inadequate.

REFERENCES

1. Junginger, H. and Werner, W., 'Phototools for Printed Circuit Board Production', *Circuit World*, **Vol. 13**, No. 4, pp. 26-32 and **Vol. 14**, No. 1, pp. 32-35, (1987).
2. Scarlett, J. A., 'Printed Circuit Boards for Microelectronics', Electrochemical Publications Ltd, Asahi House, Church Road, Port Erin, Isle of Man, British Isles (1980).
3. BS EN 60097: 1993, 'Specification for grid systems for printed circuits', British Standards Institution, London, United Kingdom (1993).
4. IEC 97: 1991, fourth edition, 'Grid systems for printed circuits', International Electrotechnical Commission, Geneva, Switzerland (1991).
5. Smith, M. D., 'Phototool Generation: The Key to a Quality Board', *Printed Circuit Fabrication*, January (1988).
6. Irving, S., 'The Dynamics of Change', *Printed Circuit Fabrication*, January (1990).
7. Scarlett, J. A., 'An Introduction to Printed Circuit Board Technology', Electrochemical Publications Ltd, Asahi House, Church Road, Port Erin, Isle of Man, British Isles (1984).
8. IPC-CM-770, 'Component Mounting Guidelines for Printed Boards', Institute for Interconnecting and Packaging Electronic Circuits, Lincolnwood, USA.
9. Hinch, S., 'SMT Component Standards Needed Now', *Circuits Manufacturing*, March (1986).

10 Lynch, J. T., 'Surface Mount — Where are the Standards?', IEEE 0569-5503/87/0000-0064 (1987).
11 'SMD Technology', Bulletin 9398 621 40011, Philips, Eindhoven, The Netherlands.
12 Boswell, D., 'Surface Mount & Mixed Technology PCB Design Guidelines', Technical Reference Publications Ltd, Asahi House, Church Road, Port Erin, Isle of Man, British Isles (1990).
13 Klein Wassink, R. J. and Verguld, M. M. F., 'Bridge-free Wave Soldering of SMDs', *Brazing & Soldering*, No. 9, pp. 24-27, Autumn (1985).
14 Klein Wassink, R. J. and Verguld, M. M. F., 'Manufacturing Techniques for Surface Mounted Assemblies', Electrochemical Publications Ltd, Asahi House, Church Road, Port Erin, Isle of Man, British Isles (1995).

Chapter 10

CAD/CAM

R. D. BATTELL
Computamation Systems Ltd, Leighton Buzzard, UK

10.1 INTRODUCTION

The vast majority of all but the most elementary printed circuit boards and flexible assemblies are designed using a *C*omputer *A*ided *D*esign (CAD) system. *C*omputer *A*ided *M*anufacture (CAM) is still primarily the province of larger organisations, although many board manufacturers use CAM to process work provided in machine-readable form by small customers.

This chapter will survey the needs, options and equipment types available now and in the foreseeable future, primarily for CAD, but also for electronics-specific aspects of CAM.

Figure 10.1 gives an outline of the operations required to take a design from circuit definition to the final PCB, including the process of applying modifications.

10.2 INPUT FORMATS

Input of the required interconnections is the first of the three major interactions between the user and the CAD system, and the one which must adapt to the form in which the design currently exists.

10.2.1 Pin List

The simplest input format is the 'pin list'. This consists of a set of device identifications and pin numbers which are to be connected. Such a list can be readily compiled by hand from existing schematic drawings which include pin numbers and component legends. This method of input is often preferred by sub-contract design bureaux. Almost all electronics schematic drawing packages can produce a format of this type. Simple 2-D drawing packages cannot analyse what is drawn in this way — the drawing has to be properly structured as devices, interconnections, etc., with which the software can cope.

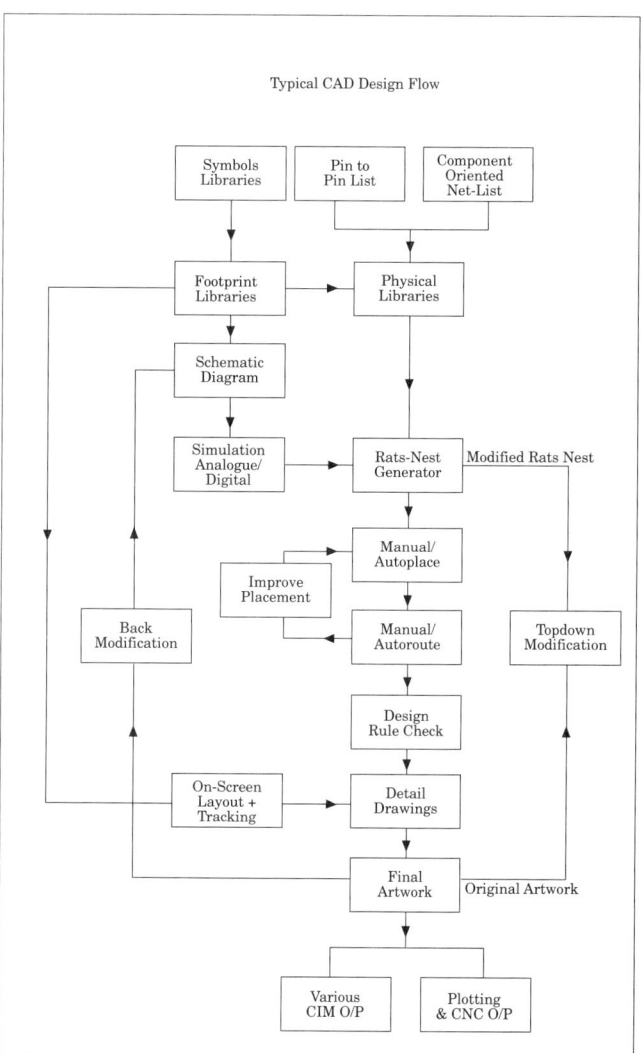

Fig. 10.1 Typical operation of a PCB design system.

10.2.2 Schematic Entry

However, when most people think of CAD for electronics, they think of schematic entry — drawing the circuit on-screen, and thereafter having the interconnections maintained by the system. An example of a small schematic is given in Figure 10.2 (this and other illustrations in this chapter are from the Vutrax PCB design system). This technique avoids many of the potential causes of error in the preparation of a PCB. Each device type must have associated with it two distinct sets of information, as indicated in the following text (although in some systems the distinction is allowed to become blurred).

CAD/CAM 441

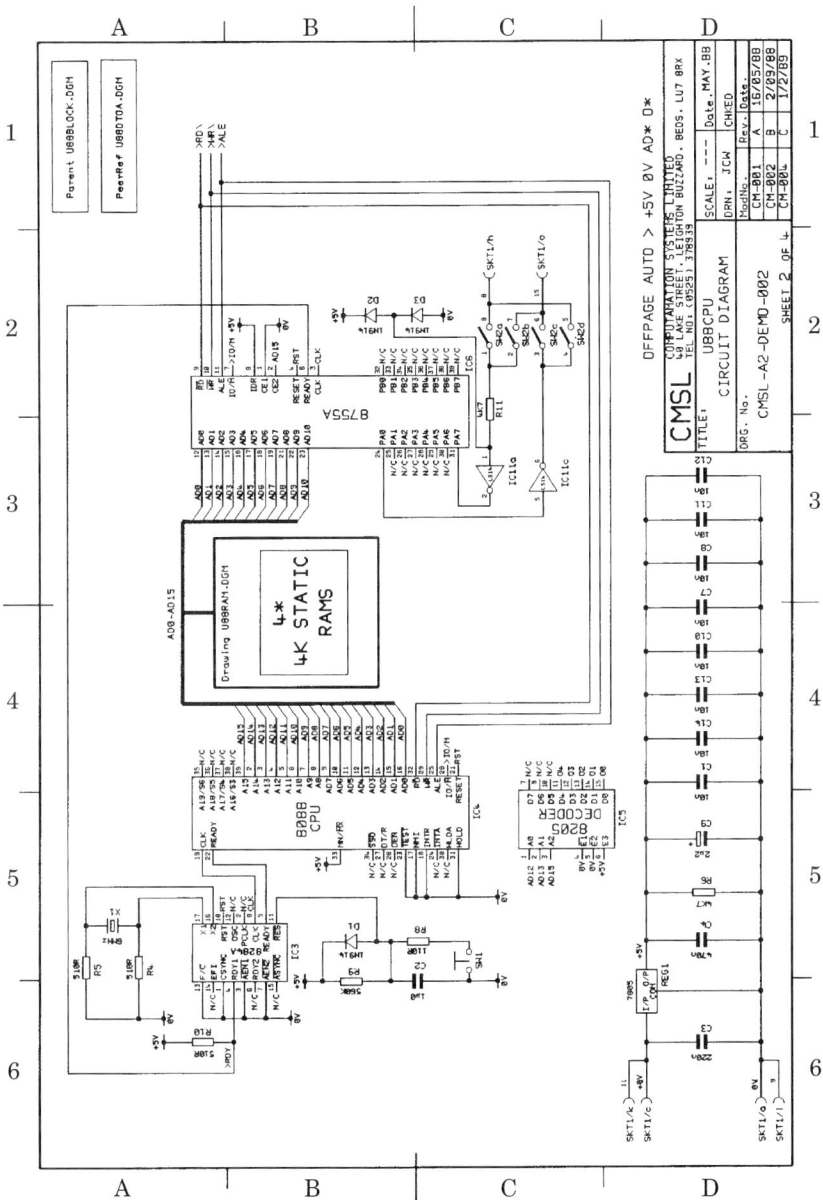

Fig. 10.2 A typical small schematic diagram forming part of a set.

(i) The *visual schematic symbol* is what is given on the circuit diagram. Some systems assign pin numbers on this shape, while others assign functional names to the pins and assign pin numbers later, providing more flexibility in the use of gates within packages and different packaging types. In the latter type of system, the pin numbers are

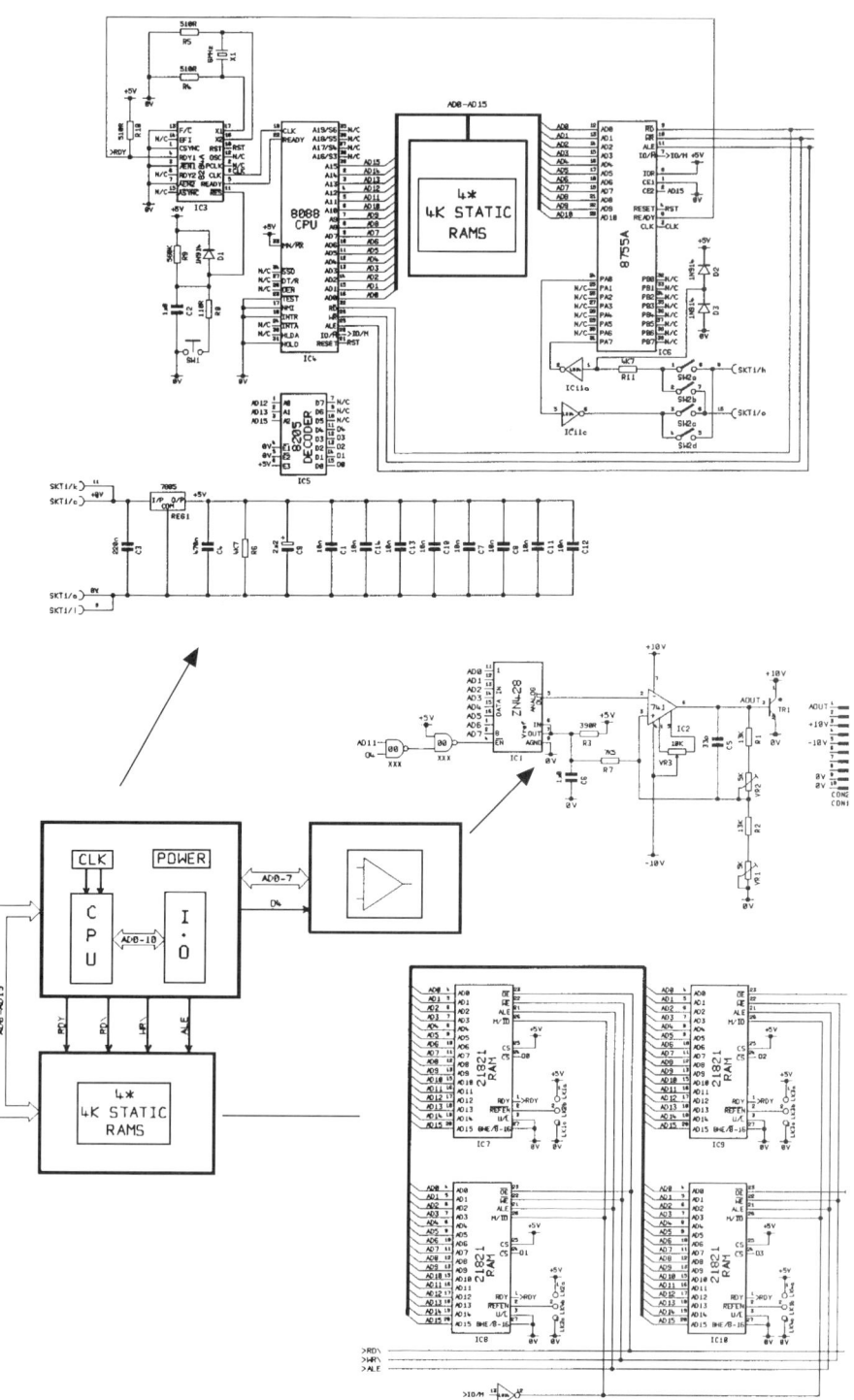

Fig. 10.3 A set of separate drawings in a hierarchy.

automatically inserted near the end of the job once they have been allocated.
(ii) The *physical footprint* is what is given on the final printed board. It will generally include the exact pad shapes, plated-through hole (PTH) sizes or surface mount nodes (the size for a given pad position possibly varying on front, back and inner layers), and the exact location and identity of the pads.

More advanced systems should also allow:

— specification of the initial size, position and orientation of the screen printing symbol and its label;
— provision of the actual device outline for use during manual or automatic placement (including automatic placement machine chuck size);
— specification of the height above the board surface for use in visualisation and for adhering to profiling limits during automatic placement;
— various density-determining aids such as density histograms and percentage of area occupied by components.

10.2.2.1 STYLE OF SCHEMATIC ENTRY

The style of schematic entry varies widely, many systems offering convenience features such as automatic squaring, automatic generation of busses, and the like. While these can be of service (especially for demonstrations), their usefulness in real drawing may be limited by lack of space and the need for unusual connection of busses (often only discovered during routing, where RAM chips are better connected in other than standard 1:1 forms).

10.2.2.2 HIERARCHICAL DESIGN

Within the topic of schematic input, there is the issue of hierarchical design. In its simplest form hierarchical design consists of drawings in a 'tree' structure starting at a single root drawing (usually a series of functional blocks), each referring to another drawing containing more detail, and so on *ad infinitum*. In principle, this hierarchy could start at an equipment rack and pass down through backplanes, other harnesses, connectors, PBs, PALs and hybrids, etc. — all for a simple design. Such a level of abstraction demands formal interfaces between the sections. In practice, only enormously expensive systems offer anything like this level of overall design. Most systems for personal computers (PCs) restrict their sphere of activity to one phase at any time, although more flexible PC-based systems can often be used independently for different aspects of the design, bringing all levels of hierarchy together. An illustration of how a set of drawings can form a single design derived from a block diagram appears in Figure 10.3.

10.2.2.3 LIBRARIES

All serious PCB CAD systems come with libraries of various commonly used parts, particularly including standard logic. These can be a good starting point but, in practice, almost all users have to prepare libraries specific to their industry and company standards. The level of control over the appearance of the schematic symbols varies enormously — in some, one may draw anything one likes (identifying the connection points by means of some specifier), while others predetermine a box shape and labelling, with arbitrary shapes limited to small symbols. Complex transformers and similar parts can be difficult to represent unless free format drawing is allowed.

With screen stencils for legends, etc., appearance is a separate issue not to be confused with schematics. Again, capabilities vary widely. Ensure that the labelling on the finished layout can be moved, resized and re-orientated — a standard set of positions is never sufficient.

10.2.3 No Format Supplied

Many customers may have the printed board layout prepared manually, or by CAD systems from vendors who will not publish their data formats. These can be adopted into some CAD systems by one of the following routes:

(i) analysing a photoplot data file: such a file will often be on magnetic tape (the pre-personal computer era data interchange standard), so that an agency may be required at least to convert the data to appropriate media;
(ii) manual digitising of an existing production master: depending on size and accuracy required, such a digitiser may cost a few hundred dollars or tens of thousands;
(iii) automatically digitising an existing production master: the machines for this are very expensive, and agencies usually offer the best solution; output is normally in photoplot data format — see (i).

These existing PB capture techniques vary in usefulness according to how well the CAD system can subsequently associate the tracks, pads and vias on various layers to make intelligent signals rather than arbitrary pads and tracks. Even so, modifications are awkward at best. Designs for major rework are best analysed into a pin list and reprocessed from scratch with a modern system.

10.3 SIMULATION

With a finished set of schematics the next stage may be simulation of the circuit. Simulation generally falls into two categories, as below.

(i) *Digital.* All components are treated as logic devices and the circuit is analysed for behaviour with time (generally measured in clock ticks). Timing diagrams are the form of output most commonly produced by the exercise. A major difficulty with digital simulation is the increasing use of high integration components. At the low end it may be feasible to define the relevant parts of a PAL, but software emulation of a microprocessor plus program in EPROM is daunting. High level simulation products permit 'real' devices to be integrated into the simulator, but even here difficulties exist related to memory refresh and minimum clock speed for which the simulated environment cannot even provide an approach.

(ii) *Analogue.* Here there is more choice in the method of analysis, reflecting the wide range of critical issues involved in analogue designs. The basic choice is between analysing in the time and frequency domains, plus analyses of the effects of power supply noise and the like. The ability to determine the tolerancing of components is frequently an important characteristic of such simulations and is one of the main reasons for performing them. At extreme frequencies, the effect of parasitic coupling between the various tracks and components can be very difficult to quantify and may require the design of the analogue section to be fully routed before it can be analysed.

From this it should be clear that simulation, even on very powerful systems, is far from the ideal of whole board testing. Generally, specific areas of the circuit are simulated in isolation, and only those parts known to present potential problems are simulated at all. Mixed digital/analogue simulation is still in its infancy. Ideally, simulation should be able to prove that a design works with all its worst case component tolerances. While sections can be tested in this way, the overall design often can not. Apart from the complexity, and the mix of analogue and digital technologies, such effects as inter-track capacitance, track self-inductance, and signal skew due to varying track lengths are very difficult to accommodate.

In integrated circuit (IC) design, the enormous effort required to achieve thorough simulation is an absolutely essential outcome of the huge expense of producing prototypes and the difficulty of fault-tracing in them once they have been produced. The experience of the author's firm has been that, for most designs of PCB, most customers consider that the cheapest and easiest simulation is to build one or more prototype boards, and to test these using marginal power supply and timings in order to ensure adequate tolerancing. Simulation becomes worthwhile in isolated sections of circuit which are known to pose potential problems and which will be difficult to fault-trace in the final form of the circuit because of its nature.

10.3.1 PAL, PLA and ASIC

These all present the same basic problems to the PCB CAD system, namely:

(i) How are they represented in the schematic?
(ii) How are the programming tapes, masks, etc., generated?
(iii) How are they simulated?

Some CAD systems already provide support for inclusion of PALs and PLAs drawn as logic in schematics, and then analyse these to produce JEDEC fuse blowing data. Manufacturers of programmable devices can always supply special-purpose CAD systems for PCs which are able to accept logic descriptions as simple schematics or Boolean equations. Also included can be simulations of the workings of the proposed device and, finally, production of the programming data either for a programmer or for mask generation (the latter performed by specialist equipment at the IC manufacturer). These are more easily kept up to date as regards new devices than a third party product can be. Unfortunately, there is no standardisation to ease the hassle of changing from device to device.

10.4 PRINTED BOARD LAYOUT REQUIREMENTS

10.4.1 Capability Required

Before the phases of printed board layout are studied, the overall capabilities required should be examined. For schematics there is no practical limitation on accuracy, line widths, etc. — as long as the final result looks acceptable and can be plotted to any desired size, it will be satisfactory. The same does not hold for board layouts.

10.4.1.1 CO-ORDINATE SYSTEM

The PCB industry now works in both metric and imperial units. Although data sheets are invariably metric, the actual pin spacings in dual-in-line packages (DIP), surface mount DIP and many other footprints are still essentially based on a 0.1 in. or 0.05 in. grid, see Chapters 3 and 4. To comply with the needs and with company conventions, the user must be able to input a mix of co-ordinate types to the library and to represent the overall layout in either type of co-ordinate.

10.4.1.2 RESOLUTION

At one time a system grid of 0.050 in. was considered the norm, fitting well with the DIP 0.1 in. grid and typical single track between pads, see Chapter 9, 9.3.1 onwards. With the advent of surface mount technology a resolution of 0.001 in. (0.025 mm) is the minimum acceptable. Although this may be the basic resolution, most systems provide the convenience feature of a 'snap' grid at some multiple of the basic resolution. When generating footprints graphically, such grids can be a hazard — what looks 'OK' may not in fact fit the device on a finished board — the intuitively simple drawn footprint requires more attention than it often receives.

10.4.1.3 LAYERS

This term means different things in different CAD systems. In some the term is restricted to layers containing interconnection traces, with legend screen, solder resist, pastes, and powerplanes as special-purpose features. Other systems assign numbered layers to whatever is desired, possibly not finally assigning the order until a later stage (the actual allocation is often only important for surface mount devices, edge connector finger pads, and surface or buried vias). Free format systems are often more flexible about including special artwork on solder resists, splitting powerplanes or even running tracks within them, and having alternative legend screens (e.g., for the real board and for documentation) or multicoloured screens. There should be extensive controls both in the interactive graphics and at the plotting phase to decide what is visible and how it is represented.

10.4.1.4 TRACK/TRACE WIDTHS

The 0.015 in. (0.38 mm) norm for track/trace widths has been steadily reducing, and 0.004 in. (0.10 mm) and finer traces are now not unusual on dense digital PBs. Wide tracks may be required for power supplies. Most CAD systems provide sensible options in this area — but it is advisable to check.

10.4.1.5 PAD SHAPES

This is a very complex area handled with varying degrees of competence by different systems. Variations in surface mount requirements (because of the lack of imposed standards) mean that any reasonable surface mount library will require dozens of different pad shapes and sizes. Even the humble PTH component pad and through-hole via may vary in size from layer to layer, for example, being larger on the solder side for improved solderability. The specification of a pad should allow any shape on any desired layer and specification of different solder resist and solder cream shapes for each side (a total of four shapes altogether).

A wide variety of pad shapes should be available. A few systems allow arbitrary shapes but, in practice, a selection including all reasonable sizes of squares, rectangles, disks, rectangles with rounded ends, annular rings, thermal break rings (also known as heat traps), and drill marks is sufficient. Post-processing or explicit drawn construction can be used for occasional strange requirements. Figure 10.4 illustrates uses of some of a variety of pad shapes (compare with Figure 9.5).

As board density has increased, optimising vias has become more important. Many systems still insist that vias must pass through all routing layers. However, vias can also be *blind* (starting on one side but stopping part of the way through), and *buried* (between inner layers only). Because in practice multilayer boards are often made as laminates of double-sided boards, buried vias appear mainly on boards with six or more layers. Blind and buried vias present special problems for CNC drill tape generation — these will be discussed later.

10.4.2 Placement

Placement is one of the most critical phases in PCB layout and the one where the experienced draughtsperson can make a major impact on quality. Most PCB CAD systems provide an autoplacement facility which is intended to optimise component position (and sometimes orientation) to maximise routing success. Unfortunately, the theoretical models for placement are far removed from the real world, and most autoplacement facilities provide what is at best a general arrangement, aimed at minimising total connection

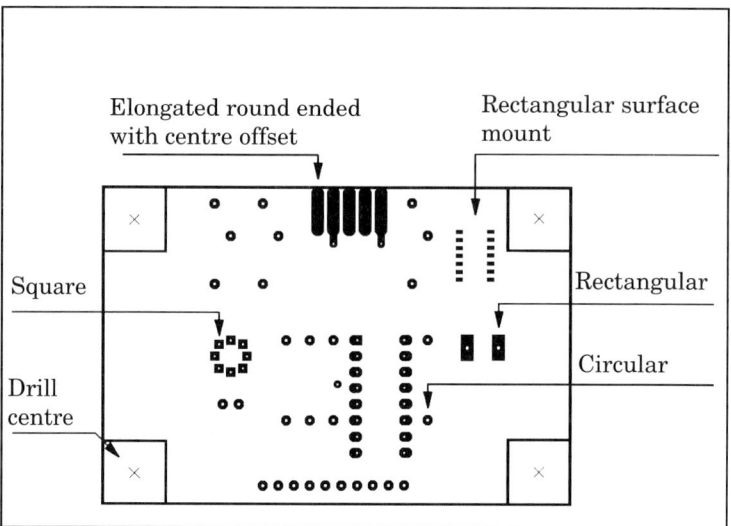

Fig. 10.4 Examples of various pad types.

length. The majority of placement routines permit the user to place critical parts first (connectors, for example, where the position is often a functional requirement). If the major components, including busses, are manually placed first, autoplacement can often make an excellent job of the tedious process of placement of minor components — indeed, the patience of the autoplacement program may lead to a better result than the user would achieve!

10.4.2.1 AUTOPLACEMENT

More advanced autoplacement routines have strategies for recognising bussed components, giving weight to accurately aligning the connections, and for tackling the thorny problem of associating power rail decoupling with specific components or in component groups. Of course, decoupling placement can never be handled by a general placement routine, as the components would all end up in arbitrary groups in order to minimise the routing between them — quite the reverse of the functional intent. Handling of power rail decoupling is an important aspect to consider when evaluating a system.

Many designs contain critical areas where short routing and/or minimising crosstalk is important. Getting this right involves the user's understanding of the design to ensure that autoplacement and autorouting will behave appropriately. All systems should, as a last resort, allow the critical part to be manually placed and then routed in such a manner that the system will not alter.

Most experienced designers working on high density layouts consider a background autorouter as *part* of the placement strategy. A proposed layout, possibly produced by autoplacement, is given to the autorouter for a first attempt at routing. Signals initially unrouted provide a direct indication of problems with the layout which can be rectified by placement before the routing is attempted again. This cycle can be repeated until a usable placement has evolved.

10.4.3 Routing

10.4.3.1 MANUAL ROUTING

The very smallest systems offer routing only using an interactive graphics system (hereafter referred to as 'manual routing'). Even at this level, PB CAD can be a great advantage. A properly designed manual routing system will include a large number of convenience features for laying traces on user-selectable grids so that correct clearances are automatically obtained for the standard track width and pad sizes. It should be possible to change layers of existing routing, move traces, insert and remove vias, and change the networking interconnection pattern, all with ease. The routing system should also provide a high degree of protection from accidental change of the required interconnect pattern, without actually preventing deliberate modification. In multilayer work, there must be the possibility of viewing any group of layers and legends desired in a chosen set of colours, with unrouted connections invisible or highlighted as required, and vias displayed so that they are easily distinguished from any potentially confusing pads of similar style.

It is important to be able to highlight all the traces and unrouted connections for a particular signal — seeing a complex signal in its entirety can substantially clarify the routing options for it.

Given all these features, an experienced designer can prepare a routed layout far more quickly than by manual taping. However, the importance of manual layout does not end with small systems without autorouting. In many larger systems, important signals are either pre-routed or modified after autorouting. Many users always route power rails manually. Designs such as power supplies and hybrids almost never autoroute satisfactorily, and it is important that even the highest level system should have a powerful and flexible interactive routing capability.

10.4.3.2 AUTOROUTE STYLES

Autorouters vary enormously in both capability and methods of use. They range from interactive routers integrated with an interactive graphics system, which route signals on demand, to highly sophisticated 'rip-up and retry' routers which can normally complete the routing task without user interaction.

While the interactive form has much initial appeal, it defeats one of the principal aims of autorouters — namely, to free the designer to do something more useful than routine track layout. Most advanced users make exclusive use of background routers, varying placement strategies, layer allocation, design rules, etc., and then running the router to gauge the quality of the layout attainable, as discussed earlier. With multiprogramming workstations, such runs may take place simultaneously with other work. Users of more modest systems can run these routing trials overnight.

The remainder of this discussion will confine itself to stand-alone background routers, examined under a series of headings.

(i) *Layers.* Very few modern routers come near to making a reasonable job of routing on a single layer board. Two layer routing is the standard for low-cost routers, using the two sides for predominantly horizontal and predominantly vertical routes. More sophisticated routers can handle any practical number of layers, usually routing all layers simultaneously. (Powerplanes are not usually considered as routing layers — see (vi)(c) below.)

(ii) *RAM routing.* The routing of the typical RAM bank (or any other aligned bus structure) requires special recognition of the structure, and routing in a systematic way. This is because successful routing requires precise alignment of the track kinks over a potentially large area, and that all tracks use the same up/down sense and the same on-grid or between-grid path for the same alignments. Most RAM routers use only 45° routes. Some close-packed structures (e.g., DIL packs at 0.1 in. (2.54 mm) spacing and 0.012 in. (0.3 mm) tracks) need angles other than this. On many jobs, a router which can tackle the RAM and other bussed areas will already complete 80% of the task.

(iii) *Gridding.* All routers are gridded inasmuch as they work within the resolution of their system. However, many routers demand the use of a grid a great deal coarser than this. This is required by the overall routing scheme where the board is mapped on to a series of grid points stored by the program. The router then explores its grid points to find routes. This is the much-vaunted 'maze' router, although in modern routers the original scheme has been modified almost out of existence. The technique has many advantages, particularly if the grid, trace width and clearance are related so that the router can just look for uncommitted grid points.

One negative aspect is that the gridding often prevents the tightest possible packing. More seriously, the grid required for successful operation with surface mount parts having a wide variety of pad pitches may have to be very fine indeed, and its use may be precluded either by router constraints (e.g., grid not finer than track width plus clearance) or because the memory required to store the grid points escalates enormously with fine grids.

These problems have led to the introduction of gridless routers. The data organisation required for such routers is complex, and the routers therefore do not operate as speedily as do gridded routers on work which allows their use.

(iv) *Angles routing.* Autorouters usually introduce routing at 45° (sometimes to excess, in the view of many users). Such angles are essential for preventing blockages round component pads and for RAM routing. A few routers can route at other angles (gridded routers are particularly tied to the 45° angle because of the natural alignment of grid points). A number of proprietary studies are said to show that use of other angles can improve routing performance but, clearly, such matters are dependent on the characteristics of the board.

(v) *Optimisations.* Good routers provide numerous optimisations, often as one or more separate passes after all signals are routed. Levels of sophistication range from attempts to make local changes to attempts to rip out major areas of routing.

Via removal is the top priority — vias cost money to drill.

Closely packed tracks, and tracks close to pads and vias, are likely causes of manufacturing defects. Where it is impossible to increase clearances, this is advantageous.

Narrow tracks are also likely causes of defects, and can be widened where clearances permit. Users often think of this the other way round — as 'necking' through narrow gaps. This is more likely to be useful on low density boards manufactured by cheaper processes.

(vi) *Power supplies.* Power supply routing makes special demands on the router because of the need to maintain low impedance, minimise power supply noise, and provide screened areas. The following three basic tactics are usually employed:

(a) Route the power supplies as a standard signal, except that the trace width is larger than standard. For CMOS circuits and similar designs with low current drain, this may be satisfactory.

(b) Grid the power supply, and tap power from the nearest rail for each device. While this scheme worked well for aligned DIL packages, it works less well with surface mounted and high density boards.

(c) Use powerplanes. These are whole layers of copper, usually inner PB layers, which include isolation for drilled pin and via holes, and connection for required power connections. Obtaining both isolation and connection is complicated by the manufacturing process in the following ways:

Isolation can be accomplished by having no copper around the drilled hole (i.e., a 'disk' of non-copper). However, this can lead to delamination and adhesive creep, so an annular ring of isolation, Figure 10.5, is normally used.

Connection can be accomplished by simply drilling through the powerplane copper. However, this can cause the powerplane to act as a thermal sink during soldering. To prevent this, a thermal break ring, Figure 10.5, can be used.

The four combinations of possibilities are all used for different manufacturing purposes, and some manufacturers promote their own variant with amazing verve! An illustration of two combinations of these options is given in Figure 10.5. Such illustrations are both plotted and illustrated as negatives for reasons of practicality — white areas become copper in the product.

Occasionally, an ability to place one or more signal tracks on a powerplane layer will avoid the introduction of additional layers. This is possible on some systems by drawing an outline of the required track and vias. Automatic handling of such constructs is very hazardous — it is very easy to isolate an area of the plane or introduce high impedance paths for adjacent parts of the circuit.

Another much more common variant of the powerplane is the 'split' powerplane. A typical instance is a design using both analogue and digital earths (and, possibly, supply rails as well). If the design is laid out in a well compartmented manner, the powerplane can be split into two (or more)

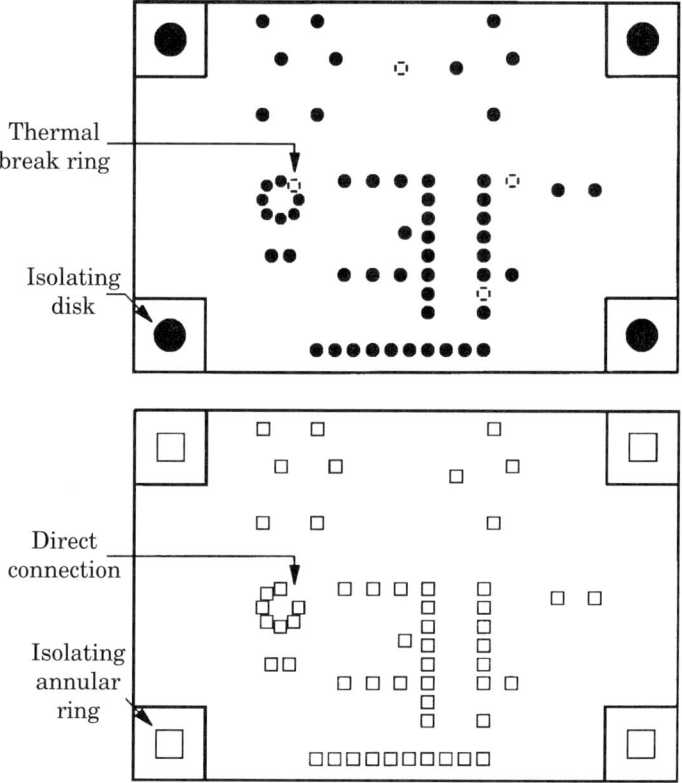

Fig. 10.5 Two powerplane connection/isolation styles.

areas. The powerplane generation feature must provide clear indications as to which connections are intended to connect to which portion of the powerplane. Dividing up the powerplane and checking for correctness is almost always a manual process which can be made quite straightforward by suitable colour coding of the powerplane connection vias.

Groundplanes differ substantially from powerplanes in that they share the layers with other signals, the latter being the dominant content of the layer. Some designs, particularly in radio frequency work, require all unused area to be filled with either solid or cross-hatched grounded copper. This is usually a specialised routing function, possible only after routing is complete.

Figure 10.6 illustrates one layer of a small printed board which incorporates a partial groundplane at the top left.

10.4.4 Design Checking

Automated design checking is an important aspect of any PB CAD system. Most references to design checking relate to checking of clearances and some systems claim to do this 'on-line'. For 'on-line' checking to be worthwhile, it must *prevent* mistakes rather than just providing a one-off warning. This means that, in practice, such design checking is almost always disabled by the user because it blocks the natural process of modification (which usually

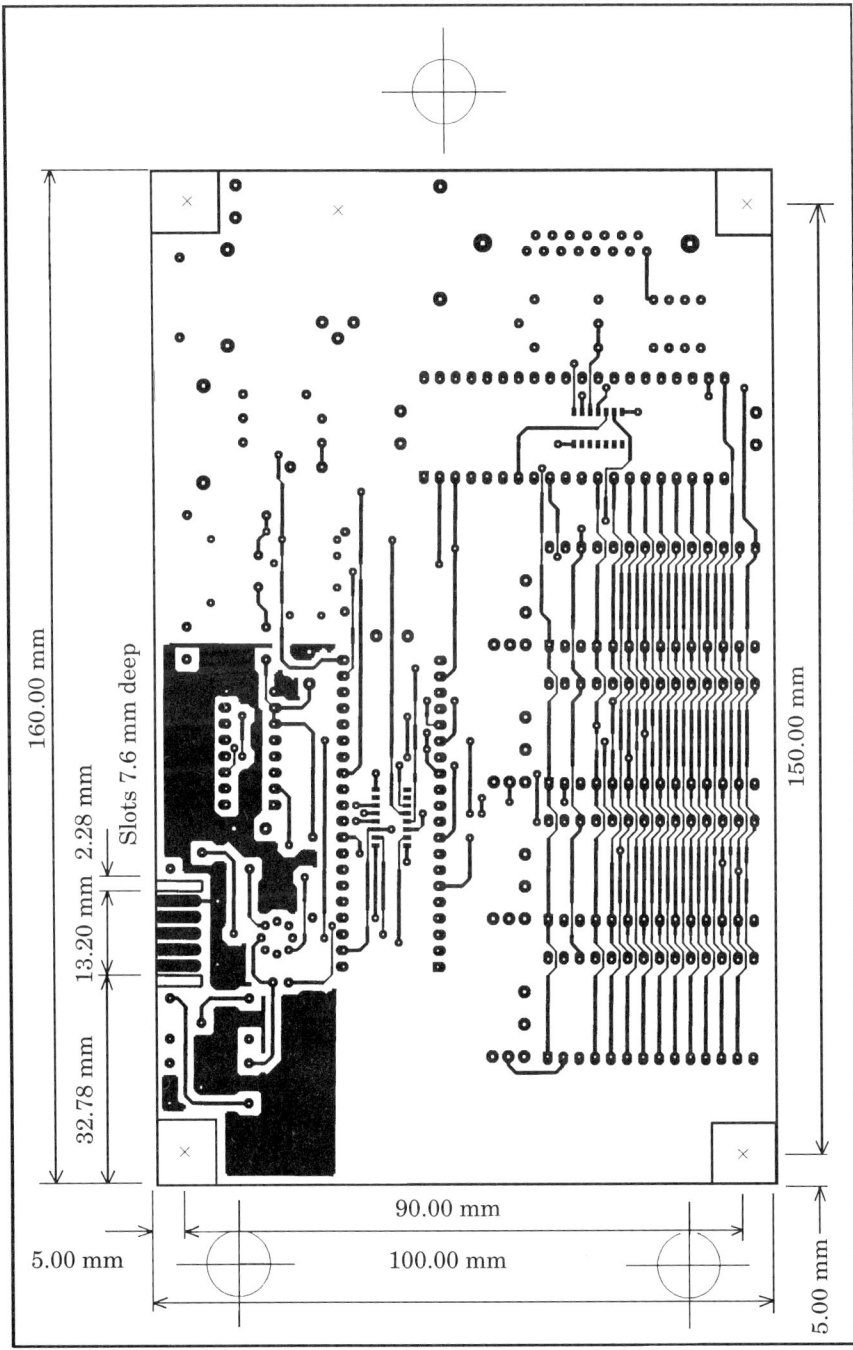

Fig. 10.6 A typical PB check plot incorporating dimensioning and a partial groundplane.

involves temporarily creating crossed tracks and tracks too close to vias that will be moved or removed subsequently).

However, proper design checking involves far more than just clearance checking and should include the following:

(i) Comparison of the original rats-nest, pin list or whatever, with the finished routed board, the check to include pad shapes, layer usage and so on.
(ii) Checking that each signal pin is actually interconnected by tracks of whatever minimum width is specified.
(iii) Checking that only the nominated tracking layers are used.
(iv) Ensuring that minimum via or pad size is not violated.
(v) Verifying that drilled items leave sufficient copper around the hole.
(vi) Clearance checks. The rules for required clearances are potentially immensely complex, involving different clearances between tracks, tracks and pads, and pads, and with potential differences in these between different signals and on different layers. Signal-sensitive clearances may involve signal group clearance (e.g., mains power traces may require 3 mm clearance from each other, but 10 mm to any other signal).

A thorough implementation of this phase of the check is extremely computationally intensive. Many systems simplify the checks and execute them during manual routing (trusting the autorouting phase to be accurate). Most large scale systems perform this phase of design checking as a background activity.

10.4.5 Modifications

Modifications to the design, once placement and routing are under way, seem to be inevitable and in some cases are actually driven from developments during the layout process, resulting, for example, from component availability, size constraints, or the opportunity to save packages by utilising spare functions of partly used components.

The second area of modification is when finished designs (either built as prototypes or already in production) need to be modified to a new function without discarding the investment already made in layout, documentation, test and other existing procedures.

Accurate and systematic handling of such modifications is the key to preventing escalation of the costs. Modification can be applied from both ends of the system:

(i) *Top-down.* Changes are made to the schematic or pin list and are required to appear on the PCB rats-nest or the partly or fully routed PB. The former case is usually trivial — regeneration of the rats-nest using existing placements is the most straightforward approach. The case where useful, or all, routing is already complete requires minimum disturbance to the existing routing. Some of the more advanced CAD systems can provide this, often taking a rats-nest for the newly required interconnection and comparing this with the routed board, applying minimum changes as necessary.

CAD/CAM

(ii) *Bottom-up.* Changes are made to the routed PB layout and are required to be reflected back on to the schematic. Some changes to the PCB will consist of exchanging pins with identical function and the schematic changes may be purely annotational or require obvious connection changes. Other changes may require removal or introduction of device symbols — these changes are usually listed for the user to make manually — the day when automatic schematic layout incorporates an understanding of circuit design sufficient to produce good schematics is still in the future.

Both modification schemes require an important extra — the cyclic design cycle (i.e., endless modifications of one or other type) and a means of checking that the modifications are correctly applied so that the schematic and the printed board are in exact correspondence. Service manuals frequently indicate the common *absence* of this feature!

10.4.6 Schematic Annotation

Most PCB CAD systems now present the initial circuit in the form of schematic diagrams. Simplistic systems allocate component names and gate groupings in the schematic. This removes any need for subsequent insertion of such information. More advanced systems defer these decisions until ratsnest generation or even later, for example, by user-instigated rearrangement of gates. Apart from improving the layout (by avoiding arbitrary groupings of gates, etc. in the schematic before the layout constraints are known), this scheme permits the allocation of parts identification on the basis of board layout rather than the schematic. The advantages during initial rectification and in-service repair are obvious. However, this scheme relies on automatic annotation and modification of the schematic to include both the final device identification label and pin number labels. For complex multi-sheet designs the advantages of automatic allocation can become very significant.

Figure 10.2, forming part of the discussion of schematic entry, includes such annotation produced from the finalised PCB layout.

10.4.7 Automatic Test

Automatic testing is an area of little consequence to the specialist designer and of great importance to the designer of boards intended for mass production.

The test machine has to be able to determine 'correct' responses. Three approaches are in general use:

(i) Using the output from a whole board simulator. During diagnostic phases the testing machine can make reference to the design and simulator, and, potentially, discover the exact cause of the fault. This approach is basically limited to all-digital boards and is found only in extremely expensive systems.
(ii) Manual encoding of (i) above. The effort involved limits this approach to relatively simple designs in large scale production (e.g., the automotive and domestic markets).

(iii) Learning the characteristics of a known good prototype board. The test programmer generally has to choose the points on the PCB to monitor signals, decide where to inject test input signals, and decide also on an input pattern which will exercise the circuit adequately. The testing machine then learns the responses expected from the known good board, with timing and level tolerancing suitable for the design. In this scheme the isolation of the fault is usually limited to listing of certain test nodes as being marginal, indeterminately incorrect, or 'stuck'. The machine will report these faults in terms of test point co-ordinates or numbers — it may be important that both a drawing of the finished layout and the schematics can be easily updated by the CAD system to indicate the position of these test points.

10.4.8 Mechanical Drawing

The degree of mechanical drawing offered by a PCB CAD system can vary from none to technical drawing capabilities rivalling those of low end specialised mechanical systems. In addition to the cost and familiarity advantages of a combined system, many other practical advantages are to be gained from having the following features:

(i) automatic dimensioning of board outlines to provide production documentation;
(ii) preparation of complex board profiles with curved sections, complex cutouts and edges at arbitrary angles with precise lengths;
(iii) hatching and filling of selected areas;
(iv) drawing of heatsinks and other mechanical details where they can be directly checked against the constraints.

Areas in which virtually no PCB CAD system can be of help are true 3-D representation (although some systems will accommodate isometric and oblique axes), and shading, stress analysis, etc. In these areas, being able to port the mechanical aspects of the PCB to a specialist CAD system (e.g., via the DXF format discussed in 10.7) can be of great value.

10.5 OUTPUTS

Output from a CAD system can take many forms, and requirements of cost, speed and accuracy, and level of user control vary greatly according to the information.

10.5.1 Schematics

For schematics, the primary requirements are those of legibility, speed, and control of the information included. The latter should provide the ability to leave out device part numbers (leaving in only the identification, including or excluding pin number annotation) and signal names, and to include or exclude various notes. Using these controls makes it possible to maintain a single schematic drawing file (known to be correct because the board was derived from it) and, from this, generate internal design and test documentation, field service documentation, and so on.

For some users, colour is an important factor in schematic output, while others will be happy with a clear black-on-white drawing, plus, perhaps, one colour for highlighting. Similarly, some users require A1 or even A0 drawings, while others prefer a high-definition small scale plot (e.g., from a laser printer) as being more practical for service documentation.

10.5.2 Layouts

PCB layouts, also, have to be produced on paper. Similar controls on outputs can have similar advantages. In this case, major savings are possible if monochrome output can be produced with half-tones (e.g., to plot a component layout against half-tone copper traces). Some systems can accomplish this with laser or matrix printers. If it is possible to choose the appearance separately of each layer, via, PTH pad and surface mount pad (including hiding any or all of them), then virtually any style of document can be produced. Figure 10.6 shows one side of a typical small PCB, incorporating dimensional and other information.

Non-graphic outputs from the PCB layout data are also important. There should be a parts list, the format of which can be adjusted to suit whatever stock control or database system the user requires, together with information on component placement and orientation which can be extracted for use by automatic component insertion/onsertion machines. As a last resort, the graphics format should be available in some sort of textual form (usually specially produced on request), so that any desired information can be extracted (even if this involves some effort).

10.5.3 Photoplotting

The primary form of graphic output for the PCB is for the production of the printed board itself. Still in use, but falling out of favour, is the production of 2:1 scale paper or film artworks, which are then photoreduced to produce manufacturing masters. The reason for the decline of this technique is the general availability and reasonable price of photoplotting services from specialist bureaux. Photoplotters work by 'drawing' directly on to photographic film with a light beam. In the traditional photoplotter, tracks are drawn by moving an aperture consisting of either a fixed size disk or an annular ring, and pads are drawn by photoflashing predetermined aperture shapes. Support of traditional photoplotters requires rather sophisticated software to group and optimise operations so that the slow changes of aperture are minimised and limitations in the number and type of apertures available are compensated for. Modern equipment tends to use high resolution raster scan devices called 'laser plotters', where the drawing is basically performed in a matrix memory and then drawn on to the film by a laser. Although laser plotters can accept pre-rastered images, most are currently used to accept a traditional laser plot data format and generate the image from this. The most common 'standard' for photoplotting data is based on the protocols used by Gerber Scientific Instruments photoplotters and most good CAD systems can output this format. Although the format is formally designated for magnetic tape in ASCII or EBCDIC[1] code, most bureaux now accept the data in ASCII on IBM-PC format disks. The format is quite variable in a number of aspects — check the details of co-ordinate system, code and precision before committing yourself to a job. Many bureaux save the user the trouble, however, accepting the CAD system's design files and generating the intermediate data format in-house. This means that they must have purchased at least part of the design system in question — check with them on formats, software release numbers, and acceptable floppy disk formats.

Plotting of printed boards places far greater constraints on the plotting process than schematics or documentation drawings. Obviously plotting accuracy is paramount, for controlling local detail to maintain track widths and spacings, and for controlling overall dimensions where multilayer plots have to register to accuracies of typically 0.001 in. (0.025 mm). There are other constraints however — it is no longer possible to be casual about many of the details of plotting control, such as:

(i) how ends and joins of tracks are handled (square (butting) end and rounded ends are among the possibilities, but raster generated images can include others) — the CAD system's design rule checker must either know the details of the rounding, or always check to worst case (assumed to be rounded ends);
(ii) whether or not tracks need to stop short of PTH pad centres in order to leave white space;
(iii) how much curvature, if any, is permitted at the corners of square or rectangular pads (where these have to be drawn by construction);
(iv) how to draw pads on photoplotters with fixed aperture sets which do not include the exact shapes required;
(v) which layers to mirror image, and whether or not text on the layer is also mirrored;
(vi) which groups of layers to plot together for special effects;
(vii) how to generate resist artworks, solder cream artworks, screen printing artworks (possibly for both sides with mirror image control), powerplanes and drilling drawings.

If the chosen CAD system cannot handle these matters as part of its basic design, fudging the artwork to achieve the right effect is likely to be both time consuming and error prone. The ability to produce proof plots to see the effects of the controls (but on a matrix or pen plotter prior to commitment to photoplotting) can represent a major saving in both time and cost.

Very few board design stations include a photoplotter as part of their equipment — the starting cost of $30K-40K for the minimum practical machine climbs steadily with performance. Most CAD systems have attached to them one or more of three general types of plotter.

10.5.3.1 PEN PLOTTERS

These range from simple A4 paper size single pen types to A1 or even huge A0 machines with multiple pen carriages. With good software, a multiple pen carriage can contain pens of the same colour but different widths which the

CAD system can choose to save drawing time. 'Blocking' pens (drawing wide one way and narrow the other) are not normally of much use in electronics CAD work.

The major drawback with pen plotters is accuracy. Inaccuracies can stem from:

(i) precise static reproducibility of pen position, particularly when approaching a co-ordinate from different directions;
(ii) drawing width of the pen;
(iii) overshoot;
(iv) registration of different pens.

Only the very highest quality plotters are adequate for 2:1 PB artwork generation. Almost any reasonable plotter using fibre pens can produce acceptable schematic diagrams or PCB documentation.

A lesser issue is that of ink flow — capillary pens in plotters are notorious for temporarily (or permanently) stopping ink flow, thinning or pooling ink. If a plotter is to be used for precision work, it is best to have a demonstration in which the actual pens required are used to perform a sample of the critical work — and no excuses for using fibre pens! Some plotters with pressurised pens and good autocapping can cope — most can not. If plotting on film is a requirement, remember that ink density can be a critical factor on film, and often requires a different pen tip (e.g., crosspoint nib).

10.5.3.2 MATRIX PLOTTERS

Matrix plotters are basically standard computer printer peripherals offering individual control of the needles. The industry has mostly adopted variations of the 'Epson' protocol for this purpose — do check that the printer and CAD system are compatible. Typical theoretical resolution for 24 needle matrix printers is 180 dots/inch, with 360 dots/inch on 'up-market' models. Dot placement is at least this good, so such plotters are capable of remarkably good work. Absolute accuracy is unusual, but the dimensions are usually very reproducible. Software with precise scaling adjustment can be calibrated to produce precisely dimensioned results. Reproducibility in the paper feed direction is the most critical factor — for accurate work a tractor feed and edge-punched paper designed for plotting are essential.

Most matrix printers use a standard impact ribbon and obtaining a dense image for transparencies is just not practicable. A number of photographic agencies can produce adequate 2:1 reduced PCBs from top lit paper plots provided that the software can use multiple passes to achieve reasonable density on good quality paper. The maximum plot size is limited in one dimension by the carriage width, typically 13.2 in. (335 mm) maximum (which can accommodate a double Eurocard at 2:1). Clever software can produce strip plots for large widths, but this is obviously unsatisfactory and is completely unacceptable for printed board artwork.

Colour plots are really only practicable with ink-jet versions of matrix printers. The accuracy and resolution of these is otherwise disappointing and they are suitable only for documentary and proofing purposes — where they serve rather well.

10.5.3.3 LASER PRINTERS

Laser printers are a form of matrix plotter, but offer typically 300 to 600 dots/inch (dpi) resolution and a dense image on paper. Their primary use is to provide high speed schematics, documentation and check plots. Do not imagine that, if a printer specifies eight pages per minute, this is the speed at which graphics are plotted — sending the approximately 1 Mbyte of data for the image through the standard parallel interface can take a minute in its own right. However, it is still usually much quicker than pen or standard matrix plotting.

Plotting on to transparencies is possible, although it is difficult to achieve the density required for PB artwork. Further, the paper size limitation (typically A4) and the resolution at 1:1 make PB production only a marginal possibility (which is nevertheless being used in undemanding cases). The figures for this chapter were all produced on an ordinary desk-top A4 laser printer at 300 dpi.

10.6 COMPUTER AIDED MANUFACTURING (CAM)

The PCB design cycle inherently includes many aspects of computer aided manufacturing — the preparation of the trace layout masters is an obvious first step. Areas more explicitly related to CAM include:

(i) *C*omputer *N*umerical *C*ontrol (CNC) machines. These are commonly used for drilling and profiling. The two 'standard' character codes used for programming CNC machines are EIA and ASCII. The situation is confused, however, by the fact that drill machine manufacturers employ different codes to command the machine to carry out its tasks. The most commonly used of these, developed by Excellon, is, fortunately, readily translated to drive most other machines.

Blind and buried vias require that the CNC generation software can be controlled to generate separate sets of drilling instructions for different sections of the multilayer sandwich. For efficiency, it should be possible to produce drilling instructions for holes drilled through all layers, plus separate sets of instructions for additional holes for blind or buried vias.

(ii) Pick-and-place machines. These really offer no standards at all, see Chapter 7. The CAD system has to provide, at a minimum, tabulated information that can be 'massaged' into a format acceptable to the machine. The difficulties of fully automating the design process of preparing mixed component bandoliers and related tasks are so peculiar to a given machine and product type that any such design tools have to be tailored to the specific requirements.

(iii) Board profiling machines. These are more straightforward to program, but in the context of PBs the task of manual preparation is usually so trivial (or the cards are of a standardised size) that there is little need to develop a program at the design stage.

(iv) Areas such as stock movement, ordering and control. These are traditionally handled by database systems. The requirement here is to obtain from the CAD system a parts and accessories list in a format acceptable for input to the database system. Translation of electronic part numbers into company stock numbers may be a requirement.

10.7 INTERFACES

The interfaces discussed here are those points of contact between the PCB CAD system components themselves and other software which needs to accept or provide information used by the PCB CAD system.

One major factor to consider is whether or not the formats used within the CAD system itself are published. The standard graphics format is usually highly encoded for compact storage and reading at high speed, but some systems offer an ASCII (visible text) version of this and publish its format. Such published data can allow the user to extract all sorts of unusual data for special requirements.

Interfacing different graphics systems can be a major problem — the graphics inevitably reflect the philosophy of the system design, supporting the special types of objects needed by the particular CAD system. The most common graphical interchange format is DXF,[2] an ASCII (i.e., readable text) format of graphical data which can be generated and used by most mechanical CAD systems, including its originator AutoCAD. Whilst this is strictly a proprietary standard, the widespread use of AutoDesk products has rendered this format almost universal. Unfortunately, DXF is primarily geared towards 2-D and 3-D drawing, and needs various tricks to provide features such as pads, vias, track necking, etc. However, interchange of basic graphical data using this technique is useful for the design of housings, panels and other mechanical adjuncts. Graphical conversion can be useful, but do not expect converted schematics to retain any machine-analysable meaning, or a PCB to be easy to modify.

As has been seen, almost every PCB CAD system can generate 'Gerber' format photoplot data. Some systems can also read such data and convert them to the internal data representation for subsequent modification and re-plotting. However, data loaded in this way inevitably lose information — do not consider such links as adequate for more than capturing existing designs on to the CAD system.

A number of 'universal' interchange formats exist that are specifically designed for schematic and PCB design systems. Some of these do little more than pass a sophisticated net-list, although this may be useful in its own right. More sophisticated formats are determined by the style of software and, in practice, are standards introduced by small cartels of companies.

Other electronics CAD proprietary formats are in use for interchange, the main ones being those originated by the Futurenet schematic entry systems. These are often used for interchange between disparate schematic entry and PCB design systems. Such interfacing severely limits the possibilities for automatic annotation and application of PCB modifications — a properly integrated schematic and PCB system is usually a better option.

All except the smallest users or design bureaux will require integration of their CAD systems with their company stock control and ordering system. This can be a major difficulty, requiring software work to generate files from the CAD system in a form appropriate for the database. Some systems allow some sort of free-format parts information which can be extracted for direct input to the database system. It is worth investigating the database input format options in advance of choosing the CAD system in order to assess the work required to provide the link between the two — even if this work is contracted out, the customer still pays.

10.8 THE PROCESSING PLATFORM

The choice of computing machinery for PCB CAD falls into a number of related options for hardware and operating systems.

10.8.1 The Personal Computer (PC)

The 'industry standard' personal computer, the PC, is a machine broadly derived from the design of the original IBM PC. This standard has now diverged into a number of areas but most are highly compatible with carefully constructed application programs.

The processors used in these machines cover a wide range of performance although they all support a 'lowest common denominator' set of machine instructions. Unfortunately, the need to function across a wide range of equipment leads to most software only using the minimum set. High performance machines of this type now normally include numeric co-processors for efficient handling of floating point numbers which can produce substantial improvements for CAD performance. Machines without this capability should be avoided.

The operating environments available on PCs vary very widely. The base PC standard is still Microsoft DOS, also known as MS-DOS. DOS (disk operating system) does not provide proper multiprogramming capabilities beyond a simple background printing facility. However, third party products extend DOS to do background communications, pop-up desktop tools etc. Products like DESQview on 80386 and later machines extend DOS to perform true multiprogramming — ideal for background routing.

Microsoft Windows is a presentation layer running on top of DOS that provides a graphical user interface and co-operative multiprogramming. Different applications share the processor to varying degrees of success, though a well written suite of software can be made to co-operate very well. More advanced operating systems use pre-emptive multitasking which guarantees more controllable sharing of resources. The key advantage of Windows is the uniformity of the graphical user interface (GUI). This provides a consistent look and feel across all of its applications by using a uniform style of menu bars, pop-up menus, icons, sizeable overlapping windows etc. This uniformity helps the learning phase but can be over-estimated in complex fields such as electronics where there is no substitute for understanding what is going on.

Marketing considerations now mean that hardware manufacturers are obliged to provide Windows drivers for their hardware, allowing the user to mix and match equipment virtually at will when the applications are all proper Windows implementations. Windows programs invariably run more slowly than optimal DOS equivalents because of the large overhead in GUI environments. However, the steadily increasing power of processors together with use of intelligent graphics cards makes this a steadily reducing factor.

Operating systems such as IBM's OS/2 and Microsoft's NT are broadly based at very high end workstations and for file servers on networks. They both make large demands on system resources. The fact that these operating systems provide Microsoft Windows compatible environments points to the future of user interfaces. Their penetration into the market is limited and it remains to be seen if one of these or a development thereof will become dominant, or whether developments of the lower scale operating environments like Windows will become the future standard for workstations.

UNIX[3] is well established as a multi-user and multiprogramming environment aimed primarily at the technical user. On PCs it comes under a number of guises (e.g., AIX, XENIX, SCO UNIX and AT&T UNIX[4]).

The main difference between buying software for DOS/Windows and for UNIX is in the way it can be packaged. DOS with the IBM PC provides a single (fairly) consistent interface that a single product can satisfy, this leading to the 'shrink-wrap' box style of software marketing. UNIX is currently customised for each machine type. Each version of UNIX requires different application software, making it impractical for any one vendor to satisfy the whole market.

However, the advantages of standardisation are clear, and the various UNIX vendors are working towards a binary applications standard for 80386/80486/Pentium UNIX, as well as for RISC (*R*educed *I*nstruction *S*et *C*omputer) processor designs. (RISC processors are highly optimised by doing one or more fundamental operations in each clock cycle, but have led to a number of incompatible designs.)

One of the major innovations in standardising UNIX has come from the Massachusetts Institute of Technology (MIT) in the form of the X-Windows system which provides a consistent interface to whatever graphics facility the machine has. A number of PCB CAD products are now based on it. Specialised 'X' protocol terminals are available to extend a single computer's use to a number of users.

CAD places great demands on processor performance and requires close coupling of processor and graphics for good performance and for CAD purposes the multi-user aspects of UNIX are of primary consequence only in that they allow extensive networking and use of a shared database on file servers. The inherent UNIX multiprogramming environment using pre-emptive scheduling (with priority given to the interactive user in this instance) is the optimal way to handle background tasks such as routing.

UNIX is also important as the common operating system used by specialised (non-PC compatible) workstations.

10.8.2 Memory Needs

Random access memory (RAM) requirements for various configurations of operating system, applications functions and number of users are highly dependent on the particular software in use. RAM is comparatively cheap and has a significant influence on the performance of routing and multi-programming. For Windows, less than 8 Mbytes should be avoided. For UNIX, NT and OS/2, the sky is the limit!

One or more fixed disks (also called 'hard' disks) will be present on any serious system except possibly a workstation on a high speed network that accesses a disk on another machine, normally known as the 'file server'. The size of disk required for convenient operation of DOS based systems will be 80 Mbytes or more depending on how many design jobs are held on it at any one time and the number of libraries stored. For UNIX systems, a disk capacity in excess of 300 Mbytes is the norm.

Hard disks are very reliable. Because of the long period between failures, it is easy to forget that they do fail and, when this occurs, it is common for all of the data to be permanently lost. Common sense dictates that the contents of the hard disk should be regularly copied to off-line storage. Although much maligned, floppy disks will serve for DOS and Windows machines. The floppy disk has the advantage that you do not need any specialised hardware on the machine on which you wish to restore the file. Selective back-ups (using date stamps or 'changed' attributes) can limit back-up time to two or three minutes a day, with an occasional full system archive when the set of disks becomes rather large.

If your patience will not run to archiving on to floppies, then the natural alternative is a cartridge tape streamer. These can record hundreds of Megabytes on to a single tape without attention while you have lunch.

10.8.3 Data Transfer

Ensure that the PC you purchase can write disks acceptable to the processing house. If not, there are a number of transfer facilities available (which do not have to be PCB specific) but there is an obvious delay and irritation factor.

Transmitting files to your processing house using a modem obviously avoids disk format problems. However, modem communication introduces a new set of potential compatibility problems in the areas of modem standards, link speeds, protocols and the like.

10.8.4 Machine Architecture

The 'PC' architecture may have one of three basic extension busses for the addition of local area networks, communication ports, some types of mouse and similar items:

(i) ISA (Industry Standard Architecture) appears in both 8 and 16 bit versions, the latter often referred to as the 'AT' bus because of its first appearance in the IBM PC/AT.
(ii) EISA (Extended ISA) is a non-IBM standard which accepts ISA cards, but provides extended capabilities for EISA specific cards.
(iii) IBM's MCA (Micro-channel Architecture), now licensed to a number of vendors including Olivetti, uses a different card form factor and connector type but targets the same performance requirements as EISA.

Furthermore, most machines now support local busses, these being more intimately connected high speed busses for memory and critical controller interfacing (e.g., memory and graphics cards). The PCI is the current market leader, but other forms will undoubtedly emerge to match ever higher performance requirements.

For CAD the critical bus is that to which the graphics controller is attached. The best option is to purchase a configured system that makes special hardware and software provision for high performance vector drawing under the target operating system. The extension bus becomes an issue only in terms of required extensions for comparatively low performance items such as printers, back-up devices, CD-ROM, local networks, modems and so on. In this respect ISA and EISA are equivalent.

10.8.5 Graphics

Graphics on PCs have to some extent become simplified by the loss of many incompatible standards through which the PC architecture has evolved. Most modern high performance machines now fall into one of the following categories:

(i) VGA (Video Graphics Array): 640 x 480 with 16 colours is the base standard for modern machines, and is well specified and standardised. It is barely adequate for printed board layout but usable for schematics. Much software that can use specific higher resolution capabilities (but not the one you have!) will work with this standard on any graphics card.

(ii) Super-VGA: Typically 800 x 600, 1024 x 768 or 1240 x 1024, with 16 or 256 colours, these are cards broadly based on VGA (and invariably supporting plain VGA mode) but with proprietary extensions to include higher resolutions. At the higher resolution end it is important to choose an implementation with a graphics accelerator such as the S3 chip if the additional pixels are not to affect graphics drawing speed. If you are using DOS, ensure that the graphics card is compatible with drivers available for your chosen software. For ergonomic reasons, check that the graphics card and monitor combination can support high scan rate non-interlaced operation at the required resolution.

10.8.6 Networks and Workstations

Large organisations may wish to network their machines. Such networking allows a central base of libraries, access to files between members of a team working on a project, and shared access to printers, plotters, etc.

Networks are not necessarily built using obviously compatible equipment — often personal workstations on PCs with DOS are supplemented by file servers running UNIX. The optimal mix has to be decided in conjunction with a network vendor. Simply sharing printers and plotters can utilise very simple and low cost networks that need little planning.

10.8.7 Processor Performance

When comparing processor performance, the days of being able to look at the clock speed and know which machine is faster are long gone. Processor performance is now a result of:

(i) Clock speed — but some processors use the clock differently from others, so that direct comparison can be very misleading.
(ii) Memory cycle time — but if there is a cache how much does it matter?
(iii) Cache — how fast? how wide? how big? what algorithm? — all have effects that are very difficult to judge.
(iv) Processor data width — generally, the wider the better given that other factors are equal.
(v) Claimed MIPS (*M*illion *I*nstructions *p*er *S*econd) — but what do the instructions actually consist of? — a *C*omplex *I*nstruction *S*et *C*omputer (CISC — such as used in PCs) processor may perform 3 or 4 RISC instructions as one instruction.
(vi) Claimed MFLOPS (*M*illion *F*loating *P*oint *O*perations *p*er *S*econd) — but what proportion of operations in the application is being considered?

Even modern, high level benchmarks can be affected by compiler optimisation and the like, as much as or more than by the processor. In UNIX and other virtual memory systems (or if the system overlays memory), disk performance may be a critical issue. Only the CAD supplier is likely to know what the critical performance issues are for the software.

10.9 MAN-MACHINE INTERACTION

Much CAD work uses specialised input devices to provide a fast and intuitive interface.

The typical input device is the 'mouse' — a small device which slides over the desk or on a special pad to input cursor movement. One to three buttons on the mouse action select operations or mark positions. The mouse is a relative movement device — picking it up and moving it to a convenient place has no effect on the cursor position. Two basic standards exist for mice — Microsoft Mouse protocol and Mouse Systems protocol. Make sure the CAD system and the mouse are compatible.

The digitiser is a mouse-like device (usually called a 'puck') which is used on an active tablet. It transmits the position of the puck to the computer, together with the position of one to twelve buttons. Such tablets can be used to provide access to printed menus placed on the tablet, or for inputting existing drawings (e.g., PCB layouts) using absolute positions. If these features are wanted, check that the CAD software can support them — in particular, digitising of existing drawings is not commonly supported by PC-based CAD packages. Digitisers can cost from a few hundred dollars to tens of thousands of dollars, according to size and accuracy. If there is no real need to digitise existing drawings, a mouse will be found more convenient. There are many protocols used by digitisers, and many digitisers can be switched to a number of them. Do not buy a digitiser without seeing it work with the available software.

The much-maligned keyboard is still the best means of providing text input, although some digitiser manufacturers will try to convince the user that selecting characters on a tablet is easier. It is often also the quickest way for experienced users to enter commands during graphical editing. Menus are important for learning and operation of unusual features, but routine operations can often be invoked by single keystrokes.

Menus fall into a number of styles. They can be presented on a separate command dialogue screen, in reserved spaces on the graphics screen, pulled-down or popped-over graphics, or in printed forms on graphics tablets. On-screen menus are generally more helpful than tablet menus because they are context-sensitive — they can offer only currently valid and useful operations, together with explanations. Some systems use text names for selections and others use the currently fashionable icons. Icons in PCB CAD tend to become overloaded. There are too many options in a fully functional PCB layout package to have easily distinguishable icons for all of them. These are all very personal matters, and the menu system demonstrated may not be the only one which a particular CAD system can use. Ask to see the options.

10.10   CHOOSING A CAD SYSTEM

Most first-time CAD buyers, and virtually all small companies, will choose a PC-based PCB CAD system — the cost advantages and the ability to use the computing equipment for a wide range of subsidiary functions make this a natural choice. Higher power platforms and networked systems have their advantages, but both hardware and software come at a premium price related to the comparatively limited market for each variant.

Choose the software first — the precise configuration of the computer hardware best suited to the CAD package varies from machine to machine. Obviously, price is a factor in the choice of a CAD package, but remember to cost the invisibles — learning time, and the effort to reproduce the drawings and libraries, should a change to a more functional system be necessary later. A number of systems are available in modular form so that specialised modules can be added as required. If the chosen software is available on both PCs and more complex hardware, with compatible file structures, options are left open for the future.

Look at the track record of the supplier — many PCB CAD systems now appearing on the market are so cheap that proper support is out of the question, and an unfortunately large number of such systems disappear from the market very quickly, together with the customer back-up. Some suppliers are merely dealers or importers who are only interested in shifting boxes and can provide very little by way of support. Ensure that the package chosen is well established, and supported by the photoplotting and/or board manufacturing agencies — these people know which packages are really successful and appreciated by their users.

NOTES

1   EBCDIC — This stands for *extended binary coded decimal interchange code*, a character coding used by early IBM machines and some current mainframes. The majority of PCs use ASCII and some CNC machines (e.g., certain drilling and routing machines) use EIA. Essentially, these are different arbitrary numbers for representing characters.

2   DXF — This is the name given by AutoDesk to their *data exchange format*. As with proprietary standards in general, there is no guarantee of long-term stability of such formats. That is, it may be found that the format of an updated version of a given CAD

software has been changed to such an extent that the immediately preceding version cannot read output from the later version. This can cause real problems in converting files to the later, more powerful software.

3 UNIX is a real time disk operating system originally developed by Bell Laboratories which is now licensed for use on many machines, including PCs — provided the disk has at least 200 Mbytes capacity. SCO UNIX is the *de facto* standard for PCs, sold by Santa Cruz Operation, the primary PC UNIX vendor.

4 Each licensed version of UNIX varies from the others by virtue of different extensions, binary formats and by different processor instruction sets, where applicable. The original licensing did not allow the licensee to use the name UNIX. This led to a whole range of peculiar names for UNIX in different variants. While the licence limitation has now stopped, some vendors carry on with their own names. AIX is IBM's UNIX variant and XENIX was coined by Microsoft.

Chapter 11

THERMAL MANAGEMENT ASPECTS

D. J. DEAN
Pearl Publications Management Ltd and Thermal Associates,
Reading, UK

11.1 INTRODUCTION

This chapter considers some of the effects caused by temperature, and describes methods of taking them into account during the circuit design stage.

The generation of heat and the associated rise of temperatures are unavoidable by-products of the operation of every electronic device and circuit. Increased temperatures are undesirable, with a reduction in reliability and impaired performance often resulting. Part of thermal design consists of examining these temperature increases and deciding which construction and layout alterations will result in optimum changes of temperature. Temperature prediction has therefore become a part of system design.

In electronic thermal design too much time, effort and resources are often expended on attempting to obtain temperature predictions to a degree of precision far beyond that which is needed. This chapter is restricted to methods whereby this effort can be limited to what is required to achieve the desired levels of system performance and reliability.

11.1.1 Thermal Design

Except for a few well known special cases, the hottest devices have the highest failure rates. Reduction of temperatures will therefore normally increase reliability. If a certain system, or one that is very similar, works, then improvement in heat flow paths will reduce temperatures and almost always extend the effective life of the most sensitive devices. The degree of correspondence between the input and output performance of a device and/or circuit and its required specification is usually also temperature-dependent.

Before embarking on a thermal design programme, it should be noted that temperature-dependent failure mechanisms are not the only factors which cause system malfunction. As a general rule the number of failures directly attributable to thermal effects can be taken to be a third of the total. They are concentrated in components operating at the higher temperatures; hence as component and power density increase so does the proportion of thermal failures. At lower temperatures systems may fail because of non-thermal effects long before they would fail as a result of temperature. If the number of failures due to elevated temperatures is a small proportion of the total, other areas of the design should be examined first.

11.1.1.1 THE INFLUENCE OF COMPONENTS AND COOLING METHODS ON TEMPERATURE

Operating temperatures are influenced by the physical make-up and cooling methods of the system. The following techniques will reduce temperatures and generally improve reliability:

1. Increasing thermal conductivity. This is achieved by, for example:
    (a) using metal-cored boards instead of standard epoxide boards;
    (b) using thermal ladders and planes.
2. Increasing heat extraction areas. This is achieved by, for example:
    (a) using thermally bonded clamps along the edge of boards instead of just relying on the edge connector;
    (b) removing heat from the circuit board surface, i.e., convection cooling and rear-mounted heat sinks.
3. Increasing efficiency of heat flow paths. This is achieved by, for example:
    (a) increasing copper thickness and width;
    (b) using shortened heat flow paths;
    (c) using power devices with incorporated heatsink pads.
4. Increasing the effective heat flow area. This is achieved by, for example:
    (a) using larger contact areas, such as larger area packages deliberately constructed to have a smaller internal thermal resistance;
    (b) using heat spreaders;
    (c) using thinly spread heat conducting compounds (see 11.4.4).
5. Placing sensitive and high-power devices close to heat extraction points; note that sensitive devices can be affected by the proximity of high-power devices; hence, power distribution is important (see 6).
6. Distributing power input. For example:
    (a) where the heat loss is predominantly from the surface, the more uniform the distribution of power-dissipating elements the lower the temperature;
    (b) where edge-cooling predominates, 5 applies;
    (c) in most cases a combination of these approaches is most effective.

Note: The greater the density of power input, the higher the temperature reached. As packages become smaller, the packing density also tends to increase along with the resultant board temperatures. This increase is usually partially mitigated by a reduction in the temperature differences associated with the shorter thermal paths of the small packages. This is especially the case where all the heat is effectively lost by convection from the surface or to a rear-mounted heat sink.

Where these techniques can be employed with little or no extra cost, they should be incorporated. Part of thermal design lies in balancing the cost of implementing improvements with the benefits gained.

11.1.1.2 THE RÔLE OF TEMPERATURE PREDICTION

An electronics system's performance and many of its failure modes are temperature-dependent.

Except in a few cases involving moisture, failure rates increase with temperature. The power output of an active device also normally increases with higher temperatures, further affecting both performance and reliability.

Modest temperature reductions can result in dramatic improvements in device reliability, and temperature prediction is used to obtain cost-effective improvements in reliability and performance. Temperature prediction is therefore an important constituent, but not the purpose, of thermal design. The effort put into temperature evaluation is often wasted, but can be minimised by considering which aspects of such predictions enable reliability and performance improvements to be made.

Temperatures can be reached at which systems will not operate at all, and at lower operational temperatures systems may fail due to non-thermal effects long before they would fail as a result of temperature. Within the range of temperatures at which significant thermal failures occur, it is important that the designer be able to determine either that a change of design will result in a certain increase in reliability (hence saving a given amount of money), or that a new design will have an optimum combination of performance, reliability and cost.

11.1.1.3 HOW FAR THERMAL DESIGN?

Thermal design enables both reliability and performance effects to be taken into account at the circuit design stage.

Sound thermal design means:

(i) choosing the best route to achieve optimum thermally related reliability and performance taking the cost into account ;
(ii) validating instinctive thermal design decisions;
(iii) knowing when diminishing returns render further improvements to thermal design uneconomical.

Thermal design and its implementation costs money; hence the level of design appropriate for a given system is always a trade-off between design and implementation costs on one hand, and the likely costs of not including the thermal design on the other. Examples of the costs that can occur as a result of not including thermal design are: the actual cost of servicing the thermally induced failures, the value that can be placed on reduced space and power requirements, and reduced profits due to lost sales.

11.1.2 Format of this Chapter

This chapter caters for several different situations:

(i) the designer new to thermal design with little or no background experience of the subject;
(ii) the designer with an established piece of equipment who wishes to reduce its temperature failure rates or improve its performance;
(iii) the designer developing a new system.

The fundamental mechanisms of heat transfer are covered in 11.2.1, while 11.2.2 discusses the effect of temperature on performance and reliability. Only the designer new to thermal problems need read 11.2.1, whereas all should read 11.2.2 which describes the underlying basis used to develop the approaches used in this chapter. The way in which thermal design is tackled in this chapter is introduced in 11.2.3, and the principles applied in 11.2.4.

A range of printed circuit board constructions is available and, in many cases, the different constructions can give rise to non-obvious thermal properties. How these can be readily evaluated are described in 11.3.

The designer who has to make thermal design decisions only very occasionally, or who needs to make them away from his home base, requires some reasonable estimates of temperature to assist with the making of these decisions. These estimates are obtained by the 'First Look' thermal design methods outlined in 11.4. Reviewing a design using 'first look' estimates is relatively inexpensive, and allows the designer to determine whether the temperatures on the boards are unacceptably high or satisfactorily cool. A simple review of this kind is often sufficient to suggest substantial improvements in layout.

Techniques for the designer whose involvement in thermal design work would not justify investment in formal temperature prediction methods, but who requires an increased precision for his decisions are provided in 11.5.

Where thermal design is an infrequent but continuing requirement, a small investment of time or money in thermal design tools such as desk-top computer programs will be well justified. These programs, which can give the designer added confidence in his own decisions, are described in 11.6. Care has to be taken in the selection of programs, as methods suitable for many engineering systems do not necessarily give the correct answers when applied to small heat sources. Some factors which should be considered when selecting programs are given in 11.6.1.

Some of the special features involved in predicting temperatures arising on circuit boards undergoing convection cooling are described in 11.7. Other thermal aspects, including thermal mismatch and phase-change cooling, are dealt with in 11.8.

For ease of reference, edge-connected printed circuit boards are dealt with first, then rear-cooled boards, and finally packages.

11.2 HEAT TRANSFER MECHANISMS AND TEMPERATURE EFFECTS

Heat flows from a higher to a lower temperature. Whatever the mechanism of heat removal from the generating components, it occurs in association with elevated temperatures. Without the elevated temperatures, the heat could not be dissipated, but it is these same elevated temperatures which affect the reliability and performance of the system.

11.2.1 Mechanisms of Heat Transfer

There are three mechanisms of heat transfer: conduction, convection and radiation. Heat can also be absorbed by a phase change. Detailed descriptions of the mechanisms of heat transfer are given in most books on heat transfer, and Reference 1 includes a chapter on the subject. Therefore, only a brief description of the different mechanisms is given here.

Many heat transfer and dissipation rates change with temperature. Most readers will find the methods contained in this chapter adequate without needing to consider the effects of temperature on physical properties. Where these effects have to be taken into account, detailed coverage is available in Reference 2.

11.2.1.1 CONDUCTION

Conduction involves the transfer of kinetic energy from one molecule, atom or electron to another. It is the dominant mechanism for heat transfer within solids.

The rate of heat transfer by conduction is proportional to the temperature gradient, the cross-sectional area and a heat transfer coefficient. This coefficient is referred to as the 'thermal conductivity, K'.

11.2.1.2 CONVECTION

Convection occurs only in fluids and is often the prime means of heat transfer in such materials. The mechanism involves the transfer of heat by mixing of fluids. When the mixing is due solely to the density differences which result from different temperatures within the fluid, the mechanism is referred to as 'natural convection'. When the mixing is caused by artificially induced fluid motion, it is referred to as 'forced convection'.

In most electronics thermal design calculations, it is taken that the heat transfer from a solid to a fluid is proportional to the temperature difference between the solid surface and the fluid, the area, and a heat transfer coefficient, h. The relevance of this approach to the temperatures occurring in electronics systems is discussed in 11.7.2.

11.2.1.3 RADIATION

Radiation is the conveyance of heat by electromagnetic transmission. It is the only means of heat transfer between bodies separated by vacuum.

Because of the relatively small temperature differences usually found between circuit boards, radiation does not enter into most thermal design calculations. If a significant effect is present, it is usually sufficient to treat radiation in the same manner as convection.[3]

11.2.1.4 PHASE-CHANGE HEAT ABSORPTION

Phase-change heat absorption is due to the latent heat involved in changes of state and occurs in nucleate boiling and heat pipes. These techniques are described in 11.8.2.2 and 11.8.3.

11.2.2 Difficulties Involved in Evaluating the Effects of Temperature on Reliability and Performance

Ideally, it would be possible to obtain quantitative relationships of temperature to the reliability and performance of devices. In such a situation, and assuming it was possible to predict the actual temperatures of a system, the designer would have all the information required to make optimum design decisions as regards cost and thermal effects. Unfortunately, in most production situations factors exist which preclude the possibility of obtaining a set of complete and accurate data on which to base thermal design decisions. These factors are:

(i) In electronics systems the relationship of temperature to reliability and/or performance is not accurately known.
   Investigations into temperature-dependent performance and failure effects are normally performed in conditions of controlled environment, giving a uniform temperature. Because the temperatures that exist in actual electronics systems are spatially variable, the temperature relationships so obtained can only be treated as qualitative.
(ii) The temperature of the regions which most frequently fail cannot be accurately known.
   (a) Most thermal design computer programs give approximate solutions in relation to approximate models. When prediction of the temperature distributions arising from small heat sources is attempted, the rapid changes in temperature gradient with position and the very low thermal capacity of the sources can lead to cumulative errors. As a result there are several examples in the literature, based on computer-generated temperature predictions, which simple physical considerations clearly show to be wrong.
   (b) Temperature predictions are often based on measurements made using thermocouples or semiconductor components.
   It is well known that thermocouples can drain heat from small sources, thus giving artificially low temperature readings. The same effect can occur when thermocouples are used to read the temperatures of poor thermal conductors such as printed boards.
   Some transistor chips are specially designed for making thermal measurements and include a junction devoted to temperature indication. The temperature-measuring area is remote from the heat-generating area when compared with the thickness of the silicon. There is, therefore, always a significant decrease in temperature between the power-dissipating and temperature-measuring junctions. When devices are operated alternately to dissipate heat and to measure temperature, a delay of between 50 and 100 µs follows

removal of the heating voltage while a constant current input is established so that temperature measurements can be made. As the effective time constant of an active device is typically 10 μs[4] about 90% of device cooling occurs during the measurement delay time.
(iii) Electronics systems often exhibit wide variations in thermal performance between apparently identical units.

Whenever components are assembled, variations in the actual contact areas occur. Similarly, when parts are bonded there are likely to be different distributions of voids. Also lead length variations can effect turbulence and hence the distribution of air as it flows through the system. Specific points of heat generation are not always clearly defined and hot spots can move during operation.[5] In general, wide variations of thermal performance can exist between apparently identical units.

The documented data on the effect of temperature on failure rate and performance are therefore qualitative: a precise forecast of temperature can lead to only an approximate value for reliability.

It is often stated that germanium devices can operate up to about 70°C, while silicon junctions can withstand 150 to 200°C. These values are, however, qualitative, and the actual values can be much higher. This is shown by the many failures in silicon devices which involve the melting of aluminium-silicon eutectic at 577°C, sometimes over quite a large area. Some estimate of the actual temperatures can be obtained by calculation, but even these are not precise because of the variations in effective junction area that can occur even during operation.[5]

Fortunately, thermal design decisions made using qualitative values are valid in most situations, as the values being compared have been produced by similar methods.

Geometry has a considerable influence on temperatures: for example, design decisions tend to be optimistic for smaller effective junction areas and pessimistic for larger ones.

Because it is not possible to predict exactly the effects of high temperatures, it is not necessary to attempt to predict the temperatures with high accuracy, especially as this can be wasteful both of time and resources.

11.2.3 Approach to Thermal Design

The traditional approach to thermal design has been to attempt to calculate temperatures and then compare the advantages of the supposed temperature with the cost of production of the different systems. Temperature prediction methods are becoming increasingly prominent in the literature, yet because of inadequacies inherent in the thermal data the resultant forecasts can be totally misleading. The decisions made on such a basis are sometimes actually the reverse of what would have been correct. The tendency to push forward with more complex prediction methods has therefore to be resisted.

Despite the shortcomings both of thermal data and of temperature prediction, designers have to make decisions regarding electronics system construction and assembly. The designer needs to decide:

— In what way are the different temperature-dependent properties, or thermal effects, important?
— What aspects of temperature data are relevant to the different thermal effects?
— How can the required temperature data be most easily predicted?

The decisions required of thermal designers in relation to electronics system construction and assembly cover the areas both of reliability and of performance.

11.2.4 Designing for Reliability and Performance

As operating temperatures increase, the mean time between thermally induced failures decreases, until eventually the system will not operate. The first requirement of any system is that it functions, all other aspects being secondary. A major function of thermal design is therefore to ensure that new systems will work.

When a system is in production, or has been designed such that it will function, a further aspect of thermal design arises — that of optimising reliability with respect to cost.

The author has found that the most effective way of approaching thermal design is to start by posing two simple questions:

Will the proposed system operate?
Can the proposed (or existing) system be improved?

Reducing the circuit designer's requirements to two limited questions simplifies the temperature prediction techniques needed to support thermally related design decisions.

In effect, the crux of thermal design is to devise by any means a system that operates, and then improve it by comparison.

11.2.4.1 WILL THE SYSTEM OPERATE?

Many systems will operate over quite a wide temperature range, but they are all subject to a maximum operating temperature at which failure is certain.

All systems have limiting thermal characteristics which can be evaluated experimentally. Consider, for example, a silicon solid-state device. At 577°C the silicon-aluminium eutectic melts and tracks lose their integrity. Also, at about 515°C, before the eutectic point with aluminium is reached, silicon becomes intrinsic: the electrical characteristics change and the system will no longer meet its design requirements. Other effects can occur at lower temperatures, but they may sometimes be controlled by device and system design.

The temperature at which failure is certain to occur can therefore be specified, but it is not possible to predict accurately maximum operating temperatures. For this reason, in the vast majority of cases, the decision as to whether a design will work or not is based on past experience. Where this is not possible some other approach has to be used.

To be confident that the system will function, the designer has to know that there is no part which will undergo a temperature in excess of its maximum operating temperature. There is no requirement to know the actual temperatures of the operating system, only that they are not too high.

The most common means of deciding whether temperatures are too high is the worst-case forecast. In such an analysis, exaggerated extremes of all the various temperature contributions are added together. The designer then knows that his system will never produce the resulting sum. When the maximum operating temperature exceeds the prediction, the circuit will function. In practice, however, worst-case evaluations can lead to the rejection of good designs.

A less extreme approach can be used where the individual contributions are not necessarily greater than the real values but where it is certain that the summation is greater than the maximum possible in the actual system. Such a method involves the calculation of upper temperature limits which can be obtained by the techniques described in this chapter.

11.2.4.2 CAN THE SYSTEM BE IMPROVED?

In spite of the fact that failure rates cannot be precisely related to actual temperatures, decisions on the economics of design variations require changes in reliability to be forecast. These decisions are often most profitably made by predicting the temperature variations resulting from design changes rather than by attempting to forecast exact temperatures. Provided the accuracy of prediction is sufficient for design purposes, the method used should be chosen for its ease of use and interpretation.

The following general rule is very helpful in forecasting the effect on reliability of small changes in temperature.

Within the range of temperatures in which electronics systems operate, failure rates double for every increase of between 10 and 15°C.[6]

It is for this reason that a reduction of peak temperatures at the expense of an increase in the lower temperatures can often improve overall reliability. Besides being temperature-related, stress-induced failures can be further aggravated by the number of temperature cycles, see 11.8.1.

11.2.4.3 RELATIONSHIP BETWEEN TEMPERATURE AND PERFORMANCE

Performance (i.e., how well a device and/or circuit input and output matches the required specification) also changes with temperature. A circuit whose design has been based on room temperature data may not meet its specifications under higher operating temperatures. For instance, electrical resistance can be affected by temperature both directly and indirectly via thermal expansion mismatch. This, in turn, can result in stress and deformation affecting performance. For the reasons given at the beginning of 11.2.2 it is not possible to obtain a precise relationship between actual temperature and performance. Because it is not possible to forecast performance accurately, approximate predictions of temperature are normally sufficient for making performance-related design decisions.

As was the case with reliability, it is possible to predict that a given change in temperature will result in a certain change in performance with far more precision than can be achieved in forecasting the actual performance.

As resistance values and individual junction outputs vary with temperature, the matching of balanced circuits can also be temperature-dependent.

When considering the performance of balanced circuits, where the output differences should be minimised, it is again a knowledge of how close the temperatures of the different components are to each other, and not a knowledge of the actual temperatures, that is important.

Raising the temperature causes device output to increase. As well as having an effect on the performance, this increases the power dissipation and hence again the temperature, further reducing reliability.

11.3 PRINTED BOARD CONSTRUCTION AND EXAMPLES

The effect that thermal properties have on the temperatures which arise in a system is referred to in 11.2.1. Before temperature predictions can be made, the thermal properties of the components have to be specified. This is especially so in the case of printed boards, where there is a wide variation in the thermal performance of the materials used in their construction. The first step towards printed board temperature prediction has therefore to be the evaluation of effective thermal properties.

Six particular board constructions will be used to illustrate the determination of the effective thermal properties of the four basic types of circuit board. These same six boards are then used in 11.4 and 11.5 to demonstrate the application of the different temperature prediction techniques.

11.3.1 Types of Printed Board

The particular type of circuit board construction affects the distribution of heat losses and therefore the influence of the different heat transfer coefficients.

Although there is some overlap in their analysis and properties, printed boards can be treated thermally as being of four distinct basic types according to their construction:

(i) poor thermal conductivity board having high conductivity tracks bonded to it — for example, the conventional printed board;
(ii) high thermal conductivity materials having a poor conductivity surface, see Volume 2, Sections 1 and 2, for example, copper-cored boards;
(iii) poor thermal conductivity board material having deliberate heat pathways bonded to it — for example, thermal planes and ladders;
(iv) single-material circuit boards — for example, ceramic or glass substrates.

All four types are, strictly speaking, printed boards, but common usage has restricted that description to the first three types. The fourth type is now described as a 'hybrid integrated circuit', see Chapter 3. For convenience, this chapter uses 'circuit board' (or 'printed board') as the global description of all four types.

11.3.1.1 A COMBINATION OF CIRCUIT BOARD TYPES

One combination of circuit board types has considerable thermal potential.
In hierarchical interconnection technology (HIT),[7] a single circuit board ('daughter board') has a number of smaller sub-circuit boards ('child boards') mounted in contact frames and held in position by a metallic clamping frame, Figure 11.1. With this system, the clamping frames can be thermally

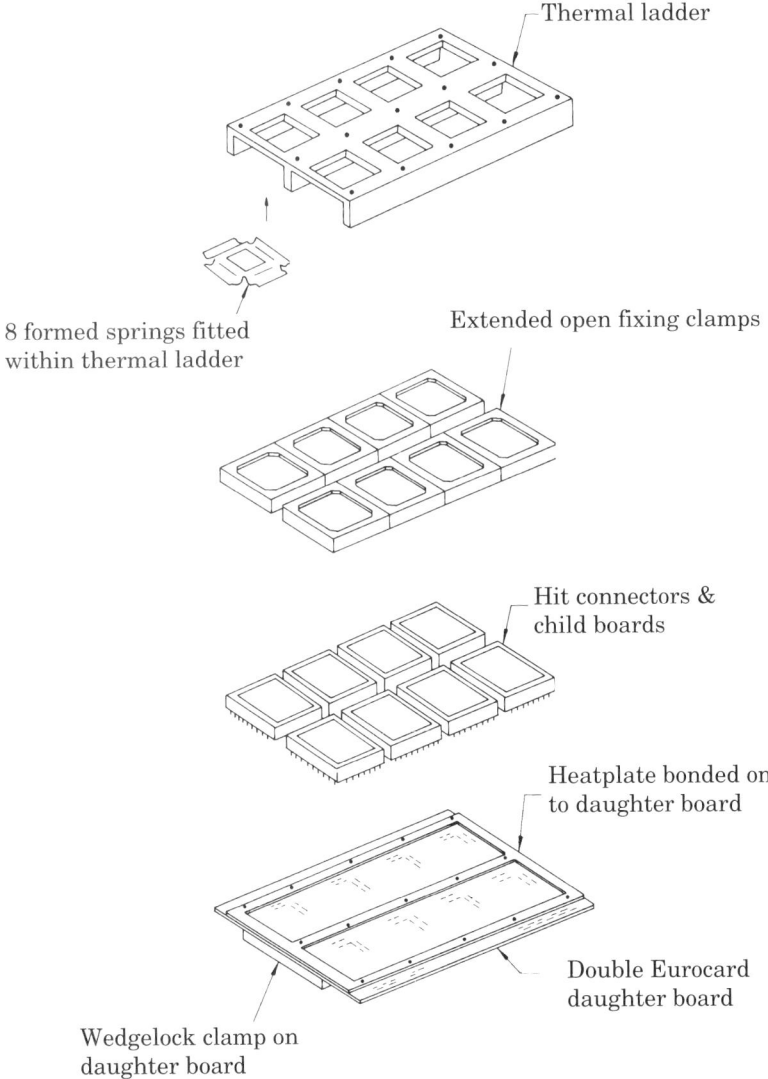

Fig. 11.1 Hierarchical interconnection technology (HIT).

interconnected, producing a thermal ladder. In addition, the sub-circuit boards which carry the high-power devices could be metal-cored. (It is often necessary to incorporate air flow deflectors to ensure adequate cooling of non-cored boards.)

11.3.2 Circuit Board Thermal Properties

Before temperature predictions can be made for conventional printed boards, it is necessary to specify the relevant thermal properties. The thermal properties of greatest importance are the thermal conductivity and the apparent rate of surface heat loss.

11.3.2.1 THE THERMAL CONDUCTIVITY OF BASIC EPOXIDE PRINTED BOARDS

Conventional printed boards have a low value of thermal conductivity. This often means that convection losses become the dominant mode of heat transfer.

The metallic layers are designed mainly for electrical performance and not, as is the intention with ladder conductors or metal-cored boards, to optimise the board's thermal conductivity. The effective lateral thermal conductivity of a printed board can range from that obtained by having no conducting tracks (bulk epoxide conductivity), to that where the copper is a continuous sheet. Typically, this effective thermal conductivity can range from 0.002 to 0.20 W/°Ccm, that is, varying by two orders of magnitude.

Two very effective visual assessment methods are available which give quick estimates of the thermal conductivity of a basic epoxide printed board.

The estimates provide averages over a relatively large area of circuit board containing several separate conducting tracks. The deviation of these estimates from the real surface averages is usually within the reproducibility of the other heat transfer characteristics.

One method uses an empirical formula based upon a combination of experimental data and mathematical analysis. The other involves visually matching the circuit board in question with a section of a chart produced from the results of computer analysis applied to a set of circuit boards having copper-clad areas ranging from no circuitry to a solid conducting sheet or 100% circuitry. (These visual means of assessment are in good agreement and detailed descriptions with comparisons are contained in Reference 8.)

(a) The empirical evaluation method (Dean's Empirical Formula taken from Reference 8) is:

$$K_b = K_{cu}(t_{cu}/t_{ep})\beta^2 + K_{ep}$$

where K_b is the effective thermal conductivity of the board
           K_{cu} is the thermal conductivity of copper
           K_{ep} is the thermal conductivity of the board's insulating material
           t_{cu} is the thickness of the copper
           t_{ep} is the thickness of the board's insulating material
and β is a visually assessed measure of the fraction of the board that is copper-covered.

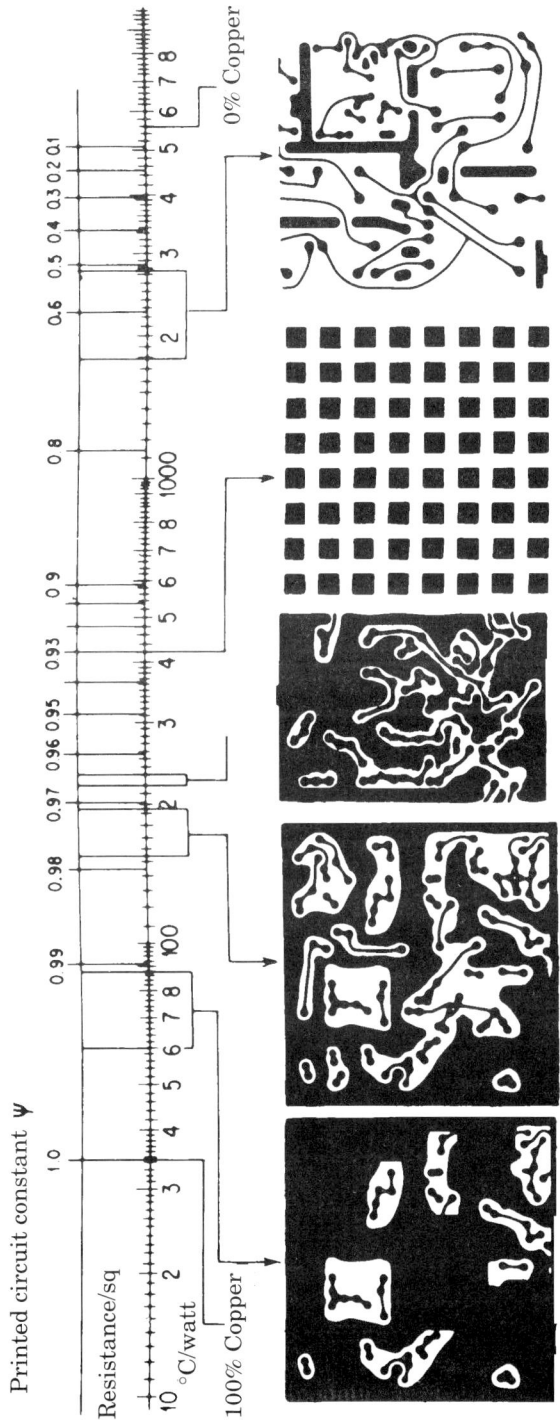

Fig. 11.2 Printed circuit board thermal resistance. (Courtesy of P. Dickerson, Reference 9)

When the board is predominantly convection-cooled, as assumed in this chapter, then ß is the fraction covered. When there is an appreciable amount of conduction heat loss, the value of ß is obtained from:

$$\beta^2 = W \cdot L^3$$

where W is the linear fraction of copper normal to the main heat flow and L is the linear fraction of copper parallel to the main heat flow.

Values of both W and L are assessed visually.

(b) The second visual assessment method involves the use of a chart developed by Dickerson[9] and reproduced here with permission as Figure 11.2. The chart provides the approximate bulk thermal resistance per square as a function of conductor configuration for any given circuit board. In order to use this figure one must compare the printed circuit board to be analysed with those shown in the figure and hence obtain an approximate value for thermal resistance per square.

The resistance per square numbers in Figure 11.2 are valid only for printed boards having a conductor thermal resistance per square of 35°C/W and a board thermal resistance per square of 5600°C/W. A thermal resistance per square for a printed board having different properties can be estimated using the following procedure:

(i) Read printed circuit board constant ψ from Figure 11.2.
(ii) Substitute into the following equation

$$Z_{sb} = Z_{seq} - \psi(Z_{seq} - Z_{scu})$$

where Z_{sb} is the average board resistance per square
Z_{seq} is the resistance per square of the insulating board
Z_{scu} is the resistance per square of the conductor material
and ψ is the printed circuit board constant.

Newcomers to thermal design often prefer to use the chart rather than the empirical formula but after a time they will come to prefer the formula because of the chart's limitations. When using the chart, care must be taken with a circuit board around 25% of which is covered by circuitry, as the layout is not typical of circuit boards and, in this case, the empirical formula is strongly recommended.[8]

11.3.2.2 EFFECTIVE THERMAL CONDUCTIVITY OF PRINTED CIRCUIT BOARDS HAVING A HIGH THERMAL CONDUCTIVITY CORE

The most typical of this class is the metal-cored board, for example, copper-cored epoxide boards.

With high thermal conductivity cored boards, the core acts in the same way as do alumina hybrid integrated circuit substrates, where all the conducted heat is essentially transferred within the basic alumina and the effective thermal conductivity is that of the bulk material. In the case of cored boards there is usually little contribution to the lateral heat flow from the epoxide. The level of this contribution can be checked using the empirical techniques of 11.3.2.1. If the surface material's thermal conductivity is significant, then it is combined linearly in proportion to the conductivity of the core. That is:

$$\frac{KT + 2kt}{T + 2t}$$

where K is the thermal conductivity of the core
          k is the effective thermal conductivity of the surface material
          T is the thickness of the core
and 2t is the combined thickness of both of the surface material layers.

11.3.2.3   SURFACE HEAT LOSS AND GAIN FROM BOARDS HAVING A HIGH THERMAL CONDUCTIVITY CORE

With boards having a high thermal conductivity core there is generally little contribution to the lateral (or sheet) conductance by their surface materials. The vast bulk of the heat flow therefore occurs within the high conductivity core. The heat loss or gain from the surface has consequently to travel through the whole thickness of the low conductivity surface material. There are other special cases such as porcelain- or ceramic-coated steel boards where the same approach to surface heat loss applies.

Consider a small area of the surface before the heat is lost by convection. The heat transfer through the surface layers is constrained from spreading by the heat flowing through the surrounding areas. This constraint maintains essentially parallel heat flow through the surface material. In such cases the heat transfer coefficient, h, as it affects the core, is modified[10] by the factor:

$$\frac{k}{ht + k}$$

where t is the surface material thickness
and k is the surface material thermal conductivity.

The total effective heat transfer coefficient for a board with a low thermal conductivity material on both sides of a metal core is hence:

$$\frac{2h.k}{ht + k}$$

Where circuit boards receive heat from a relatively small area on their surface, the metal core acts as a rear-mounted cooling plane, and the methods described for rear cooling in 11.4.4, 11.5.3 and 11.6.2.2 can be used to evaluate the heat input temperature differences.

11.3.2.4   POOR THERMAL CONDUCTIVITY BOARDS HAVING DESIGNED HEAT PATHWAYS BONDED TO THEIR SURFACE

When thermal pathways are incorporated with circuit boards they often become the predominant means of heat transfer.

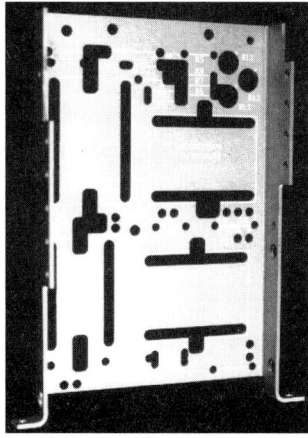

Fig. 11.3 Example of a thermal plane (heat sink).

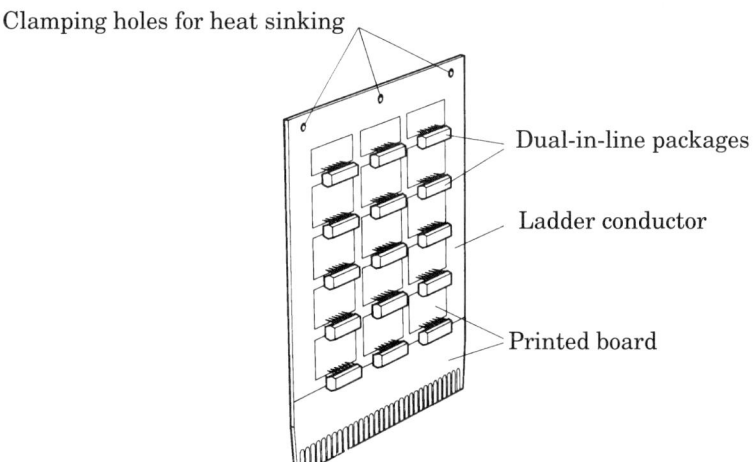

Fig. 11.4 Ladder conductor on a printed board.

Where the pathway is formed by a thermal plane and hence covers a very significant part of the surface, Figure 11.3,[11] the effective thermal conductivity of the combination may be obtained from the empirical formula in 11.3.2.1.

Where the predominant thermal pathways are discrete and sparse (for instance, a thermal ladder as in Figure 11.4), the pathways may be treated as an interconnecting mat of thermal resistances running parallel to the board itself, Figure 11.5.

The prediction of temperatures arising from the use of ladder conductors and some thermal planes is made using standard circuit analysis and an electrothermal analogue.[12] The thermal resistance is obtained from:

$$\text{Thermal Resistance} = \frac{\text{Temperature Difference (cf. voltage difference)}}{\text{Rate of Heat Flow (cf. current)}}$$

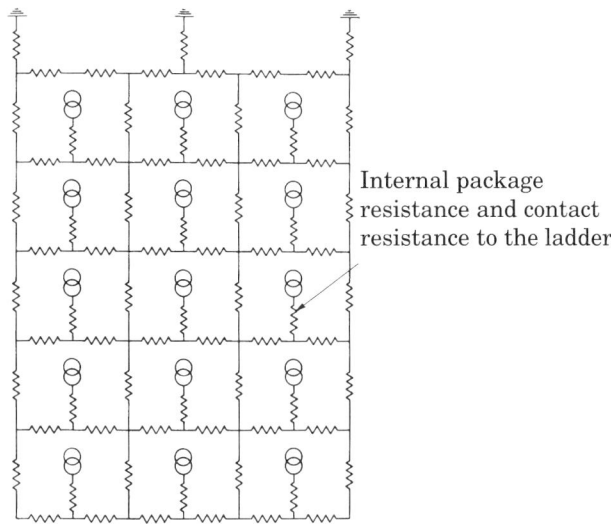

Fig. 11.5 Thermal resistance equivalent circuit.

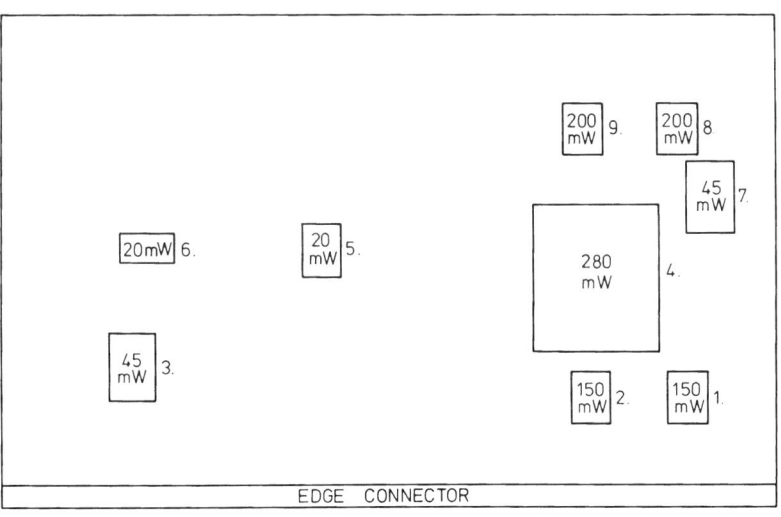

Fig. 11.6 Example layout and power levels.

11.3.3 Circuit Board Examples for Illustrating the Estimation of Effective Thermal Properties

A single layout, Figure 11.6, will be used to illustrate the different principles. The dimensions of the components and the circuit board will vary in proportion, depending on the particular type of board. The board is 10 units of length deep (L), 16 units wide (W) and has a thickness t. The dissipation is indicated in milliwatts in the figure and totals 1.11 watts (Q).

Six types of construction are used as demonstration examples. The first five are derivatives of conventional epoxide circuit boards and the last is a thick film hybrid integrated circuit, see Chapter 3, made with an alumina ceramic substrate.

The thermal conductivity of the materials used are as follows:

    Epoxide board: 0.0025 W/°Ccm
    Copper: 3.8 W/°Ccm
    Alumina: 0.25 W/°Ccm

The heat transfer coefficient for natural air convection cooling is taken as being 0.001 W/°Ccm^2 and for forced air cooling 0.01 W/°Ccm^2. These values are discussed in 11.7.2.

The first five circuit board examples are 6 cm long and 10 cm wide, while in the sixth example the exposed part of the alumina substrate is 2.4 cm long and 4 cm wide.

Type A

A 0.15 cm thick epoxide board with a linear fraction of 0.2 covered with 0.007 cm thick (2 oz) copper on one side.

From the empirical formula of 11.3.2.1, the effective thermal conductivity is:

$$3.8 \left(\frac{0.007}{0.15} \right) 0.2^4 + 0.0025 = 0.003 \text{ W/°Ccm}$$

Type B

A similar board to Type A but with a linear fraction of 0.45 covered with copper on both sides. The effective thermal conductivity is then:

    0.018 W/°Ccm

Type C

Again, a similar board to Type A but with a copper thermal plane 0.15 cm thick mounted on one side. The thermal plane is effectively continuous in the predominant direction of heat flow and covers a linear fraction of one third across the width of the board. As the plane is continuous in the direction of predominant heat flow, the effective thermal conductivity is the area fraction of copper times the fraction of copper in the combined thickness times its thermal conductivity, that is:

$$\frac{3.8}{3 \times 2}$$

$$= 0.63 \text{ W/°Ccm}$$

The method described at the end of 11.3.2.1 (a) is used to take into account a predominant flow direction when the copper is not continuous in the line of predominant heat flow.

The epoxide board inhibits surface heat loss from the lower face and the heat transfer coefficient of this face is modified, see 11.3.2.3, by a factor of:

$$\frac{K}{ht+K} = \frac{0.0025}{0.15h+0.0025}$$

$$= \frac{1}{60h+1}$$

where K is the thermal conductivity of the epoxide board (0.0025 W/°Ccm)
      t is the thickness of the epoxide board (0.15 cm)
and   h is the surface heat transfer coefficient.

In the case of natural convection, where h is 0.001 W/°Ccm², the equation reduces the surface cooling from the epoxide side by a factor of 1.06 and at the same time increases the amount of heat spreading within the board and hence the relative effect of the board's thermal conductivity.

In the case of forced convection, where h is 0.01 W/°Ccm², the factor is 1.6.

The difference in these factors is due to the fact that, in the case of natural convection, the temperature drop across the epoxide boards is small compared with that between the surface and the cooling fluid. It is more significant in the case of forced convection.

Type D

The next case is the same as that in Type C except that the thermal plane covers two thirds of the area of the board.

In this case, the effective thermal conductivity is

    1.27 W/°Ccm

The epoxide board on the lower face inhibits the surface heat loss in exactly the same way as in the case of Type C.

Type E

This is a 0.2 cm thick board consisting of a 0.1 cm copper core with 0.05 cm epoxide boards on both sides. The thermal conductivity of this board is effectively half that of copper, that is:

    1.9 W/°Ccm

The epoxide boards on the sides do, however, inhibit the surface heat loss and the heat transfer coefficient is modified, see Section 11.3.2.3, by a factor of:

$$\frac{K}{ht+K} = \frac{0.0025}{0.05h+0.0025}$$

$$= \frac{1}{20h+1}$$

In the case of natural convection, this reduces the surface cooling from both sides by a factor of 1.02 and at the same time increases the relative effect of the thermal conductivity as for Type B.

In the case of forced convection the factor is 1.2.

Type F

This is an alumina substrate of thickness 0.06 cm.

The substrate is homogeneous and there is no added impedance or conduction in any direction. In this case, there is no modification of the heat transfer coefficients.

11.4 'FIRST LOOK' METHODS OF CIRCUIT BOARD THERMAL DESIGN

Thermal design involves making decisions about electronics system construction and assembly. Before going into detail, consider a hypothetical circuit board with a tidy-looking but rather fine line circuit.

If high operating temperatures are expected, the first action could be to increase the copper density to the maximum possible, taking into account possible solder bridging. At the same time the copper thickness could be increased to the limits set by production and delamination problems.

Using average thermal conductivity values, the circuit board with the increased copper would be 40°C cooler at the most sensitive component.

There are now two designs — the original one, and the new one set at production and physical limits. The production processes are similar and the new design would involve little, if any, extra cost. As the change in temperature difference is the maximum that can be obtained within the production limits, there would be no advantage in choosing a design resulting in a smaller reduction in temperature difference.

Using the 10 to 15°C rule of 11.2.4.2, the improvement gained by using the new design will be between 8- and 20-fold, and probably, as a result of secondary effects, at least 10-fold.

A very detailed temperature calculation, taking into account all the different track distributions, may perhaps give a 55°C difference. Ignoring the problems of cumulative errors and lack of confidence in such an analysis, the estimate of improved reliability is still going to be similar to that which would be obtained using an all-over average thermal conductivity and will have just as much precision. There is therefore no advantage in the detailed analysis, and also the result may be totally erroneous.

If a cored circuit board or a thermal plane were to be used, the temperature variations would again be step changes and the arguments used for the earlier increased copper example would still apply.

The accuracy of the temperature prediction is not the important feature — only its relevance to the making of design decisions.

The same single layout of components is used for all the different types of board construction and combination of cooling modes. The particular example used to illustrate the methods described in this chapter is not typical, as there is an extreme imbalance in the power distribution. This non-typical example has been chosen to enable the reader to see the effect of ignoring the lack of uniform dissipation compared with making an intuitive subdivision of the problem.

11.4.1 'First Look' Temperature Prediction for Circuit Boards without Surface Heat Loss

When making a 'first look' analysis of a circuit board without surface cooling, a base temperature is obtained by taking the heat generation to be averaged over the whole effective surface of the board. This board base temperature evaluation can be performed at two different investigation levels. The effect of component size is then added to this value to give the predicted temperatures for contact areas.

The type of board considered here is one where all the heat is lost by conduction to an edge-mounted heat sink, Figure 11.6.

11.4.1.1 CONCENTRATION OF HEAT GENERATION AT THE MEAN DISTANCE FROM THE HEAT SINK

The first estimate of board temperatures is made by assuming that the temperature that would arise if all the heat were concentrated at the mean distance from the heat sink ($L/2$) is typical of the temperatures which arise from uniform distribution.

At distance x from the heat sink, the temperature θ is hence:

$$\theta = \frac{\dot{Q}.x}{KtW} \text{ from the heat sink to a distance } \frac{L}{2}$$

and

$$\theta = \frac{\dot{Q}.L}{2KtW} \text{ between distances } \frac{L}{2} \text{ and } L$$

where K is the circuit board's effective thermal conductivity
L is the length of the circuit board
W is the width of the circuit board
and t is the circuit board thickness.

The maximum error in this approach is found at the mid-distance and, by comparison with the refined estimate that follows in 11.4.1.2, is 33%.

Table 11.1 tabulates, alongside the designation of 'Whole Circuit Board', the general board temperatures obtained by this first estimate.

It is apparent that considerable imbalance exists in the heat source distribution. It would not be unreasonable to treat the problem as two separate circuit boards, the right hand one containing heat sources 1, 2, 4, 7, 8 and 9, and the left hand one components 3, 5 and 6. Without any strict guidelines, a reasonable dividing line would appear to be that alongside heat source 5. This results in two circuit boards, one of width $9/16$ths of the original and with a power input of 1.025 watts and one of $7/16$ths of the original with a power input of 0.085 watts. The temperatures obtained for these separate boards are also contained in Table 11.1, listed alongside the designation 'Divided Circuit Board'.

Table 11.1
First Estimate of Conduction-only Cooled Circuit Board Base Temperatures in Degrees Centigrade

Component	Circuit Board	A	B	C	D	E	F
1	Whole	262.08	43.68	0.62	0.31	0.31	7.86
	Divided	430.25	71.71	1.02	0.51	0.51	12.91
2	Whole	262.08	43.68	0.62	0.31	0.31	7.86
	Divided	430.25	71.71	1.02	0.51	0.51	12.91
3	Whole	354.58	59.10	0.84	0.42	0.42	10.64
	Divided	62.06	10.34	0.15	0.07	0.07	1.86
4	Whole	632.08	105.35	1.50	0.75	0.75	18.96
	Divided	1037.65	172.94	2.46	1.23	1.23	31.13
5	Whole	724.58	120.76	1.72	0.86	0.86	21.74
	Divided	126.83	21.14	0.30	0.15	0.15	3.80
6	Whole	724.58	120.76	1.72	0.86	0.86	21.74
	Divided	126.83	21.14	0.30	0.15	0.15	3.80
7	Whole	740.00	123.33	1.75	0.88	0.88	22.20
	Divided	1214.81	202.47	2.88	1.44	1.44	36.44
8	Whole	740.00	123.33	1.75	0.88	0.88	22.20
	Divided	1214.81	202.47	2.88	1.44	1.44	36.44
9	Whole	740.00	123.33	1.75	0.88	0.88	22.20
	Divided	1214.81	202.47	2.88	1.44	1.44	36.44

11.4.1.2 UNIFORM DISTRIBUTION OF HEAT GENERATION WITH NO SURFACE HEAT LOSS

A more refined first look estimate of the general board temperature is obtained from an exact evaluation of the temperatures when the heat is generated uniformly.

For a uniformly heated board cooled by conduction to an edge connector or heat sink, the temperature at distance x from the edge connector can be demonstrated to be:

$$\theta = \frac{\dot{Q}}{KtW}\left(x - \frac{x^2}{2L}\right)$$

Provided the heat-generating components are reasonably uniformly distributed, this equation gives useful general or base temperatures for the circuit board.

Table 11.2 sets out the results of this more refined general board temperature estimate in the same format as Table 11.1.

Table 11.2

Refined Estimate of Conduction-only Cooled Circuit Board Base Temperatures in Degrees Centigrade

Component	Circuit Board	A	B	C	D	E	F
1	Whole	238.88	39.81	0.57	0.28	0.28	7.17
	Divided	392.15	65.36	0.93	0.46	0.46	11.76
2	Whole	238.88	39.81	0.57	0.28	0.28	7.17
	Divided	392.15	65.36	0.93	0.46	0.46	11.76
3	Whole	312.11	52.02	0.74	0.37	0.37	9.36
	Divided	54.63	9.10	0.13	0.06	0.06	1.64
4	Whole	497.11	82.85	1.18	0.59	0.59	14.91
	Divided	816.07	136.01	1.93	0.97	0.97	24.48
5	Whole	547.21	91.20	1.30	0.65	0.65	16.42
	Divided	95.78	15.96	0.23	0.11	0.11	2.87
6	Whole	547.21	91.20	1.30	0.65	0.65	16.42
	Divided	95.78	15.96	0.23	0.11	0.11	2.87
7	Whole	624.05	104.01	1.48	0.74	0.74	18.72
	Divided	1024.47	170.75	2.43	1.21	1.21	30.73
8	Whole	693.75	115.63	1.64	0.82	0.82	20.81
	Divided	1138.89	189.81	2.70	1.35	1.35	34.17
9	Whole	693.75	115.63	1.64	0.82	0.82	20.81
	Divided	1138.89	189.81	2.70	1.35	1.35	34.17

11.4.1.3 EFFECT OF COMPONENT SIZE

Having obtained a first look base temperature of the board by one or other of the two methods described in 11.4.1.1 and 11.4.1.2, the effect of component size has to be estimated. The first step is to treat each component as being equivalent to a disc or combination of discs.[13] As the choice of disc size is for design comparison, several criteria are acceptable provided the same criterion is chosen for each component.

The size of each disc can be related to the corresponding component in one of a number of ways, but the choice must be the same for each component. Examples are:

The disc can be totally inscribed within the component.
The disc can circumscribe the component.
The disc can have the same surface area as the component.

With a long rectangular component, the central disc should follow the chosen criteria, while the remainder of the discs can be of any convenient size which illustrates the component's geometry.

The heat spreads from each disc into an area of the board referred to as its 'effective area' (Ae).

The ratio of each area to the total area of the board is the same as the ratio of the heat dissipated by the particular heat source to the heat dissipated by the whole board.

The first look approach represents this effective area by a disc of the same area to which it is taken that the heat spreads.

This results in the temperature distribution for such conduction-only cooled areas being:

$$\theta = \frac{\text{Component Power}}{2\pi \times \text{Conductivity} \times \text{Thickness}} \times \text{Ln}\sqrt{\frac{\text{Board Area}}{\text{Component Area}} \times \frac{\text{Component Power}}{\text{Total Power}}}$$

That is:

$$\theta c = qf \cdot \text{Ln}\sqrt{(Ar)}$$

where qf is the component power factor $= \dfrac{\dot{q}}{2\pi kt}$

Ae is the effective area $= \dfrac{LW\dot{q}}{\dot{Q}}$

Ac is the component area $= lw$

and Ar is the area ratio,

which is the effective area divided by the component's own area, that is:

$$= \frac{Ae}{Ac}$$

$$= \frac{LW\dot{q}}{lw\dot{Q}}$$

Table 11.3 lists, in the same format as the previous two tables, the temperature difference generated by the heat spreading.

11.4.1.4 ESTIMATE OF COMPONENT AREA TEMPERATURES

The component area temperature is the temperature of the circuit board adjacent to the component, it does not include the temperature differences which exist within the device and package. The component area temperatures are obtained by adding to the values in Table 11.3 those in Tables 11.1 and 11.2 to give 'first' and 'refined' area estimates. These are set out in Table 11.4

Table 11.3

Centigrade Temperature Differences due to Conduction Heat Spreading in the Immediate Proximity of the Heat Dissipating Components

Component	Circuit Board	A	B	C	D	E	F
1	Whole	86.37	14.39	0.20	0.10	0.10	2.59
	Divided	73.22	12.20	0.17	0.09	0.09	2.20
2	Whole	86.37	14.39	0.20	0.10	0.10	2.59
	Divided	73.22	12.20	0.17	0.09	0.09	2.20
3	Whole	11.88	1.98	0.03	0.01	0.01	0.36
	Divided	25.75	4.29	0.06	0.03	0.03	0.77
4	Whole	83.33	13.89	0.20	0.10	0.10	2.50
	Divided	58.79	9.80	0.14	0.07	0.07	1.76
5	Whole	4.39	0.73	0.01	0.01	0.01	0.13
	Divided	11.44	1.91	0.03	0.01	0.01	0.34
6	Whole	5.41	0.90	0.01	0.01	0.01	0.16
	Divided	12.46	2.08	0.03	0.01	0.01	0.37
7	Whole	11.88	1.98	0.03	0.01	0.01	0.36
	Divided	7.93	1.32	0.02	0.01	0.01	0.24
8	Whole	125.33	20.89	0.30	0.15	0.15	3.76
	Divided	107.80	17.97	0.26	0.13	0.13	3.23
9	Whole	125.33	20.89	0.30	0.15	0.15	3.76
	Divided	107.80	17.97	0.26	0.13	0.13	3.23

for the initial estimate and in Table 11.5 for the more refined estimate. For comparison, the actual temperatures as obtained by the method described in Reference 14 and further discussed in 11.6.2 are included.

11.4.2 'First Look' Temperature Prediction for Surface-cooled Circuit Boards

As in the case of conduction-only cooling, the assumption when making a first look analysis of circuit boards with surface heat loss is that the heat generation is averaged over the whole effective surface of the board.

In many air-cooled circuit boards the vast bulk of the generated heat is lost from the board surface, with very little lost by conduction to the edge connector. In such a case the mean temperature required to sustain heat loss from the surface is used as the general base temperature. Thermal conduction does, however, play a part, because heat has to travel from the heat source until it has spread sufficiently for there to be enough surface area for complete dissipation to occur at the board's mean temperature. This heat spreading temperature difference in the area of the heat sources is discussed in 11.4.2.3.

Table 11.4

Conduction Cooled, Component Area Temperatures — First Estimate — in Degrees Centigrade

Component	Circuit Board	A	B	C	D	E	F
1	Whole	348.45	58.08	0.83	0.41	0.41	10.45
	Divided	503.47	83.91	1.19	0.60	0.60	15.10
	Actual	470.58	78.43	1.11	0.56	0.56	14.12
2	Whole	348.45	58.08	0.83	0.41	0.41	10.45
	Divided	503.47	83.91	1.19	0.60	0.60	15.10
	Actual	455.07	75.84	1.08	0.54	0.54	13.65
3	Whole	366.46	61.08	0.87	0.43	0.43	10.99
	Divided	87.81	14.63	0.21	0.10	0.10	2.63
	Actual	192.09	32.01	0.45	0.23	0.23	5.76
4	Whole	715.41	119.24	1.69	0.85	0.85	21.46
	Divided	1096.44	182.74	2.60	1.30	1.30	32.89
	Actual	794.14	132.36	1.88	0.94	0.94	23.82
5	Whole	728.97	121.50	1.73	0.86	0.86	21.87
	Divided	138.27	23.04	0.33	0.16	0.16	4.15
	Actual	419.52	69.92	0.99	0.50	0.50	12.59
6	Whole	729.99	121.67	1.73	0.86	0.86	21.90
	Divided	139.29	23.21	0.33	0.16	0.16	4.18
	Actual	318.63	53.11	0.75	0.38	0.38	9.56
7	Whole	751.88	125.31	1.78	0.89	0.89	22.56
	Divided	1222.75	203.79	2.90	1.45	1.45	36.68
	Actual	998.60	166.43	2.37	1.18	1.18	29.96
8	Whole	865.33	144.22	2.05	1.02	1.02	25.96
	Divided	1322.62	220.44	3.13	1.57	1.57	39.68
	Actual	1165.80	194.30	2.76	1.38	1.38	34.97
9	Whole	865.33	144.22	2.05	1.02	1.02	25.96
	Divided	1322.62	220.44	3.13	1.57	1.57	39.68
	Actual	1073.88	178.98	2.54	1.27	1.27	32.22

11.4.2.1 SIMPLIFICATION OF TEMPERATURE PROFILE APPROACH

Before using the temperature which would exist should all the heat generated be lost from the surface, it has to be decided if this is an appropriate value for design purposes. This decision is made by estimating what proportion of the board loses almost none of its heat to the edge connector.

A model is used where the circuit board is divided into two. The part remote from the edge connector is taken to be at a constant temperature, sufficient that all the heat generated within that strip is lost from the surface.

The strip closest to the edge connector is taken to be at temperatures which change linearly between its maximum at the temperature of the other strip to that of the edge connector. The surface heat loss from this strip is taken to be that which would occur if it had a uniform temperature the same as that of the mean. It is also assumed that all the remaining heat generated within

Table 11.5

Conduction Cooled, Component Area Temperatures — Refined Estimate — in Degrees Centigrade

Component	Circuit Board	A	B	C	D	E	F
1	Whole	325.25	54.21	0.77	0.39	0.39	9.76
	Divided	465.37	77.56	1.10	0.55	0.55	13.96
	Actual	470.58	78.43	1.11	0.56	0.56	14.12
2	Whole	325.25	54.21	0.77	0.39	0.39	9.76
	Divided	465.37	77.56	1.10	0.55	0.55	13.96
	Actual	455.07	75.84	1.08	0.54	0.54	13.65
3	Whole	323.98	54.00	0.77	0.38	0.38	9.72
	Divided	80.37	13.40	0.19	0.10	0.10	2.41
	Actual	192.09	32.01	0.45	0.23	0.23	5.76
4	Whole	580.44	96.74	1.37	0.69	0.69	17.41
	Divided	874.86	145.81	2.07	1.04	1.04	26.25
	Actual	794.14	132.36	1.88	0.94	0.94	23.82
5	Whole	551.60	91.93	1.31	0.65	0.65	16.55
	Divided	107.22	17.87	0.25	0.13	0.13	3.22
	Actual	419.52	69.92	0.99	0.50	0.50	12.59
6	Whole	552.62	92.10	1.31	0.65	0.65	16.58
	Divided	108.24	18.04	0.26	0.13	0.13	3.25
	Actual	318.63	53.11	0.75	0.38	0.38	9.56
7	Whole	635.93	105.99	1.51	0.75	0.75	19.08
	Divided	1032.40	172.07	2.45	1.22	1.22	30.97
	Actual	998.60	166.43	2.37	1.18	1.18	29.96
8	Whole	819.08	136.51	1.94	0.97	0.97	24.57
	Divided	1246.69	207.78	2.95	1.48	1.48	37.40
	Actual	1165.80	194.30	2.76	1.38	1.38	34.97
9	Whole	819.08	136.51	1.94	0.97	0.97	24.57
	Divided	1246.69	207.78	2.95	1.48	1.48	37.40
	Actual	1073.88	178.98	2.54	1.27	1.27	32.22

the strip is transmitted across half the width of the strip, that is, all the heat is assumed to travel the mean distance.

Judgements as to how the board behaves are based on the width of this latter strip.

If the heat is being lost from each face at a rate of h units of heat per unit area per unit of excess temperature, the temperature in the region where all the generated heat is lost will be:

$$\frac{\dot{Q}}{2hLW}$$

Taking the width of the strip closest to the edge connector as being x, x is given by:

$$x^2 = 4\left(\frac{Kt}{2h}\right)$$

$2h/Kt$ reappears frequently in thermal design problems and is represented as λ. Hence,

$$x = \frac{2}{\sqrt{\lambda}}$$

The exact evaluation for uniform heating described in 11.4.2.2 demonstrates that the temperature reached at this distance is actually over 85% of the maximum value where all the heat is being lost from the surface. Hence, if x is a small part of the length of the board then, for practical purposes, it can be taken that the whole of the board is at the maximum temperature of

$$\frac{\dot{Q}}{2hLW}$$

If the distance x is not small compared with the length of the board, then it is taken that there is a linear temperature gradient from zero excess temperature at the plugged-in edge to the maximum temperature over the distance x.

Table 11.6 gives the maximum temperature which would exist if all the heat were lost by convection and the distance into the board at which surface heat losses are effectively complete.

Table 11.6

Centigrade Temperature Difference of the Board over Ambient if all the Heat were to be lost by Convection along with an Estimate of the Probable Distance at which this would occur

	Circuit Board	A	B	C	D	E	F
Natural convection	Whole	9.25	9.25	9.54	9.54	9.44	57.81
	Left side	1.62	1.62	1.67	1.67	1.65	10.12
	Right side	15.19	15.19	15.65	15.65	15.50	94.91
Distance at which situation occurs		0.95	2.32	19.79	27.99	27.85	5.48
Forced convection	Whole	0.93	0.93	1.14	1.14	1.11	5.78
	Left side	0.16	0.16	0.20	0.20	0.19	1.01
	Right side	1.52	1.52	1.87	1.87	1.82	9.49
Distance at which situation occurs		0.30	0.73	6.84	9.67	9.55	1.73

Using this approach, Tables 11.7 and 11.8 give the resulting first estimates of general board base temperatures.

Table 11.7

First Estimate of Natural Convection Cooled Circuit Board Base Temperatures in Degrees Centigrade

Component	Circuit Board	A	B	C	D	E	F
1	Whole	9.25	4.23	0.51	0.36	0.36	4.49
	Divided	15.19	6.94	0.84	0.59	0.59	7.36
2	Whole	9.25	4.23	0.51	0.36	0.36	4.49
	Divided	15.19	6.94	0.84	0.59	0.59	7.36
3	Whole	9.25	5.72	0.69	0.49	0.49	6.07
	Divided	1.62	1.00	0.12	0.09	0.09	1.06
4	Whole	9.25	9.25	1.23	0.87	0.87	10.82
	Divided	15.19	15.19	2.03	1.43	1.43	17.76
5	Whole	9.25	9.25	1.42	1.00	1.00	12.40
	Divided	1.62	1.62	0.25	0.18	0.17	2.17
6	Whole	9.25	9.25	1.42	1.00	1.00	12.40
	Divided	1.62	1.62	0.25	0.18	0.17	2.17
7	Whole	9.25	9.25	1.75	1.23	1.23	15.30
	Divided	15.19	15.19	2.87	2.03	2.02	25.13
8	Whole	9.25	9.25	2.17	1.53	1.53	19.00
	Divided	15.19	15.19	3.56	2.52	2.50	31.19
9	Whole	9.25	9.25	2.17	1.53	1.53	19.00
	Divided	15.19	15.19	3.56	2.52	2.50	31.19

Table 11.8

First Estimate of Forced Convection Cooled Circuit Board Base Temperatures in Degrees Centigrade

Component	Circuit Board	A	B	C	D	E	F
1	Whole	0.93	1.34	0.18	0.13	0.12	1.42
	Divided	1.52	2.20	0.29	0.21	0.20	2.33
2	Whole	0.93	1.34	0.18	0.13	0.12	1.42
	Divided	1.52	2.20	0.29	0.21	0.20	2.33
3	Whole	0.93	1.81	0.24	0.17	0.17	1.92
	Divided	0.16	0.32	0.04	0.03	0.03	0.34
4	Whole	0.93	0.93	0.43	0.30	0.30	3.42
	Divided	1.52	1.52	0.70	0.50	0.49	5.62
5	Whole	0.93	0.93	0.49	0.35	0.34	3.92
	Divided	0.16	0.16	0.09	0.06	0.06	0.69
6	Whole	0.93	0.93	0.49	0.35	0.34	3.92
	Divided	0.16	0.16	0.09	0.06	0.06	0.69
7	Whole	0.93	0.93	0.60	0.43	0.42	4.84
	Divided	1.52	1.52	0.99	0.70	0.69	7.95
8	Whole	0.93	0.93	0.75	0.53	0.52	6.01
	Divided	1.52	1.52	1.23	0.87	0.86	9.86
9	Whole	0.93	0.93	0.75	0.53	0.52	6.01
	Divided	1.52	1.52	1.23	0.87	0.86	9.86

11.4.2.2 EXACT EVALUATION OF UNIFORMLY HEATED CIRCUIT BOARD WITH SURFACE HEAT LOSS

As with the board cooled only by conduction, a refined first look estimate of the general board temperature is obtained from an exact evaluation of the temperatures when the heat is generated uniformly.

For a uniformly heated board with the edge connector at the cooling fluid temperature, the temperature at distance x from the edge connector can be shown to be:

$$\theta = \frac{\dot{Q}}{2hLW}\left(1 - \frac{\text{Exp}(x\sqrt{\lambda}) + \text{Exp}(2L\sqrt{\lambda} - x\sqrt{\lambda})}{1 + \text{Exp}(2L\sqrt{\lambda})}\right)$$

where λ is $2h/Kt$ as in 11.4.2.1.

Tables 11.9 and 11.10 use this evaluation to give refined estimates of the general base temperatures for the six different types of circuit board described in 11.3.3, cooled by both natural and forced convection respectively.

Table 11.9

Refined Estimate of Natural Convection Cooled Circuit Board Base Temperatures in Degrees Centigrade

Component	Circuit Board	A	B	C	D	E	F
1	Whole	8.27	5.54	0.50	0.27	0.27	5.65
	Divided	13.57	9.10	0.82	0.44	0.44	9.27
2	Whole	8.27	5.54	0.50	0.27	0.27	5.65
	Divided	13.57	9.10	0.82	0.44	0.44	9.27
3	Whole	8.80	6.56	0.65	0.35	0.35	7.34
	Divided	1.54	1.15	0.11	0.06	0.06	1.28
4	Whole	9.21	8.23	1.03	0.55	0.55	11.51
	Divided	15.12	13.51	1.69	0.90	0.90	18.89
5	Whole	9.23	8.51	1.13	0.60	0.60	12.61
	Divided	1.62	1.49	0.20	0.11	0.11	2.21
6	Whole	9.23	8.51	1.13	0.60	0.60	12.61
	Divided	1.62	1.49	0.20	0.11	0.11	2.21
7	Whole	9.25	8.83	1.29	0.69	0.69	14.29
	Divided	15.18	14.50	2.11	1.13	1.13	23.47
8	Whole	9.25	9.04	1.43	0.76	0.76	15.80
	Divided	15.18	14.85	2.34	1.25	1.25	25.94
9	Whole	9.25	9.04	1.43	0.76	0.76	15.80
	Divided	15.18	14.85	2.34	1.25	1.25	25.94

Figure 11.7 illustrates, for each of the six circuit board types, the ratio of temperatures along the board to that which would exist if all the heat were surface-dissipated at the rate of 0.001 W/°Ccm². This figure shows that the

Table 11.10

Refined Estimate of Forced Convection Cooled Circuit Board Base Temperatures in Degrees Centigrade

Component	Circuit Board	A	B	C	D	E	F
1	Whole	0.92	0.87	0.28	0.19	0.18	2.22
	Divided	1.52	1.43	0.46	0.30	0.30	3.64
2	Whole	0.92	0.87	0.28	0.19	0.18	2.22
	Divided	1.52	1.43	0.46	0.30	0.30	3.64
3	Whole	0.92	0.91	0.36	0.24	0.24	2.77
	Divided	0.16	0.16	0.06	0.04	0.04	0.49
4	Whole	0.92	0.92	0.55	0.37	0.37	3.94
	Divided	1.52	1.52	0.90	0.61	0.60	6.48
5	Whole	0.92	0.92	0.59	0.40	0.40	4.21
	Divided	0.16	0.16	0.10	0.07	0.07	0.74
6	Whole	0.92	0.92	0.59	0.40	0.40	4.21
	Divided	0.16	0.16	0.10	0.07	0.07	0.74
7	Whole	0.92	0.92	0.66	0.46	0.45	4.58
	Divided	1.52	1.52	1.08	0.75	0.74	7.52
8	Whole	0.93	0.92	0.72	0.50	0.50	4.88
	Divided	1.52	1.52	1.18	0.82	0.81	8.01
9	Whole	0.93	0.92	0.72	0.50	0.50	4.88
	Divided	1.52	1.52	1.18	0.82	0.81	8.01

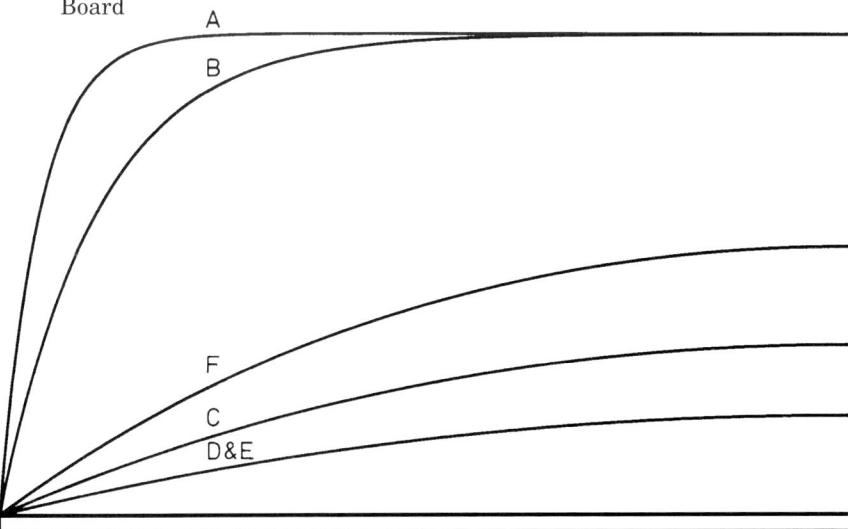

Fig. 11.7 Ratio of temperatures for uniform input with natural convection.

rate of surface heat loss becomes the predominant factor for poor conductivity boards.

By coincidence the effective thermal conductivity of circuit board types D and E is equal. In addition, the apparent surface heat loss is essentially the same for both board constructions. This results in almost identical temperature values and it is unnecessary to illustrate both.

Figure 11.8 illustrates the same temperature ratios as Figure 11.7 but this time with forced convection taken as being a rate of heat loss of 0.01 W/°Ccm². The surface heat loss is increasing its significance for the good thermal conductance boards, as is the blanketing effect of the epoxide on the cored board.

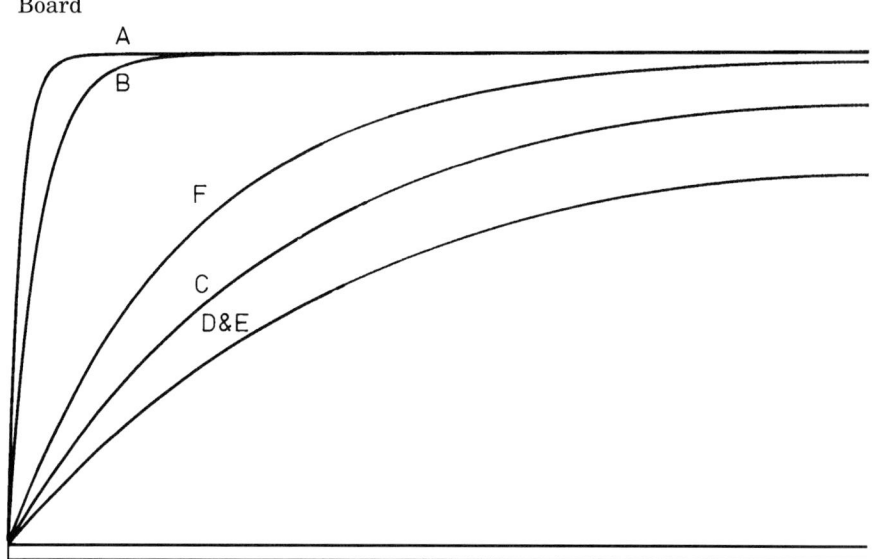

Fig. 11.8 Ratio of temperatures for uniform input with forced convection.

The curves in Figures 11.7 and 11.8 are the result of uniform heating and constitute an approximate average base value for the board temperature. In the immediate proximity of the devices there is the heat spreading effect and, within this area, the board conductivity has a greater effect on the temperatures than surface heat loss.

11.4.2.3 EFFECT OF COMPONENT SIZE UNDER SURFACE HEAT LOSS CONDITIONS

Having obtained a first look base temperature in 11.4.2.1 and 11.4.2.2, the effect of component size has to be estimated. As in 11.4.1.3, this is achieved by treating each component as being equivalent to a disc or combination of discs.[13]

For first look evaluations the heat spreading temperature is that used for the conduction-only case, see 11.4.1.3, but modified to take into account the increased rate of surface heat loss. The heat spreading temperature for conduction-only cooling is given by:

$$\theta_c = qf.\text{Ln}(Ar)$$

where θ_c, qf and Ar are the heat spreading temperature which would arise from conduction-only cooling, the component power factor and the area ratio, as described in 11.4.1.3.

The heat then spreads from this disc into an area of a board until the temperatures are lost in the mean value of the board.

The ratio of each area to the total area of the board is the same as the ratio of the heat dissipated by the particular heat source to the heat dissipated by the whole board. The temperature of the area where this heat spreading is taking place is greater than the board base temperature and this excess temperature has a mean value of θ_m, obtained by integrating over the surface the spreading temperature values:

$$\theta_m = \frac{\dot{q}}{4\pi kt}\left(1 - \frac{lw\dot{Q}}{LW\dot{q}}\right)$$

where L is the circuit board length
       l is the heat source length
       W is the circuit board width
       w is the heat source width
       $\dot{Q}$ is the total heat dissipation for the whole board
and $\dot{q}$ is the dissipation of the heat source.

Hence the mean temperature can be written as:

$$\theta_m = \frac{qf}{2}\left(1 - \frac{1}{Ar}\right)$$

Additional heat is lost from the surface of this disc over and above that already taken into account by evaluation of the board base temperature.

The first estimate of the increased heat loss from the surface (qe) would be the effective area (Ae of 11.4.1.3) times the mean temperature θ_m times twice the heat transfer coefficient 2h (to take into account both sides of the board); that is:

$$qe = 2.Ae.h.\theta_m$$

The actual spreading temperature is therefore modified in the ratio of the component power level plus increased heat loss to the actual component power level.

This gives a spreading temperature difference of:

$$\theta_s = \frac{\dot{q}.\theta_c}{\dot{q} + qe}$$

where θ_c is the heat spreading temperature for conduction-only cooling.

The steps in this analysis are all simple and easily programmed in a short calculator or computer program.

Tables 11.11 and 11.12 tabulate the temperatures due to spreading obtained by this approach for both natural and forced convection.

Table 11.11

Centigrade Temperature Differences due to Spreading in the Immediate Proximity of Heat Dissipating Components with Natural Convection

Component	Circuit Board	A	B	C	D	E	F
1	Whole	23.02	9.87	0.20	0.10	0.10	2.56
	Divided	27.85	9.60	0.17	0.09	0.09	2.18
2	Whole	23.02	9.87	0.20	0.10	0.10	2.56
	Divided	27.85	9.60	0.17	0.09	0.09	2.18
3	Whole	7.37	1.80	0.03	0.01	0.01	0.36
	Divided	4.51	2.40	0.06	0.03	0.03	0.76
4	Whole	16.24	8.23	0.20	0.10	0.10	2.45
	Divided	20.78	7.51	0.14	0.07	0.07	1.75
5	Whole	3.58	0.70	0.01	0.01	0.01	0.13
	Divided	3.73	1.42	0.03	0.01	0.01	0.34
6	Whole	4.24	0.86	0.01	0.01	0.01	0.16
	Divided	4.02	1.54	0.03	0.01	0.01	0.37
7	Whole	7.37	1.80	0.03	0.01	0.01	0.36
	Divided	6.51	1.28	0.02	0.01	0.01	0.24
8	Whole	26.61	12.91	0.29	0.15	0.15	3.69
	Divided	33.55	13.13	0.25	0.13	0.13	3.20
9	Whole	26.61	12.91	0.29	0.15	0.15	3.69
	Divided	33.55	13.13	0.25	0.13	0.13	3.20

11.4.2.4 ESTIMATE OF COMPONENT AREA TEMPERATURES FOR SURFACE COOLING

As in 11.4.1.4, the component area temperatures are obtained by adding to the values in Table 11.11 those in Tables 11.7 and 11.9, and by adding to the values in Table 11.12 those in Tables 11.8 and 11.10, to give first and refined area estimates. In the case of natural convection, these are set out in Table 11.13 for the initial estimate and in Table 11.14 for the refined estimate. In the case of forced convection, these are set out in Table 11.15 for the initial estimate and in Table 11.16 for the refined estimate. The actual temperatures, again obtained by the method described later in 11.6.2, are included for comparison.

11.4.3 General Features of Temperature Prediction Using 'First Look' Methods

First look temperature predictions give considerable guidance to circuit designers. Basing designs on such a first look evaluation may not result in the optimum design, but will avoid most foolish decisions. Figures 11.9 to 11.11 show the way trends are clearly indicated and the sort of scatter from the optimum that can be expected when using the first look approach.

Table 11.12

Centigrade Temperature Differences due to Spreading in the Immediate Proximity of Heat Dissipating Components with Forced Convection

Component	Circuit Board	A	B	C	D	E	F
1	Whole	3.03	2.58	0.19	0.10	0.10	2.29
	Divided	4.24	3.29	0.17	0.09	0.09	2.04
2	Whole	3.03	2.58	0.19	0.10	0.10	2.29
	Divided	4.24	3.29	0.17	0.09	0.09	2.04
3	Whole	1.67	0.98	0.03	0.01	0.01	0.35
	Divided	0.53	0.48	0.06	0.03	0.03	0.63
4	Whole	1.97	1.76	0.18	0.09	0.09	2.09
	Divided	3.05	2.42	0.13	0.07	0.07	1.62
5	Whole	1.34	0.53	0.01	0.01	0.01	0.13
	Divided	0.53	0.43	0.03	0.01	0.01	0.31
6	Whole	1.44	0.62	0.01	0.01	0.01	0.16
	Divided	0.57	0.46	0.03	0.01	0.01	0.34
7	Whole	1.67	0.98	0.03	0.01	0.01	0.35
	Divided	2.50	0.97	0.02	0.01	0.01	0.24
8	Whole	3.29	2.91	0.28	0.14	0.14	3.19
	Divided	4.66	3.83	0.24	0.12	0.12	2.92
9	Whole	3.29	2.91	0.28	0.14	0.14	3.19
	Divided	4.66	3.83	0.24	0.12	0.12	2.92

Figure 11.9 illustrates all the first look evaluations for the different heat sources, board constructions and cooling combinations plotted and compared with the actual ascending temperature values.

Figure 11.10 is a similar comparison for the refined evaluation applied to a divided circuit board. Figure 11.11 is the same plot, on the same scales, for just the hottest heat-dissipating component.

Gross errors in circuits with balanced components can be avoided by using first look approaches. Where designs for balanced circuits are being finalised, relative temperatures become the critical parameter and the slightly more advanced techniques described in 11.5.1 are used.

11.4.3.1 AREA OF SPARSE POWER DISSIPATION

In the portion of the substrate where the power input is more concentrated, the divided substrate gives a good representation of the actual temperatures. This concentration of power input results in a higher temperature in this region than in the sparsely powered portion of the circuit board. Even if there was no power input in the sparsely powered region there would be an appreciable increase of temperature due to heat flow from the more concentrated area. The subdivision into two boards prevents representation of this contribution to the sparsely powered area of the substrate.

This effect appears significant in terms of the low temperatures of the sparsely populated area of the substrate, but is small in relation to the

Table 11.13

Natural Convection, Component Area Temperatures — First Estimate — in Degrees Centigrade

Component	Circuit Board	A	B	C	D	E	F
1	Whole	32.27	14.10	0.72	0.46	0.46	7.04
	Divided	43.04	16.54	1.01	0.68	0.68	9.54
	Actual	63.42	23.24	1.03	0.54	0.54	12.17
2	Whole	32.27	14.10	0.72	0.46	0.46	7.04
	Divided	43.04	16.54	1.01	0.68	0.68	9.54
	Actual	66.37	24.22	1.00	0.52	0.52	11.80
3	Whole	16.62	7.52	0.72	0.50	0.50	6.42
	Divided	6.12	3.40	0.18	0.12	0.12	1.82
	Actual	14.17	4.78	0.39	0.21	0.21	4.20
4	Whole	25.49	17.48	1.43	0.97	0.97	13.27
	Divided	35.96	22.69	2.17	1.50	1.50	19.51
	Actual	38.98	24.21	1.71	0.89	0.89	19.69
5	Whole	12.83	9.95	1.43	1.01	1.00	12.53
	Divided	5.35	3.04	0.27	0.19	0.19	2.51
	Actual	7.97	3.89	0.84	0.46	0.46	9.15
6	Whole	13.49	10.11	1.43	1.01	1.00	12.56
	Divided	5.64	3.16	0.28	0.19	0.19	2.54
	Actual	9.15	3.54	0.63	0.34	0.34	6.61
7	Whole	16.62	11.05	1.77	1.25	1.24	15.66
	Divided	21.70	16.46	2.89	2.04	2.03	25.36
	Actual	25.88	21.46	2.13	1.12	1.12	24.34
8	Whole	35.86	22.16	2.46	1.68	1.67	22.69
	Divided	48.74	28.31	3.81	2.64	2.63	34.39
	Actual	85.74	35.30	2.49	1.31	1.31	28.70
9	Whole	35.86	22.16	2.46	1.68	1.67	22.69
	Divided	48.74	28.31	3.81	2.64	2.63	34.39
	Actual	84.51	33.52	2.29	1.20	1.20	26.28

temperatures reached in the concentrated heat source region. Because the influence of the low-temperature components on the overall reliability is usually small, the difference in actual and estimated temperatures should not affect design decisions, but it can weaken the confidence of inexperienced designers.

11.4.3.2 POOR THERMAL CONDUCTIVITY CIRCUIT BOARDS

Temperature estimates for poor thermal conductivity circuit board materials tend to be optimistic under circumstances of surface cooling. When a sequence of temperature values is being assembled, the order of estimated

Table 11.14

Natural Convection, Component Area Temperatures — Refined Estimate — in Degrees Centigrade

Component	Circuit Board	A	B	C	D	E	F
1	Whole	31.28	15.41	0.70	0.37	0.37	8.21
	Divided	41.42	18.70	0.99	0.52	0.52	11.45
	Actual	63.42	23.24	1.03	0.54	0.54	12.17
2	Whole	31.28	15.41	0.70	0.37	0.37	8.21
	Divided	41.42	18.70	0.99	0.52	0.52	11.45
	Actual	66.37	24.22	1.00	0.52	0.52	11.80
3	Whole	16.18	8.36	0.68	0.36	0.36	7.69
	Divided	6.05	3.55	0.17	0.09	0.09	2.04
	Actual	14.17	4.78	0.39	0.21	0.21	4.20
4	Whole	25.45	16.45	1.23	0.65	0.65	13.96
	Divided	35.89	21.02	1.83	0.97	0.97	20.64
	Actual	38.98	24.21	1.71	0.89	0.89	19.69
5	Whole	12.81	9.21	1.14	0.61	0.61	12.75
	Divided	5.35	2.91	0.23	0.12	0.12	2.55
	Actual	7.97	3.89	0.84	0.46	0.46	9.15
6	Whole	13.47	9.37	1.14	0.61	0.61	12.78
	Divided	5.64	3.03	0.23	0.12	0.12	2.58
	Actual	9.15	3.54	0.63	0.34	0.34	6.61
7	Whole	16.62	10.63	1.32	0.70	0.70	14.65
	Divided	21.69	15.78	2.13	1.14	1.14	23.70
	Actual	25.88	21.46	2.13	1.12	1.12	24.34
8	Whole	35.86	21.95	1.72	0.91	0.91	19.49
	Divided	48.73	27.97	2.60	1.38	1.38	29.14
	Actual	85.74	35.30	2.49	1.31	1.31	28.70
9	Whole	35.86	21.95	1.72	0.91	0.91	19.49
	Divided	48.73	27.97	2.60	1.38	1.38	29.14
	Actual	84.51	33.52	2.29	1.20	1.20	26.28

temperatures is sometimes the reverse of the actual values. In practical situations this reversal between apparent and actual temperature difference normally only becomes significant when boards of widely varying thermal conductivity are being compared. The problem is also generally limited to those areas of the poor conductivity boards where the distance from the edge at which the heat loss is entirely by convection is small compared with the distance of components to the same edge.

In such cases, the slightly more advanced technique described in 11.5.1.2 for balanced circuits can also be used, see 11.5.2.2.

Table 11.15
Natural Convection, Component Area Temperatures — First Estimate — in Degrees Centigrade

Component	Circuit Board	A	B	C	D	E	F
1	Whole	3.95	3.91	0.37	0.22	0.22	3.71
	Divided	5.75	5.48	0.46	0.29	0.29	4.37
	Actual	17.56	8.29	0.73	0.43	0.43	6.84
2	Whole	3.95	3.91	0.37	0.22	0.22	3.71
	Divided	5.75	5.48	0.46	0.29	0.29	4.37
	Actual	17.81	8.56	0.71	0.42	0.42	6.85
3	Whole	2.60	2.79	0.27	0.18	0.18	2.27
	Divided	4.57	0.80	0.10	0.06	0.06	0.97
	Actual	4.94	1.84	0.19	0.13	0.13	1.45
4	Whole	2.89	2.69	0.61	0.40	0.39	5.51
	Divided	4.57	3.94	0.83	0.56	0.56	7.24
	Actual	4.94	4.26	1.07	0.67	0.67	9.03
5	Whole	2.27	1.46	0.50	0.35	0.35	4.05
	Divided	0.69	0.59	0.11	0.07	0.07	1.00
	Actual	2.34	1.07	0.38	0.28	0.28	2.29
6	Whole	2.36	1.54	0.50	0.35	0.35	4.08
	Divided	0.73	0.62	0.11	0.07	0.07	1.03
	Actual	2.79	1.21	0.25	0.19	0.19	1.52
7	Whole	2.60	1.91	0.63	0.44	0.44	5.19
	Divided	4.02	2.49	1.01	0.71	0.71	8.18
	Actual	3.57	2.73	1.26	0.81	0.81	9.78
8	Whole	4.21	3.83	1.03	0.67	0.67	9.20
	Divided	6.18	5.35	1.47	0.99	0.98	12.79
	Actual	23.46	11.14	1.54	0.97	0.96	12.70
9	Whole	4.21	3.83	1.03	0.67	0.67	9.20
	Divided	6.18	5.35	1.47	0.99	0.98	12.79
	Actual	23.42	11.02	1.40	0.89	0.88	11.64

11.4.4 'First Look' Temperature Prediction for Circuit Boards Cooled by a Rear-mounted Heat Sink

The first look at the temperatures arising on a circuit board cooled by a rear-mounted heat sink assumes that the heat passes straight through the board. The heat drop across the circuit board is then added to either the manufacturer's quoted value for the heat sink or, if this is not available, to a calculated value depending upon the heat sink effective area.[15] That is:

$$\theta = \frac{\text{Total Power}}{\text{Heat Transfer Coefficient} \times \text{Effective Area}}$$

The temperature drop across a circuit board is obtained by the summation of the temperature drops across each of the component layers (i.e., one layer in the case of circuit board Types A, B and F, two layers in the case of Types

Table 11.16

Forced Convection, Component Area Temperatures — Refined Estimate — in Degrees Centigrade

Component	Circuit Board	A	B	C	D	E	F
1	Whole	3.95	3.45	0.48	0.29	0.28	4.51
	Divided	5.75	4.72	0.63	0.39	0.39	5.68
	Actual	17.56	8.29	0.73	0.43	0.43	6.84
2	Whole	3.95	3.45	0.48	0.29	0.28	4.51
	Divided	5.75	4.72	0.63	0.39	0.39	5.68
	Actual	17.81	8.56	0.71	0.42	0.42	6.85
3	Whole	2.60	1.89	0.39	0.25	0.25	3.12
	Divided	0.70	0.64	0.12	0.07	0.07	1.12
	Actual	3.46	1.84	0.19	0.13	0.13	1.45
4	Whole	2.89	2.69	0.73	0.47	0.46	6.03
	Divided	4.57	3.94	1.03	0.68	0.67	8.10
	Actual	4.94	4.26	1.07	0.67	0.67	9.03
5	Whole	2.27	1.46	0.60	0.41	0.41	4.34
	Divided	0.69	0.59	0.13	0.08	0.08	1.05
	Actual	2.34	1.07	0.38	0.28	0.28	2.29
6	Whole	2.36	1.54	0.60	0.41	0.41	4.37
	Divided	0.73	0.62	0.13	0.09	0.08	1.08
	Actual	2.79	1.21	0.25	0.19	0.19	1.52
7	Whole	2.60	1.91	0.69	0.47	0.47	4.93
	Divided	4.02	2.49	1.10	0.76	0.75	7.76
	Actual	3.57	2.73	1.26	0.81	0.81	9.78
8	Whole	4.21	3.83	1.00	0.64	0.64	8.07
	Divided	6.18	5.35	1.42	0.95	0.94	10.94
	Actual	23.46	11.14	1.54	0.97	0.96	12.70
9	Whole	4.21	3.83	1.00	0.64	0.64	8.07
	Divided	6.18	5.35	1.42	0.95	0.94	10.94
	Actual	23.42	11.02	1.40	0.89	0.88	11.64

C and D, and three layers in the case of Type E. The resultant temperature drop is given by:

$$\theta = \frac{\dot{q}}{A} \sum_{1}^{\text{Layers}} \left(\frac{t}{k}\right)_n$$

where $\dot{q}$ is the power of the given component
$(t/k)_n$ is the thickness divided by the thermal conductivity for each layer in turn
and A is the contact area of the given component.

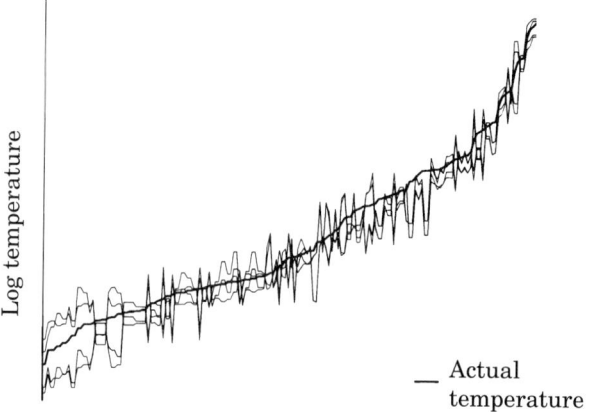

Fig. 11.9 First look temperature prediction.

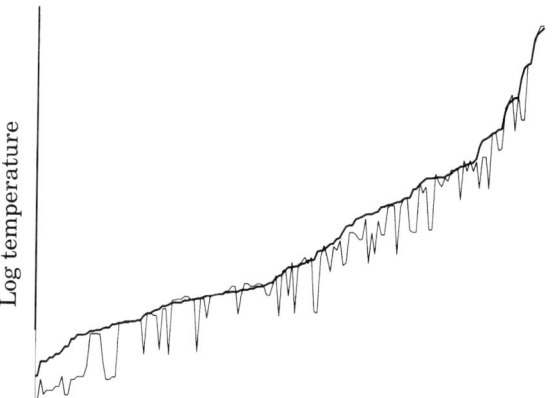

Fig. 11.10 Refined first look temperature prediction.

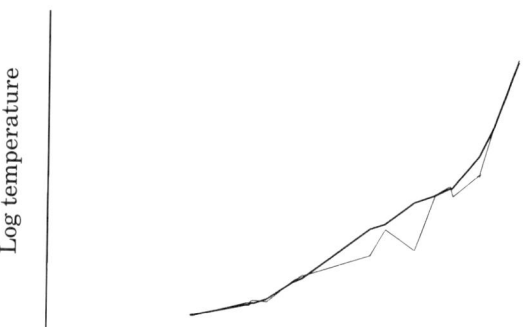

Fig. 11.11 Refined first look temperatures for component No. 8.

Sometimes the contact area is the sum of the footprints of the leads and at other times it is the area of the package. If it is the area of the package then, if glue is used, the glue becomes an additional layer. Should thermal conducting paste be used its contribution to the temperature difference is taken into account. Assuming only a thin smear, a reasonable value for the temperature drop across conducting paste is:[16]

$0.65°C/Wcm^2$

11.4.5 'First Look' Temperature Values for Components

The circuit board designer usually has little influence on detailed package or device design and in such situations either 'rule of thumb' or manufacturers' quoted values of thermal resistance are sufficiently accurate for a first look evaluation. Table 11.17 lists some rule of thumb values for some common package types.

Table 11.17

Thermal Resistance Values for Typical Package Types (°C/W)

TO-3 Steel header (Chapter 3, 3.6.1)	2.5
TO-5 Steel header (Chapter 3, 3.6.1)	150
Plastic dual-in-line (Chapter 3, 3.9.5)	120
Plastic small outline (SOT-23) (Chapter 4, 4.4.2)	100
Plastic small outline (SOT-89) (Chapter 4, 4.4.3)	30
Plastic leaded chip carriers (Chapter 4, 4.4.11)	75*
Leadless ceramic chip carriers (Chapter 4, 4.4.13)	25*
TO-220 Tabbed plastic package (Chapter 3, 3.6.1)	4**

* Values based on 0.16 cm square device; larger devices result in a very much reduced thermal impedance, see 11.5.4.
** When mounted directly on a heat sink, the value can be reduced to 1°C/W.

To all the above values, the thermal resistance from the junction to the device bonding pad has to be added. This is critically dependent upon bonding method and junction and device sizes. However, as most circuit board design decisions will involve the same or similar devices, then a value of 50°C/W will suffice for most applications.

Manufacturers sometimes quote relationships between wattage and maximum case operating temperature. This information is often of more practical use to circuit board designers than claimed maximum junction temperatures, since it is based upon the manufacturer's experience of failure rates for a particular family of devices. In circuit design the power requirement is normally known, hence a maximum case temperature is recommended.

More reliance can be placed on these recommendations than on maximum junction temperatures. This is because case temperatures are not affected by the true junction temperature and the generation of the data has already taken into account the effect of device size. If the power rating remains the same, then a reduction in case temperature of 15°C will also lead to a reduction in junction temperature of about the same, hence doubling the reliability. Because most manufacturers use the same test methods, the relationship is also useful, although not perfect, for making design judgements between suppliers. There are some dangers in this last statement and personal experience of the bias in different manufacturers' data should be taken into account.

11.5 MORE ACCURATE METHODS OF ESTIMATING THE TEMPERATURES REACHED IN CIRCUIT BOARDS

First look estimates provide considerable guidance and can prevent most of the gross and costly mistakes apparent in many systems, but they do have limitations.

The uniform power dissipation assumption of the first look approach is simplistic and cannot differentiate between two identical sources at the same distance from the cold edges. In addition, except in the very general aspect of the contribution to the overall power density, the approach does not demonstrate the influence that components have on each other when in close proximity. This is of special importance both with bare devices on alumina substrates and with the smaller sizes of surface mounted packages on printed circuit boards.

The inability to differentiate particular effects is even more limiting when evaluating the performance of balanced circuits. When component size effects are being evaluated, negligible interaction is assumed between the different components in balanced circuits. Where such interaction exists, circuit boards with and without surface cooling require more detailed evaluations. The effects of interaction on balanced circuits and on excess temperature are dealt with separately.

11.5.1 Balanced Circuits

When the performance of balanced circuits is being examined it is relative properties, such as output, that have to be considered. Frequently these require the same temperature and it is a knowledge of how close the temperatures of the different components are to each other, and not a knowledge of the actual temperatures, that is important. The critical design criterion is therefore the relative temperature difference between two components. Because of the inherent difficulties in maintaining the same temperature in components at a distance from each other, it is desirable to mount those components which require the same operating temperature close to each other.

A technique is therefore required for the prediction of the temperature difference between components in the immediate vicinity of each other. Provided that the relative influence of the various components is not changed, the choice of model of their actual geometrical properties is not inviolable. It is therefore acceptable to use discs to represent rectangular components.

The methods consist in calculating the differences in the effect of the heat sources in the immediate vicinity. By way of example, the temperature difference between components 8 and 9 is evaluated, taking into account the adjacent components 4 and 7. As a refinement, component 4 is divided into four equal sub-components. By representing the components by circular sources, the evaluations depend only on their distances.

The effect of adjacent boundaries is represented by mirror images of the various components, Figure 11.12.[17]

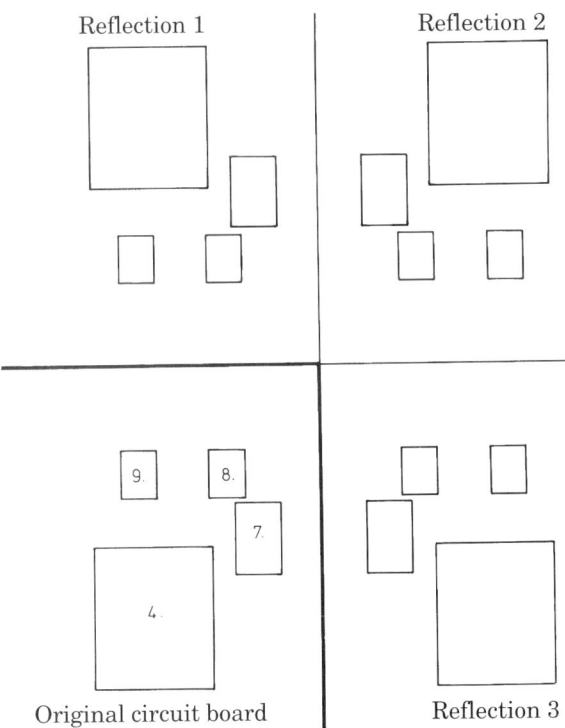

Fig. 11.12 Original circuit board and its reflections.

Calculation methods for balanced circuits mounted on conduction-only cooled boards and those also cooled from the surface are explained in 11.5.1.1 and 11.5.1.2 respectively.

11.5.1.1 BALANCED CIRCUITS MOUNTED ON CONDUCTION-ONLY COOLED CIRCUIT BOARDS

In balanced circuits, the critical components are normally situated close to each other. Remote heat-dissipating components will hence have a similar thermal influence on each of the critical components and the effect of the remote components can therefore be ignored. Equally, where the effect of the critical components on each other is the same, those effects do not require evaluation.

When listing results for convenience of systematic evaluation, such cancelling evaluations are sometimes included but even then very crude predictions are often used as the errors are also self-cancelling.

Representing the heat sources by circular heat sources, the temperature difference between two different radii is given as in 11.4.1.3, by:

$$\theta_c = qf.Ln(A/C)$$

where qf is the component power factor = $\dfrac{\dot{q}}{2\pi kt}$

A is the larger radius
and C is the smaller radius.

If boundaries are represented by mirror-image heat sources, the larger radii can be taken as being very large and the difference between the logarithms of two such radii is very small. Therefore, no significant reduction in accuracy results if the temperature contributions are defined as being:

$$\theta_c = \text{Constant} - qf.Ln(C)$$

where the Constant = $qf.Ln(A)$.

In the case of balanced circuits it is the difference that is important and there will be an equal number of identical constants included in the temperatures of all the critical components which can be cancelled out. The temperature difference between these components is hence obtained from the difference in the respective sums of:

$-qf.Ln(C)$

Table 11.18 tabulates the sums of the respective logarithms obtained by this method for a Type B circuit board. For convenience, cancelling values are included and the self-contribution of the critical components is taken as being the temperature at their boundary, a radius of 0.25 cm. The resultant temperature difference is 19.5°C compared with the actual value of 15.3°C. Both the estimated value and the actual value lead to the same conclusion — namely, that these temperatures are too high and some form of thermal bonding is required for these components.

In the case of a Type F alumina substrate, the respective values would be 3.5 and 2.8°C, which is within the variations that can exist in devices and is hence close enough for most design purposes.

Table 11.18

Distances, Logarithms and Conduction Temperature Differences for a Type B Circuit Board

Component	Reflection	0	1	2	3	Temperature Summation
8	Distance	0.25	3.00	4.07	2.75	-25.0815
	Logarithms	1.39	-1.10	-1.40	-1.01	
7	Distance	0.98	3.90	4.51	2.47	-9.9499
	Logarithms	0.02	-1.36	-1.51	-0.91	
9	Distance	1.25	3.25	5.00	4.00	-51.8435
	Logarithms	-0.22	-1.18	-1.61	-1.39	
4A	Distance	2.80	5.61	6.83	4.79	-25.7512
	Logarithms	-1.03	-1.72	-1.92	-1.57	
4B	Distance	2.52	5.47	6.40	4.16	-24.3664
	Logarithms	-0.92	-1.70	-1.86	-1.43	
4C	Distance	1.99	4.65	6.06	4.37	-22.6924
	Logarithms	-0.69	-1.54	-1.80	-1.47	
4D	Distance	1.57	4.48	5.58	3.67	-20.4969
	Logarithms	-0.45	-1.50	-1.72	-1.30	
					Total	-180.1818
9	Distance	0.25	3.00	6.05	5.25	-37.3726
	Logarithms	1.39	-1.10	-1.80	-1.66	
7	Distance	1.90	4.23	5.26	3.67	-13.3803
	Logarithms	-0.64	-1.44	-1.66	-1.30	
8	Distance	1.25	3.25	5.00	4.00	-51.8435
	Logarithms	-0.22	-1.18	-1.61	-1.39	
4A	Distance	2.44	5.44	7.65	5.90	-26.3885
	Logarithms	-0.89	-1.69	-2.03	-1.78	
4B	Distance	2.52	5.47	7.14	5.23	-25.7568
	Logarithms	-0.92	-1.70	-1.97	-1.65	
4C	Distance	1.44	4.44	6.97	5.56	-22.7565
	Logarithms	-0.37	-1.49	-1.94	-1.72	
4D	Distance	1.57	4.48	6.41	4.84	-22.2188
	Logarithms	-0.45	-1.50	-1.86	-1.58	
					Total	-199.7171
				Temperature difference		19.5353

11.5.1.2 BALANCED CIRCUITS MOUNTED ON SURFACE-COOLED CIRCUIT BOARDS

As in 11.5.1.1, with balanced circuits the critical components are normally situated close to each other, but in this case remote heat-dissipating components have little or no thermal influence and their effect can be ignored. The basis of the approach to surface-cooled circuit boards is therefore the same as for conduction cooling, except that the resultant equation is rather more complex.

$$\theta_c = D.I_o\left(r\sqrt{\lambda}\right) + E.K_o\left(r\sqrt{\lambda}\right)$$

where $I_o(r\sqrt{\lambda})$ and $K_o(r\sqrt{\lambda})$ are modified Bessel functions.
λ is, as defined in 11.4.2.1, equal to $2h/kt$.
D and E are integration constants obtained from the boundary conditions

$$\left(\frac{d\theta}{dr}\right)_A = 0 \quad \left(\frac{d\theta}{dr}\right)_B = \frac{qf}{B}$$

where A represents the size of the board
and   B the component radius.

Fortunately it is possible to combine the critical parts of the equation and these have been tabulated as DCMBs (Dean's Combination of Modified Bessel Functions) in Reference 18. The simplified expression is then:

$$\theta_c = \frac{qf}{B\sqrt{\lambda}} \text{ DCMB}\left(A\sqrt{\lambda},\ B\sqrt{\lambda},\ C\sqrt{\lambda}\right) \quad \text{(Reference 19)}$$

Assuming that boundaries are represented by mirror-image heat sources, the largest radius quoted in each of the respective tables is the radius A as, by the time this distance has been reached, all the heat has effectively been lost from the surface.

Table 11.19 tabulates the sums of the respective DCMBs for a Type B circuit board cooled by natural convection. Table 11.20 similarly tabulates the sums of the respective DCMBs when a Type B circuit board is being cooled by forced convection. As in 11.5.1.1, the cancelling values are included for convenience. The self-contribution by the critical components is obtained by making radius C equal to radius B. The resultant temperature difference is, for natural convection, 2.9°C compared with the actual value of 1.8°C and, for forced convection, 0.28°C compared with the actual value of 0.12°C. Both estimated values differ from the actual by less than the variations usually found between different packaged devices of the same type and are therefore adequate for most design purposes.

The formula requires the division by B so it is convenient that the interpolation of DCMBs for different values of B is by dividing by B first.

11.5.2 Excess Temperature Estimation

Several methods exist that enable temperatures in excess of some ambient value to be estimated with more accuracy than is the case with the first look methods.

As in 11.5.1, this is sub-divided into two sub-headings: conduction-only cooling, and cooling including surface heat loss.

11.5.2.1 A GRAPHICAL METHOD OF TEMPERATURE PREDICTION FOR CONDUCTION-COOLED CIRCUIT BOARDS

There exists an approximate graphical technique for predicting temperatures arising on conduction-cooled circuit boards which generates

Table 11.19

Distances times Root Lambda, DCMBs and Temperature Summations for a Type B Board, Natural Convection

Component	Reflection	0	1	2	3	Temperature Summation
8	C√λ	0.20	2.58	3.50	2.37	
	DCMBs	0.3670	0.0243	0.0057	0.0330	25.3468
7	C√λ	0.84	3.36	3.88	2.13	
	DCMBs	0.1127	0.0061	0.0044	0.0427	2.2019
9	C√λ	1.08	2.80	4.30	3.44	
	DCMBs	0.0855	0.0155	0.0031	0.0059	6.4834
4A	C√λ	2.41	4.83	5.87	4.12	
	DCMBs	0.0314	0.0014	0.0006	0.0036	0.7616
4B	C√λ	2.17	4.71	5.51	3.58	
	DCMBs	0.0412	0.0017	0.0007	0.0054	1.0114
4C	C√λ	1.71	4.00	5.21	3.76	
	DCMBs	0.0597	0.0041	0.0007	0.0048	1.4294
4D	C√λ	1.35	3.86	4.80	3.16	
	DCMBs	0.0744	0.0045	0.0015	0.0068	1.7984
					Total	39.0329
9	C√λ	0.20	2.58	5.20	4.52	
	DCMBs	0.3670	0.0243	0.0001	0.0024	23.2127
7	C√λ	1.64	3.64	4.53	3.16	
	DCMBs	0.0627	0.0052	0.0023	0.0068	1.0224
8	C√λ	1.08	2.80	4.30	3.44	
	DCMBs	0.0855	0.0155	0.0031	0.0059	6.4834
4A	C√λ	2.10	4.68	6.58	5.08	
	DCMBs	0.0439	0.0018	0.0004	0.0008	0.9666
4B	C√λ	2.17	4.71	6.14	4.50	
	DCMBs	0.0412	0.0017	0.0005	0.0024	0.9465
4C	C√λ	1.24	3.82	6.00	4.79	
	DCMBs	0.0788	0.0046	0.0005	0.0015	1.7623
4D	C√λ	1.35	3.86	5.52	4.17	
	DCMBs	0.0744	0.0045	0.0007	0.0035	1.7142
					Total	36.1083
					Temperature difference	2.9247

an understanding of heat flow while at the same time giving surprisingly accurate temperature values. These temperature predictions are better than those obtained using the method of 11.4.1, and the method is therefore an extremely useful design tool.

The technique is a pencil-and-paper method and is used for producing an all-over temperature map of a circuit board containing several heat sources.

A thermal map of a circuit board is drawn by plotting a series of isotherms and flow lines. An isotherm or temperature contour is a line of constant temperature, that is, a line of zero temperature gradient. A flow line is a line across which there is no net flow of thermal energy at any point. Hence, there

Table 11.20

Distances times Root Lambda, DCMBs and Temperature Summations for a Type B Board, Forced Convection

Component	Reflection	0	1	2	3	Temperature Summation
8	C√λ	0.80	8.16	11.80	7.48	
	DCMB	0.6560	0.0002	0.0000	0.0009	9.6837
7	C√λ	2.66	10.61	12.28	6.73	
	DCMB	0.1168	0.0000	0.0000	0.0020	0.3939
9	C√λ	3.40	8.85	13.61	10.89	
	DCMB	0.0339	0.0001	0.0000	0.0000	0.5013
4A	C√λ	7.62	15.26	18.58	13.04	
	DCMB	0.0007	0.0000	0.0000	0.0000	0.0039
4B	C√λ	6.85	14.90	17.42	11.33	
	DCMB	0.0019	0.0000	0.0000	0.0000	0.0095
4C	C√λ	5.41	12.64	16.49	11.89	
	DCMB	0.0039	0.0000	0.0000	0.0000	0.0201
4D	C√λ	4.27	12.20	15.17	9.98	
	DCMB	0.0180	0.0000	0.0000	0.0000	0.0929
					Total	10.7053
9	C√λ	0.80	8.16	16.46	14.29	
	DCMB	0.6560	0.0002	0.0000	0.0000	9.6699
7	C√λ	5.17	11.50	14.33	9.98	
	DCMB	0.0043	0.0000	0.0000	0.0000	0.0141
8	C√λ	3.40	8.85	13.61	10.89	
	DCMB	0.0339	0.0001	0.0000	0.0000	0.5013
4A	C√λ	6.64	14.80	20.81	16.06	
	DCMB	0.0021	0.0000	0.0000	0.0000	0.0111
4B	C√λ	6.85	14.90	19.43	14.23	
	DCMB	0.0019	0.0000	0.0000	0.0000	0.0095
4C	C√λ	3.93	12.08	18.97	15.14	
	DCMB	0.0242	0.0000	0.0000	0.0000	0.1250
4D	C√λ	4.27	12.20	17.44	13.18	
	DCMB	0.0180	0.0000	0.0000	0.0000	0.0929
					Total	10.4238
				Temperature difference		0.2815

is no temperature gradient normal to flow lines, and isotherms must intersect flow lines at right angles. The fundamental basis for the graphical method is the principle of thermal resistance per square. In electronics it is common practice to describe the electrical resistivity of a sheet material in terms of ohms per square. Heat flow may be considered in exactly the same way.

In the graphical technique, the sequence of construction involves the sketching of isotherms and lines of heat flow such that the figures formed by the intersection of the two sets of lines will approximate to curvilinear squares. These have many properties in common with rectilinear squares, especially that of having a thermal resistivity independent of size. The isotherms and flow lines are re-drawn alternately until the majority of shapes formed approximate to curvilinear squares and the few deformed 'squares' remaining are well scattered in both position and direction.

Figure 11.13 is the analysis for the example layout and took half an hour to complete. Each heat flow path carries the same quantity of heat $\dot{q}$, approximately 0.075 watts. With the alumina substrate circuit board (Type F) the temperature difference between each temperature contour is given by

$$\frac{\dot{q}}{kt} \text{ and equals } \frac{0.075}{0.015} = 5°C$$

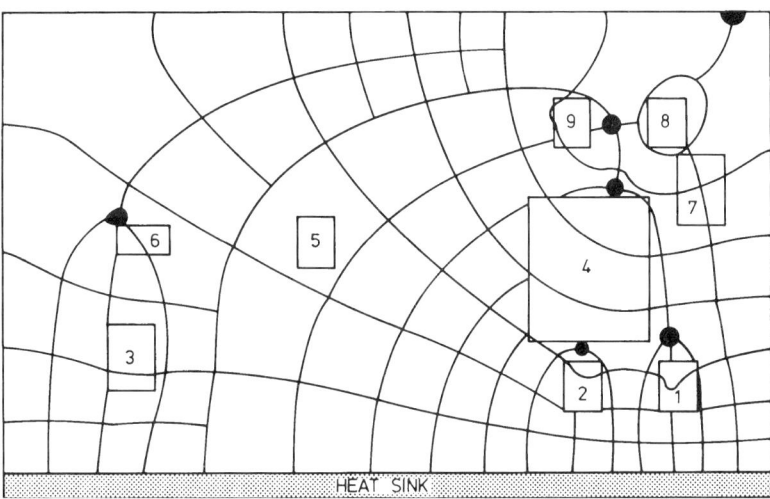

Fig. 11.13 Graphical solution of example problem.

The temperatures of the different components are compared with accurate values in Table 11.21.

A step by step description of the method is given in Reference 20.

Table 11.21

Comparison of Graphical Temperature Values and Accurate Values (°C)

Component	1	2	3	4	5	6	7	8	9
Graphical	15	13	6	23	12	9	30	36	31
Accurate values	14.1	13.7	5.8	23.8	12.6	9.6	30.0	35.0	32.2

11.5.2.2 TEMPERATURE ESTIMATES FOR SURFACE-COOLED CIRCUIT BOARDS

Whereas in a conduction-cooled board it was necessary to evaluate the contribution to temperature from all the components, this is not so in a board with significant surface cooling. Because all the heat produced by remote components is lost before it reaches the point of evaluation, the temperature contribution from remote components can be ignored.

The temperatures obtained in Tables 11.19 and 11.20 can therefore be taken as being the estimated temperatures. The resultant temperatures obtained for components 8 and 9 are, for natural convection, respectively, 39.0 and 36.1°C compared with the actual values of 35.3 and 33.5°C. For forced convection they are 10.7 and 10.4°C compared with the actual values of 11.1 and 11.0°C. Similar evaluations for a circuit board of Type A result in temperatures, for natural convection, of, respectively, 83.5 and 79.6°C compared with the actual values of 85.7 and 84.5°C; for forced convection they are 28.9 and 28.8°C compared with the actual values of 23.5 and 23.4°C.

These estimated values are adequate for most design purposes. The ones for component 8 and board Types A and B are used in Figure 11.14 to replace the first look estimates used in Figure 11.11.

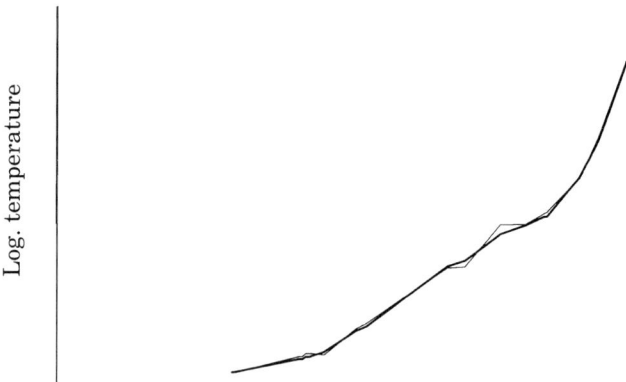

Fig. 11.14 Temperatures for component No. 8 with corrections for board types A and B.

11.5.3 Estimates of Temperatures Arising on Circuit Boards Cooled by a Rear-mounted Heat Sink

Where the contacts on the surface are large compared with the circuit board thickness, the heat flow through the board to a rear-mounted heat sink is essentially parallel, as explained in 11.3.2.3. The method given in 11.4.4 may be used to obtain the actual temperature differences.

As the dimensions of the contact reduce, the effect of heat-spreading becomes more significant. A useful estimate of the temperatures which then arise is obtained by integrating the temperature difference which would occur if the heat flow were uniform across each cross-section of a 45 degree model.[21] For a rectangular heat source:

$$\theta = \frac{\dot{Q}}{2k(a-b)} \, \text{Ln} \left(\frac{a(2t+b)}{b(2t+a)} \right)$$

where $\dot{Q}$ is the source power
  k is the circuit board thermal conductivity
  t is the circuit board thickness
  a is the heat source width
and b is the heat source length.

For a square source where width equals length:

$$\theta = \frac{\dot{Q}}{k} \left(\frac{t}{a(2t+a)} \right)$$

11.5.4 Temperature Values for Components

In most cases the circuit designer may have a choice of different devices, but has no influence on the detailed design of the actual packages. In such situations, the manufacturer's quoted values or the typical values of Table 11.17 provide enough data with which to compare components for design purposes. Detailed design of packages has been covered in Reference 22.

11.6 DESK-COMPUTER-SUPPORTED CIRCUIT BOARD THERMAL DESIGN

One of the analyses described in 11.4 and 11.5 will take between ten and thirty minutes to carry out, the graphical method taking the longest. Each analysis involves making some judgements which, although easy for those who are experienced, are likely to cause a casual thermal designer to hesitate. Many more stages are possible in approximate temperature prediction, but their use tends to become increasingly uneconomical because of the time involved.

If only slightly more accurate results are needed than those obtained by the methods so far discussed and if predictions of temperatures are required frequently, then a computer-supported method is recommended.

11.6.1 Defects of Computer Packages when used for Thermal Design

Many circuit designers are tempted to rely on computer-supported methods, but there are pitfalls in the use of computer-supported predictions. They can appear deceptively simple, since one simply keys in the data and out comes the required design information. Although the method may give correct results for many geometries, the method may not be applicable to the designer's particular problem.

The potential for making unsafe design decisions is a result of many factors, but originates with the most commonly used methods of temperature prediction and especially finite-element and finite-difference methods.

There are four basic approaches to temperature prediction:

(i) exact mathematical techniques applied to the actual problem;
(ii) approximate mathematical techniques applied to the actual problem;
(iii) approximate mathematical techniques applied to an approximation of the problem;
(iv) exact mathematical techniques applied to an approximation of the problem.

Although the first approach is clearly ideal, its use is limited to a few very simple situations. The last approach is generally best for the electronics engineer because he is able to make a judgement as to the suitability of the model (and hence that of the method) based on physical flow rather than mathematics. By comparison, when the third approach is used, for example, in finite element-regimes, the interaction between both approximations has to be taken into account.

The methods described in 11.6.2 pertain to the last basic approach, but first some of the dangers inherent in most computer programs will be amplified.

Most thermal design computer packages are based on these methods and, when predicting the temperatures that result from the existence of small heat sources, difficulties arise which can lead to cumulative errors.

Caution must therefore be exercised when using any computer thermal design package. Although the designer now has available suites of programs for quick and easy design, much is in danger of being lost. Something that is certainly being lost is the understanding of the real thermal processes which comes with experience in traditional 'at the desk' design methods. This appreciation of heat flow is being replaced by a sense of remoteness; when a computer program is used as a 'black box', the designer acquires no 'feel' or understanding of what the calculations really mean. Most computer packages are themselves based upon approximate analytical methods and require understanding to avoid the pitfalls inherent in their use. The experienced designer can take into account these pitfalls by modelling the problem, using an appreciation of heat flow to assess the validity of the results. Engineers new to thermal problems need to develop this appreciation of heat flow before blindly accepting computer predictions.

The use of finite-element and finite-difference methods of thermal analysis has resulted in many examples in the literature of designs which on physical grounds alone are clearly wrong. This is a direct result of the designer having no clear idea of the relationship of the model and the solution to the actual situation. A further problem lies in the difficulty of obtaining accurate temperatures for small components or poor thermal conductors. This sometimes results in wrong decisions which appear to be experimentally confirmed, and a false sense of confidence may be instilled.

11.6.2 Two Reliable Computer-assisted Methods

Methods which avoid these pitfalls are described in 11.6.2.1 and 11.6.2.2. Computer packages utilising these methods are not only speedy and informative but allow a considerable degree of effective interaction.

11.6.2.1 A METHOD FOR EDGE-COOLED CIRCUIT BOARDS

The method used to obtain the actual temperature values quoted in the earlier tables of temperature prediction is a computer technique which gives an exact mathematical solution to an approximation of the problem. The temperature at a point (x,y) arising from a heat source of length l and width w mounted at position (x_1, y_1) on a circuit board of length L, width W, thickness t, thermal conductivity k, and undergoing surface cooling of h from both sides is given by:

$$\theta = \frac{\dot{Q}}{l\text{wkt}} \sum_{m=1}^{m=\infty} \frac{8 \sin\left(\frac{(2m-1)\pi x_1}{2L}\right) \cdot \sin\left(\frac{(2m-1)\pi l}{4L}\right) \cdot \sin\left(\frac{(2m-1)\pi x}{2L}\right)}{(2m-1)\pi}$$

$$\left[\frac{W}{\left(\frac{(2m-1)^2 \pi^2}{4L^2} + \frac{2h}{kt}\right)} + \sum_{n=1}^{n=\infty} \frac{4\cos\left(\frac{n\pi y_2}{W}\right) \cdot \sin\left(\frac{n\pi w}{2W}\right) \cdot \cos\left(\frac{n\pi y}{W}\right)}{\left(\frac{(2m-1)^2 \pi^2}{4L^2} + \frac{n^2 \pi^2}{W^2} + \frac{2h}{kt}\right)(n\pi)} \right]$$

The technique is fully documented in Reference 14. It provides the designer with an exact prediction of the temperatures which would arise from a power input representation and it is the power input that is approximately modelled, not the circuit board. The ability of the engineer to make a visual judgement as to the appropriateness of the power input allows the evaluation time to be reduced from several hours to a few minutes.

Many programs employing this form of analysis start with a basic number of terms which can sometimes be modified by the designer on loading. The program then illustrates the form of the power input representation for the first heat source in its first direction, at the same time requesting a decision on acceptance or refinement. This procedure is repeated in the other direction and then for each source in turn.

The layout in Figure 11.6 is again used as an illustration. Figures 11.15 to 11.19 illustrate the power input representation for different numbers of Fourier terms used. Using a standard 16-bit desk computer, the times taken and the greatest error reached for each of these numbers of terms are given in Table 11.22.

The two hottest heat sources, 8 and 9, always have an error equal, or close to, the greatest error.

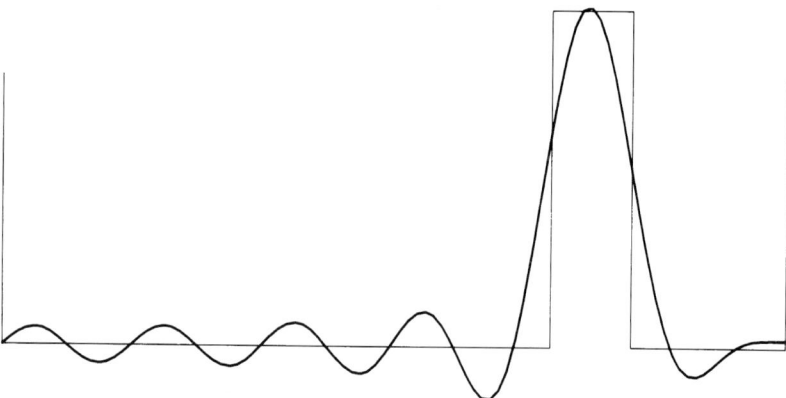

Fig. 11.15 Power input for source No. 8 with 12 terms.

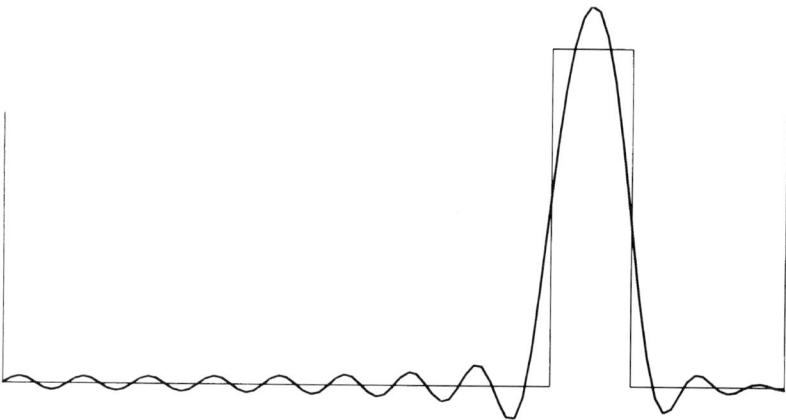

Fig. 11.16 Power input for source No. 8 with 24 terms.

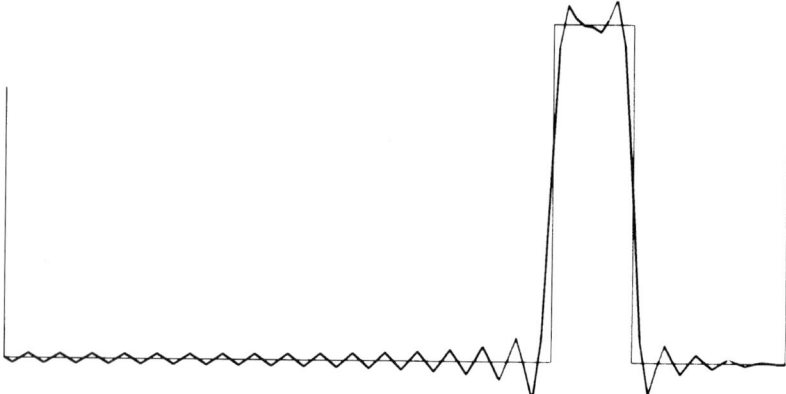

Fig. 11.17 Power input for source No. 8 with 48 terms.

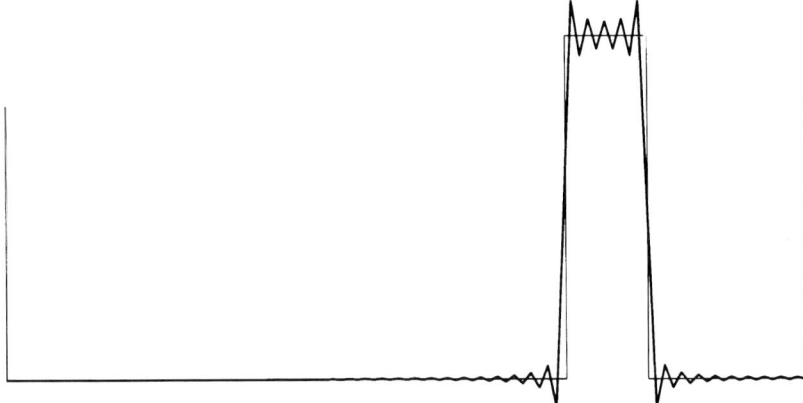

Fig. 11.18 Power input for source No. 8 with 96 terms.

Fig. 11.19 Power input for source No. 8 with 192 terms.

Table 11.22

Errors and Times for Different Numbers of Terms in the Evaluation
(Hewlett Packard 300 Standard — Basic 5.0)

Terms	12	24	48	96	192
Time	1 min. 58 s	4 min. 26 s	13 min. 49 s	48 min. 27 s	3 h. 0 min. 59 s
Greatest error	3.9%	0.80%	0.12%	0.008%	0.001%
Error in right hand corner	0.03%	0.0005%	3×10^{-5} %	5×10^{-6}%	3×10^{-7}%

As indicated by the values for the error in the right hand corner, which has a similar temperature to heat source 9, it is the size of the heat source which governs the number of terms required for a given accuracy. An example of a large heat source is 4, which, with only 12 terms, has an accuracy of 0.10%

and a power input form as shown in Figure 11.20. In Table 11.22 the same number of terms was used for all the sources. In practice the number of terms chosen from the respective power input displays varies, being dependent upon the component's size and power contribution.

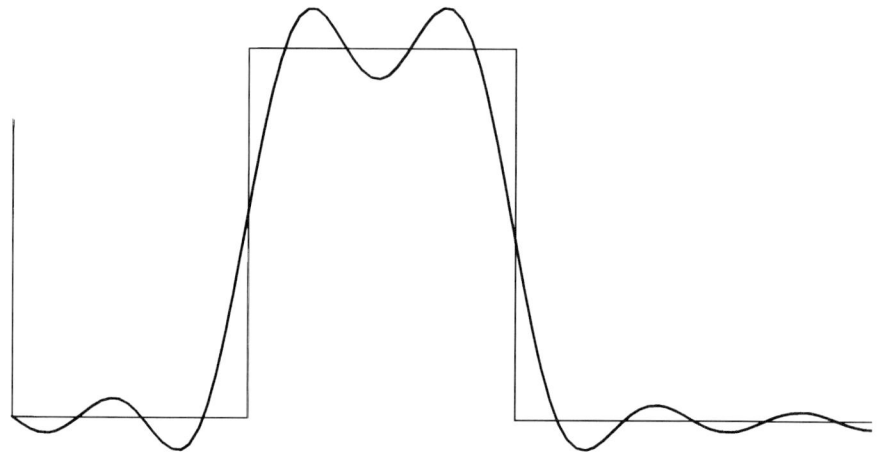

Fig. 11.20 Power input for source No. 4 with 12 terms.

For a set number of terms, the change in power representation when moving a source is significant only when the size is changed. The power representation for comparing circuit layouts is therefore the same, and the deviation from the optimum in design decisions is a second order effect. That is, an accuracy of 90% in temperature prediction results in a design confidence of about 98%.

In the particular example examined, having 9 heat sources, the addition of one source would take one ninth of the base time for evaluation and moving one source two ninths (the subtraction of one and the addition of one). This allows the system to be used in effect as a real-time design aid.

In general, having decided on a provisional layout, the effect of moving a component to a new position takes about 2/n of the original total time to evaluate. This time can be further reduced by using separate storage of the different component contributions. Finite-element and finite-difference evaluations normally require complete re-evaluation for even the smallest change.

As it is relatively easy for the designer to make a sound judgement on the appropriateness of power input representation, not only does he receive the results of the evaluation in just a few minutes, he also has absolute confidence in their relevance to his circuit design problem.

There is therefore no question as to the basis of the information which he is using to make the design decisions.

The method can also be used to evaluate the effects of transient operation.

11.6.2.2 A METHOD FOR REAR-COOLED CIRCUIT BOARDS

In 11.5.2 a 45 degree model was used to predict the effect of heat spreading with rear-cooled components.

The errors in the 45 degree model predictions can be as great as 30%.[23] However, it is not so much the size of the error that is the major weakness in the 45 degree model as the problems that arise from variations in the accuracy of temperature prediction between components with very small geometry changes.

There are other techniques available for rear-cooling temperature prediction, the multiple reflection method being the most useful.[24] This method requires some rather lengthy, but straightforward, equations which can be loaded into a programmable pocket calculator or personal computer. Once the program has been loaded, the designer has at his fingertips a versatile tool which gives quick results to a predetermined accuracy for both steady-state and transient situations.

By way of example, consider the prediction of the temperatures which would arise from a 0.1 by 0.05 cm heat source on a 0.15 cm circuit board cooled by a rear-mounted heat sink.

Using the multiple reflection program in a Hewlett Packard HP41C pocket calculator, the solution to better than 0.006% was obtained in 39 seconds. Because the program actively requests the respective inputs, the loading of the data is slow, accounting for 20 seconds of this time. With the 45 degree method, after detailed planning of the keying sequence, memorising the keying inputs and a lot of practice, the evaluation took 45 seconds. The 45 degree result was only 0.8077 of the accurate result, i.e., 20% in error.

Once the multiple reflection method's program has been stored on cards, or on disk in the case of a computer, the user has a rapid and accurate prediction method where he knows he is within the accuracy band he requires. The method is time-saving, geometrically versatile and allows transient effects to be predicted. The method can also be applied to multi-layered problems by systematic modelling.

11.7 CONVECTION COOLING

Convection cooling depends upon fluid flow and temperature difference. A surface is deemed to undergo convection cooling when there is a net transfer of heat to an in-contact fluid. Convection occurs only in fluids and involves the movement and mixing of the fluid. The heat is carried away from the proximity of the surface by a continual exchange of the bulk of the fluid.

11.7.1 The Mechanisms of Convection

In convection cooling the heat is being conveyed mechanically and is dependent upon the motion of the fluid which is affected by two mechanisms:

- (i) *Conduction*. From a surface across the boundary to the fluid's molecules and between the molecules themselves both when stationary and when mixing.

(ii) *Motion.* The heat energy absorbed by the individual molecules is transferred as the fluid particles move from one area to another.

There are two distinct types of fluid flow, although during a transition period they can exist in adjacent regions:

(i) *Laminar Flow.* Here the fluid moves in layers, each molecule following a smooth path. The movement is continuous and the individual molecules tend to remain in an orderly sequence. This type of streamlined flow is described as being 'laminar'.
(ii) *Turbulent Flow.* Here the fluid molecules move in a randomly disorganised manner. There is considerable exchange of molecules normal to the mean flow and, in general, the heat transfer by conduction is negligible. This type of randomly disorganised flow is described as being 'turbulent'. Except for a very thin layer of fluid near the surface, wherein laminar flow continues to exist, turbulent flow results in complete loss of separate streamlines.

The rate of fluid movement also influences the heat removal and depends on the type of force which is causing the movement. There are two types of force:

(i) *Natural.* Here the movement of the fluid is a result of hotter fluid being less dense and therefore rising. Natural convection between boards packaged in conventional card files drops by 25% between sea level and 15,000 feet of altitude, and effectively ceases entirely at slightly above 30,000 feet.
(ii) *Forced.* Here the fluid is propelled by mechanical means; in electronics this is normally a fan. In practice forced convection is about ten times as effective a cooling mode as natural convection.

11.7.1.1 DESIGN FEATURES OF AIR-COOLED SYSTEMS

Convection is governed by, among other things, the rate of change of those air molecules that are in close proximity to the circuit boards. The inhibiting of the movement of such molecules greatly reduces the heat transfer rate. The relative significance of the factors that influence molecular exchange close to the boundary differ depending on which type of force is causing convection. They are therefore discussed under the appropriate heading in the following text.

A — Natural Convection

Because with natural convection the addition of external components such as fans is not required, the method is effectively a fail-safe system. The lack of fans does not necessarily result in a saving in space, as larger channels are required with natural convection to allow the build-up of the momentum required to produce a significant rate of air change in the vicinity of the boards.

The momentum is gravity-induced and is affected by the ease with which air can be supplied and removed from a given position. The fresh colder air has the higher density and it is the buoyancy given by the expansion on heating which provides the force to cause it to flow. The inlet for the air should therefore be at the lowest possible position, and the outlet at the top.

Boards should be mounted vertically and air inlet slots should be above and below the board files. The worst design would be a horizontal board where the air has to flow across the board to the centre with very little motivation.

Resistors are less sensitive to temperature than solid-state devices, hence placing some high-dissipation resistors at the top of a circuit board can sometimes improve the cooling and reliability of the rest of the board.

B — Forced Convection

The pressure differences within an enclosure cooled by forced convection are obtained by external means such as fans and therefore the aspect of the circuit boards has less influence than is the case with natural convection. The flow pattern is more amenable to manipulation by the designer where the air flow is pressure-fed. In pressure feeding, the air is given momentum on entering the container and this momentum allows some control of flow distribution. Methods of evaluating the flow are the same as those described for cabinet cooling in 11.7.4.

11.7.2 Accuracy of Heat Transfer Coefficients

The equations that describe convective heat transfer belong to the most difficult classes in theoretical physics. Traditional convection cooling evaluations are therefore based upon empirical data obtained with regular and uniformly heated geometric shapes. However, electronics assemblies are rarely perfect planes, spheres or cylinders and adopting this approach adds to the possible sources of error. Fortunately for the thermal designer, heat loss by convection is closely proportional to the mean temperature difference for the whole of the surface area. Often with convection-cooled circuit boards as much as 80% of the temperature difference occurs over only 20% of the surface area. This phenomenon, which is due to heat having to spread by conduction before adequate surface area is available for the heat to be lost by convection, can result in a substantial proportion of the maximum temperature difference being within the circuit board itself. The effect is of considerable assistance to the circuit designer because it reduces the significance of variations in the rate of heat transfer coefficients.

This reduction of required accuracy applies with the majority of components. However, the sensitivity of component temperature to variation in heat transfer coefficient increases as the temperature contribution from other components increases. A special example of such a component is a large screen-printed resistor, where the hottest region in the centre can be considered as a separate component surrounded by other components, all constituting separate parts of the same resistor.

Components which receive a considerable temperature contribution from other components are cooler than the components which supply the heat. Therefore the cooler components are not usually of critical importance in reliability terms and their greater sensitivity to changes in coefficients does not influence the sensitivity of the whole system's reliability. For components having the highest temperatures, sensitivity to heat transfer coefficient variations increases at a much slower rate than the temperatures. As a result, the use of values of an order of magnitude, such as 0.001 W/°Ccm² for natural convection and 0.01 W/°Ccm² for forced convection, is usually sufficient for design purposes.

When the effectiveness of convection cooling is being considered, it is convenient to compare the temperatures which occur as fractions of some temperature. The temperature used is that which would exist if there was no surface cooling and all the heat was lost by conduction to the board edge.

Figure 11.21 illustrates the effect of heat transfer coefficients on the fraction of conduction-only temperatures for circuit board Type B, taking each component package to have an internal thermal resistance of 50°C/W. To give a linear comparison, temperatures inversely proportional to the heat transfer coefficient are also illustrated. These take 0.0001 W/°Ccm² as a base value.

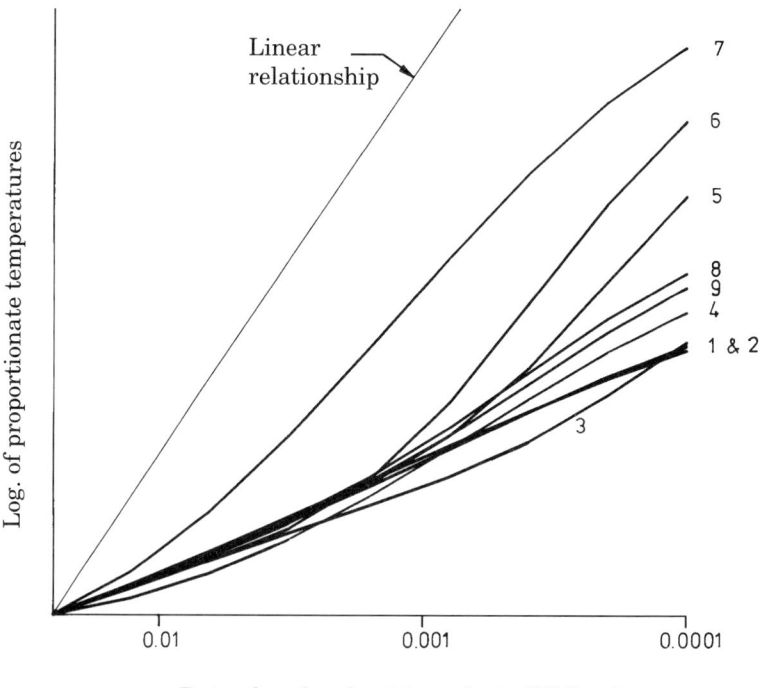

Fig. 11.21 Illustration of reduced sensitivity to heat transfer coefficient.

All the components demonstrate a much less than linear relationship with the heat transfer coefficient. Components 5, 6 and 7 show the slightly greater sensitivity caused by a significant temperature contribution from the other components.

11.7.2.1 INFLUENCE OF CHANGES IN AIR TEMPERATURE AND EDGE-CONNECTOR EFFICIENCY

The influence on convection heat transfer coefficients of changes in air temperature, of excess board temperatures and of variations in edge connector conductances does not normally need to be taken into account for early design decisions. They are therefore not included in this chapter, but methods are described in Reference 25 for designers who have to consider large variations in air temperature.

11.7.3 Packages with an Attached Heat Sink

It can sometimes be more economical to dissipate the heat generated directly to the cooling air via an attached heat sink than to improve the thermal pathway within the circuit board. There are many forms of attachable heat sink, which may be clamped on, bolted on or built in. A very varied selection of heat sinks is commercially available from a large number of suppliers.

For almost all circuit design decisions a straightforward analysis is sufficient:

A — The temperature difference from the heat sink to the cooling air is obtained by dividing power dissipated by the surface area of the heat sink available for dissipation and by the appropriate heat transfer coefficient (0.001 W/°Ccm^2 for natural convection and 0.01 W/°Ccm^2 for forced convection).

B — The temperature difference between the package and the heat sink is obtained from the thermal resistance of the contact. With the use of thermal grease, this will be approximately 0.65°C/Wcm^2. Other interfaces are covered in detail in Reference 26.

C — The final temperature difference is that from the junction to the surface of the package.

These respective temperature differences are then added to give the overall temperature of the device.

In an increasing number of packaged devices the heat sink is part of the package design. Some of these packages have heat sinks moulded in and others have facilities for bolting on heat sinks. In all these cases, the manufacturer's data are usually sufficient for design decisions.

The thermal resistance of the package, but not the device, is often reduced to almost negligible proportions when the device bonding area constitutes part of the same piece of copper as the heat sink mounting tab.

In almost all these cases, by far the major part of the temperature difference is that between the heat sink and the cooling air.

11.7.3.1 THE USE OF DIE-CAST ZINC HEAT SINKS

Zinc die casting is a cheap and versatile method of manufacturing heat sinks. In addition to the standard types normally available in aluminium, more efficient but complex-shaped heat sinks, such as ones having flat-faced pins, can be produced cheaply and easily by zinc die casting.[27] The thermal conductivity of die-cast zinc, although very close to that of cast aluminium, is only about half that of extruded aluminium. The thermal designer has therefore to decide if die-cast zinc is going to have any effect on his system's reliability.

Take the example in Reference 28, where a 0.1-cm square silicon device 0.01 cm thick has a 0.01-cm square heat generating area; this device is mounted on a 1-cm square alumina substrate 0.1 cm thick, in turn mounted on a 0.4-cm thick steel TO-3 header. The header is mounted on a heat sink of 150-cm dissipating area cooled by natural convection. The temperature drop per watt across each component is given in Table 11.23.

Table 11.23
Control of Aluminium and Zinc Heat Sinks

Material	Temperature Difference °C/W
Silicon device	45
Alumina substrate	18
Steel header	2.4
Heat sinking compound	0.25
Spreading within heat sink	
Aluminium	0.25
Zinc	0.5
Convection thermal resistance	6

The temperature penalty for using cast zinc instead of extruded aluminium is only 0.25 out of a total of 72°C. All the advantages of the cheap versatility of cast zinc can therefore be utilised in this case, and in practice most others, with very slight thermal cost.

11.7.4 Cooling within the Cabinet

The system container or cabinet provides the ambient temperature surrounding the circuit board. Of far greater significance to circuit board cooling than small variations in theoretical heat transfer coefficients is the distribution of air flow within the cabinet.

The temperature differences in the cabinet are probably the most difficult to predict. However, these temperatures are the most easily measured. This is fortunate, as all the prediction methods are empirically based on temperatures that have been measured in similar arrangements.

When selecting a cabinet, reliance should be placed as far as possible on mean temperatures obtained from past experience, or supplied by cabinet and fan manufacturers. The greatest problem in predicting temperatures is the large variations which can occur in local cooling. It is important to know how to control these variations in order to improve circuit board ambient temperature and reliability.

Back pressure and sheltered areas can cause problems in cabinet cooling by leaving large regions without any significant exchange of fluid. Figure 11.22 illustrates the air flow into an empty cabinet. Ignoring momentum effects, the air flows down the pressure gradients and Figures 11.23 and 11.24 illustrate the resultant pressure profile. It may be helpful to visualise how the flow will distribute itself by imagining skiing down the pressure gradients, personal momentum taking one beyond the regions of significant slope, towards the corners for instance, and sometimes even up slight hills.

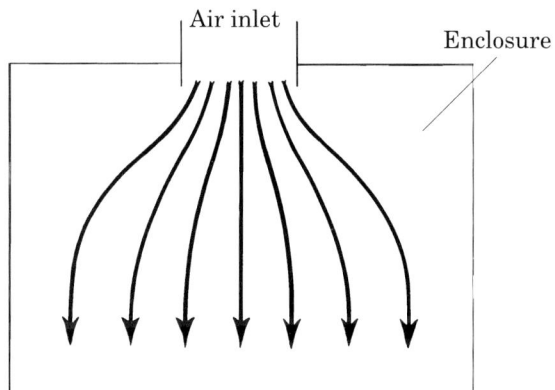

Fig. 11.22 Air distribution in an enclosure.

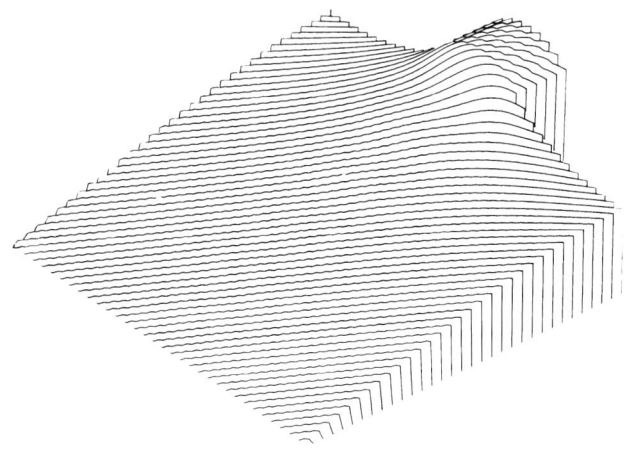

Fig. 11.23 Air pressures in an empty cabinet.

Within areas where the pressure gradients are small, significant quantities of air reaching such a region are dependent on momentum and deflection, the deflection being from physical components or other gas molecules in turbulence.

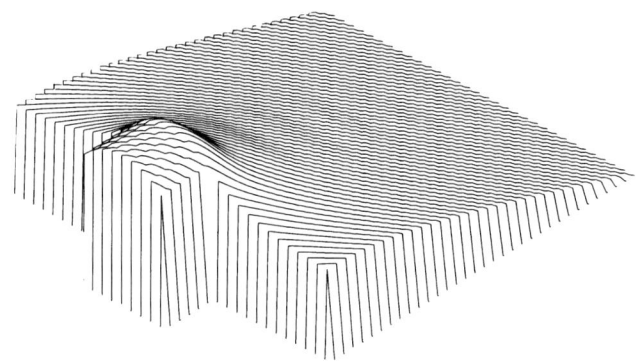

Fig. 11.24 Air pressures in an empty cabinet from the air inlet.

A near optimum distribution of fluid flow can be achieved. Consider how table tennis balls would move around the various obstructions. First visualise the effect of throwing a continuous stream of table tennis balls into a corner. The balls will bounce back, inhibiting the flow of further balls and causing 'back pressure'. The flow should be encouraged to curve away from such areas. Now visualise the effect of throwing a continuous stream of table tennis balls square on to the ground. The balls would bounce back, slowing down the projected balls. They would also build up a thick 'boundary layer' of balls, preventing subsequent balls from approaching the ground, Figure 11.25. However, if the table tennis balls were thrown at an angle they would help to sweep the boundary layer sideways and their momentum would tend to take them through any remaining few balls so that they would actually reach the ground, Figure 11.26. In a real situation, the stream of fluid would ideally impinge on the whole surface of the board. Hence, the fluid flow should be not only deflected towards sheltered areas, but should also be allowed to sweep away without building up unnecessary back pressure which would reduce the overall flow. Curving of heat sinks and the use of deflectors assist in this control of air flow. In summary, air flow should be gently deflected and coaxed to where it is required, and not left to find its own way.

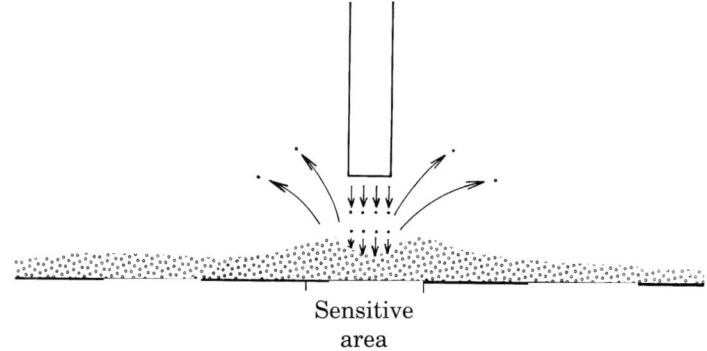

Fig. 11.25 Table tennis balls impinging normally.

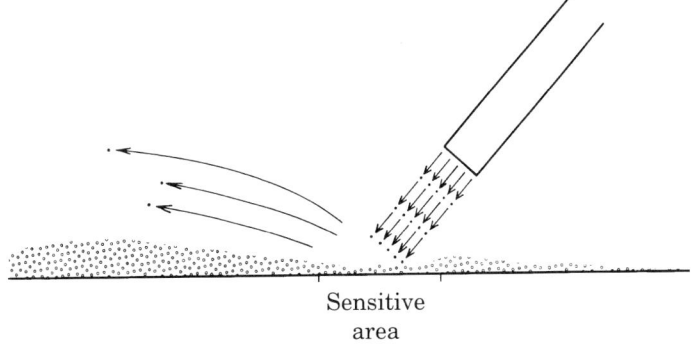

Fig. 11.26 Table tennis balls impinging at an angle.

In the case of forced convection cooling, increased turbulence and hence heat transfer can be induced in sensitive regions. Unfortunately, turbulence causes back pressure and reduces momentum so all the implications of these effects must be taken into account throughout the cabinet. For instance, using the table tennis ball analogy, the effect can be visualised by assuming a small wall in such a region which causes bouncing back with its associated back pressure.

As discussed in 11.7.1.1, cooling by buoyancy-induced flow is affected by the lie of the circuit boards. This factor must be included during the final design stage.

In many systems the space available for air flow is limited, resulting in further back pressure. The effect of this problem can be reduced by graduating the available flow paths relative to the air flow, that is increasing the size of the flow path relative to the flow, Figure 11.27.

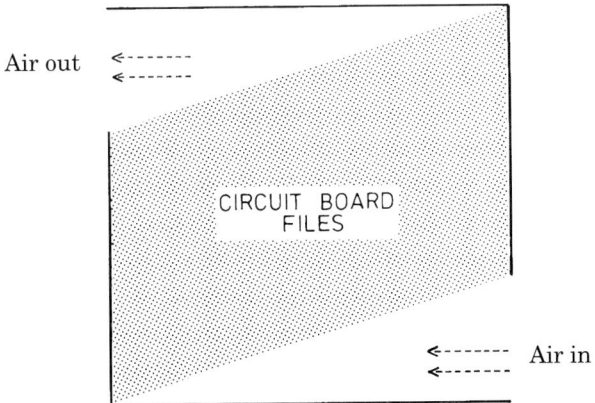

Fig. 11.27 Graduation of flow paths.

To go further with cabinet temperature prediction requires the generation of experimental data particular to the cabinet and its internal layout, and is therefore beyond the scope of this chapter.

11.8 OTHER ASPECTS OF THERMAL DESIGN

11.8.1 Thermal Expansion Mismatch

Failure or performance degradation may be a result of the strain caused by thermal expansion mismatch. This strain occurs when materials with differing thermal expansions are bonded or annealed at temperatures other than the temperature of operation or storage. The resultant stress should be considered concurrently with other thermal design evaluations and manifests itself in two fields, electrical and mechanical.

The electrical manifestations include piezo-resistance effects in precision resistor arrays on silicon devices mounted in Kovar and alumina packages.

Mechanical effects consist of bowing (or flexing) and cracking. Bowing often results in microcracking in regions of high stress concentrations. This microcracking can lead to the conductor material migrating through insulants. Under humid conditions and with a few volts (8 V) applied across about 0.003 cm of porous insulation, silver will electrically short in hours and gold in weeks.

Stress-induced failures are also temperature-related but are often further aggravated by the number of temperature excursions.

11.8.1.1 SOFT SOLDER JOINT EMBRITTLEMENT

Soft solder joints are often used to relieve thermal expansion mismatch and also to reduce the concentration of thermal-expansion-induced stress. With solder-bonded devices, cycling results in refining of the grain size, which strengthens the joint and can result in brittle fracture. In addition, although in typical circuit operation the solder is often annealed by being held at slightly elevated temperatures, these same raised temperatures can result in increased brittleness of thick-film conductors due to the formation of intermetallic, see Section 5.

11.8.1.2 COMPONENT FEATURES

One benefit provided by copper lead frames in plastic leaded chip carriers is their flexibility. During circuit operation, when elevated temperatures cause the circuit board to expand, the leads flex slightly and absorb the mechanical stress produced.

The leads provide a small air gap between the bottom of the package and the circuit board. As there is normally only air conduction in this region it is an advantage to fill the gap with thermally conducting paste.

The contact pads of leadless ceramic chip carriers do not have any flexibility to absorb the stress. In addition, the thermal expansion of the ceramic is very much less than that of the conventional printed circuit board, so the potential for significant stress is greater.

The solution is to use a printed board which has approximately the same coefficient of expansion as that of the ceramic. Various materials have been used, one such being copper-Invar-copper (Invar is a 64% iron, 36% nickel alloy). Temperature predictions are obtained by the same methods as are employed for cored boards.

11.8.2 Liquid Cooling

11.8.2.1 CONVECTION COOLING USING LIQUIDS

Convection cooling using liquids is about ten times as effective as forced air cooling. All the comments made previously regarding natural convection cooling apply to design for liquid cooling.

Although most packages can be described as being 'hermetic' in that they are impervious to mercury, many are not proof against water vapour. Unless there are deliberate or occasional accidental leaks, welded or soldered metal cans and glass or solder-sealed ceramic packages are impervious to significant water ingress. For other package types, and unprotected devices, special inert fluorocarbons have to be used to prevent device failure.

Liquids require containment and this becomes the main design problem with liquid convection cooling.

11.8.2.2 PHASE-CHANGE COOLING BY NUCLEATE BOILING

Nucleate boiling is a very efficient means of device cooling, about five times as effective as convection cooling with liquids. Except with bare devices, the temperature drop from surface to the fluid is a negligible part of the total temperature difference and can be ignored.

The containment problem is greater than that for convection cooling, as the container not only has to prevent spillage but also facilitate venting or condensing as well as coping with the pressure generated internally.

11.8.3 Heat Pipes

Artificial conductors or heat pipes to assist in heat extraction have been used in electronics systems in a limited number of situations.

The heat pipe is a particular example of nucleate boiling. In its simplest form, a heat pipe consists of an air-tight cylinder lined with a wick material which is saturated with a volatile fluid. When heat is applied to one end of the pipe, the working fluid is vaporised and carries the heat to the cool or condensing end of the pipe. The condensed fluid is drawn back by capillary action along the wick to the hot end of the pipe. As heat pipes have no moving parts and are completely self-contained, theoretically they have high reliability. The heat-carrying capability is also high: for example, a pipe 60 cm long of 1.25 cm diameter can handle 200 watts at 100°C with a 0.5°C temperature differential.

To maintain the capillary forces, the surface film needs to be continuous. This film can be broken by acceleration, severe jarring, or burn-out due to the heat input exceeding the extraction rate. Without the capillary forces the heat pipe ceases to operate and has to be reprimed, which can often be difficult in operational conditions. Some so-called 'heat pipes' use gravity to maintain a fluid pool in the area of the heat input and are really thermo-syphons.

A further problem arises because the vapour within the heat pipe will condense at any position where the temperature is lower than the dew point. Instead of extracting the heat from the system, the heat pipe can hence dump all its heat in the vicinity of some very sensitive device.

The heat pipe does have its place in electronics cooling but it is not a universal solution and care must be taken in its use.

11.9 POSTSCRIPT

Thermal design is an important aspect of circuit board design. However, the geometries must be simplified before it is possible to begin to make the temperature predictions required as a basis for design decisions. The need to generate data appropriate for safe design decisions must be the overriding factor when the geometries are being simplified. All simplifications should therefore be tailored primarily to the needs of the designer, not the computer.

An application of thermal common sense can result in reductions in temperatures and improvements in reliability and performance. For instance, temperature reductions may be achieved by increasing thermal conductivity and surface heat transfer rates, and shortening and widening heat flow paths.

Almost all levels of thermal design, when combined with thermal common sense, will normally lead to an improvement in the reliability of the system.

If thermal design problems occur more than occasionally a 'once and for all' small investment in either programming time or in the cost of purchasing pre-written packages will more than repay the designer both in time and confidence. However, because of the inherent dangers, the programs should not be of a general solve-all type. The best choice is an interactive program which allows the user to select the most appropriate routines for the different aspects of the design geometry. The choice is made on the basis of the progressive data emerging from the calculation combined with the designer's understanding of the heat flow in the actual system.

Do not be afraid of thermal design. There is nothing to be gained by complicating the means of solution to a problem by including factors which do not significantly affect the result.

REFERENCES

1 Dean, D. J., 'Thermal Design of Electronic Circuit Boards and Packages', Electrochemical Publications Ltd, Asahi House, Church Road, Port Erin, Isle of Man, British Isles (1985).
2 *ibid*, Chapter 10.
3 *ibid*, Chapter 10, 10.2.
4 *ibid*, Apendix A6.
5 Nuttall, K. I., Proceedings Internepcon 79, p. 312, Brighton, UK (1979).
6 Parker, G. W., Marconi-Elliot Report 71/76/A, D560/8/00101, January (1971).
7 Ashman, J., *IEE Electronics and Power*, p.671, September (1986).
8 Dean, D. J., 'Thermal Design of Electronic Circuit Boards and Packages', Chapter 11, 11.3.2.1.1, Electrochemical Publications Ltd, Asahi House, Church Road, Port Erin, Isle of Man, British Isles (1985). (Reproduced for convenience)
9 Dickerson, P., Proceedings Nepcon, p. 43, Long Beach (1967).
10 Dean, D. J., 'Thermal Design of Electronic Circuit Boards and Packages', Chapter 11, 11.3.3.2, Electrochemical Publications Ltd, Asahi House, Church Road, Port Erin, Isle of Man, British Isles (1985).
11 Hamilton, S., Proceedings Internepcon 87, p. 1/93, Brighton, UK (1987).
12 Dean, D. J., 'Thermal Design of Electronic Circuit Boards and Packages', Chapter 2, 2.3.1, Electrochemical Publications Ltd, Asahi House, Church Road, Port Erin, Isle of Man, British Isles (1985).

13 *ibid*, Chapter 11, 11.3.2.2.
14 *ibid*, Chapter 9, 9.3.2.
15 *ibid*, Chapter 15, 15.4.1.
16 *ibid*, Chapter 5, 5.6.1.
17 *ibid*, Chapter 2, 2.3.3.
18 *ibid*, Appendix B.
19 *ibid*, Chapter 5, 5.5.2.
20 *ibid*, Chapter 3.
21 *ibid*, Chapter 6, 6.2.1.1.
22 *ibid*, Chapters 12-14.
23 *ibid*, Chapter 6, 6.3.
24 *ibid*, Chapter 6, 6.2.3.
25 *ibid*, Chapter 15, 15.3 and 15.2.2.
26 *ibid*, Chapter 5, 5.6.
27 Bullen, S. R., Proceedings Internepcon 87, p. 108, Brighton, UK (1987).
28 Dean, D. J., 'Thermal Design of Electronic Circuit Boards and Packages', Chapter 15, 15.4.1, Electrochemical Publications Ltd, Asahi House, Church Road, Port Erin, Isle of Man, British Isles (1985).

Section 5
Soldering & Joining

Chapter 12

SOLDERING

GERT BECKER
Consultant, Sköndal, Sweden

12.1 FUNDAMENTALS OF SOLDERING

Soldering is a very forgiving method of joining most metals. This is why it has been with us for such a long time. The joints have changed from macroscopic to microscopic in size and the quantity of joints on an assembly from relatively few to large numbers. The demands placed on both the mechanical and electrical properties of the joints have increased. The development has gone from hand soldering to machine soldering, from single-sided boards to double-sided boards with plated-through holes, from SMT to fine-pitch technology. All the different kinds of connections then have to work together in the equipment manufactured. Every change has been made possible only by the improvement of the soldering process and tightening of tolerances of one or another parameter in soldering. If one attempted to solder today as was done 40 years ago, complex systems could not be manufactured, far less sold. The cost of a solder joint would be extremely high and the reliability of the products would be inadequate.

Soldering is a process based on many different technical disciplines. It is based on materials technologies — metallurgical science is important as joining metals for a knowledge of chemical science is needed to deoxidise and clean the metallic surfaces using fluxes. Furthermore, in soldering a knowledge of statistics and some understanding of stress and thermal analysis are necessary. Solder joints can be made without knowledge of these disciplines but the joints may well be inferior and not last for the expected life. For that reason, those who are seriously concerned with soldering have to cover and apply many different technologies. The fundamentals of soldering are rooted in:

— metallurgy
— fluxing
— heat considerations
— wetting of surfaces.

12.2 MAKING A SOLDER JOINT

The basics of making a solder joint sound very easy. Two parts which are to be joined together are required along with an appropriate solder. The junction of the parts must form a suitable capillary gap into which the solder can flow. It must be possible to feed the solder to the gap.

The solder has to be heated until it reaches the temperature at which it melts. In addition, the two parts to be joined must be heated to the soldering temperature so that solder can flow into the capillary and form the joint. The mating surfaces must be able to be wetted by the solder chosen and the two parts have to be kept motionless during the solidification of the solder, otherwise a crack may be formed causing an unreliable joint.

Those who are skilled in electronics assembly know from their daily work that it is not so easy in practice to make a solder joint. Each of the above conditions needs quite a lot of knowledge, thought, practical experience and careful work to make the soldering process successful. The different steps and essentials required to reach the goal of producing a solder joint are discussed in this section.

12.3 COST BREAKDOWN (THE ECONOMICS OF SOLDERING)

Soldering is a joining method which is used because of its many merits. It can be used for both small-scale production and mass production. It is versatile and it *is* easy to produce reliable joints. In addition, soldering is a low-cost joining method which may be why the economics of soldering are seldom discussed.

An analysis shows that it is wrong to neglect the cost of soldering. Doing the wrong thing in soldering usually has serious economic consequences, such as:

— products are too expensive;
— designs which have cost huge sums of money to develop are impossible to manufacture;
— claims from disgruntled and disappointed customers become very expensive; in the worst case, serious or even lethal disasters can result;
— the company earns (*and deserves*) a bad reputation;
— markets are lost.

It should not be necessary to emphasise this, but the examples which follow result from actual results achieved in an assembly business.

12.3.1 Manufacturing Costs

An analysis of the cost in US dollars to produce a solder joint in mass production gives the following, approximate, figures. The calculation is based on 1,000,000 solder joints a year, produced by wave soldering pin-in-hole boards (see 'PIH assembly' in Chapter 8). One million solder joints per year correspond to 100 printed boards per day, each board having 40 to 50 solder joints.

The costs for one solder joint are:

— Material costs (solder, flux)   0.13 cent
— Production cost (equipment, operator, etc.)   2.27 cents

from which 1 million solder joints per year cost $24,000, which must be considered cheap.

However, nothing is perfect. Assume that 0.05% of the solder joints are bad, and have to be repaired. This means that 500 solder joints have to be reworked.

Not all bad solder joints are detected at once. They are detected at different stages in the manufacturing process or even later. On average, it can be assumed that 75% of the bad solder joints are detected and repaired in the workshop at a cost of $1 per joint

  = $375

12.5% are detected and repaired at the final control point for $10 per joint

  = $625

12.5% are detected and repaired at the customer's site for $100 per joint

  = $6,250.

The assumed failure rate indicates problems in production which cost the manufacturer money. The usual problem is poor solderability which increases the soldering time to a value much greater than needed, perhaps to as much as double the norm so that the production speed in our example is halved. (This could have the immediate consequence that new machines have to be purchased.) If it is considered only that the soldering time increases from 2 to 4 s per joint, this will double the production cost. The total cost for 1 million solder joints will therefore be $(2 × 22700 + 1300 + 6250) = $52,950 of which very nearly $30,000 could have been saved by doing the job properly in the first instance.

Large companies manufacturing *billions* of solder joints per annum may be losing *millions* of dollars.

12.3.2 Materials Cost versus Failure Cost

Example One: Buying cheap termination material

Very often a choice has to be made between two materials or components. In this case, the choice for a solder tag is between German silver and brass. The German silver costs $6.60/kg and the brass $4.29/kg. The difference is a considerable saving of $2.31/kg if brass is used. So the designer chooses brass, not knowing the problems arising from its use.

Brass is easy to solder when fresh. The prerequisites for soldering brass are that the time between rolling the brass tags and soldering must be short. Transport and storage must be controlled as well as temperature and humidity. If these conditions are not met, the brass tags will lose their solderability fast. The solderability retention of German silver is better.

Let us assume that we can manufacture about 25,000 solder tags from 1 kg of either brass or German silver. 40 kg are needed to produce one million tags and the price difference will be $92. But we have just seen that poor solderability will cost us nearly $23,000 by doubling soldering time alone. By buying cheaply, we have lost $23,000.

Sometimes both German silver and brass are tinned to improve and preserve solderability. However, the designer may not be aware that even then problems can arise with the use of brass and no advantage will be gained from the cheaper purchase price. Zinc from the brass will diffuse through the tin coating, oxidise on the surface of the tin forming a non-solderable zinc oxide, and destroy the solderability.

Example Two: Buying cheap consumable tools

The Purchasing Department was very proud of having found a manufacturer selling soldering iron tips some dollars cheaper than those being used. As every company needs to keep costs down, the cheaper tips were bought. The problem was simply that the new tips did not wet properly and had to be thrown away after making some tens or, at most, hundreds of solder joints. A good soldering iron tip can last for around ten thousand solder joints. To sum up, approximately 100 soldering iron tips at $1.50 were needed to produce the same number of solder joints as just one 'expensive' tip costing $2.50, a loss of $147.50 per 10,000 solder joints.

12.3.3 Machine Cost

There are a number of soldering methods used today. As soldering machines are expensive, any purchase is, of course, carefully considered. The cost calculation can be made in many different ways, suiting one or another purpose. Sometimes even new factors come in which are not known generally and so cannot be considered. One example is the process of soldering in controlled atmospheres, see Chapter 15, 15.1.4. The machine itself is expensive and then the additional costs of gas installation and consumption may deter the prospective purchaser. The careful calculation given in Reference 1 should convince the most sceptical. The savings are greater than the difference in cost from a normal machine.[1] Another example is laser soldering. Laser soldering is considered to be slow as the joints are not made simultaneously. Normally, the laser beam makes the joints one at a time, each taking 0.2 to 2 s to manufacture. However, the failure rate decreases so that output is increased when compared with some other soldering methods.

Large cost savings can be made. In addition, due to the fact that the solder joints are checked during laser soldering, examination of the solder joints on the board is not necessary after soldering. Both cost savings may well pay for the higher cost of a laser soldering machine, see also Chapter 15, 15.8 on laser soldering and 15.15 on robot soldering.

12.3.4 Quality Costs

By doubling the cost for preventive measures from 5 to 10%, the failure cost can be reduced from 65 to 35% and inspection costs from 30 to 20%. The

Quality Costs are reduced overall to 65% of the original cost,[2] Figure 12.1 and Volume 2, Chapter 19.

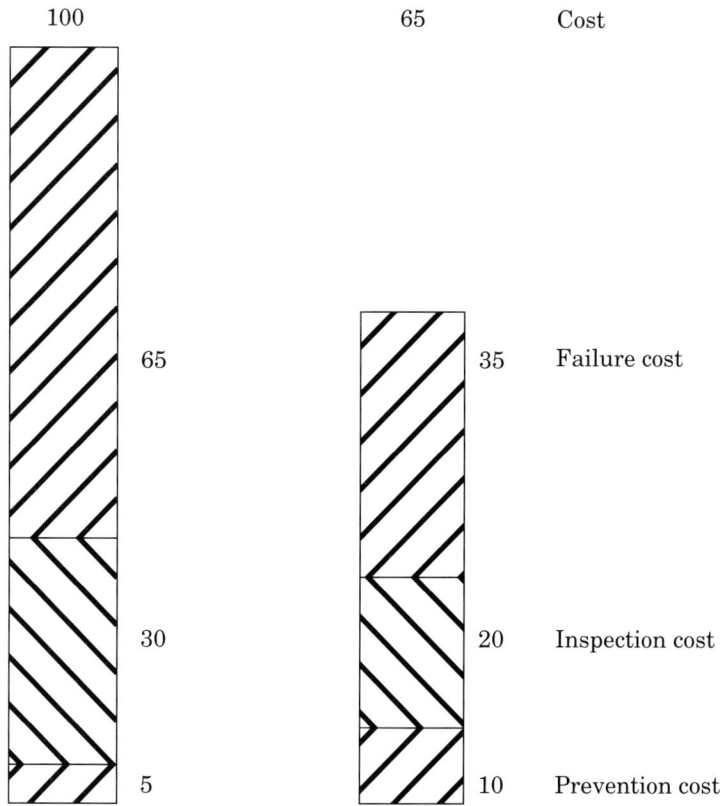

Fig. 12.1 Distribution of Quality Costs. In this case, increasing the costs for failure prevention by 5% reduced the total cost of quality to 65% of the original.

12.4 PROBLEM AREAS IN SOLDERING

Not so long ago when speaking of soldering it was usually said that soldering was an art, with this probably indicating that nothing real was known about soldering. Nowadays it can be heard that soldering is a science. This may be accepted as meaning we are on our way to knowing something about soldering. There is some truth in this. We know certain things, but there are still problem areas and areas of controversy in soldering.

The first problem is how to inform all people concerned in any way with soldering on what is known about the subject. Quite a lot of effort has been put into this task. A number of excellent books on soldering exist.[3-24] There is one journal essentially devoted to soldering.[25] Many other magazines publish articles on soldering now and then. How important soldering has become can be seen in the increasing number of publications and conferences which appear on the subject. For those having a serious interest in soldering, many sources of information are available from organisations such as ISO, IEC,

ITRI, IPC, IEEE and several others working nationally, e.g., the British BABS and the German DVS (see list of organisations).

A specific practical problem area is the solderability of the parts to be joined. Measuring methods for solderability testing are still under development. The components are becoming increasingly smaller and it has become a real problem to measure the solderability. That the solderability of components in many cases is not good enough is not only due to the ignorance of some component manufacturers, poor storage conditions and similar causes; there are other real problems involved. Components are subjected to many different stresses in production and subsequently. Electroplating and other processes are not always fully developed. Even here there is simply a lack of knowledge in certain areas.

The third major problem is to make the solder joint last. Some environments are harsh, for example, automobile electronics. The requirements for reliability and extended lifetime are becoming more severe. New developments in component and assembly/production techniques such as surface mount technology (SMT) or fine pitch technology (FPT) put extra stress on the solder joint. We simply lack enough materials data on both the components and the solders to be able to design a solder joint in the same way as an educated engineer would design a bridge.

The fourth very serious problem, particularly since the growth of environmental awareness in the late eighties and early nineties, is the cleaning of electronic assemblies after soldering. Many different ways have been, and are being, tried such as flux-less soldering, soldering with reducing gases or shielding gases, 'no-clean' fluxes and others.

12.5 PREREQUISITES FOR A SOUND, RELIABLE SOLDER JOINT

There is no doubt that many beginners in soldering are misled into believing that the soldering process itself is the most important part of the creation of a solder joint. This may well be true in the case of an amateur's soldering. It is certainly not true when a professional job must be done and a sound, reliable and long-lasting solder joint is expected, whether it be on a small scale or in mass production. There are quite a number of prerequisites which must be fulfilled to produce such a joint. The prerequisites can be divided into three phases. Each phase contains a number of points, every single one of which must be carefully considered.

The three phases are:

— the design phase,
— the preproduction phase and
— the actual production phase.

Those things to be considered in every phase indicate clearly the following points:

(i) the designer has the main responsibility for the production of a sound and reliable solder joint;
(ii) if the work which has to be carried out before soldering is done correctly,

- the solder joint will automatically become perfect and reliable;
- there is very little left to do during the actual soldering process;

(iii) it becomes obvious that, during the actual soldering process, there is no longer any chance of influencing the quality of the solder joint such that a potentially bad joint may be turned into a good one;

(iv) as a consequence,
- the soldering process becomes cheaper;
- there is no longer a need for rework;
- 100% inspection of the solder joints on the board can be reduced to inspection based on statistical sampling methods;
- the complaints disappear.

While not exhaustive, the following discussion gives greater detail of the different phases. In parallel with examining the technicalities of these requirements, the economics of their application in a given situation must be checked.

12.5.1 The Design Phase

During the design phase a number of design conditions have to be checked.

12.5.1.1 PLACEMENT OF THE SOLDER JOINT

Placement of the solder joint is normally dictated by the circuitry and the type of components used. However, a number of factors should be considered here such as thermal considerations and the practical ability to solder and to repair. Unfortunately, the effect of placement of the joint on its quality and life expectancy is generally overlooked.

12.5.1.2 THE JOINT DESIGN (DIMENSIONS, GEOMETRY AND TOLERANCES)

This raises not only the question of how the solder joint should look, which includes the dimensions and tolerances as well as geometry, but also asks how surfaces and terminations to be soldered shall be designed and what dimensions and tolerances can be applied to them. This is especially important when components having many terminations or lead-outs are used and in SMD technology.[20-24] Typical examples of bad designs for the wave soldering process are the design of the terminations for the SOT 23 and for tantalum capacitors, Figure 12.2. Due to the surface tension of the solder, it is difficult for the solder wave to reach the terminations and wet them.

The trend towards microjoining has made it important to ensure that the joint is correctly dimensioned, both electrically and mechanically. Normally the electrical dimensioning does not cause great problems whereas the mechanical dimensioning can cause trouble especially when surface mount components are involved. Unfortunately, very few guidelines have been developed and for quite a number of components none have yet been generated. This applies to the design of the component terminations as well as to the size of the pads on the boards and the fillet of solder between termination and pad. These points have considerable influence on the lifetime of the joint.

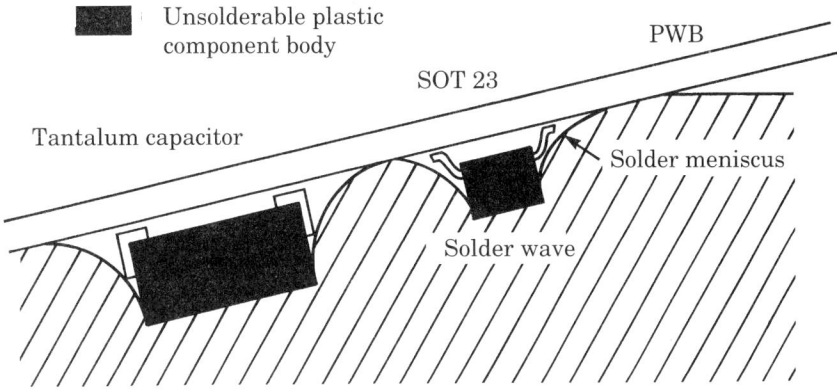

Fig. 12.2 The natural solder meniscus makes it difficult to wave solder certain SMDs.

12.5.1.3 THERMAL PROBLEMS

One of the main problems in electronics today is how to remove the heat generated in electronic components, see Section 4, Chapter 11. Another thermal management problem is how the heat needed for soldering influences the components and other materials involved in, or likely to be affected by, the soldering process.[7,26] Careful choice of the lead material can prevent overheating of a component during soldering. An important question, which has not yet been discussed, is how to provide the correct amount of heat to the solder joint during soldering so that the same metallurgical structure is obtained in all solder joints on the board.[27]

12.5.1.4 REPAIR

As early as at the design stage, the possible need to repair the solder joint has to be considered and thought has to be given to the method by which the repair will be made. This influences the placement of the component on the board and the amount of space that must be available between the components to allow for repair.

12.5.1.5 DEMANDS PLACED ON THE SOLDER JOINT IN MANUFACTURING, STORAGE, TRANSPORT AND OPERATION

A very important point to consider is to what the future solder joint will be subjected. This influences quite a number of other points. The solder joint has to withstand all the possible stresses which arise during manufacturing. These can be unexpectedly high if manufacturing problems arise.

The store has to be suitable; this means that the solder joint shall not be subjected to too high or too low a temperature and excessive humidity for too long a time. These conditions are valid also for transportation of the finished product, in this case the solder joint, to the consumer. If the stresses during storage and transport are too high, they can use up a part of the expected lifetime of the joint and thus of the product.

It is even necessary to analyse the expected stress conditions for the product in use. The main stress condition which reduces the lifetime of the solder joint is the combination of temperature swing and the time during which the joint is subjected to it. The result is known as thermal fatigue and has become especially important in surface mount technology. Other stresses such as tensile stress, shear stress, vibration, acceleration and shock can also harm the solder joint. Corrosion can be a problem which in normal cases is caused by using the wrong flux in production. Corrosion problems can even arise when special solders are used.

12.5.1.6 PROPERTIES OF THE MATERIALS USED

The designer must consider all the properties of the materials being joined which can affect the integrity of the design. Properties which are relevant to the solder joints are solderability, rate of diffusion into solder, TCE, resistance to heat and to corrosion and their compatibility with each other and the joining material.

One of the very important properties of the materials which will be joined together is solderability. If the material does not solder well, it will cause problems both in production and during the lifetime of the joint. Another important aspect is the diffusion of the different materials used, during both production and life. It is often forgotten that diffusion is occurring in the solder joint in the solid state phase, with the consequence that the properties of the joint continually undergo change. (For more detail see 12.6.5 of this chapter.)

It is necessary to know the coefficient of thermal expansion (CTE or TCE) of the materials joined together. Too large a difference in TCE values can cause cracking of the joint. Therefore, it is also necessary to study the heat flow from the component to the board in order to find out if there are any risks involved.

Certain materials or combinations of different materials are especially prone to corrosion and must be avoided. The materials which are to be joined as well as the material used in the joining process must have compatible properties.

12.5.1.7 CHOICE OF SOLDER AND FLUX

The choice of solder and flux has to be made at the design phase as it will have an impact not only on the selection of the soldering method to be used but also on the cleaning procedure to be used after soldering. Because of this, it will have a major influence on the choice of production machines.

The solder alloy will be selected after consideration of both its composition and its melting and solidification temperature(s). The flux can only be selected after assessing the metals to be joined, methods of flux application and the parameters of the fluxes available, see Chapter 13. It may be that both selections will be combined in a solder paste.

The selection of these fundamental materials has a considerable influence on the economics of the solder process.

12.5.1.8 CHOICE OF SOLDERING METHOD

Once the foregoing points have been considered, it is time to give thought to which soldering method shall be used. Here, of course, several aspects have to be observed.

Quantity is important. A low volume of production may dictate a different method of soldering from that used for large volume production. If, however, only a few joints out of many must be made by hand soldering, it can make a difference to the main method of soldering selected.

SMT boards may need a soldering technique which differs from that used for pin-in-hole boards (for equal numbers of joints).

SMT very often requires solder paste reflow soldering instead of adding solder to the joint by wave soldering.

Even different component types need different soldering methods. For instance, certain frequency crystals need to be soldered at low temperatures. It may be necessary to solder such a crystal separately after the mass soldering process.

Selection of a suitable soldering method will be made from a list similar to that given below. The methods listed are described later in this section.

- Belt soldering
- Dip soldering
- Drag soldering
- Furnace soldering
- Hand soldering
- High frequency soldering
- Hot air soldering
- Hot bar soldering
- Hot plate soldering
- Infra-red soldering
- Laser soldering
- Light soldering
- Protective gas soldering
- Reflow soldering
- Robot soldering
- Ultrasonic soldering
- Vapour phase soldering
- Wave soldering

The design phase must also consider the fact that not everything can be soldered and, therefore, alternative joining methods may have to be used such as welding, riveting, gluing or one of the many other possible joining methods.

12.5.2 Preproduction Phase

Before production starts, a number of actions are essential.

12.5.2.1 EQUIPMENT AND WORKSHOP

The first actions needed are to choose the relevant equipment, i.e., tools and machines, and to prepare the workplace. Proper design of the working place is as important as the right choice of equipment. The workplace must be properly ventilated, partly to eliminate the heat of the soldering equipment and partly to remove fumes.

To assist operators to reach maximum efficiency, there must be sufficient light and the workshop must be designed paying attention to ergonomic considerations. Furthermore, it must be, and must be kept, clean and tidy.

Instructions and specifications should be placed visibly at the work stations at which they are needed.

12.5.2.2 PERSONNEL

The importance of ergonomics to personnel efficiency has been mentioned. It is equally important that personnel be educated properly in the soldering process. Their eyesight should be checked. This is especially important for hand soldering, repair soldering and inspection.

12.5.2.3 PREPARATION FOR SOLDERING

Just as the parameters for the soldering process and post-soldering cleaning must be carefully chosen, so too must the parameters for the preparation for soldering. A common cause of failure is the mounting of the components. Either they are wrongly mounted or components are missing. Special care has to be taken when solder paste and/or glues are used. Particular attention must be given to the printing and drying of the solder paste as well as the dispensing and hardening of the glue.

12.5.2.4 PREPRODUCTION CHECKS

Before production can start, a number of checks have to be made. Both printed board and components must be checked to ensure they are as specified in type and value. Similar checks must be made on the solder, flux, solder paste and adhesive. Especially important are the checks on solderability of the boards and components.

The final surface treatment, e.g., solder mask and/or conformal coating, must be checked carefully to confirm that it fulfils the specified requirements.

12.5.2.5 STORAGE OF MATERIAL

The storage of material, in store and in the workplace, is a specialist subject. Modern computer data logging techniques really assist the use of the principle of FIFO (*First In First Out*). FIFO means that the first batch of a given material to be delivered into store shall be the first to be withdrawn for production.

Storage conditions must be reasonable. The temperature in the store should be between 10 and 25°C and the relative humidity between 30 and 70%. It is important to keep the workshop, material and equipment clean.

12.5.3 Production Phase

In this phase, the major tasks are those of inspection and process control. Inspection is needed during the production process to check specified soldering parameters, such as speed, soldering time, temperature and so on. Such checks are essential for early detection of failures in production, e.g., a broken thermocouple which can result in bad solder joints because the process temperature was wrong. Other specified checks will include control of flux density.

A major ongoing task is that of exercising *statistical process control* (SPC) on those factors which are considered to be critical for the process. SPC can lead to process stabilisation and derivation of capability indices which can be used to monitor process improvements.

12.6 METALLURGY

12.6.1 Soldering

Soldering can be defined as a process by which two metal parts are joined together by a third metal which has a melting point below those of the two metal parts to be joined. To differentiate soldering from brazing, the highest melting point for a solder has been defined to be 450°C.

In electronics, the most common metal joined is copper and the most common solder is a tin-lead solder consisting of 60% tin and 40% lead. These materials will be used to explain certain phenomena in soldering. Most of what will be described here applies also to other solders used in electronics.

In order to solder, the solder has to be heated not just to its melting point but to a temperature which is significantly higher. Normally a temperature of 50°C above the melting point of the solder is recommended. The reason is loss of heat by conduction to the workpiece and by radiation to the air. The solder cannot wet the surface of the workpiece unless the surface is at a temperature above the melting point of the solder. A flux is needed during soldering to clean the surface of all three metals thoroughly so that they can come into close atomic contact. The workpiece is usually heated by external means, e.g., by a soldering iron, by solder itself as in a wave soldering machine or by other means such as infra-red radiation. As soon as the solder has wetted, flowed and formed the joint, the soldering process is complete.

What happens during this soldering process metallurgically? When a molten solder comes into direct contact with clean copper, the copper is 'wetted'. The surface of the copper dissolves into the fluid solder and an intermetallic phase is formed which lies at the interface between the copper and the solder. Migration of the copper into the solder continues slowly in the solid state and changes the structure of the joint during its lifetime. Wetting is of paramount importance to soldering and is described in 12.7.

12.6.2 Dissolution of Metals

During soldering, the metal(s) in contact with the molten solder will dissolve in the solder at a rate which depends on a number of factors. The dissolution rate is a function of the solder composition, the temperature, the type of metal being dissolved and its surface structure. The amount which will be dissolved by a given solder is also dependent upon the time for which the materials are in contact.

The most important metals with which we are concerned are copper, nickel, gold, silver and iron.

Figure 12.3 shows how quickly copper wire is dissolved by molten solder. The wires were dipped into two solder baths, of different compositions, at temperatures of 310°C and 370°C and held there for the times indicated.

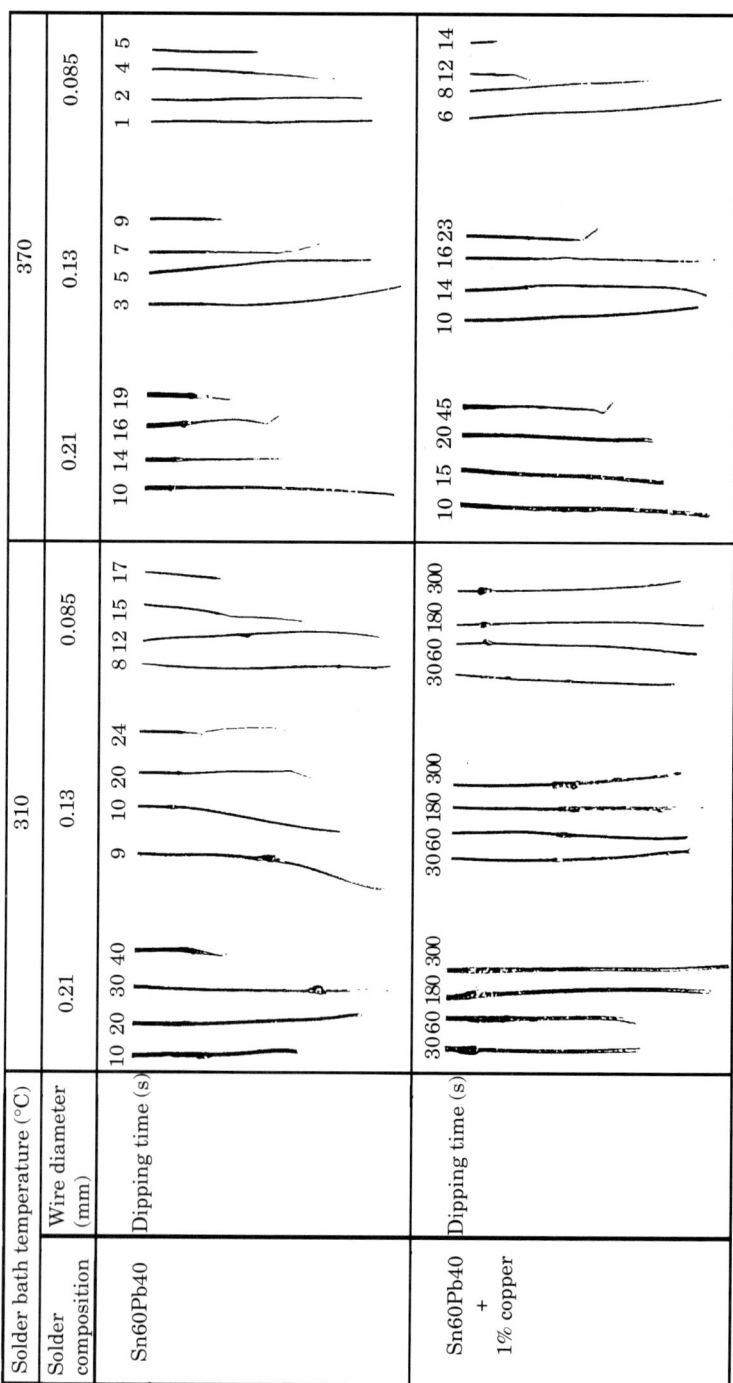

Fig. 12.3 The dissolution of copper wire in solder. The amount dissolved depends on the solder temperature, the time available for dissolution and the type of solder.

Two things can be directly established. The higher the solder bath temperature, the faster the copper wire becomes thinner and vanishes. An addition of 1% copper to the solder bath reduces considerably the speed with which the copper is dissolved in the solder. The rates of dissolution can be plotted graphically, Figure 12.4, and show that, under the same conditions, different metals have different dissolution rates.

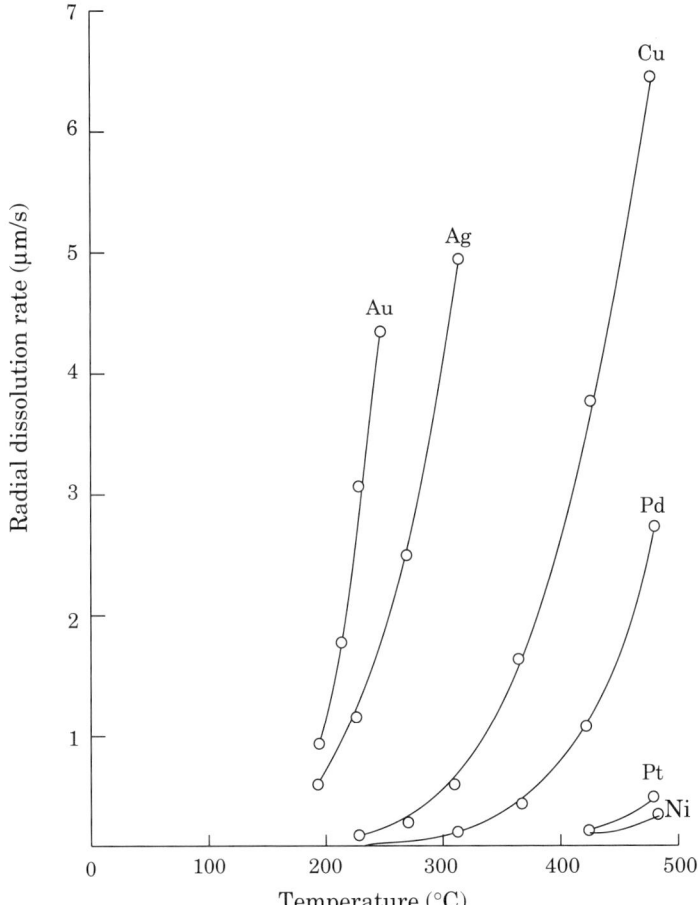

Fig. 12.4 The rate of radial dissolution of different substrate wires in liquid SnPb as a function of temperature — see Ref. 56.

Copper dissolves readily in solder (like sugar in hot coffee!). When the amount of copper in the molten solder exceeds a given level, the copper forms intermetallic crystals with the tin. These crystals can be detected easily in the solidified solder. The amount of copper which can be dissolved depends on the temperature of the melt. The relationship with pure tin solder is given in the phase diagram for tin and copper.

The dissolution of metallic terminations in solder, such as can happen with certain SMD terminations, is called 'leaching' and presents a serious problem when the termination practically vanishes.

12.6.3 Solidification of Solder

In the most usual case, solder consists of 60% tin and 40% lead (written as Sn60Pb40). This is a non-eutectic composition which means that the alloy solidifies over a range of temperature and that the microstructure is not uniform. The phase diagram (which describes the phases present at equilibrium in an alloy of two or more elements as a function of temperature and composition) for Sn60Pb40 shows that, on cooling, a small amount of lead will solidify first. When the temperature during solidification reaches 183°C — the eutectic temperature for tin and lead — both tin and lead will solidify simultaneously in the matrix of the already solidified lead crystals in a finely divided (eutectic) structure. The fineness of the eutectic structure depends on how fast the melt solidifies. If it solidifies slowly, coarse structures are obtained; if it solidifies fast, finer structures are obtained. In general, fast solidification of the solder is preferred; normally, the finer the grain size, the better the mechanical properties. Fast solidification can be promoted by fast heating to relatively low soldering temperatures. This reduces the heat input into the board and components.

12.6.4 Intermetallic Phases

As already described, copper (and most of the metals joined by soldering) dissolves in molten solder. When it goes into solution, an intermetallic layer is always formed between the copper and the solder. If the amount of copper exceeds a given level, the copper forms intermetallic crystals with tin. These can be detected by using a microscope. Long soldering times and high temperatures increase the dissolution of copper in the solder, creating more intermetallic crystals which penetrate into the solder itself. The intermetallic layer is regarded as a sign of good wetting and soldering. However, when soldering has been carried out at relatively low temperatures and for short times, the intermetallic phase may no longer be detectable with an optical microscope as it is too thin. Other metals such as German silver or nickel dissolve slowly into the solder and the intermetallic phase will thus be so thin that it is not detectable using an optical microscope. However, this does not mean that the quality of the solder joint is poor.

The soldering conditions should be kept such that as little intermetallic is formed as possible. Although an intermetallic layer is necessary for good soldering, excessive intermetallics are not wanted as they make the joint brittle. All intermetallic compounds are more or less brittle. Copper forms less brittle intermetallics than most other metals and only formations with a thickness of more than 3-5 μm seem to cause cracks. Gold intermetallics, however, are highly brittle and are a common cause of joint failure in stress or fatigue.

12.6.5 Diffusion

Even in the solid state, diffusion of the copper into the solder occurs by formation of copper-tin intermetallics. The intermetallic layer between the copper and the solder will grow with time and temperature, Figure 12.5 and Figure 12.6. This will change the structure of the solder joint significantly,

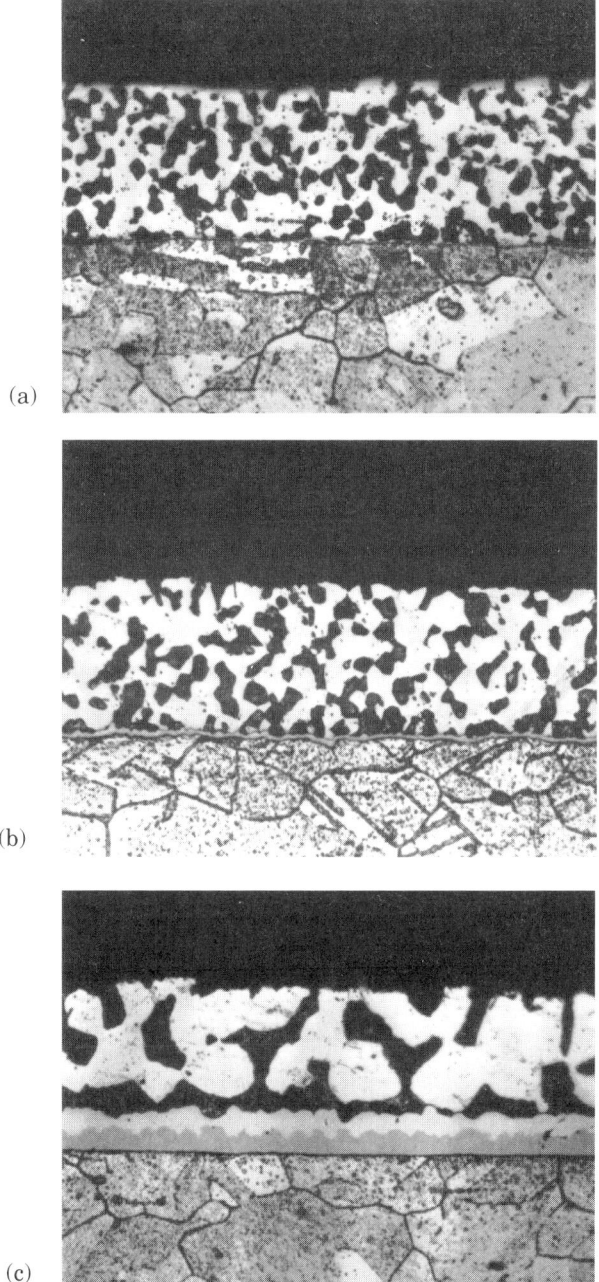

(a)

(b)

(c)

Fig. 12.5 Growth of the intermetallic layer between the basis material, copper, and solder (Sn60Pb40) in the solid state phase due to ageing. (a) Ageing 200 days at room temperature. The Cu_6Sn_5 and Cu_3Sn phases are not easy to distinguish. (b) Ageing 200 days at 70°C. (c) Ageing 200 days at 135°C. (Courtesy of the International Tin Research Institute)

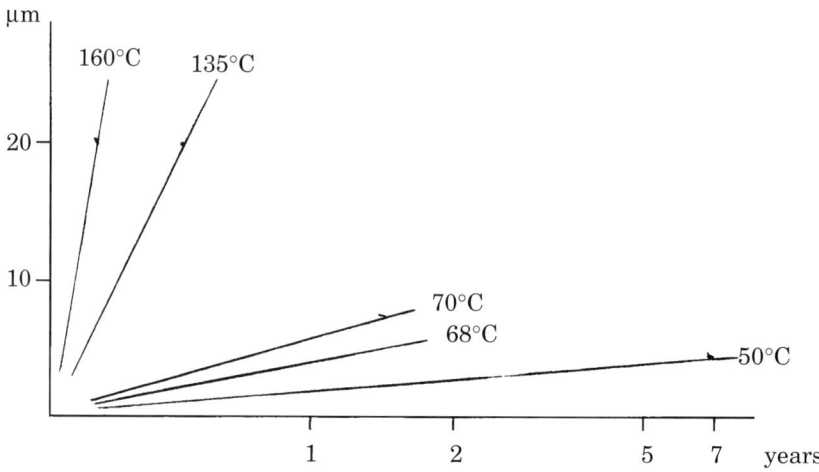

Fig. 12.6 The growth of the copper-tin intermetallic phase depends on time and temperature. The thickness of the intermetallic layer in microns, Y axis, is plotted against time, X axis. (Courtesy of L. Revay, Ericsson Telecom)

Figure 12.7, and so alter its properties. In the case of copper coated with tin or solder as a solderable layer, the solderability may be destroyed in a relatively short time by transformation of the tin or solder layer into the copper-tin intermetallic phase. This oxidises and becomes unsolderable if the layer is so thin (1 μm or less) that the intermetallic is exposed to the atmosphere. Such a thin layer may be found on the 'knees' of through-plated holes. This is one of the reasons why the solder sometimes does not flow up on to the top (component side) land of the through-plated hole. A remedy in this case would be to solder the PBs directly after manufacturing them.

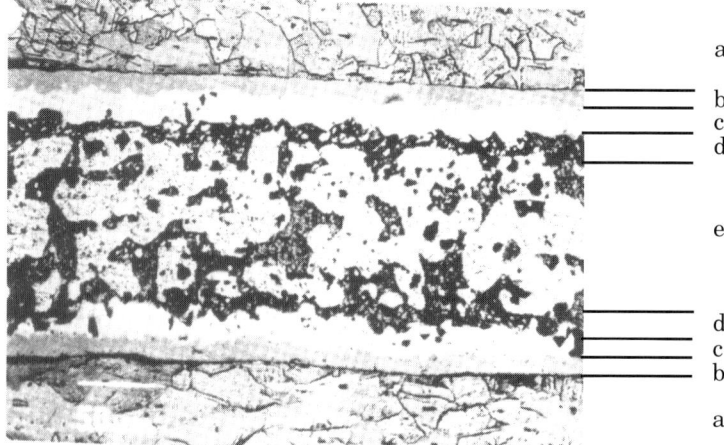

Fig. 12.7 Separation of solder constituents due to ageing sample for 200 days at 170°C. (a) Base material (copper); (b) copper/tin intermetallic compound (ε - phase Cu_3Sn); (c) copper/tin intermetallic compound (η - phase Cu_6Sn_5); (d) tin-depleted lead-rich phase; (e) solder (Sn60Pb40).

The strength properties, also, are altered by diffusion in the solid state and it is therefore important to consider the effect of the change of the material in the joint in sensitive applications.

12.7 WETTING

Wetting is the central concept in soldering. However, the reader should be warned: Good wettability may not be confused with good solderability, see 12.9.2. Without wetting there can be no soldering. Three factors are indicators of wetting:

— the spread of the solder on the surface being soldered;
— the formation of an adherent coating of the solder on the surface;
— a small contact angle between the solder and the surface.

If all three conditions are fulfilled at the same time, wetting has been achieved.

Wettability is a surface property which is influenced by the:

— melting of the material, as, for example, tin or solder;
— dissolution of the material, as, for example, gold or silver.

The phenomenon of wetting is familiar to observant people. Water on a greasy dish or metal plate does not wet. It forms droplets which roll down if the surface is tilted. If a detergent is used and the dish or metal plate is degreased, the water no longer forms droplets and, instead, spreads to make a thin layer over a large area. To achieve wetting of solder on metals a flux is needed.

Another notable phenomenon is that, when the dish is tilted, only the surplus water runs off; a thin layer of water is left. It is difficult to wipe this water away as we know from cars when we want to clean the windscreen with a slightly defective windscreen wiper. There is an adherent coating.

Figure 12.8 shows what is meant by the contact angle θ. If the contact angle is large, as in the left part of the figure, none or only a very little part of the solder is in contact with the base material. Such a droplet can easily be lifted away. If the contact angle is low, as in the right part of the figure, a much larger part of the solder is in contact with the surface. This indicates that both metals, the solder and the base metal, have been able to react with each other to form a metallurgical bond (intermetallic phase) which cannot be destroyed. It is now impossible to remove the solder by simple mechanical means.

From Figure 12.8, the relationship shown in Figure 12.9 can be derived. This can be used to analyse what happens when liquid solder wets the base metal. If the system is in equilibrium, the following equation can be written:

$$\gamma_{sl} = \gamma_{sg} - \gamma_{lg} \cdot \cos \Theta$$

where γ_{sl} = surface tension between solid phase and liquid
γ_{sg} = surface tension between solid phase and gas
γ_{lg} = surface tension between liquid and gas
Θ = wetting angle.

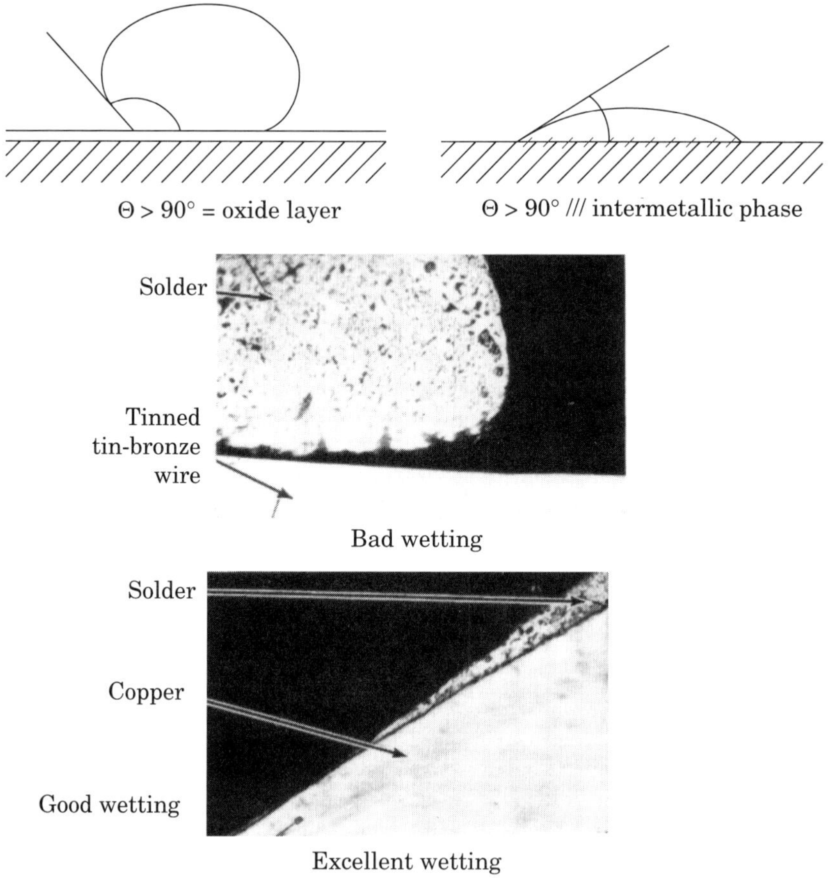

Θ > 90° = oxide layer   Θ > 90° /// intermetallic phase

Solder

Tinned
tin-bronze
wire

Bad wetting

Solder

Copper

Good wetting

Excellent wetting

Fig. 12.8 Wetting is indicated by the wetting angle. In these illustrations, good and bad wetting are shown.

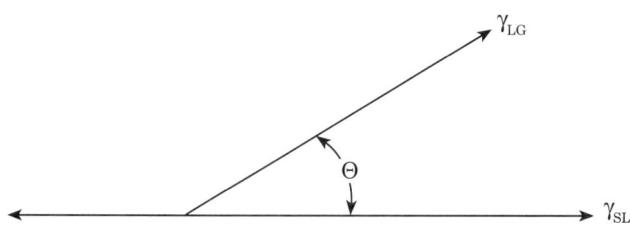

Fig. 12.9 Relationship between the different surface tensions acting and the wetting angle.

$\gamma_{SL} = \gamma_{SG} - \gamma_{LG} \cos\theta$ *where*
γ_{LG} = surface tension molten solder/flux
γ_{SL} = surface tension molten solder/base metal *and*
γ_{SG} = surface tension base metal/flux

The wetting angle can be obtained by different methods, either by direct measurement or by calculation from the wetting forces measured by a wetting balance. Sometimes the wetting angle is used as an indication of the wetting quality. A typical classification is:[8]

Wetting Quality in terms of Contact Angle

$0° < \theta \le 10°$ perfect wetting
$10° < \theta \le 20°$ excellent wetting
$20° < \theta \le 30°$ very good wetting
$30° < \theta \le 40°$ good wetting
$40° < \theta \le 55°$ adequate wetting
$55° < \theta \le 70°$ poor wetting
$70° < \theta \le 90°$ very poor wetting
$90° < \theta$ non-wetting

This classification must be considered as rather theoretical. For practical soldering, it is sufficient to differ between good, bad and non-wetting.

Good wetting should not lead the reader into believing that a good solder joint *will* be obtained. A very low wetting angle of approximately 5°, as shown in Figure 12.10, indicates perfect wetting but will not lead to a solder joint as, due to the wetting angle, the height of the solder is so low that it cannot bridge the gap between the two parts to be soldered. A wetting angle of less than 8° has never been observed by the author during measurements on solderable surfaces. Very often it is not possible to determine the real wetting angle as the design of the land for the solder joint puts a limit on how far the solder can flow, Figure 12.11(b). It must be accepted that an angle of 90° *can* occur with good wetting. If the board is tilted, at one end the solder will form a wetting angle of perhaps 120° and on the other end a wetting angle of 20°, Figure 12.11(a).

Fig. 12.10 If a good wetting angle is achieved but the gap between the land and the component lead is too large, the solder cannot bridge the gap and no solder joint can form. (Courtesy of the Swedish Institute for Metals Research)

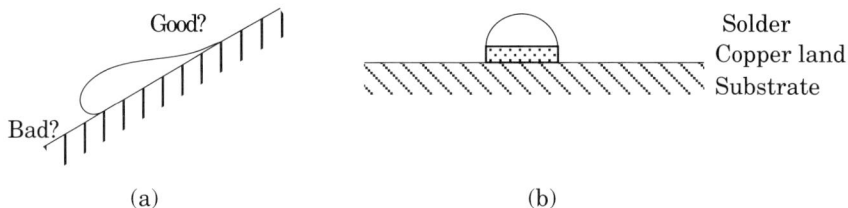

Fig. 12.11 The formation of the wetting angle can be influenced by a number of factors, two of which are shown here. In these cases, an objective judgement on the quality of the wetting cannot be made by visual observation. (a) The molten solder flows down a tilted surface and solidifies. Does the lower end of the solder indicate bad wetting and the upper end good wetting? Only a thorough metallographic investigation can give an answer. (b) The solder is limited in its flow by the restricted surface available.

Which wetting angles can be obtained depends on a number of factors, of which the most important are the material and how the material was rolled and annealed. Different wetting angles will be obtained on copper depending on the structure of the copper surface. German silver will not give as high wetting angles as copper. Different solders will give dissimilar wetting forces and different wetting angles. An acid flux will give a low wetting angle, a high solder meniscus and a low wetting force, whereas a non-activated resin gives a high wetting angle, a low solder meniscus and a high wetting force.

The question to be answered is whether wetting angle or wetting force should be regarded as the decisive measure of wetting quality. In the case of soldering, it is easy to give a practical answer. It is difficult to determine directly the wetting angle on the different shapes and sizes of component termination and it is easy to measure the wetting forces with the equipment available.

12.8 THERMAL CONSIDERATIONS

Soldering is a process which is performed at elevated temperatures and which needs a certain time. Minimum values for both temperature and time must be achieved before soldering results. It cannot be assumed that all the components to be soldered to a board can be exposed to the 'normal' soldering process, e.g., 5 s at 250°C. Some components are damaged, that is, their electrical properties change or the components are destroyed mechanically. In other words, soldering can only be carried out within a certain temperature and time range if a good solder joint is to be achieved and the components involved shall not be destroyed, Figure 12.12. Certain components, such as electrolytic or ceramic capacitors, cannot withstand the temperature shock imposed by a temperature rise from the preheating temperature of, say, 100°C to the soldering temperature of 230°C or more. Even the preheating must be selected and controlled with care.

Not only do different components have different thermal needs but dissimilar leads on a given component may need different soldering times.[28]

If a temperature-sensitive component is to be soldered, there are several ways of going about it. If there is only one sensitive component on a board, it can be soldered:

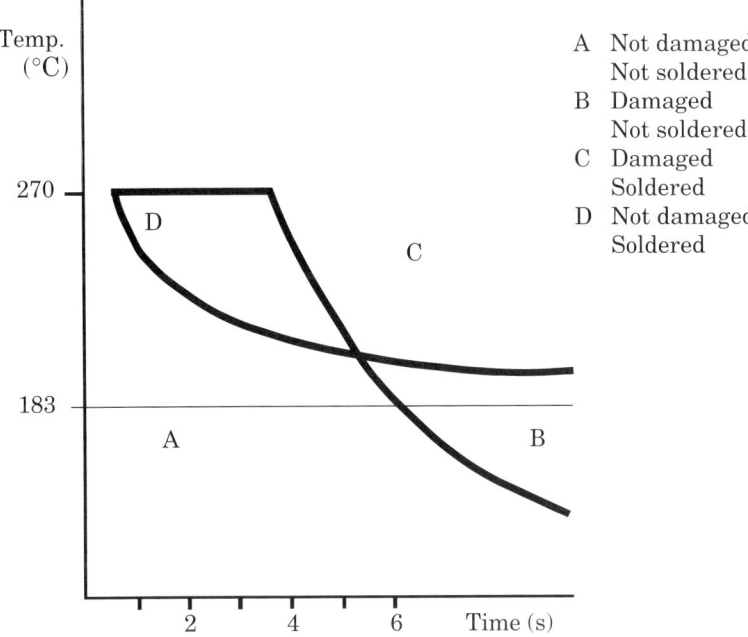

Fig. 12.12 Thermal considerations when soldering. A solder joint can only be achieved within given time and temperature limits. If the time is too short and/or the temperature too low, no solder joint can be formed. If the time is too long and the temperature too high, the component will be destroyed. The time and temperature values shown here are for the commonly used Sn60Pb40 solder.

— simultaneously with the other components;
— by extending the component lead;
— by decreasing the cross-section of the lead and/or by using a lead material with a low thermal conductivity in such a way that the temperature at the component body will never reach a dangerous level during the soldering time;
— taking precautions such as adding a heat sink to the component;
— using special methods such as laser soldering;
— using special solders such as low melting point solders.

Both laser soldering and low melting point solders can, of course, be used in mass production as well.

To attempt to avoid damaging components by increasing the preheating temperature and decreasing the soldering temperature can be risky. For a given type of electrolytic capacitor, known to be damaged at a temperature of 150°C, calculations were made in accordance with Klein Wassink.[7] It was found that:

— if the soldering temperature is lowered from 250°C to 235°C, the time interval between wetting and damaging the component decreases from 2.6 to 1.6 s;

— if the preheating temperature is increased from 80 to 100°C, the time for wetting decreases from 2.1 to 0.9 s;
— if the preheating temperature is increased to 140°C, the component lead wets immediately and the component is damaged after only 0.9 s, regardless of whether the soldering was carried out at 235 or 250°C;
— at yet other values, the component is damaged before it is soldered.

12.9 SOLDERABILITY

12.9.1 General

Approximately 70% of faulty solder joints are caused by material of poor solderability. Flux, solder, bad design or manufacturing methods and incorrect specification of the solder joint accounts for the remaining 30%.[29-31]

With figures such as these and the knowledge of the number of bad solder joints manufactured in every company every year, it is clearly seen that solderability is of utmost importance, to both the quality of the solder joint and the economics of the manufactured product.

The advantages of good solderability are so many and important that the value of solderability testing should be evident to every component manufacturer and producer of electronic equipment.

12.9.2 What is Solderability?

Solderability is a measure of the ability of a surface to be readily wetted by molten solder. It can be determined by a given test method under a specific set of conditions, for example, the solder spread test and wetting balance test.

However, solderability also defines the total suitability for industrial soldering, covering four aspects:

— the thermal demand;
— the practicability of feeding solder to the joint, Figure 12.2;
— the wettability of the joint components;
— the resistance to soldering heat.

Here it is appropriate to say what is meant by 'good' and 'bad' solderability. If thermal demand can be ignored for the moment, if a solderable material in a solderability test gives wetting time values of less than 0.8 s the material must be regarded as being of good solderability. In most cases, the material will not change its value even if it is subjected to the normal standardised ageing methods. If it does change its value (i.e., shows increased solderability values), the suspicion is great that this material can cause trouble after storage. Normally, material which gives time values between 0.8 and 2.0 s is considered to be solderable. However, the time value of such material may often increase after artificial ageing. This is an indication that it can cause trouble when soldered after natural storage. Solderability values above 2 s clearly indicate future solderability problems. These can be experienced with hand soldering but even more in sensitive machine soldering processes where failure rates of less than 1000 ppm are needed. Materials having good solderability indicate that materials have been chosen which are either stable in their properties or have been suitably processed or both. Materials having poor solderability simply indicate the wrong choice of material or faulty processing of the material.

12.9.3 The Advantages of Solderability Testing

The knowledge of solderability values helps the engineer to set the process parameters — soldering time and soldering temperature — more accurately. Material of good solderability gives the following advantages:

— a lower soldering temperature;
  this is especially important when using surface mount technology or heat-sensitive components; a low soldering temperature prevents or diminishes the risk of material damage and helps to slow down the formation of intermetallic phases and thus increases the life of the solder joint; the sensitivity of the solder joint to corrosion is decreased;
— a shorter soldering time;
  this gives increased productivity and thus lower costs; it also reduces formation of intermetallic phases;
— a smaller variation in soldering time;
  this contributes to short soldering times, uniform solder joint quality and increases both the reliability and the life of the solder joint and the assembly;
— a shorter overall production time;
  this results from shorter soldering times and reduction of rework in the workshop to a minimum;
— a flux with low activity can be used;
  this is of great importance to the electronics industry as it implies less corrosion, greater reliability and a longer life for the assembly;
— stable electrical resistance values;
  this means more reliable functioning of the circuit, especially important in medical and computer electronics; less reconditioning and repair are needed, resulting in fewer customer complaints; the risk of fire due to high resistance developing in the joint is eliminated;
— less signal noise;
— fewer repairs;
  this is a major contributor to greater reliability.

Taken together, these result in a higher level of built-in quality at a lower overall price. One might argue that solderability testing is expensive and that the equipment needed to do the testing is not cheap either. But compared with the gains — increased productivity, greater reliability and higher quality as well as reduced customer complaints — money spent on solderability testing is money well spent.

There are many points at which solderability testing can be applied in a company and the flow chart in Figure 12.13 gives some ideas and alternatives.

12.9.4 Preservation of Solderability

One of the great problems in circuit assembly is to preserve the solderability of the components and boards from deterioration until the component

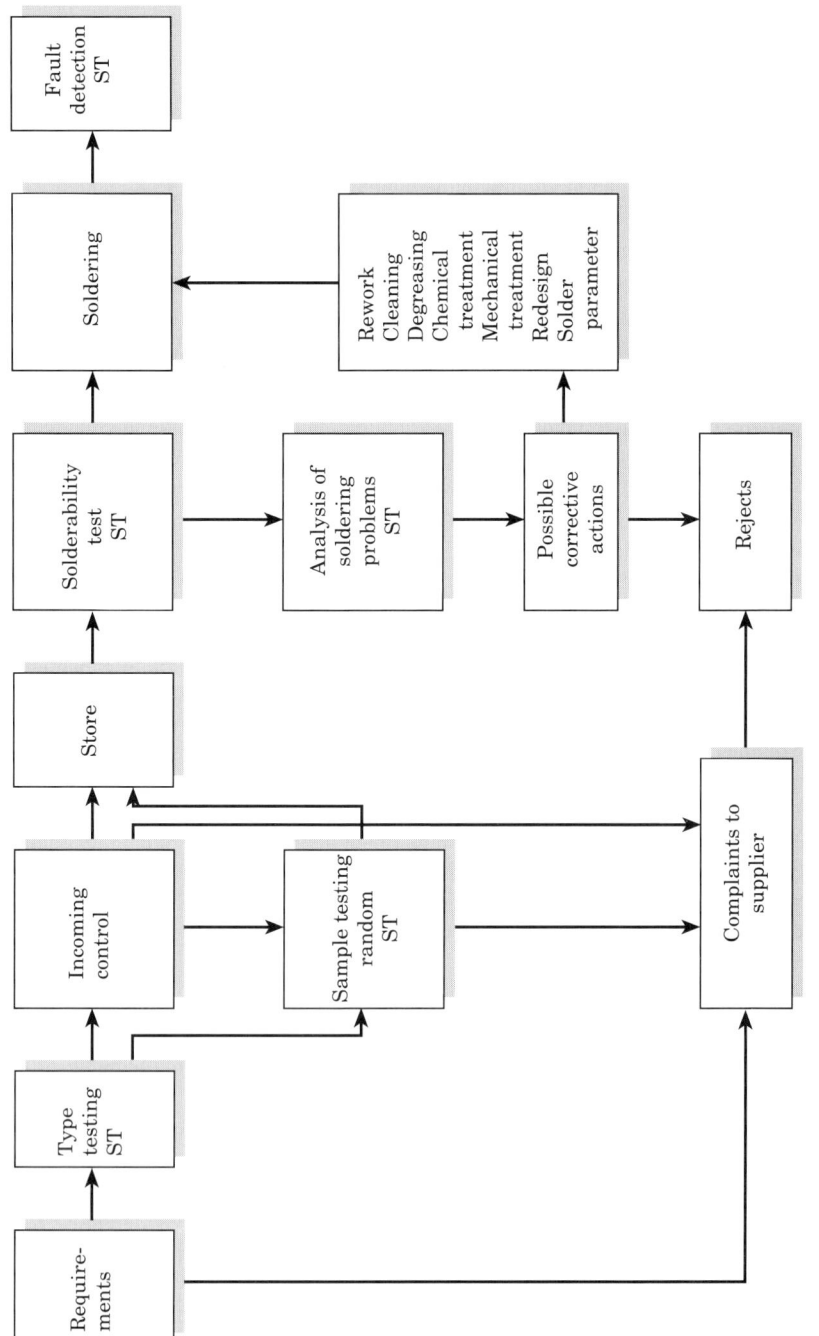

Fig. 12.13 Flow chart on the possible uses of solderability testing (ST).

terminations are joined to the board. As it is known that the solderability of material can be degraded by two phenomena, these can be counteracted. The first phenomenon is the diffusion of the different metal layers forming both the termination and the solderable layer and the second is the degradation of the surface by the surrounding air.

The air can be loaded with different contaminants such as water vapour or sulphur-containing gases. The aim is to prevent these and other agents reacting with the solderable layer. This can easily be done by enveloping the material in plastic foils or keeping the items in boxes of either plastics or metal. When using plastics, it is advisable to place a water-absorbent material in the package as the plastic foil is not totally impermeable to water vapour. The storage of solderable materials, even when fully packaged, should be at low temperatures (but above freezing point) as every chemical reaction increases with temperature. Low storage temperatures reduce the speed of diffusion.

The humidity level must be low. Any possibility of moisture condensing on to the material to be soldered must be counteracted. Next to diffusion, condensation is the main reason for destruction of solderability. It is suggested that the storage temperature should not exceed 25°C and that the humidity should be less than 70% RH. At temperatures above 30°C with a relative humidity of 90% RH, even material of good solderability can be rendered unsolderable in less than a month. Material of good solderability can be stored in a tightly closed metal can for many years without loss of solderability. It is suggested that no more than ten boards should be packed in a single plastic foil package for storage.

12.9.5 Solderability Testing Methods

Various types of solderability testing methods exist.

Qualitative subjective tests

— dip and look test;
— dewetting test;
— solder spread test;
— workshop method.

Quantitative objective tests

— wetting balance test;
— meniscometer method;
— solder globule method.

The dip test and the solder globule test are described in IEC 68-2-20 and in ANSI/J/STD 002, April 1992. The latter also describes the wetting balance test of IEC 68-2-54. IEC 68-2-44 'Guidance on Test T: Soldering', now under revision, is the most important document.

Unfortunately, the most commonly used test today is the 'dip and look' test which can give very varying results depending on who is judging and the experience of the person in question (see 12.10.3.1, Visual Inspection). This test must therefore be considered unreliable. IEC 68-2-54 (1985) is interesting in so far as it gives a list of temperatures at which different properties related to soldering shall be tested, Figure 12.14.

Property tested	Immersion conditions				
	3	2	5	30	s
	215	235	260	260	°C
Wettability	x	x			
Resistance to dewetting			x		
Resistance to soldering heat			x		
Resistance to dissolution of metallisation				x	

Fig. 12.14 Testing conditions for different properties related to soldering of SMDs. Time and temperature conditions relate to the dipping method of testing.

The most commonly used quantitative objective test today is the wetting balance test. The solder globule test should *not* be used any longer as it gives information only on wetting time and not necessarily information on the wetting quality. The wetting balance test gives a measure of wetting quality with the wetting force. Nor does the globule test give any measure of dewetting, which the wetting balance test does.

The only other test which can be recommended is the solder spread test. This is an inexpensive, fast and discriminating test. The meniscometer method which is widely used in France is well suited as an in-house method giving good results.

12.9.5.1 THE WETTING BALANCE METHOD

The component lead to be tested for solderability is dipped at a defined speed of 20±5 mm/s into the solder bath, Figure 12.15. Initially, the lead cannot solder as it has not yet reached the temperature needed for soldering to take place. As the dipping process continues, the lead forms a dimple in the solder bath — a negative meniscus. After a short time, when the lead has reached the proper soldering temperature (and has good solderability) the solder begins to flow up the lead, forming a collar of solder around the lead — the positive meniscus. The lead stays for a given time in the solder bath before being pulled out of the bath. The degree of contact between the lead and the solder can be measured by a force gauge. When the lead forms the dimple in the solder and the lead is being pushed out, a negative force value will be registered. When wetting starts and the sample is pulled into the solder, a positive meniscus is formed and positive force values are obtained.

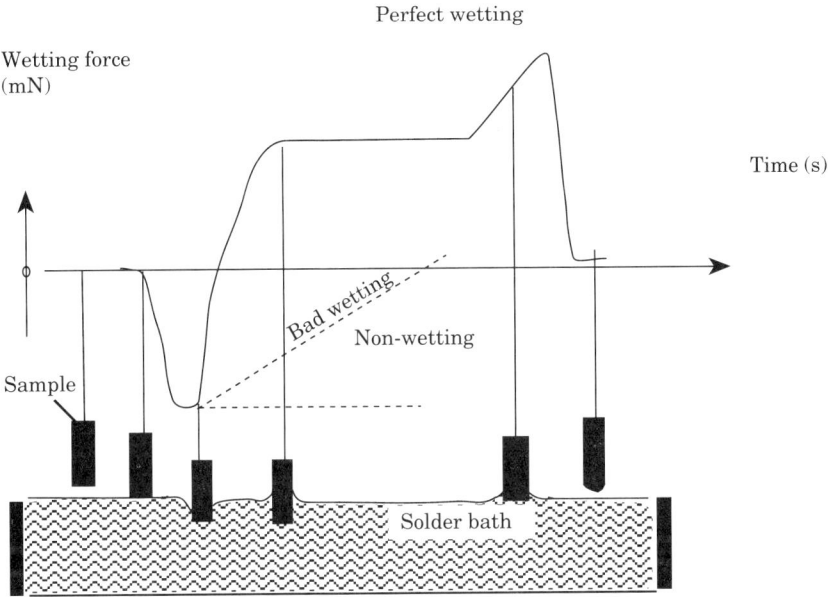

Fig. 12.15 The measuring principle of the wetting balance. The sample is dipped into the solder bath and the action of the solder on the sample is recorded as wetting force over wetting time. (This is the standardised way in which to show the curve.)

The whole wetting process can be presented as a wetting curve, that is, showing the wetting force versus the wetting time:

— the wetting force gives an indication of the quality of the wetting;
— the wetting time is useful to know for the soldering process.

To conduct this test, a rather complex equipment is needed, Figure 12.16(a). The special advantage of the wetting balance test is that leaded components and SMDs can be tested for solderability due to the ability to immerse the SMD to exactly the depth required in the molten solder, Figure 12.16(b). In testing SMDs, as shown in the figure, very often the devices are lowered into the solder bath at an angle other than 90° to the solder surface.

12.9.5.2 THE SCANNING METHOD

The scanning method is a modified wetting balance technique where the sample is dipped very slowly, preferably at 1 mm/s, into the solder bath. A major advantage, compared with all other solderability testing methods, is that this method makes it possible to measure the solderability by sequential scanning over the entire length of a component lead, Figure 12.17,[32] whereas the normal wetting balance method measures the solderability over a very small range only at the intersection with the solder bath surface. The low

566 A Comprehensive Guide to the Design and Manufacture of
Printed Board Assemblies

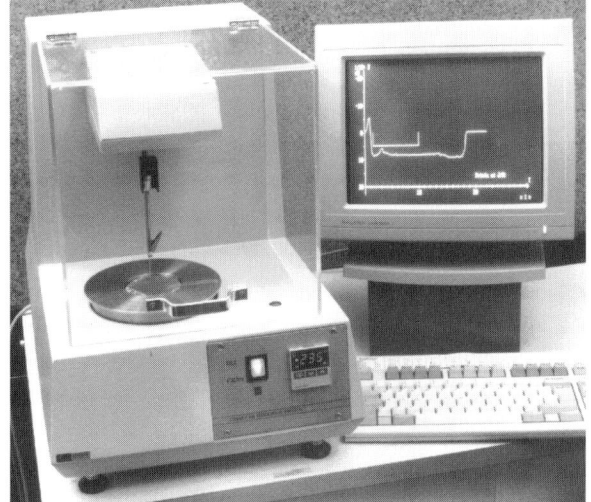

(a)

(b)

Fig. 12.16 A computerised wetting balance apparatus: (a) The curve being outside the failure area indicates that the tested sample fulfils the specified solderability requirements; (b) The SMD being tested here is lowered into the solder bath not perpendicularly but at an angle of 45° as it has short terminations.(Courtesy of Convey AB, Stockholm)

dipping speed is chosen due to the fact that, under this condition, a specimen will wet simultaneously with dipping as the place to be wetted has already reached the soldering temperature before it contacts the solder. Only at the start-up is more time needed to bring the sample to soldering temperature.

By using the scanning method, the solderability over the entire length of a termination can be determined. With this method it is even possible to determine, for a given component, how long it takes for the soldering process to start. This is important information for practical soldering. Other information which can be gained (important for the quality of the solder joint) is how much the solderability of the component lead adjacent to the component body is influenced by the cooling of the component body which is acting as a heat sink, Figure 12.18.

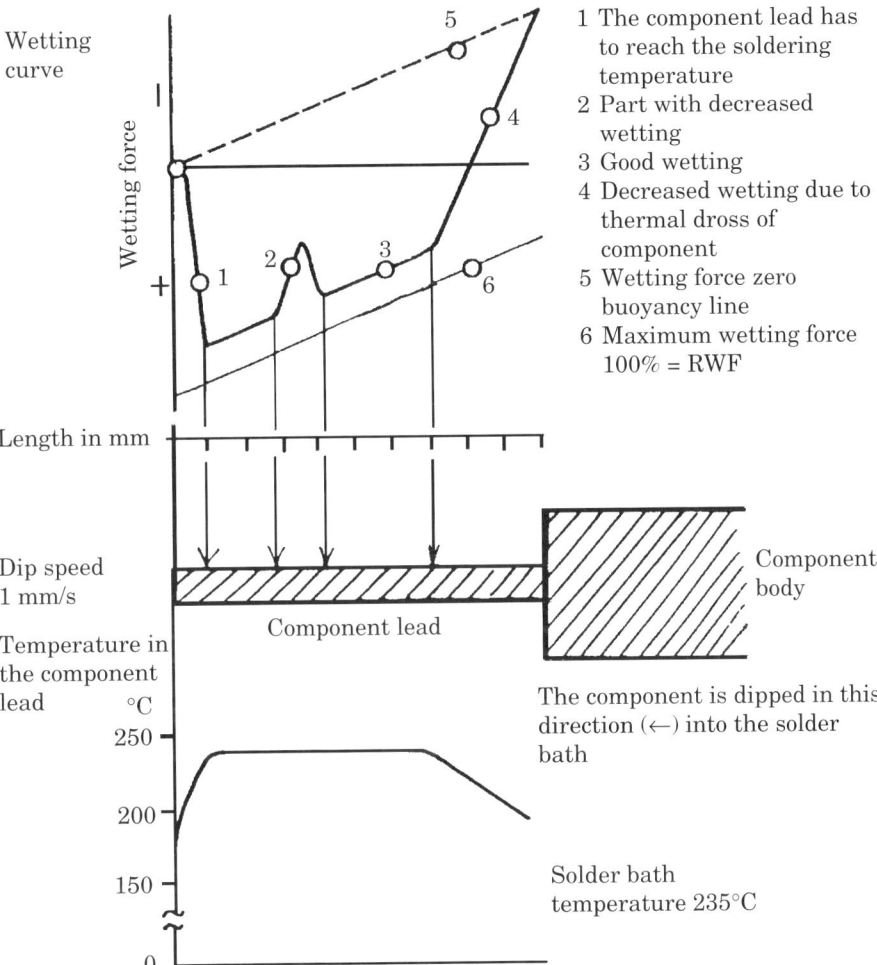

Fig. 12.17 The scanning solderability test method.

12.9.5.3 THE WORKSHOP METHOD

The principle of the wetting balance method can be used in the workshop to give a better indication of the solderability of component leads than can be obtained by the dip and look method.

Even though this method does not give figures, it will give a reliable idea on whether a component termination is solderable or not. For this purpose, the termination of the component is dipped slowly into a solder bath. (A simple little solder pot will suffice provided the temperature of the solder can be checked.) At first the component termination will cause a depression in the solder surface. This depression is easily visible by eye — compare with Figure 12.15. If the termination solderability is really bad, the depression will remain. If the termination is easily solderable, the solder will flow up the termination readily and an upward meniscus will be formed. How soon the

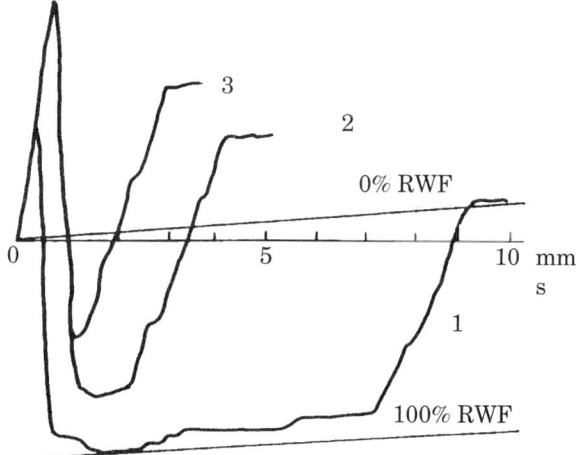

Sample No.	Wire diameter 0.6 mm Wire length mm
1	10
2	5
3	3.6

Component body:
Al Ø 7.2 x 25 mm

Fig. 12.18 Thermal demand measured with the scanning wetting balance method. 0% RWF indicates no wetting, 100% means perfect wetting. From the diagram it can be seen how long it takes before the wetting of the component lead starts (in the case of curves 2 and 3, 1 s) and at which point of the component lead the wetting action decreases (for curve 1, it is 2.5 mm before the component body).

solder starts flowing upwards, and how fast it flows, gives an indication of the solderability of the sample.

It is also important to observe the appearance of the solder as the component termination is pulled out of the bath. If it looks tenacious, and if the solder surface is not oxidised, this too indicates good solderability. If the solder jumps off the lead, it may be an indication of bad solderability or dewetting.

The dipping speed should be between 1 and 4 mm/s with a dipping time of about 5 s. This is a speed which, after some training, can be achieved by hand. The operator must be trained also in the observation of how the meniscus is formed at the termination. These are things which can be done easily in a workshop.

12.9.5.4 THE SOLDER GLOBULE METHOD

This method is worthy of description here as it was the first objective method that was widely used and gave numerical values which could be treated statistically. Even though it has been replaced by the wetting balance method, the globule method has its merits in so far as many solderability studies, which are frequently referenced, have been carried out using this method.

A defined amount of solder is heated to 235°C on an iron rod. The sample, usually a component wire, is dipped into the solder, Figure 12.19. The time measurement starts when the wire touches the solder, and when the solder passes over the wire the time measurement is finished. A number of tests, at least 10 and preferably 50, are carried out. The measured values are arranged in sequence and plotted on a logarithmic scale with time on the X-axis and the probability on the Y-axis, Figure 12.20. Any deviation from a straight line plot or a low inclination of the curve indicates a problem with the solderability. The steeper the inclination of the curve, the smaller the spread of results and, as a consequence, the better the material. The curve is extrapolated to the 99.99% value as shown in Figure 12.20 and, depending on the time value achieved for this probability level and its spread, the sample is accepted or rejected, Figure 12.21. As a limit for the time value, 1 s is suggested. Of course, suitable values other than that suggested can be chosen. Internationally, a time value of 2 s is widely accepted for normal electronic components. The exact procedure for evaluation of the measured values is described in Reference 33.

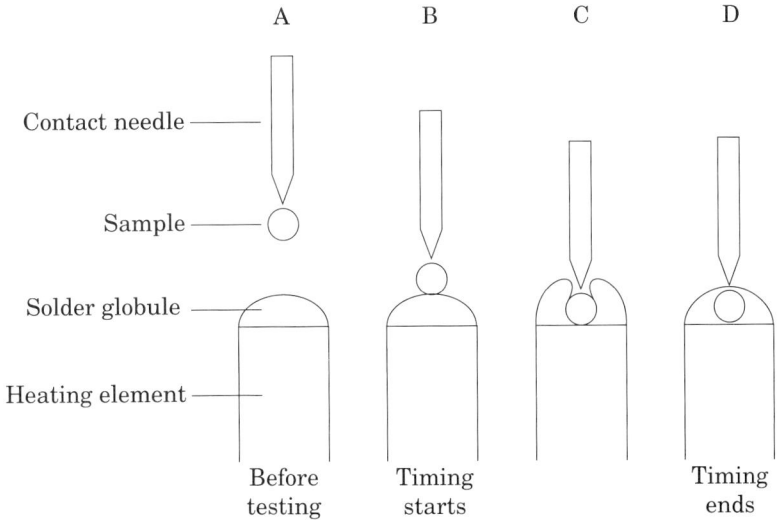

Fig. 12.19 The solder globule method. A defined amount of solder is melted on an iron pin (A). The time measurement starts when the component wire touches the solder globule (B) and ends when the globule closes over the wire and touches the contact needle (D).

12.9.5.5 THE SPREAD TEST

This is a simple, useful and discriminating test. It is not possible to test component leads with this test but it will show how well fluxes and solders work on given basis materials. According to the German standard DIN 8527,

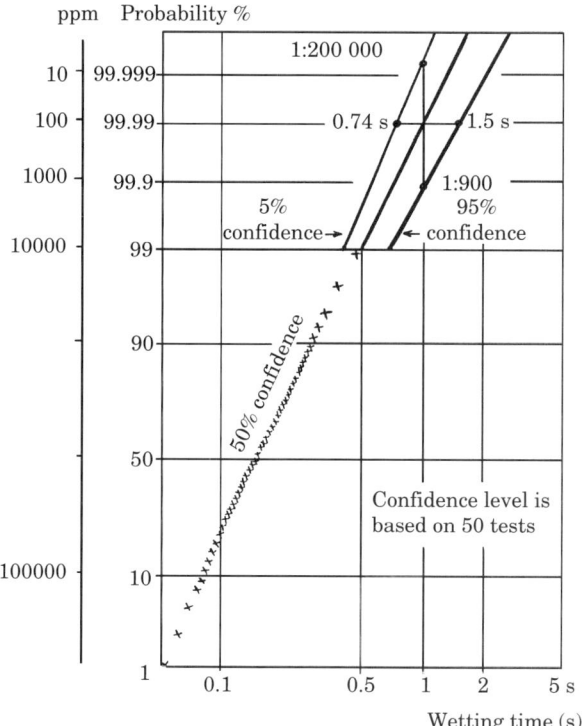

Fig. 12.20 The spread of test results given by the solder globule method.

Sheet 1, the test is carried out as follows. To compare fluxes and solders, brass sheets of size 0.5 x 40 x 40 mm are carefully degreased, etched and cleaned. 500 milligrams of solder wire of diameter 1.5 mm is wound into spiral form and pressed together to 10% of the wire size. 0.025 ml of the flux under test is dropped on to the brass sheet and the solder spiral laid on the fluxed surface. The sheet is lowered on to a solder bath surface at a temperature of 200 to 300°C. After a dwell of 5 s, the brass sheet is carefully lifted off. When the solder has solidified, the spread of the solder on the (top of the) brass sheet is measured in mm^2. The standard states that ten measurements are to be made.

The spread test is valuable as, with the spread of the solder, it gives information which the wetting balance method cannot give.

12.9.5.6 TESTING OF PRINTED BOARDS

A solderability testing method for printed boards which is relatively simple and objective, like the wetting balance method for component terminations, does not yet exist. For this reason, the dip and look method is used in different variations. The suggested IEC method is the so-called

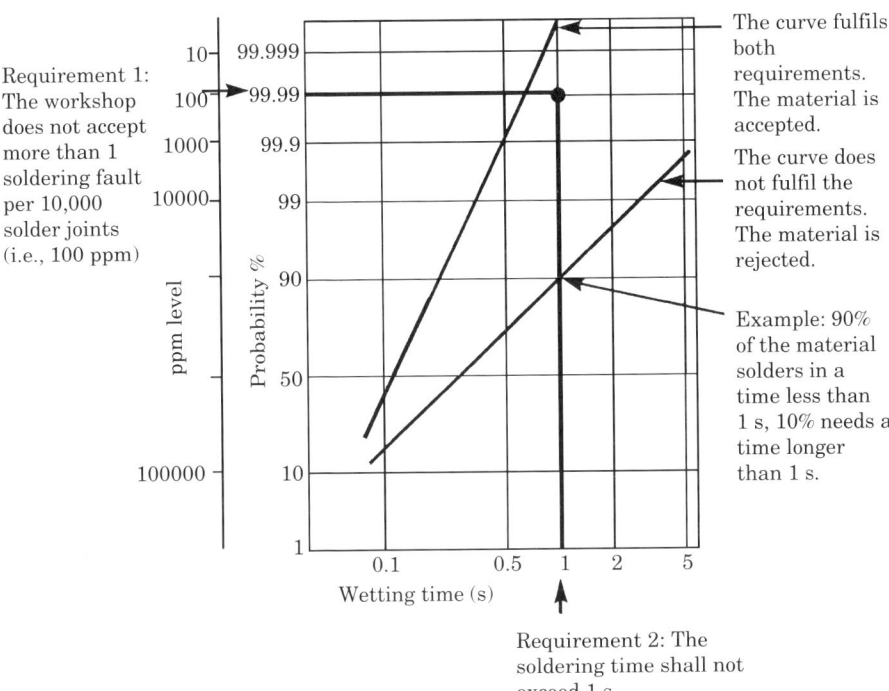

Fig. 12.21 Analysing solderability test results: Accept — Reject criteria.

rotary dip test. In this method, test coupons are brought into contact in a defined way, for a defined time, with a solder bath (held at the specified temperature). The degree of wetting is assessed visually. This is not very satisfactory. Therefore, wherever practicable, a number of samples should be run through the production soldering machine under production conditions and judged visually.

A better idea can be obtained of how readily solderable the boards are if the boards are soldered using different, increasing, times and judging the results afterwards. By doing this, it is relatively easy to establish the minimum soldering time which gives acceptable wetting.

The solderability of plated-through holes can be measured with the solder globule method, Figure 12.22. To avoid false measurements, the timing needle should be placed only over the land and never over the middle of the hole.[34] The values obtained can be treated statistically. This method is circumstantial, takes time and does not give information on the solderability of the leads on the board. It is, nevertheless, sometimes necessary as it gives valuable information on the production process and problems, Figure 12.23.

One way of testing the solderability of lands or conductors on boards which has gained much importance with the use of SMDs is the following. At the design stage, one or more lands are planned on the board for solderability testing. The land or conductor should have a length of approximately 20 mm and a width of about 2 mm. The land is carefully pulled off the board for solderability testing. The strip is bent together with the copper on the outside and a wetting balance test, preferably the scanning test, carried out.

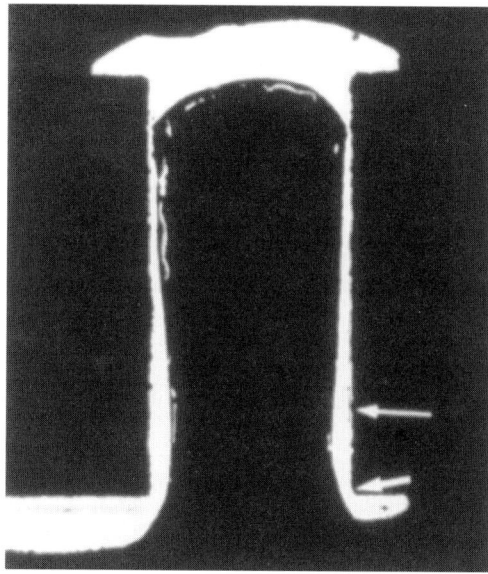

Fig. 12.22 Measuring the solderability of plated-through holes with the solder globule method.

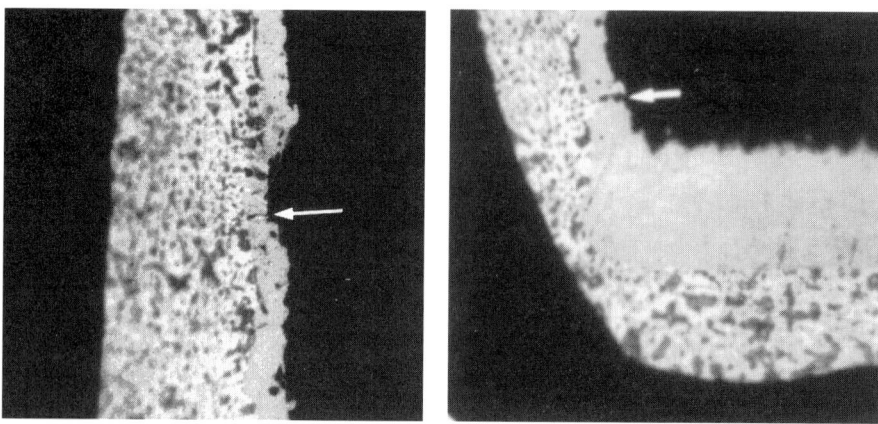

Fig. 12.23 Defects in a plated-through hole which are suspected to cause long soldering times. Measurement was made using the solder globule method.

12.9.5.7 ARTIFICIAL AGEING METHODS

Ageing methods have been developed in order to test whether or not a material is still solderable after storage for a half-year, one year or more.

Material to be tested for solderability should always be tested first in the unaged or as-received condition. When the material meets the requirements specified for that condition, the tests can be carried out on specimens in the aged condition. By so doing, changes in solderability can be detected which can give a warning that the material in question may not be solderable after the intended natural storage period. It will be a warning only because there is no strict correlation between natural and artificial ageing of solderable surfaces. The choice of ageing method is dependent on the type of ageing expected, for example, diffusion in the material (which actually accounts for most cases of degraded solderability), changes on the surface of the material, or both. If it is known that pollutants such as sulphur dioxide, hydrogen sulphide or other agents may be present to cause ageing, special tests have to be designed.

The following, typical, methods all give different results:

A   *Ageing*

The samples are stored for at least 2 but at the most 24 hours in normal room atmosphere. This 'ageing' is used as a recovery period for material which has been aged before solderability testing. In the case of printed boards, baking at $105 \pm 5°C$ for 1-2 hours is recommended to remove moisture and other volatiles.

B   *As-received condition*

The as-received condition is the condition in which the components are delivered to the purchaser or user. Normally, the ageing history from manufacture to receipt is unknown.

C   *Humidity test — Storage for 10 days at 40°C and 90-95% relative humidity*

This corresponds to test CA of IEC Publication 68-2-3 (1969). For 'tinned' material, this ageing treatment corresponds to about six months' storage in normal room atmosphere.

D   *Steam ageing*

While the 10 day humidity test equates to normal storage for about six months, often a shorter test is needed. For this purpose, the 4 hour steam test is used. In this test the samples are suspended at a height of 25 to 40 mm above boiling water for 4 hours.

E   *Dry heat*

The specimens under test are stored for 16 hours at 155°C. This test is designed to reveal diffusion problems, that is, to show if there is a diffusion barrier and, if so, if the barrier is sufficiently thick and impervious.

12.9.5.8 ANALYSING SOLDERABILITY RESULTS

A   *The solder globule method*

The solder globule method was developed as the first quantitative and objective technique of measuring solderability. Statistical methods were developed which allowed analysis of the results measured. A number of facts were learned from these statistical analyses:

— The material of a given batch does not all solder in the same unique time.
— The time values measured as an indication of solderability have a spread and, usually, follow a normal distribution.
— Any deviation from the normal distribution indicates a soldering problem.
— If the curve obtained slopes at a low angle, it indicates there are soldering problems. This explains why most of the material from a given batch can be soldered easily, while the remainder of that batch is unsolderable, Figure 12.21.
— Conclusions can be drawn from a small number of tests as to how well a large number of joints will solder. In other words, the expected failure rate (in ppm) and the spread of results can be determined from relatively few tests, Figure 12.20.
— It is possible to extrapolate values from a relatively small number of tests (>10).
— It is not possible to draw any conclusions about solderability from a very small number of tests, that is from two or three tests. The reason for this is the previously mentioned spread in the measured values.
— To be able to draw sound conclusions, at least 10 tests have to be carried out and 50 tests will give still more cogent results.
— Conclusions can be drawn from solderability testing which are valid for practical soldering.

B   *The wetting balance*

When developing the wetting balance test, it was a very long time before the wetting balance curve was sufficiently understood to establish the points of interest, Figure 12. 24.

Today, there are computer programs which automatically evaluate wetting balance curves. The values obtained for time and wetting force from a number of curves can then be evaluated statistically in similar manner to that for the solder globule method, giving conclusions as indicated in 12.9.5.8A.

In Figure 12.25, a method quite often used today is shown. Even here, 10 to 50 tests should be carried out. Normally, the curves will be drawn either by a computer or by a chart recorder and can, therefore, be treated statistically.

As a requirement for solderable material, the following guidelines are given:

> Leaded components:
> Within 2 s a wetting force of 300 mN/m should be reached.

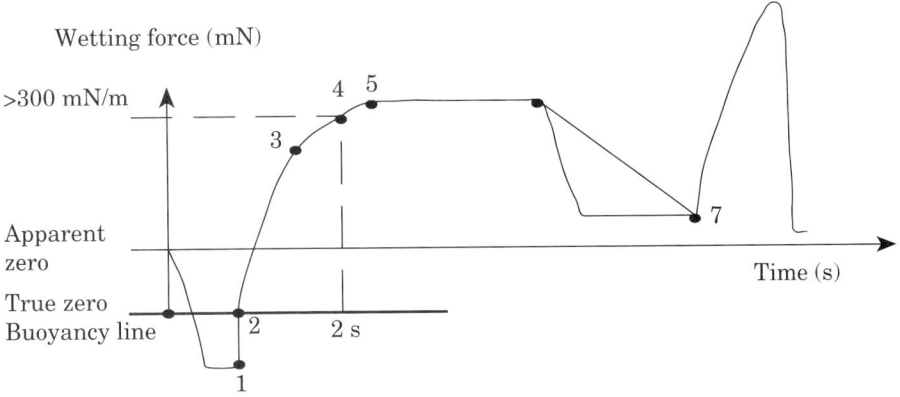

T1 Time to induce wetting
T2 Time to achieve a wetting angle of 90°
T3 Time to achieve a specified wetting force, e.g., 300 mN/m
T7 Measuring time
F4 Wetting force after a specified time, e.g., 2 s
F5 Maximum wetting force
F7 Stability of wetting, e.g., by onset of dewetting F7/F5

Fig. 12.24 Evaluation of the wetting balance curve. (T = time, F = force) Points of interest are shown.

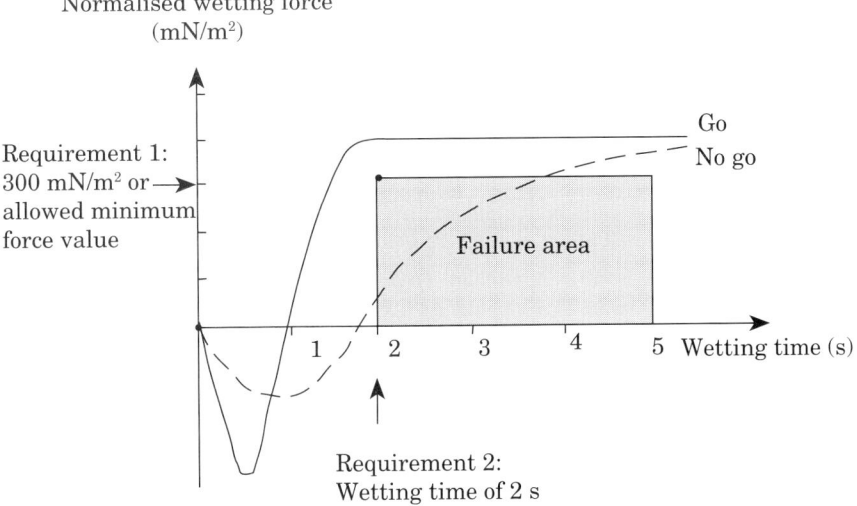

Fig. 12.25 Evaluation of wetting balance curves in a go/no go test.

Surface mounted components:
No general value can be given for the wetting force as the value depends largely on the materials used and the design of the device. In many cases, the height of the termination does not allow the full meniscus rise so that a reasonable wetting force cannot be reached. Because of this, in most cases for SMDs, only the force value in mN can be noted. For leaded SMDs a value of 250 mN/m at 235°C can be obtained and, for non-leaded devices, 150 mN/m.[32]

12.9.5.9 SOLDERABILITY TESTING *VERSUS* ACTUAL PERFORMANCE

The value of solderability testing has been proven over the last 25 years, first with the solder globule procedure and then with the wetting balance method. During that time it was observed that, in a very few cases (less than 10%), the measured results did not agree with the soldering reality. That is, a material which failed dismally in the test performed well during soldering or vice versa. No explanation has yet been found for such phenomena, but, nevertheless, these occurrences have been used as reasons for saying that solderability testing is of little value. However, these occurrences have never really been investigated. It can be assumed that the cause for a number of such happenings could have been found in design problems, thermal problems or process parameters such as flux, preheating or type of soldering process.

12.9.6 Correcting Bad Solderability

Unfortunately, workshops still encounter soldering problems. As already mentioned, the majority of these arise because of poor solderability. The puzzle is how to solve the problems posed. Before any countermeasures can be taken, an analysis must be made in order to find the cause of the problem. If the cause is found, action to resolve the problem can be suggested, Figure 12.26. Sadly, only a few possible actions can be offered. Re-tinning is, unfortunately, seldom a real solution. Figure 12.27 shows how little is gained in solderability by a simple re-tinning process. In this particular instance it was possible to achieve nearly perfect solderability using ultrasonic tinning (but do not forget the flux!!). In most cases material of poor solderability has to be rejected, but the workshop still has the problem and must deliver. The only way out, however unrealistic it may sound, is mechanical cleaning of the surface, e.g., by abrasion. While in certain cases chemical cleaning is possible, one has to face the problem that the component may even be destroyed. It is essential to be able to provide thorough cleaning after treatment with the agents used and it is necessary to work with highly poisonous chemicals.

These difficulties highlight the importance of correct preparation for soldering (see 12.5) especially for high volume and 'just in time' production.

When cracks develop in solder joints, they are mainly caused by wrong material combinations or thermal cycling. It is often worthwhile to remake the solder joint, by re-soldering, since this can get the equipment working again for quite a while.

Countermeasures Imply Analysis		
Problem	Cause	Action
Unsolderable	Cu Sn phase	Scraping, reject
Unsolderable	Zn oxides	Scraping, reject
Difficult to solder	Cu oxides	Retinning, reject
Difficult to solder	Organic material	Retinning, reject
Corrosion	Basic lead carbonate	Scraping, reject
Corrosion	Sulphidation	Chemical cleaning
Crack	Au Sn phase	Remove gold, resolder
Crack	Thermal cycling	Resolder

Fig. 12.26 Examples of solderability problems, their cause and actions to resolve them.

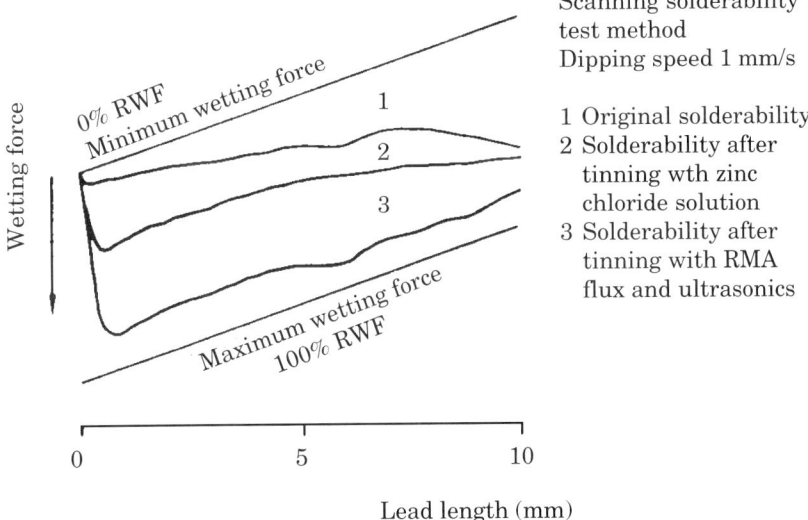

Fig. 12.27 Re-tinning material of poor solderability.

12.10 QUALITY AND RELIABILITY

The solder joint has to fulfil several functions. Therefore, it is necessary to produce a solder joint in such a way that it can fulfil these functions. If it does perform these tasks, the solder joint has the right quality.

What are these functions? The primary function is to conduct electricity from one end of the joint to the other. The other essential is that it shall do so for a given time under given circumstances of mechanical, thermal and/or chemical stresses. In the case of SMD technology, the thermomechanical stresses are the most serious ones. The time function must not be neglected either. A solder joint which has to last for 20 or more years must be designed much more carefully than a joint which has to last only one year.

It is a myth, however, that you can make a reliable joint, which will last only one year, by using inferior materials, qualities or processes with the idea of saving money. The intention in making a solder joint is to achieve a metallurgical bond. This metallurgical bond cannot be obtained unless a number of conditions are fulfilled. If they are met, the bond which has to last for only one year could easily endure for five years or more. As demonstrated in 12.3.2, money is best saved by using proper material and, by so doing, reducing the number of faults by improving the process and the quality at the same time.

One example will be given to show the way in which the quality can be improved by using the right material. Tests have been carried out to establish the relation between solderability ageing and the contact resistance between the lead and the solder in a joint.[35] One material was employed in two different conditions. When first tested, 99.99% of the material used showed soldering time values of 0.86 s or lower. The material was then aged at 155°C for 15 hours to give the second condition. After this treatment, 99.99% of the material showed soldering times of up to 5 s. After soldering the wire in both conditions, the joints were aged and then subjected to contact resistance measurements at 100 mA dc. The result is shown in Figure 12.28.

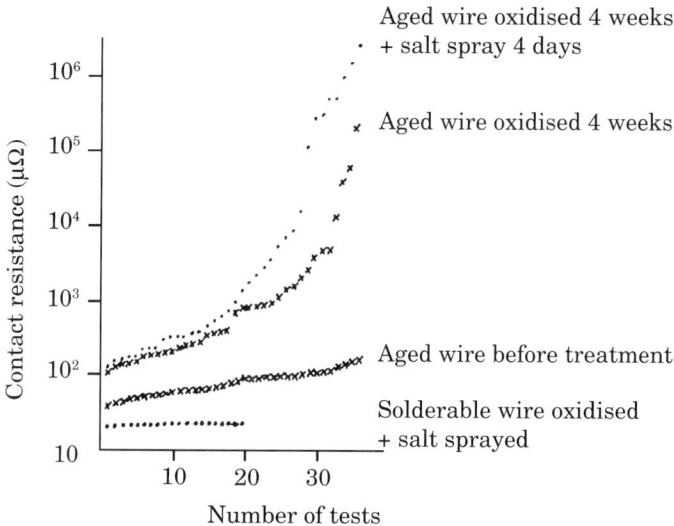

Fig. 12.28 Results of electrical contact resistance measurements *versus* solderability.

The wire having good solderability did not change in contact resistance after 4 weeks' oxidation and salt spraying for 4 days. The joints made with the wire having bad solderability showed, before any ageing, resistance values which were considerably higher than those of the joints made with the wire of good solderability after ageing. When the joints made with the wire of poor solderability were aged and salt sprayed, the resistance values increased from about 100 to over 10^6 micro-ohms. The motivation for conducting these measurements was electrically malfunctioning electronic systems.

Manko[5] claims that material having bad solderability gives solder joints of low strength.

12.10.1 Education, Information, Training

It is essential to train all personnel concerned in one way or another with soldering. This must include not only those personnel directly concerned with soldering such as soldering machine operators, quality inspectors or those who solder with a soldering iron, but also persons such as designers and managers. Both design staff and management have to make decisions which will influence the ability to make a solder joint, affect the attainment of better quality and avoid unnecessary costs. The information given to each group of persons must, of course, be specific to their needs.

12.10.1.1 MANAGEMENT

Soldering is considered to be a simple and low-cost technique. This may be the reason why the importance of this technique for cost, function and quality of the manufactured equipment is seldom recognised by management.

Quoting from a paper presented by Robert L. Moore, supervisory engineer and quality assurance staff consultant, Naval Weapons Center, China Lake, California, "One of the biggest problems associated with producing a quality product is management attitude. Management attitude is one of the major cost drivers... Eighty-five per cent of quality systems faults are related to management attitude and 15 percent are attributable to the man-and-machine relationship".

This is the reason why management must be informed on both cost and quality consequences when decisions have to be made. A decision made to save money may well turn out to cost much more than ever expected (see 12.3 et seq.).

12.10.1.2 DESIGNERS

It is obvious that the designer must be informed on *all* aspects of the interconnection technique as well as the soldering technique. As shown in 12.5, a very large number of decisions have to be made by him if the ultimate goal of manufacturing a cheap and reliable solder joint is to be achieved. Thus the designer has to be provided with all necessary information and be taught that even a solder joint needs care to be properly designed. One very important source of information will arise from intimate co-operation with the workshop.

12.10.1.3 SOLDERING OPERATORS AND INSPECTORS

Before any training for hand soldering operators can be started, the operators must be checked for their eyesight and their basic manual skill.

As well as inspectors, all soldering operators have to learn what is needed to form a solder joint. They have to be informed on the solder joint itself, on the components and boards, the solder, the function of the flux and the heat problems as well as the tools. They must learn what solderability is and, if required, be given the facts about cleaning. They must not only be able to judge the quality of a manufactured solder joint but even know the most common cause of bad solder joints and what corrective actions can or must be taken. This implies, of course, that practical training alone is not enough; a certain amount of theory is needed as well if the result is to be responsible, observant operators.

12.10.2 Statistical Considerations

Mass production has resulted in increased demands for high production yields, the necessity of minimising the number of complaints and the need to increase the quality and reliability of solder joints. As a consequence, statistics has become a vital tool in soldering technology. A popular phrase has been coined to highlight this — 'zero defect soldering'.

12.10.2.1 THE PROBLEM OF THE GREAT NUMBER OF SOLDER JOINTS

This problem probably became urgent for the first time when the black and white TV was replaced by colour TV. At that time, quite frequently the customer came back to the shop with the new colour TV set — because the TV did not work properly. A large TV manufacturer at that time published the fact that twenty percent of the TV sets the company produced did not work on the final check.[36] What had gone wrong? More components were needed for the colour television set than for the black and white one. Both the quality of the components and the quality of the solder joints remained the same as it had been for the black and white TV.

The manufacturers apparently did not observe the simple statistical law that the function of an equipment depends on the number of components in the equipment and the probability that they will work, which is expressed in the following equation:

$$Y = \left(1 - \frac{R}{1,000,000}\right)^N$$

where Y = the yield
       R = the probability of component malfunctioning
       N = number of components.

This equation can be presented in different types of graphics, one of which is shown in Figure 12.29. For the sake of simplicity, it is assumed that all components were perfect and all faults depended solely on the solder joints. A further assumption made is that a black and white TV at this time contained 100 solder joints and the new colour TV 1000 joints. It can easily be seen in Figure 12.29 that, if 99.9% of the 100 solder joints work properly (1000 ppm defective), 90% of the black and white TV sets will work satisfactorily. If the same quality level is maintained for colour televisions containing the larger number of 1000 solder joints, only 37% of the sets will work properly and can be sold. The production yield has decreased dramatically to 37%. What can be done? The only possible course of action is to increase the number of correct, working, solder joints from 99.9% to 99.99% (100 ppm defective). After this corrective action the yield will come back to 90%.

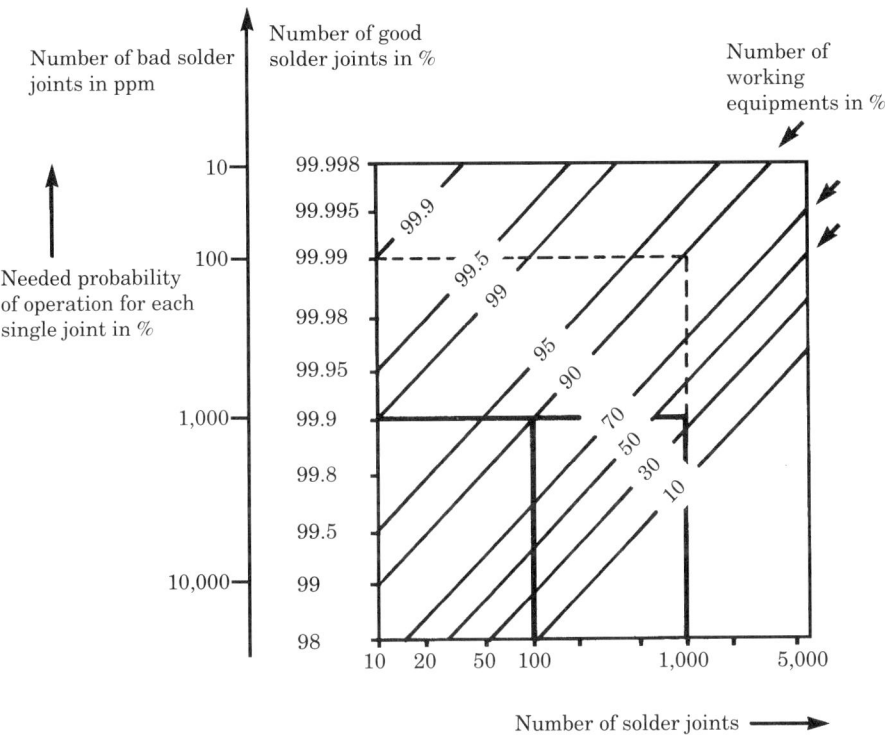

Fig. 12.29 The effect of increasing numbers of joints on production yields — the importance of good solderability.

12.10.2.2   HOW EXACT IS A MEASURED VALUE?

Nowadays a quality level of 100 ppm is no longer good enough. One printed board can have many thousands of solder joints and an electronic system or a complex module can contain many boards. Because of this, the quality level must now be better than 1 ppm.

There are several reasons for the value being so low. When we speak of solder joint quality, we speak of different things. An inspector is satisfied when the joint looks as it is specified to look. Another person wants good solderability, a third wants the solder joint to last for the specified lifetime. All these and more properties are included in the word 'quality.' All the different requirements for one solder joint are multiplied by their respective failure rate and this requires in the end a lower total failure rate (ppm value) for the solder joint.

Another reason lies in statistics. If a property is measured, such as the wetting time of component leads using the solder globule method, and a number of measurements are carried out, a number of different values for the wetting time will be obtained. The measured values are arranged in sequence and plotted into a diagram, Figure 12.20. On starting, it is seldom known what type of statistical distribution the measurements will follow. Assume that the values follow a logarithmical normal distribution as they do if the material is of good solderability, such as shown in Figure 12.20. In the end it will be seen that all the measured values gather along a line. If the measurements are repeated with a component lot from the same batch, another line will be obtained. However, instead of doing that it can be calculated and determined with a specific level of confidence how large the overall spread of all possible measurements will be. This is done in Figure 12.20 and the lines of confidence established give the possible spread of results. From the spread the following can be read. While a value of 100 ppm can be obtained for a wetting time of 1 s, one may be lucky and achieve a value of 10 ppm *but* one must be prepared for a value of 1000 ppm as well. In the example, the spread in the yield — or quality level — ranges over two orders of magnitude. If one wants to be sure not to exceed a quality level of 10 ppm, the material must be improved so much that the mean value lies below 1 ppm.

If such low values have to be realised — and it has come to that point in today's production — this has a number of consequences on the whole way of working. Visual inspection is no longer possible and meaningful. One has to build in quality instead of inspecting out defects. The quality level is no longer produced solely on the workshop floor — the decision to build in the needed level of quality has to be made by the managers and designers.

12.10.2.3   SPC — STATISTICAL PROCESS CONTROL

As was seen in 12.10.2.1, the more parts that are involved, the greater the problems become and that is true for processes as well. What is more complex than the manufacture of an integrated circuit, the production of a printed circuit or a complete electronic system? A special system has been developed to assist in reducing the number of problems in production and thus the number of faults on the unit produced, namely SPC.

The aim of SPC is to detect the faults, to remove them and to get the process under control by reducing the variations in the respective process parameters. To this end, SPC uses the following tools:

— *Pareto diagram* lists the most apparent faults in descending order of frequency of occurrence or cost, as required, Figure 12.30;
— *Fish bone diagram* helps to find possible causes of failures and thus to reduce the ppm value, Figure 12.31. Figure 12.32 shows the fish bone diagram for the failure analysis of the reflow soldering process;[37]
— *Scatter diagram*;
— *Control chart*; when the failure is found and corrected, the changes must be followed up; the control chart is set up with upper and lower limits calculated (usually set at ± 3 sigma) for the variables to find special failures and systematic variations; from the control chart further curves can be derived such as the:

— *Histogram*, Figure 12.33, or *normal distribution diagram*, Figure 12.20; the histogram can be used to govern the process to the specified quality which in modern mass production often is ± 6 sigma.

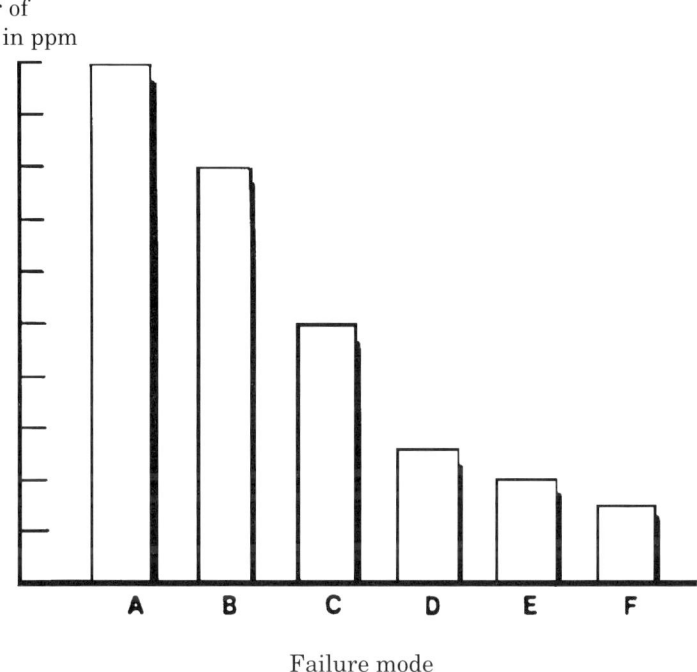

Fig. 12.30 A Pareto diagram.

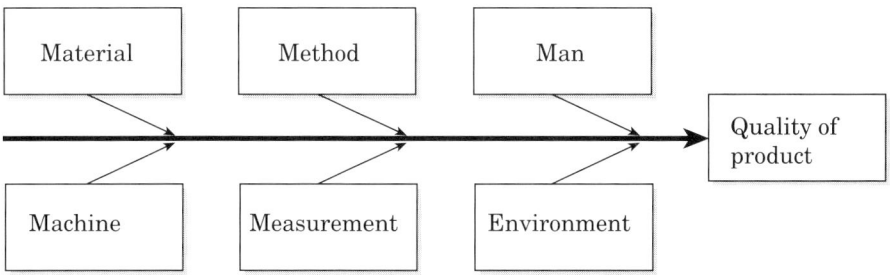

Fig. 12.31 A 'Fish bone' diagram.

If the largest or costliest failure, according to the Pareto diagram, is eliminated, work can then begin to eradicate the next problem.

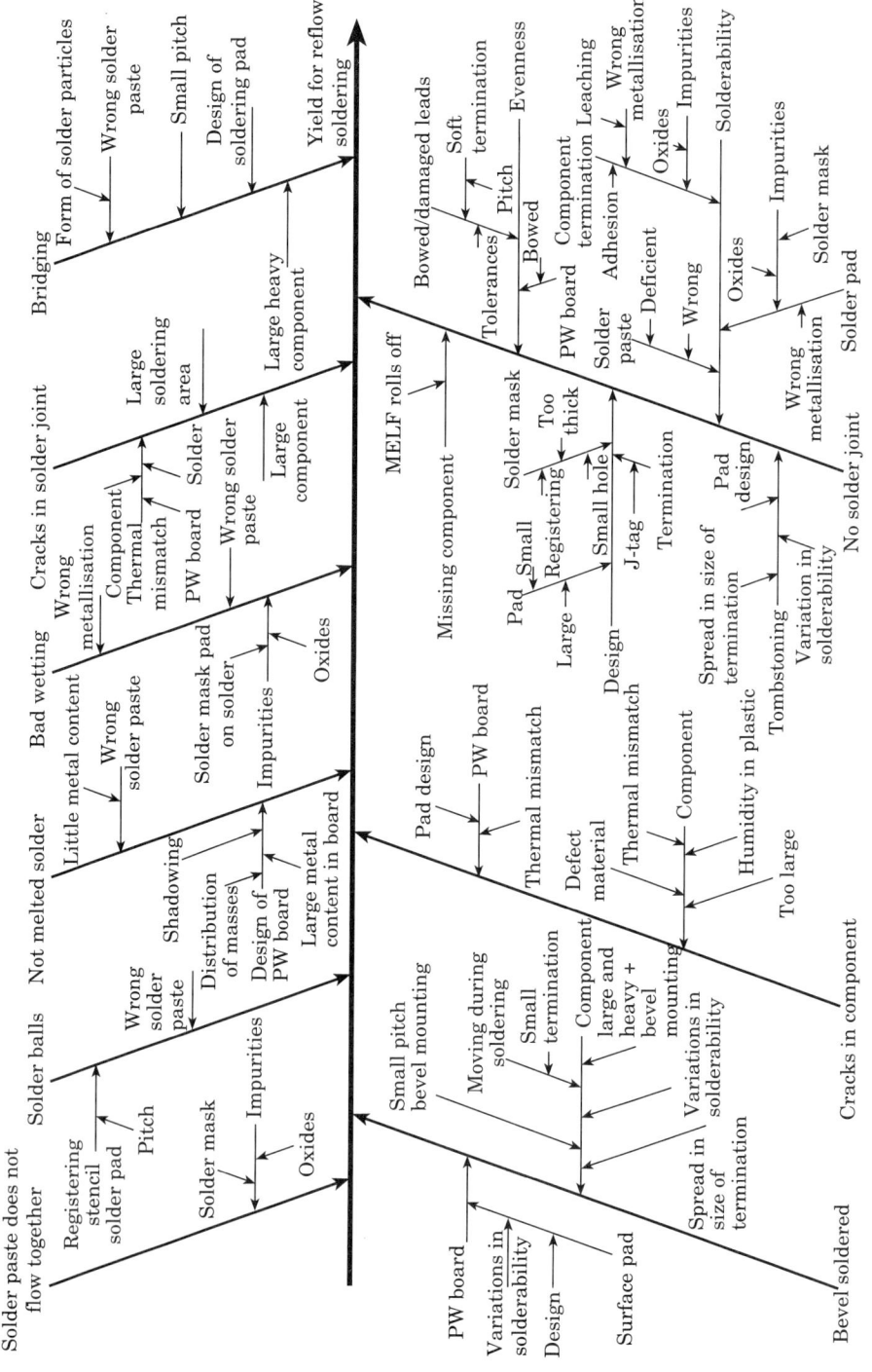

Fig. 12.32 Fish bone diagram for failure analysis of the reflow soldering process. (Courtesy of the Swedish Institute for Production Research)

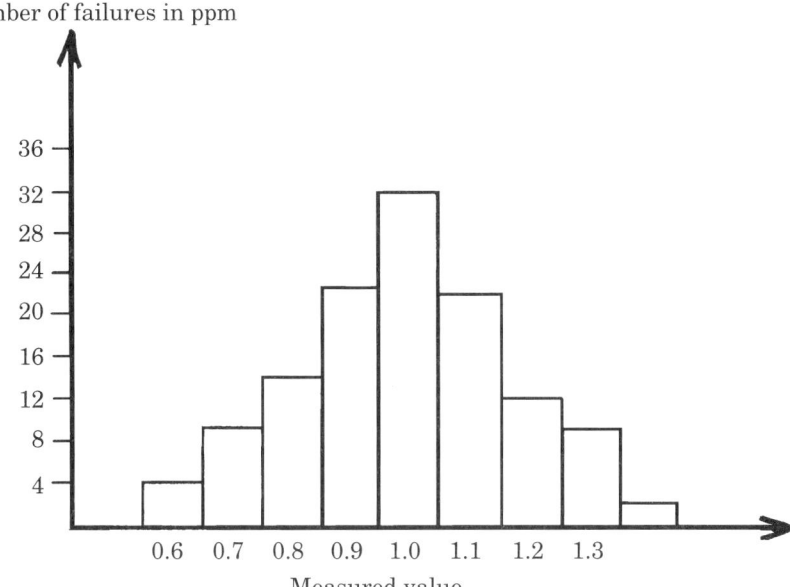

Fig. 12.33 Histogram showing a 'normal' failure distribution.

In Figure 12.34 an example is given which shows how rapidly the SMT process can be improved by the use of SPC.[38] Other experiences gained with SPC methodology show that problems with soldering are not necessarily caused by the soldering process but can be produced by the board design and that these by far surpass other problems.[38,39]

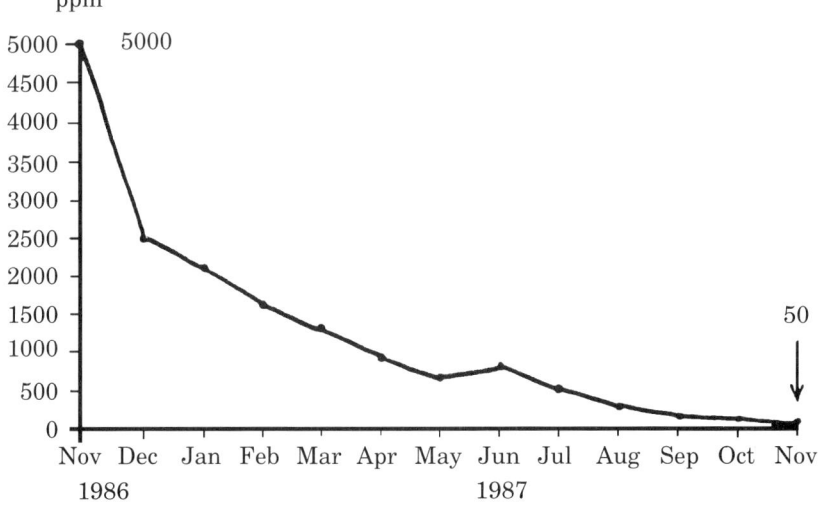

Fig. 12.34 A result of the use of SPC in surface mounting technology. After two years of using SPC, the failure rate had reduced from 5,000 ppm to 50 ppm. (Courtesy of Expira)

SPC is not new but its use has been made much easier today by the modern computer.

12.10.3 Inspection of Solder Joints

When a good solder joint is made it is not obvious that this solder joint is a faultless and reliable one, even if all the points made in 12.5 have been considered. Visual inspection, which until quite recently was the predominant method, can check only the exterior appearance. It cannot check the ultimate characteristic for the solder joint quality, the bond between the solder and the materials which have been joined; nor can visual inspection find all the faults hiding in the solder joint such as cracks and pores. During all the many years that soldering has been used, a certain experience of how a good solder joint should look has developed. Rules have been generated for the visual inspection of solder joints. However, several developments have made reliance on visual inspection hazardous:

— more and more solder joints are manufactured by practically untrained people;
— it has become virtually impossible to check the huge amount of solder joints manufactured visually due to human incapability to work with the necessary exactness and alertness for hours;
— the requirements placed on the solder joints have increased considerably; a specified failure rate of 100 ppm implies that 1 defective solder joint must be detected among 10,000 solder joints!
— even the possibility of checking solder joints visually disappears with the advent of new component types such as SMDs.

Because of this, the question of which non-destructive testing method can be used for checking the quality of the solder joint has arisen. If suitable non-destructive testing methods have the capability to be automated and computerised, which they have, totally new possibilities arise. It will be possible to list the faults, count them, find where in the process the faults arise and improve the processes.

Nearly all the non-destructive testing methods considered for the examination of solder joints have been used for many decades. So their potential uses are well known. It may be said straightaway that none of the common methods such as radiography or ultrasonics is suitable on its own for inspecting the quality of a solder joint. The method which is best suited is thermal inspection, which governs the soldering process, or combinations of methods such as optical inspection coupled with radiographic methods.[40]

Whatever the inspection method(s), it is important to realise that the boards and the solder joints must be designed for the chosen methods, first to enable inspection and, second, to get the most out of it.[41]

12.10.3.1 VISUAL INSPECTION

To detect soldering faults, criteria which can be followed are essential. The only simple ones which exist are the appearance of the solder joint and the ability of the solder joint to fulfil its electrical function. In the absence of other objective standards, visual inspection of the appearance has been used to decide if a solder joint is good or not. It is true that the appearance of a solder joint can reveal a soldering fault but it is equally true that a solder joint which looks 'good' can be responsible for the electrical malfunction of an equipment. It is also true that a solder joint which does not fulfil all visual requirements may function just as well as a solder joint which looks perfect. The conclusion is that visual judgement is an inadequate measure of the quality of a solder joint. What is acceptable for one type of solder joint may not be acceptable for another type of solder joint. General rules for the appearance of a perfect solder joint may not apply to a special design.

Visual inspection requires a magnification of at least 5 times and at the most 30 times, specially trained personnel and pictures which show the types of faults as well as the acceptance and rejection standards which are needed, for example, in the case of excessive or insufficient solder.

To make life more difficult, it is disputed whether a number of 'faults' are faults or not. A well-known instance of this is the cold solder joint which according to definition has a dull and disturbed surface. Such a surface can be obtained when using solder containing bismuth and an operator not familiar with this appearance will reject such a joint as faulty. Probably quite a number of perfect solder joints will be rejected — and repaired. Keller[42] has coined the expression of cosmetic faults, suggesting that too many solder joints are rejected and repaired unnecessarily. This not only costs money but can even give inferior solder joints after repair. It is not known how many rejected joints really needed to be rejected — but it *is* known how incompatible are the opinions of different inspectors, Figure 12.35.[43]

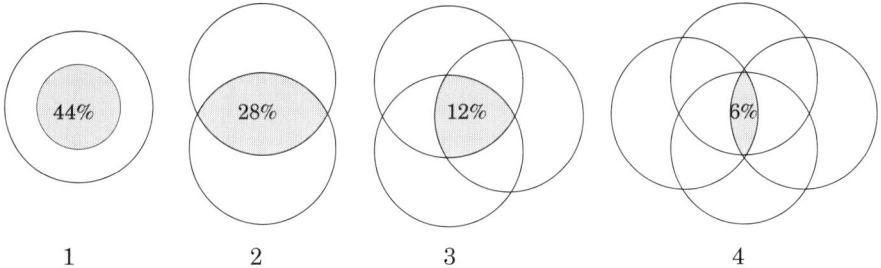

1 — a single inspector judged the same part twice (on separate occasions) — with only 44% agreement.
2 — two inspectors judged the same part in isolation with 28% agreement.
3 — three inspectors judged the same part on separate occasions with 12% agreement.
4 — when four inspectors examined and judged the same board assembly (in isolation), the level of agreement fell to only 6%.

Fig. 12.35 Agreement of results of visual inspection by different inspectors of solder joints on assembled printed boards.

12.10.3.2 AUTOMATED OPTICAL INSPECTION (AOI)

Automated optical inspection was developed to replace manual visual inspection as a method of judging the outside of a solder joint. Provided that the right characteristics are entered in the computer with which the image of the work is compared, automated visual inspection can be of tremendous help. Unlike even the best of inspection personnel, such a system does not become tired and does not change the basis for its judgement. Of course, automated optical inspection is not only able to detect deviations in the appearance of a solder joint but after programming will 'see' other solder faults, such as non-wetting and bridging, and even other production faults like missing or misplaced components.

With automated optical inspection it is possible not only to detect faults but, by using SPC, to see faults developing in the processes and to correct them in time. Different methods of AOI exist such as the 2-D (dimensional) and 3-D methods, stereo viewing and others. These can both prevent faults and improve the solder joint quality as shown in Figure 12.36. Here solder paste prints are checked for amount of solder and solder paste height.[44] This is one of the problems in surface mount technology. All such methods are constantly being upgraded to increase sensitivity and versatility and are expensive. However, in view of the cost of bad or defective solder joints, or other visible faults, it is worthwhile keeping up-to-date with these equipments.

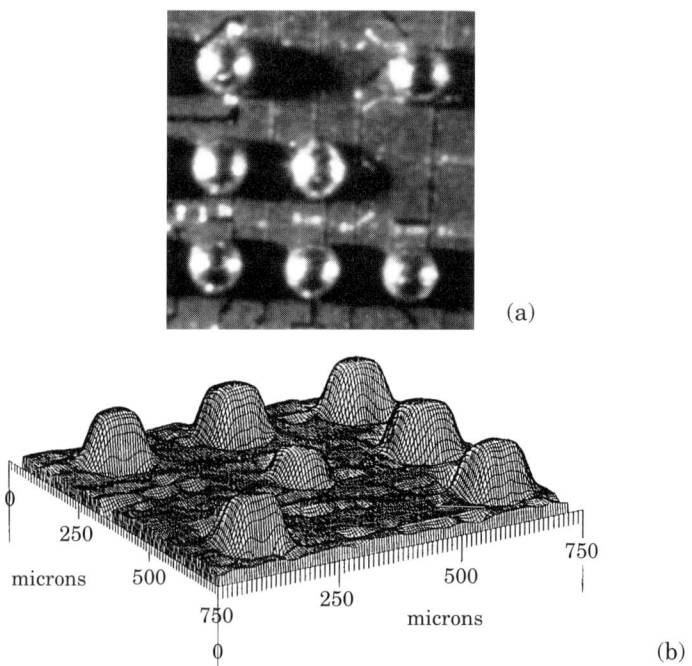

Fig. 12.36 The result of 3-D laser profile inspection. (Courtesy of Elliot Scientific Ltd)
(a) This shows the visual image of the solder. It is very difficult to detect any difference in the amount of solder in each mound.
(b) This shows the 3-D laser profile measurement. It is clearly obvious that different amounts of solder paste have been printed on to the board. It can be anticipated that the central solder mound may provide difficulties in forming a joint. If a joint is formed, it may be of lower reliability than the others.

12.10.3.3 X-RAY INSPECTION

For the inspection of solder joints fine focus X-ray equipment is needed. The primary sign of quality, a good bond between the solder and the metal to be soldered, cannot be detected by radiography, but other defects such as solder balls between the leads, bridges between leads, holes in the solder joint and cracks parallel to the X-rays can easily be seen. Cracks perpendicular to the X-rays cannot be detected. In some cases it is possible to compare the height of different solder joints. The X-ray pictures can only be judged correctly by experienced operators.

A special and interesting variant of the X-ray technique is the scanned-beam laminography method.[45] With this, the same faults can be detected as with the normal X-ray method but it is very fast. In scanned-beam laminography, images of X-ray slices or laminae are taken parallel to the mounting plane of the board. For a double-sided, surface mounted board, four laminae are viewed. This takes less than 2 s for a board of size 125 x 150 mm. Interpretation can be handled simply and easily by a computer.

X-ray equipment is expensive, especially if facilities for tilting, automatic X-ray inspection and other options are used.

12.10.3.4 ULTRASONIC INSPECTION

In principle, it should be possible to test the bond of a solder joint in an electronics assembly with ultrasonics, as differences in density, elasticity and sound speed and, thus, changes in material and phases, as well as cracks and pores, can be detected with this method. However, the practicality of the method in this application cannot yet be proven as no suitable equipment is on the market. The problem with normal equipment is the difficulty of coupling the transducer to the irregular shape of the solder joints in electronics. Larger solder joints of regular design have been tested successfully with ultrasonics.

The acoustic microscope, which can measure down to very small dimensions, down to some ten micrometres, suffers from the problem that it cannot penetrate to the depth necessary to reach the bonding plane of a normal solder joint, due to the frequency used in such a microscope. It is difficult enough to comment on the ultrasonic images of larger subjects of simple geometry having only one or two materials involved. It is much more difficult to judge ultrasonic images of solder joints on boards which have complex geometry and invariably involve two or more materials and phases. The analysis of such images needs much training and experience.

12.10.3.5 THERMAL INSPECTION

The spread of heat in a body or the rise and distribution of temperature can be used to assess a number of parameters. Examples of these are the uniformity of the material, the heat capacity, rate of heat conduction and the presence of obstacles to the spread of heat, such as cracks.

These properties are used in the thermal inspection of solder joints, in which the principle is the analysis of the infra-red radiation from the solder joint.

There are different methods of thermal inspection. In one technique, a complete, processed board is heated either externally or by applying a voltage to the circuit. The thermal camera images the temperature distribution over the board. Hot spots can be detected instantly. If the thermal signal of a good solder joint is known, a comparison can be made with other solder joints.

Another method which has been publicised by Vanzetti[46] measures the heat radiation from a solder joint when it is heated by a laser beam to about 60°C and then allowed to cool down. The values obtained are then compared with the values of a perfect solder joint. The comparison is made with the help of a computer. Work has been carried out[8,41,47] which shows the advantages and disadvantages of this system. Even though the Vanzetti method does not detect all faults (or all types of fault), it is a useful adjunct to visual inspection. It is worthwhile to compare the costs for the equipment with that for visual inspection if mass production and/or critical equipment is involved.

MBB (Messerschmitt Bölkow Blohm) has developed this system independently for direct use during soldering. Here the thermal signal from the solder joint, which is produced using a laser, is employed to govern the soldering process. This means that the quality of the solder joint is checked directly in the process of making the joint. This gives a clear advantage over the Vanzetti process as no extra process and extra time are needed.[48] Every solder joint is provided with the exact amount of heat required to produce the joint. (The amount of heat needed is determined by experiment.) If, therefore, during manufacture a given solder joint requires more heat or less heat, this is an indication of a faulty solder joint. Using this method, the material will never be overheated and destroyed. The system is, of course, computerised and automated. While such a system seems to be expensive, when the many advantages it gives are analysed, 15.9, it is comparable in price with other soldering systems.

12.10.4 Soldering Defects

12.10.4.1 WHAT IS A SOLDERING DEFECT?

Apparently it is impossible to avoid obtaining faulty solder joints when soldering.

There is general agreement that most of the soldering 'faults' *are* faults. In certain cases, however, a joint can be considered as faulty or not depending on the circumstances and, perhaps, on the underlying knowledge of the material used, the design and the process. An example of this is given by Figure 12.37 and Figure 12.38 which both show a solder joint with a rough surface. Because of this, it is necessary for every company to specify and describe what faulty solder joints are, which faults are to be considered as serious faults, which are minor faults and which may be considered only cosmetic.

It is not possible to give a general description such as 'the solder joint shall be bright and shiny with a wetting angle of less than 30°'. This is certainly correct for most solder joints but there are solder joints which never have a bright and shiny surface or never could achieve a wetting angle of less than 30° yet must be considered as perfect solder joints, Figure 12.38 and Figure

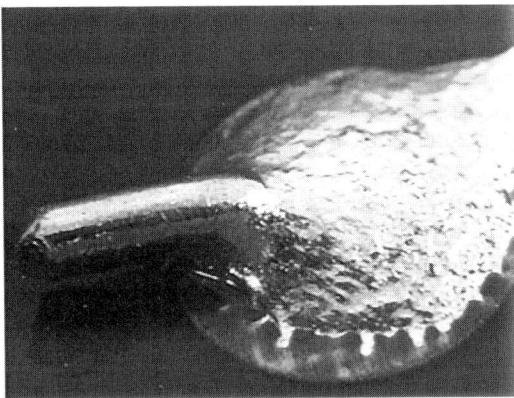

Fig. 12.37 A solder joint with a rough surface. The quality of this solder joint is doubtful as the exact reason for the rough surface is not known. This joint has to be rejected.

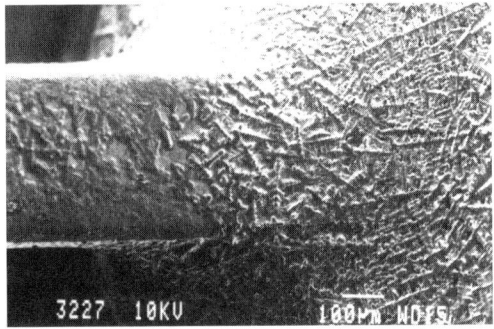

Fig. 12.38 A solder joint with a matt, rough surface. The quality of this joint is good as the reason for the matt, rough surface is known. Bismuth has been added to the solder to assist inspection. The surface finish is created by primary lead dendrites which are easily formed when bismuth is present in the solder.

12.11 (a), (b). For reasons such as these, quality criteria must be worked out for every specific type of solder joint.

A good basis for this work can be found in terms and definitions created by the International Institute of Welding (IIW), the International Tin Research Institute (ITRI),[49] the Institute for Interconnecting and Packaging Electronic Circuits (IPC)[50] and a few private enterprises[51] which produce pictures and slides that are available at low cost. It will be found that, while in the descriptions of the faults the appearance of the solder joint itself is well described, the name given to the fault does not always point to the cause of the fault.

12.10.4.2 SOLDERING DEFECTS AND FAILURE RATE

According to References 29, 30 and 31, the distribution of soldering defects is as follows (values in %):

Literature	30	29	31
Year	1972	1975	1990
			SMD Technology
Solder		10	
Flux		25	
Solderability	60	70	70
Bridging	100		
Physical damage	20		
Miscellaneous	10		25
Boards			50
Components			25

Even though the different authors had dissimilar views on soldering problems, it is apparent that solderability is the main problem. The problem exists not just for one company or even country and is not limited to a specific soldering technique. It is a problem which has been in existence for a long time and is not yet solved.

It is seldom that a soldering defect appears alone. The normal combination is a bad design or manufacturing process in combination with a solderability problem. Generally, the design and manufacturing problem is much more severe than the solderability problem. It often happens that, when the design problem is resolved, the solderability problem becomes apparent.

By eliminating the different kinds of faults, the number of faulty solder joints has been reduced to 100 ppm from 1000 ppm. Improvement of the solderability then pushes the number of faults below 100 ppm, down into the ppb area.[52] However, examples are known where the failure rate due to bad solderability was in the range of some 10,000 ppm. In such cases, the solderability problem had to be solved before the design or manufacturing problems became apparent. Faults can be typical of a design or a manufacturing process. In such a case, it is very easy to achieve a low failure rate.

12.10.4.3 SOLDER-FILLED THROUGH-PLATED HOLES OR NOT?

This is a question to which the answer is based mostly on opinion and in very few cases on facts. In other words, a solder-filled hole is very often requested where there is actually no need for it. Only in very few cases of extreme stresses on the solder joint is the filling of a through-plated hole with solder prompted. This case, however, has another bearing on quality. Originally the plated-through hole was only a conductor connecting both sides of a printed board and this has nothing to do with soldering. However, when the board was soldered the holes were sometimes filled and sometimes not. Failure to fill was often caused by insufficient quality of the plated holes, such as holes in the copper wall and circumferential cracking of the hole. The idea of filling was then that the solder should bridge such faults.

There are several other natural reasons why it can be difficult to fill plated-through holes with solder. Verbeek[26] has shown that it can be very difficult to reach the necessary soldering temperature in the component lead at the top of the plated-through hole, Figure 12.39, even if the material has good solderability. For example, if the temperature in the solder bath is 250°C, the solder at the upper side of the hole will have a temperature of 200°C or less. It is also known that the ease of soldering copper or solder-clad copper of good solderability decreases sharply below 235°C, Figure 12.40.[53] A study of the filling behaviour of plated-through holes shows that the filling is governed by the heat in the material and not by the capillary forces in the hole. These two effects combine to give an oscillating solder 'pillar' in the hole which cannot reasonably be governed by the soldering process. Whether the holes are nicely filled or not is totally dependent on where the upper level of this solder pillar is when the solder pillar is disconnected from the solder source. This is one of the reasons why, on one and the same board, both filled and partly filled holes appear.

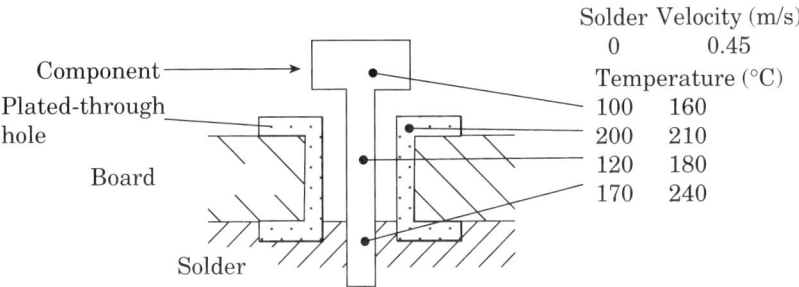

Fig. 12.39 Temperature distribution in a plated-through hole and component after a 2 s contact time with a solder bath at 250°C.

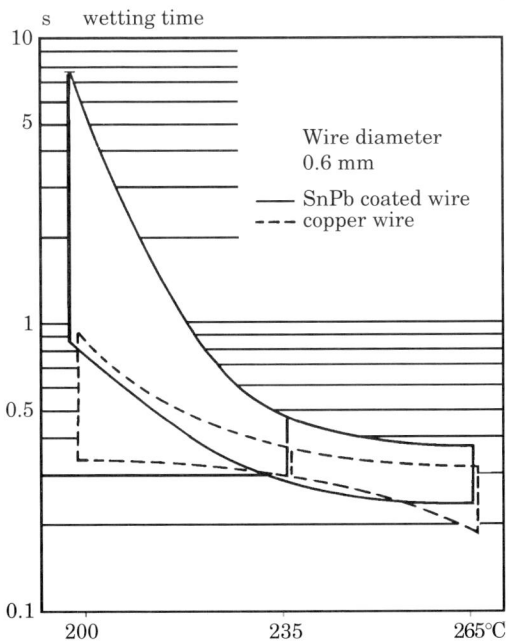

Fig. 12.40 The variation of wetting time with temperature.

Unfilled holes should be accepted under the following conditions:

— if it can be proven that a sound electrical connection is obtained; this requires a plating technique which ensures that the wall of the hole is covered with an adherent and ductile copper deposit;
— if the strength problems which arise in a hole with a wire (component termination) are carefully considered;
— if the corrosion risk in the hole is investigated and found to be negligible;
— if the solderability of the holes is found to be good; the solderability here is a sign of the plating process quality and if the solderability is good this is a sign that there are no cracks or holes in the hole wall;
— if the problems which can be caused by any insulation lacquer are analysed.

12.10.4.4 DEFECT TERMINOLOGY

Soldering defects can be serious, for example cracks or leaching, or they can be solely cosmetic. Soldering defects can be more frequent for one soldering process than for another. The definitions which follow are taken from either the IEC or the IPC. In cases where the author believed the definitions were inappropriate or inadequate, other definitions were suggested or changes were made to the IEC/IPC text.

Blowholes	These are discontinuities, holes or sharp depressions in the solder joint caused, e.g., by vaporisation of entrapped chemicals, flux and/or moisture. Studies by Lea[54] show that there is no evidence that blowholes cause electrical failure, cracks or corrosion. Therefore, they are not soldering defects but cosmetic faults and indicate process problems which must be taken seriously.
Bridging	The formation of an unwanted conductive path between two conductors.
Cold solder joint	A joint in which the solder usually has a rough, dull, wrinkled or piled-up appearance and shows signs of imperfect flow or wetting action, i.e., abrupt lines of demarcation of the solder and high contact angle between the solder and the surfaces being joined. Another definition could be a solder connection exhibiting poor wetting due to too low a soldering temperature and/or too short a soldering time.
Dewetting	A condition that results when molten solder coats a surface and then recedes to leave irregularly shaped mounds of solder separated by areas which are covered with a thin layer of solder and with the basis metal *not* exposed, Figure 12.41.
Disturbed solder joint	An unsatisfactory bond resulting from relative motion between the joint surfaces during solidification of the solder.

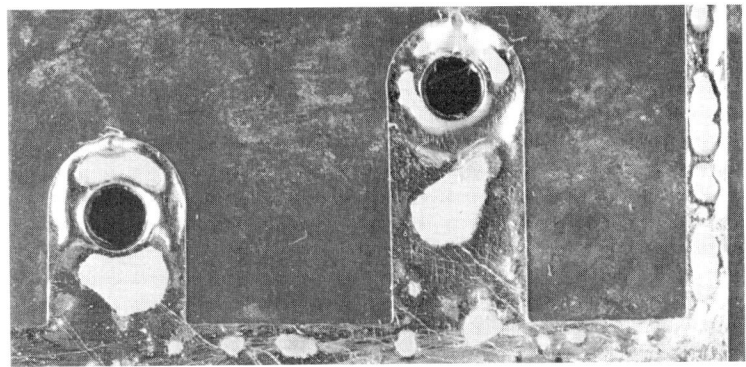

Fig. 12.41 Dewetting of a copper surface.

Dross — The granular material resulting from oxidation on the surface of molten solder consisting of solder and contaminants enveloped by a solder alloy metal oxide skin.

Dry solder joint — A solder connection exhibiting a greyish, gritty porous appearance due to inadequate cleaning, wrong flux or excess impurities in the solder.

Excessive solder — A soldered joint with so much solder that the outlines of the connected parts are not visible and the quality cannot be judged, Figure 12.42 and Figure 12.43.

Fig. 12.42 Solder joints with different amounts of solder.

Faulty solder joint — Collective name for all kinds of bad solder joints such as cold solder joint, dewetting, dry solder joint, excessive solder, icicle, insufficient solder, non-wetting, solder deficit, solder sputter, voids etc.

Granular solder — A solder connection exhibiting a rough dull surface usually caused by an inappropriate cooling rate, Figure 12.37.

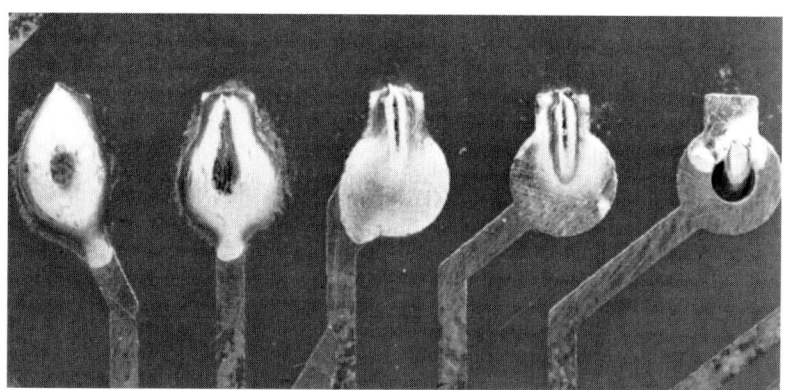

Fig. 12.43 Solder joints with different amounts of solder.

Heating fault	Soldering defect caused by a wrong soldering temperature and/or soldering time.
Icicle (or stalactite)	A cone-shaped peak or sharp point of solder, usually formed by premature cooling and solidification of solder upon removal of the heat source.
Insufficient solder	see Solder deficit.
Lack of heating	Soldering defect caused by too low a soldering temperature and/or too short a soldering time.
Leaching	The dissolution of the termination metallisation by molten solder.
Non-wetting	The material has not been wetted by the molten solder due to one or another of a number of causes such as failure in the soldering process, contamination of the surface to be soldered, shadow effects caused by soldering fixtures, too low soldering temperature or too short soldering time. The wetting angle is much greater than 90°.
Opens	These are open circuits primarily caused by insufficient solder or solder paste deposition and/or non-coplanar component leads.
Residues	Any visual or measurable form of contamination on the workpiece, device or assembly related to processing (plating salts, resists, flux, etc.).
Sink mark	Depression in soldering surface appearing during solidification.
Solder balls	Small spheres of solder adhering to the substrate or solder resist.
Solder deficit	A solder joint with too little solder, so that the mechanical strength is jeopardised, Figures 12.42, 12.43 and 12.44.
Solder slivers	Portions of solder plating overhang on conductor edges which are partially or completely detached.
Solder spatter/splatter	Unwanted particles of solder adhering to the surface of the printed board.

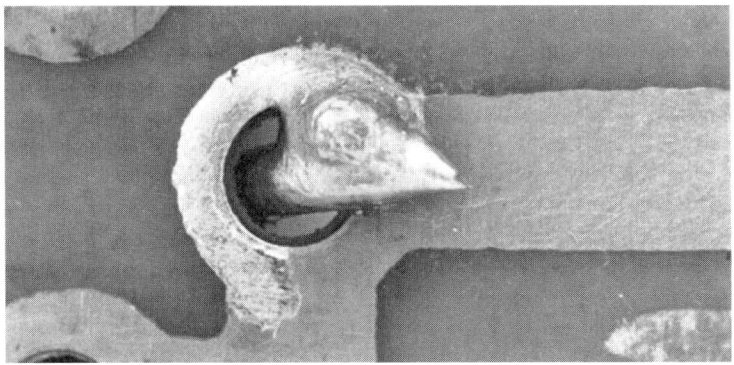

Fig. 12.44 A solder joint with insufficient solder.

Solder wicking	This is the preferential wetting of solder on either the component termination or the pad. See also under Wicking.
Tombstoning	This is an effect occurring with chip components where one end of the chip rises due to differential wetting on the ends of the chip. This is caused by different heat flow, different wetting properties and tensile forces due to surface tension. According to Klein Wassink[6] there are 14 causes of tombstoning.
Webbing	A continuous film or curtain of solder parallel to but not necessarily adhering to a surface or between separate sections of circuitry that should be free of solder.
Wetting defect	A soldering defect caused by the fact that solder has not wetted the joint completely because the solderability of the surface, the solder or the flux are not optimised, Figure 12.45.

Fig. 12.45 A solder joint with a wetting fault on the component lead.

Wicking	The flow (usually undesirable) of molten solder, usually along the strands and under the insulation of stranded lead wires, away from the immediate joint area.

12.10.5 Expert Systems in Soldering

Within computer technology, an area has developed which is termed artificial intelligence. A result of the research in this field is the so-called expert system. This system is a new medium whereby one or more experts can convey their knowledge to a larger group of people via a computer in which their knowledge is stored. Hitherto, expert systems have not been used to a large extent in industry.

As soldering has developed in the last 25-30 years from an art to a technology, from a technique where little real knowledge existed to a complex subject where knowledge of many different disciplines is needed, some companies have looked into the possibilities of the expert system for making the accumulated knowledge more easily accessible and to widen the range of this knowledge. It is relatively difficult to build such systems and therefore a potential user has to be critical and must investigate to see which system fills his needs.

Taking an example, the expert system 'waves', developed by STC (now part of Northern Telecom) in Britain, indicates with its name that the system is not thought to be useful for iron or vapour phase soldering. This system is intended to help the operator of a wave soldering machine and other people concerned with wave soldering to find out what may have caused failures on a soldered printed circuit board after wave soldering. It is interesting to note that a thorough analysis can be carried out within 15 minutes by a person untrained in the field of soldering. Probably a soldering expert could sort out most of the failures in a shorter time but certainly would miss some of the possible causes for a failure as no one person can be as logical and consequential as a properly programmed computer.

Troubleshooting charts became common in soldering workshops after Leonida[10] published probably the first one. A revised version is given in Figure 12.46. It is interesting to compare this with an expert system. The expert system allows the user to find complex causes for a failure which is not possible with a simple chart. Not only are actions to correct the faults suggested but the results of the actions are checked to determine if consequential failures can occur. The system is more dynamic and includes logic and probability considerations, and sets out a method of working. In other expert systems for soldering which have been developed, graphics have been integrated into the program, making it easier to see and explain the different types of soldering faults. It should be noted that such a program requires a VGA or SVGA monitor whereas the 'waves' program does not.

Future expert systems will be connected directly to soldering machines and will control them when a soldering fault is detected. It will take time for really comprehensive expert systems to reach the market. The reason for this is that there is, still, a great lack of knowledge in the field of soldering and, especially, a lack of data for the materials used in soldering.

Fault	Cause
Bad or no wetting	1, 2, 3, 11, 12, 13, 14, 16, 19, 20, 24, 25, 29, 30
Blowholes/empty holes	3, 4, 7, 8, 15, 17, 24, 26, 27, 29
Bridges	1, 3, 11, 12, 13, 14, 16, 21, 23, 24, 25, 27, 29, 30, 31
Cold solder joints	1, 3, 5, 16, 24, 27, 29
Dewetting	1, 3, 8, 20, 24, 27
Dull joints	9, 16, 25, 27
Excess flux on board	3, 7, 14, 18, 19, 24
Excess solder	2, 3, 6, 7, 10, 14, 18, 19, 24, 27, 31
Excessive dross generated	10, 25, 26
Excessive solder drag-out	14, 25, 26
Grainy solder	5, 6, 9, 16, 25, 26, 27, 28
Icicles	1, 3, 5, 8, 11, 12, 13, 14, 16, 24, 25, 26, 27, 29, 31
Insufficient flow through PTH	1, 3, 8, 10, 13, 16, 19, 24, 27, 29
Insufficient solder	4
Lifted conductor	22, 26, 28
Missing solder	1, 3, 10, 12, 13, 18, 19, 24, 29, 31
Pinholes	3, 4, 7, 8, 13, 15, 17, 24
Solder splatter	5, 7, 12, 13, 18, 21, 26, 27, 30
Solder balls	7, 10, 11, 15, 22, 24, 30
Voids	2, 3, 8, 13, 14, 15, 24, 27
Webbing	9, 11, 13, 14, 20, 22, 30
White residues	20, 30

Key to cause:

1. Bad solderability
2. Board not secured in fixture and/or does not move correctly on conveyor
3. Conveyor speed too high
4. Conveyor speed too low
5. Conveyor vibration
6. Copper or gold impurities
7. Density of flux too high
8. Density of flux too low
9. Dross in solder
10. Erratic solder wave
11. Flux contamination
12. Flux no longer active
13. Flux not in contact with board
14. Fluxer uneven
15. Gassing caused by cracks or voids in plated hole
16. Heat capacity of joint
17. Hole to lead diameter ratio too large
18. Incorrect depth or incorrect dipping of board
19. Incorrect fixture
20. Incorrect flux
21. Incorrect pattern design
22. Lamination problem
23. Oxides on solder wave
24. Preheat too low
25. Solder contamination
26. Solder temperature too high
27. Solder temperature too low
28. Soldering time too long
29. Soldering time too short
30. Unsuitable solder resist or insufficient cure
31. Unsuitable wave shape/height

Fig. 12.46 Troubleshooting chart for wave soldering.

It may be concluded that people in the workshop, or others who have no special knowledge in the field of soldering, can already make great use of an expert system. Even though such systems may still seem expensive, they must be considered worth the price when compared with the cost of not knowing — interruptions in production, bad quality and much rework.

12.11 MOUNTING METHODS

Today the principal component mounting methods are still the pin-in-hole ('conventional') and the surface mounting techniques, see Section 3.

As not all components are available for one or other technique, it is sometimes necessary to mix both mounting techniques, the 'mixprint' or mixed technology board — see Chapter 7 and Figures 7.3 and 7.4.

The fact that two very different types of components can be mounted on the same board must not blur the fact that each technique has its own implications for production and that the requirements placed on the soldering technique are much higher for the surface mounting method than for the hole mounting method. These may be summarised as follows:

— Soldering provides an electrical connection in both cases.
— While, in most cases, the requirements for safety against corrosion have been moderate in pin-in-hole technology, they have become much more important for SMT.
— In pin-in-hole technology, components were considered to be held by the lead going through the hole — nobody faced the fact that, in the majority of cases, actually the solder in the hole kept the component in place; it is obvious that in SM technology the component is held on the board by the solder.
— The leads of the leaded components can be used to provide stress relief for the joints; this facility is not available for the (non-leaded) surface mounting component and, suddenly, the fatigue properties of the joint became important.
— The amount of solder for each joint was reduced in the SMD by a factor of about 10.
— Where the number of wired components on a printed board were relatively low, the number of components increased dramatically when using SMDs.

All the changes or new requirements arising from the developments in component mounting technology led to the need of much increased reliability of the joints.

12.12 CLASSIFICATION OF ASSEMBLIES

Many different kinds of products are manufactured which include electronics, such as medical equipment, toys, space hardware and vehicles, TVs and computers. The list is almost endless. It is obvious that these different products have different requirements for both lifetime and reliability. To make it easier both for the vendor to choose the right quality level and for the user to know what type of product he will get, the IPC has created three assembly classes which can be found in American National Standard ANSI/ J-STD-001: Requirements for Soldered Electrical and Electronic Assemblies, published in April, 1992. The three classes are:

Class 1: General Electronic Products
Includes consumer products, some computer and computer peripherals, and hardware suitable for applications where the major requirement is function of the completed assembly.

Class 2: Dedicated Service Electronic Products
Includes communication equipment, sophisticated business machines and instruments, where high performance and extended life are required and for which uninterrupted service is desired but not critical. Typically the end-use environment would not cause failures.

Class 3: High Performance Electronic Products
Includes equipment for commercial and military products where continued performance or performance on demand is critical. Equipment downtime cannot be tolerated, end-use environment may be uncommonly harsh and the system must function when required, such as life support systems and critical weapons systems.

Of course, it is not always easy to determine to which class a certain equipment belongs and there may be overlap between classes. Therefore, the user has the responsibility to determine the class into which his electronics interconnection falls. These classes are used, for example, for the characterisation of the fluxes used, see 13.5.

To establish the fatigue stress service life of solder joints in surface mount electronics, Engelmaier[55] suggests nine different levels of severity starting from Consumer products and going through Computer, Telecommunications, Commercial Aircraft to the last and most severe one — Automotive under hood (i.e., in the engine compartment).

12.13 HEALTH AND SAFETY

When soldering, work is carried out with agents which may be detrimental to health such as solders and fluxes or which may cause a safety risk such as certain solvents used. Regulations vary from country to country, therefore it is necessary to check these with the appropriate health and other suitable authorities.

12.13.1 Solders

When using solders, it must be borne in mind that solder is composed of tin and lead. Lead and lead oxide are poisons which can, after a long period, affect the health seriously. However, the health risk can be avoided by ensuring that those who work with solder:

— wash their hands before taking any meal;
— are not allowed to eat, to drink, to chew chewing gum or tobacco, or to smoke at their workplace.

A separate place has to be provided for such purposes with washroom facilities close to the location.

All processes connected with soldering must be checked to determine if solder dust or fumes can be generated by the process. Such processes have to be eliminated or safeguarded. One example of such a process is the removal of the dross formed during the soldering process or during tinning. Another example is the cleaning of screens or stencils used for applying solder pastes by blowing through with compressed air. This must be avoided as the small solder particles from the paste are dispersed into the air and can cause lead poisoning when inhaled or ingested. Soldering operators should wear protective working garments, not personal clothes, as solder particles can settle in the clothes and subsequently contaminate foodstuffs, etc. Personnel working with solder in any form should be subject to regular health checks.

A number of solders have cadmium as a constituent. There is no reason to use a cadmium-loaded solder for its melting point as another, less hazardous, solder melting in the required region can always be found. However, it may be difficult to replace a cadmium solder if a defined property such as strength is desired. With the possibilities opened up by laser soldering in mind, this last argument should no longer be valid.

12.13.2 Fluxes

When mass soldering printed board assemblies, large amounts of solvents are released from the flux. Ventilation is therefore necessary. People who are allergic to flux fumes (many are, especially to colophony which can cause difficulties in breathing and, in the long term, asthma) must be excluded from any location where soldering is carried out.

Nearly all fluxes contain activators, some of which are a health risk and so must be excluded — despite their good activation properties. Some fluxes contain very aggressive substances which, if they get into the eyes or on to the skin, must be washed away immediately with copious amounts of clean water. Hydrazine compounds and α- and β-napthylamine are carcinogenic. Many amines cause dermatitis.

The supplier of a flux must, on request, produce a certificate of analysis for the purchasing company's health or safety officer who must inform those responsible for the soldering activity which, if any, precautions must be taken.

12.13.3 Soldering Equipment

Smaller soldering machines are usually placed on benches. Care must be taken to ensure the machines are mounted in a stable manner on rigid benches free from any movement. The space under the solder pot or tank shall be empty or of metal. The pot must not be allowed to contact plastics or wood which can outgas or catch fire. Even the larger soldering machines can catch fire but, fortunately, this very seldom happens. The usual causes are solvents, flux dripping on the preheater or oils.

For health and safety reasons, soldering machines must never be operated unless the ventilation system is functioning. The ventilation system must meet the relevant safety regulations for avoiding fire and explosions and it must continue to work for a sufficient time after the soldering work is finished to extract all gases and fumes.

Molten solder is hot and personnel working directly with the molten metal should wear eyeglasses and protective gloves and clothing. If dross has to be removed from the machine, the operator doing it should wear a mask or breathing equipment. The dross must be kept in a closed container.

Because of the sizeable amounts of heat given out by soldering equipment, it may be necessary to control the air temperature in the working area to ensure operator comfort and meet the relevant regulations for working environments.

12.14 TERMS AND DEFINITIONS

Every human activity needs its own special expressions in order to define clearly what is meant. If the same expressions are not used, it is difficult to communicate. In the field of soldering, special expressions are used which are taken in many instances from different scientific disciplines. We do not all have the same background and, for that reason, need to define the major special expressions in the field of soldering. The following list of the most frequently used terms is divided into different groups. Whenever possible, definitions given by the IEC, ISO or IPC are used.

1 General definitions related to soldering
2 Terms related to soldering
3 Time — Temperature terms
4 Terms related to joint form and size
5 Terms related to solder joints
6 Material to be joined
7 Flux terms
8 Solder terms
9 Soldering aid material
10 Soldering methods and processes
11 Quality terms
12 Soldering defects.

12.14.1 General Definitions Related to Soldering

Ageing   The change of a property with time.
Accelerated ageing   A specific set of environmental conditions and/or chemical treatments of the test sample to simulate long-term environmental exposure in a relatively short period of time.
Capillary force   This is a result of the surface tensions of the liquid solder and the basis material. It causes the molten solder to be drawn into the joint gap.
Contamination   Examples of contamination are an undesirable surface film or embedded foreign matter associated

	with a joint surface. The contaminants can lower the functional performance of characteristics such as solderability or can cause corrosion.
Corrosion	The deterioration of a material by chemical or electrochemical reaction with its environment.
Diffusion	This is a migration of particles, e.g., atoms, between two materials in contact with each other under certain conditions of temperature, pressure, etc. The properties of a soldered joint are changed by diffusion.
Diffusion layer	Layer in material which has assumed a particular composition by diffusion.
Insulation resistance	The electrical resistance of an insulating material determined under specified environmental and electrical conditions between a pair of contacts, conductors or grounding devices in various combinations. This is one of the most important parameters in determining the electronic reliability of the soldering process.
Intermetallic phase	By diffusion of metal atoms into another metal, intermetallic phases are formed if a defined concentration ratio between the two metals is reached.
Reliability	This is a general term describing the performance/lifetime characteristics of an electronic device or a solder joint.

12.14.2 Terms Related to Soldering, Figure 12.47

Eutectic	An alloy which melts and sets at a fixed temperature. There is no melting range (setting range or pasty range).
Inert atmosphere	An inert atmosphere is comprised of a gas, e.g., nitrogen, argon, etc., which does not react with either the workpiece or the solder during the joining process. It is used to prevent oxidation of the workpiece surfaces.
Liquidus temperature	The liquidus temperature of a material is the temperature at which it becomes completely molten.
Leaching	The loss of metallisation from a termination by its solution in molten solder.
Melting range	The melting range of a solder is the temperature interval between its solidus and liquidus temperature.
Solidus temperature	The solidus temperature of a solder is the temperature at which it begins to melt.

12.14.3 Time-temperature Terms, Figure 12.48

Cooling time	The cooling time is the time span during which the solder cools down from the soldering temperature

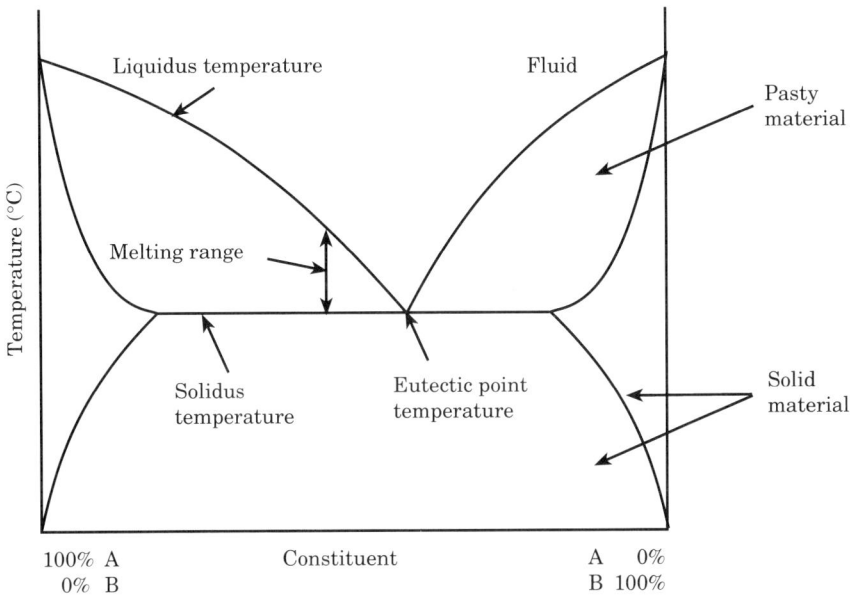

Fig. 12.47 A phase diagram as used in metallurgy.

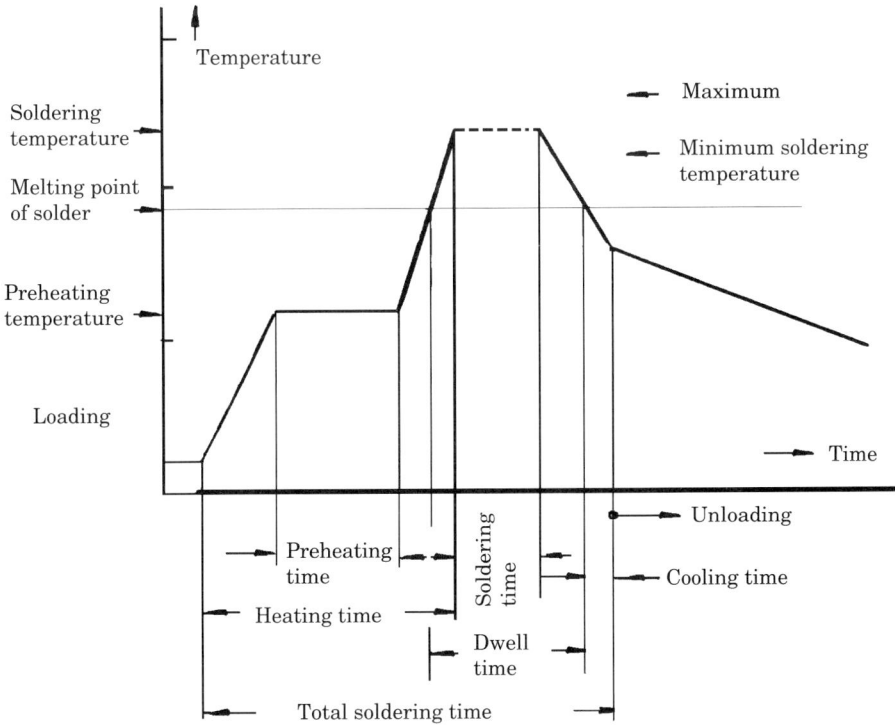

Fig. 12.48 Time and temperature terms characteristic for soldering.

	to a temperature below the melting point of the solder at which the soldered board can be handled.
Dipping speed limit	Maximum dipping speed at which a sound solder joint is obtained.
Dwell time	Time during which the parts to be soldered are in contact with the molten solder.
Heating time	The heating time is the time span during which the soldering temperature is attained. It is determined by the solderability, the heat capacity and heat dissipation characteristics of the workpiece to be soldered.
Maximum soldering temperature	Temperature above which soldering is not allowed. The upper limit of the soldering temperature is determined by the solderability of the objects to be soldered, the rate of deterioration of the solder at a given temperature, flux used and sensitivity to heat of the basis material and components to be soldered.
Minimum soldering temperature	This is the lowest temperature at which soldering is possible. This lower limit of the soldering temperature is determined by the solderability of the surfaces being soldered, their heat capacity and the liquidus temperature of the solder.
Soldering temperature	The soldering temperature is the temperature prevailing at the joint during the soldering operation.
Soldering time	The time required for a surface to be wetted or for a joint to be completed under specific conditions.

12.14.4   Terms Related to Joint Form and Size

Butt joint	Joining of two workpieces end to end (e.g., rods), a non-preferred solder joint design.
Fillet	In a solder connection, the configuration of the solder in the joint after solidification.
Interfacial connection	A conductor which connects conductive patterns on opposite sides of a PB.
Joint gap	The joint gap or joint thickness is the distance between the surfaces of the components required to be joined by soldering. When tubular components are required to be joined, it is the radial clearance at the gap at soldering temperature which should be specified and not the diametrical dimensions of the parts. The joint gap is generally different from the initial clearance.
Lap joint	The joining of two parallel overlapping surfaces (e.g., sheet metal) by soldering. This is the preferred design for solder joints.
Soldered connection	An electrical/mechanical connection which employs solder for joining two or more metals.

12.14.5 Terms Related to Solder Joints

Clinched lead — A component lead which extends through a hole and is bent over towards the conductor pattern.

Comb pattern — A set of interdigitated arrays of uniformly spaced conductors (combs) placed on a dielectric (printed board substrate), which is used to determine the influence of the soldering process (primarily the effect of flux and flux residues) on the reliability of printed circuits.

Eyelet — A short metal tube inserted in a terminal or a hole in a printed board to provide mechanical support for component leads or electrical connection between the two faces.

Land — The termination of a conductor (track) on a printed board to which component terminations are soldered to form an electrical connection.

Pad — see land

Solder side — The side of a printed board on which the soldering is conducted. (The term is not always applicable to SM printed boards as both sides may carry soldered joints.)

Solder tag — A metal part designed for connection of a conductor.

Stress relief — The formed portion of a component lead or wire providing excess length to minimise stress between solder terminations.

Terminal — A tie point device used for making electrical connection with flexible wires. Basic styles of terminals are bifurcated, hook, perforated or pierced, solder cup, turret and straight post.

Thermal shunt — A device with good heat dissipation characteristics used to conduct heat away from a component being soldered to prevent heat damage of adjacent components.

12.14.6 Materials to be Joined, Figure 12.49

Basis (base) metal (material) — The various materials to be joined by soldering.

Diffusion barrier or layer — Layer of metal to prevent diffusion between basis material and solderable layer.

Protective lacquer — Layer of lacquer applied on printed board assemblies after soldering is completed.

Substrate — The supporting dielectric material upon which the elements of a circuit are deposited or installed.

Tinning — Tinning is a general expression applied to the preparation of a surface for soldering by coating with either a tin-lead alloy or with pure tin.

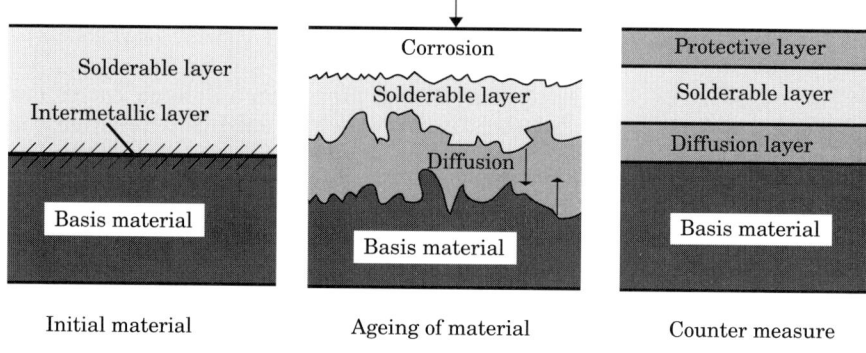

Fig. 12.49 Material terms.

12.14.7 Flux Terms

Acid flux — Term referring to an aqueous solution of inorganic halide or acid.

Activated rosin flux — A solution of colophony in an organic solvent containing small amounts of an organic halide or acid (an 'activator') to improve wetting.

Activator — A substance promoting flux activity usually selected from organic and inorganic acids, amines and certain thermally unstable organic halogen-containing compounds and amine hydrohalides.

Aqueous flux — General term for a water based or water soluble flux of inorganic halide or organic acid type.

Colophony — A natural resin obtained as the residue after removal of turpentine from the oleo-resin of the pine tree. It consists mainly of abietic acid and related resin acids, the remainder being acid esters.

Flux — A chemically and physically active substance which facilitates wetting of the surface by molten solder by removing the oxide or other surface films from the basis metal and the solder. It also protects the surfaces from re-oxidation during soldering and reduces the solder/base metal surface tension. A flux is necessary to complete the soldering process.

Flux activity — The ability of the flux to remove contamination and chemically clean the surface being soldered.

Flux efficiency — The ability of the flux to promote wetting of surfaces by molten solder within a specific time.

Fluxing — Applying flux to a metal surface to be soldered by spraying, dipping, foam applicator, brushing and so on.

Flux paste — A flux made in the form of a paste.

Flux residues	Surface contaminants found after soldering resulting from the use of soldering fluxes.
Halide content	The halide content is the ratio of the mass of (free) halide to the mass of the flux solids, expressed in mass percent of free chloride ions.
Inorganic flux	A general term denoting, typically, aqueous flux of various inorganic acids and halides used in general industry, e.g., tinning of steel plates using zinc chloride/ammonium chloride/hydrochloric acid flux.
Ionic/Ionisable contaminants	Process residues, such as flux activators, finger prints, etching and plating salts, etc., that exist as ions and increase the electrical conductivity when dissolved in water. Their presence on a product may promote electrolytic corrosion and thus reduce the product reliability.
Non-activated flux	A soldering flux made of natural or synthetic resins with no activators added to increase activity. Non-activated flux residues in the solid state, e.g., after soldering, cause no risk of corrosion.
Organic acid flux	A non-specific term for a flux not containing colophony or inorganic salts.

12.14.8 Solder Terms

Cored solder	see 'flux-cored solder wire'
Dross	Oxides and other contaminants which form on the surface of molten solder.
Flux-cored solder wire	Solder wire or preforms containing one or more cavities filled with flux which is released as it melts.
High purity solder	A non-specific term commonly used to refer to the solder alloys with low levels of impurities. Such solders are preferred by the electrical and electronics industries.
Preform	Solder shapes (e.g., disc washers, foil), often of specified volume, formed to fit a joint. The preform is placed on the joint prior to heating to avoid the necessity of applying solder during heating. Preforms may contain flux cores or be flux coated.
Solder (soft solder)	A metal or an alloy of two or more metals (usually tin and lead but other metals may be present) of a melting range below 450°C used to join metallic surfaces. Solder can be supplied as, e.g., wire, sheet, bars or preforms.
Solder paste or solder cream	A mixture of solder powder, flux and special agents added to facilitate application of the paste by printing methods or from a syringe.
Wipe solder	This is a soft solder with a relatively wide melting range and pasty consistency, enabling it to be wiped on to the object being soldered during the soldering process.

12.14.9 Soldering Aid Material

Solderable insulated wire
An insulated conductor which can be soldered without stripping the insulation prior to soldering.

Soldering oil
Liquid formulations for use in oil intermix wave soldering equipment as solder pot covering on still and wave solder pots to eliminate solder dross formation.

Solder resist
A non-wettable and heat-resistant coating applied to prevent the contact of underlying circuitry and molten solder. Openings are left at the lands to be soldered. Its use improves soldering efficiency, reduces the effect of aggressive environments on the printed circuit and prevents electrical shorts between the conductor lines and/or electronic components placed over them.

Stop-off media
These are materials which can be applied either to the components to be soldered or to the surface of supporting jigs in order to prevent the wetting or spread of material on to the areas masked.

12.14.10 Soldering Methods and Processes

These terms are defined in Chapter 15 where the different soldering processes are described.

12.14.11 Quality Terms

Contact angle
In general, this is the angle enclosed between two planes, tangential to a liquid surface and a solid-liquid interface at their intersection. In particular, it is the contact angle of liquid solder in contact with a solid metal surface, Figures 12.8 and 12.11.

Dewetting
A condition which results when molten solder has coated a surface and then receded, leaving irregularly shaped mounds of solder separated by areas covered with a thin solder film. The basis metal is not exposed, Figure 12.41.

Meniscus
The contour of the solder surface shaped as a result of surface tension forces acting during the wetting process.

Resistance to soldering heat
Property of a component to be soldered indicating the effects of soldering heat and thermal stresses associated with that heat upon change of component characteristics.

Solder spread
A test in which the area of spread of a specified weight of solder placed on a fluxed and heated metallic surface is determined. This test can be

	used to compare or control the solderability of flat surfaces, solders and the efficiency of fluxes.
Solderability	A measure of the ability of a surface to be wetted by molten solder. This can be determined by a given test method under a given set of conditions — see, for example, the solder spread test and the wetting balance test. Solderability is a property defining the total suitability of an item for industrial soldering, covering three aspects — thermal demand, wettability and resistance to soldering heat.
Surface tension, interfacial	Natural physical forces acting at the liquid/solid/flux interface enabling spreading of the liquid solder across the surface of the base metal.
Thermal demand	The characteristic of a component to be soldered which indicates the ability of the solder joint area to be heated to the desired temperature within the time available for the soldering operation.
Wetting	In soldering practice, wetting is the irreversible spreading of a molten solder on the surface of the base metal, resulting in the formation of an adherent coating of solder on the surface. A small contact angle is indicative of good wetting, Figure 12.8.
Wetting balance	An instrument used in the measurement of wetting performance and in solderability testing. The balance is able to measure the change of wetting force as a function of time during the immersion of the test sample into the solder bath, Figures 12.15 and 12.16.

12.14.12 Soldering Defects

Soldering defects are described and discussed in 12.10.4, the terms used being defined in 12.10.4.4.

REFERENCES

1. Holm, T., 'Controlled Atmospheres: Soldering of Printed Circuit Boards', Report REPM 93378, AGA AB, 181 81 Lidingö, Sweden (1993).
2. Kendrick, 'Quality in the Electrical Component Industry', *Metals and Materials*, pp. 118-120, April (1968).
3. Thwaites, C. J., 'Soldering Technology — A Decade of Developments,' International Tin Research Institute, Publication No. 644 (1984).
4. Manko, H. H., 'Solders and Soldering', 2nd Edition, McGraw Hill (1979).
5. Manko, H. H., 'Soldering Handbook for Printed Circuits and Surface Mounting', Van Nostrand Reinhold (1986).
6. Klein Wassink, R. J., 'Soldering in Electronics,' Electrochemical Publications Ltd, Asahi House, Church Road, Port Erin, Isle of Man, British Isles (1984).
7. Klein Wassink, R. J., 'Soldering in Electronics', Second Edition, Electrochemical Publications Ltd, Asahi House, Church Road, Port Erin, Isle of Man, British Isles (1989).
8. Lea, C., 'A Scientific Guide to Surface Mount Technology', Electrochemical Publications

Ltd, Asahi House, Church Road, Port Erin, Isle of Man, British Isles (1988).
9. Woodgate, R. W., 'Handbook of Machine Soldering', John Wiley & Sons (1983).
10. Leonida, G., 'Handbook of Printed Circuit Design, Manufacture, Components & Assembly', Electrochemical Publications Ltd, Asahi House, Church Road, Port Erin, Isle of Man, British Isles (1981).
11. Thwaites, C. J., 'Soft Soldering Handbook', International Tin Research Institute, Publication No. 533.
12. Allen, B. M., 'Soldering Handbook', Iliffe, London (1969).
13. Lüder, E., 'Handbuch der Löttechnik', Verlag Technik, Berlin (1952).
14. Wuich, W., 'Löten', Vogel Verlag, Würzburg (1972).
15. Lenz, E., 'Automatisiertes Löten elektronischer Baugruppen', Siemens AG (1985).
16. Lambert, L. P., 'Soldering for Electronic Assemblies', Marcel Dekker Inc., New York, Basel (1988).
17. Müller, W., 'Lote', Deutscher Verlag für Schweisstechnik, Düsseldorf (1990).
18. DeVore, J. A. and Westerman, J., 'Solder and Solder Joint Properties Handbook', General Electric Company, Syracuse, Technical Information Series R80ELS028, September (1980).
19. NN, 'Solder Alloy Data, Mechanical Properties of Solders and Soldered Joints', International Tin Research Institute, Publication No. 656.
20. NN, 'SMD-Technology', Philips Electronic Components and Materials, Philips Export BV, No. 9398 621 40011 (1986).
21. Maiwald, D., 'Soldering in SMD Technology', Siemens AG.
22. Mullen, J., 'How to Use Surface Mount Technology', Texas Instruments (1984).
23. Strauss, R. S. 'Surface Mount Technology', Butterworth Heinemann (1994).
24. Danielsson, H., 'Surface Mount Technology with Fine Pitch Components: the Manufacturing Issues', Chapman & Hall, London (1994).
25. *Soldering & Surface Mount Technology*, Wela Publications Ltd, Port Erin, Isle of Man, British Isles.
26. Verbeek, H. J., 'A Model to Evaluate Design and Application of Components regarding Heat Transfer during Soldering', International Conference on Soldering, Brazing and Welding in Electronics, Munich, DVS Bericht 40, pp. 35-39, November (1976).
27. Nylén, M. and Norgren, S., 'Temperature Variations in Soldering and their Influence on Microstructure and Strength of Solder Joints', *Soldering & Surface Mount Technology*, No. 5, pp. 15-20, June (1990).
28. Steen, H., 'Quantitative Solderability Measurement of Surface Mounting Components using the Wetting Balance', Solderability Seminar of the Swedish Institute for Metals Research, Steningevik, March (1991).
29. Mathes, H. and Gamalski, J., 'Praktische Anwendung des Weichlötens', *Der Praktiker*, Deutscher Verlag für Schweisstechnik GmbH, Düsseldorf, No. 9 (1975).
30. Keller, J. D., 'A Review of Soldering Technology and Related Areas', Soldering Seminar of Hollis Engineering Inc., April (1972).
31. Woodgate, R., *Printed Circuit Assembly*, June (1988) and private communication (1990).
32. Becker, G., 'Scanning the Solderability of a Surface', *Welding Journal*, pp.202s-206s, October (1981).
33. NN, 'Solderability Testing of Component Terminations', Mekanresultat 72014 A, Sveriges Mekanförbund, 114 85 Stockholm, Sweden, December (1972).
34. Becker, G., 'New Method Measures Solderability of Plated-Through Holes', *Insulation/Circuits*, pp. 73-78, December (1967).
35. Becker, G., 'Solder Joint Quality — Ageing — Solderability, Modelling of Environmental Effects on Electrical and General Engineering Equipment', 4th International Symposium CSSR, Liblice, October (1975).
36. Sdp, '270 000 Farbfernseher jährlich aus einem Werk' (270 000 Colour TVs per year from one factory), *VDI-Nachrichten*, No. 39. p. 16, September (1970).
37. Uddèn, M., 'Yield and Causes of Failures in Surface Mount Production,' IVF Resultat 90626, Mekanförbundets Förlag Box 5506, 114 85 Stockholm, October (1990).

38 Södersved, H., paper given at the STF meeting in Stockholm, September (1988).
39 Coleman, J. R., 'SPC for PCBs', *Assembly Engineering*, No. 12, pp. 23-24 (1988).
40 Chander, K., 'Automated Solder Joint Inspection in a CCAPS Environment', Proceedings 13th Annual Electronics Manufacturing Seminar, Naval Weapons Center, China Lake, pp. 221-240 (1989).
41 Thompson, D. and Stroebel, T., 'Designing Electronics for Automated Inspection', *Circuit World*, **Vol. 14**, No. 4, pp. 13-20 (1988).
42 Keller, J., 'Can the US afford the "Cosmetic look" in Solder Joints?', *Assembly Engineering*, **Vol. 16**, No. 10, pp. 32-35 and No. 11, pp. 38-41 (1973).
43 Lilley, F., Hobson, C. A. and Konkash, H., 'Visual Inspection of Surface Mounted Solder Joints', *Circuit World*, **Vol. 16**, No. 4, pp. 13-15 (1990).
44 Elliot, '3-D Laser Profile Inspection', BABS Seminar, London, September (1990).
45 Adams, J., 'Scanned-Beam Laminography breaks Through the 3-D Barrier', Proceedings Nepcon West '89, Cahners Exposition Group, **Vol. 1**, pp. 112-116 (1989)
46 Vanzetti, R., Traub, A. C. and Richard, A. A., 'Laser Inspection of Solder Joints', *Brazing & Soldering*, No. 2, pp. 34-37, Spring (1982).
47 Chen, S., '3-D Machine Vision for Solder Joint Inspection and Process Control', Proceedings Nepcon West '89, pp. 93-101 (1989).
48 MBB Laserlöten.
49 'Surface Mount Devices. A photographic guide of profiles and defects encountered during the soldering of surface mount devices', ITRI Publication No. 700.
50 IPC has a number of specialised publications and articles available.
51 'SMD Workmanship Standard', EPF, Oslo, Norway (1994).
52 Becker, G., 'European Solderability Control', *Brazing & Soldering*, No. 5, pp. 41-44, Autumn (1983).
53 Allen, B. M. and Becker, G., 'The Effect of Impurities in Soft Solder for Electrical Purposes', Proceedings InterNepcon, II, pp. 52-56 (1970).
54 Lea, C., 'The Effect of Blow Holes in Soldered PTH Assemblies', *Circuits Assembly*, pp. 38-47, November (1990).
55 Engelmaier, W., 'The Use Environments of Electronic Assemblies and their Impact on Surface Mount Solder Attachment Reliability', IPC, Smart VI Proceedings, **Vol. 2**, pp. 129-145 (1990).
56 Bader, W. G., 'Dissolution and Formation of Intermetallics in the Soldering Process', Physical Metallurgy of Metal Joining, AIME Symposium, St. Louis, Editors Kossolvsky and Glicksman (1980).

Chapter 13

MATERIALS USED IN SOLDERING

GERT BECKER
Consultant, Sköndal, Sweden

13.1 GENERAL

Quite a number of materials are involved in the formation of a solder joint. Essential, of course, are the parts which are to be joined together, the solder and the flux. In addition to these, other materials may be needed such as solder pastes and glues. After the soldering process, if the boards have to be cleaned, then cleaners are required. All of these materials have special properties which must be controlled if a satisfactory result is to be obtained.

13.2 BASE MATERIALS

Only a few metals are readily solderable with the fluxes used in electronic production. These are copper, copper alloys such as brass, tin bronze and German silver, silver and gold. All other metals, typified by beryllium bronze, Alloy 42, Kovar, steel, stainless steel and aluminium, or non-metals, such as ceramics and the different plastics used in printed boards, need either special fluxes or a surface treatment, for example, coating with a solderable material. Even glass for optical communications is coated for mounting purposes.

13.3 SURFACE TREATMENTS

Without surface treatments, modern electronics would not be possible. However, as the majority of engineers who design or manufacture electronic circuits and equipments are not directly involved with surface treatment, the technology is not credited with its true value, either in its direct form or in its consequences.

A surface treatment is normally used to generate a material which the designer needs. The material so created must fulfil the required function with a defined life and reliability. It must be possible to manufacture this material for a price which the customer is prepared to pay. When it is realised that a surface treatment in electronics is seldom performed solely to reduce cost, it is not surprising that there are many different surface treatments available to resolve specific problems.

Surface treatment is carried out for many different reasons. Sometimes it has to fulfil several different functions at the same time. The main function in the present case is to facilitate a contact function, whether it be a mechanical contact, a weld or a solder joint. Protection against corrosion is a function needed from most finishes.

Just as for any other technique, a general basic knowledge and a number of preconditions are essential to be able to solve problems and achieve good results. Surface treatment needs *good* specialists and, unfortunately, there are only a few specialists in the field of surface treatments for use in electronics.

13.3.1 Methods of Surface Treatment

When speaking of surface treatment, one thinks immediately of chemical or electrolytic processes. When dealing with finishes for soldering, hot tinning in which the base material is dipped into molten tin or tin-lead springs easily to mind, but there *are* other processes available. For example, the CVD (chemical vapour deposition) technique, the PVD (physical vapour deposition) procedure and ion implantation techniques which are used mainly in 'chip' component production can be employed in other applications.

13.3.2 Mechanism of Bonding

The way in which one material binds to another determines the quality of the bond and the properties connected with it. There are four different types of bonding:

— *atomic or molecular forces*; these give very good bonds; an example is the soldering of cadmium on aluminium;
— *formation of alloys*; this happens when brazing copper with brass;
— *chemical reaction*; this is typified by the intermetallic phases obtained when soldering with tin or tin-lead on copper;
— *mechanical anchoring*; in this case, surface roughness is used to fasten or 'key' another material to the surface; an example is in the electroplating of plastics where the substrate may be 'roughened' by mechanical or chemical means.

If problems arise, to find the correct solution it is very important to determine which type of bonding mechanism is involved.

13.3.3 Layer Structure

Obtaining good plating is difficult. Besides the technical problems of the plating process itself there are the problems of achieving:

— a good bond;
— a proper intermediate layer;
— a suitable surface layer.

It is obvious, therefore, that it will not normally be sufficient just to deposit a substance on to the basis material in order to achieve the required property. The selection of both substance and deposition process has to be made very carefully.

13.3.3.1 THE BOND

The characteristics which promote bonding between the functional layer and the basis material may not be good enough and must be improved. In the extreme case, bond promotion may be non-existent as in the instance of plastics where a solderable layer cannot fasten directly on to the surface. A weak bond may not withstand the stresses caused by temperature, time, corrosion, manufacturing and usage. In such instances, an intermediate layer which bonds well to both the basis material and the functional layer may be the solution. This technique of applying an intermediate layer is used not only in joining by soldering but also for contacts, in welding and in SMD technology.

It is quite common to plate materials of poor solderability first with nickel and then with tin, the tin forming the solderable layer. If, after the nickel plating process, the material is stored, nickel oxides form on the nickel surface to give a 'passive' layer. Nevertheless, it is possible to plate this surface with tin but, unless the nickel is activated prior to the tinning process, the bond between the passive layer and the tin will be bad. Initially, the soldering action takes place on the tinned surface. As the tin is only of the order of a few micrometres thick it is soon dissolved and the solder has to bond to the nickel surface. If the nickel surface is oxidised, no bond can be formed and dewetting occurs. This phenomenon can easily be detected and followed in the wetting balance test, as shown in Figure 13.1 for an Olin Alloy which was not activated before tin plating.

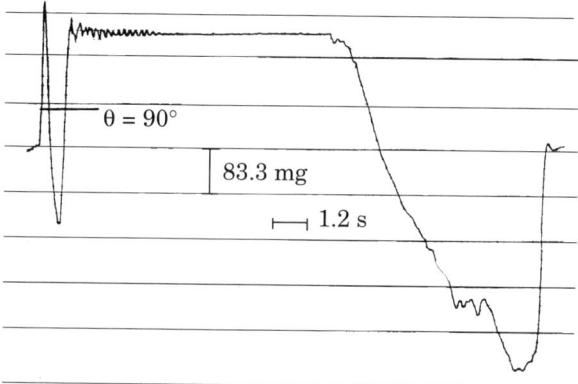

Fig. 13.1 Wetting balance curve showing total dewetting. The basic material surface was intentionally not activated before electrolytic tinning and, because of this, the tin was totally dissolved into the solder bath. Conditions were:

    Solder   Sn60Pb40
    Soldering temperature   235°C
    Flux   IEC, activated

13.3.3.2 DIFFUSION BARRIERS

In other cases, an intermediate layer is needed to prevent the diffusion of harmful constituents from one layer into the other. In such instances, therefore, the intermediate layer is called the diffusion barrier. An example is the diffusion of zinc from the basis material brass into the solderable layer which may be tin. The zinc moves through the tin and forms unsolderable zinc oxide on the tin surface. Without a diffusion barrier, usually of nickel, between the brass and the solder, the good soldering properties of the solder will be negated.

13.3.3.3 PROTECTIVE LAYERS

Very often the material which is to be soldered has to be stored for a time. During this period the solderable layer is exposed to the surrounding air, which may lead to reaction with the functional layer to form oxides, compounds of sulphur or other corrosion products. This 'ageing' of the solderable surface will, in most cases, degrade the solderability. To prevent or at least retard this ageing, it may be advisable to cover the surface of the functional layer, e.g., with a chromate coating.

13.3.3.4 MATERIALS BUILD-UP FOR SOLDERING

By looking at each material from base material to the protective layer, the build-up of material shown in Figure 13.2 is obtained:

— basis material;
— intermediate layer (i.e., the diffusion barrier);
— functional layer (i.e., the solderable layer);
— protective layer (to prevent tarnishing of the surface(s) to be soldered).

Base Material	Typical Intermediate Layer (diffusion barrier)	Typical Functional Layer (solderable layer)	Typical Protective Layer

Ceramics
Glasses
Metals
  Aluminium
  Copper
  Copper alloys
  Iron alloys
Plastics
	Copper
	Nickel
	Tungsten
		Copper
		Gold
		Nickel
		Palladium
		Silver
		Tin
		Tin-lead
			Benzotriazole
			Chromate layer
			Lacquers

Usual thickness (µm)	4	0.4 - 8	See text, 13.3.3.4

Fig. 13.2 Materials build-up for soldering purposes.

Usually the materials and thicknesses shown in Figure 13.2 are used. The thickness of the protective layer will vary from mono-molecular in the case of chromate and benzotriazole films to 25 µm or more for lacquers.

13.3.4 Solderable Coatings

As mentioned previously, there are different reasons for choosing different materials to form a solderable coating. By looking at their behaviour during soldering, they can be divided into materials which are wetted but not completely dissolved by solder, materials which are wetted and *are* completely dissolved by solder and, finally, those which are displaced by solder such as a rosin lacquer. To the first group belong thick noble metal layers such as gold and silver. In the second group are found thin noble metal layers as well as tin and solder.

13.3.4.1 ELECTROLYTIC TINNING

Electrodeposition of tin or tin-lead, often referred to as 'electrolytic tinning', is widely used to prepare material for soldering. Wire and lead frames for component leads are mostly hot tinned but electrolytically tinned wire and lead frames are used as well. Printed board terminations are frequently electroplated although hot tinning processes such as hot air solder levelling are widely used.

As always in the case of soldering, the term 'tinning' is not restricted just to the element tin but can mean tin-lead in various compositions. The different compositions again can be deposited over each other in order to achieve specific results, for example:

— to prevent diffusion;
— to provide a hard surface for better machine feeding;
— to improve the ageing properties.

If copper is *tin* electroplated, it is advisable first to deposit a layer of nickel, approximately 4 µm thick, to avoid formation of tin whiskers. Tin directly deposited on to copper can also be prevented from forming whiskers by fusing (melting) the tin.

Copper, itself, can sometimes be used as a diffusion barrier, for example, when using zinc-containing alloys.

A tin, or tin-lead, electroplating process is recommended in cases where close control of thickness is needed for specific purposes. Close tolerances and uniform layer thickness are very difficult to obtain with a hot tinning process.

Certain points of importance should at least be mentioned here. At the beginning of any electrolytic process, thorough, uniform degreasing of the material to be plated is necessary as only a clean surface can be activated satisfactorily. The electrodeposited layers will adhere without causing problems in the subsequent soldering process *only* if the surface has been adequately activated. All process steps needed to achieve the desired surface should be carried out in a continuous process. If this cannot be done, an intermediate surface activation or 'pickling' stage will be necessary in order to activate the surface to be treated.

As will be appreciated from reading other chapters in this book, in particular Volume 2, Chapter 6, Metal Deposition, much knowledge and experience are required in order to control electroplating solutions and processes properly.

13.3.4.2 HOT TINNING

Hot tinning is considered to be an easy process. However, when tinning is carried out to prepare material for soldering in electronics, one of the requirements is controlled thickness of the tin or tin-lead layer. In practice, three different thicknesses are often used, a very thin one of 0.5 to 2 µm, a medium thickness of 4 to 8 µm and a thick one of more than 50 µm.

All these deposit thicknesses can be obtained by electrodeposition but hot tinning will give only the very thin and medium deposits. The thickness of 4-8 µm gives excellent solderability, even after long-term storage, and is needed for normal soldering purposes. A thickness of more than 50 µm is sometimes required in reflow soldering processes.

The parameters which influence the hot tinning process are the temperature, the tinning speed, the flux and the solder. Figure 13.3 shows that both the temperature and the tinning speed have the greatest influence on the coating thickness achieved. The influence of different solders on coating thickness is marginal and that of fluxes is negligible.

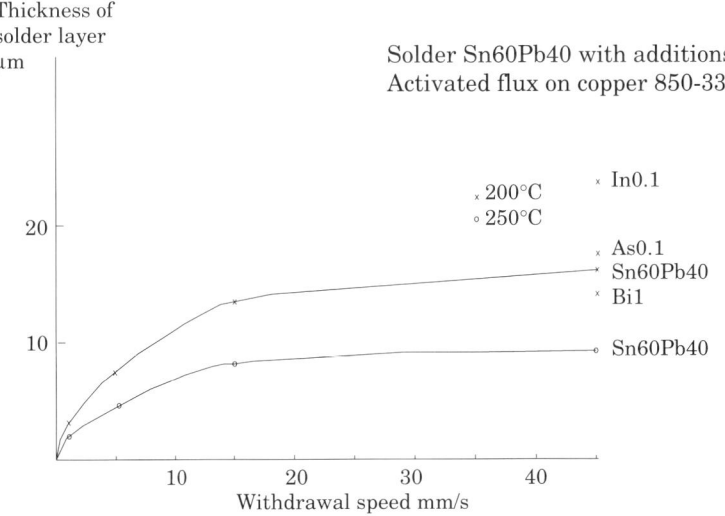

Fig. 13.3 Factors influencing the thickness of the solder layer when 'tinning' by a dipping process.

Hot tinning processes are usually highly mechanised and a number of different methods have been developed to achieve the solder layer thickness required. There are machines which operate in the horizontal as well as in the vertical direction. Such machines work properly provided adequately trained people run the process and maintain the machine.

Hexacon[1] has developed a simple, fast, hot tinning process which is suitable for tinning lands on SMT boards in low volume production, rework, and similar situations. This method is advantageous in cases where the cost for making stencils and screen printing with solder paste is considered excessive. A roller mounted on a soldering iron is tinned in a small solder bath and the tin on the roller transferred to the lands on the board, Figure 13.4. A very uniform solder thickness is achieved. As no bridging occurs, this method can be employed even for fine pitch surface mount boards.

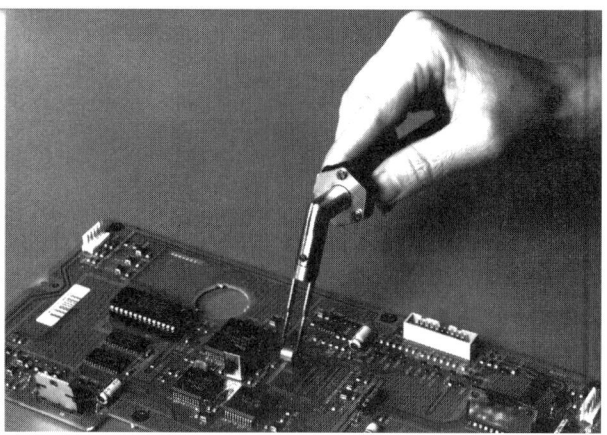

Fig. 13.4 With the Hexacon SoldeRoll™ tool, SMD mounting lands can be roller-solder coated in small-scale production or even after the boards are part assembled.
(Courtesy of Hexacon Electric Company)

13.3.4.3 NICKEL

Nickel is not considered to be an easily soldered metal. Despite this, electroplated nickel plays an important part in soldering as a diffusion barrier of thickness 1-4 µm. A deposit thickness of 4 µm is normally recommended as this thickness really guarantees that diffusion will be prevented. A 1 µm thickness is inadequate for inhibition of tin whisker growth, Figure 13.5. Nickel is also frequently used where it is important to hold close tolerances, as in the case of punching or injection moulding tools.

As mentioned in 13.3.3.1, it is most important to make and keep the nickel surface active during the plating process. A passive or inactive nickel surface allows no bonding to the functional (solderable) layer and will cause the serious soldering problem known as 'dewetting'. If dewetting occurs only in small areas, the dewetted areas can be covered by solder and hidden, and such areas reduce the fatigue life of a solder joint.

13.3.4.4 GOLD

A. The Use of Gold in Electronics

The major uses for gold in the electronics industry are:

Materials Used in Soldering 621

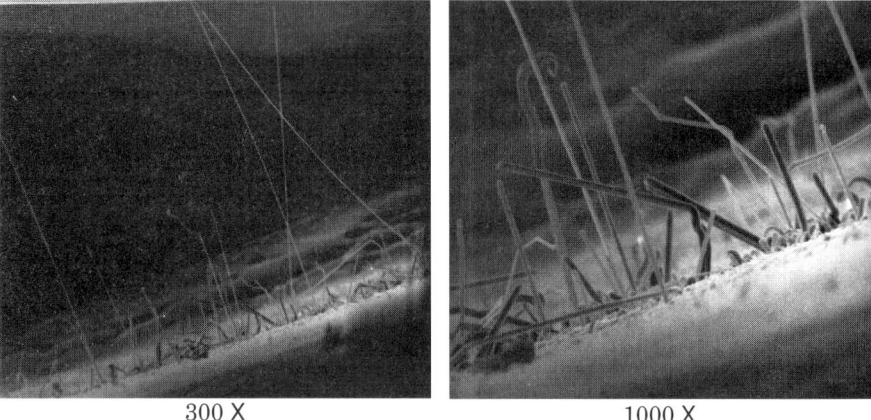

300 X   1000 X

Fig. 13.5 Whiskers on a solder tag coated with unalloyed tin. The whiskers could grow as the diffusion barrier did not meet the requirement of 4 μm Ni but had a Ni thickness of only 1 μm. The tin layer was 10 μm thick.

— to provide a safe and stable electrical contact in detachable contacts, e.g., edge connectors;
— to provide a suitable surface for the diffusion welding process in electronic devices.

While gold should not be used as a soldering surface because of the problems created, described in 13.3.4.4 B, it *is* frequently used as such because:

(a) To make the internal connections in a component often requires the use of gold. As it is very difficult to plate only the tiny spots actually needed for the internal connections, normally the whole surface of the lead frame or carrier is plated. To plate only the spots which are needed would create extra costs for the component manufacturers which they wish to avoid.
(b) On occasion, when manufacturing components or in surface mount technology, problems in tooling or in thicknesses or tolerances arise which cannot be solved by using a tinning process as the tin, or tin-lead, melts at a relatively low temperature and does not form a flat surface. In such a case, a material must be chosen which does not melt at such a low temperature, will withstand the different time-temperature conditions arising during component manufacture and is easily soldered after all processing.
(c) Where a consistently flat, solderable surface is required, as in SMT, the use of an immersion gold finish over electroless nickel has become widely favoured. However, the gold thickness is extremely low and in corrosive conditions solderability can be severely impaired, see Volume 2, Chapter 6. The intending user must remember that a joint may look adequate but may be totally unreliable due to hidden defects and must therefore keep tight control of processes, storage and all other factors which may affect the joint integrity.

B. The Problems with Gold in Soldering

Gold is really a problem in soldering.[2-7] Gold dissolves in molten solder and diffuses easily into the tin-lead, even in the solid state. Problems similar to those described in 13.3.3.1 and 13.3.4.3 can arise. If the ratio of gold to tin in the solder joint is appropriate, brittle gold-tin intermetallics are formed. These brittle intermetallics can endanger the mechanical stability of the joint and may under certain types of stress cause the solder joint to crack, Figure 13.6.

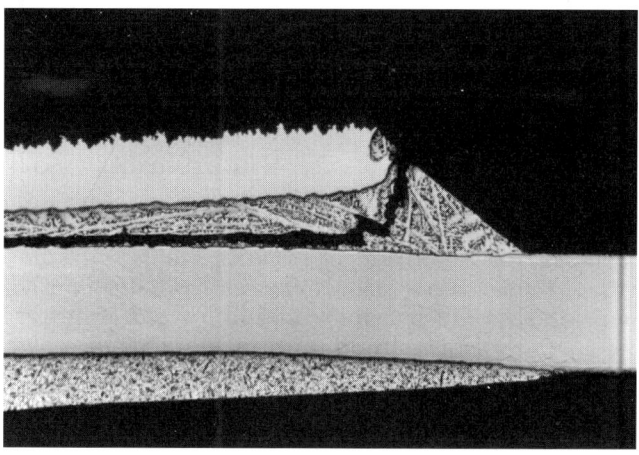

Fig. 13.6 Cracking in a solder joint. The gold on the terminal was unprofessionally removed. Gold was left on the terminal surface and, during soldering, formed the brittle, long white gold/tin intermetallic compounds which lead to fracture.

If the life of the joint is not ended prematurely by the gold-tin intermetallics formed, it may be ended by the following process.

A solder joint is a living product: it ages, it responds to stresses with a change in its metallurgical structure. The diffusion of gold into solder takes place not only when the solder is molten but also in the solid state, albeit more slowly as the rate of the diffusion processes is dependent on temperature. In due course the gold layer will be totally diffused into the joint. If there has not been a good bond between the gold and the nickel layer, due to the nickel layer not being sufficiently active when the gold was deposited, the component lead can be pulled easily out of the 'joint'.

Shock and thermal fatigue are especially dangerous for a joint containing brittle intermetallics and these are exactly the main types of stress on solder joints in electronics. The fatigue life of a solder joint containing 4% w/w gold is bad, the joint surviving for only 19 cycles, under defined conditions, compared with 37 cycles for an Sn60Pb40 solder loaded with 3.6% Ag (silver). In Figures 13.7 to 13.11, Beester[2] relates the values for different types of stress to the amount of gold in the solder. It can be seen that, when the gold reaches a level of 4% in the solder, the impact strength values decrease dramatically. This is also true for ductility values.

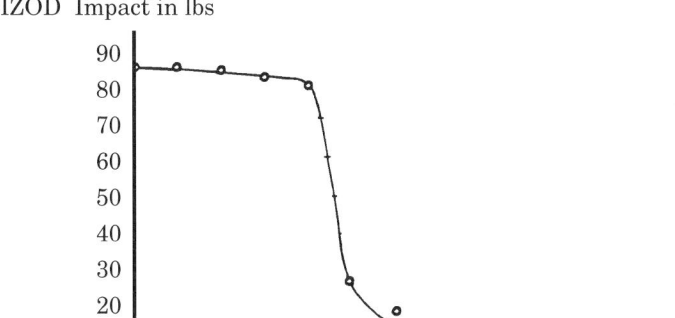

Fig. 13.7 The impact strength of Sn63Pb37 solder as a function of the gold content, Ref. 2.

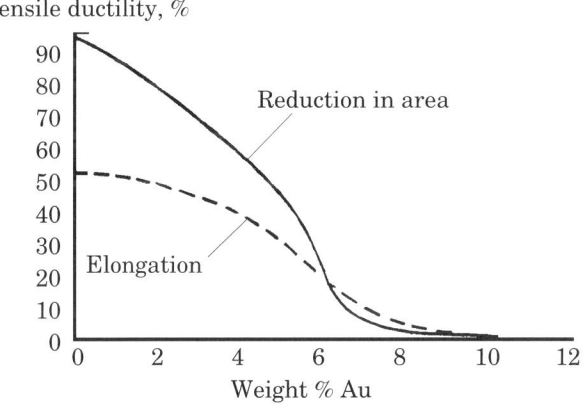

Fig. 13.8 The ductility of Sn63Pb37 solder as a function of the gold content, Ref. 2.

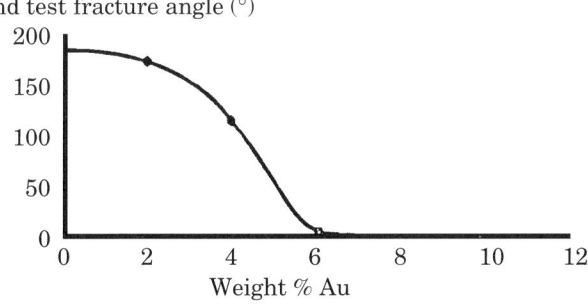

Fig. 13.9 The bend properties of Sn63Pb37 solder as a function of the gold content, Ref. 2.

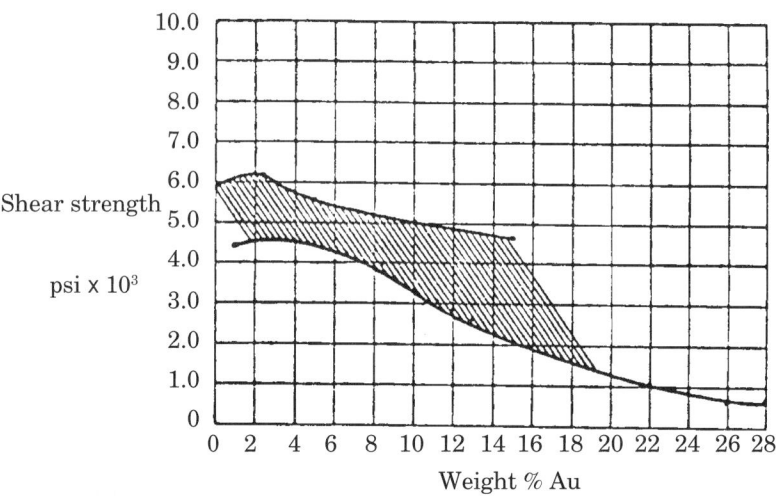

Fig. 13.10 The shear strength of Sn63Pb37 solder as a function of gold content, Ref. 2.

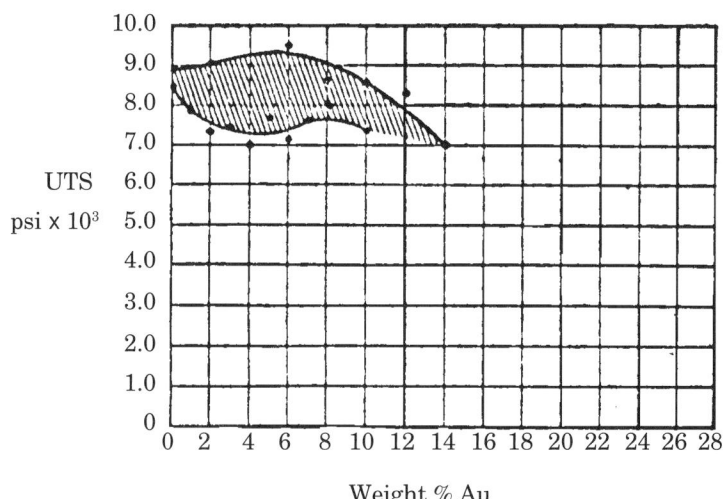

Fig. 13.11 The pull strength of Sn63Pb37 solder as a function of gold content, Ref. 2.

C. The Distribution of Gold in a Solder Joint

Beester's work suggests that less than 4% Au in solder does no harm. This may be valid only under very special conditions:

(a) that the gold really is distributed uniformly throughout the solder joint;
(b) that the gold is really dissolved in the tin.

The reality is, however, that the gold dissolves from the plated deposit into the solder. Close to the deposit there will be a very high concentration of gold which continually decreases to lower values as one moves further from the solid deposit into the joint, Figure 13.12. There will be a very high gold concentration in a zone close to the gold layer in which the cracking of the gold-tin crystals starts and propagates through the whole solder joint.

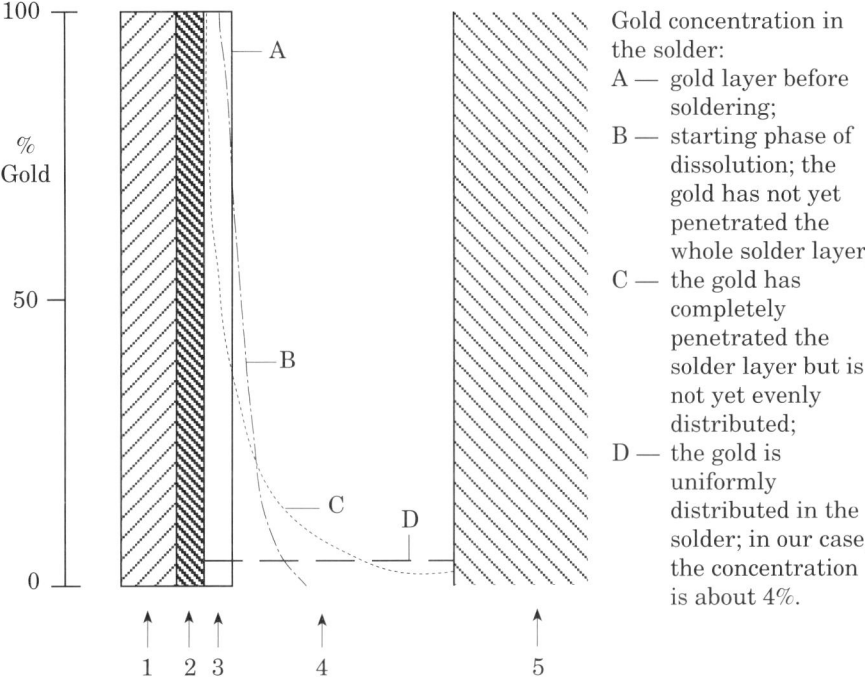

Fig. 13.12 Model for the distribution of gold in a solder joint after dwelling at a given temperature for a given time. The curves showing the gold percentage depend on the gold layer thickness as well as on the amount of solder.

Like copper and silver, gold dissolves readily in solder, but at a much faster rate, Figure 12.4 and Figure 13.13. While it is possible to reduce the dissolution rate of copper or silver by adding these elements to solder, this is not possible with gold. With gold, more than 10% has to be added to the solder and such a solder cannot be used in normal soldering processes.

The results show that gold, if used, should be plated in thicknesses well below 0.5 µm. When electrodeposited, a relatively thick gold layer is used to ensure freedom from porosity, although improved processes can solve this conflicting problem by giving pore-free deposits at lower thicknesses.

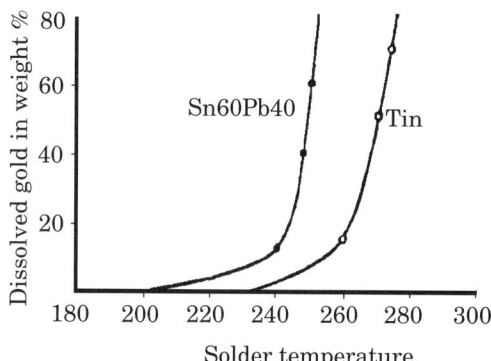

Fig. 13.13 Dissolution of fine gold wire in molten tin and molten solder as a function of the temperature, Ref. 5.

D. How Can Gold be Soldered?

There is an argument that if the gold has been plated in a very thin layer only (<0.5 µm) it will be totally dissolved in the solder. This statement must be carefully investigated from one case to the next. Plating can never be carried out exactly. Depending on the circumstance, the deposits may have large variations in their thickness.

For a given gold thickness a defined amount of solder is needed to dissolve the gold so that not more than the desired gold concentration will be found in the solder. This can be calculated and, in Figure 13.14, the relationship can be seen between gold layer thickness, the amount of solder and the concentration level of the gold in the solder. The assumption is that the gold is totally dissolved. It can be anticipated that gold-tin crystals can be found in the solder joint below the level where the first gold-tin crystals could form according to the gold-tin phase diagram.

It is possible to reduce the dissolution rate for gold at the same soldering temperature by using a solder other than Sn60Pb40. Such a solder could be pure lead or a eutectic lead-silver alloy. With lead, gold forms an intermetallic with a melting point of 215°C. It is also possible to solder at a lower temperature by using an indium-containing solder. A special flux is needed and the corrosion property of such a joint should be studied carefully. Intermetallics cannot be avoided and their properties are not fully known.

Two cases must be considered:

— *soldering of gold surfaces in a wave soldering machine.* The consequence is that the gold is washed away into the solder pot. If the soldering time is long, the plated gold will disappear completely and the joint may contain very little gold.
— *soldering of gold using a reflow method.* Reflowing means that only a defined amount of solder is available in which the gold will be dissolved. Here the amount of gold in the solder can be calculated depending on the thickness of the gold layer, the surface which is soldered and the amount of solder used, Figure 13.14.

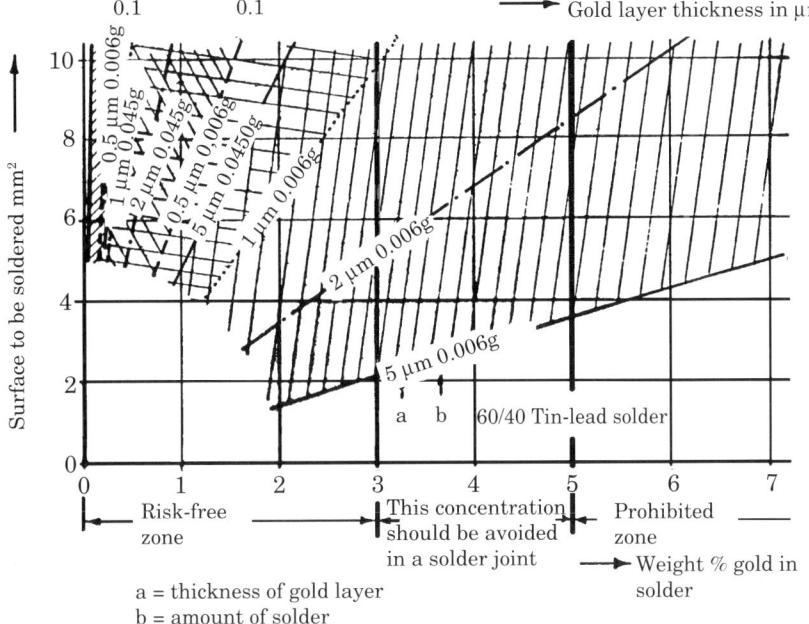

Fig. 13.14 The calculated amount of gold in a solder joint after complete dissolution of the gold in the solder.

E. Gold Plating (for fuller details see Volume 2, Chapter 6)

Gold plating is rarely carried out directly on a base metal, as gold diffuses quickly into most of the common metals. To prevent this, a diffusion barrier is normally deposited between the gold and the conductor being plated. The diffusion layer very often is nickel with a thickness of at least 1μm or, more commonly, 4 μm.

F. Solderability of Gold Plating

The solderability of pure gold is good. Sometimes it is argued that gold plating preserves the solderability of the component lead. This is true but there are a number of other coatings which perform the same task at lower cost.

Different types of electrolytic gold give different solderability. This depends on the different additives in the electrolyte which are necessary to achieve special properties of the gold. For example, cobalt, an additive to control hardness, in the gold layer degrades solderability. Additives for enhancing the brightness of the deposit can give the same effect.

G. How to De-gold

If one has to use components with gold plated component leads and really control the quality of the product, only one recommendation can be given by the author. De-gold the leads. It is essential to check with the component supplier that the time/temperature of immersion in solder will not adversely affect the component before proceeding as follows:

— Flux the component leads with a flux used in production.
— Dip the leads into a solder bath held at 250°C to 260°C for 5 seconds. The solder should be of the 60 tin 40 lead type. Pure tin can also be used.
— Pull the leads out of the solder bath slowly so that the solder can run off the surface and repeat the dipping operation in a second solder pot.

The reason for the two baths is that the first accumulates the gold and the second coats the leads with new fresh solder. A larger area must be de-golded than required for soldering. If this is not done, cracking of the gold at the edge of the solder joint can lead to a crack in the joint, Figure 13.15.

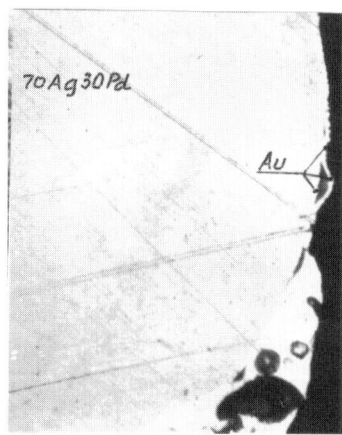

Fig. 13.15 Cracking of surface layer due to bad plating. The contact is silver, gold plated. When the contact was resistance welded, the gold layer cracked due to the welding heat. Similar phenomena have been observed on tin plated surfaces after soldering.

13.3.4.5 SILVER, SILVER-PALLADIUM PLATING

Electroplated silver is not normally used as a solderable finish. There are, however, a number of components, e.g., chip capacitors, which are silver or silver-palladium plated or coated in the manufacturing process, see Chapter 4.

Silver should never be plated directly on copper in a thickness of less than 1 μm as it diffuses fast into the copper and the solderability will deteriorate quite rapidly. A nickel anti-diffusion layer is recommended or a silver thickness of 5 μm or more. Fresh silver is readily solderable. The solderability will be impaired by sulphur in the air, forming silver sulphide on the surface. The deterioration of silver surfaces by air can be prevented as described in Chapter 12, 12.9.4, and in 13.3.3.3.

The dissolution rate of palladium in solders is lower than that of silver and it is sometimes used for terminations on components which must be soldered. It may be used as an overcoat for silver or co-deposited with nickel. Palladium is a catalyst for many organic reactions and can promote polymerisation of organic substances with which it is in contact to produce an insulating coating which also degrades solderability.

13.3.4.6 PROBLEMS IN PLATING

Numerous problems can arise in plating, many of them well known, others totally unexpected. Some of the problems affecting soldering are:

(a) Porosity is an all-too-common defect in plated deposits. There are a number of causes, such as too thin a deposit, an unsuitable bath has been used, or the process is out of control. The material beneath the layer can diffuse through the pores, form a film on the outer surface of the deposit and react with air-borne contaminants. One example of this is silver which reacts with sulphur in the air to form silver sulphide which is not solderable and has a high contact resistance.
(b) Another problem arises when a nickel diffusion barrier is too thin and cannot stop the migration of zinc from a brass base material into the solderable layer. Zinc can then accumulate on the surface of the solderable layer and form zinc oxide or other compounds which are not solderable.
(c) Unsuitable electrolytes can cause problems other than porosity. Most electrolytes contain inorganic and/or organic additives to control brightness, throwing power, deposition rate or deposit properties such as hardness. The effect of cobalt in a gold deposit is cited in 13.3.4.4 F.
(d) Poor adhesion of the surface layer may cause great problems in certain cases. Lack of adhesion can be detected on the wetting balance as described in 13.3.3.1, The Bond. Alternatively, poor adhesion of a layer can be shown by heating the material to a suitable temperature and examining for blisters and/or cracking, Figure 13.15.

13.3.4.7 POINTS TO BE CHECKED

Things to be checked on materials and processes used for coatings depend on the stresses expected in manufacture and service and on failure modes known to be common in a specific process. Some of the more important properties which have to be checked in soldering applications are solderability, especially after ageing, bond between layers, thickness of the layers, possible effects of and resistance to corrosion, ductility and porosity.

13.4 SOLDER FOR ELECTRONIC PURPOSES

13.4.1 General

Solder may be defined as a metal or an alloy of two or more metals (usually tin and lead but other metals may be present) of melting range below 450°C used to join metallic surfaces. Solder can be supplied as, e.g., wire, sheet, bars, preforms, pastes and so on.

The temperature of 450°C is set to distinguish (soft) soldering from brazing (hard soldering).

The most common solder for electrical and electronic purposes is composed of 60% tin and 40% lead. The reason for this popularity is that it has a relatively low melting range of 183°C to 191°C and is relatively cheap due to the lead constituent. Sometimes there are other elements in the solder either added intentionally or unintentionally. The elements added intentionally are alloying elements added to give the solder specific properties. The elements added unintentionally are impurities.

13.4.2 Properties of Solder

Quite a number of properties of solder are of interest when soldering and others for the behaviour of the solder joint during its life. For 'normal' solders of the type Sn60Pb40 quite a depth of knowledge has been gathered but for alloys differing from this basic type astonishingly little information or values are available.

Properties of solder influencing the soldering process are primarily the melting point or melting range, the oxidation behaviour, the surface tension, the reaction with fluxes, the composition of the solder and its impurities, and those characteristics affecting the dissolution of metals in the solder.[8]

The main property influencing the solder joint is the strength in all its various forms such as shear strength, pull strength, creep strength and fatigue with low cycle fatigue. Other important properties include the diffusion behaviour,[9] corrosion,[10,11] thermal and electrical conductivity,[12] and special phenomena such as whiskers or tin pest.

Another property, highlighted especially in the last few years, is the environmental suitability of solder containing lead. In the early nineties, much consideration was given in the United States and elsewhere to the replacement of lead-rich solders with other solder types such as Sn95Pb5 or quaternary solders. No substitute has been readily found and it may be a long time before a true replacement is established.

13.4.3 Requirements Specified for Solders

The first thing specified for a solder is the melting point, normally given by the composition of the solder. The composition of solders, including many which are not used in electronic production, are given in ISO standard ISO 9453 and in References 13 and 14. In the ISO document, forms of delivery, sampling and analysis as well as marking, labelling and packaging are also described.

For soldering electronics normally a solder based on 60% tin and 40% lead is specified. In wave soldering machines, however, quite often an alloy based on 62% tin and 38% lead is used, which is close to the eutectic alloy. The reason for this is that tin is used up in the wave soldering machine by oxidation.

The soldering properties of a solder can be determined best by using either the wetting balance method or the spread test. The behaviour of the surface during heating can indicate if difficulties can be expected, e.g., sputtering or, at the larger quantities in the solder bath, small explosions. When the solder is melted the rate of oxidation can give an indication of whether or not the solder delivered was properly manufactured. The trouble is that a chemical analysis alone does not always tell whether the solder obtained is good or not.

Other important properties specified can include the melting point or the melting area, as well as physical properties such as strength, resistivity and resistance to corrosion.

Eutectic solders are only used where their specific properties, e.g., the composition or the fact that no melting range exists, are essential. In normal soldering processes, it is easier to solder with a non-eutectic solder which has a melting range.

13.4.4 Low Melting Point Solders

Sometimes a melting point of 183°C is too high. Heat-sensitive components may be damaged, or the necessary soldering temperatures cannot be achieved in multilayers or a fuse would not be able to function. In such cases, low melting solders are needed.

There are more than 40 alloys which could be used as a solder in the melting point range of 16°C to 183°C. Such alloys are mainly composed of two or more metals selected from tin, lead, bismuth, cadmium, indium with smaller additions of copper, zinc, silver and gallium. Quite a number of these alloys need special fluxes for proper soldering. Normal fluxes are designed for soldering at temperatures of 200-280°C and for solders based on tin-lead and, therefore, do not work satisfactorily at lower temperatures with other solders.

In many cases, solders with indium are preferred for use as indium has excellent wetting and good strength properties. However, indium is expensive. Additionally, indium forms intermetallics fast[15] and some of the indium solders, e.g., In90Ag10 and In70Pb30, corrode. For such reasons, indium solders should be used in electronic production only with special precautions.[16] The problem of corrosion has led to indium solders being replaced for multilayer applications with an Sn42Bi58 solder.

The problem with low melting point solders is that very little is known about their strength properties, such as creep, fatigue, elongation and properties such as diffusion, corrosion, intermetallics (and their brittleness) and electrical conductivity. Care is needed with bismuth-containing solders as Bi forms a ternary eutectic with lead which melts at 93°C. As many component leads and printed boards are tinned with Sn60Pb40, the joint will be ruined when soldering is carried out with the low melting point solder Sn42Bi58 — which has a melting point of 138°C. (Incidentally, this solder is brittle.)

Actually, so little is known of most low — *and high* — melting point solders that different literature sources give different values for the melting point or range of the same solder composition.

13.4.5 High Melting Point Solders

Again because normal solders do not fulfil the necessary requirements, the use of other solder alloys with a higher melting point has been found necessary. In microwave technology, a solder with 70% gold and 30% tin is preferred as it has low electrical resistivity. Even with the high melting point solders, very little information can be found on the different properties of the solder. Here, also, it is sometimes found that special fluxes have to be used to guarantee a good soldering result.

13.4.6 Step Soldering

Sometimes in an electronic design a number of joints have to be made in sequence and very close to each other. In such cases the possibilities offered by high and low melting point solders with their different melting points can be put to good use. Suitable solders are selected which have a difference in melting point of at least 30°C (or, better, 50°C) and, if possible, are of eutectic composition, since such a solder has a sharply defined melting point and not a melting range. The first joint is made with a high melting point solder and the others are made successively with a solder of a lower melting point.

In Table 13.1 the solders most commonly used in electronics are shown, ranging from low to high melting point. Cadmium solders are excluded due to their health risks.

13.4.7 Impurities in Solder

Although a considerable amount of work has been done to investigate the effect of impurities in solder, these are not a real problem in practical soldering. It is estimated that only 3% of soldering problems are related to the solder, see Chapter 12, 12.10.4.2.

Impurities in solder may have two main sources. The first is the manufacturer of the solder. It is possible that the origin of the impurities is in the raw materials used or arises from accidental contamination after refining and during manufacture. Occurrences of such contamination must be considered to be very rare. The national or international solder standards give advice on the amount of a given impurity allowed in solder purchased to the relevant standard, Table 13.3.

The second main source is the user of the solder. The solder at the user is applied either in the form of wire when using a soldering iron, or as solder paste or from the solder baths of soldering machines. Only in the last case is there a real chance that the solder can be polluted. It should be noted that the values for impurities in a solder bath in use can be set at a higher level than those in the material which is purchased, Table 13.3.

The materials being soldered, such as the copper of the terminations on the printed board or a surface finish, e.g., gold or silver, react with the solder, dissolve into the solder bath and so pollute it. A component may fall into the solder bath and contaminate it. If the board holder is made of aluminium, an active flux can start dissolving the aluminium into the solder bath. In very high volume soldering, not even a very active flux is needed to promote attack on the aluminium.

Impurities in solder can influence the soldering properties in three different ways,[17] Table 13.2:

(a) they can influence the solidification and the structure of the solidified solder by changing the liquidus and solidus temperature and forming ternary eutectics and primary intermetallic phases;

Table 13.1

Composition of the Solders Most Commonly Used in Electronics

Solidus[1] °C	Liquidus[2] °C	Composition (% by weight)					
		Tin (Sn)	Lead (Pb)	Bismuth (Bi)	Indium (In)	Silver (Ag)	Other
57	Eutectic[3]	12	18	49	21		
96	Eutectic	16	32	52			
95	110	15	40	45			
110	130	50	25	25			
117	Eutectic	48			52		
124	Eutectic		43.5	56.5			
143	Eutectic	43		57			
157					100		
178	190	62	36			2	
183	Eutectic	63	37				
183	191	60	40				
183	215	50	49				Cu 1
183	255	70	30				
163	194	50	47	3			
221	Eutectic	96.5				3.5	
221	240	95				5	
232		100					
236	243	95					Sb 5
252	289	30					Au 70
268	290	10	88			2	
280	Eutectic	20					Au 80
280	305	8	92				
296	301	5	93			2	
311	314	5	95				
320	325	2	98				
363	Eutectic	92.5			5	2.5	

Notes:
1 The *solidus* is the temperature at which an alloy becomes completely solid on cooling.
2 The *liquidus* is the temperature at which an alloy becomes completely liquid on heating.
3 An alloy having the same melting/solidification temperature is a *eutectic* alloy and the temperature at which solidification/melting occurs is the *eutectic* point.

(b) they can change the oxidation behaviour;
(c) they can influence the surface properties.

Table 13.2

Summary of Impurity Effects

Property Affected by Contamination	Characteristic Defect	Impurity Name	Symbol
Liquidus temperature	Primary intermetallics cause grittiness, reduce solder fluidity and hinder wetting	Arsenic	As
		Copper	Cu
		Gold	Au
		Iron	Fe
		Nickel	Ni
		Silver	Ag
		Sulphur	S
Oxidation behaviour	Increased drossing, icicles and bridging	Aluminium	Al
		Cadmium	Cd
		Zinc	Zn
Surface energies	Decreased areas of spread, icicles and bridging	Antimony	Sb
		Bismuth	Bi
		Cadmium	Cd
		Phosphorus	P

Table 13.3

Limits for Impurities in Solder (% by weight)

Impurity		Fresh Solder QQS, ASTM, DIN, BS	Working Solder Bath ANSI/J-STD-002 and 003
Aluminium	Al	0.001-0.005	0.006
Arsenic	As	0.01-0.03	0.030
Cadmium	Cd	0.005	0.005
Copper	Cu	0.05-0.08	0.300
Iron	Fe	0.02	0.020
Antimony	Sb	0.12-0.5	0.500
Zinc	Zn	0.003-0.005	0.005
Silver	Ag		0.100
Gold	Au		0.200
Bismuth	Bi	0.1-0.25	0.250
Nickel	Ni		0.010
Phosphorus	P		

Notes:
1. The tin content of the working solder bath shall be maintained at ±1% of the nominal alloy being used.
2. The total aluminium, cadmium, copper, gold and zinc contaminants shall not exceed 0.4%.

The correlation between the visual appearance of the solidified solder surface, the impurities in solder and solder joint quality is bad. However, it can be shown that the solidified solder surface is a very sensitive indicator which reacts long before critical impurity values in the solder bath are reached.[18] Therefore, it should not be necessary in normal circumstances to carry out a check on the impurities at regular intervals.

Considering the different ways by which a solder can become polluted with impurities, the following materials are of primary interest.

Metallic materials: copper, gold, silver, iron, nickel, aluminium, cadmium, zinc, bismuth, antimony.
Non-metallic materials: oxygen, sulphur, phosphorus.

For certain impurities, a state of equilibrium can be reached between dissolving in the impurities and removing them by drag-out with the solder which forms the joint. This balance is dependent on the size of the solder bath, the temperature of the solder bath (which determines the dissolution speed), the rate of replenishment with fresh solder and, of course, the dissolution properties of the relevant metal with reference to the solder constituents. This is why some solder baths never reach the maximum permitted values for certain elements which can be found in the specified, valid standard.

13.4.7.1 COPPER

Copper in solder is normally considered an impurity. However, there are solders to which copper is added intentionally. This is done in the case where soldering iron bits made of copper are used. Here a copper addition of 1.4 to 3% to the solder will prevent or slow down dissolution of the copper of the tip into the solder, thus increasing the useful life of the tip. It has been observed that copper in the solder improves the wetting of the solder to the soldering iron tip. Copper in solder is also used for certain mass soldering processes for a similar reason, that is, to avoid the copper of enamelled copper wire, which has to be soldered at temperatures over 320°C, being dissolved totally into the solder, jeopardising the connection.

Copper is an impurity in the case of normal printed board soldering. In standards for working baths, a value of 0.3% is given which is not to be exceeded. This is a good rule which should be followed in all normal cases. However, experience has shown that a content up to 0.5% Cu in the solder pot does not harm the quality when coarse solder joints are being produced on boards with wide spacing of the conductors. In practice, copper-tin crystals are sometimes found in soldered joints even when the solder bath contains 0.2% Cu or less. This can happen if mixing in the bath is not very effective (e.g., in drag soldering machines). The copper dissolved from the board is not homogeneously distributed and forms Cu_6Sn_5 crystals which float on the surface and are then incorporated in the solder joint before there is time for them to be dissolved.

In normal solder baths which operate at the usual soldering temperatures of 230°C to 250°C, copper in the solder is seldom a problem as the amount of copper in the bath stabilises at about 0.3%.

As the copper-tin compound is heavier than solder, this can be used to remove copper from the solder bath. The temperature of the solder bath is lowered to 200°C, and the compound bailed from the bottom with a scoop. Precautions must be taken by the operator as the solder is still a very hot liquid: the operator must work with eye glasses, special gloves and protective clothes.

13.4.7.2 GOLD

Gold as an impurity never comes from the supplier of the solder. Gold is introduced into the solder by the soldering process when soldering on gold surfaces. Gold is dissolved rapidly by tin-lead solder and the properties of the solder will change very quickly. The solder rapidly loses its normal ability to spread.

With increasing temperature, an increasing amount of gold dissolves in the solder. In theory, 15% gold can be dissolved in 60/40 tin-lead solder at 250°C. The figure of 15%, however, is never reached in a normally operating solder bath. Solder is dragged out of the solder bath to form a joint and is replaced with fresh solder. This process stabilises the amount of gold in the bath, and that of any other impurity, at a much lower level.

As gold in a solder bath makes the solder sluggish at concentration levels above 0.5%, the temptation is to raise the solder bath temperature because this effect then disappears. However, the raised temperature enables the bath to dissolve more gold, making the solder sluggish again and inviting a further rise in temperature. This could end in the proverbial vicious circle! Raising the solder bath temperature to 'cure' this kind of problem should *never* be allowed.

13.4.7.3 SILVER

Silver easily dissolves in a solder bath and can accumulate to relatively high values. A maximum of 5% has been measured. As silver up to an amount of 2% does no harm either to the soldering process or to the solder joint, it is not normally necessary to check the silver content in a production bath. Depending on the crystallisation of the silver-tin crystals, silver can improve the joint strength slightly. This is the reason why silver is sometimes added intentionally to the solder. However, it is difficult to ensure the correct crystallisation in a normal soldering process on a board with many components of different thermal behaviours, as the temperature distribution on a board cannot yet be controlled during soldering and cooling. Therefore, it is not yet possible to use the strengthening properties of silver in real soldering.

The ease with which silver dissolves can become dangerous when soldering ceramic components using silver coatings as terminations. The silver dissolves and the termination disappears.

13.4.7.4 IRON

Solder pots and certain component terminations are made of iron alloys. Therefore, iron is found in solder baths but at very low concentration levels, as the solubility of iron in solder is low unless the temperature is very high (>400°C). For this reason, iron is not a problem in normal soft soldering.

Problems may arise with iron if the solder pot is new or if it has been damaged in some way. The action to be taken is to keep the bath surface free of dross.

13.4.7.5 NICKEL

Nickel does not create any problem as a contaminant in solder as the solubility of nickel in tin-lead is very low.

13.4.7.6 BISMUTH

About 3% of bismuth is sometimes added to solder to make the solder joint surface matt and easy to inspect. This is a considerable advantage to the inspection personnel and savings can be made in the inspection process. A surface which is tinned with a bismuth-containing solder withstands ageing better. An addition of bismuth to the solder does not degrade the important properties of solder such as solderability, strength, electrical conductivity or corrosion resistance.[19]

However, unless specified, bismuth is an impurity and, if a bismuth solder is used by a company, the scrap material must be kept separated from the normal solder scrap. This ensures a better price for the normal solder scrap as bismuth is considered to be a contaminant. See also 13.4.4, Low Melting Point Solders.

13.4.7.7 ANTIMONY

Antimony is already found in tin ore and has to be eliminated to give an antimony-free solder. The refining process is comparatively expensive and it is therefore a cost advantage to leave the antimony in the solder.

It is said that antimony is necessary to avoid the destruction of a solder joint by tin pest initiated by low temperatures,[20] but it is also known that small amounts of bismuth and lead have the same effect. (This may be the reason why in the last 30 years no case of a solder joint failure due to tin pest has been reported in the Scandinavian countries where solder connections are used at temperatures lower than -30°C.)

As a contaminant, up to 0.3% antimony seems to improve wetting but larger additions degrade the wetting slowly. Contamination problems with antimony are minor because it is not likely to be introduced by accident into a solder bath.

13.4.7.8 ARSENIC

Even though limits for arsenic as an impurity are specified in standards, arsenic as an impurity in solder baths is virtually unknown. Arsenic in small amounts (0.03 to 0.05%) is known to cause dewetting on brass.

13.4.7.9 ALUMINIUM, ZINC AND CADMIUM

If tools or components made of aluminium, zinc or cadmium are used intensively in contact with solder as soldering jigs, skimming or other tools, contamination of the solder can result. Tough oxide films are formed on the surface of the solder which are difficult to penetrate. The colour of zinc oxide and aluminium oxide is white, that of cadmium brownish red. The oxides formed rise to the surface and are skimmed off so that aluminium, zinc and cadmium are seldom found in large concentration levels in solder. Depending on the soldering task, aluminium can be tolerated up to higher levels than specifications indicate.[18]

If brass parts are soldered in a solder bath, the zinc of the brass dissolves into the solder. The zinc oxidises and forms a skin on the solder bath which prevents proper soldering. Therefore, a limit for zinc in a solder bath has been set at 0.005%.

Screws, rivets or other fabrication parts may be used on printed boards. Such parts are often made of aluminium or steel protected against corrosion with zinc or cadmium. Experience has shown that, if the zinc coated parts are chromated or treated with a protective lacquer, they can be used and soldered together with the components in a wave soldering machine. Even aluminium rivets do not cause problems due to:

(a) drag-out of the solder when forming solder joints and replenishment of the bath with new fresh material;
(b) the fact that the oxides rise to the surface of the solder and are skimmed off.

However, it should be said that use of aluminium rivets or the other materials indicated can only be made without problems if personnel are properly trained and instructed, and the solder bath is continuously checked for aluminium and/or zinc contamination.

Cadmium is highly toxic and should therefore not be used in solders or for surface coatings. In a number of countries the use of cadmium is not allowed, see Volume 2, Chapter 18.

13.4.7.10 OXYGEN

Normally oxygen is not found *in* solder but only on the solder bath surface in the form of a black dross of metastable tin oxides. This dross can cause soldering defects such as bridging and webbing. Any soldering machine driven by a pump — the wave soldering machine is the best example — feeds oxygen in smaller or larger quantities into the solder bath where it reacts with the tin and forms dross. The dross does not stay long in the solder bath but proceeds quite quickly to the surface where it can easily be removed. In well maintained soldering machines, the risk of getting dross into a solder joint is zero. Dross in agitated machines should be removed regularly.

13.4.7.11 SULPHUR

Sulphur as an impurity has been reported to the author only in connection with soldering iron tips where it destroys the wettability on the iron tip, see Chapter 14, 14.2.3. A concentration of 0.03% of sulphur in the solder is dangerous. The analysis of the small amounts of sulphur in question is difficult. Fortunately, a good grade of solder will not contain more than a few parts per million of sulphur.

13.4.7.12 PHOSPHORUS

Phosphorus is used in solders for static as well as wave solder baths to replace the tin oxide on the solder surface with tin phosphide. The tin phosphide is more easily penetrated by the workpiece in a dip soldering operation. A solder containing phosphorus forms less dross than a normal solder and the dross is powdery. As the phosphorus is fed into the solder by a copper-phosphorus alloy, such special solders also contain copper.

13.4.7.13 DROSS

More than 90% of the dross on a normal solder bath is non-oxidised tin-lead and, at a maximum, 10% of the dross is oxidised material, mainly tin oxide.[21] Drossing leads to tin being dragged out of the solder bath and it has to be replaced regularly in critical applications. This is the reason why, in critical soldering tasks where an agitated solder bath is used, a solder alloy of 63% tin and 37% lead is preferred.

As the dross contains only a small amount of oxides and a great deal of solder, it should not be thrown away as it can be sold to a refinery. Some companies recover the solder from the dross and reuse it. The dross is heated and melted with highly activated flux. A considerable amount of money will be saved. The formation of dross can be reduced dramatically by using protective gases, such as nitrogen. This requires specially designed machines which are readily available.

13.5 FLUXES

13.5.1 Definition of Fluxes

The flux has the following tasks:

— to dissolve any final solder lacquer;
— to clean the surfaces to be soldered from oxides and other impurities;
— to keep the surfaces to be soldered clean during the heating process until the solder is able to wet the metals involved and form the joint;
— to complete the soldering process in a proper way, by helping to prevent bridging, webbing and the formation of icicles and solder balls.

The main constituents of a flux are:

— the carrier material;
— the activator;
— the solvent;
— the wetting agent.

13.5.2 General

Normally, the flux must clean copper, tinned surfaces and the solder itself. For that purpose, resin fluxes are generally sufficient. However, there are other metals which have to be soldered or tinned, e.g., brass, German silver, Alloy 42, nickel surfaces, beryllium bronze, etc. Due to the alloying constituents in these metals other oxides are formed on the surface[22] which are difficult to remove with normal resin fluxes. Sometimes the coating, which for soldering purposes is usually tin or tin-lead, reacts with the metal underneath, the base material, say copper, and forms a compound which may be very difficult to solder after a long time in bad storage conditions.

When mass soldering came into use, the main fluxes used were rosin based with or without an activator. These fluxes had one advantage — they could be left on the solder joint because they encapsulated the activator used and no corrosion occurred. However, more and more different types of components were used and the component manufacturers did not seem to care that their components had to be soldered because the component leads were often of inferior solderability.

At this point two developments started. One was the development of solderability testing methods in order to overcome the problem by introducing materials of better solderability. The other was aimed at developing new fluxes, using very active fluxes which of necessity were corrosive. Such fluxes had to be cleaned off. An example is the SA (which stands for *synthesised activity*) type flux, the residues of which remain liquid after soldering.

Sometimes the term 'corrosive' or 'non-corrosive' flux is used. This must be understood in the right context. Fluxes are *always* corrosive, otherwise they cannot fulfil their task. The only difference between the two types is the temperature at which they become active and whether the flux residues are corrosive or not. For example, non-activated rosin based fluxes become active above 80°C and their residues are non-corrosive at room temperature.

Probably as a result of the fact that a more active flux gives better soldering, an instinctive belief has developed that it is the flux which is responsible for better or worse wetting. However, this is not true: the wetting is independent of the flux and depends only on the forces which develop on the surface of the solder and the material to be soldered. The flux only cleans material from the surface to be soldered which would otherwise hinder the action of such forces, that is, the production of wetting. If the materials are clean, they cannot wet better or worse. The flux in many respects acts as a catalyst in a chemical process.[23]

13.5.3 Classification of and Tests on Fluxes

Due to the large developments in soldering and considerable changes in fluxes, the American QQ-S-571E Federal specification was revised and the ANSI/J-STD-004, April 1992, Requirements for soldering fluxes, forms a part of that revision. This standard is based on the experience of both American and international experts and American and international standards in soldering.

Fluxes can be classified according to ISO 9454-1, see Table 13.4. In the same document, guidance is given for the use of test methods for the various flux types. The fluxes may be tested for:

— non-volatile matter (or the solid content) — needed to determine the halide content and acid value;
— acid value — the acid value together with the halide content gives an indication of both the flux activity and corrosiveness;
— conductivity of water extract — to show if ionic material(s) are in the flux;
— copper mirror — this test is a corrosion test and the mirror must be properly prepared; correct judgement of the results is critical;
— ionic residues;
— halide content (by different methods) — in early specifications the halide content was required as it is an indication of the corrosiveness of the flux;
— free acid in resin;
— zinc content;
— ammonia content;
— flux efficiency (by different methods);
— ease of removal of residues;
— steel tube corrosion test;
— printed circuit surface resistance — this test is very important, as it gives an indication of the reliability and the life of the assembled board; the normally used comb pattern can also be printed on the board directly under SMDs in order to check whether the cleaning under the components has been successful or not;
— flux sputtering test;
— dryness test (tackiness).

Table 13.4
Classification of Soft Soldering Fluxes according to their Major Ingredients

Flux Type	Flux Basis	Flux Activation	Flux Form
1 Resin	1 Colophony rosin	1 No activator added	A Liquid
		2 Halide activated*	B Solid
	2 Non-colophony (resin)	3 Non-halide activated	C Paste
2 Organic	1 Water-soluble		
	2 Non-water-soluble		
3 Inorganic	1 Salts	1 With ammonium chloride 2 Without ammonium chloride	
	2 Acids	1 Phosphoric acid 2 Other acids	
	3 Alkalis	1 Amines and/or ammonia	

*Other activators may be present.

Table 13.5
IPC Flux Activity Classification Tests

Flux and Flux Residue Activity Level	Copper Mirror	Corrosion Resistance Tests	
		Silver Chromate	Corrosion
	(Section 5.3.3.1)*	(Section 4.4.3.2) or Halide Test (Section 4.4.3.3)	(Section 4.4.3.4)
L[†]	No evidence of complete copper removal (no white background visible).	For Class 1 and 2 less than 0.5% halides (based on flux solids). For Class 3 must pass the silver chromate paper test.	No evidence of corrosion. If a blue/green peripheral border is visible, the flux must pass the silver chromate paper test.
M	Partial or complete copper removal allowed.	<2% halides	Minor corrosion is acceptable. The flux must pass the halide requirement.
H	Complete copper mirror removal allowed.	>2% halides allowed	Major corrosion evident.

*Section numbers refer to IPC-SF-818, *General Requirements for Electronic Soldering Fluxes*
[†]L = low or no activity: M = moderate activity; H = high activity.

Table 13.6
IPC Surface Insulation Resistance Requirements for Fluxes

Flux and Flux Residue Activity Level	Assembly Class Test Conditions, ANSI/J-STD 004		
	1	2	3
	50°C 90% RH 4 days	50°C 90% RH 7 days	85°C 90+% RH 7 days
L	100 Megohm Residues not cleaned (N) or cleaned (C)	100 Megohm Residues not cleaned (N) or cleaned (C)	100 Megohm Residues not cleaned (N) or cleaned (C)
M	100 Megohm Residues not cleaned (N) or cleaned (C)	100 Megohm Residues not cleaned (N) or cleaned(C)	100 Megohm Residues not cleaned (N) or cleaned(C)
H	100 Megohm Residues cleaned (C)	100 Megohm Residues cleaned (C)	100 Megohm Residues cleaned (C)

L = low or no activity; M = moderate activity; H = high activity.

Even though numerous test methods are still under consideration or only at the draft stage, many of the tests have been used widely for years. Examples of these are the wetting balance test for testing flux efficacy and the surface insulation resistance (SIR) test.

The IPC Flux Activity Classification Test is of interest. It classifies flux types by corrosion tests which they pass or not, Table 13.5. In addition, surface insulation resistance requirements are specified for the different flux types, Table 13.6. These classifications are linked to the ANSI classification of assemblies described in Chapter 12, 12.12.

Other aspects which are important for practical soldering are:

— the density of the liquid flux;
— what solvent is used in the flux — the solvent in a working flux evaporates and has to be replaced; also, the flux residues may have to be cleaned off after soldering and the type of solvent determines the type of cleaner to be used for this purpose;
— the dry content;
— an infra-red diagram is a good aid to identifying whether or not the flux is the same from one delivery to the next;
— a health and safety declaration from the supplier is a requirement in most countries, see Volume 2, Chapter 18;
— some fluxes which may appear safe to leave on the solder joint may after a period of use decompose and become powder; this can contaminate the assembly and cause malfunction.

Other points which must not be forgotten are:

— the smell, especially when soldering with a soldering iron;
— the development of fumes (and their removal);
— the impact on lacquers and solder resist; not only may they be dissolved but they can cause webbing and other detrimental effects; in other words, the flux must not only be chosen with regard to metallic materials, corrosion and so forth but also with regard to the non-metallic materials involved in the soldering process, such as the board material, lacquers used and component materials.

It is not sufficient simply to make one or other measurement(s) when choosing a flux. A high halide content or acid value does not automatically mean that the flux is more active or corrosive, nor does a low value imply the contrary.[10] Once all the tests needed to select a flux have been carried out, a production test is essential to determine whether this flux really can be used or not. Only a test in your own production will show the number of occurrences of icicles, bridging, webbing and other typical solder faults, which will certainly rule out some of the fluxes which showed good values in all the other tests.

The choice of the flux type generally does not influence the soldering process line in important points. A given type of soldering machine can be used for different types of flux. However, there can be differences in which flux applicators are suitable and which are not for the type of flux chosen.

When using fluxes special care should be taken when changing from one flux vendor to another. Fluxes tested and supplied to the same specification may behave very differently. Thus it is advisable to carry out tests first to see if a change can be made without incurring any problems. Even fluxes supplied by the same vendor under the same code can vary significantly.

13.5.4 Resin Fluxes

In the case of resin fluxes, the resin is usually dissolved in alcohol. The alcohol will also dissolve any solder lacquer on the board. The resin itself contains weak acids capable of removing oxides. An activator reinforces the ability of the flux to remove oxides.

ANSI/J-STD-004 gives the following designations for resin-based fluxes which are widely used:

— R: Rosin flux with only pure colophony in a proper solvent
— RMA: Rosin, mildly activated
— RA: Rosin, activated. Only to be used with appropriate cleaning.

13.5.5 Colophony (Rosin)

As colophony is a natural product from various species of pine trees, its properties vary due to the species of pine tree the sap comes from, where the pine tree grows and in which year the sap was harvested.

The variation of the properties is due to the fact that colophony is a mixture of a number of natural acids, the main one of which is abietic acid. This explains why colophony can behave differently during soldering unless special care is taken by the flux manufacturer to achieve the same quality of flux. The resin acids can be changed intentionally or unintentionally to give other properties to the colophony. Polymerisation, for example, can both improve and decrease the soldering properties depending on the soldering temperature used.

As it is not necessary to remove colophony from the soldered joint, it protects the solder joint in the same way that a lacquer will protect it and increases the lifetime of the joint if it is subjected to thermal cycling.[24]

13.5.6 Activated Rosin Fluxes

Strictly speaking, the term *activated rosin flux* is wrong. It is not the rosin, or for that matter, other agents in fluxes, which is activated; the activator itself works directly on the surface to be soldered. Quite commonly used activators are dimethylammonium chloride (DMA-HCl) and diethylammonium chloride (DEA-HCl), but other salts such as amines or pyridine as hydrochloride or hydrobromide are used, as are organic acids. The amount of activator in the flux is usually expressed as per cent by weight of the chloride ion Cl^- of the colophony content. When acid fluxes are used, the acid value has to be determined. As a flux is often a blend of different agents and activators, the acid value together with the halide content gives an indication of both flux efficiency and corrosiveness.

Corrosion at an early stage can be observed as a discoloration on the site of the flux after soldering. It should be noted, however, that a halo of green or blue-green coloured rosin is due to the solution of copper oxide and the coloured copper complexes which form as a result. This is not corrosion and has nothing to do with corrosion but is a normal reaction of tarnished copper with flux.

In the telecommunications and computer industries, the requirements for long life and high reliability of the products require the use of a non-corrosive flux in soldering. Such a flux must fulfil the following specification:

— The acid value of the dry content must not exceed 170 mg KOH/g colophony.
— The weight percent of the chloride must not exceed 0.5% on the colophony content.
— The flux residues must not change the surface insulation resistance.
— No salts in the form of white or green-blue corrosion products are allowed on the copper in the neighbourhood of the joint after corrosion testing for 30 days, in 95% relative humidity at 40°C.
— It may sometimes be specified that the conductivity measured on a water extract shall not be greater than 22 µS.

13.5.7 Halogen-containing Fluxes and Corrosion

Halogen-containing fluxes are considered as especially dangerous due to the way in which corrosion proceeds. When soldering with a flux containing an organic halide or an inorganic salt such as zinc chloride, ammonium chloride or similar, the compound decomposes (usually just below the soldering temperature) and hydrochloric acid is formed. The hydrochloric acid reacts with the copper oxide on the copper surface to form copper chloride, water and hydrogen. The copper chloride can then react at room temperature with sulphur dioxide, carbonic acid or carbon dioxide, which are always present in the air, especially in industrial areas, to produce hydrochloric acid again. A new cycle begins and, in the worst case, the copper lead is completely corroded away. How fast this cycle will be depends on many factors including the type of flux, the amount of flux, the temperature and so on.

This process is the reason why only approved fluxes may be allowed in a workshop. It is a good reason, also, for ensuring that the material to be soldered has good solderability. Otherwise, a poorly trained or irresponsible operator might decide to solve the solderability problems on his own by using more effective fluxes. Here it must be especially emphasised that residues formed by the reaction products of certain inorganic halides such as zinc chloride or zinc ammonium chloride cannot be washed away by cold water alone.

Often component leads are of gold- or tin-plated Kovar. Many fluxes cause degradation by stress corrosion cracking on such material — which is not always visible. To avoid this problem, a number of requirements must be fulfilled. For details of these requirements, see Reference 10.

13.5.8 Water-soluble Fluxes

Water-soluble fluxes are fluxes whose residues after soldering can be dissolved in water and washed away. Removal of the residues is necessary as the water-soluble fluxes contain aggressive agents to improve the efficiency of the flux when soldering. These agents in normal circumstances can cause corrosion and have to be cleaned off. Water-soluble fluxes do not necessarily contain water, the solvent more often than not being alcohol or polyethylene glycol. Water is not actually a good solvent in this instance as its thermal properties do not compare favourably with those of alcohol and glycol and create problems such as sputtering when soldering.

13.5.9 Gaseous Fluxes and Soldering without Fluxes

For many years, engineers have dreamed of eliminating the problems generated by the use of fluxes in soldering printed board assemblies by rendering the use of a flux unnecessary. However, nobody can ignore the fact that, before soldering, oxides are formed on metallic surfaces under normal atmospheric conditions and that these oxides have to be removed to promote soldering. To this end, fluxes are necessary.

From welding processes it is known that oxides can be removed by 'forming gas' which consists of nitrogen with 5 to 10% hydrogen. Therefore, it is quite natural to investigate the possibility of using such an oxide reducing gas for the purpose of soldering. Figure 13.16 shows that copper, lead and nickel oxides can be reduced by hydrogen as their curves, which indicate the energy needed to form their oxides, lie above the curve for forming water. To reduce tin oxide, more energy is required than to form water and thus tin oxides, under normal conditions, can be reduced by hydrogen only above a temperature of 600°C. There is another aspect concerning hydrogen and flux to be considered. In a controlled oxygen-free atmosphere the oxidation of flux is eliminated. An addition of hydrogen is even more efficient in promoting volatilising, possibly by lowering the vaporising temperature through hydrogenation of hydrocarbons.

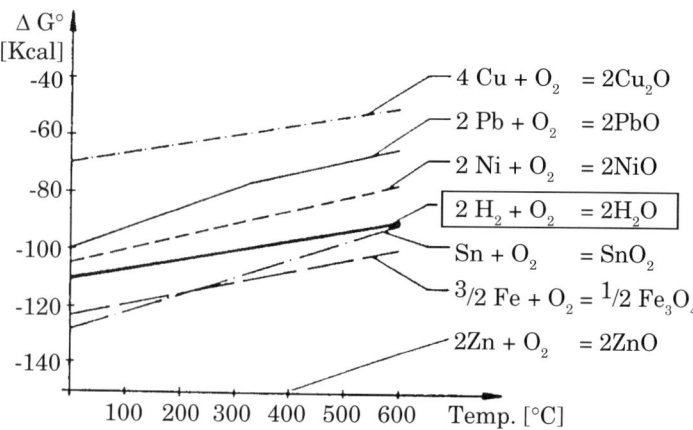

Fig. 13.16 Thermal conditions for reducing metal oxides. Copper, lead and nickel oxides can be easily reduced by hydrogen whereas tin oxide can be reduced by hydrogen only at temperatures above 600°C.

The consequence is obvious. Solder contains 60% tin and 40% lead. As tin oxides are easily formed and therefore are found on the surface of solder, soldering of printed boards cannot be carried out using hydrogen as a flux. The temperature needed to reduce the oxide will destroy the board *and* the components. Even the usefulness of hydrogen in a shielding or protective gas must be questioned.

From Figure 13.16, another conclusion can be drawn. If a solder based mainly on lead can be used for soldering (which would need soldering temperatures of 300°C and more), soldering on copper or nickel with reducing forming gas is possible. This technique is used in the production of components where materials are used that are not damaged at such process temperatures.[25]

With the development of modern soldering machines which are more or less totally enclosed, especially infra-red and vapour phase machines and those using inert gas such as nitrogen, soldering without flux was tried. It was quickly found that it is not possible to make a reliable solder joint at the normally used soldering temperatures of 250°C and lower and that a flux was really necessary. However, by using inert gases, oxidation of the material during the heating period could be prevented, enabling less active fluxes to be used. This also reduces considerably the amount of dross formed in wave soldering machines with corresponding economic gains. It was then only a short step to explore the use of organic acids such as formic acid, which at low temperatures is liquid and at higher temperatures is gaseous. Unfortunately, formic acid reacts with the lead of the solder and forms lead formate which goes directly from the gaseous state to the solid state in the form of very fine particles. This happens at every point where the temperature is below 200°C, that is, on the boards, the fixtures, the conveyor and so on. Lead formate is poisonous. For such reasons, soldering is now conducted under a protective gas with mild fluxes.

13.5.10 Low Solid Content Flux, No Residue Flux

The development in electronics of smaller components, the necessity to check the circuitry with test pins and the knowledge that it is very difficult to clean under components have led to a search for fluxes which do not hinder testing, are totally used up after soldering and which do not leave any residues which could corrode. Some suppliers claim that the goal of 'No residue flux' has been reached: others claim that their 'low solid content flux' is safe and/or easy to clean from an assembly. The development of fluxes is continuously ongoing.

13.5.11 Cleaning off Fluxes and Flux Residues

Flux residues are cleaned from assemblies because they:

— may cause corrosion; not all fluxes are non-corrosive at the working temperature of the board and joints;
— assist migration and reduce the insulating resistance between conductors;
— insulate contact surfaces needed for testing the board assembly;
— may dry and powder, creating particles which could fall on to contacts and cause insulation problems;
— prevent protective lacquers and coatings adhering properly to surfaces;

— may collect dust during operation which, in humid conditions, then causes crosstalk, short circuits and corrosion on the board;
— may assist mould growth;
— may trap solder sputter;
— degrade the appearance of the assembly; in some cases, a clean board surface is a good selling point (and may be demanded by the customer).

13.6 SOLDER PASTES

13.6.1 General

A solder paste is a suspension of solder powder in a flux vehicle. Agents are added which are necessary to prevent separation of the metal powder from the flux and to assist the application of the paste. In general, a solder paste consists of about 90% by weight of solder powder, 5% resin, 4% solvents and 1% additives.[26]

In the ISO/TC44/SC12 document of 1989-11-28 concerning solder pastes, a classification system for solder pastes is suggested. The document classifies powder size and pastes in Tables 1 and 2, and uses the flux classification of ISO 9454-1 and the solder alloy classification of ISO 9543.

Further test methods are proposed.

The most commonly used solder paste for SMT is based on the alloy Sn62Pb36Ag2, but Sn60Pb40 and Sn63Pb37 alloys are also widely used. In addition to the normal requirements for a flux, fluxes for solder pastes have additions which influence the paste rheology and thus the printability or dispensability of the paste. They must prevent evaporation of the solvent during storage, or during the work with the paste, and contribute to good slump properties of the paste.

When printing solder paste, air is worked into the paste. The air oxidises the solder and the amount of flux in the paste is usually so limited that it cannot take care of non-intended extra oxidation. This is the major reason why used solder paste must not be saved and reused.

When soldering with solder paste, the heating must be carefully controlled. There must be sufficient time to drive off any moisture in the paste and thus avoid explosions which can spread solder particles over the board. On the other hand, the heating must not be so long that unnecessary oxides are formed on the surfaces to be soldered.

13.6.2 Solder Balling

Solder balling can result from several different causes, not all of which are yet known. One reason is insufficient cleaning of the solder surface by the flux or, put simply, bad solderability of the solder paste. Another cause is insufficient heating of the solder paste. The results of these failure modes appear directly at the joint. Formation of solder balls around the solder joint can be due to the flow properties of the paste and, when the solder itself melts, the small balls formed cannot flow back and recombine with the mass of solder. If the solder paste has been stored in such a way that moisture in the air was absorbed by the flux, spattering can occur. If the stencil is not cleaned properly after a few prints, smearing can occur and solder particles can be transferred to the board and remelted in the soldering process.

The ease of removal of solder balls depends upon the type of solder and organic coating on the board.

13.6.3 Viscosity

The viscosity of the solder paste must be matched to the method by which the paste is applied. For pin transfer, the recommended values are 100,000 to 200,000 cPs whereas for screen printing 700,000 to 800,000 cPs paste is used with still higher viscosity values being employed for fine pitch technology.

13.6.4 Slump

'Slump' describes that property of a solder paste which has to be controlled to keep the contours stable both after printing and during preheating. This is necessary for printing pastes on solder lands which are located close to one another, especially in fine pitch technology, to avoid bridges and solder balls. Too much slump can result in solder displacement.

13.6.5 Corrosion of Residues

As the cleaning of an SM board after assembly and soldering is difficult to ensure, the ultimate aim must be to use fluxes whose residues do not have to be removed.

13.6.6 Assessment of Shape and Size of Solder Powder Particles

The solder particle size, the size distribution, and the shape of the particles all have an influence on the application (printing or dispensing) of the solder paste and the soldering properties. The solder particles should be spherical, not irregular or needle-shaped, Figure 13.17. Such shapes impair the

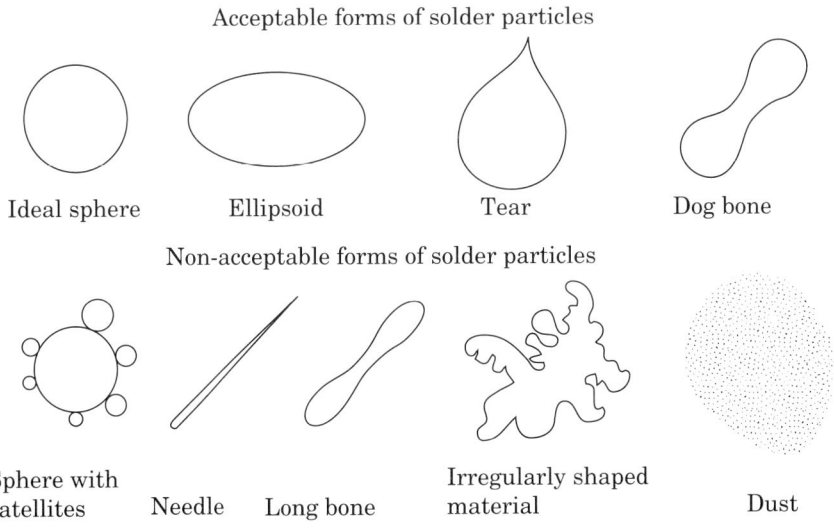

Fig. 13.17 The shape of solder powder particles. (Courtesy of Demetron)

printability of the paste. The particle size should be suitable for the pad or pitch size used, Table 13.7, and should follow a normal distribution with defined tolerances to avoid very fine particle sizes in the paste. The smaller the solder land, the smaller the solder particle size has to be. The result is that the surface area of the solder (which will be oxidised) increases, putting greater demands on the flux which has to remove these oxides.

Table 13.7
Size Distribution of Solder Particles in Powders

Termination Pitch		Recommended Particle Size	
mil*	mm	Class	μm
50	1.25	2	44-74
30	0.75	3	35-44
20	0.5	4	25-35

*1 mil = 0.001 in. = 1 thousandth of an inch.

13.6.7 Storage of Solder Pastes

It has been normal practice to store solder pastes in a refrigerator although recently solder pastes have been developed where this is not a requirement. Before using a paste which has been stored at lower temperatures, it is essential to allow the paste to reach room temperature in the sealed package for two reasons. The first is that the viscosity is temperature dependent: the solder paste must reach the right temperature. The second is that condensation of water onto the solder paste, which is detrimental to the paste, is avoided by this procedure.

13.7 PROPERTIES OF SOLDER INFLUENCING THE SOLDER JOINT

13.7.1 Cracks in the Solder Joint

A solder joint in electronics does not crack abruptly; the crack develops slowly. The crack starts very early when about one tenth to one third of the normal lifetime is used up. The solder joint still works satisfactorily for quite a while after the crack has started. The normal case, however, is that an increase in the resistivity can be measured when the crack has become large enough to reduce conductivity. The crack grows until it forms a complete crack or until it reaches a zone in the solder joint with such low stresses that it cannot propagate further.

Solder joints have been known to work correctly even after a total crack has formed. This is explained by fresh metal surfaces lying close to each other and forming electrically conducting bridges. Such a crack can open for a very short time (microseconds) and close again. The 'open' time becomes longer and longer until the surfaces oxidise and the solder joint becomes an open-circuit. When the cracks open for times of only microseconds, the working of digital instruments may be impaired, the instrument showing an intermittent fault. Normally, a crack causing a short, intermittent break in electrical continuity cannot be detected in an equipment but the consequences can be very serious, e.g., in a computer. The values published for fatigue of solder joints refer to either a total crack or a decrease of the electrical resistance of 10% and, to the author's knowledge, no values exist which refer to such very short intermittent breaks.

13.7.2 Mechanisms of Solder Joint Failure

There are three mechanisms which cause a solder joint to fail:

A. Pure Mechanical Stress

Mechanical stresses may be low and cyclic, e.g., imposed vibrations, or high and occurring only a few times, e.g., when checking a board after production, on test or insertion into a magazine. When a board is bent, the solder joints can be subjected to very high strains. The strains can be so large that the solder joint will crack after only a few bends, Figure 13.18. In most cases, vibration is relatively harmless as the forces are low and the frequency is high. Frequency of vibration and joint life are related to each other.

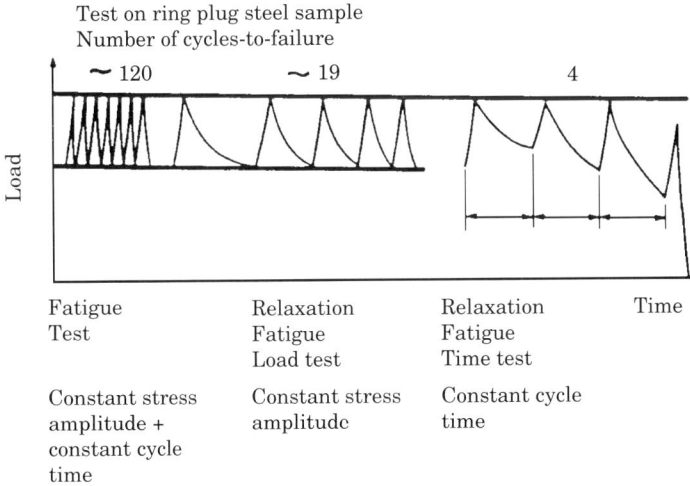

Fig. 13.18 Different types of mechanical fatigue stress. Depending on the type of fatigue stress applied, a solder joint can endure a large or small number of cycles. In the case of stress relaxation cycling, which, unwittingly, may be applied in production of the assembly, the solder joint may survive only a small number of cycles.

Accelerations and stresses applied in use can be so large that a solder joint will be destroyed. This has been observed in mining equipment, rockets and during earthquakes. In such cases, the design has to be made in such a way that the forces cannot be transferred to the solder joint. Alternatively, the solder joint may be designed so that it can endure such forces. If this cannot be done, other joining methods will have to be employed.

B. Thermal Stress

A joint is subjected to different temperatures during its life. This both accelerates diffusion and causes changes in the crystal structure which gives the solder joint properties other than those the solder itself had before soldering. Such changes occur slowly but may become substantial after a long time. The higher the temperature to which the solder joint is subjected, the faster such changes take place. The crystals become coarse, Figure 13.19.

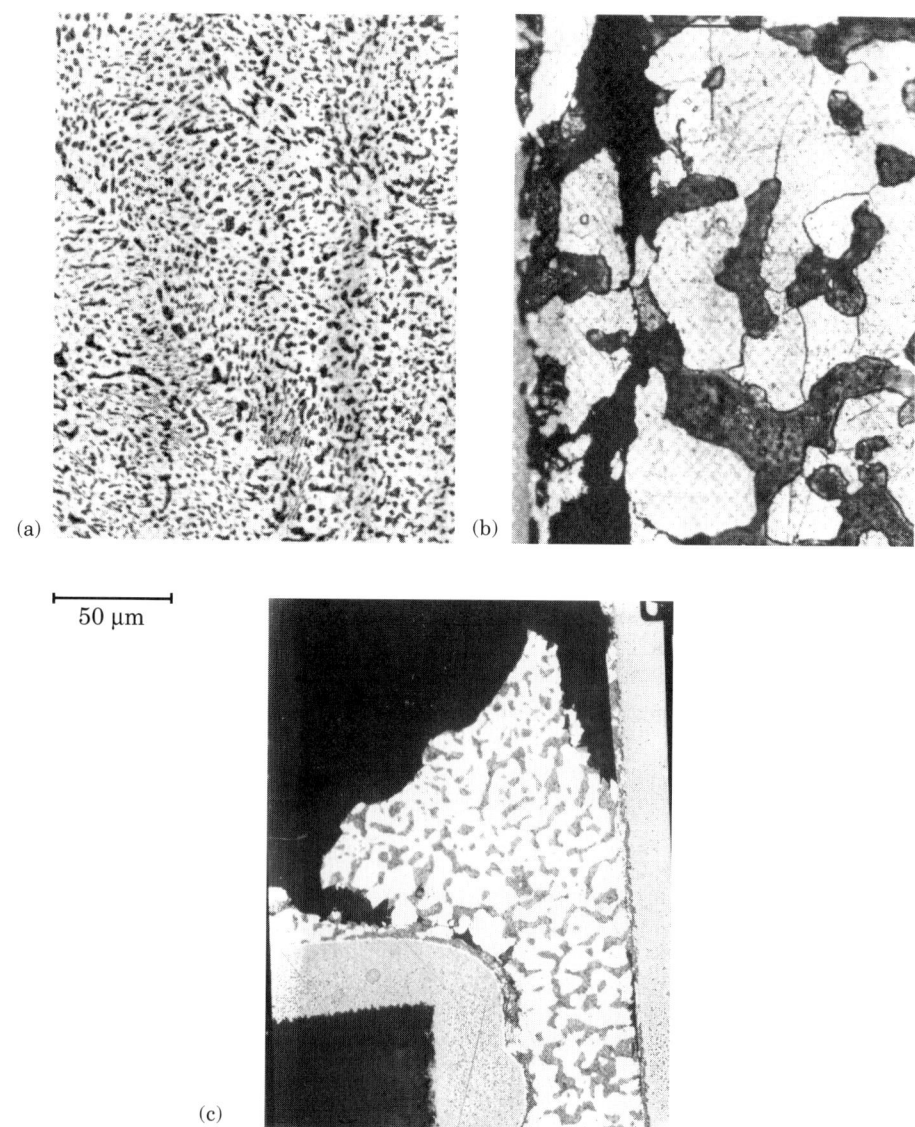

Fig. 13.19 Coarsening of the solder joint structure due to thermal cycling, with the development of a crack as a consequence.

The diffusion layer between the basis material and the solder increases in thickness at the expense of the tin. The lead is correspondingly enriched and concentrated in the solder joint, see Chapter 12, Figure 12.6.

C. Thermomechanical Stress

The first effect is due to the fact that a tin crystal has a different length in two axes, Figure 13.20, so that the expansion is different during heating. As different types of crystals with different coefficients of thermal expansion are lying in different directions in the joint, they work against each other under heating and cooling and, in time, destroy the structure of the joint without any external forces being involved, Figure 13.21.

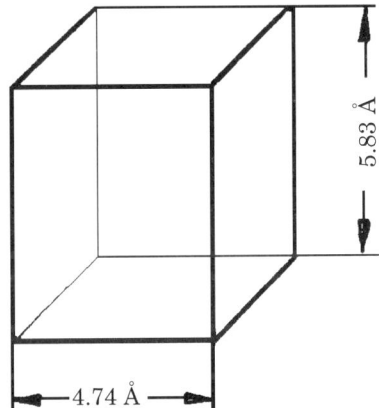

Fig. 13.20 The crystal structure of tin.

The second effect is the diffusion of, for example, impurities to the grain boundaries. Yet another effect is described later in 13.7.4.

13.7.3   The Different Types of Strength

It is rare that the tensile strength is of importance for a solder joint. Such a case arises in butt joints which are seldom used. Most solder joints are exposed to shear forces either in the creep or, more importantly, in the fatigue mode. A solder joint subjected to pure shear creep will rupture sooner or later depending on the load and temperature to which the joint is subjected.

The following illustrates the phenomenon of creep. Printed boards must be fastened in one way or another in a cabinet or an instrument. One of the common methods is to use holes in the board and screws for fastening. Sometimes the screw connection is used also to establish an electrical function, such as grounding. If the copper terminations on the board are plated with tin or solder, the land at the hole for the fastener is normally plated as well. When the board is fastened with a screw, the screw head is pressed on to the tin layer and a good, tight contact is established. After a while, with no other stress applied to the connection, the screw loosens and the electrical contact is jeopardised. The reason is simply that the solder has

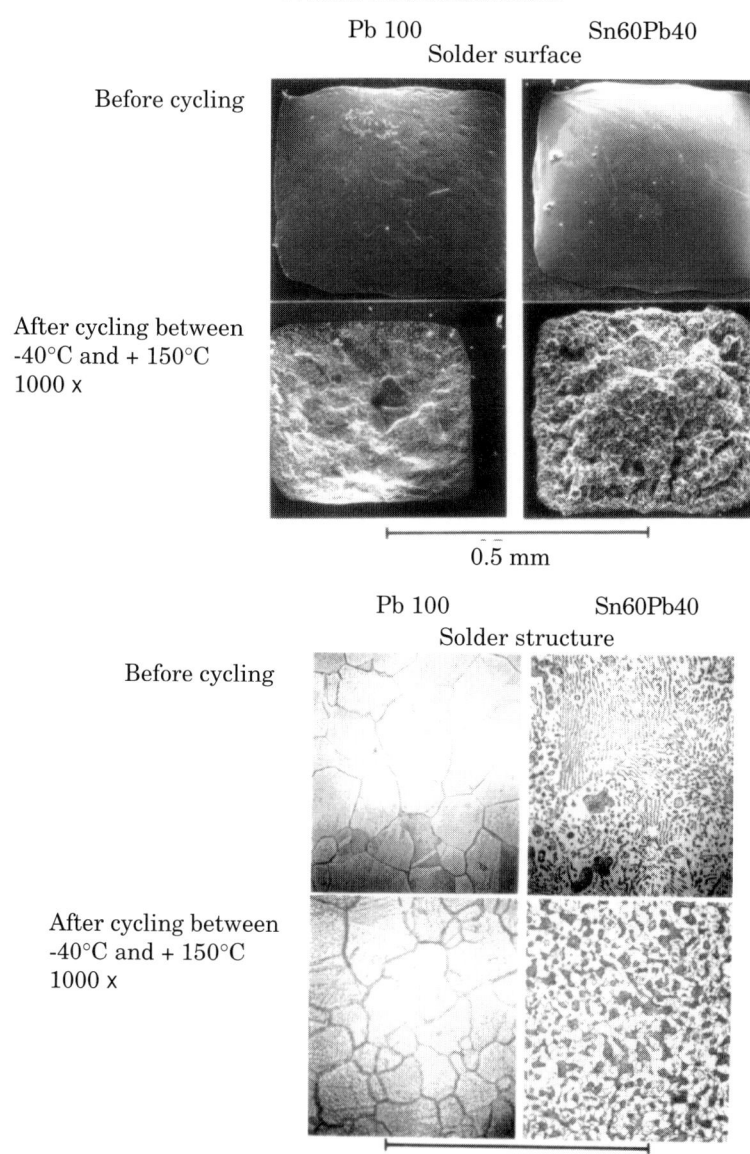

Fig. 13.21 Self-destruction of solder by thermal cycling. (Courtesy Dr Leibfried, Robert Bosch AG)

crept away — relaxed, Figure 13.22, from under the screw head as it could not withstand the pressure. This is the reason why a copper land must never be tinned if a safe mechanical and electrical contact is to be established between it and a screw.

How tensile strength data for solder can be used to calculate fatigue properties is not yet known. However, a relationship between the creep strength and the fatigue strength has been established.[27,28] Some creep

Materials Used in Soldering

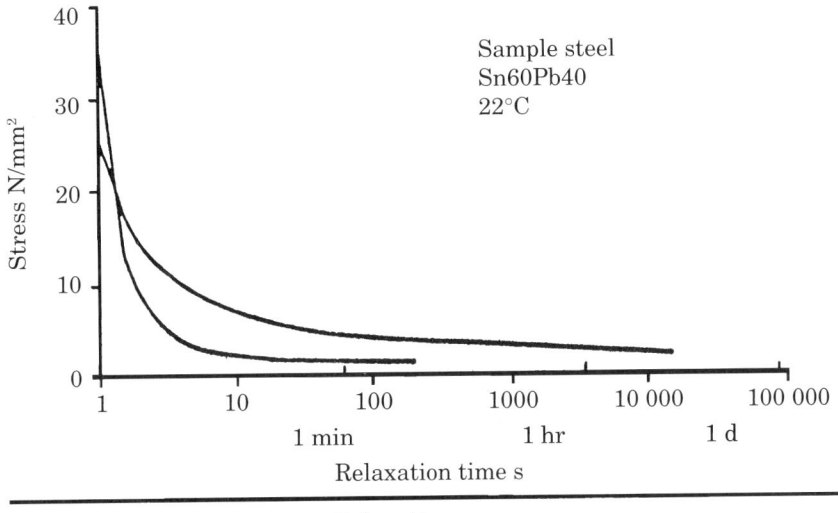

Fig. 13.22 Stress relaxation of solder in a ring plug sample.

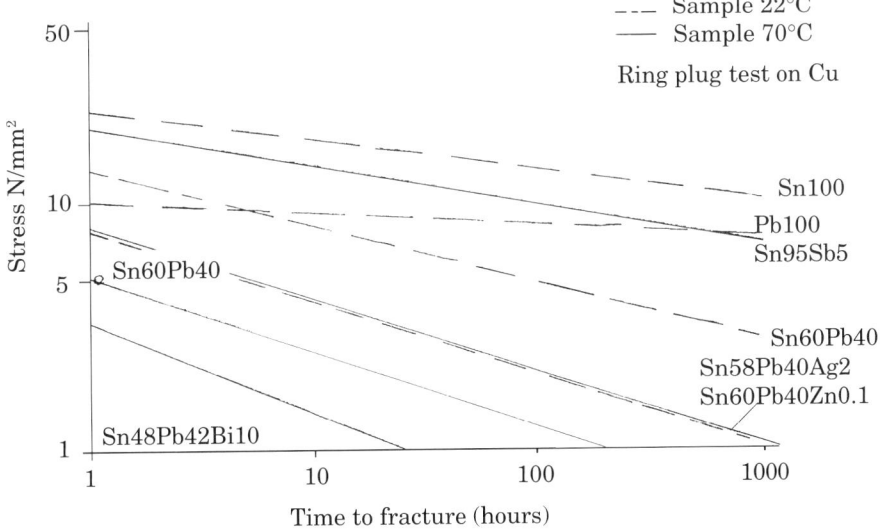

Fig. 13.23 Creep curves for different types of solder.

strength data are shown in Figure 13.23. From the data, it becomes obvious that the difference in their values is not large. Further creep strength data can be found in References 29, 30 and 31.

The order of strength value for the different types of stress to which a joint can be subjected is:[27, 29-32]

— *shear stress* for solder in a freshly made solder joint is 35-40 N/mm^2;
— *creep strength* after 10,000 hours (417 days) is 3.5 N/mm^2;

— *fatigue strength* after 10,000 cycles with two cycles/day (13.7 years) is 0.3 N/mm^2.

A SOLDER JOINT USED IN AN ELECTRONIC EQUIPMENT WORKING AT 70°C — which is quite normal — CAN BE COMPARED TO A BRIDGE WHICH HAS TO OPERATE AT 1,000°C.

Nobody would design and use such a bridge without taking all the necessary precautions. However, most solder joints are designed and used without taking such precautions. The primary reasons for this are very simple:

(a) It is not fully understood why a solder joint breaks;
(b) the necessary engineering data are not available, e.g., in the form of materials data such as creep, fatigue, structure and diffusion data, the impact of form and size, data on long-term effects and so on.

13.7.4 Fatigue

The fatigue of solder joints is not a new problem, as one might assume. The problem was already known in the 1920s to radio amateurs and repair shops, as may be illustrated by a radio valve with a connection to the anode on the top of the valve. The soldered connection at that point normally worked correctly for five to ten years after which electrical noise appeared and increased in volume. The experienced radio amateur removed the solder and re-soldered the joint. The author is pleased to be able to pass on this knowledge due to the interest in technical matters of his retired colleague, Hans Sköld.

The use of surface mount components has promoted a growth in interest of fatigue problems in solders. Heat is generated in a component, in the present example an SMD, when it is in use. The heat causes the component to expand, Figure 13.24. The heat dissipated through conduction flows through the solder joint to the board and heats the board. The board material has a different coefficient of expansion from that of the chip component. This thermal mismatch and the delay in the transport of the heat cause relative movement between the chip and the board. The movement occurs not only in the X- Y-direction but also in the Z-direction with very different values. This gives an irregular pattern which can be seen clearly in Figure 13.25. The pattern does not repeat from one heating to the next of the component and the board with the consequence that the same solder joint will be subjected to very different stresses during its life. This is why a purely mathematical treatment of the stresses in a solder joint can, at the most, give only a rough estimate. In Figure 13.25 it can be seen that the edges of the same component are subjected to very different stresses. The solder joint is the weakest link in this design and is stressed by the repeated heating. This leads eventually to a fatigue crack.

Solder joints in most working electronics equipments are subjected to a temperature of 50 to 100°C. The strength of solders decreases with increased temperature. This limits the choice of solder primarily to 60/40 tin-lead

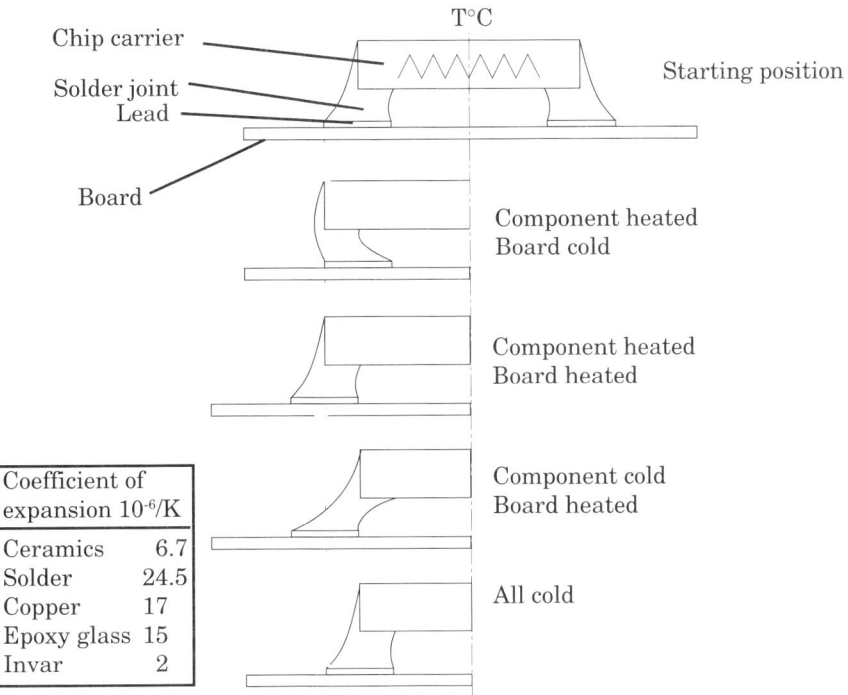

Fig. 13.24 Stressing of a solder joint caused by the heat generated in a component and the mismatch of the coefficient of expansion of the materials involved.

Fig. 13.25 The deformation of a board and the mounted components can be determined using laser interferometric methods. In this case the chip carriers are heated internally. The approximate distance from one dark line to the next is 0.3 μm.

solders. It should be noted that, although the melting point of eutectic tin-lead solder is 183°C, the working limit for the strength seems to be 120°C.

13.7.4.1 LIFETIME PREDICTION

A solder joint in general has a limited lifetime. Therefore it is advisable to analyse the working conditions of a solder joint to check whether or not there are conditions which could jeopardise its life in an equipment. Two examples of such analysis are given, the first being a telephone exchange and the second the electronics in the engine compartment of a motor vehicle.

Example One — Telephone Exchange

Figure 13.26 shows to what stresses a solder joint in a telephone exchange could be subjected. The greatest stresses actually occur during production, storage and transport although the number of stress cycles is low. The high number of cycles during operation at low stresses is remarkable. From this figure, it cannot be determined whether or not the number of cycles and the

		ΔT	Cycles
1	During Production		10
	Cooling after soldering		
	From 183°C to 20°C	163	
	Repeated at repairs		
2	Storage and Transport		
	Thermal cycling between -40°C and +80°C	120	500
3	Station at work		
	Day and night cycling		
	Full power during daytime. 80°C at PWB		
	Switch off during night 20°C	60	15,000
4	Operating, Telephone call		
	On/off cycling of one component. All others at average temperature: component 100°C, PWB 60°C	40	350,000
	One cycle/hour		

Fig. 13.26 Cases of thermal cycling used in stress analysis for solder joints in a telephone exchange.

stress are sufficiently low to guarantee the desired lifetime of 40 years. If the values are plotted in a Coffin-Manson diagram using the values of Wild[32] or Hagge[33] for the slope of the curve, it becomes obvious that there is a risk that the solder joints will crack before the exchange has reached the end of its life, Figure 13.27. A more exact calculation can be performed to confirm this by using Miner's rule, Figure 13.28, which states that at failure the sum of the ratios of the actual number of cycles at each strain to the number of cycles to fracture at that strain is one. It is possible to draw the following conclusions about what can be done to increase joint life:

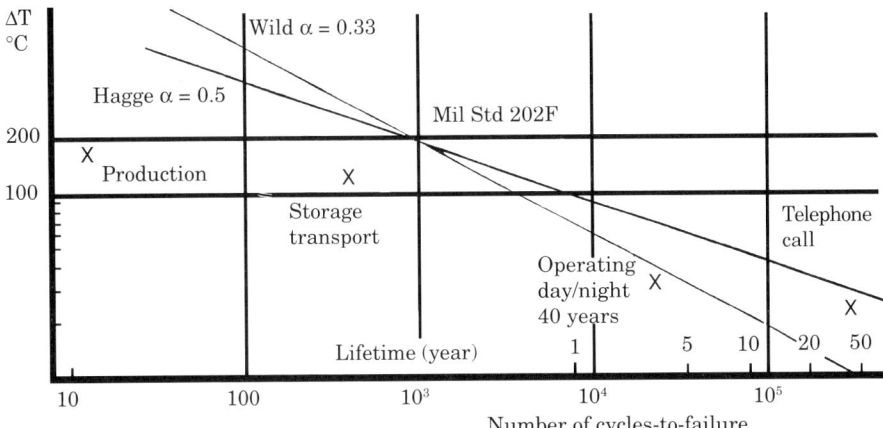

Fig. 13.27 Coffin-Manson diagram for stress analysis of a solder joint in a telephone exchange.

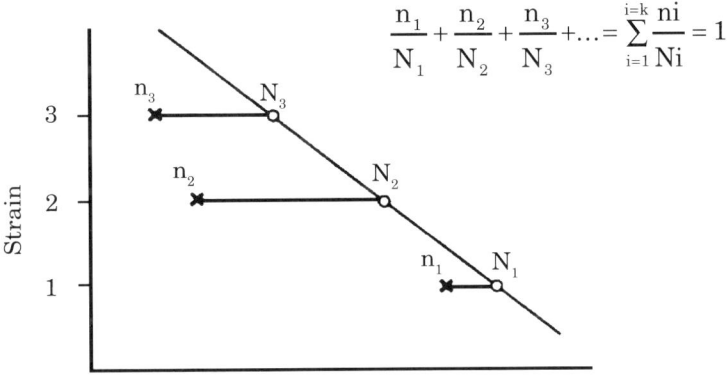

$$\frac{n_1}{N_1} + \frac{n_2}{N_2} + \frac{n_3}{N_3} + \ldots = \sum_{i=1}^{i=k} \frac{n_i}{N_i} = 1$$

Number of cycles to fracture N_f
n_1 number of used cycles at strain σ_1
N_1 number of cycles to fracture at strain σ_1

Fig. 13.28 Miner's Rule.

1 Improve the method/control of production to reduce the number of repair cycles.
2 Improve the storage conditions regarding both the temperature and the time to start the operation of the exchange.
3 Do not switch off the exchange during the night, to avoid any temperature swing.

Example Two — Automobile Electronics in the Engine Compartment

Assume the following conditions. The expected life of a motor car shall be:

15 years *or*
300,000 kilometres' driving *or*
5,000 hours of operation.

The stresses on the electronics in the engine compartment are very severe and of very different types:

1   Fatigue

   (a) High frequency fatigue $f > 10^3$ Hz.
   (b) Normal fatigue.
   (c) Low cycle fatigue (LCF) $f < 10^{-3}$ Hz.

2   Vibration
3   Shock
4   Temperature -40°C to +120°C
5   Corrosion

If each of these different stresses is considered, it becomes obvious that, with great probability, fatigue will be the most serious problem. An analysis of the working conditions of the electronics under the hood with regard to thermal cycling gives the values in Table 13.8, assuming two cycles each day, one for short-distance and one for long-distance driving.

Table 13.8

Analysis of Working Conditions under an Automobile Bonnet

Time of the Year	Number of Days	Distance Travelled	Temperature (°C)		
			Minimum	Maximum	Difference
Winter	90	Short	-40	+60	100
		Long	-40	+120	160
Spring	94	Short	+15	+60	45
		Long	+15	+120	105
Summer	90	Short	+25	+60	35
		Long	+25	+120	95
Autumn	91	Short	+15	+60	45
		Long	+15	+120	105

The temperature difference to which the solder joint is subjected is obtained from the assumed or measured minimum temperature when starting the engine and the maximum temperature which the engine will have reached at the end of the journey. A new table is established, Table 13.9, where the data for the temperature differences are arranged in sequence from the highest to the lowest value.

Column three gives the number of cycles when driving for one year. The fourth column gives the number of cycles expected during the engine's life of

15 years. Normally, to test the solder joints for 15 years under realistic conditions is impractical and it is of interest to determine how many cycles

Table 13.9

Comparison of the Number of Cycles

Season / Distance	ΔT °C	Number of Cycles		
		Driving for 1 Year	Driving for 15 Years	Cycles which must be Obtained in Test
Winter/Long	160	90	1350	3916
Spring/Autumn/Long	105	185	2775	3467
Winter/Short	100	90	1350	1530
Summer/Long	95	90	1350	1380
Spring/Autumn/Short	45	180	2700	636
Summer/Short	35	90	1350	187
Total			10,875	11,116

must be run in a test to give the same effect as the calculated number of cycles for the life in the actual case. For this purpose, the Coffin-Manson equation can be re-formulated as:

$$N_A \times \Delta T_A^2 = N_T \times \Delta T_T^2$$

where N is the number of cycles to failure
ΔT is the temperature difference, in the test case 100°C
index A stands for actual
index T stands for test

The value for N_T is calculated and given in the last column of Table 13.9. Having these values, they can be compared with the values in Table 13.10 which gives the number of cycles which solder joints of different qualities can endure.

It can easily be seen that a solder joint in an automobile engine compartment will not survive even a tenth of the required life. The number of cycles both in the actual case and under test conditions exceeds by far the number of cycles to which a solder joint can be subjected without a failure. In this

Table 13.10

Low Cycle Fatigue Quality Assessment

Number of Cycles-to-Failure	Assessed Quality
<400	Bad
400-600	Reasonable
>600-1000	Good

example the actual life cannot be simulated by a test which takes a shorter time. The consequence of this investigation is that solder joints are unsuitable for use in an automobile engine compartment and either the electronics must be removed from the engine compartment or the components must be connected by other methods or the whole design for the equipment must be altered.

Note also that other factors such as diffusion, spread in the quality of the solder joint as well as variations in manufacture, the statistical spread, vibration, shock and corrosion have not been taken into account.

13.7.4.2 TESTING OF FATIGUE PROPERTIES

Quite a number of electronic equipments have to work safely for a very long time, from 10 years up to 40 years being common. It is not possible to have test runs of this length and the time for fatigue testing a solder joint must be shortened considerably. For this purpose, test methods have been worked out by increasing the temperature range during test as much as possible based on the fact that most solder joints are in practice stressed within only a limited temperature range. The Coffin-Manson relation is used which states that the number of cycles to fracture is inversely proportional to strain, Figure 13.27. Miner's rule is used to convert the figures measured into a value for the expected lifetime, in this case the number of cycles to failure, Figure 13.28. There are several problems involved:

(a) With one test at a defined strain, a number of cycles to fracture is obtained. How much the solder joint will change with time is not known but it is known there will be a change due to diffusion and increasing grain size. Therefore, it is questionable whether or not the extrapolation to the desired lifetime gives the correct value if a linear extrapolation is used, Figure 13.29.
(b) Very little is known about the spread of the lifetime at the point of the stipulated life, Figure 13.29.
(c) Different types of cycle give different results, Figure 13.30[34]. They may even be found in the same equipment. In Figure 13.30, the fatigue curve is divided into two parts. The left part, with long cycling times, indicates that both creep and relaxation occur, which gives a low number of cycles to fracture. The right part, with short cycling times, shows only a small variation in the lifetime but the number of cycles to failure is large. This part of the diagram indicates that it is possible to predict the lifetime of a solder joint subjected to short cycle times with reasonable precision by cycling at a faster rate. The method used today, MIL-STD-202F, method 107D, with a cycle between -55°C and +125°C with half an hour at each temperature, lies precisely in the critical area where the transition from slow cycling to fast cycling occurs, giving very different results. Thus, depending on the aim, two different types of cycle should be used for predicting the lifetime of a joint — a very fast one and a very slow one.

13.7.4.3 MEASURES TO INCREASE THE LIFETIME OF A SOLDER JOINT

A solder joint has only a given lifetime but there are certain measures which can be taken to increase this lifetime. When designing with respect to

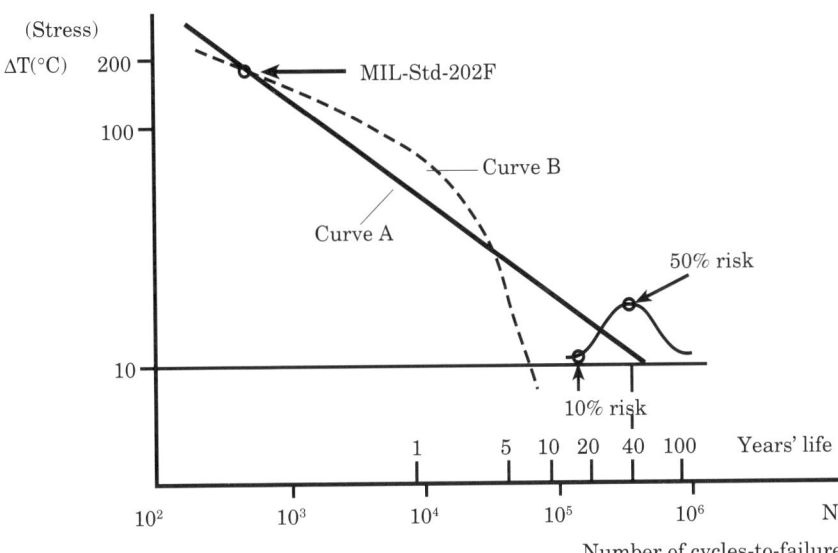

Fig. 13.29 Effect of change in solder properties on lifetime of solder joints. Curve A: Extrapolation from measured lifetime according to MIL-STD-202F. Curve B: Effect of possible changes of solder properties due to diffusion and increase of grain size.

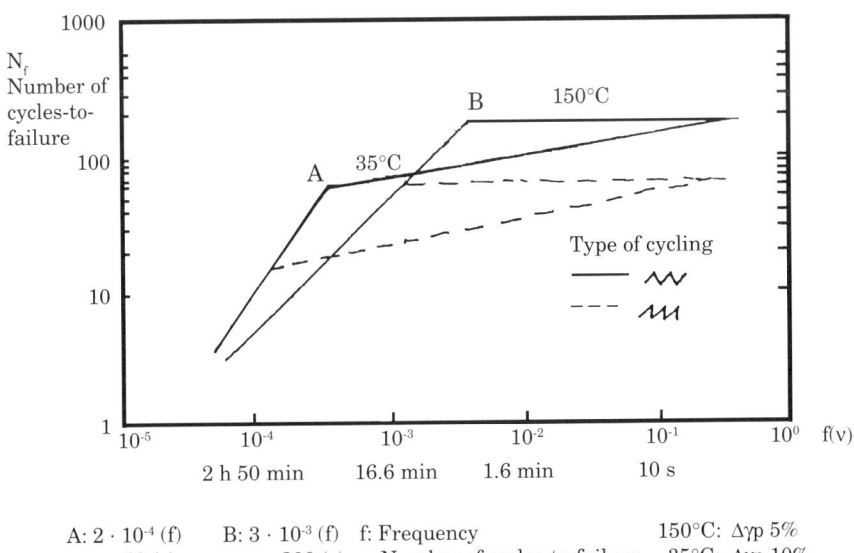

A: $2 \cdot 10^{-4}$ (f)   B: $3 \cdot 10^{-3}$ (f)   f: Frequency   150°C: $\Delta\gamma p$ 5%
   60 (c)   200 (c)   c: Number of cycles-to-failure   35°C: $\Delta\gamma p$ 10%

Fig. 13.30 The influence of stress type on fatigue life, Ref. 34.

the fatigue of solder joints, a number of things can be done to produce improvements. The first is to carry out an analysis of the stresses to find out if the current design is sufficiently good or not. If it is not, other methods of joining must be considered, such as welding. To this end, the designer must try to establish the following facts:

1. The highest working temperature of the solder joint. If it is higher than 120°C, there is a definite risk of the joint having a short life.
2. Fast or slow temperature cycling — slow cycling gives a short life, Figure 13.18 and Figure 13.30.
3. The number of cycles expected — if only a few cycles are expected, the stresses can be high and long cycle times can be allowed. If a large number of cycles are required, the stresses in the solder joint must be minimised.

The distribution of the heat-generating components on the board is important, in order to achieve as low thermal stresses as possible on every solder joint. For this purpose, special computer programs have been written.

If the design is *just* adequate or at the limit, a number of measures can be taken to extend the life to a safer value, always bearing in mind that the improvements are only marginal:

(a) Considering the coefficient of thermal expansion, another type of printed board can be chosen such as one made using a restraining core material like copper-clad Invar sheet, IEC 1249-7-1. In such boards, the coefficient of expansion in the X and Y directions can be adjusted to that of the most important components.
(b) Consider the choice of material for the terminations of the components. Alloy 42 has a very low coefficient of thermal expansion compared with both solder and copper. A joint on Alloy 42 is more likely to crack than one on a copper-based material, Figure 13.31.
(c) Study the types of intermetallics formed during soldering. Keep the thickness of intermetallic layers as small as possible by using short soldering times and low soldering temperatures.
(d) Avoid gold in the solder joint. This can be done by correct design, de-golding or proper processing.
(e) Use short soldering times, a low soldering temperature and fast solidification in order to ensure a small grain size in the solder joint. This will gain a number of cycles. The choice of soldering method is important.
(f) Consider the use of high melting point solders. This is a real possibility as such solders can be used on normal board material by employing laser soldering without degrading either the board material or the component.
(g) Investigate the possibility of having a solder joint working temperature of 20°C or less.
(h) Aim to have a constant higher solder joint temperature with small temperature fluctuations (±5°C) rather than a low general temperature with large temperature swings (±20°C or more).
(i) Design in such a way for surface mount components that high solder

joints are obtained (higher than 50 µm — the higher, the better).
(j) Keep close tolerances on the component terminations to avoid the need for excess solder to 'bridge' the gap between the board conductor and the component termination. The height of the solder joints all around the component must be the same with a close tolerance if the life is not to be shortened.
(k) Try to arrange stress reliefs by, e.g., using suitably formed terminations or chip carriers.
(l) Ensure that the solderability of the parts to be soldered is excellent. When speaking of solderability, one normally thinks only of its importance for the soldering process. Only a few are aware that good solderability is necessary for correct electrical function of the solder joint and that bad solderability is a contributory factor to a short life for a solder joint.[35]

Fig. 13.31 Crack in a solder joint caused by thermal cycling and materials with different coefficients of thermal expansion. The terminals are of Alloy 42 and are soldered using Sn60Pb40 solder.

13.7.5 Tin Pest

In all the years that the author has worked with soldering in Scandinavia, not one single case of tin pest on solder, solder joints or electroplated tin has come to his attention. Only one case of tin pest on pure (99.75%) tin bars is known in Scandinavia. It was reported by Mr Egon Ekberg from the Boliden Bersjö Company, a Scandinavian solder manufacturer. In the sixties he was invited to inspect a strategic reserve of tin in a mountain store in the north of Sweden. The origin of the tin was unknown and it was observed that the room where the tin was stored was quite humid. The bars, while showing attack, were not completely disintegrated.

Tin pest, often called tin disease, is the transformation of ß-phase or white tin into α-phase or grey tin. While the transformation temperature is 13°C, the transformation into grey tin usually occurs at temperatures well below 0°C. Grey tin is a powder and the transformation can only be reversed by melting. Transformation to the a-phase can be suppressed by addition of 0.1% of antimony[20] or 0.5% of bismuth.

13.7.6 Whiskers

In the 1950s, a new phenomenon became known to, in particular, the telecommunications industry — tin whiskers. The problems caused by these single-crystal metallic growths increased during the 1960s and 1970s with the increase in the use of electroplated tin as a finish. Whiskers are thin needles, almost faultless single crystals, Figure 13.5, which grow from the surface layer of the tin. They can be so long that they cause short circuits between the leads. Even now very little is known about them. However, electroplaters have learned how the electrolytes must be controlled in order to avoid them or to ensure that, if they develop, they do so in such a size and form that they cannot cause any harm. As whiskers have not been observed on hot-tinned surfaces, one rule for avoiding them is to melt ('fuse') the electrolytically tinned surface. Another rule is to avoid plating directly on to copper and brass, by using a diffusion barrier of nickel, making certain that a sufficiently thick nickel layer is used, Figure 13.5.

REFERENCES

1. Hexacon Electrical Company, brochure on SoldeRoll.
2. Beester, M. H., 'Metallurgical Aspects of Soldering Gold and Gold Plating', Proceedings International Electronic Packaging and Production Conference, Internepcon, Brighton, UK, October (1968).
3. Ross, W. M. and Lee, E. A. R., 'The Embrittlement Problem of Gold in Soldered Joints', Proceedings International Electronic Packaging and Production Conference, Internepcon, Brighton, UK, October (1968).
4. Foster, F. G., 'Embrittlement of Solder by Gold from Plated Surfaces', Bell Telephone Systems Monograph 4552.
5. Harmsen, U. and Meyer, C.-L., 'Über Weichlötungen an Gold', *Zeitschrift für Metallkunde*, **Vol. 56**, No. 4, pp. 234-239 (1965).
6. Alpha Metals, Technical Bulletin No. 6a - 56.
7. Whitfield, I. and Cubbin, A. J., 'Experimental Observations on the Effect of Gold and Palladium on Soldered Joints', *ATE Journal*, January (1965).
8. Bader, W. G., 'Dissolution and Formation of Intermetallics in the Soldering Process', in 'Physical Metallurgy of Metal Joining', edited by Kossowsky and Glicksman, AIME Symposium, St Louis (1980).
9. Creydt, M., 'Diffusion in galvanisch aufgebrachten Schichten und Weichloten bei Temperaturen zwischen 23° und 212°C', Thesis, ETH Zürich (1971).
10. Dunn, B. D. and Chandler, C., 'The Corrosion Effect of Solders, Fluxes and Handling of Some Electronic Material', *Welding Journal*, October (1980).
11. Becker, G., 'Impurities in Solder — Their Impact on Solderability and Corrosion', *Soldering & Surface Mount Technology*, No. 7, pp. 24-31, February (1991).
12. Reichenecker, W. J., 'Effect of Long Term Temperature Ageing on the Electrical Resistance of Soldered Copper Joints', *Welding Journal, Welding Research Supplement*, pp. 290-294, October (1983).

13 Fasching, G. M., 'Weichlöten, Lote und Flussmittel in der Elektrotechnik', *Feinwerktechnik*, **Vol. 73**, No. 12, pp. 519-528 (1969).
14 Müller, 'Lote', Deutscher Verlag für Schweisstechnik, Düsseldorf (1990).
15 Romig Jr, A. D., Yost, F. G. and Hlava, P. F., 'Intermetallic Layer Growth in Cu/SnIn Solder Joints', in 'Microbeam Analysis', San Francisco Press, Inc., San Francisco, pp. 87-92 (1984).
16 Kossowsky, R., Pearson, R. C. and Christovich, L. C., 'Corrosion of Indium Base Solders', Proceedings Annual Reliability Physics Symposium, **Vol. 16**, pp. 200-206, (1978).
17 Steen, H., 'The Effect of Impurity Elements on the Soldering Properties of Eutectic and Near-eutectic Tin-lead Solder', Swedish Institute for Metals Research, Report No. IM-2031 (1985).
18 Becker, G., 'Impurities in Solder: How They Influence Solderability and Strength of PCB Plated-through Holes', *Insulation Circuits*, March (1982).
19 Steen, H., 'The Effect of Bismuth Additions on the Properties of Tin-lead Solder', Swedish Institute for Metals Research, Report No. 2032, May (1985).
20 Thwaites, C. J., 'Antimony in Soft Solders — A Review of the Effects and Its Use in the Soldering Industry', *Brazing & Soldering*, No. 11, pp. 22-26, Autumn (1986).
21 Steen, H., 'The Oxidation of Molten Tin-lead Solder', Swedish Institute for Metals Research, Report No. IM-1610, October (1981).
22 Becker, G., Biverstedt, A. and Tolvgård A., 'The Surfactant Flux — A New Flux for the Ozone Age', *Hybrid Circuits*, No. 16, pp. 66-68, May (1988).
23 Manko, H., *Welding Journal*, No. 1 (1967).
24 Brox, B. and Livh, C., 'Ytskyddets inverkan på lödfogars tålighet mot temperaturväxling' ('The Influence of Surface Protection on the Durability to Temperature Cycling'), The Swedish Institute for Production Engineering Research, Report 1988-12-23.
25 Leibfried, W., 'Löten ohne Flussmittel' ('Soldering without Fluxes'), in 'Widerstandsschweissen und Mikrofügeverfahren', Fachbuchreihe Schweisstechnik, Deutscher Verband für Schweisstechnik, **Vol. 62**, pp. 141-150.
26 Hwang, J. S., 'Solder Paste in Electronics Packaging', Van Nostrand Reinhold (1989).
27 Koch, H. and Wasserbäch, W., 'Untersuchungen an Weichlötverbindungen als Beitrag zur Entwicklung geeigneter Prüfverfahren für die statische und dynamische Festigkeit', *Schweissen und Schneiden*, **Vol. 12**, No. 1, p.2 (1960).
28 Lundberg, L. and Sandström, R., 'Application of Low Cycle Fatigue Data on Thermal Fatigue Cracking', *Scandinavian Journal of Metallurgy*, No. 11, p. 85 (1982).
29 DeVore, J. A. and Westerman, J., 'Solder and Solder Joint Properties Handbook', General Electric Company, Syracuse, Technical Information Series R80ELS028, September (1980).
30 NN, 'Solder Alloy Data, Mechanical Properties of Solders and Solder Joints', International Tin Research Institute, Publication No. 656.
31 Becker, G., 'Testing and Results Related to the Mechanical Strength of Solder Joints', IPC TP 288, Proceedings IPC Fall Meeting, September (1979).
32 Wild, R. N., 'Fatigue Properties of Solder Joints', *Welding Journal*, November (1972).
33 Hagge, J. K., 'Predicting Fatigue Life of Leadless Chip Carriers using Manson-Coffin's Equations', Proceedings IEPS Conference (1982).
34 Solomon, H. D., 'The Influence of Hold Time and Fatigue Cycle Wave Shape on the Low Cycle Fatigue of 60/40 Solder', Proceedings 38th Electronic Components Conference, IEEE, pp. 7-13, May (1988).
35 Manko, H. H., 'Soldering Handbook for Printed Circuits and Surface Mounting', Van Nostrand Reinhold (1986).

Chapter 14

MANUAL SOLDERING

GERT BECKER
Consultant, Sköndal, Sweden

14.1 MANUAL SOLDERING

Before the invention of mass soldering processes, solder joints were made with a soldering iron. Today we are accustomed to electrically heated soldering irons but in earlier times it was quite usual to heat the soldering iron bit with a flame. It may have been thought that the development of mass soldering processes would render the soldering iron redundant but this was not the case.

On the contrary, the use of soldering irons is steadily increasing. The reason is simply that the expansion of electronics has meant an increase both in the number of solder joints which have to be repaired or touched up after mass soldering and in the number of components which have to be attached to the board by hand soldering for one or another reason, see Chapter 5. In addition, one must not forget small companies which manufacture electronics but cannot afford mass soldering machines and those whose hobby is electronics.

14.2 THE SOLDERING IRON

In modern electronics production, electrically heated irons are used exclusively. The reason is simple — it is much easier to control the temperature of the soldering iron bit using an electrical heating element than using a flame torch. The need to control temperature more closely as soldering tasks become more difficult has led to the replacement of 'uncontrolled' electrically heated irons with 'controlled' irons.

An uncontrolled iron is one on which a predetermined maximum temperature is given by the design of the iron. A controlled iron, on the other hand, is one on which different maximum temperatures can be set by different means.

14.2.1 How Does a Soldering Iron Work?

According to Figure 14.1, it takes a given time SW for the soldering iron tip to reach the maximum temperature TH. When this temperature has been reached, the soldering operation can start. Each solder joint consumes a certain amount of heat, corresponding to TW during SH. When the solder joint has been made, the iron is lifted from the joint and the temperature of the tip can recover in time period SM.

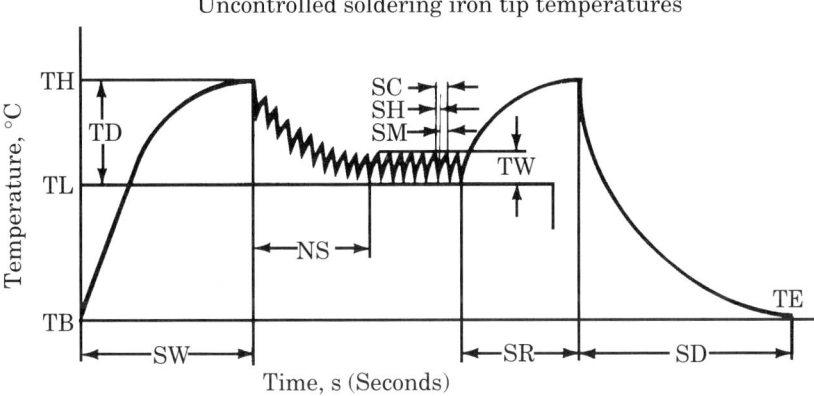

Fig. 14.1 This graph contains terms for describing the tip temperature changes, plotted against time, which occur during a sequence of similar soldering tasks. Typical of electronics assembly, the soldering starts and ends with a cold iron. (Reprinted from *Assembly Engineering*, October/November (1975), by permission of the publishers, Hitchcock Publishing Co., (©) 1975. All rights reserved.)

Once a number of joints, NS, have been made successively, the tip temperature stabilises at a lower temperature TL. When the work is finished and the electricity supply is switched off, it takes a certain time for the iron to cool to room temperature, SD.

The diagram for a controlled iron is similar. Such an iron has a steeper heating curve, Figure 14.2. The temperature regulator starts working when a preset temperature is reached and fluctuates with a certain amplitude TA and frequency SF. When soldering, the tip temperature drops as it does for uncontrolled irons, but stabilises at a lower temperature after only a few joints have been made. The temperature difference TD is smaller for a controlled iron.

These diagrams are important. They explain some of the problems an operator may experience with the iron. For example, a soldering iron needs a defined length of time to heat and should never be used until that heating time has elapsed. Another problem is that a soldering iron has to be chosen in any production having regard both to TH and to TL. If these factors are not taken into consideration, the joints may be overheated and the material

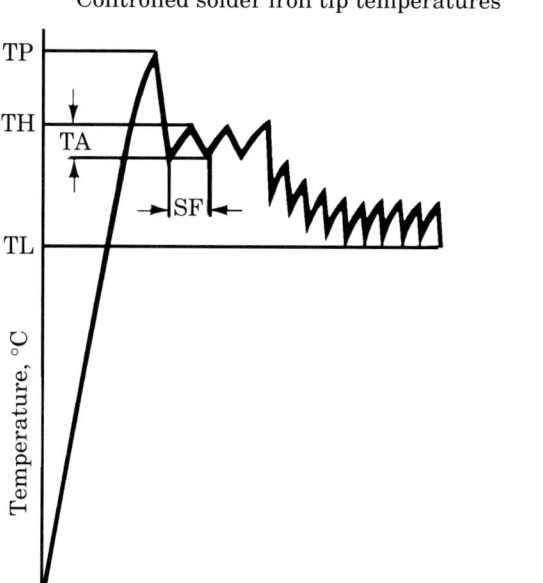

Fig. 14.2 This graph contains the additional terms required to describe the tip temperature changes of a controlled iron during a sequence of soldering tasks, starting with a cold iron. Some terms common to both graphs are not carried over from Fig. 14.1.

destroyed. Alternatively, unreliable joints may result because the soldering temperature was too low to achieve a proper metallurgical bond, Figure 14.3.

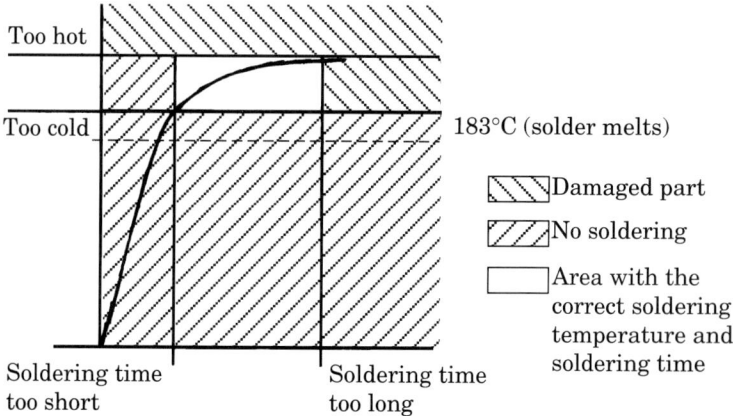

Fig. 14.3 The effect of temperature and time on soldering with 63:37 SnPb solder.

14.2.2 The Soldering Iron Tip

As the design and size of a soldering iron tip (or bit) affect the quality of the solder joint, it is worthwhile defining the different parts of the tip as shown in Figure 14.4.

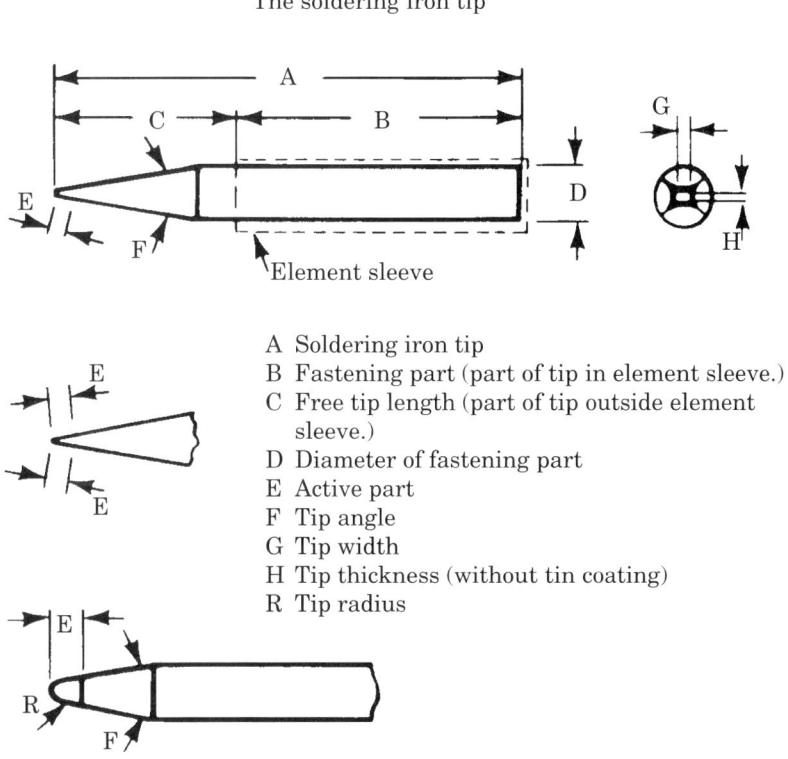

Fig. 14.4 In the interests of uniformity (a basic requirement of an international standard) a proposal was made for dimensioning and describing soldering iron tips as shown here.

In the past, soldering iron tips made from copper were common. These have been replaced by copper tips plated with iron. With the exception of the active part E, the iron is commonly coated with chromium or nickel to prevent the solder wetting the tip as well as to prevent oxidation of the fastening part B. The reason for this is that iron dissolves much more slowly in solder than does copper, Figure 14.5. The life of iron-plated tips is much longer than that of plain copper tips. The tip efficiency and working capacity is independent of the plating thickness and even an aged iron-plated tip works as efficiently as a new iron-plated tip. If the plating is of good quality, the end of the tip can withstand bending through 90° without the plated layer separating from the bend.

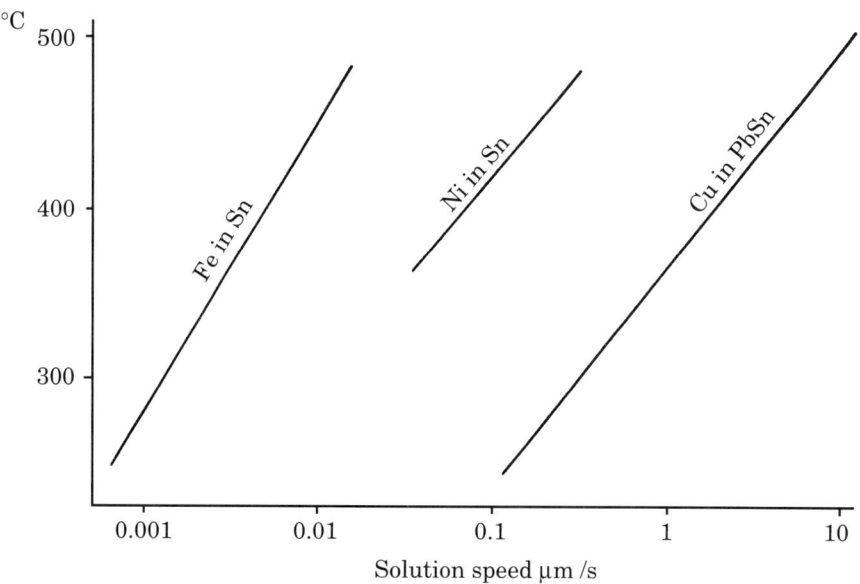

Fig. 14.5 Rate of solution of iron, nickel and copper in tin (and tin-lead).

Soldering iron tips must be carefully designed to suit their task. The form of the iron must not make it difficult for the operator to watch the soldering process. The tip must be able to reach the point at which the solder joint is to be made. In addition, the active portion of the tip must be of sufficient cross-section to guarantee the heat flow necessary to form the joint. The fact that there are many difficult and sometimes conflicting requirements means that there are many different types of soldering iron tip, Figure 14.6.

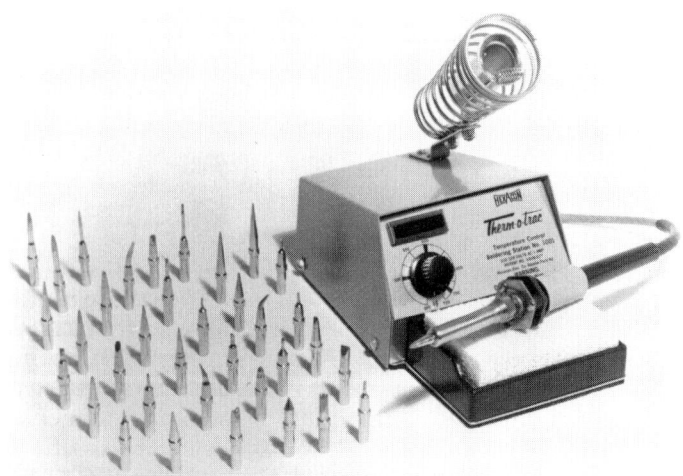

Fig. 14.6 Many different soldering iron tips are needed to meet the varying technical requirements arising in the production of solder joints. (Courtesy of Hexacon Electric Company)

14.2.3 The Wear of the Soldering Iron Tip

A soldering iron is normally constructed so that the tip may be changed. While different shapes of tip are available for most irons, a change of shape is not the main reason for the desirability of being able to change the tip. Tips are changed because they wear out. Depending on the type of work, a tip will last from a few hours to several weeks. Obviously, then, for companies using a soldering iron as a production tool, the cost lies in the tip and not in the soldering iron. The spread in the lifetime of a soldering iron used for one type of work only is surprisingly wide, ranging from several hundred to around 10,000 joints.

As mentioned, soldering iron tips used for production are made from copper plated with iron. This iron coating is brittle and the brittleness can cause cracks through which the molten solder can destroy the soldering iron, Figure 14.7. For this reason, the tip must always be handled with care and should be annealed after plating. An annealing coloration on a new tip should be taken as an indication of good quality. Quite often the iron is plated with nickel or chromium to avoid oxidation (and thus changes in the dimensions). The chromium also prevents solder from flowing on to parts of the tip where it is not wanted, so that the active part, Figure 14.4, cannot be chromium plated. It should be noted that it is bad practice to have a nickel layer only over the iron plating, as nickel dissolves more rapidly in molten solder than does iron, Figure 14.5.

During soldering, solder is added to the tip and will dissolve some of the iron coating. Fresh solder dissolves the iron more quickly than does solder which is loaded with iron. Solder should therefore be wiped off the iron only when absolutely necessary. This decreases the dissolution process and increases the life of the tip. It will help also if the soldering iron tip is maintained at the lowest temperature compatible with the work being carried out. The higher the temperature, the more iron will be dissolved by the solder in a given time, Figure 14.5, with a consequent reduction in tip life. Besides cases where a high tip temperature is essential, high temperatures may be used because of poor solderability of the material.

Sometimes the solder may not wet the active part of the tip. It is known that, both in flux and solder, sulphur in quantities as low as 1 ppm causes dewetting of tips.[1] Chromium, plated as a protection for the tip, can cause dewetting even although it has not been applied to the active part of the tip. It has been observed that tips which wet badly will wet properly if a solder containing 1-1.5% of copper is used.

14.2.4 Heat Flow from the Iron to the Workpiece

To produce a satisfactory joint, the soldering iron tip must be at the right temperature — but what is the right temperature? In the case of a printed circuit board, the general rule is that 260°C is the maximum temperature and 5 to 10 seconds the maximum soldering time if destruction of the adhesive bond between the copper foil and the laminate is to be avoided.

To achieve transport of heat, there must be a difference in temperature between the tip and the joint being made. This is why the temperature of the workpiece never reaches the temperature of the tip, even after prolonged

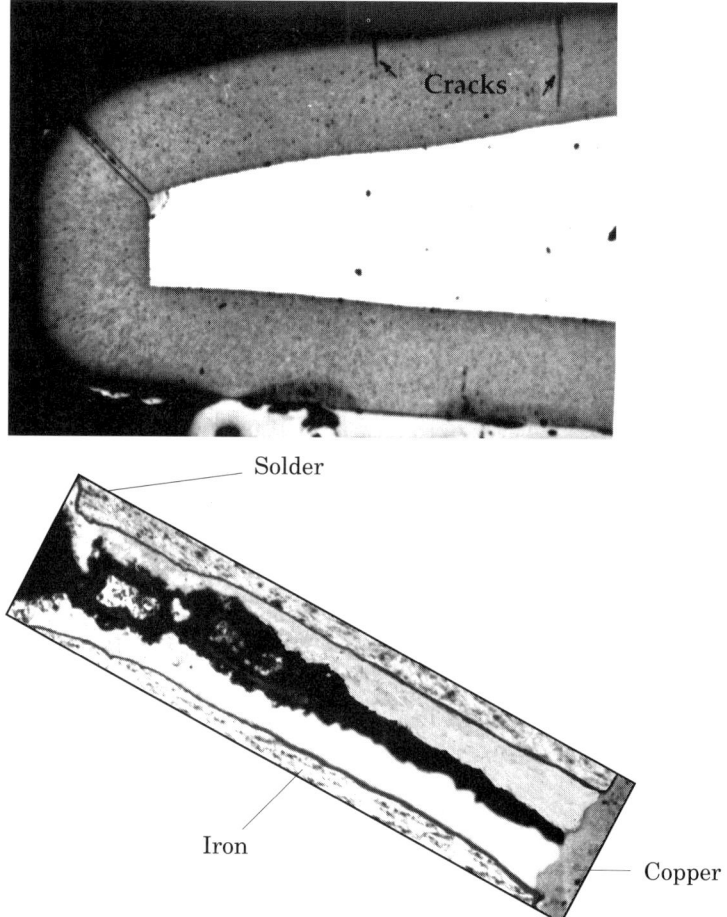

Fig. 14.7 Detail (x 500) from the edge of a soldering iron tip. The solder has widened this crack and started to dissolve the copper core of the tip.

contact. Heat losses always occur through radiation and conduction, and temperature differences of 5°C to 30°C have been noted for different types of soldering work. Oxides both on the workpiece and on the soldering iron tip can act as a heatsink and prevent the workpiece from being sufficiently heated. The tip should therefore be cleaned and tinned during the work as often as is necessary.[2]

The parts to be joined *must* be heated to a temperature at which they can be wetted by the solder. The flux and the solder must be heated to their melting points and the melting processes require extra heat input. In PCB soldering, the heat taken from the tip has been shown to reduce tip temperature by 40°C to 70°C under normal conditions, the tip temperature dropping approximately 20°C as the first solder joint is made. So, taking 260°C as the maximum allowable temperature for soldering to a PCB and adding the heat losses of the iron, solder and parts to be joined, the correct maximum tip temperature, T_{max} (or TH, see Figures 14.1 and 14.2), is 350°C.

However, there is also a lowest allowable tip temperature, T_{min} (TL). This is obtained by taking the melting point of the solder, which for eutectic SnPb is 183°C, and adding all the heat losses which have to be taken into consideration. Another factor to be considered when establishing TL is the solderability of the material. For easily soldered material, TL may be 210°C to 230°C. If, on the other hand, enamelled wire is to be soldered, a TL of at least 320°C must be reached.

Soldering iron tips have to work between clearly defined minimum and maximum temperatures if reliable and economical solder joints are to be produced, Figure 14.3, and soldering time t_{min} must not be too short, as no soldering will be achieved. Similarly, soldering time t_{max} must not be too long, as the materials involved can be damaged.

14.2.5 Measuring the Tip Temperature

There are a number of instruments on the market with which the temperature of the soldering iron tip can be measured. In principle, these are millivoltmeters but measuring with them gives limited reliability, and it is advisable to calibrate such instruments. With the standard instruments, it is not practicable to plot the temperature of the soldering iron tip during soldering operations.

This can, however, be accomplished using a thermocouple with a wire diameter of 0.3 mm. The wire is welded or brazed on to the tip at a distance of 3±1 mm from the end of the tip. By connecting the thermocouple to a calibrated millivoltmeter and a recorder, the temperature variations of the tip under actual production conditions in the workshop can be followed. Figure 14.8 shows all the instruments necessary for such measurements combined in one unit. The same instrument will also check the iron for electrostatic discharge (ESD) safety as well as for voltage leakage.

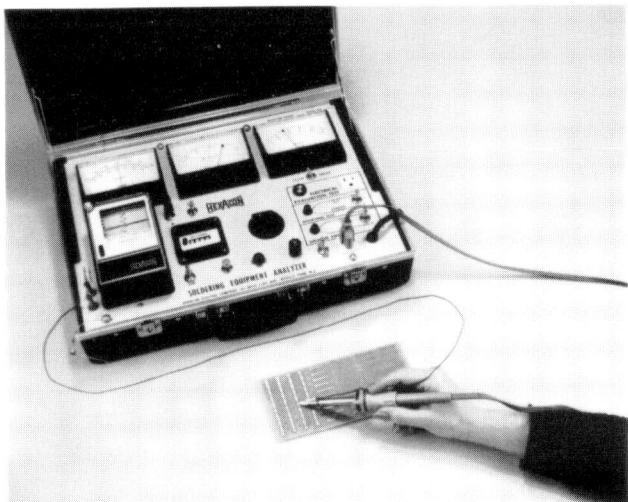

Fig. 14.8 An instrument to measure the tip temperature of a soldering iron under production conditions. (Courtesy of Hexacon Electric Company)

14.2.6 The Rating of the Soldering Iron

The process of soldering with a soldering iron can be adequately described in terms of temperature and time. However, to attain the required temperatures, an iron of the correct wattage is needed, as solder joints are of different sizes and require different amounts of heat for satisfactory working. The working capacity of a soldering iron is therefore indicated by the rated wattage. Unfortunately, this wattage, the input wattage, indicates only the amount of power the iron consumes and not how much heat the iron can transfer to the workpiece through the tip, Figure 14.9.

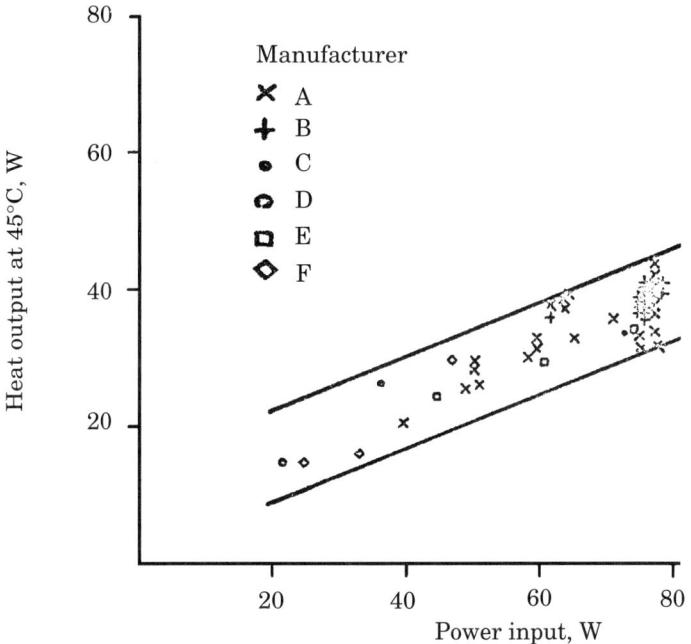

Fig. 14.9 Heat outputs *vs* power inputs of the soldering irons of six different manufacturers.

14.2.7 Tip Temperature and Heat Output from the Tip

If making a solder joint requires a given amount of energy, it may happen that one soldering iron will do the work perfectly while the next soldering iron from the same batch will not.[3] The reason is simply the spread in the behaviour of soldering irons, demonstrated in Figure 14.9. Another important point is that a slight change in the form or dimension of the soldering iron tip will affect the heat output of the iron and hence the quality of the solder joint, Figure 14.10.

In electronics soldering, both the soldering iron and the tip must be strictly defined. Moving the soldering iron tip in the heating element sleeve of the soldering iron or changing the free tip length must not be permitted. As Figures 14.11 and 14.12 show, if either of these occur, very different tip temperatures will be obtained.

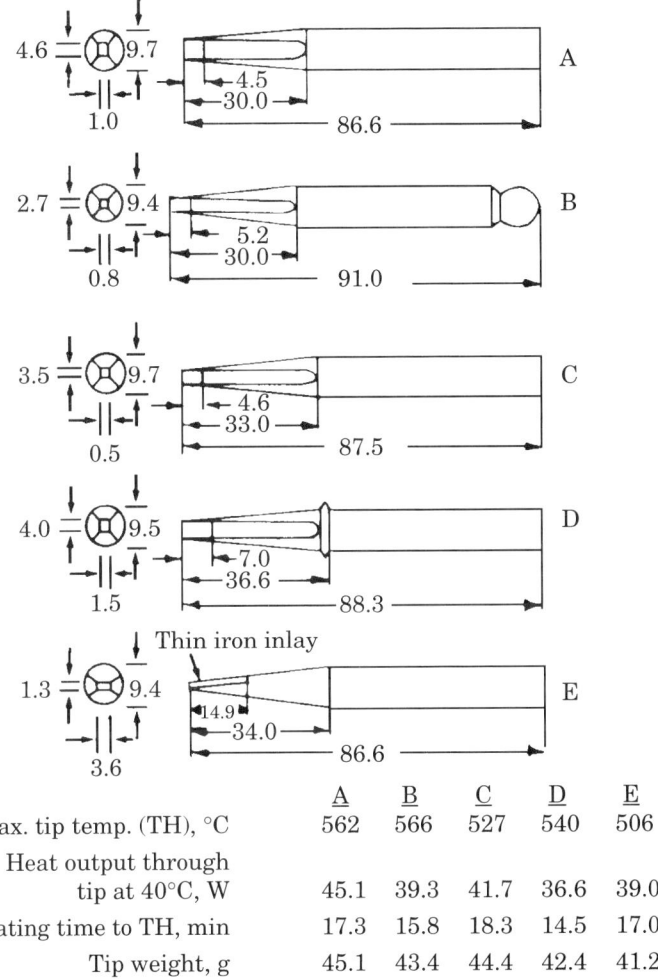

Fig. 14.10 Influence of tip design on the tip temperature and heat output of a 92 W (input) soldering iron.

14.2.8 Criteria for a Soldering Iron

The essential points on which a soldering iron may be judged are as follows:

— it should be light, with the weight of the element, handle and flex balanced to each other for ease of working;
— the handle should not become warmer than 40°C;
— it should fulfil specified time/temperature requirements;
— it must be approved to national and/or international standards for electrical safety.

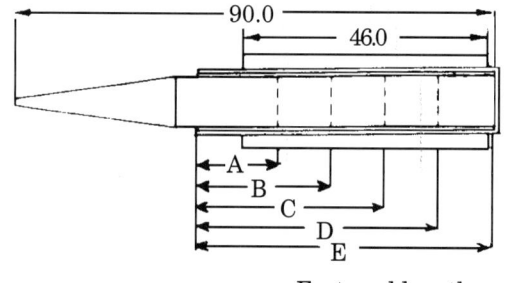

Fastened length, mm

	16.0 (A)	26.0 (B)	36.0 (C)	46.0 (D)	56.0 (E)
Max. tip temp. (TH), °C	145	502	517	530	560
Heating time to TH (SW), min	15.0	15.0	14.0	15.5	16.0
Heat output (tip) at 45° C, W	25.2	27.8	37.8	41.3	40.2

Fig. 14.11 Influence of fastened length on tip temperature and heat output of 70 W (input) soldering iron.

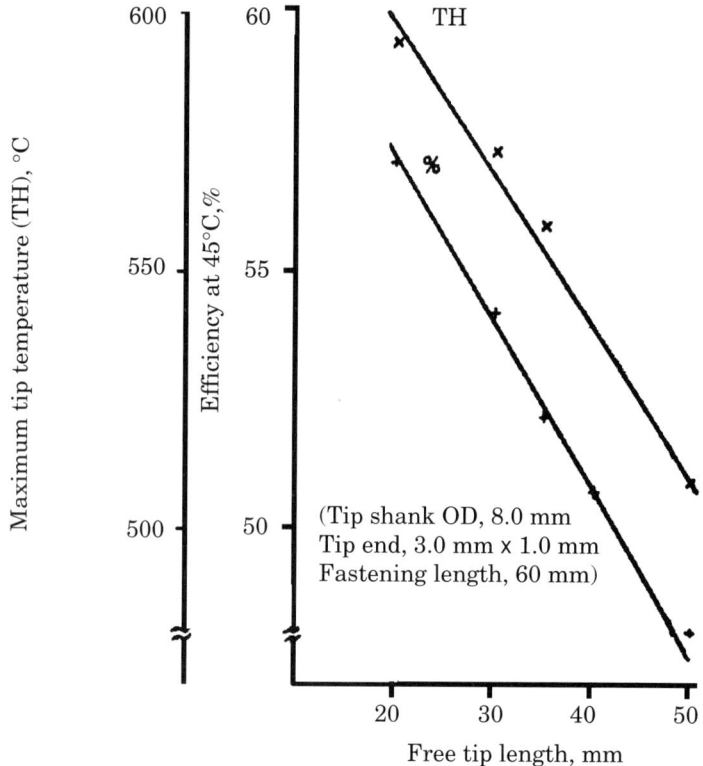

Fig. 14.12 Influence of free tip length on tip temperature and efficiency of 77 W (input) soldering iron.

14.2.9 Soldering Iron Support

The support for the soldering iron must be designed to prevent burns, fire and excessive cooling of the soldering iron tip. The active part of the tip must not be contaminated by the support and the support should be provided with a tip cleaner.

14.2.10 'Tin Surplus' Collector

The tin surplus collector, Figures 14.13 and 14.14, is used during repair desoldering to remove solder from the joint. In order to do this, the solder to be removed is heated until it melts. When molten, the solder is sucked away by the tin surplus collector.

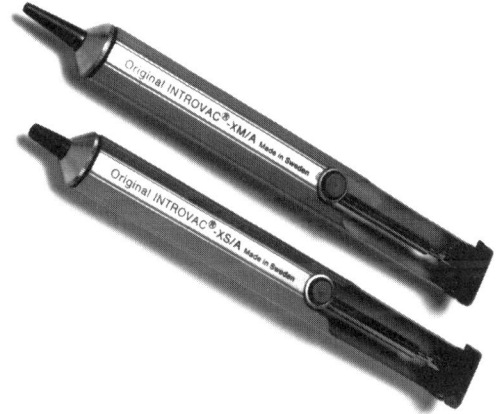

Fig. 14.13 'Tin surplus' collectors. (Courtesy Ahiko)

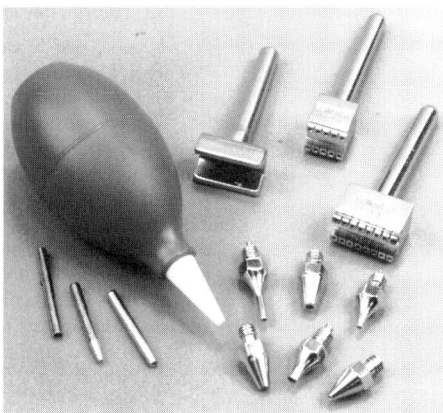

Fig. 14.14 Specially designed tips for repair of different types of component. In the picture a solder sucker, which can be used generally, can be seen. (Courtesy of Hexacon Electric Company)

The tin surplus collector should have an in-built mechanism to eliminate recoil and improve the quality of the work. Manual collectors must be designed in such a way that the knob is guarded and cannot accidentally hurt the eyes. The nozzle should be easily changed.

14.3 SOLDERING PRINTED BOARD ASSEMBLIES

14.3.1 Boards, Components and Solder Requirements

The solderability of boards, and of components and other parts to be attached to the boards, must meet any specified requirements, for example, of ANSI/J—STD-002, ANSI/J—STD-003 or the agreed IEC equivalents[4,5]. The solder wire diameter is chosen according to the size of the joint to be made. Both the solder wire used and the flux must meet relevant specification(s) for composition and purity.

Very thin copper wire, with a diameter of less than 0.08 mm, must always be soldered using a copper-loaded solder to avoid dissolution of the wire into the molten solder. In this case, the solder joint will become slightly matt in appearance.

14.3.2 Operator and Working Conditions

The soldering operator must have good eyesight and have completed a training course for the types of joint to be made, see Volume 2, Chapter 21. The operator must also be aware of the specified requirements for mounting and soldering components to the printed boards so that he/she is able to check the work done and ensure that it is of good quality.

The workplace should be well ventilated/extracted so that fumes emitted during soldering are removed and do not inconvenience the operator. When required, the workbench must be provided with antistatic mats and straps. Lighting is very important and should be a minimum of 600 lux at the workplace, without glare, dazzle and/or shadowing of the work.

Precautions must be taken to ensure the absence from the workplace of any materials which might contaminate or cause deterioration of the soldering work. For that reason, no handcreams, lotions or food and drink may be permitted in the area.

14.3.3 Soldering Conditions

The soldering iron must be earthed and, unless a controlled iron is to be used, the maximum temperature and wattage of the iron must be defined. The size and type of soldering tip must be specified, together with the free length of the tip (C in Figure 14.4).

A recommended range for the soldering time to be allowed should be given for operator guidance. When parts which require especially high temperatures are to be soldered (such as enamelled wires), or if parts such as multilayer boards need unusually high heat flow, irons and/or tips suitable for the task must be chosen. When heat-sensitive components are to be soldered, a heatsink must be attached to the lead between the soldering iron and the body of the component to minimise the risk of heat damage to the component.

14.3.4 Making the Joint

It is essential to ensure that the surfaces to be soldered are not touched by anyone's fingers. The active part of the soldering iron is placed on the part of the joint to be soldered which has the greatest thermal mass. There must be good contact between the active part of the iron, the copper conductor on the board and the component termination.

To improve the heat flow from the soldering iron to the joint, it is of advantage to melt a small amount of solder on to the active part of the iron touching the workpiece. The solder wire is then moved to the opposite side of the workpiece, touching both the land and the component termination, Figure 14.15. If conditions are correct, the solder will flow well and cover all parts, forming the joint.

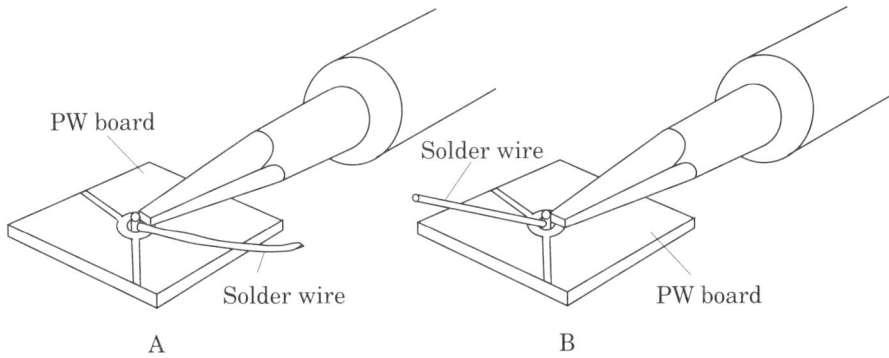

Fig. 14.15 The drawing shows in A how to start the soldering process and in B how to continue. To achieve better heat conduction, it is allowed in phase A to touch the tip with the solder wire. To obtain a correct solder joint, the solder wire must be moved as shown in B.

Once the joint has formed, lift the soldering iron away quickly but with care. The solder in the joint must solidify without any disturbance. For this reason it is not permissible to hold components or other parts by hand during the soldering operation.

14.4 REWORK AND REPAIR

A common sight in factories is that of personnel touching up solder joints after machine soldering. Finding soldering faults is difficult, and there is an inevitable tendency to rework a joint which may not need it rather than miss reworking a joint which does need it. It is not practicable for the operator carrying out the rework to analyse the fault and so choose the correct method of repair.

Can a badly designed solder joint really be improved by touching up? For instance, on a joint made to brass, zinc oxide may have formed. The only correct solution is to remove the oxide and solder again. Why should we expect a solder joint to become faultless on reheating with a soldering iron when the soldering machine could not produce a good joint in the first place? It is generally agreed that touching up is in fact more likely to degrade a joint than to improve it. This is particularly true if the joint was a good one to begin with. Therefore, if a solder joint fault has to be corrected, it is necessary to determine the cause of the problem and tackle that cause. Solder joint repair should be permitted only as a temporary expedient and not as a part of the production process.

Nothing is perfect. Rework and repair cannot be avoided, they can only be kept at a low level. The two items which normally undergo repair are conductors on the board and solder joints. Special care must be taken when repairing — there are methods to be developed and specifications to be followed.[6-10]

14.4.1 Repair of Conductors on the Board

Repairing a conductor on a board can be a quite simple matter. A copper wire of suitable diameter and length can be laid over the damaged track and soldered to it. But before doing this, consider whether or not the repair will adversely affect the electrical performance of the circuit.

In some cases it is necessary to glue long wires, as well as the ends of the damaged conductors, on to the board. When dealing with larger production volumes and higher demands on finishes, a metal ribbon or tape can be welded over the damaged portion of track. For the welding, a highly defined process must be worked out and, of course, the operators should be trained and certified. The welding machine and the material used must be of a high standard if the work is to be reliably performed. For an outline of the various types of welding which can be used, see Chapter 16.

Not all circuit configurations can be repaired by welding, but most can be repaired by soldering. It is advisable to define the types of damage to conductors which are permitted to be repaired, and most customer/supplier contracts and specifications will do this.

14.4.2 Repair of Solder Joints

If a solder joint has to be repaired, it is essential to determine the cause of the fault. If, for example, the basic cause is poor solderability, it is pointless merely to reheat the solder joint. The component must be removed, and the mating surfaces cleaned down to the basis metal and retinned. The component can then be put back into place and new solder joints made.

If a new component is to be used, it is easier to remove the old one by cutting the latter off its leads and desoldering and removing the leads one at a time. In other cases, for example when the solder joint is broken as a result of thermal fatigue (a phenomenon often seen on the joints of power components in elderly television sets), melting the solder in the joint restores the joint structure and the set will work happily for many more years.

The methods used to repair PCB assemblies are very similar to normal production techniques. The same attention has to be given, for example, to using an iron which is neither too hot (this could damage both component and board) nor too cold (this could cause 'cold' solder joints). If non-corrosive fluxes are not used, special cleaning may have to be provided.

When desoldering components, an antistatic solder sucker must be used with specially designed tips for different types of components, Figure 14.14. An alternative to a solder sucker is a fluxed copper wire braid which sucks or 'wicks' up the solder, Figure 14.16. The normal rules for component mounting must be followed, taking care while repairing one joint not to damage another joint or component. If no clean gloves are available, boards should be handled by the edge only, to avoid corrosion.

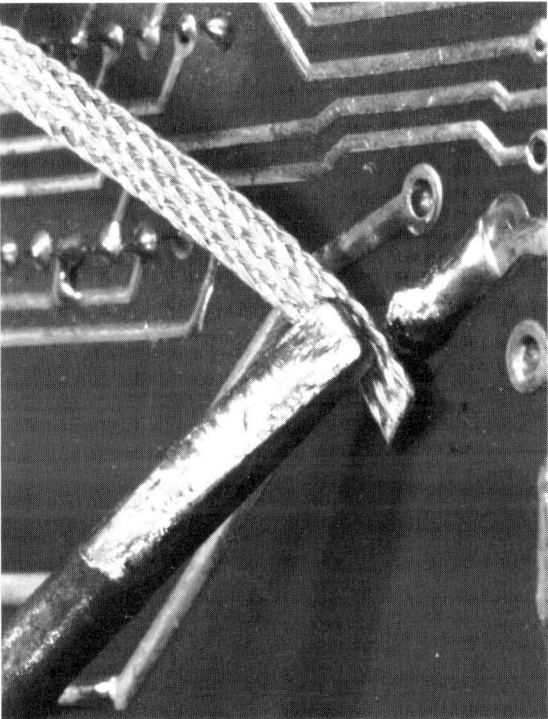

Fig. 14.16 This figure shows the start of the repair of a soldered joint. The soldering area is being cleaned of excess solder by sucking it up with Solder Wick® a proprietary fluxed copper braid, before resoldering the joint.

A number of different tools and workstations are used for the heating, removal, retinning and replacement of component(s). The basic tools are a soldering iron, hot gas soldering equipment (the most common tool for multi-leaded components) and infra-red heaters or heater blocks.

REFERENCES

1 NN, 'AWS Conference Features the Solder Connection', *Assembly Engineering*, pp. 36-39, June (1972).
2 Mehl, R. R., 'A Clean Tip, Well Tinned, Solders Best', *Assembly Engineering*, pp. 38-40, June (1973).
3 Becker, G., 'Specifying and Measuring the Working Capacities of Soldering Irons and Tips', *Assembly Engineering,* October and November (1975).

4 ANSI/J-STD-002, 'Solderability Tests for Component Leads, Terminations, Lugs, Terminals and Wires', April (1992).
5 ANSI/J-STD-003, 'Solderability Tests for Printed Boards', April (1992).
6 Verguld, M. M. and Leenaerts, M. H. W., 'Repair of Printed Circuit Boards Carrying Surface Mount Components', *Circuit World*, **Vol. 14,** No. 2, pp. 11-15, January (1988).
7 Bechtold, J., 'Printed Circuit Rework, Repair and Cleaning', Bench Briefs, Service Information from Hewlett Packard, July-October (1982).
8 Hunn, N., 'Getting Results from Hot Gas Rework', *Circuits Assembly*, **Vol. 1**, No. 2, pp. 21-26, November (1990).
9 Dow, S. and Helton, D., 'The Use of Collimated Infrared Light', *Circuits Assembly,* **Vol. 1**, No. 2, pp. 28-30, November (1990).
10 D'Andretti, L., 'Essentials of Fine Pitch Repair', *Circuits Assembly*, **Vol. 1**, No. 2, pp. 32-37, November (1990).

Chapter 15

MASS SOLDERING

GERT BECKER
Consultant, Sköndal, Sweden

15.1   GENERAL CONSIDERATIONS

Mass soldering in electronics was made possible with the invention of the wave soldering method by Barnes, Elliot and Strauss in 1956.[1] Other methods, such as drag soldering, soon followed. In the sixties, ICs in flat packages, see Chapter 3, 3.9.5, were mounted on the printed board in surface mounting fashion by soldering using a resistance welder where the electrodes acted as thermodes. (A plastic bodied version of the flatpack is still used in SMT, see Chapter 4, 4.4.9.) Soldering of large boards led to the invention of the condensation or vapour phase method. It was a considerable time before the laser was developed to the point where it could be used without problems for soldering printed board assemblies.

The soldering method used is selected according to the soldering task required. New tasks generate new soldering methods. Soldering of printed boards with components mounted in holes or plated-through holes is successfully achieved with wave soldering or drag soldering equipment. The advent of surface mount technology accelerated the use of vapour phase soldering and IR soldering. A need for high reliability soldering, added to a requirement for use of solders which have a higher melting point than the usual Sn60Pb40 solder, promoted laser soldering.

Mass soldering has developed gradually and has shown an extraordinary ability to adjust to different requirements and overcome great problems in production. The continuing evolution of the different processes, especially of the wave soldering process, gives rise to the suspicion that none of the existing processes has come to an end of its development and that further evolution can be expected. This is why the basics only of the different methods will be described. What is valid today may not be valid tomorrow. A good example is the development of vapour phase soldering (VPS) which came into existence because of an urgent production need, quickly became a recognised and widely used method and almost vanished totally as suddenly as it came into existence due to the environmental problems with CFCs (chlorofluorocarbons), which have been banned by the Montreal Protocol, see Volume 2, Chapter 17. Today VPS is experiencing a renaissance due to the development of less environmentally harmful heat transfer media.

15.1.1 Heat Supply

The heat can be transferred to the solder joint by:

— conduction;
— radiation;
— convection.

The three forms of heat transfer are listed in decreasing order of efficiency, conduction processes being the most effective and convection the least.

15.1.2 Temperature Profiles

In every soldering process the solder joint and the surrounding material are heated and, after soldering, cooled down. This gives a specific temperature profile which depends on the soldering process used and the materials involved. There are many different soldering processes in use, each one having its own principal temperature profile which can be altered slightly according to the need. It can be quite confusing to see the many different temperature profiles in the literature, every one proving either that the given profile is best or that improvements are required.

A soldering machine has a given length. This more or less establishes the time for the total soldering process. The actual soldering time (the time at soldering temperature) is determined by the soldering method. In the wave soldering process, the material is heated very fast to the soldering temperature by the liquid solder which, by its intimate contact with the material, gives good conduction of heat, Figure 15.1.[2] It takes much longer for either infra-red heating or condensation heating to heat the material to soldering temperature. The longer the material is at elevated temperature, the more heat is stored in the material. The heat must be dissipated when the soldering process is ended and the board, or whatever it is, cools down to ambient temperature. The amount of heat stored determines how long the cooling process will take. From a study of Figure 15.1 it can be seen that wave soldered boards will cool faster than others. This is — seen from the viewpoint of the solder joint — an advantage, see Chapter 13. *Note*: Normally no forced cooling is used at the outlet of the soldering machine.

To solder, a certain temperature is needed which, for the common tin-lead solder, is above 183°C. The normal soldering temperature is 250°C. However, some components cannot withstand such a high soldering temperature so a maximum temperature of 210-220°C is chosen.

There are a number of reasons why preheating is needed:

(i) to evaporate superfluous flux;
(ii) to activate the flux;
(iii) to give the flux the time to do its job;
(iv) to heat sensitive components carefully, so that they do not get damaged.

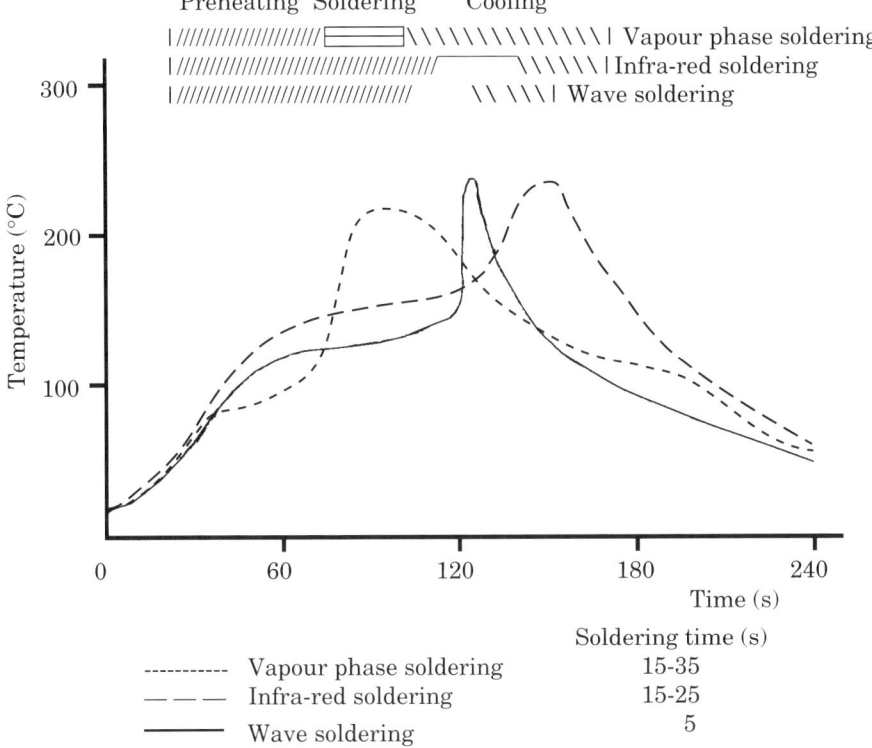

Fig. 15.1 Typical time-temperature curves for different soldering processes.

All these different factors determine the preheating temperature and the preheating time. They also have an influence on how the transition from the preheating temperature to the soldering temperature should be controlled. In some cases this is crucial for a good soldering result.

Consideration of the foregoing may show why it is necessary to study temperature profiles or to establish one's own.

15.1.3 Reflow Soldering

Today the term reflow soldering is widely used. Depending on the context, reflow soldering can mean different things. Very often it is an indication that a solder paste or preform is used for the soldering process. However, to make a solder joint, tinned surfaces can be reflowed. Many soldering methods are reflow processes, for example, the vapour phase process or the infra-red soldering method. Other soldering methods such as laser soldering can be carried out either as a reflow soldering process or as soldering with a flux-cored wire. Thus the term reflow soldering does not define which soldering process is used. The only common characteristic is that no solder is added *during* the soldering process.

Often there is no freedom of choice between the wave soldering process and reflow soldering processes. However, when there *is* freedom to choose the method, it must be remembered that each method will have metallurgical consequences.[3]

15.1.4 Controlled Atmosphere Soldering

In the case of *controlled atmosphere* soldering, it is easy to become confused about what the term means. The process may use *protective* or *reactive* gases, it may be called *hot air* soldering, *inert* or *nitrogen* soldering and other terms depending on the gas or mixture of gases applied and the precise purpose.

A protective atmosphere is an inert atmosphere which contains a gas which is disinclined to react with the process environment constituents. Typical of such gases are nitrogen and the 'noble' or chemically inert gases — argon, helium, krypton, neon and xenon. A reactive gas is either *oxidising*, containing gases such as oxygen, carbon dioxide or water vapour, or *reducing*, containing gases like hydrogen or carbon monoxide. The intention is to minimise or eliminate the requirement for activator and flux by creating a suitable environment for soldering, see also Chapter 13, 13.5.9.

In addition, different soldering methods are applied of which infra-red, forced circulation atmosphere soldering and wave soldering are the most important ones. In the case of wave soldering, a fast-flowing soldering wave is oxide-free at an oxygen level of 50 ppm or lower. A wave flowing with normal velocity demands an oxygen level of less than 15 ppm to remain oxide-free. The interest in nitrogen soldering has increased greatly since 1992 due to its many virtues which are thoroughly described in a report by T. Holm.[4] The main effect of nitrogen is that it hinders the formation of new oxides, which leads to many advantages. One of these is that soldering times become shorter. Another is that the amount of residues left when using low-solid RMA pastes is reduced in nitrogen soldering and the flux residues are more easily cleaned off. This is especially important in the case of surface mounting and fine pitch technologies. Other advantages are listed in Table 15.1. The use of nitrogen is also economically feasible despite greater cost for equipment. The most remarkable saving is the reduction in operating cost due to reduction in dross. Dross formation decreases by more than 85%. Keeping the oxygen level below the permitted level can be accomplished by using sealed chambers, drawing a vacuum and purging with the protective atmosphere until the required oxygen level is reached. It is even practicable to modify an older wave soldering machine with a retrofit package into a 'nitrogen machine'. The retrofit can be a tunnel, a hood or a special solution such as the Electrovert CoN$_2$tour.

Sometimes it is claimed that no flux is needed when using controlled atmosphere soldering but some kind of fluxing *is* needed. Figure 15.2 shows the wetting balance curve of copper which was tinned with the aid of flux, curve 1. The material soldered excellently. The freshly tinned copper specimen was dipped again into the same solder bath after only 15 seconds hanging in the air over the solder bath. This time the specimen was not fluxed and curve 2 shows that the sample did not solder properly. It is, however, true that a less active flux giving lower amounts of residues can be used. It is important that nitrogen (or other protective gas) is available at the point where the board leaves the wave. Flux is often already washed away and

Table 15.1
The Most Commonly Reported Advantages of Nitrogen Soldering

Benefit	Soldering Performance and Quality	Economy	Safety and Environment
Reduced amount of solder — almost 100% reduction of dross in wave soldering	Yes	Yes	Yes
Milder and less flux gives less residues, less cleaning — for no-clean fluxes, elimination of cleaning	Yes	Yes	Yes
Less fire hazard in the equipment			Yes
Reduced rework — reduced number of defects	Yes	Yes	
Improved quality and reproducibility (shiny joints, no board discoloration, no white haze, no influence of ambient air conditions, e.g., humidity)	Yes	Yes	
Fewer testing problems thanks to less residues	Yes	Yes	
Reduced soldering machine cleaning and maintenance	Yes	Yes	
Improved solderability — capability of soldering fine pitch components	Yes	Yes	
Cleanliness of the copper foil of the printed board is maintained and multiple reflowing is possible	Yes	Yes	
Improved productivity by high forced convection (reflow)	Yes	Yes	

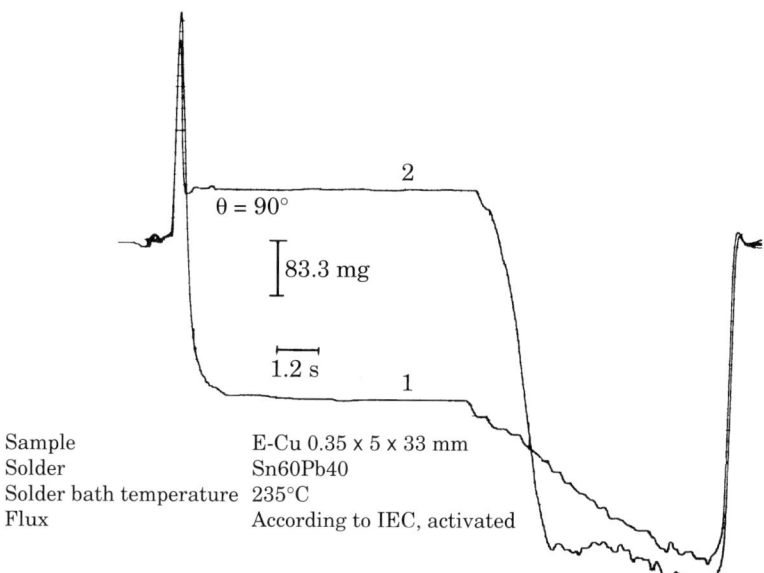

Fig. 15.2 The wetting balance curves show that flux is needed when soldering. Curve 1 shows perfect wetting when 'tinning' a copper surface. Flux was used for this operation. Curve 2 shows non-wetting. The tinned sample, curve 1, was allowed to dwell for only 15 seconds over the solder bath before being dipped again into the bath. On this second dip, no flux was applied.

cannot protect the board at this critical point, causing bridging, icicles and white haze. Nitrogen can give the missing protection. Another most welcome advantage is that the process windows are widened allowing the use of higher or lower temperatures, shorter or longer soldering times, slower or faster gas velocities, smaller or larger pitches and so on. The cost savings, both in reflow and wave soldering, are remarkable. Especially important is that the amount of repair and rework drops considerably due to the improvement in the defect level. Even machine maintenance becomes easier and cheaper.

15.2 THE MASS SOLDERING MACHINE

In Figure 15.3, the most commonly used type of mass soldering machine, the wave soldering machine, is described. Parts of the description are also applicable to other types of mass soldering machines. Such a machine consists essentially of a conveyor which links the main stations required together. The main stations are:

— the loading station;
— the fluxing station;
— the preheating station;
— the soldering station;
— the cleaning station;
— the unloading station.

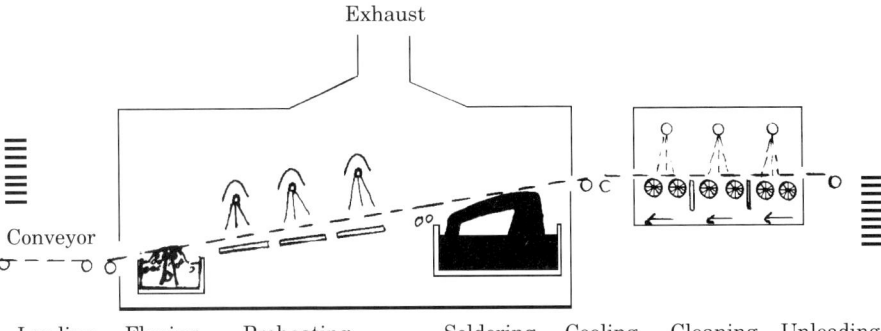

Fig. 15.3 A mass soldering machine. The figure shows the major stations required.

Each station has its own characteristics and needs auxiliaries such as its own conveyor system, specific controls, and aids without which the effectiveness and reproducibility of the soldering process would not be obtained.

15.2.1 The Soldering Fixture

The board assembly to be soldered in a soldering machine has in most cases to be mounted into a fixture. The fixture is placed on the conveyor which transports the board to the different stations of the soldering machine.

There is a variety of fixtures from the simple one with fingers into which the board is mounted manually, to more complex ones which shield the board, its contacts and edges from contamination by the flux and solder. Some fixtures can control the soldering machine. Others can be automatically separated from the soldered and cleaned board.

A fixture is precision engineered and must be handled with care. It must be checked at regular intervals for deformation, warpage and wear so that it can fulfil its task which is to dip the board to exactly the right depth into the flux and the solder bath. If the more complex fixtures are wrongly designed, they can create a shadow effect in the solder wave which causes the shadowed areas on the board not to be soldered. Fixtures with too tight tolerances can cause the board to bow, with non-uniform soldering as a result. The fixture should allow for large boards to be supported suitably to prevent bowing due to their own weight. Fixtures are built of non-wettable material such as stainless steel or heat resistant plastics. Aluminium is not recommended although it is sometimes used with a plastics coating. The fixtures should be cleaned at regular intervals.

15.2.2 The Conveyor

The conveyor links all the stations together and transports the board assembly from one station to the next. Every station in the soldering line may have its own conveyor but each of them must be able to transport the type of fixture used.

The conveyor must be rigid and precisely adjustable. It is the conveyor which determines the precision of the contact between the board and the flux and solder wave and controls the quality of the solder joint (see 15.2.5). The contact depth has a tolerance of only some tenths of a millimetre. If, for example, the conveyor is mounted lopsided, it may happen that only part of the board is soldered. The conveyor must travel smoothly, without vibration. To achieve this, the conveyor has to be maintained and serviced carefully. In modern equipment, the speed of the conveyor can be adjusted to the optimum for the different stations.

The angle at which the board meets the solder wave is important for the soldering result. Therefore, it is recommended that the conveyor is mounted at an angle of 6 to 8 degrees to the horizontal in the direction of travel, Figure 15.3. It should be possible, however, to adjust the angle from 0 to 8 degrees.

An emergency stop should be available along the soldering line.

15.2.3   The Fluxing Station

As fluxing is an important part of the soldering process, it is not surprising that much work has been done on the fluxing station with the consequence that many different types exist.

As fluxes and fluxing are difficult issues, a flux manufacturer must be able to advise the user on how to use the flux in the intended fluxing station. For example, fluxes with a solid content lower than 10% are more difficult to use in a foam flux applicator if special foam control agents are not added. The so-called low residue fluxes come into this category.

There are certain requirements for a fluxing station. It must be adjustable in the vertical direction. As most of the fluxes contain solvents which evaporate, necessary means must be taken to avoid or reduce excessive evaporation. It is necessary to check the density of the flux at least every two hours, see 15.2.3.6, and to add solvent if needed. Checking and replenishment with solvent(s) can be done automatically.

The parts of the flux applicator which come into contact with the flux should be manufactured of plastic such as PVC or polypropylene and *not* of metal. The reason is that most of the fluxes are corrosive and some of them can attack even stainless steel. Even 'non-corrosive' fluxes can attack metals under certain conditions and, by the chemical reactions, alter the behaviour of the flux. The flux applicator should be designed in such a way that it can be cleaned easily. The applicators should be maintained properly and cleaned regularly to avoid trouble, especially if the machine is shut down for some time. When cleaning, the type of flux used should be considered to make sure that the correct solvents are used. (Flux may become thick and sticky and is then difficult to remove.)

If the fluxing station is fed with compressed air, the following aspects should be carefully considered. As the properties of fluxes can be changed by additions of water, oils and other ingredients, it is absolutely necessary to keep the compressed air dry and clean. If necessary, both a dust filter and a water separator will have to be built in at the compressor intake and an oil filter fitted on the compressor outlet to remove any oil which may come from the compressor into the air system. This point should be especially checked *after* the soldering machine is assembled and *before* it is commissioned and put into use.

After fluxing, excessive flux is often wiped off. This is normally done with a brush or an air knife which is mounted on the exit side of the fluxing nozzles. The brush properties are important. If the bristles are too stiff, the brush may displace the mounted components. If too soft, it may fill with flux so that it cannot carry out the task of wiping off excess flux and ensuring a thin, continuous film of flux on the board. Because of this, an air knife is usually preferred.

It is necessary to obtain a continuous and even film of flux on the surface to be soldered. Too much flux on certain points will lead to insufficient drying in the preheater and cause soldering failures. On the other hand, parts which are not fluxed will not solder, giving the type of failure called webbing.

It is a good rule to change the flux at regular intervals, at least every three weeks. The reason is that the composition of the flux changes during use. The copper content increases, oxides, dust and grease, both from the board and the machine, collect in the flux and the humidity of the air which is blown in degrades the flux properties. Too much water in the flux will cause solder spatter and splashes, see Chapter 13.

15.2.3.1 BRUSH FLUXING AND DIP FLUXING

In small scale production, when no complete soldering machine is available, the flux can be applied with a brush, which is dipped into the flux container, or the board can be dipped into flux held in a suitable bath. It is difficult by such methods to obtain an even layer of flux over the solder side of a board and thus a consistent soldering result. This method is also used when different fluxes are being tested for their suitability. Dip fluxing can be very effective in mass soldering electronics assemblies other than printed board assemblies.

15.2.3.2 ROTARY BRUSH FLUXING

This fluxing method was mainly used in the manufacturing of single-sided board assemblies. It is a simple method. A brush is rotated in a container with flux, takes up a certain amount of flux and transfers it to the underside of the board. The components have to be restrained, otherwise they can be pressed out of the hole by the brush. (There are several methods of restraining components, for example, see Chapter 5, 5.2.) When the soldering process is finished for the day, the brush must be cleaned carefully and stored in a flux solvent.

15.2.3.3 WAVE FLUXING

The wave fluxer is very simple in its design and operation. It uses the principle of the wave soldering machine. It supplies more liquid flux to the board than any other method and is therefore especially suitable for densely packed circuitry. When using this method for plated-through hole assemblies, it should be observed that the height of the wave with respect to the board is critical. If the height is not set correctly, either the flux will flow over the top of the component side of the board or the solder side of the board will only be partly wetted. The flux can penetrate through the holes into unsealed components such as relays. This flux cannot be removed and can cause corrosion and malfunction in due course.

Surplus flux must be removed from the board, either with brushes or an air knife, as it can fall in droplets on to the preheater and burn there. Alternatively, it can cause trouble in the soldering process itself. The height of the flux wave should be easily controllable and adjustable during the fluxing process so that the board is fluxed with just the right amount of flux. A general rule is that the top of the flux wave should be about 2 mm above the printed board. It is of advantage if the flux wave starts and stops automatically.

The wave fluxer has been proven to be unsuitable for SMT soldering as too much flux can be trapped between the board and the component. This can adversely affect the soldering process and cannot be reliably removed after soldering.

15.2.3.4 FOAM FLUXING

In all probability the most commonly used method of flux application, especially for boards with plated-through holes, is foam fluxing. The foam is generated in the liquid flux by blowing air through a porous ceramic or plastic tube to form bubbles of flux. These bubbles of foam are forced through the nozzle of the fluxer so that they come into contact with the underside of the board. The bursting bubbles not only coat the board with an even, continuous layer of the flux but help the flux to penetrate into plated-through holes. The height of the foam is 5 to 12 mm above the nozzle top which does not, therefore, allow long component lead protrusions on the solder side of the board as used in the solder-cut-solder process. In such a case, a flux wave or a spray fluxer is recommended.

Even though a thin and consistent layer of flux is formed on the board by a foam fluxer, it is still necessary to remove surplus flux using a brush or air knife. It is important that the air pressure, which should be below 0.3 atmosphere, is adjusted precisely to obtain both the correct height of the foam and the right size of the bubbles in the foam. These determine the amount of flux supplied to the board whereas the dwell time and conveyor speed have virtually no effect. Small amounts of oil in the flux decrease the height of the foam which can be obtained and hot fixtures can cause the foam to collapse.

15.2.3.5 SPRAY FLUXING

Spray fluxers are used only for special purposes. The reason is that it is difficult to keep such fluxers clean for correct work. The cases in which spray fluxing is used are:

— in a soldering process where long leads have to be used;
— when only that amount of flux which is absolutely necessary has to be used to avoid special problems during soldering or due to the flux/component relationship (contamination of open components on the component side of the board and in SMD technology);

— when component movement must be avoided;
— in surface mount technology.

The spray can be generated either by:

— compressed air using nozzles;
— pumping the flux directly through the nozzles;
— blowing air against a screen the pores of which are filled with flux, Figure 15.4;
— using rotating brushes.

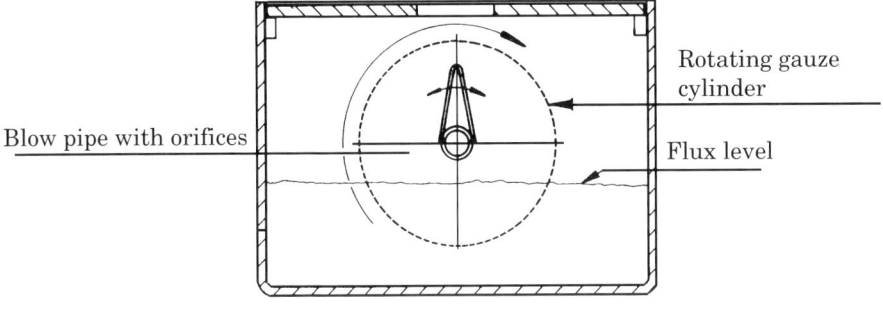

$$\text{Flux layer thickness (mm)} = \frac{24\,n}{v}$$

n = drum speed (RPM)
v = conveyor speed (m/min)

Fig. 15.4 The principle of a spray flux application machine for application of thin flux layers. (Courtesy of Soltec Ltd)

15.2.3.6 DENSITY CONTROL

The density (specific gravity) of the flux has a considerable influence on the soldering result. In general, flux manufacturers give advice on the density to be used for their fluxes in the soldering process. In most instances, there will be no problem in using the suggested density values but, under certain conditions, the recommended figures should be used with caution as the flux manufacturer may not know all the soldering conditions that exist in your specific case. A slight change in the density *could* solve a particular problem experienced. The density of a flux for a soldering machine should not be specified on the basis of manufacturers' data only. It is the soldering result that counts.

As the density of the flux is important to soldering quality, it should be checked regularly. This can be done either with a hydrometer or with equipment which will measure the density automatically and continuously, Figure 15.5. Such equipment will compensate for changes due to temperature when the machine is started. By evaporation of the alcohol the flux is chilled, the solid content, and thus the density, increases and is brought back to the

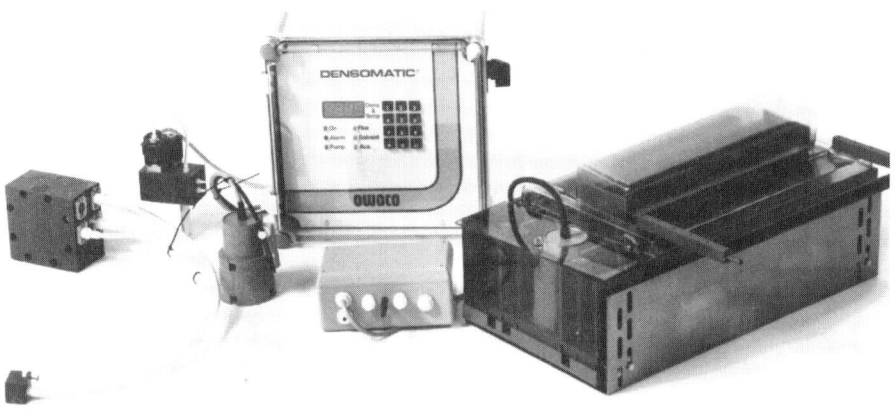

Fig. 15.5 The Densomatic fluxing unit for printed boards in automated soldering machines. It controls the even distribution of flux as well as the accuracy of the solid content of the flux by measuring the specific gravity and adjusting as necessary by adding solvent(s). The flux and solvents are stored in the unit in such quantities that refilling is unnecessary during operation. (Courtesy of Owoco AB)

right level by adding fresh alcohol. When the machine is switched off, or the flux in the fluxing station warms, the density is decreased and the flux flows over into a store. It is difficult to return to the correct density in such a case. Therefore, it is recommended that the soldering machine is not switched off unnecessarily. It is difficult to measure the density of a low solid content flux as most of the flux is alcohol which absorbs water and ruins the measurement. Flux manufacturers should be asked for advice on how to measure the concentration of their particular low residue fluxes on the production line. It should be noted, however, that at the point at which the water concentration affects the density noticeably, the flux should have been replaced a long time previously as it will not be working correctly. The increase of the density due to the water must be considered as an ultimate signal of the need to change the flux.

When the density is too low, the board will not be wetted properly and solder splashes may occur. When the density is too high, 'fat' solder joints (joints with excessive solder) and bridges will appear.

15.2.4 The Preheating Station

Preheating is performed for several reasons. Most fluxes contain large amounts of solvents, alcohol or water which act as a diluent and transfer medium. The solvents are not needed in the actual soldering process. Therefore, they have to be removed. The simplest way is to heat the flux and evaporate the solvents before soldering. This saves considerable time in the soldering process when compared with evaporating the flux first during the soldering process. Preheating to eliminate the solvent avoids a number of soldering faults such as solder spatter, poor wetting, pores or blowholes in the joint, bridging and webbing. In normal cases, preheating up to 80 to 100°C on the solder side of the board is recommended. By preheating, less heat is 'stolen' from the molten solder and so the soldering speed is increased.

The type of solvent used is one of the factors which determine the temperature and time of preheating.

The activator in a flux is usually inactive at room temperature. It becomes active at an elevated temperature and needs some time at this, or a still higher, temperature to break down the oxides and clean the surfaces to be soldered. It is obvious, therefore, that the activator also influences the time and temperature of preheating.

When soldering, the temperature has to be raised from ambient to above 200°C. This is normally done within seconds or even fractions of a second. This can create a shock to the items involved, with cracks in component bodies as a consequence. Preheating at a lower temperature for a longer time reduces this shock considerably. If there is any risk for the parts to be soldered, the preheating temperature and time must be carefully evaluated. If this is not done, there could be the risk that a component will be destroyed before it is soldered. In such cases the only solution is either to solder at a lower temperature or for a very short time, which a laser can provide.

The preheating parameters also depend on the type of board, with large or small heatsinks (copper area, components), the heat sensitivity of the components used and the rate of dissolution of the SMD terminations in the solder bath. After consideration of all these factors, the correct preheat temperature should be chosen with regard to the speed of the conveyor belt, the length of the preheating zone, the distance between the heating element and the board, and the ventilation. All these factors are machine related and difficult to regulate absolutely. In most cases, for example, it is necessary to change the conveyor speed if the preheating temperature has been altered or *vice versa*.

When using resin fluxes, the effect of the heating can be checked by touching the board lightly with a finger to see if it feels slightly tacky. A tacky surface implies that good soldering will result. If it feels dry, non-soldered surfaces can be expected; if it feels wet, this will result in solder splashes. However, this rule is not necessarily valid when fluxes with low solid contents are used. In such cases the board will feel dry. A sign of insufficient drying is a whistling noise when the board touches the solder wave. The whistle is caused by the solvent evaporating.

For surface mount technology, preheating on both sides of the board is usually required and separately adjustable and controllable heaters for top and bottom should be provided.

As new fluxes are developed, more efficient preheaters are required. Preheating may be performed using hot air or infra-red radiation generated in different ways such as hotplates, lamps, resistance rods and infra-red panels, all of which can be set to defined times and temperatures. Infra-red is very efficient for heating the board assemblies while hot air is needed to assist evaporation and transport the evaporated solvents to the exhaust system. So, in most cases, the infra-red radiation is used to heat the air as well. Normally the heaters are mounted in sections transverse to the transport direction of the board to facilitate heat regulation.

The main difference between the different heating systems is the thermal mass of the heater, how efficiently the energy is used, the response time and speed with which the boards can be heated and, of course, the price.

It must be remembered that it is very important when using alcoholic solvents to follow the necessary safety and security measures in order to prevent fire or explosions. See also Volume 2, Chapter 18.

15.2.5 The Soldering Station

The soldering station is essentially a soldering pot which is heated by electrical heaters. Solder pots for mass soldering normally have a large solder volume which takes some hours to heat to the soldering temperature. Therefore, when purchasing a soldering machine, the required heating time should be specified by the supplier of the machine. A timer which switches the power on at the right time to ensure that the machine is at operating temperature at the beginning of the shift should be available.

The solder pot must be adjustable in the vertical direction. Access to the solder pot should be easy for daily maintenance and the design of the machine must be such that it is easy to keep the machine clean.

The solder pot has to be rather large in size to store enough solder to provide a stable temperature within close tolerances for the wave and thus a consistently good soldering result. The wave is rather thin in cross-section and the solder in the wave is easily chilled. The heat loss is much greater than that of a static bath. This is one of the key reasons for preheating the assembled boards before passing them over the wave. A mechanical or magnetic pump is mounted in the soldering pot to feed the solder to the nozzle. Arrangements are made to minimise oxidation of the solder when falling back into the solder pot.

The solder pot is normally made of cast iron or stainless steel. Neither metal dissolves into the solder due to the oxides which are formed on them. Therefore, pots must never be cleaned with metal brushes or sharp scrapers to prevent the destruction of the oxide films.

The electrical heaters have a finite lifetime and it is necessary to check the temperature of the solder regularly. To facilitate changing, it is most practical to use a design of solder pot where the heaters either are placed into tubes which are welded into the pot or are immersion type heaters. If the heaters have to be changed, it is not then necessary to drain the solder from the pot. However, as solder pots filled with solder are very heavy and solder may have to be tapped off for one reason or another, the larger solder pots generally have a suitable outlet for the (molten) solder. External heaters are useful in so far as the pot is easier to clean.

The normal temperatures for soldering boards are in the range of 230 to 250°C with a tolerance of ±5°C. Obviously, when using other types of solder, lower or higher temperatures are employed. Wave soldering machines have been built which are able to go up to 450°C. The original reason for these was to make soldering of enamelled wire possible without first stripping the enamel from the wire. Heating of the joint to the soldering temperature is done by the solder and takes a finite time. If this time is too short because the conveyor moves too fast, the soldering temperature will not be reached and a bad joint will be produced. If the conveyor moves too slowly, the board and/

or the components can be damaged by the heat. In normal cases, depending on the thermal masses of the board and the components, the soldering time will be between 2 and 7 seconds.

Besides the fact that flux must be present on the board when it leaves the solder bath in order to allow the solder to run off the board easily, the conveyor speed is important. If the conveyor speed is too high, the solder has no time to run off and bridges and icicles will be formed. When the conveyor speed is too low, too much solder can run off resulting in the formation of joints with insufficient solder. When the joint leaves the bath the solder is still liquid. If vibrations occur at this stage before the joint has solidified, the metallurgical structure of the solder joint can be damaged resulting in an unreliable solder joint.

The mechanical pump consists of an electric motor above the solder surface connected by a shaft to the pump lying in the solder bath. Incorrect design of the shaft can generate tin oxides and draw them down into the solder bath. The magnetic pump can provide a much higher wave than the mechanical pump and can force solder into narrow spaces.

The shape of the nozzle is critical for the shape of the wave which controls the first and last contact of the board with the solder — which determines the soldering result. Much effort has gone into improvements to the form of the nozzle.

Certain components, such as the SOT-23, see Chapter 4, are difficult to solder in a wave soldering machine. Modifications have been introduced to the wave which generate additional waveforms on the wave, produce more than one wave and/or generate more pressure so that the solder is able to flow higher and penetrate better. For SMDs, two waves have been coupled together.

In some soldering machines an extra nozzle is mounted immediately after the soldering nozzle. Hot air is pumped through this nozzle towards the board. The pressure of the hot air destroys any bridges which have formed. In addition, it pushes away solder which has not bonded properly to the surfaces so that, in this respect, the solder joints are tested for their quality directly after soldering.

In all mass soldering machines the fluxing station, the preheating station and the soldering station are built together in one unit. They are totally enclosed and fitted with an exhaust system, mainly for health and safety reasons, such as:

— to remove solvents which must be evaporated and other noxious fumes;
— to guard the operator from the hot solder pot and solder splashes.

15.2.5.1 SOLDER REPLACEMENT

The solder which is dragged out of the solder pot to form a solder joint must be replaced. The level of the solder pot must, therefore, be checked regularly in order to achieve a trouble-free soldering process, especially in mass soldering. This is valid for both static and dynamic solder baths. An indicator should be used which shows the height of the bath level, when the minimum value is reached and gives a warning signal. In automated mass production, where much solder is dragged out of the pot, the most practical method of replenishment is to arrange a system which keeps a constant solder level in the pot. There are quite a number of such systems available and solder bars can be bought prepared for this purpose. Such bars are automatically fed and dipped into the solder bath under the control of a level sensor.

If the solder level is too low in a pump-driven bath, the dross will be pulled down into the pump. This will cause a gritty appearance of the solder joint and, in the longer term, will damage the pump.

If solder has to be replaced in a solder bath, never blend different makes. Each, by itself, may solder without problems but trouble can arise when mixing them. If mixing is considered essential, e.g., to save cost, make a small experimental mixture first and study the behaviour of the solder both during melting and on the test run. In normal cases, remove the solder from the pot totally before filling with material of the same type from a different supplier. The cost of reject solder joints due to a solder fault, stopped production and investigation of the cause of the problem far exceeds the cost of the discarded solder — which can in any case be sold for reclaim.

When a solder pot has to be filled for the first time, ensure that the heater is covered with solder, otherwise the heater can become too hot and will be damaged. Be careful! Air bubbles may form between solder rods which are laid on the bottom of the pot and erupt. Protective clothing and eyeglasses *must* be worn. Before the first tests or production runs are made, the fluid solder must be stabilised by keeping it above the melting point for at least 3 hours.

15.2.6 The Cleaning Station

When a board assembly has to be cleaned after soldering, and there are a number of reasons for doing this, the cleaning should be done immediately after soldering when it is still easy to remove the flux residues. This is the reason for modern machines in mass production having a cleaning station fitted directly after the soldering station.

The cleaning process is complex and is described in Volume 2, Chapter 15.

15.2.7 Process Control Aids

15.2.7.1 TEMPERATURE INDICATION

Temperature-indicating stickers are available in the range from about 40 to 600°C and are, therefore, suitable for indicating temperatures in soldering. The accuracy is claimed to be ±1°C. The stated temperatures are indicated by a change of the indicator colour from a pastel shade to solid black. The change is irreversible. The heat-sensitive elements are hermetically sealed in high temperature-resistant plastics. They can be easily mounted on almost any clean, dry surface. They resist exposure to solvents and reducing atmospheres and can be immersed in both flux and liquid solder.

15.2.7.2 SOLDERING PROCESS TESTING

In order to achieve consistent soldering results, the soldering parameters should be checked regularly. It is not always necessary to use complex instruments. For example, the temperature in a solder bath can be checked with a mercury-in-glass thermometer. This needs, of course, some care on the part of the operator so that he will not be burnt or otherwise hurt (and the thermometer does not get broken!). The next step would be a thermocouple, which requires knowledge on how such a thermocouple is correctly adjusted and applied.

The soldering time can be measured easily by watching the solder flow up in the holes of a printed board and timing when the process ends with a stop watch. A translucent plate can be used for this purpose, a transparent plate of heat-resistant glass being ideal, Figure 15.6. With this it is not only possible to check the flux wave, the amount of flux applied to the board and the preheating, but also to observe how the flux acts in contact with the solder wave. The contact of the solder wave with the board can be studied over the entire width and length of the wave. It is easy to see if any adjustments are necessary.

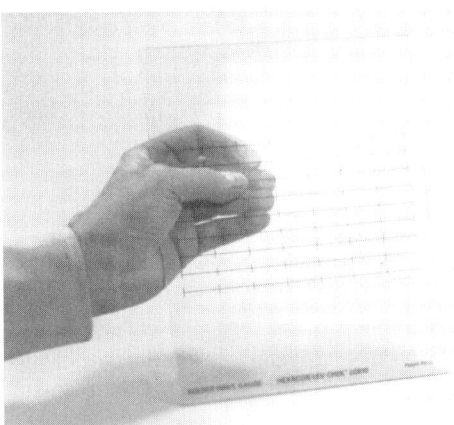

Fig. 15.6 A transparent, heat resistant, glass gauge for adjusting the solder wave. The Lev Chek Solder Wave Gauge permits topside observation of wave-to-board contact to be made on a calibrated grid for easy and accurate adjustment of the transport system of wave soldering machines. (Courtesy of Hexacon Electric Company)

Calculation of the soldering time is simply performed by measuring the width of the wave on the plate and dividing this by the conveyor speed.

There are even electronic testers on the market which make all these measurements easy. This is especially important in mass production. The test equipment is mounted on a board and transported through the soldering machine. Thermocouples are connected to the board to be measured, Figure 15.7. The data registered are date and time, continuous registering of the temperature, conveyor speed, soldering temperature, soldering time and the height of the solder wave if applicable. All values are logged and can be printed out as proof of the correct setting of the machine.

15.2.8 The Maintenance of a Soldering Machine

To obtain good soldering results, a number of parameters of the soldering machine have to be checked regularly, e.g., hourly, daily or monthly as

Fig. 15.7 Datapaq Tracker. During wave soldering, this instrument registers automatically time, temperatures, speed and height of the solder wave. (Courtesy of Datapaq Ltd)

deemed necessary. This avoids not only the costly repair of solder joints but also searching for the cause of the problems, which can take hours. The manufacturer's recommendations for maintenance should be followed. If the machine is used only now and again, the suggested interval will have to be altered accordingly. The majority of checks listed can be carried out either with simple equipment or with continually registering automated sensors which feed the values measured into a computer. The computer controls the necessary actions.

Certain functions, however, can be checked only by eye, such as proper functioning of the fluxer or disturbance of the solder wave.

A. Hourly or more than once per day:

— The size of the bubbles in the foam and the surface of the solder wave should be checked visually as often as possible;
— flux density at least each second hour;
— air pressure twice per day, at the beginning of the shift and after lunch.

B. Daily:

— Correct functioning of preheating once per day at the beginning of the shift;
— correct soldering temperature;
— wave height and the parallelism of the wave to the board;
— correct speed of the conveyor.

Note:
1. All those things which can be measured directly, such as the temperature or the speed, should be measured at least once per day in order to detect malfunction of built-in measuring and indicating systems.
2. The solder dross should be removed whenever it appears necessary, but at least at the end of the working day.
3. It should be self-evident that the machine should be left in a clean condition at the end of the day.

C. Weekly:

— All flux and solder which is not in its proper place shall be cleaned off the machine;
— the fluxing unit shall be cleaned and washed out;
— both the preheater and the solder pot heater(s) shall be checked for the proper electrical values.

D. Monthly:

— The transport system shall be checked carefully for wear, maintained and adjusted where necessary;
— moving parts shall be lubricated as required;
— depending on usage, the solder pot, pump, pump chamber and nozzles shall be cleaned, all flux residues and solder dross being removed; each part shall be inspected for wear or other problems.

Note: As the pumping system and solder pot form the heart of a wave soldering machine and must work properly to give the shape of solder wave required, it should be obvious that it is necessary to inspect and clean these at regular intervals. Care must be taken not to scratch the surfaces with hard tools so that the protective oxides are damaged.

15.2.9 Buying a Soldering Machine

Before a machine is selected for purchase, the conditions necessary for operation of the soldering machine in the intended workshop should be investigated. Things to be checked include the space available, the ventilation, the availability of clean and dry compressed air and a stable electrical system without voltage drop. This can prevent many problems arising during installation and operation of the soldering system since the soldering machine can be fitted with any necessary equipment from the start.

At the earliest possible point in the selection process, the company Health and Safety Officer should be involved.

The properties of the machine to be bought should be agreed on a precisely written specification. The specification should include all necessary details as just described, with exact requirements specified for speeds, temperatures, dimensions and relevant tolerances. Care must be taken to ensure that there is enough space and that the machine is designed for easy maintenance and cleaning, especially the fluxing unit and solder pot. The controls needed and the manner in which they will function must be described. It is helpful to detail the tasks the machine will be expected to execute and with which safety and other standards the machine will be required to comply. The spare part position, service and maintenance requirements, handbooks, delivery time and guarantee also must be specified. Adequate training on the machine should be included in the requirements.

15.3 DIP SOLDERING

Dip soldering is a general soldering process, rarely used in the manufacture of printed board assemblies. However, it has its merits, as it is a very simple process which can be of help in small-scale production or in solving certain soldering problems in electronics cheaply. The method can even be mechanised and used in mass production when appropriate.

Robots are used in conjunction with solder pots to remove gold from component terminations and then tin them on a mass production scale. The robot gives the precision needed in the de-golding and tinning process for the dipping speed, dipping depth and the position of the component in space.

It is easy to arrange precise temperature control for a solder pot. However, as the size of the normal solder pot is relatively small, it can be used for producing solder joints only on small volumes of small printed board assemblies. With the development of robots, see Chapters 6 and 7, it became easier to handle and to dip small sized populated boards in a proper manner.

15.4 DRAG SOLDERING

The drag soldering machine consists in principle of a static solder pot and a conveyor system. The populated board is placed in a soldering jig and transported by a conveyor to the solder bath, lowered until it floats on the bath, Figure 15.8, and is then dragged over the bath. After that transport distance which corresponds to the soldering time desired, the board is lifted from the bath. A scraper is fitted immediately in front of the board to scrape the oxides from the solder bath surface so that the board will come into contact only with a clean, oxide-free solder surface. The bath can be built into an in-line system as described in 15.2.

Fig. 15.8 The principle of drag soldering (not to scale).

The board size which can be soldered is limited by the length and width of the solder bath. Compared with the wave soldering machine, a much smaller quantity of solder is needed in the solder pot. This has the advantage that considerably less solder has to be replaced if the solder bath is seriously contaminated. This, of course, costs much less. For difficult soldering jobs it is possible to arrange a temperature gradient in the drag solder bath in the direction of the board transportation. The solder is not moved as in the wave soldering machine, therefore much less oxidation occurs. A disadvantage of the system is that gas bubbles coming from the flux can be trapped under the board and prevent wetting.

When mounting boards for drag soldering into a fixture, the board should not be placed right at the front of the fixture. The front is lifted up from the solder bath more quickly than the rear, with the result of more icicles and bridging. The first position should be left empty even if it will reduce production capacity. The result will be much less rework.

For many years there has been an ongoing discussion about which machine is the better soldering machine, the drag soldering machine or the wave soldering machine. For general tasks and if the solderability of the material is good, there is actually no difference between the two machines. However, in certain cases there could be a difference, an advantage for one or the other machine. In the case of material of inferior solderability, the washing-off action of the wave soldering machine could achieve more acceptable solder joints. The development of printed boards to produce denser circuits and finer lines has promoted the development of wave soldering machines rather than drag soldering machines.

SMDs are difficult to solder with a drag soldering machine as gas from the flux is easily trapped in the complex geometry between the solder, component and board and will prevent soldering. The difference in the hole filling process between the drag and the wave soldering machine gives the drag soldering machine an advantage for multilayer boards.

15.5 WAVE SOLDERING

Since the invention of this process by Barnes, Elliot and Strauss in 1956, this process has become widely used and much developed. Because of its basic simplicity, it is one of the most widely employed mass soldering processes.

Solder is heated in a solder pot. When molten, the solder is pumped through a nozzle and forms a wave while falling back into the solder pot, Figure 15.9. The object to be soldered, in most cases a populated printed board, is taken in such a way over the wave that only the parts to be soldered come into contact with the solder to form the solder joint.

This simple principle has been improved upon steadily over the years. The wave shape and the nozzle have undergone a number of changes. Fluxing and preheating stations were added at the beginning and a cleaning station at the end, Figure 15.3. The wave soldering system became conveyorised, computerised and each of the individual stages was examined critically, improved and reduced to its essentials. The wave soldering line has also been adjusted to the needs of surface mount technology. More recently, the whole soldering machine was enclosed to enable the use of controlled atmospheres.

All these developments have made a complex sophisticated machine, such as shown in Figure 15.10, from the old simple wave soldering machine. Compared with the older machines, the newer ones give highly improved soldering results as well as higher solder joint quality. Almost certainly, the pattern of continuous development has not yet ended. Such development is

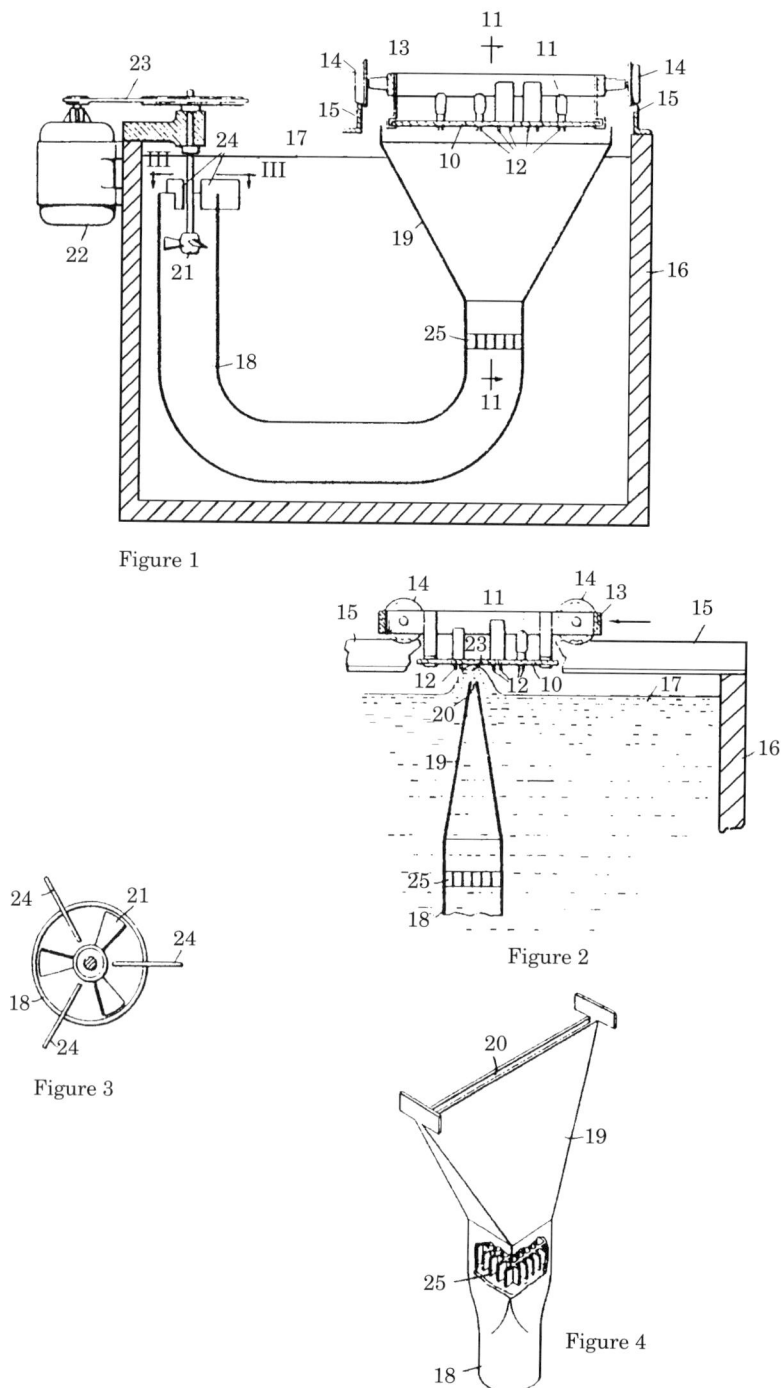

Fig. 15.9 The first patent drawing for the wave soldering principle. British Patent 798 701. Fry's Metals, 3.10.1956.

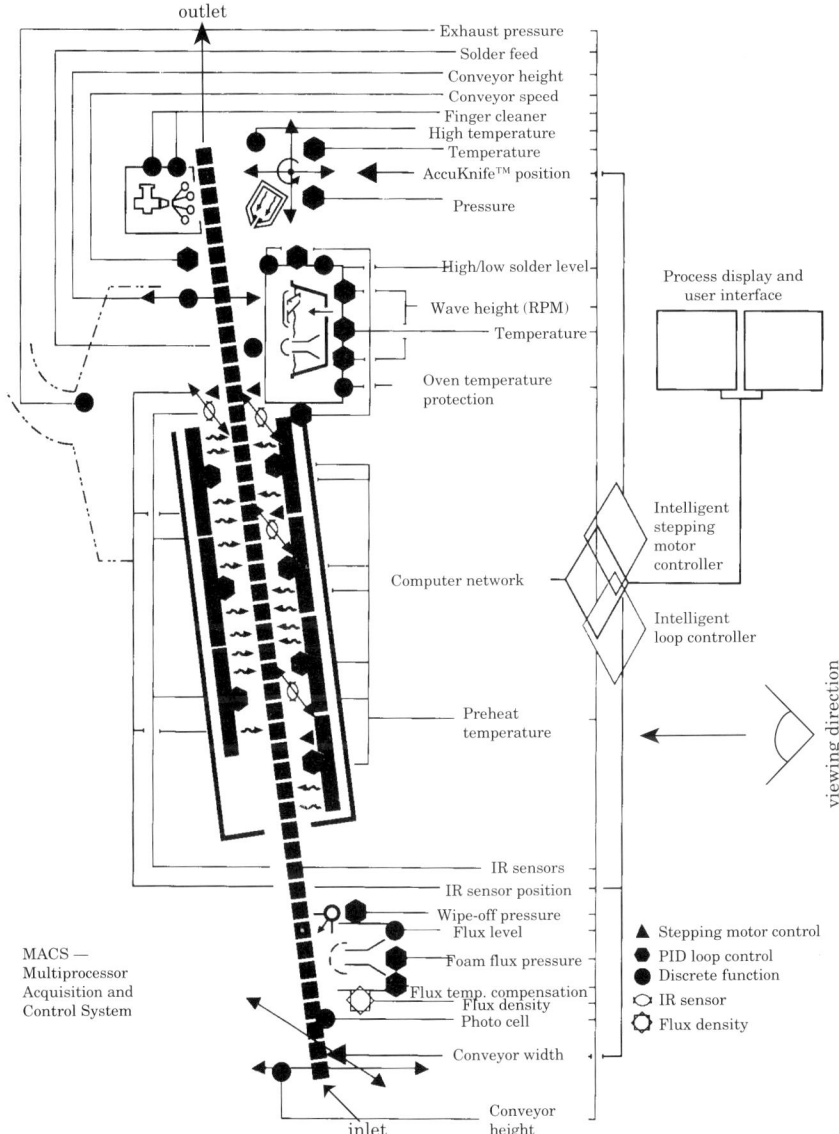

Fig. 15.10 A computerised soldering machine. (Courtesy of Hollis)

characteristic for all other types of soldering machine. (The individual parts of most modern soldering machines are described in 15.2.)

There are certain requirements which are specific to a wave soldering machine. The soldering wave should be started automatically some seconds before the board meets the wave (but there should be built-in means of preventing the solder pump starting when the solder temperature is under a preset limit). This reduces the dross formation and also helps to keep a tighter temperature tolerance.

As soldering is a complicated process, and is carried out on sophisticated machines, all operators must be given thorough training.

15.5.1 The Wave Form and the Nozzles

The main problems with wave soldering are excessive solder, bridging and icicles. These defects are caused by:

— insufficient heating of the parts so that excessive amounts of solder can be gathered on the areas to be soldered; *or*
— the solder joint leaving the wave too fast, giving the solder no time to drain back to the wave; excessive solder is left on the board and bridging occurs; *or*
— the solder on the joint and the solder which is draining back cool too quickly and solidify before the surface can act properly, forming icicles and bridges.

These phenomena were counteracted by special designs of the nozzles which form the wave shape as well as by taking the board at a low angle over the wave. Correct heating was achieved by keeping the board in longer contact with the wave, e.g., by lengthening the wave presented to the board, giving the wave an asymmetrical form, Figure 15.11. To avoid bridges and icicles, the wave was extended on the board exit side and the conveyor was angled towards the wave. The conveyor angle, in conjunction with the correct speed of the board, makes it possible for the excess solder to flow back into the wave. The extension on the board exit side means that the board is still heated by the solder although that particular part of the board is no longer in contact with the solder. This prevents premature solidification of the excess solder on the board which thus can flow back on the board to the wave. The surface tension of the wave helps the drainage of the solder from the board.

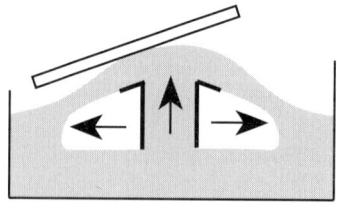   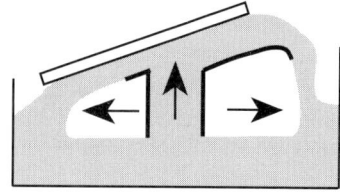

Symmetrical wave   Asymmetrical wave

Fig. 15.11 Different types of soldering wave. The wave form is shaped by plates. The plates can be adjusted to alter the form of the wave and the length of the solder path presented to the board. This influences the soldering time and hence the soldering properties.

The different types of nozzles with their ability to shape the wave, in conjunction with horizontal or angled conveyors, offer a wide variety of methods for solving different soldering tasks. Such methods should be considered carefully as they influence the economics of soldering. A careful choice is especially necessary when different soldering tasks have to be conducted on the same machine. Therefore, when selecting a machine, it is also necessary to check whether or not other nozzle types can be used on the chosen machine and if a change of nozzle can be made when the solder pot is full.

15.5.2 Wave Soldering Machines for SMT

Wave soldering machines which are able to solder SMDs can, of course, solder hole-mounted components as well. This is a necessity as very often mixed technologies are used, see Chapter 7, 7.1.3 and 7.1.4. Machines have been developed specifically to solder.

— chips, SO components, chip carriers and other SMDs, see Chapter 4;
— densely packed surface mount assemblies.

In both cases, the solder cannot always make contact with the joints when normal soldering machines are used. Flux can be entrapped between closely spaced components and bridging between them easily happens. See also Chapter 7.

15.5.2.1 THE DOUBLE WAVE SOLDERING MACHINE

The double wave soldering machine is a product of surface mount technology. The distance between the waves should not be larger than 110 mm to prevent solidification of the solder on the board. This normally gives a contact time of the board with the solder of 4 to 6 s. Each wave should have its own pump and be separately adjustable. The first wave is a turbulent wave which may cause icicles and bridges which are then removed by the second, smooth, wave, Figure 15.12. Different specific solutions are based on this principle.

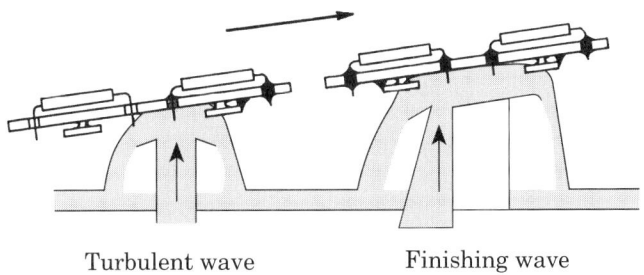

Turbulent wave   Finishing wave

Fig. 15.12 The dual wave principle of wave soldering.

15.5.2.2 THE PULSED WAVE SOLDERING MACHINE

Pulses or vibrations ease the solder flow into both smaller capillaries and deeper spaces. They are generated in different ways for different machines at the point where the board enters the wave. On the exit side of the wave the board finds the same conditions as those in a normal wave soldering machine, to remove excessive solder and bridges and give a clean solder joint. As only one wave is used, a shorter soldering time results.

Electrovert generates the vibration with a bar which is moved up and down in the solder bath at 50 Hz, Figure 15.13, while Soltec employs a rotating hexagonal bar with adjustable speed, Figure 15.14.

15.5.3 Oil in the Wave

Oils are used in wave soldering machines to lower the surface tension of the solder on the exit side of a wave, mainly on symmetrical waves, in order

Fig. 15.13 The Electrovert pulsed solder wave in a Europak wave soldering machine. Note the fingers in the upper part of the picture. These are for mounting the populated printed board. (Courtesy of Electrovert)

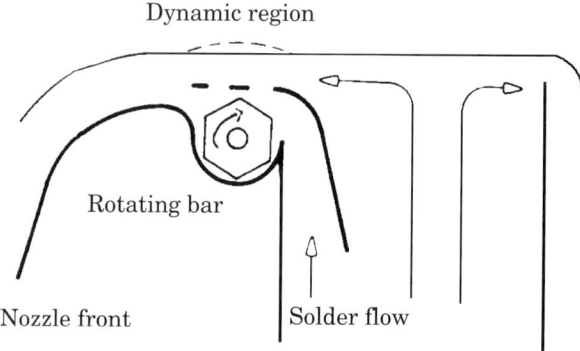

Fig. 15.14 The Soltec principle for producing a pulsed wave — 'Smart Wave'.

to overcome the problems of bridging and icicles. At the same time, oil reduces the oxidation on the solder wave and solder bath surface. The oil will deteriorate with time despite its high temperature stability and must, therefore, be replaced at regular intervals (which may be after one day to one week of use).

A disadvantage with oil is that the boards have to be cleaned, but this can be done simultaneously with the cleaning of the flux residues. In addition, the use of oil is an extra cost and the use of soldering machines with oil must be carefully investigated to determine their cost-effectiveness. The decisive factor for this should be the cost of rework necessary on the soldered boards calculated with and without oil.

15.6 INFRA-RED SOLDERING

For quite a number of years infra-red soldering has been used in the production of hybrid circuits. However, this method was seldom employed in soldering normal printed board assemblies. A change came when surface mount technology was introduced generally and especially when the crisis arose in vapour phase soldering due to the use of chlorinated fluorocarbons and their detrimental effect on the environment, see Volume 2, Chapter 17.

As the name implies, infra-red soldering uses infra-red radiation in the range of less than 1 µm to 6 µm to heat the board, the solder and the component. One characteristic of infra-red soldering is the long soldering time. Another is that surfaces and colours of board and components can be especially susceptible to the infra-red wavelength used. Both together can jeopardise the board as well as the solder joint.

It should be noted that certain chemicals have a high absorption rate in the infra-red spectrum. This could harmfully reduce the effectiveness of the heat transport by radiation. Infra-red soldering is a process which needs much care at the development stage to eliminate all problems but, when the engineer does his homework properly, it is a reliable process which works well.

Long soldering times cannot be considered to be good for the board, the components or the solder joint. The cooling rate for the solder joint is low, due to the stored energy in the board and components. This causes a coarse grain size in the solder joint with the consequence that one of the important properties of a solder joint, the low cycle fatigue resistance, is negatively influenced.

15.7 CONVECTION AND FORCED CONVECTION SOLDERING

The manufacturers of infra-red soldering machines have carried out a lot of work to develop suitable radiators, to use the right wavelength, to avoid the problem that only the surface of the objects will be heated and to speed up the process. In reality, the greater part of the soldering heat now comes from the atmosphere which is heated by infra-red and only a minor part comes from direct infra-red heating (convection soldering). This gives more uniform heating. As convection may not be enough, it is reinforced by fans. For practical reasons, the machines are more or less enclosed which makes the use of protective gases, such as nitrogen (which increases the heat transfer capacity), quite convenient.

712 A Comprehensive Guide to the Design and Manufacture of Printed Board Assemblies

15.8 VAPOUR PHASE SOLDERING OR CONDENSATION SOLDERING

In the vapour phase process, a liquid is heated until it vaporises. The parts to be soldered are dipped into the vapour which condenses on them while the parts have a temperature below the vaporisation point of the liquid, Figure 15.15. As the vapour condenses, energy is released and heats the part to the temperature at which soldering becomes possible. This implies, of course, that the vaporisation temperature of the liquid lies above the soldering temperature. As the energy comes from the condensation of the vapour, the process is often called *condensation soldering*.

The parts to be soldered must be dipped into the vapour. Both the flux and the solder have to be pre-placed at the site of the solder joint. This can be done in the form of solder preforms or solder paste. Thus, vapour phase soldering is a *reflow soldering* method.

This process has a number of advantages. It is never possible to overheat and damage the material involved in the soldering process. As the vapour excludes the air, and hence the oxygen, the metals to be soldered cannot be oxidised during the heating process. So, provided the parts have good solderability as they should have, a very weak flux can be used. When the machine is properly designed, large amounts of heat can be transferred to the

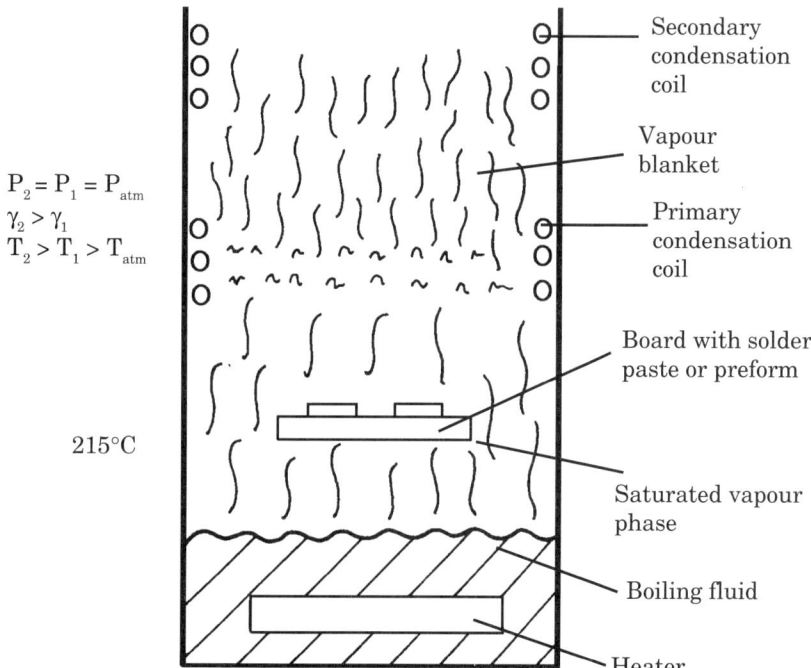

Fig. 15.15 The principle of vapour phase soldering. The pressure (P), density (γ) and temperature (T) conditions for a working vapour phase process are shown.

parts to be soldered and very large and very heat-consuming parts can be soldered.

The heat distribution across the workpiece is very uniform. If the parts to be soldered consume more energy than can be provided by the vapour, the vapour phase zone can collapse and no soldering is possible. The parts to be soldered should be designed in such a way that the condensed fluid does not stay at those parts where the temperature has to be raised as, in this case, the liquid is a thermal insulator. Care must be taken when designing the fixtures which carry the boards. Different sections can cause distortion due to differential heating and expansion of the sections.

For a number of years vapour phase soldering machines were close to being banned from soldering workshops. The reason was as follows. The first vapour phase machines were batch machines as shown in Figure 15.15, more or less open to the atmosphere at the top. The recommended fluid then, and now, for soldering is a completely fluorinated organic compound with a boiling point of 215°C. The trade name from 3M is Fluorinert liquid FC-70 and, from Galden, LS 215. As this compound is very expensive and evaporates during soldering, a cheaper fluid with a lower boiling point was included in the tank to form a blanket above the FC-70 to hinder evaporation. This fluid belongs to the family of chemicals known as chlorofluorocarbons, or CFCs, which was banned in 1987 by 31 nations in the Montreal Protocol as CFCs destroy the protective ozone layer surrounding the world. (Ozone functions as a shield for living beings against harmful ultra-violet (UV) radiation, see Volume 2, Chapter 17.)

The two types of fluid must not be confused as they are different chemical types with different properties, one of which is banned (the CFC) and the other not.

The CFC problem has led to three major developments. The first was to enclose the batch machines and equip them with a vapour recovery system (VRS) for the evaporated fluorocarbon fluid. The second was to develop batch machines, Figure 15.16, which were considerably improved by using infrared for preheating, reducing the otherwise common solder balling, tombstoning and solder wicking. The provision of a rapid cool-down system eliminated condensation of water and the consequent corrosion problems. The third development was aimed at replacing the secondary fluid used as a vapour blanket. The search for a replacement was successful and the old type of batch machine can be used again.

Fluids with different boiling points, e.g., 174°C, 215°C and 253°C, are now available. This widens the range of solders which can be used and spurred the development of newer batch and in-line machines for use with a single fluid. Features included are palletised conveyors, IR preheat, vapour phase reflow, forced convection cooling and vapour recovery,[5] Figure 15.17. Such machines overcome another problem of the older machines in which moisture from the air condenses on the cooling coils and combines there to form hydrochloric or hydrofluoric acids, depending on the chemicals available, causing corrosion of the coils and the tank.

Fluorocarbons have a number of unique properties.[6] Besides those already named, they are electrical insulators, they do not react chemically unless grossly overheated and, therefore, electronic components made of virtually any material can be soldered without being affected by the fluids (although,

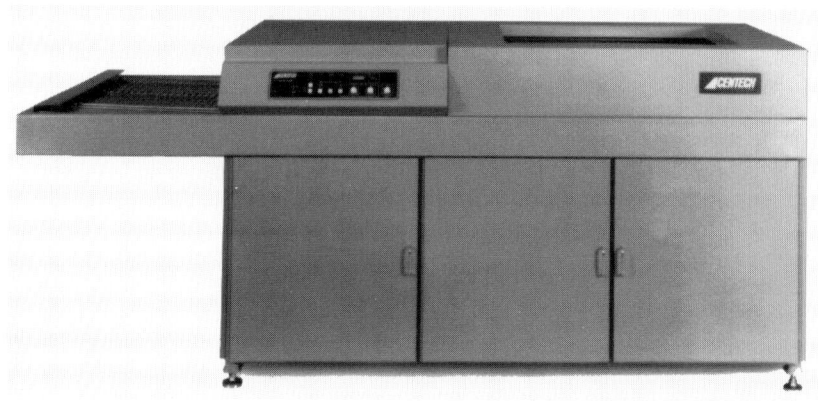

Fig. 15.16 A batch machine for vapour phase soldering. (Courtesy of Centech Corporation)

as said before, they may be affected by the temperature!). Fluorocarbons are 17 times heavier than air, they are non-flammable and non-explosive and do not leave any residue on the workpiece after soldering.

Reference 7 summarises the differences between VPS (vapour phase soldering) and IR (infra-red soldering), the two important mass *reflow* soldering methods, as follows:

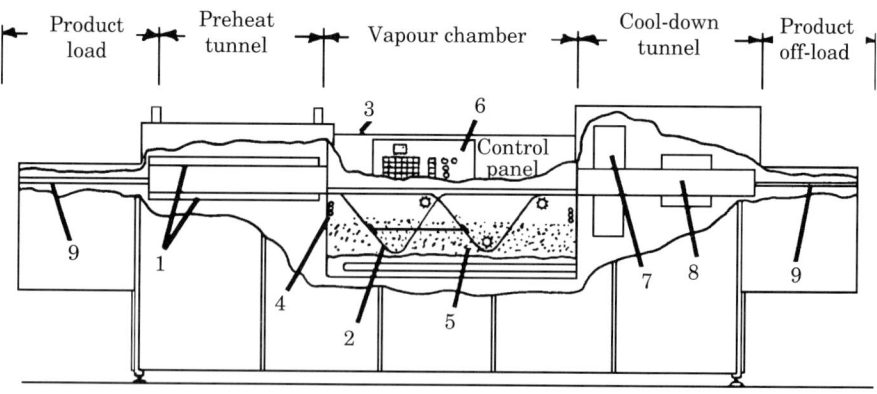

Key: 1 Ceramic preheaters
2 Arrangement for horizontal board transport
3 Window for visual inspection
4 Vapour containment (single fluid)
5 Saturated vapour
6 Control panel
7 Rapid cool-down section
8 Product reflow tunnel
9 Entry and exit pallet areas

Fig. 15.17 An in-line vapour phase soldering machine. (Courtesy of Centech Corporation)

Factor	Infra-red Soldering	Vapour Phase Soldering
Production throughput	Good	Excellent
Yield	Comparable	Comparable
Preheat	Included	Recommended
Process control	Fair	Excellent
Operating temperature	Adjustable	Fixed
Temperature uniformity	Poor	Excellent
Maximum temperature	Unknown	Fixed
Fail-safe operation	No	Yes
Versatility	Fair	Excellent
Sensitivity to materials	High	None
Ease of setting operating data	Moderate	Faster
Operating costs	Low	Higher
Capital costs	Comparable	Comparable
Operating environment	Air or inert gas	Anaerobic
Maintenance	Low	Required
Flux residue cleaning	Poor	Excellent

15.9 LASER (See also Chapter 12)

In laser soldering the energy of a focused laser beam is used to heat the solder joint and to melt the solder. Laser soldering is very fast. Normally, it manufactures the solder joints sequentially and not simultaneously as is done in, e.g., a wave soldering machine. Laser soldering is the latest of the soldering methods. Laser soldering was first used in production at the end of the seventies. The growth of laser soldering has been inhibited by several problems. First, a laser soldering equipment is considered to be relatively expensive. Secondly, laser soldering is not considered to be a mass production method. Thirdly, a number of technical problems arose of which the most important was that the printed board could be burnt when soldering with a laser. These problems will be discussed later.

15.9.1 Types of Laser used for Soldering

Two types of laser are used for soldering today, the CO_2 laser which operates at a wavelength of 10.6 µm and the YAG laser which operates at 1.06 µm. Metal absorbs the laser energy efficiently at a wavelength of 1.06 and less efficiently at 10.6 µm. It is just the opposite with plastics.

This has the following implications. More energy is needed to heat metals with a CO_2 laser than with a YAG laser. When using a CO_2 laser for soldering, the risk of burning the plastic insulation and the printed boards is much greater than when using a YAG laser. This is the reason why YAG lasers in general are preferred to CO_2 lasers for soldering.

As not much heat is needed for laser soldering and the soldering is carried out at low temperatures, the Nd:YAG laser is more suitable. A Nd:YAG laser is even more compact than a CO_2 laser. For soldering purposes in electronics it is also important that the Nd:YAG laser beam can be focused to finer spots

(< 50 µm) than a CO_2 laser beam. The optical system can be more easily adjusted for special geometrical demands. The possible use of fibre optical cables is of great advantage to production as it increases the three-dimensional flexibility and allows the same work — that is not only on the same component terminations but also on all terminations having the same thermal requirements — to be performed at different places. This makes a semi-simultaneous soldering machine out of the former sequential soldering machine.

15.9.2 Pulsed and CW Lasers

There is another property of lasers which is important for soldering. There are pulsed lasers and CW (continuous wave) lasers. Pulsed lasers give their energy in the form of pulses in a defined short time of parts of milliseconds to some ten milliseconds. If the energy needed for making the solder joint is known, a pulsed laser can be used. The normal case, however, is that the energy needed varies from joint to joint on the board. This variation is not only (and this is important) due to the different thermal masses and heat conductivity of every single joint, but also due to the varying solderability of the materials which are to form the joint. This is the reason why, for soldering of components on a board, a controlled laser should be used.

15.9.3 Controlled Laser for Soldering

Due to the complexity of the soldering process as described earlier, only on-line control of the soldering process can guarantee a perfect solder joint without damaging other materials involved.[8] This can be done by measuring the temperature on the solder joint during soldering, Figure 15.18. The infra-red heat radiation is gauged with an infra-red detector. When soldering, the joint and the solder are heated at a constant rate. When the solder melts, energy is consumed. The wetting action of the solder increases the heat flow from the solder to the material connected to the joint. This causes a non-linearity in the temperature rise. After a while, when conditions have stabilised, the laser energy again raises the temperature at a constant rate. Thus the laser soldering process can be controlled on the basis of the non-linear temperature rise. This makes it possible to feed precisely the right amount of heat to the solder joint. It will be neither overheated nor underheated.

A controlled laser normally uses the CW principle as the duration of the pulse length is easily controlled and adjusted to the needs. However, the duration of the laser radiation can be perfectly governed even with a pulsed laser.

15.9.4 The Heat Flow in Metals Radiated by Laser

The general opinion of a laser is that the beam generates considerable heat and thus destroys sensitive materials. This is not true as the desired temperature in the material can be adjusted precisely by controlling the time the laser beam radiates on a spot. This is the reason why the laser is the most

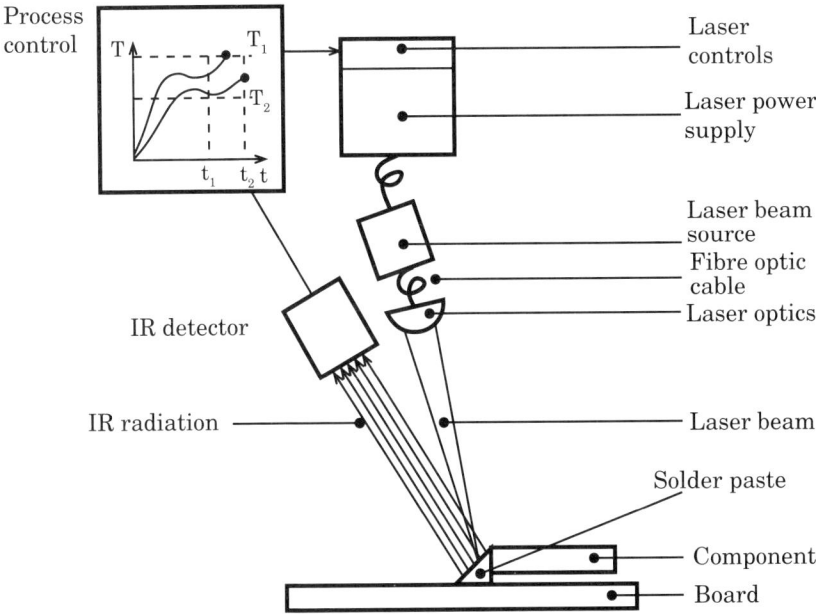

Fig. 15.18 Measuring the temperature on the solder joint surface during laser soldering.

suitable method of soldering heat-sensitive components. The beam can be adjusted such that there is sufficient energy for soldering, but, due to the short time the laser is on, the total energy fed into the joint is so small that it is not enough to heat and so damage the component.

This is demonstrated by Figure 15.19. A piece of metal (German silver) with a thickness of 0.3 mm has been heated with a laser pulse and a bell-shaped area has been melted. Next to the melted zone, a small grainy zone can be seen which indicates how deeply the heat has penetrated the original material and transformed (recrystallised) it. Beyond that zone, the rolled structure of the original material can be seen. The melting point of the material is 1020°C, the recrystallisation temperature about 300°C and the width of the grainy zone is 0.03 mm. From this it becomes clear that the temperature gradient is very steep from the melted to the uninfluenced zone, a property which is of great advantage in soldering.

15.9.5 The Laser Solder Joint

Depending on the size of the solder joint, a laser solder joint can be produced in 5 to 2000 ms. No more than the energy necessary is supplied to form the solder joint which gives the following metallurgical advantages:

— the joints are more uniform compared with those formed by manual soldering;
— the structure of the solder is very fine as the solidification is fast;
— the intermetallic phases which are formed are much thinner.

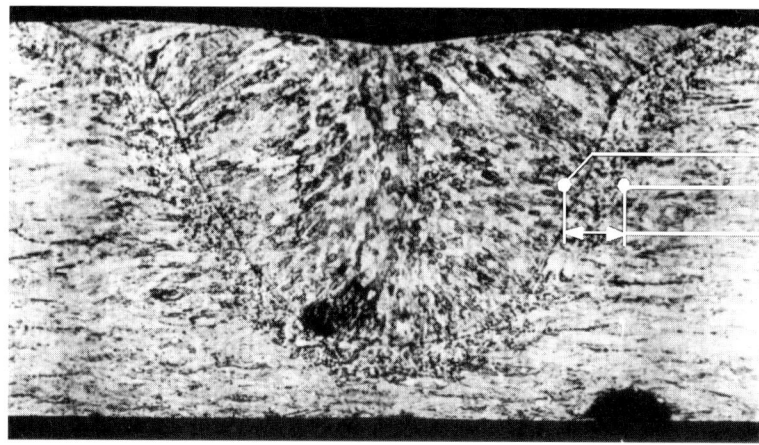

Fig. 15.19 Temperature distribution in a piece of metal after melting with a YAG laser and solidification.

This results in a joint with better mechanical strength.

With respect to low cycle fatigue, a significant increase of the life has been reported.[8] It may be, however, that this improvement should be regarded as being marginal, considering how the structure, and hence the properties, of a solder joint change with time.[9,10] It is a fact that the overall quality of a board wholly soldered by controlled laser is better, as all the joints have the same structure and there is not such wide variation in the structure as can be found on normally soldered boards.

15.9.6 Maximising the Advantages of Laser Soldering

Laser soldering allows the use of solders with special properties as, for example, higher soldering temperatures can be permitted. It even allows the use of solders with different solder compositions on one board to meet different demands on the solder joint, such as, for example, good low cycle fatigue properties or good electrical conductivity in high frequency equipment. It must be noted, however, that for laser soldering good solderability is a necessity.

As a laser beam is controlled by electronics, it is clear that a laser soldering system can be perfectly robotised and governed in all aspects by a computer. This gives high precision resulting in a high yield and excellent quality. With a robot it is possible to position the laser beam with an accuracy of ±0.08 mm. The position of every single solder joint, the energy needed to make a specific solder joint, the position of the laser and the optical conditions can be fed into the computer.

Laser soldering is only possible where the laser beam can hit a surface, e.g., solder joints under a chip carrier cannot be made. It is, however, possible to make solder joints at castellations which stretch further under the component. By the use of fibre optics, a number of components can be soldered simultaneously. By these and other possibilities, the speed of a laser soldering apparatus can be increased substantially.

15.9.7 Areas for Laser Soldering

Laser soldering was first used to solder chip components on to a board. The next step was to solder component leads of components having a very high thermal capacity with a laser instead of a large soldering iron. Then there followed attempts to solder computer boards of normal design and layout. One problem was always apparent, that a number of boards were burnt. There was no controlled laser to effect output so the yield was not at an ptimum.

Now the laser is used in a wide field of applications. It is used from soldering component leads in plated-through holes, Figure 15.20, and soldering of surface mount components to soldering of TAB assemblies. The board material ranges from glass fibre epoxide to polyimide film to ceramics.

Solder wire is used as well as solder paste, Figure 15.20. The normal fluxes can be used as well as controlled atmospheres. There is no limit to the choice of solder alloy. This means in practice that, instead of using low melting solders for soldering sensitive components, high melting solders such as Au80Sn20 with a melting point of 280°C can be used in order to obtain a solder joint with special properties.

Laser soldering has made possible soldering tasks which could not be executed only a few years ago, such as soldering thermally sensitive components with a solder of high melting point to a polyimide board. Some of the most difficult designs have become possible, e.g., soldering high heat dissipating components on to a thin board which sits on an aluminium cooler.

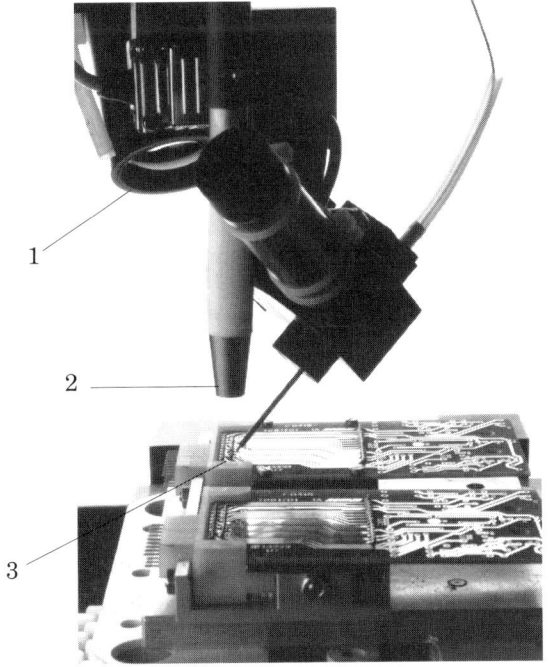

1. Infra-red optics measure the temperature on the solder joint.
2. YAG-laser optics.
3. Solder wire.

Fig. 15.20 Controlled laser-soldering equipment. Laser soldering a printed board with plated-through holes using solder wire.

It must be remembered, however, that most of these soldering tasks can be carried out with reproducible success only with a controlled laser.

15.9.8 The Economics of Laser Soldering

The following calculations are based on Reference 11. They are interesting in that they show that, even in 1989, laser soldering could compete commercially with other soldering methods under given conditions, without taking the technical advantages into account.

A controlled laser soldering equipment cost (US) $300,000 (1989). The cost, including equipment amortisation over 5 years, was 90 $/h when used 100 h/month and 54 $/h when used 200 h/month. In calculating the cost for other machine soldering methods (wave, IR, VP), the following assumptions were made: conveyor speed 1 m/minute and a PBA/0.5 m, which amounts to 120 PBA/h. With 250 solder joints/board, this gives 120 ms/solder joint and, with 500 solder joints/board, 60 ms/solder joint. To produce a solder joint by laser takes about 60-120 ms. If two beams are used simultaneously, which is possible by beam splitting with fibre optics, the time will be 30-60 ms. This means that, under similar conditions, the economics favour the laser above 250 (or 500 as the case may be) solder joints/PBA. Under certain conditions the controlled laser can even compete with hand soldering. The minimum number of PBA/lot lies here at 10,000 with 30 solder joints/PBA or more than 300,000 solder joints.

An important point in evaluating cost is that a solder joint produced with a controlled laser does not need to be inspected after soldering, as the in-line control itself gives a protocol which determines whether a solder joint has passed or not. A considerable amount of money can be saved by discarding the visual inspection of solder joints. These savings should be taken into account when calculating the costs for a laser soldering machine.

15.10 LIGHT SOLDERING

Actually it is not the visible light but the infra-red part of the spectrum which gives the energy to heat the solder joint. Light soldering in principle is infra-red soldering. The infra-red part is used in many connections.[12,13]

Soldering with visible light is possible, but very rarely used in the production of printed board assemblies. For this purpose halogen lamps can be used which have a special reflector to focus the light beam. The characteristic of visible light is that it is a blend of numerous wavelengths and does not, as in the case of the laser, consist of just one defined wavelength. This makes it difficult to use visible light for the soldering of printed boards as a wavelength of about 10 µm is absorbed by plastics with the consequence that the board material can be damaged. To avoid this, screens can be used to shield the board, Figure 15.21. The light radiates through holes in the shield on to the solder joint.[14] Another problem with lamps is that the beam cannot be focused to a sufficiently small spot since the filament of the lamp, where the light and heat are generated, has a finite size which cannot be made smaller by optical means. With light it is possible to solder joints which are enclosed in glass.

Heating with light is fast. Therefore, there is a risk of overheating and precautions must be taken if this could create a problem.

15.11 HOT BAR SOLDERING

A hot bar is pressed against the pieces to be soldered. When the parts are pressed together, the bar is heated electrically by using either the resistance of the bar itself or by a resistance coil inside the bar. (If the resistance of the bar is to be used, molybdenum, nickel plated, or Hasteloy Alloy X can be used with success.) After a predetermined time the electricity is switched off, the

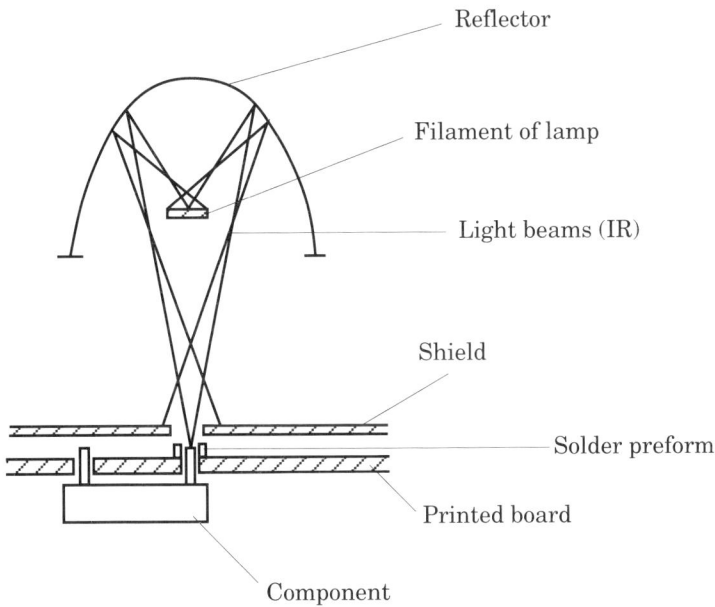

Fig. 15.21 Soldering printed board assemblies using high intensity light.

bar cools to below the melting point of the solder and the solder joint solidifies. This is the point at which the bar can be lifted off.

In the early days of the hot bar soldering method, small resistance welding machines were used. Machines have since been developed which are specially adjusted to the required time, force, power and control needs.

With the hot bar, mainly flat leads are soldered together. It is not really a mass soldering method as only a small number of connections can be made simultaneously. The soldering time is some seconds and the bar cannot be lifted off the solder joint before the solder has solidified which takes a further few seconds.

There are a number of points which must be taken into account when this method is used. The bar itself has to be designed specially so that the temperature is uniform along the length of the bar. It must be sufficiently stiff so that the pressure will be distributed uniformly along the length of the bar. The parallelism between the bar, the workpiece and all the leads which are to be soldered together must be good. A finite pressure is needed to ensure good heat flow from the bar to the workpiece. This pressure must not be so large that the leads are deformed or the board material under the leads will be cracked or otherwise damaged. If a number of leads is to be soldered, quite a high total pressure is reached. If reflow soldering is to be executed on a tinned surface, the tinning thickness should be at least 20 μm. In certain cases, only a thickness of 50 to 80 μm will give a good result. However, good results have been obtained even with very thin layers. The problem here is that the material must be soldered within days or weeks of processing if soldering problems are to be avoided. Good solderability is a necessity, as in all soldering tasks.

The placement of the thermocouple on the bar is important as false temperatures can be easily obtained. The temperature variations on the bar should be not greater than 10°C. The weld of the thermocouple must be protected against mechanical stresses.

15.12 HOTPLATE SOLDERING

A heated plate, in the simplest case a cooking plate or 'hotplate', can be used for special types of soldering. Normal printed boards cannot be soldered on such a plate but surface mounted components on ceramic tiles can. The temperature can be controlled very simply and cheaply. If this is done, the soldering process will be under reasonably good control.

15.13 BELT SOLDERING

Belt soldering can be seen as an extension of hotplate soldering. A belt moving over the hotplate increases the soldering possibilities up to large-scale production. In some cases, belt soldering can be combined with infrared soldering and/or with a furnace or tunnel oven in order to use protective gases. As with many other soldering methods, numerous combinations can be used in order to optimise the soldering process.

One problem which may arise with the belt is that the heat conductivity from the hotplate to the ceramic tile will no longer be direct and the desired temperature may not be achieved on the tile. In addition, the belt is subjected to wear which may give rise to further problems.

Taken as a whole, when properly set up and controlled, belt soldering is a simple and reliable method for the right purpose.

15.14 HOT GAS SOLDERING

Here, hot gas soldering must not be confused with protective gas soldering, see Chapter 13, 13.5.9, and 15.7. Hot gas soldering is known mostly for its employment in the repair shop for surface mount components, see Chapter 14, 14.4.2. Hot gas soldering has also found application in the soldering of large backplanes and in cases where other methods cannot deal with large boards. It is typical that, in both cases, special equipment is designed for the purpose, for example, see Chapter 7, 7.5.1.

15.15 FURNACE SOLDERING

In this method, hot gas or infra-red heating is used or a combination of both. Belt soldering can be used in conjunction with a furnace. If furnace soldering is not applied in combination with other soldering methods in mass production, it may be applied in cases when boards are too large for the normal soldering equipment on hand and a furnace is available, or it may be used for special electronic soldering tasks.

15.16 ROBOT SOLDERING

A major advantage of robot soldering is that a really slow point-by-point soldering process is speeded up by using a robot and thus gains the characteristics of a mass soldering process. Processes which are speeded up in this way are hand soldering, tinning processes (de-golding) and laser soldering. Robot soldering is not limited to these but can be used for any sequential soldering process.

The reasons for using robot soldering are manifold. In small-scale production usually hand soldering is used. Sometimes, however, it is difficult to define what small-scale production is. If a product is manufactured in large quantities but in a great variety of types, each variant can be achieved only by a separate soldering process, while that part of the product common to *all* variants can be made in a true mass soldering process.[15] In normal cases, the separate soldering process will be performed by hand soldering with a soldering iron — but there is another solution. The soldering iron can be handled by a robot which is controlled by a computer. All the points to be soldered are programmed with the help of the teach-in technique, together with the soldering temperature required, the soldering time and amount of solder (which is wire fed) for each specific solder joint. Compared with hand soldering, the quality of the solder joints is improved, the output of printed board assemblies can be doubled and, by this, the investment of, say, $97,700 for the robot equipment needed can be paid back in 1.5 years.

This technique can be applied in similar cases, e.g., when heat sensitive components must be soldered separately at soldering temperatures lower than the norm and when components or material cannot be subjected to aggressive fluxes or to the cleaning process which is normally used for the board. In addition, it can be used on double-sided boards containing a number of components mounted in plated-through holes.

Robot techniques are also useful in tinning processes for small component terminations as it is possible to dip the termination into the bath with great precision at the desired angles for the defined times. They are excellently suited to de-golding component leads in order to improve the quality and the lifetime of the joint by avoiding the formation of brittle gold-tin intermetallics.

Finally, laser soldering can be performed without a robot, but a robot is the natural complement to a laser as the laser energy and the dwell time also have to be computer controlled.

15.17 ULTRASONIC (US) SOLDERING

An ultrasonic generator is placed either in the soldering iron or in the soldering bath, as appropriate, in order to improve the soldering result. Ultrasonic soldering is rarely used in normal soldering such as soldering of printed boards.

The application of ultrasonics is sometimes suggested as a solution for solderability problems. If such a suggestion should be necessary in a mass production operation, one is facing fundamental problems which should be resolved in the proper way by analysing the problem, determining the correct solution and using it.

Ultrasonic soldering is used as a solution in the instances where particular wetting difficulties in soldering exist. This may be the case, for example, in aluminium soldering, tinning special gold electrodeposits or Alloy 42, German silver, beryllium bronze and other metals. It is, nevertheless, a mistaken belief that a good wetting result can be obtained without using flux. When fluxes are used, a very good and adherent coating of tin or solder can be achieved on the metals mentioned with the aid of ultrasonics.

15.18 HIGH FREQUENCY (HF) SOLDERING

As the name indicates, high frequencies are used to generate the heat in the parts to be soldered. High frequency soldering is an *induction* soldering process. This method is not used in the manufacture of printed board assemblies but can be used for soldering or preparing electronic components for soldering. The problems lie in the correct design of the coils which are used and in selecting the right frequency, which is not always very easy. When using high frequency, it is easy to overheat the solder joints. In other words, it is difficult to stop the input of heat at the right point on the heating curve.

REFERENCES

1 Barnes, A. F. C., Elliott, V. B. and Strauss, R. S., British Patent No. 798 701, 'Improvements relating to Soldering Components to Printed Circuits', 3 October (1956).
2 'An Introduction to Surface Mounting Techniques', Thomson Customer Application Laboratory (1989).
3 Strauss, R. S. 'Wave Soldering versus Reflow Soldering, Metallurgical Consequences of the Choice between the Two Methods', *Brazing & Soldering*, No. 14, pp. 5-12, Spring (1988).
4 Holm, T., 'Controlled Atmospheres: Soldering of Printed Circuit Boards', AGA AB, Report REPM 93378 (1993).
5 Linman, D., 'Vapour Phase: The Third Generation', *Circuits Manufacturing*, pp. 26-28, March (1989).
6 Lea, C., 'Heat Transfer Fluids for Vapour Phase Soldering — An Appraisal', *Soldering & Surface Mount Technology*, No. 1, pp. 23-32, February (1989).
7 Headrik, L. G. and Ruffing, J. F., 'Vapour Phase and IR — Each has Reflow Soldering Advantages', 3M Industrial Chemical Products Division, St Paul, MN.
8 Möller, W., Knödler, D. and Vaihinger, K. U., 'Laser-Mikrolöten mit Temperatur und Zeitsteuerung', *Opto Elektronik Magazin*, No. 8 (1988).
9 Quillam, R. M. and Flicos, D. M., 'Reliable Solder Joints', BABS Seminar 'The Metallurgy of Soldered Joints in Electronic Assemblies', London, May (1991).
10 Harris, P. G., Chaggar, K. S. and Whitmore, M. A., 'The Effect of Ageing on the Microstructure

of 60:40 Tin-lead Solders', *Soldering & Surface Mount Technology*, No. 7, pp. 20-23, February (1991).
11 Ringle, H. and Vaihinger, K. U., 'Präzisionsinstrument. Wann Laserlöten heute schon wirtschaftlich ist', *Elektronikpraxis*, No. 15, pp. 32-37, August (1989).
12 Dow, S. and Helton, D., 'The Use of Collimated Infrared Light', *Circuits Assembly*, **Vol. 1**, No. 2, pp. 28-30, November (1990).
13 'Optical Soldering Promotes Automated Packaging', *Electronics*, pp. 90-94, 22 March (1963).
14 Isert, H., 'Weichlöten an durchkontaktierten Leiterplatten mit Hilfe fokussierter Infrarotstrahlen', *Feinwerktechnik*, **Vol. 73**, No. 4, pp. 185-191 (1969).
15 Ebensberger, 'Lötkolben in Roboter Hand', *Flexible Automation Betriebstechnik*, No. 3 (1987).

Chapter 16

MICROJOINING METHODS

GERT BECKER
Consultant, Sköndal, Sweden

16.1 INTRODUCTION

In this chapter, other methods of forming 'permanent' interconnections between electronics components, their packages and printed boards are discussed. Many of the methods outlined come from microcircuit fabrication, microjoining methods being adapted for use in printed board technology.

In the search to solve the problems of connecting chip components, see Chapters 3 and 4, with the package in a reliable and, especially, cheap manner, many different ways are being tried of using and adapting well-known techniques. The costs of microjoining techniques as a part of the total cost of a product ranges from 35 to 75%. The number of such joints made worldwide per year was estimated in the late eighties to be 10^{14}, see Reference 1. The trend for microjoining in the mid-nineties is for tape automated bonding (TAB) to be used more and wire bonding less, although wire bonding is important in HF technology. In the future probably flip-chip technology will replace TAB technology.

One general aspect of the following joining methods is that they are not so universally used in industry as the methods described in the preceding chapters in Section 5. Even though certain of them are used in mass production, they are used only under certain circumstances, e.g., wire bonding mainly in the semiconductor industry and tape automated bonding primarily in the watch and calculator industries. The flip-chip, beam lead and the chip-on-board (COB) methods are not even widely used in special industries but rather in certain companies which have developed these methods to meet their specific needs.

The methods mentioned here give impetus to the further development of electronic technology towards smaller, lighter, cheaper, faster and environmentally more acceptable circuit realisations than achievable with the older electronic technology, Table 16.1.[2] Their use will spread with continuing miniaturisation and the need for greater reliability, Table 16.2,[2] (which is achieved by avoiding unnecessary joints), as well as with the need

Table 16.1

A Comparison of Typical Printed Boards with MCM

	PB	MCM
Size of substrate	Large	Small
Weight/surface area	Smaller	Larger
Number of layers	6 to 8	2 to 4
Percentage of surface used (%)	5	75
Lead tightness (cm/cm^2) [Note 1]	30	500
Number of joints needed to mount the board	100	16
Power consumption (W)	16	10
Heat conduction	Lower	Higher
Clock frequency	Lower	Higher
Number of processes to manufacture	Many	Few
Bath volume (e.g., for electroplating)	Large	Small
Investment required	Smaller	Larger

Note 1: 'Lead tightness' is the length of conductor in centimetres which can be placed on a surface of 1 square centimetre.

Table 16.2

The Reliability of Different Joining Methods

	PTH	FPT	Wire Bond	TAB	Flip-chip
Failures (ppm)	100-10,000	200	20-100	10	2-5

for shorter leads between the components due to the higher clock frequencies used in many circuits.

It is characteristic of these connection methods that the different needs both for bonding and cooling, which is a very serious problem, see Chapter 11, have developed many different mounting techniques with different names.

16.2   THE PRINCIPLES OF METALLURGICAL JOINING METHODS

This covers those methods described in which a metallurgical bond is formed between the parts being joined.

Every method has its pros and cons. Naturally these should be investigated thoroughly when looking for a new method or process to be used. From Table 16.3 it can be seen that, even if some processes have different names, they equate to each other in the joining method, in the equipment needed or in the materials joined to each other. For this reason, these similarities are discussed before the different methods are described.

Table 16.3

The Similarity between Different Application Methods

Process	Joining Method				Equipment			Material	
	S	R	D	US	W	T	Au	Sn	Al
Tape automated bonding (TAB)	x	x	x	-	x	x	x	x	-
Beam lead	x	x	x	-	x	x	x	x	-
Flip-chip	x	-	-	-	-	-	x	x	-
Chip-on-board (COB)	x	x	-	-	x	-	x	x	-
Multichip module (MCM)	x	x	-	x	x	-	x	-	x
Wire bonding	-	x	-	x	-	-	x	-	x
Isothermal bonding	-	-	x	-	x	x	x	x	-
Diffusion bonding	-	-	x	-	x	x	x	x	x

Key: S   = soldering method   W = use of welding equipment as energy source
R   = resistance welding   T   = use of thermode
D   = diffusion welding   Au = gold; Sn = tin; Al = aluminium
US = ultrasonic welding   x   = applicable to process; - = not applicable

The soldering or welding methods used are advanced methods derived from normal soldering and welding techniques by reducing sizes and amounts of material and using greatly improved equipment.

The metallurgy used is of utmost importance for the success and reliability of the joints or for most of the failures which occur. Moreover, the principles and metallurgy of the majority of the systems are quite alike and much can be learned from one system for application to another.

16.2.1 Soldering Methods

As soldering is a fairly inexpensive, well-known, simple, flexible and reliable method which can be used in manual production as well as mass production, it is natural to use it whenever possible. Application of the soldering material to miniaturised sizes can be made at the joint site by well-known processes such as electrolytic or evaporation processes. Equipment, processes and controls for soldering methods like infra-red soldering, thermode soldering, hot bar soldering and others, see Chapter 15, had to be developed to give better regulated ones, suitable for the materials used and sizes needed. To avoid oxidation of the solder, it is necessary in most processes to use a controlled nitrogen atmosphere or, if the soldering is carried out at temperatures above 300°C, a reducing atmosphere.

16.2.2 Welding Methods

With these, there is a slight confusion in terminology. Quite a number of welding methods have been improved and applied as microjoining techniques which were given very different names by different authors. The term 'bonding' is frequently used, e.g., as in 'solid state bonding' to differentiate diffusion welding from melt (fusion) welding methods. To clarify this for the uninitiated, as an example, diffusion bonding is not a soldering process but a welding process. For the comparison of different applications, the different bonding methods are referred back to the correct welding term. However, to avoid confusing the reader, the term 'bonding' is used in 16.3 as it appears commonly in the literature.

Different welding methods are used, such as resistance, diffusion, ultrasonic and explosive welding. A common factor of the welding methods used is that, when required, the welds very often can be repaired relatively easily by using a precision chisel to cut away the previous material.

16.2.2.1 RESISTANCE WELDING

Different types of electrodes can be used, an example being shown in Figure 16.1.

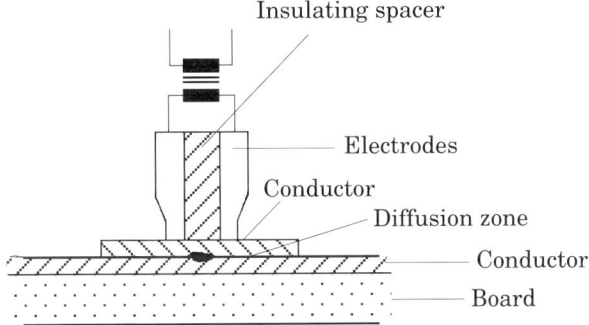

Fig. 16.1 Parallel gap welding electrodes as used in the TAB, beam lead and isothermal techniques.

The material must not be dirty so that the welding voltage can break through the surface. On the other hand, if the material is too clean the resistance necessary to generate the heat for welding cannot be built up between the parts. Figure 16.2³ demonstrates how the different contact resistances are distributed on a material. This, of course, influences the result of the welding. This is one of the reasons why gold is preferred in interconnection techniques as it has a quite uniform contact resistance. The uniformity and strength of the joint formed can depend on the direction of the electric current (polarity, Peltier effect) and a bad joint can result.

16.2.2.2 DIFFUSION WELDING

This is a metallurgical process in which parts of the same or different suitable metal(s) are joined by pressing them together and subjecting them to a temperature below the melting point of the metal(s) for a given time. The connection is established by diffusion of the atoms at the contact points. Methods such as wire bonding, tape automated bonding, beam lead technology, methods of repairing leads on printed boards, see Chapter 14, 14.4.1, isothermal soldering, the SLID process, see 16.3.5, impulse bonding and others belong to this class. Thermocompression wire bonding is a special

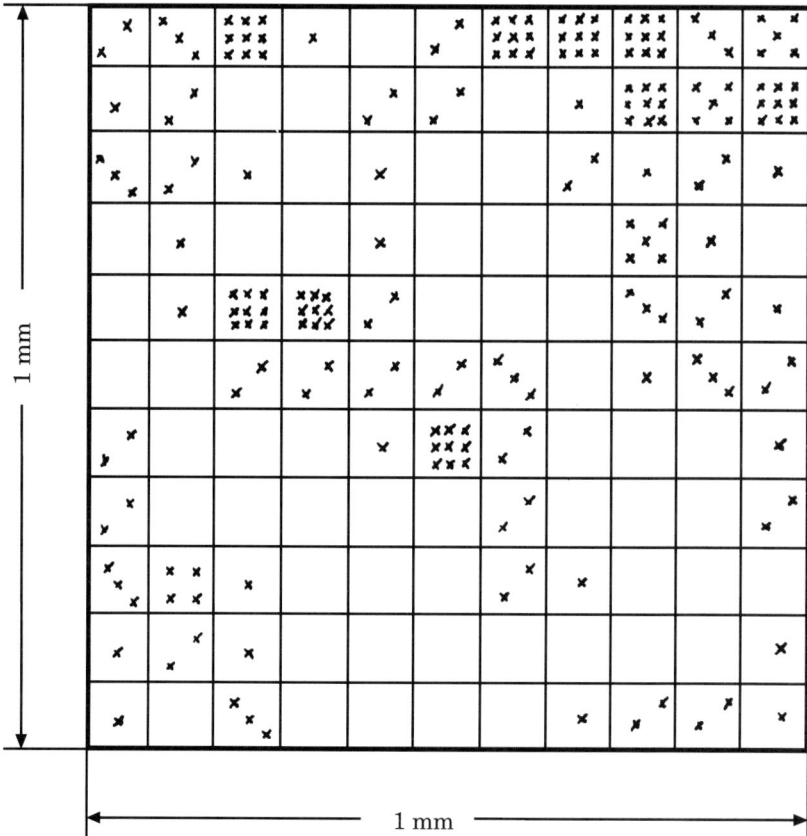

Fig. 16.2 The distribution of the contact resistance on a piece of copper.[3] Both the contact resistance and the voltage required to break through the insulating layer can vary widely in very small areas. This explains the differing results which can be obtained in both resistance and diffusion welding.
Nine crosses correspond to values >1 kΩ.
Zero crosses correspond to values <1 mΩ.

form of diffusion welding, mainly using a gold wire connected to gold surfaces. The joining method in both the wire wrap and explosion welding procedures belongs to the cold pressure welding class and thus to the diffusion welding methods. Often the electrodes used are in the form of two closely spaced fingers separated by an insulating material, 'parallel gap electrodes', Figure 16.1.

The material used in this joining method is preferably gold. Copper as a material has been tried but with no greater success. The tough copper oxide layer needs a deformation of 30-40% and a copper sulphide layer (also tough) is difficult to destroy with thermocompression bonding at 525°C. Reference 4 reports successful copper ball/wedge bonding to copper thick film in an atmosphere of $N_2/10\%H_2$. It is important that cooling down of the formed joint takes place under pressure. The joints are significantly stronger than those with gold.

Every thermocompression bonding method implies a risk of damaging the bond between the copper lead and the board. The fibreglass in reinforced boards can be cracked.

16.2.2.3 ULTRASONIC (US) WELDING

The welding process is a transverse rubbing action and needs specially designed electrodes which must be adjusted carefully. Normally no heat is needed or generated. US welding is especially suitable for joining metals which have brittle oxides, such as aluminium, or no oxides at all. It is used to join aluminium to aluminium or aluminium wires to gold surfaces. To give uniform and high joint strength the wire used must be free from lubricants and the material must be homogeneous. The aluminium diffuses faster into gold than *vice versa*, forming voids in the gold which isolate the gold/aluminium joint from the surrounding metallurgical structure.[5] Thermal cycling causes cracks in the brittle compound of gold and aluminium. Plastic outgassing induces wire bond failures.[6] US bonding is a slow process and can be used in one direction only on the chip.

16.2.3 The Equipment

For successful microjoining, it is of great importance that the equipment is not *just* suitable for the purpose. When a welding machine is used as a power source for welding or soldering, the choice of an alternating current, a direct current or a capacitor (or 'condenser') discharge machine with precise controls must be considered carefully.[7] In a number of cases the power must be switched off, to allow the joint to become solid, before the thermode leaves the joint. Since first used for joining in microelectronics, the machines have been developed considerably and adjusted to the special requirements.

A constant welding force during the welding process is of utmost importance for a good welding result. Therefore it must be checked that the electrode is easily movable and so can apply a constant welding force. This movement can be used to switch off the energy at the right moment when the right quality and correct height of the weld are reached.

The geometry of the surface must fulfil certain conditions, both for soldering and welding processes, otherwise it may, for example, be difficult to obtain reproducible results. When welding, this problem can sometimes be eliminated by using a two pulse welding machine in which the first pulse heats the parts so that they can be pressed together tightly and the second pulse is the actual welding pulse.

Another problem is that successive joints can become of inferior quality, even if everything in the cycle appears to be the same. A check of the cables may reveal that some of the copper strands are broken, thus not allowing the cable to transfer the correct amount of power. In this instance, a metallurgically sound bond on the ends of the cable should solve the problem.

It is of great importance that the surfaces to be joined lie parallel to each other. This is natural when using parallel gap welding but may be difficult when laser joining TABs. With laser joining, a number of problems mentioned here can be overcome. Especially significant is the fact that maintenance of welding electrodes is expensive and saving that will pay a great part of the cost of a laser equipment. These and other factors have a considerable influence on the selection of the different types of welding machine.

16.2.4 The Materials

16.2.4.1 MATERIALS TO BE JOINED

There are always two materials joined to each other making one joint. If more than two materials have to be joined or more than two types of joint need to be made simultaneously, with only one type of electrode head difficulties must be expected.

The three main metals in microelectronic joining are gold, tin, including tin-lead, and aluminium. This is due to a number of advantages. Both gold and tin are easily deposited by electrolysis or by vacuum. They are easily joined by different methods. The bond strength and electrical conductivity are reasonable. When soldering gold to tin, see Chapter 13, 13.3.4.4, normally brittle phases can be obtained. This can be avoided by using either a thick tin layer (10-20 μm) and a gold flash over copper or a thick gold layer (25 μm) and a thin tin layer (0.5 μm). Due to the layer thickness giving rise to the risk of diffusion and oxidation prior to use, it is necessary to store the material in a protective nitrogen atmosphere and to use it fast.

When using copper, especially electrolytically deposited copper, plated with tin, the disadvantage is the formation of brittle copper-tin crystals. This can be avoided by depositing a nickel barrier coat. Gold wires used are in the range of 12 to 500 μm in diameter and are alloyed with small amounts of beryllium-copper to control grain growth during welding or with magnesium and silicon to give an alloy which will withstand short-term exposure to a temperature of 300°C during welding and long-term exposure at temperatures up to 150°C without loss of strength. One failure mode, which is characteristic for welding methods, is cracking in the material brought about by the large deformation and sharp notches caused by the electrodes.

Aluminium is mainly used for its good properties in ultrasonic welding and for its superior electrical conductivity. Pure aluminium wire, or an alloy of aluminium and silicon, is used in ultrasonic welding with a wire diameter of 25 μm. In special applications wire diameters up to 500 μm are used. Aluminium in ultrasonic welding has the disadvantage that only a fixed direction of welding motion is possible. Failures of aluminium-silicon wire are mostly due to work hardening at the heel of the wire.

16.2.4.2 CHIPS, DICE OR DIE (SEE ALSO CHAPTERS 3 AND 4)

For all joining methods involving active chip components, the problem is to buy tested, un-packaged chips or wafers with a certified yield. Depending on the manufacturing process of the active part of the bonding pad, different metal layers are laid on to the chip forming the bump needed for the connection, Figure 16.3. The size normally is 125 μm but can be made larger or smaller, as can the insulation distance between the bumps.

The bump can have a height of 15-75 μm when the outer layer is of gold or 50-100 μm when made of Sn90Pb10. The corresponding part of the bearer

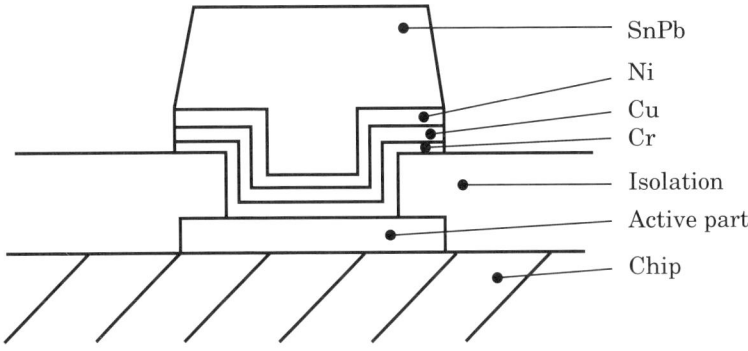

Fig. 16.3 A possible design for a solder bump on a chip component. See also Fig. 4.35.

material (board of different kind, tape, etc.), to which the chip is fastened, is usually coated with the same material, often with a different thickness. A large number of bump materials and thickness values have been evaluated to determine those which give the best adhesion to the silicon and the best bump quality for joining. Examples of layer materials are:

    titanium — palladium — gold (Ti-Pd-Au)
    titanium — copper — nickel — gold (Ti-Cu-Ni-Au)
    aluminium — chromium — nickel —copper (Al-Cr-Ni-Cu)
    titanium — tungsten — gold (Ti-W-Au).

16.2.4.3 ELECTRODES AND THERMODES

Electrodes are used in melt or fusion welding processes. When they are used in diffusion welding processes, they are called *thermodes*. The choice of the materials for the electrode or thermode is important. The heat transfer properties are important in the case of certain TAB and beam lead bonding applications. Inconel and diamond-tipped Inconel 718 thermodes have been specially developed for device bonding.[8]

One thing which must be checked regularly is the condition of the electrode or thermode. The surface of an electrode can change its electrical properties by alloying with the material to be welded or soldered, thus causing problems. Because of this, it is fitting to choose the electrode material not necessarily with regard to the basis material to be joined, e.g., copper, but rather with regard to the surface layer, which could be tin. Alloying can be, if not avoided, considerably retarded by chilling the contact surface of the electrode to around 0°C. If such alloying *is* detected, the process has run out of control and must be stopped until the correct contact surface is restored. While using the electrode/thermode, it can be deformed, de-adjusted or it will wear. The surfaces of the electrodes should be checked regularly to ensure they are still parallel to each other, are not worn, burnt or otherwise defective. The pressure between the electrode and its counterpart has to be checked regularly and adjusted if necessary. The wrong measurement may mean that the electrodes slip in the electrode holder.

16.2.5 Process Control

Today the requirements for both quality and reliability of joints exceed those of only ten years ago by several powers of ten. This is a consequence of yield problems in the workshop arising for highly complex circuitries. The problems are simply economic problems. A yield problem can mean that the product cannot be manufactured or, at least, not in the given environments with the equipment available. As materials become more sophisticated and joints smaller and smaller, good quality and reliability can only be achieved by analysing the preconditions and following the laws nature has given us. Especially in microjoining, it does not pay to ignore problems in an attempt to be faster or cheaper. What has been said for the soldering process in Chapter 12 (e.g., 12.3, 12.5, 12.10) is also valid here, maybe even more so.

The material, the equipment, the process and even the failure modes must be properly specified to achieve good results and the operators must be well trained.

The soldering process is forgiving to a certain extent. Microjoining processes are not. The quality of wire wrap joints is good partly because of the tight checks required — which should be self-evident. If such specified checks were the normal case in soldering, the failure rate of solder joints would decrease dramatically to extremely small values. Here the workshop has a really serious responsibility to monitor and measure changes in the process and to use all means, including statistical analysis and other methods, in order to detect problems, see Volume 2, Chapter 19.

The way to avoid problems is to use a reliable process and to keep the process stable. Visual inspection allows operators only to sort out obvious failures. There is no method available to measure bond failures, except for certain electrical measurements, and none at all for detecting diminished bond reliability. Here it may be truly said that the quality has to be built into the product and cannot be tested or inspected into the product.

For a lack of other methods, quality is often tested using methods which are quite questionable. For wire bonds the joint is frequently tested using the pull test. A certain percentage of the rupture value (which is often determined only by experiment) shall be achieved as a pass value. The empirical rupture value very often has not been checked against theoretically possible values. If such a pull test is carried out with success, who can swear that the joint has not been destroyed by the test?

One of the most common failure modes in the different welding processes is fracture initiated by sharp transitions in the welded material, essentially caused by the electrode/thermode.

16.3 DIFFERENT JOINING METHODS

16.3.1 Wire Bonding

Wire bonding is derived from and closely related to semiconductor packaging. In this technique very small dimensions have been used for many years. A great advantage is that materials used for the manufacture and packaging of the chips, such as gold, are quite suitable for the wire bonding methods.

Wire bonding is a welding method, more specifically a diffusion welding method, where gold or aluminium wire is used. If the wire was welded perpendicular to the surface, the strength of the weld would be insufficient. Therefore, two different methods have been introduced, each having its own advantages and disadvantages, ball bonding and wedge bonding, Figure 16.4.[9]

In ball bonding the fine wire is fed through a capillary and a ball is formed by melting the end of the wire with a gas flame or electrical spark. The ball is pressed against the land and, normally, heated by a pulsed current to form the connection. As shown in Figure 16.4(a) the other end of the connection is wedge bonded using the side of the capillary. The wire is then broken off or flame cut and a ball is formed for the next operation.

Wedge bonding is illustrated in Figure 16.4(b). A hardened shaped tool (e.g., of tungsten carbide) is used to apply pressure and heat to the wire to form the first joint. The wire is then released by the clamp and the tool tracked back to form the second joint. Finally, the wire is re-clamped and pulled back to break the wire free while the joint is still under pressure from the bonding tool.

An alternative method which may be employed when using aluminium wire is cold pressure welding. In this case ultrasonic welding is used. Sometimes the ultrasonic electrodes are heated. The chips and/or substrates may also be heated, either alone or in conjunction with the electrodes.

Ball bonding gives the greater strength and allows more lead flexing. The wire bonding technique covers the whole field from small-scale, semi-automatic production to highly automated mass production. A wedge bond will probably fail before a ball bond and a gold/aluminium wedge type bond will fail more often than an aluminium to aluminium bond. It is interesting to note that, according to Fitch,[10] the degradation of the wire bond starts from the first cycle.

Defective aluminium wire bonds account for a high percentage of hybrid microcircuit failures.[11] On one IC, such as used in a dual-in-line package (DIP), relatively few bonds are made. As such ICs are relatively low-cost, it is possible to test the ICs thoroughly and throw away the faulty ones. This leads to a very high bonding quality level in those components. Compared with TAB, yield losses are obtained when using wire bonding for circuit chips with more than 100 bond pads.

16.3.2 Tape Automated Bonding (TAB)

Tape automated bonding is a relatively old method, see Chapter 4, 4.4.16. Factors which have inhibited wider use are the cost of the materials and tools involved (which could only be recovered on real mass production) and a lack of suitable components for all the possible applications. In addition, the TAB film was not standardised and there was no great need for the method. TAB thus only found wide application where naked chips were to be connected to a board without using DIPs or in DIP fabrication itself.

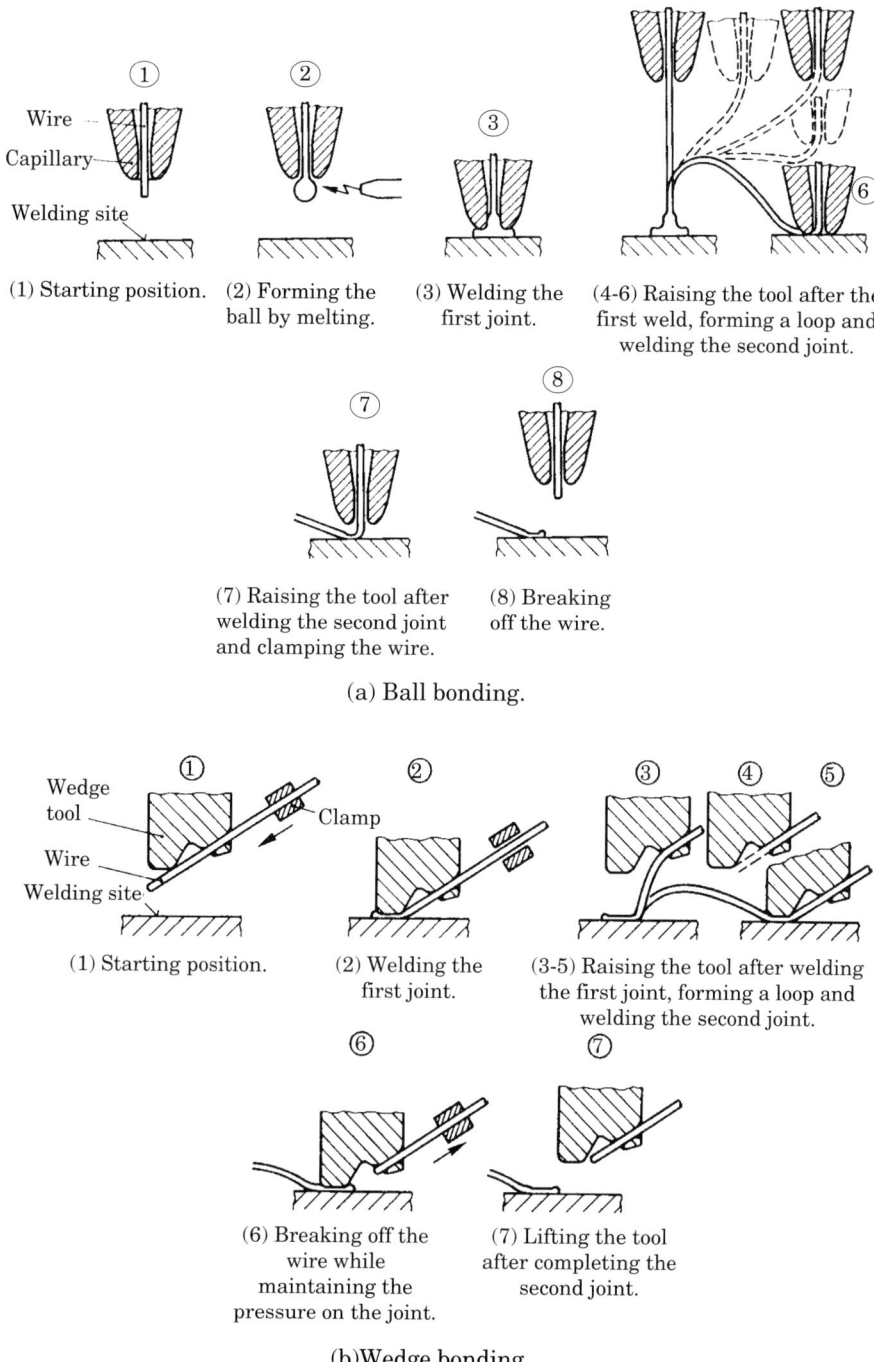

Fig. 16.4 Ball and wedge bonding methods.

The principle of tape automated bonding is shown in Figure 16.5. In the TAB method a tape is used bearing inner leads to which the circuit chip is connected ('ILB' — *Inner Lead Bonding*). After this it is possible to test the electrical function of the chips. The chip is then connected to the substrate, which may be ceramic for hybrids or thick film circuits, glass, ceramic or fused silica for thin film circuits or glass/epoxide for, e.g., SMT, using the outer leads ('OLB' — *Outer Lead Bonding*). This explains why different joining materials and methods can be chosen. The TAB component can be encapsulated or 'glob topped', see Chapter 4, 4.4.16, with polyimide, epoxy or silicone plastics at any stage after testing.

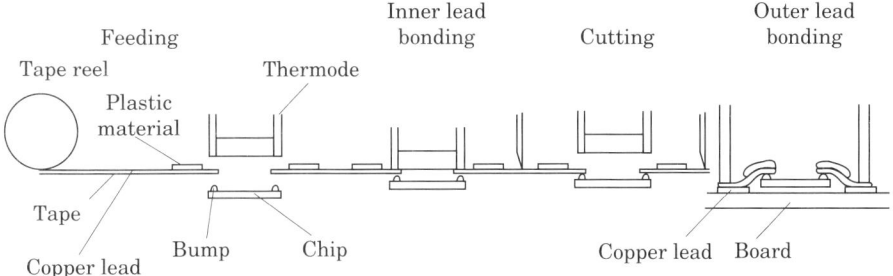

Fig. 16.5 The principle of tape automated bonding.

The material used in the joining process depends on the components to be mounted on the tape, on the tape material and on the substrate to which the tape leads are to be joined. Rolled copper, half hard, with a gold flash is recommended for the tape. The plating most used on both inner and outer leads is Au-Au or Au-Sn. The high quality of the joint is due largely to the high melting point of the joint. However, Sn-Sn or solder is also used. It is a high quality advantage that no other material is involved in the formation of the joint. No flux, with all its disadvantages as in soldering, is needed. Only when solder is used for the outer lead bonding (15-20 µm on the copper lead) can the joint strength be inferior. The board material has to be preheated up to 70-130°C and the flux has to be carefully cleaned off. The solderability can be doubtful if the tape is not stored in an nitrogen atmosphere and used up in a short time.

The materials may be joined by soldering or by the different diffusion welding methods with a hot bar or thermode, see Chapter 15, 15.11. The hot bar has the disadvantage that it may crack large integrated circuits with many terminations, as the pressure needed increases with the number of bumps. Because of this, diffusion welding, as well as soldering, has been tried with a laser, Figure 16.6,[12] which does not need physical contact for joining.

The leads of the tape can be formed in different ways to bridge height differences or to compensate for the tensions on the bond induced by the heat developed in the circuit chip. The tape can be made of different insulating plastic materials, e.g., polyimide (Kapton) and polyester (Mylar), polyester being used for temporary packages. Different electrical conducting materials (e.g., different types of copper with different coatings) and different numbers of layers (e.g., one or two layers of copper on plastic film) may be used. The different plastics and adhesives, and their different thicknesses, e.g., for two

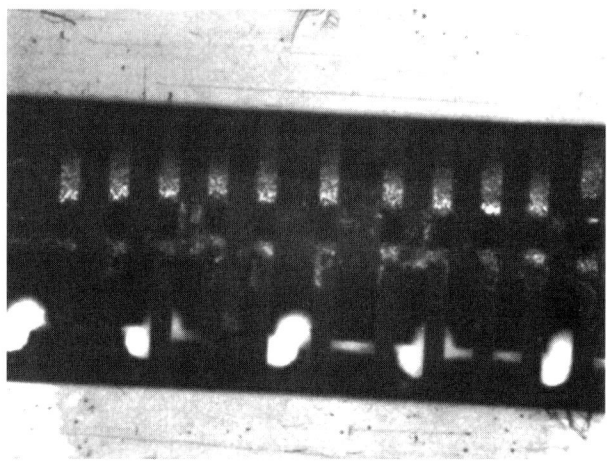

Fig. 16.6 TAB leads soldered using a laser.[12] Soldering tinned copper foil to gold-plated ceramics. Lead thickness 35 μm, lead width 50 μm, tinning 0.8 μm, gold plating 1.5 mm x 1.5 mm.

and three layer film, can cause different water absorption and give rise to problems with insulation resistance or corrosion. The design possibilities are numerous.

The copper leads will recrystallise when bonded but this does not cause any problems. Whiskers can be difficult to detect but can be avoided by suitable choice of materials or tinning methods, see Chapter 13, 13.7.6.

Tape widths of 8, 16, 18, 20, 35 and 70 mm have been reported. TAB needs only little space for fastening to the component and the component height will be reduced to a minimum. Multilayer TAB development is going to 800 inputs/outputs per die.

16.3.3 Flip-chip

Here the die is provided with bumps as described in 16.2.4.2 and is then soldered or ultrasonically bonded to the carrier material. Because there have been many problems in reliability (due to thermal fatigue of the solder joints), in keeping the height of the solder joint constant, and in the production technique which needs high measurement accuracy, this technique has not found wider appeal. It is, however, possible that with better understanding of the thermal fatigue and material data which will become available, and better production techniques, solutions will be found for these problems.

In such a case flip-chip technology will experience a renaissance, for it will then have the advantages of simplicity, high reliability, low cost, and of using little material and space and thus being friendly to the environment. The methods used for chip attachment are ultrasonic welding, thermocompression bonding and soldering.

Automation, which would improve reliability, is only justified by large quantities. The flip-chip technique requires electrolytically plated solder pads, is difficult to repair and the joint quality cannot be inspected.

16.3.4 Beam Lead

The beam lead technique was developed to provide circuit chips with short leads by which the chips could be packaged. Dense packaging and high quality joints were produced. The leads were manufactured by electroplating during the circuit fabrication and then removed from the silicon substrate by etching away the excess silicon to leave 'cantilever beam' leads.

The beam lead method has been the model for other joining techniques; it was altered and used in different ways. Small gold-plated metal ribbons were used to provide components with longer leads, to repair leads on printed boards, Figure 16.7, and to increase the design scope of printed boards by bridging one lead to connect two non-adjacent leads.

(a) This weld can be approved. The melting zone has a well defined and uniform width.

(b) This weld cannot be approved. The welding zone is not sufficiently wide and/or uniform.

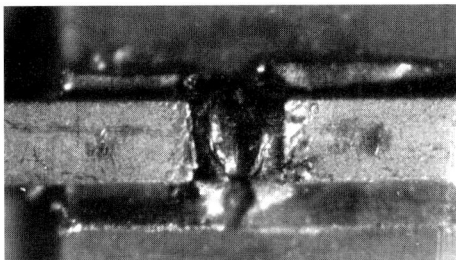

(c) This weld cannot be approved. Both ribbon and lead are destroyed.

Fig. 16.7 Repair welding of leads on printed boards using a gold-coated ribbon and a diffusion welding method.

A thorough knowledge of the method and detailed specification thereof is needed if a sound joint is to be obtained. The leads are gold coated and joined using a resistance welder and parallel gap electrodes to provide the heat necessary for the production of a diffusion joint. In the case of a gold-to-gold joint, a thermocompression joint can be produced.

16.3.5 Isothermal Soldering

There are cases where solder joints between component leads and PBs are subjected to high stresses by thermal cycling to high working temperatures causing both lifetime and reliability problems. One example is the case of high frequency component connections. These joints cannot be made with normal welding processes as the high temperatures involved would destroy the material. Therefore new soldering methods were developed which were called isothermal soldering,[13] the SLID process (*Solid Liquid Interdiffusion*),[14] or simply diffusion soldering.[15]

The result is a higher strength at a temperature higher than the melting point of the soldering material used. Isothermal soldering uses the system Sn-Cu. Cu-In, Ag-Sn and Ni-Sn have also been suggested. According to the phase diagram the bond is made at 250°C. The joint withstands creep at temperatures above 250°C. High temperature phases are formed by diffusion in the presence of a liquid in both isothermal soldering and the SLID process which uses the systems Ag-In, Au-In and Cu-In. With these, high temperature stable bonds are made when sealing flat packs or making joints at low temperatures.

16.3.6 Wire Wrap (see also Volume 2, Chapter 14, 14.1.2.1 *et seq.*)

Wire wrap is a connecting method in which two or more contact pins fastened in a PWB are connected by a wire which is wrapped four to six or seven times around them, see Volume 2, Chapter 14, Figure 14.1. Thus this method is a cold pressure diffusion welding method.[16] A number of wires can be wrapped on to one pin.[17] The contact pins are generally of gold-plated tin-bronze or a similar material and the wire either of tin-bronze or of silver-plated copper. A gastight contact is achieved in the deformed zones between the wire and the pin.

Correct choice of the insulation material is of importance. It is necessary to remove the insulation around the wire if a reliable contact is to be achieved. If the wire is pressed against other wrap posts, the insulation can be cut through with electrical malfunction as a consequence. An important matter to consider is the available space for the wire between the posts. The long-term reliability of the post (contact pin) in the plated-through hole of the PB may be a problem unless very closely controlled.

This method of wiring and jointing is often used in experimental set-ups to save time and money but it is used in the mass production of printed boards where the dimensions of the board exceed the dimension of normal soldering machines.

For wire wrapping, tight control of the tools and the methods used is essential.

16.3.7 Explosive Welding

Explosive welding does not sound as though it would be employed in printed circuit joining technology. However, this method has actually been used by Western Electric[18] to reclad defective gold contact fingers of printed boards with a gold foil. Materials bonded in this manner by Western Electric have included among others copper to nickel, stainless steel to copper and gold-plated copper to tantalum.

Potentially harmful byproducts of detonation create an environmental problem which can be solved by effective filtration.

The method can be used in other electronic applications, especially where the need arises to join metals which are considered incompatible. The material properties are not changed and the explosive force does not affect the surrounding areas. This is common when using other types of fusion bonding. Specially clean surfaces are not needed and only a relatively simple set-up is required.

16.3.8 Adhesives

In contrast to soldering and welding, joining with adhesives is carried out with organic materials at relatively low temperatures. For microjoining purposes epoxide, acrylate and cyanoacrylate based adhesives are mainly used. The adhesives are applied by dispensing, stamping or by a screening process.

Curing can be carried out in a furnace, an infra-red oven, with microwave or high frequency heating and/or with ultra-violet light. In the latter case, it should be observed that the curing action will not take place unless UV can contact the adhesive, see Chapter 7, 7.2. Adhesives, at one time mainly used to keep components in place on the boards, have for many years been used successfully with metallic fillers to make electrical joints, see Chapter 7, 7.4. The small metallic particles which make the adhesive conductive have a given form and size and are normally of silver, although gold or aluminium and certain types of carbon can be used. The thermosetting plastic binder is usually an epoxide system which also provides a protective coating over the termination.

Adhesives have been used to join components to a board for many years with good success in cases where no other joining methods were possible, where it was more feasible in production and, apparently, the requirements for lifetime and/or quality were not the highest ones. There are some doubts as to how metal particles behave in plastic material under thermal stress for a longer time and, in particular, whether or not the electrical contact between the metal particles remains uniform. For example, it is known that epoxide systems age when subjected to higher temperatures.

The main disadvantage in use seems to be that the method is slow since the adhesive system must cure. In some instances it can be useful to combine the use of two adhesives, one fast hardening adhesive, say a cyanoacrylate, to position and a stronger one, say an epoxide, which needs more time to cure fully but gives a good, structural joint.

When employing heat-curing adhesives or those which generate heat during polymerisation, it should be noted that the thermal load of the components during polymerisation may be limited.

It has been shown by Pape[19] that adhesives without metallic fillers or conductive polymers, see Volume 2, Chapter 12, 12.10, can be used to establish an electrical contact between two parts if the distance between the two parts is less than 1 µm and the surface roughness is approximately 100 nm. The connection shows low sensitivity to cracking and has high mechanical stability with low contact resistances. The bonded connections can be subjected to 127°C and 80% RH without failure.

16.3.9 Chip-on-board (COB) and Multichip Module (MCM)

Both of these may be viewed as a component, where a number of bare ICs and chips are mounted on a substrate. In the case of COB, the substrate is normally an organic one, such as epoxide/glass or polyimide, whereas in the case of MCM the substrate can be a high-density multilayer or it can be ceramic, silicon or even metal. These small packages, reminiscent of hybrid circuits, are then assembled on a normal PB. Connecting the IC/chip pads to the substrate conductors is done either by wire bonding, TAB or the flip-chip method. The ICs are then covered with a glob topping, 16.3.2, or other protective coating.

This gives a number of advantages such as saving space and minimising the area of printed boards needed. In addition, the board is lighter, less material is needed and the production line becomes more environmentally friendly due to reduction in the number of joints (less solder), fewer processes required, decreased power consumption and so on, Table 16.1. Higher clock frequencies can be used due to the shorter path lengths.

The major difficulty is in obtaining tested wafers or chips with certified yields from the supplier. The yield of the chips on delivered wafers is critically important and a poor yield can ruin a project. The connection can only be tested electrically, and that only if the circuit is not too complex.

Multichip modules are often encapsulated, hermetically or non-hermetically. Certain designs of MCM are called high density packages (HDPs) or high density interconnect technology (HDI). Constant developments are taking place in MCM technology, a major spur being to extend the use of MCM into the traditionally high volume, low cost commercial and industrial markets. One such development uses a highly innovative plastic encapsulation approach to form the package and protect the components.[20]

16.3.10 Board Wiring Techniques

16.3.10.1 IMPULSE BONDED WIRING

Impulse bonded wiring was developed some years ago in order to fabricate boards which were fast to produce, as dense as multilayer boards were,[21] and cheap to redesign to meet changing needs of the customer (or until the development was stabilised). The primary advantage of boards made in this way is that they can be produced in smaller lots.

In this technique all components are installed and soldered in place on to a specially made board. The interconnecting wiring is the last step in the process. Wiring and bonding are on the same side of the board which gives a relatively thin board. The high purity nickel wires are diffusion welded (by impulse bonding) to a 150 µm thick nickel land on the board which has a connection to the plated-through hole. The bonding layer on the nickel land to which the weld is made is of copper 3 µm thick.

This system was derived from the Stitchwire technique. A similar technique, but using polyurethane-insulated copper wire which is soldered to the land, was developed by Augat.

16.3.10.2 STITCHWIRE

This is a semi-automatic system of point-to-point interconnection in which a Teflon-insulated nickel wire is welded to either a gold-plated stainless steel pin or a specially prepared land on a proprietary laminate clad with a stainless steel/composite foil. Both stitchwire and impulse bonding are described and illustrated in Volume 2, Chapter 14, 14.1.2.

16.3.10.3 MULTIWIRE AND MICROWIRE

Multiwire and Microwire are fully automated interconnection systems which are described and illustrated in detail in Volume 2, Chapter 14 .

REFERENCES

1 NN, 'Leistungsfähige Chips fordern neue Einbautechniken', VDI-Nachrichten, 06-16 (1989).
2 Tolvgård, A., Private communication.
3 Becker, G., 'Finsvetsning inom Teleindustrien', *Svetsen*, Årg. 37, No.1, pp. 60-73, January (1978).
4 Winkle, R. V. and Cater, S. R., 'Copper Ball/Wedge Bonding to Copper Thick Film', *Hybrid Circuits*, No. 15, pp. 13-17, January (1988).
5 Wirsing, C. E., 'An Ultrasonic Bond, Monometallic, Gold Interconnection Technique for Integrated Packages', IBM Manssas, Virginia.
6 Thomas, R. E., Winchell, V., James, K. and Scharr, T., 'Plastic Outgassing Induced Wire Bond Failure', Semiconductor Research and Development Laboratories, Motorola Semiconductor Group, 5005 E. McDowell Road, Phoenix, Arizona 85008.
7 Vroomans, H. M., 'Mikrofügen von Metallen', Weld-Equip, Helmond, Netherlands.
8 Kershner, R. C. and Panousis, N. T., 'Diamond-tipped and Other New Thermodes for Device Bonding', *IEEE Transactions*, **Vol. CHMT-2**, No. 3, pp. 283-288, September (1979).
9 Lindner, K., 'Drahtbonden — Verfahren, Werkstoffe, Anwendungen', *Verbindungstechnik in der Elektronik*, H.2, pp. 77-81, Juni (1989).
10 Fitch, W. T., 'The Degradation of Bonding Wires and Sealing Glasses with Extended Thermal Cycling', Motorola, Inc., SPD, P.O. Box 20 906 M 402, Phoenix, Arizona 85 208.
11 Himmel, R. P. and Pratt, I. H., 'Analysis of Hybrid Microcircuit Failure Modes', ISHM Proceedings, Vancouver, pp. 347-352 (1976).
12 Becker, G., 'From Soldering Iron to Laser', *Hybrid Circuits*, No. 12, pp. 22-27, January (1987).
13 Bader, S., Gust, W. and Hieber, H., 'Entstehung und Aufbau isotherm erstarrter Schichtverbindungen', *Verbindungstechnik in der Elektronik*, Fellbach, pp. 227-231, Februar (1988).
14 Bernstein, L., 'Semiconductor Joining by the Solid-Liquid-Interdiffusion (SLID) Process', *Journal of the Electrochemical Society*, pp. 1282-1288, December (1966).

15 Humpston, G. and Jacobson, D. M., 'New Diffusion Soldering Processes for Electrical Applications', BABS Seminar 'The Metallurgy of Soldered Joints in Electronic Assemblies', London, May (1991).
16 'Solderless Wrapped Connections', *The Bell Telephone System Technical Journal*, **Vol. 32**, pp. 523-591, May (1953).
17 Potochnik, J., 'Solderless Wire Wrapping, Fast, Reliable, and Economical', *Assembly Engineering*, pp. 28-31, August, and pp. 40-44, September (1977).
18 Cranston, N. H., Machusak, D. A. and Skinkle, M. E., 'Small Area, High Energy Bonding', *Western Electric*, pp. 26-35, October (1978).
19 Pape, K., 'Elektrische Kontakte, hergestellt mit nichtgefüllten Klebstoffen', *Elektronik Produktion and Prüftechnik*, pp. 40-44, Mai (1989).
20 Fillion, R., Wojnarowski, R., Gorcyzca, T., Wildi, E. and Cole, H., 'Plastic Encapsulated MCM Technology for High Volume, Low Cost Electronics', *Circuit World*, **Vol. 21**, No. 2, pp. 28-31, January (1995).
21 Schulz, F. G., French, D. C., Criscenzo, B. E. and Klotz, P. T., 'Impulse-bonded Wiring is Economical Alternative to Multilayer Boards', *Electronics*, pp. 75-81, March 6 (1975).

Acknowledgement

It is obvious that a book on any technical subject is based on many years' experience and knowledge, not only that of the author but of many other people. Therefore, at this point, I would like to thank all those who helped me to learn about soldering. I would like to mention in particular Mr John Lind who opened my eyes to this special technique, Mr Stig Berglund who helped me with statistical analysis, Dr Hector Steen with whom I had many fruitful discussions concerning metallurgy and Mr Arne Biverstedt who taught me about the chemistry of fluxes. I much appreciated the help of the last two gentlemen when writing certain chapters of this book.

I also want to thank the many companies who helped with both pictures and much valuable information. In addition, as the study of soldering is not only theoretical but very practical, I am grateful to many observant and careful operators in the workshop.

My special thanks go to the Ericsson Company which gave me the opportunity to specialise in this field and allowed me to present the knowledge I have gained in the company for the benefit of others.

Gert Becker

Author Index

(See also separate Subject Index)

Adams, J. 613
Allen, B. M. 612, 613
Appleton, Sir Edward 29
Ashman, J. 536

Bachman, W. S. 29
Bader, S. 743
Bader, W. G. 613, 666
Bandyopadhyay, N. 327
Barnes, A. F. C. 724
Bauer, B.B. 29
Bechtold, J. 684
Becker, G. 170, 612, 613, 666, 667, 683, 743
Beester, M. H. 666
Bernstein, L. 743
Beuhler, E. 4
Biverstedt, A. 667
Block, J. 28
Boswell, D. 170, 438
Bowers, B. P. 29
Britton, P. L. 28
Brody, M. 28
Brown, W. C. 29
Brox, B. 667
Buckley, D. 265
Bullen, S. R. 537
Burr, R. P. 28

Cannizzaro 28
Carr, L. H. A. 29
Cater, S. R. 743
Chaggar, K. S. 724
Chander, K. 613
Chandler, C. 666
Chang, J. J. 30
Chen, S. 613
Chopko, D. A. 28
Christovich, L. C. 667
Chu, T. Y. 28

Clark, E. 170
Clayton, P. A. 383
Cobine, J. D. 29
Cole, H. 744
Coleman, J. R. 613
Coursey, P. R. 29
Cox, D. 28
Cranston, N. H. 744
Creydt, M. 666
Criscenzo, B. E. 744
Cubbin, A. J. 666

D'Andretti, L. 684
Daniels, R. 265
Danielsson, H. 612
Darnell, P. S. 29
De Fodor, E. 30
Dean, D. J. 536, 537
DeVore, J. A. 612, 667
Dickerson, P. 536
Dow, S. 684, 725
Duffeck, E. F. 28
Dummer, G. W. A. 4
Dunn, B. D. 666

Ebensberger 725
Eisler, Dr P. 27, 28
Elliot 613
Elliott, V. B. 724
Engelmaier, W. 613
Esaki, L. 30
Everitt, W. L. 29

Fasching, G. M. 667
Fillion, R. 744
Fitch, W. T. 743
Flicos, D. M. 724
Flood, J. E. 29
Footner, P. K. 208

Foster, F. G. 666
French, D. C. 744

Gamalski, J. 612
Garrow, R. 30
Gerstberger, H. 28
Gleason, J. 170
Glicksman, R. 29
Goldmark, P. C. 29
Goldstine, H. H. 29
Gorcyzca, T. 744
Gunn, J. B. 30
Gust, W. 743
Gustin, J. 265

Hagge, J. K. 667
Hamilton, S. 536
Harmsen, U. 666
Harria, N. 327
Harris, P. G. 724
Haskard, M. R. 100
Haven, R. 170
Headrik, L. G. 724
Helton, D. 684, 725
Hieber, H. 743
Himmel, R. P. 743
Hinch, S. 437
Hlava, P. F. 667
Hobson, C. A. 613
Hockham, G. A. 30
Holm, T. 611, 724
Holmes, P. J. 100
Hsu, F. S. L. 4
Humpston, G. 744
Hunn, H, 684
Hwang, J. S. 667

Irving, S. 437
Isert, H. 725

Jacobson, D. M. 744
James, K. 743
Jansen, C. J. 28
Johnson, J. 30
Jones, M. D. 208
Jonger, H. 28
Junginger, H. 437

Kao, K. C. 30
Keller, J. 613
Keller, J. D. 612
Kelly, M. 29

Kendrick 611
Keogh, R. J. 28
Kershner, R. C. 743
Kirchner, G. 327
Klein Wassink, R. J. 438, 611
Klein, J. G. F. 30
Klotz, P. T. 744
Knödler, D. 724
Koch, H. 667
Konkash, H. 613
Kossowsky, R. 667
Kunzler, J. E. 4

Lambert, L. P. 612
Landis, D. 170
Lassen, C. 30
Lau, J. 170
Lea, C. 170, 208, 327, 611, 613, 724
Lee, E. A. R. 666
Leenaerts, M. H. W. 684
Leibfried, W. 667
Lenz, E. 612
Leonida, G. 612
Lilley, F. 613
Lindner, K. 743
Linman, D. 724
Livh, C. 667
Loasby, R. G. 100
Lozier, G. S. 29
Lubin, G. 28
Lüder, E. 612
Luke, D. A. 28
Lundberg, L. 667
Lynch, J. T. 438

Machusak, D. A. 744
Macklen, E. D. 100
MacRae, A. U. 29
Maerz, M. 30
Maiwald, D. 612
Mallendorf, J. C. 28
Manko, H. H. 30, 611, 667
Marczi, M. 327
Marnott, P. J. 28
Marsh, H. G. 28
Marsten, J. 29
Mathes, H. 612
McLean, D. A. 29
Mehl, R. R. 683
Messner, G. 28, 30
Meyer, C-L. 666
Mimura, S. 170

Miremadi, J. 170
Möller, W. 724
Morehouse, C. K. 29
Motteley 29
Mullen, J. 612
Müller, W. 612, 667

Norgren, S. 612
Notman, J. 170
Nuttall, K. I. 536
Nylen, M. 612

Olson, H. F. 29
Ottoboni, S. 170

Panousis, N. T. 743
Pape, K. 744
Park, C. H. 327
Parker, G. W. 536
Parry, S. 30
Pearne, N. 28, 170
Pearson, R. C. 667
Pfahl, R. C. Jr. 28
Pitt, K. E. G. 100
Postma, L. 28
Potochnik, J. 744
Power, F. S. 29
Pratt, I. H. 743
Prichard, D. J. 208

Quillam, R. M. 724

Reichenecker, W. J. 666
Reynolds, R. 170
Richard, A. A. 613
Richards, B. P. 208
Rider, D. K. 28
Ringle, H. 725
Romig Jr., A. D. 667
Ross, W. M. 666
Rothschild, B. 28
Ruffing, J. F. 724

Sanders, D. 28
Sandström, R, 667
Sargrove, J. A. 27
Scarlett, J. A. 437
Scharr, T. 743
Schlaback, T. D. 28
Schneble, F. W. 28
Schnorr, D. P. 27
Schulz, F. G. 744

Siegel, E. S. 208
Sinnadurai, F. N. 170
Skinkle, M. E. 744
Smith, M. D. 437
Socolovsky, A. 30
Södersved, H. 613
Solomon, H. D. 667
Staller, J. 28
Steen, H. 612, 667
Strauss, R. S. 612, 724
Stroebel, T. 613
Swann, W. F. G. 29
Swigget, R. L. 28

Thomas, R. E. 743
Thompson, D. 613
Thompson, M. K. 28
Thwaites, C. J. 611, 612, 667
Tolvgård, A. 667, 743
Traub, A. C. 613
Trovato, R. A. 327
Tucker, Dr W. B 28
Tyler, R. L. 170

Uddèn, M. 612
Ueltzen, K. 170

Vaihinger, K. U. 724, 725
Vanzetti, R. 613
Verbeek, H. J. 612
Verguld, M. M. F. 438, 684
Vroomans, H. M. 743

Wasserbäch, W. 667
Werner, W. 437
Wernick, J. H. 4
Westerman, J. 612, 667
Wheeler, N. D. 29
Whitfield, I. 666
Whitmore, M. A. 724
Wild, R. N. 667
Wildi, E. 744
Winchell, V. 743
Winkle, R. V. 743
Wirsing, C. E. 743
Wojnarowski, R. 744
Wood, J. C. 29
Woodgate, R. W. 28, 612
Woods, B. 28
Wuich, W. 612

Yost, F. G. 667

Subject Index

(See also separate Author Index)

abietic acid 608, 644
adders 83
additive process
  definition 329
  development 8, 9
adhesion
  of electroless copper 9
adhesive tape packaging SMCs 159
adhesives
  conductive adhesives 286
  for microjoining 741
  for surface mounting 272-284
  heat-curing 323
  manual application in SMA 177
alumina
  for heat dissipation 108
  for manufacture of chip resistors 113
  inherently gives better tolerances 420
  use for LCCC fabrication 148
aluminium
  as a heatsink material 530
  as an impurity in solder 637
  as back-up material for drilling 12
  for base and emitter contacts 23
  for primary cells 25
  for wire bonds 66, 156
  in aluminium-paper or plastic capacitors 43
  in glass capacitors 45
  stick carriers for leaded components 221
  use in microjoining 732
aluminium electrolytic capacitor
  PIH — construction and use 46, 47
  surface mounting — construction and use 120
  vertical and leadless for SMA 128
amines 641
ammonium persulphate 11

ammopack 218
analogue ICs
  applications 80
  definition 77
annotation
  definition 330
ANSI/J-STD-001 601
ANSI/J-STD-002 563, 680
ANSI/J-STD-003 680
ANSI/J-STD-004 640, 644
antimony 71, 634, 637
  as an impurity in solder 637
arsenic 56, 71, 634, 637
  as an impurity in solder 637
artwork
  'English style' 387
  automatic film production 396-400
  definition of artwork master 330
  laser printers 460
  manual layout 384-393
  matrix plotters 459
  pen plotters 458
  use in creating placement program 302
ASCII 237, 467
ASICs 15, 86, 87
  definition and use 86
assembly
  classification by product use 601
  division of components for assembly 32
  increase reliability by using SMCs 106
  types of printed board assembly 103
assembly, automatic, PIH
  advantages and disadvantages of off-line sequencing 240
  automatic insertion machines 222
  availability of equipment 211

Subject Index

board handling 262-265
cutting and clinching leads 235
economics 209
in-line sequencing 228
insertion verification on-line 236
inspection of PBAs 259-262
optical verification systems 234
packaging components for assembly 215-222
programming insertion machines 242-245
rotating heads and tables 233
sequencing of axial components 224
simple aids and pantographs 212, 213
transfer lines 214
assembly, automatic, SMA
a typical pick-and-place machine 291
adhesives for mounting components 272-284
ancillary equipment 322-324
application head, automatic mounting machine 297-304
board loading and positioning 292
component feeding 293-297
conductive adhesives 286
control of automatic placement equipment 304-307
effect of method used on cost of mounting SMCs 290
evolution of pick-and-place equipment 288
integrated assembly lines 324
multiple arm pick-and-place (in-line) machines 315
packages designed with automation in mind 167
parallel, simultaneous or multi mounting machines 316-321
parallel-sequential machines 321
pick-and-place (sequential) machines 309-315
robots 324
selection of equipment 326
solder pastes 285
table-top machines 308
testing components during placement 300, 301
turret heads 303
types of surface mount assembly 266-272
assembly, manual
cleaning after assembly 178
component sequencing 199
creating operator interest 173
hand soldering PBAs 680
inspection of PBAs 206-208
lamp displays as an aid 202
layout memorisation 195
lead trimming 174
minimum equipment requirements for SMA 173
mounting SMCs 177
operator requirements 172
problems with SMCs 176
screen printing as an aid 200
see also 'soldering irons'
sequence assembly 197
simple tools and board holder 173
slide projection as an aid 202
stage assembly 196, 197
the laser scanning system 204, 205
the LED as an aid 203
use of optical fibre light guides 204
uses of 171
astable multivibrators 83
automated optical inspection (AOI) of solder joints 587
automatic assembly of conventional components
see 'assembly, automatic, PIH'
automatic assembly of surface mount components
see 'assembly, automatic, SMA'
axial components
adverse effects of vertical mounting 343
classification of insertion machines 238
equipment for forming leads 187, 190
insertion pitch, specification 342
insertion span, calculation 415-417
lead taping for assembly 215
sequence for manual assembly 175
sequencing for automatic assembly 224
shapes for preformed leads 181-186

backplanes
soldered 348
wire wrapped 347
ball bonding 66, 136, 735, 736
base material
definition 330
base metal
copper 330
beam leads 739
belt soldering 722
BGA — ball grid array 151
BI-MOS 75
bipolar junction transistor (BJT) 58, 75

bismuth
  as an additive to solder 637
  as an impurity in solder 634, 637
  as ruthenate in thick film inks 97
  in solder to aid inspection 591
  solders, composition 633
blowholes 184
bond strength 275
bonding
  components to the substrate 274
  improving the solder bond 616
  the four types of bond between materials 615
Boolean operators
  the basis of digital electronics 78
brass
  for fabrication of square pin inserts 223
  solderability 540
breakdown voltage 25, 42, 53
bridging 186, 286, 370, 592, 594, 634, 711
  definition 594
British Standards Institution (BSI)
  BS 3934 50
  BS 415 383
  BS 4727 329
  BS 6221: Part 22 382
  BS 6221: Part 23 382
  BS 6221: Part 3 382
  BS EN 60097 331
brush and dip fluxing 693
bulk packaging SMCs 157

CAD
  capability needed for printed board layout 446-449
  choosing a system 467
  circuit simulation 444
  design checking ability 452
  development of computer-aided design 14, 15
  hardware and operating systems 462
  outputs for computer aided manufacturing (CAM) 460
  outputs from the system 456-460
  routing conductors 449-452
  schematic entry 440-444
  software for control of inserters 242
  software for control of placement machines 302
  system interfacers 461
  the man/machine interface 466
  the pin list 439
cadmium
  as an impurity in solder 634, 637
  the nickel-cadmium cell 20
  the Weston cadmium cell 19
capacitors
  characteristics 42
  chip 115
  manufacturing technology (non-electrolytic) 43
  marking 49
  MELF 108
  microchip 118
  surface mounting trimming capacitors 129
  uses 40
  variable 49
capacitors, electrolytic
  aluminium 46
  construction 46
  marking 49
  orientation on the board 360
  tantalum 47
  uses 46
CCD 76, 77, 86
CENELEC Electronic Components Committee (CECC)
  CECC 00 802 113
  CECC 50 000 51
ceramic capacitors 45, 110, 219
charge coupled devices (CCDs) 76, 77, 86
chip components 112
  'chip coil' inductors 126
  capacitors 115
  capacitors — manufacture 115
  inductors 119
  microchip capacitors 118
  placement on the board 366, 368
  resistor 112
  resistors — manufacture 113
chip-on-board (COB) 155, 742
chromic acid 13
classification
  of assemblies by product use 601
  of axial component insertion machines 238
  of basic types of PIH inserter 222
  of capacitors by dielectric used 118
  of electronic components 32
  of fluxes to ISO 9454-1 641
  of integrated circuits 74
  of printed boards 334-340
  of solder wetting quality 557
  of surface mount components 104
  of surface mount placement machines 308
cleaning
  cleaning on-line after soldering 700

expect difficulty with SMAs 266
manual assemblies 178
removal of fluxes and flux residues 647
clinching 185, 186, 235, 236, 246, 248, 253, 416
CMOS 76
co-ordinatograph 388
coatings
   for soldering 618-629
component area
   in temperature prediction 492-495, 502-507
   PB-to-component area ratios 343
component holes
   definition 330
component side
   definition 333
components
   'chip-foil' capacitors 120
   capacitors 40
   Cerachip ceramic capacitors 110
   ceramic resonators 111
   chips 112
   comparison of conventional and surface mount technologies 161, 162, 164, 165-169
   conventional or traditional 31
   diodes 50
   diodes, SMT 133
   hybrid integrated circuits 94
   integrated circuits or ICs 74
   MELFs — advantages and disadvantages 106
   mini-MELF and micro-MELF 110
   odd shaped 172
   optoelectronic devices 69
   resistors 32
   SM connectors 131
   surface mounting component (SMC) defined 103
   surface mounting inductors 125
   transistors 58
computer aided manufacture (CAM) 460
   factory automation easier with SMT machines 308
computers
   application of digital ICs 82
   control of dispensing trays in manual assembly 205
   early development of 17
   for control of PIH insertion machines 237
   for control of SMA machines 292, 305, 307, 308
   introduction of the IBM 360 series 102

replacement of mini-computers by workstation hardware 15
the role of the adder in numerical computers 83
use in 'expert' systems 598
use in thermal analysis/design of circuits 519-525
use of add-ons 351
use of the RAM in computers 85
condensation soldering 712-715
conduction
   by electrons 2, 3
   heat supply in soldering 686
   improving heat conduction in hand soldering 681
   of heat — MCM versus PB 727
   thermal, definition 473
   thermal, estimating board temperatures 490-495, 512-514
conductive adhesives 286, 287
conductive pattern 330
   definition 330
conductor
   branched — definition 330
   current carrying capacity 378
   definition 330
   dimensioning conductors 373-380
   spacing — definition 330
   thermal 473
   thermal ladder 484
connectors
   selection of type 348
   contact resistance 578, 730
   contact time 593
contamination
   difficult to remove on SMAs 266
   effects of impurities in solder 634
   effects on wave soldering 599
   in solder, reduces reliability 177
   of insulators by polar/ionisable matter 2
   of SMT adhesives 283
   of solder by jigs and tools 638
   of soldering areas by glue 282
   of soldering areas by SMT adhesive 283
   robots are cleaner than humans! 259
controlled atmosphere soldering 688
convection
   convection soldering 711
   cooling backplane assemblies 348
   cooling using liquids 535
   heat transfer in fluids 473
   least effective for heating solder joints 686

mechanisms 525-527
copper
 a conductive filler for adhesives 287
 as an impurity in solder 635
 definition of base copper 330
 dissolution in molten solder 551
 for jumper wires 223
 for SOD leadframes 133
 in thick film inks 97
 in thin film circuits 94
 rate of solution in tin and tin-lead 672
copper-clad
 definition 330
 introduction of metal-clad laminates 11
 polyimide film used for TAB 153
copper-clad Invar, IEC 1249-7-1 664
corrosion
 due to halogen-containing fluxes 645
 prevention on die by glob topping 156
cost comparison
 German silver versus brass 541
 manual versus automatic inspection 209-211
 of methods of mounting SMCs 290
 of printed boards 344
 of soldering machines 541
counters 83
CQFP — ceramic quad flatpack 142
cracks
 can cause problems in wave soldering 599
 in solder joints 650
 lead trimming can crack joints 176
creep
 curves for different solders 655
 of SMT adhesives 283
crossover 290
cupric chloride 11
curing 274-276, 283, 284, 322-324, 742
cutting 179, 235, 245, 246, 248, 253, 390

defects
 definitions of soldering defects 594-598
 effects on conductors 374
 in board design 425
 in soldering 590-598
 of computer software for thermal design 519
 reduced by soldering in nitrogen 689
 vertical mounting and insertion defects 343
design of printed board assemblies
 backplane mounting 348
 calculating dimensions 342
 calculating gross and net area 356, 358, 359
 component map for assembly and repair 395
 conductor dimensioning 373-380
 coping with heavy components 371
 design conditions for reliable solder joints 544
 designing the board 352-373
 effect of PB construction on thermal properties 478
 estimating effective thermal properties 485
 for PIH (pin-in-hole) assembly 359-363
 for SMC (surface mount component) assembly 364-371
 hole-to-lead ratios 417
 initial assessment 340-342
 mixed SM and PIH components 370
 mother/daughter board assemblies 350
 PB master drawing 393-395
 possible outputs from CAD systems 456-460
 the 'sandwich' structure 351
 the book type connection 349
 the effect of a multiple board solution 346
 the effect of vertical mounting 343
 the effects of a single board solution 345
 thermal design 469
 thermal properties of PBs 480
 use of soldered 'add-ons' 351
 using computer aided design (CAD) 439-468
design rule
 for a 'matrix' 4-layer PB 414
 for a 50 mil grid 407
 for printed board layout 401
 semiconductor conductor width 80
desmearing hole walls 13
desoldering
 using a solder sucker 679
 using fluxed copper braid 683
 using hot gas 290
dewetting
 a measure of wetting stability 575
 definition 594
 of a copper surface 595
 on soldering iron tips 673
 shown by wetting balance curve 616
 troubleshooting guide 599

die-stamping
  first patent filed 7
digital ICs
  applications 82
  definition 78
digitising 396-398
DIN 41429 38
DIN 8527 569
diode
  definition 50
diodes
  characteristic 50
  contact diodes 56
  current regulators 54
  double-anode regulator 53
  manufacturing technology 56
  marking 58, 217
  rectifiers 51
  Schottky diodes 51
  signal diodes 51
  the Esaki diode 54
  tunnel diodes 53
  video detectors 51
  Zener 53
dip soldering 704
dip test 563, 571
DIPs 87-92, 222
  automatic insertion of 250-255
  clinching 235
  gravity feed for 231
  guide for manual insertion 195
  manual insertion of 193
  packing for shipping 222
disc capacitors
  typical forming shapes 193
double-sided boards
  definition 336
  PB-to-component area ratio 343
  relative cost of PB (schematic) 344
drag soldering 27, 704, 705
drilling 12, 390, 395, 414
dross
  composition and recovery 639
  definition 595
  reasons for excessive dross 599
  reduced formation in controlled
    atmospheres 688, 689
dual-in-line packages, see 'DIPs'
ductility 623
dull joints 599
DXF 467

EBCDIC 467
EISA (extended ISA) 464
electrodes, welding 733

electroless copper 331
  definition 331
electroless plating 8, 10
electrolytic capacitors
  41, 42, 46, 120, 122-124, 128
electrolytic copper
  definition 331
electron 2, 20, 80
electronics
  definition 1
  social impact 1
electrons
  conduction by 2, 3
encoders 83
environment
  and board design 355
  and printed boards 16
  nitrogen soldering —environmentally
    friendly 689
epoxide
  first epoxide resin produced in 1938
    10
  in conductive adhesives 287
  PB examples for estimating thermal
    properties 486-488
  printed boards, thermal conductivity
    480
  removal of epoxide smear in holes 13
  resins in SMT adhesives 275, 276
etch resist
  definition 333
  development of organic resists 11
  dry film resists introduced in 1968 11
  introduction of metal etch resists 14
etchants 11
etching
  definition 331
explosive welding 741
extraction 324

ferric chloride 11
field effect transistors (FETs)
  59-61, 70
fillers 275, 287
film resistors 20, 36, 37, 40
fine pitch 152, 169, 689
flexible printed boards
  definition 340
flip-chip bonding 738
  easier and cheaper than wire bonding
    152
flip-chip bumps 153, 733
flip-flop 83-85
  definition 83
fluorocarbons 713

fluxes
  'no residue' flux 647
  and corrosion due to halides 645
  application in mass soldering 692-696
  classification 640
  colophony 644
  density control 695
  early development 27
  flux activators 644
  functions and main constituents 639
  gaseous fluxes 646
  health hazards 602
  in solder paste 648, 649
  low solid content 647
  practical considerations 643
  reduced requirement in nitrogen
    atmospheres 688, 689
  removal of flux and flux residues 647
  selection 643
  soldering without flux 646
  tests for fluxes 640
  water-soluble 645
foam fluxing 694
footprint
  definition 331
  footprints for SMCs 428-437
forming leads
  formed shapes for axial leads 181-185
  machines for preforming 186-190
  non-axial leaded components 190-194
  objectives of preforming 180
  on an automatic inserter 245
  shapes for disc capacitors 193
furnace soldering 723
fusing 14

gallium arsenide 22, 60, 71
gates 24, 60, 76, 82, 83, 86
  definition 82
  gate arrays 86
German silver
  solderability 540
germanium 56, 77
glass capacitors 45
glob topping 154-156, 737, 742
globule test 563
glue, see 'adhesives'
gold
  as a connector finish 348
  as a metal etch resist 14
  as an impurity in solder 636
  ball bonding chip to package leads 66
  dissolution in molten solder 551
  electrodes for microchip capacitors 118
  for die attach by eutectic bonding 66

gold-tin intermetallics 622
gold-tin solder preforms 148
in conductive adhesives 287
in glass capacitors 45
in thick film inks 97
in thin film circuits 94
over nickel as a surface finish 17, 156
problems in soldering 622
removal from component leads 627
safe soldering? 626
use in contact diodes 56
use in microjoining 732
uses in electronics 620
uses in LCCC fabrication 148
graphite 13, 35, 48
grid
  definition 331
grid system
  use and selection of a grid 401-413

halogen-containing fluxes 645
harness 255
health and safety
  in soldering 601-603
heat pipes 535
heat transfer
  and hot components 358
  coefficients 527, 528
  mechanisms and temperature effects 472
  to solder joints 686
heat transfer mechanisms
  conduction 473, 686
  convection 473, 686
  phase-change heat absorption 473
  radiation 473, 686
heating
  preheating systems for soldering 697
  producing soldering defects 596
  stressing LCCCs 150
  with light 721
heatsink 57, 91, 255, 484
  thermal properties 530
hierarchical interconnection technology (HIT) 479
high frequency (HF) soldering 724
high melting point solders 631
history of electronic components 17
  0.25 µm technology arrives 26
  Alessandro Volta invents the voltaic pile 17
  Alexander Graham Bell patents the telephone 19
  Bell Telephone Laboratory announce the solid state transistor 21

Subject Index

Blathy, Deri and Zipernowski patent a power transformer 19
Boyle and Smith report on the CCD 23
British patents on electromagnetic relays in telegraphy 18
C. S. Bradley and the carbon composition resistor 19
Charles Wheatstone develops the Wheatstone bridge 18
D. G. Fitzgerald develops the rolled paper capacitor 19
D. H. Mattox invents ion plating 22
Edison patents the phonograph 19
Emile Berliner, father of disc recording 19
Esaki of Japan invents the tunnel diode 21
first gallium arsenide phosphide LED invented by Holonyak 22
first integrated circuit patent filed by J. S. Kilby 22
first moving coil loudspeaker invented by E. W. Siemens 19
flip-chip bonding is patented by Wiessenstern and Wingrove 22
Frosch develops oxide film maskants for silicon 21
Gambrell and Harris patent a carbon film resistor 20
Gaugain develops a CRT 18
Guglielmo Marconi patents wireless telegraphy 20
H. J. Round and the tetrode 20
Heinrich Hertz proved radio waves exist 20
Hitachi demonstrate a neural computer on a chip 25
IBM launch the personal computer 24
infra-red discovered by Herschel 18
J. A. Fleming patents the thermionic valve 20
J. A. Hoerni invents the planar transistor 21
J. B. Gunn develops the Gunn diode oscillator 22
J. Czochralski develops production of single crystal silicon 20
J. G. Stoney, inventor of the word 'electron' 20
J. J. Thomson demonstrates the nature of cathode rays 20
J. W. Ritter, discoverer of UV, invents the first accumulator 18
Kao and Hockham — pioneers of fibre optics 22
Knight moulds first iron cores in the UK 17
L. Clark invents the standard Clark cell 19
L. de Forest, radio pioneer and 'grandfather of television' 20
Leyden jar — the first? 5
Lombardi files a patent for ceramic capacitors 20
Luigi Galvani studies animal electricity 17
M. Bauer develops the mica capacitor 19
Max Deri, a pioneer in electricity distribution 19
McLean and Power develop the solid electrolyte capacitor 21
Michael Faraday discovers electromagnetic induction 18
Oersted — discoverer of electromagnetism 18
Ohm's law is formulated 18
Planté discovers the lead-acid battery system 18
RCA develop the MOS integrated circuit 22
Robert Noyce — a man with ideas 23
S. Darlington invents the Darlington pair 21
S. Loewe patents a sprayed metal film resistor 21
S. Teszner produces first commercial FET 21
Samuel Morse develops telegraphy 18
surface acoustic wave device patents filed by Rowan and Sittig 22
T. J. Seebeck, discoverer of thermoelectricity? 18
the flatpack invented by Yung Tao (Texas Instruments) 22
the Leclanché cell is invented 19
the Leyden jar 17
the Weston cadmium cell is patented 19
Valdemar Poulsen invents magnetic recording 20
von Kleist or von Muschenbrock, which was first? 17
William Shockley's team invents the solid state transistor 21
history of printed boards and circuits 5
A. W. Franklin patents die-stamping 7
Abner Brenner and the first practical

electroless process 9
Arlt patents process for metal patterns connected by rivets 6
Baekeland discovers phenolic resins 10
Bartels develops a rip-up and re-route autorouter for CAD 15
CAD router with variable grid as a function of density 15
Charles Ducas — an early patent 6
Dahlgreen invents new technique for clad flexible material 10
Dr Paul Eisler — 'the father of printed circuits' 6
Eric Wolfendale develops auto-interactive CAD 15
Franklin patents die stamping 7
G. Parker assists in making first printed circuit radio set 6
G. W. A. Dummer forecasts use of fibre optics in transmission 22
Hazeltine Corporation — the 'multiplanar' process 8
John Sargrove, a pioneer of automatic production equipment 6
Joseph Gerber designs first mechanical-optical photoplotter 15
Micromodule Program (US Army) — precursor of MCM? 7
Motorola — the first PTH in production 8
multilayer MCM arrives 15
Parolini patents an early method of circuit fabrication 6
Pierre Costain produces epoxide resin 10
project 'Tinkertoy' 7
Rhuyssenaers — an early method of board fabrication 6
Rudolf Strauss *et alia* invent the wave soldering machine 14
the push-and-shove router for CAD introduced by Massteck 15
history of soldering
  Barnes, Elliott and Strauss patent wave soldering 27
  first known solder joint for electrical purposes 27
  flux 27
  J. G. F. Klein — first book on soldering in 1760! 26
  the oldest metallurgical joining method known 26
hot bar soldering 721
hot gas desoldering 290
hot gas soldering 722

hotplate soldering 722
hybrid circuits
  development of manufacturing techniques 7
  packaging active components 132
  precursor of SMT 102
  thick film 97
  thin film 94
  use of chip components 112

icicles 599, 634
ICs 74, 77-82, 87, 94-96, 132, 343
IGFET 60
impulse bonded wiring 742
in-line sequencing of PIH components 228
Inconel 733
indium 71, 633
inductors 109, 119, 125, 126
infra-red soldering 711
ink 37, 97
inner lead bonding (ILB) 737
insertion machines
  'typical' axial inserters 247, 248
  'typical' inserters for radial leaded components 248-250
  automatic 222
  DIP insertion machines 250-255
  dual inserters (both axial and radial) 250
  for 'odd' or 'non-standard' packages 255
  purpose built robots 257
insertion span for axial components 415-417
insertion verification for PIH assembly 236
inspection
  automatic inspection equipment 261
  automatic visual inspection 262
  of automatically assembled PBAs 259-262
  of manually assembled PBAs 206-208
  of soldered joints 586-590
  optical recognition equipment in SMT 303
insulation 341, 642
  must be adequate between components 341
integrated circuits
  analogue ICs 77
  digital ICs 78
  level of integration 78
  linear ICs 78
  manufacturing technology 79

interfacing different graphics systems 461
intermetallics 631, 634
International Electrotechnical Commission (IEC)
 definition of LAS and LASCR 70
 definition of SCR 67
 definition of TRIAC 68
 IEC 1249-5-1 373
 IEC 1249-7-1 664
 IEC 191-1 50
 IEC 194 102, 156, 329
 IEC 249-3A 373
 IEC 326-3 382, 418
 IEC 425 37
 IEC 65 383
 IEC 68-2-20 563
 IEC 68-2-3 573
 IEC 68-2-44 563
 IEC 68-2-54 563, 564
 IEC 97 331, 359, 390, 402
International Standards Organisation (ISO)
 ISO 9000 169
 ISO 9453 630
 ISO 9454-1 641, 648
 ISO 9543 648
IPC
 IPC-CM-770 417
 IPC-D-275 382
 IPC-D-330 382
 IPC-SF-818 642
 IPC-T-50E 329
iron
 as an impurity in solder 636
 rate of solution in tin and tin-lead 672
ISA (Industry Standard Architecture) 464
isothermal soldering 740

JEDEC
 12-F 50
 77-A 50, 54, 55, 63
 RS-236-B 58
 RS-370-A 58
JFET 60, 62
jumper
 a zero-ohm resistor 115
 an illustration 329
 automatic insertion 223-225
 definition 331

labels 390
ladder conductor for heat dissipation 484
laminate 11, 344, 349

lamp displays as an aid to assembly 202
land
 definition 331
land pattern
 definition 331
lands
 dimensioning 425-428
LAS 70, 71
laser
 3-D profile inspection 588
 continuous wave (CW) lasers 716
 diodes 71
 for soldering TAb leads 737
 inferometry 657
 laser soldering 715-720
 laser trimming resistors 95
 packages for laser diodes 73
 printers 460
 pulsed lasers 716
 scanning system as an aid for assembly 204, 205
 solid state 71
 types used in soldering 716
layout
 design 353
 manual 384-393
layout memorisation for assembly 195
Lazy Susan workstation for manual assembly 198, 199
LCCC — leadless ceramic chip carrier 148
LCD 24
LDCC — leaded ceramic chip carrier 150
leadframe 79, 92
 in thick copper for high power diodes 133
LED 22, 24, 25, 71, 203
 as an assembly aid 203
legend
 definition 331
light emitting diode 22, 71
light soldering 720
linear ICs
 definition 78
liquid crystal display 24, 73
locating hole
 definition 332
logic circuits
 definition 78
loss angle 42
low melting point solders 631, 632
low volume assembly 173

magazine or cartridge packaging SMCs 158
manual assembly — see 'assembly, manual'
manual methods 171
marking 58, 201
mass production
  of records made possible 19
  using manual insertion, some tooling is essential 178
mass soldering
  advantages of nitrogen atmospheres 689
  belt soldering 722
  cleaning after soldering 700
  condensation soldering 712-715
  controlled atmosphere soldering 688
  convection and forced convection soldering 711
  conveyor 691
  dip soldering 704
  drag soldering 704
  fixtures 691
  fluxes and applicators 692-696
  furnace soldering 723
  high frequency (HF) soldering 724
  hot bar soldering 721
  hot gas soldering 722
  hotplate soldering 722
  infra-red soldering 711
  is often applicable to manual assembly 172
  laser soldering 715-720
  light soldering 720
  machines developed for SMT 709
  made possible by Strauss et alia 27
  nitrogen or inert gas soldering 688
  preheating 696
  process control and machine maintenance 700-703
  reactive gas 688
  reflow soldering 687
  robot soldering 723
  selecting a soldering machine 703
  temperature profiles 686
  the soldering machine 698-700
  ultrasonic (US) soldering 724
  vapour phase soldering 712-715
  wave soldering machine — main stations 690
master drawing 242, 244, 393-395
MCA (micro-channel architecture) 464
MCM-C
  definition 156
MCM-D
  definition 156
MCM-L
  definition 156
Meissner, W.
  Meissner effect 3
melamine 10
MELF 105-111, 368, 425, 429
  compared with mini-MELF and micro-MELF 109
  dimensions and weight 107
  marking 108
  placement on the board 366, 368
  zero-ohm resistors 108
Melinex 44
memories 84
  definition 78
  DRAM 85
  EPROM 85
  PROM 85
  RAM 85
  ROM 84
  SRAM 85
metal cored PBs 150
metallised polyester capacitors 43, 44, 120
metallurgy of microjoining methods
  — principles 727-734
metallurgy of soldering 549-554
mica capacitors 44-46
microjoining
  adhesives 741
  beam lead technology 739
  electrodes 733
  equipment for microjoining 731
  flip-chip technology 738
  materials to be joined 732
  process control 734
  thermodes 733
microprocessor
  central processing unit of all digital equipment 86
Microwire 9, 743
MIL-HDBK-217 170
MIL-STD-202 662, 663
MIL-STD-275 382
mixprint
  definition 332
  description of PBA 103
  PB-to-component area ratio 343
  production cycle 269-273
monolithic integrated circuits 74
  definition 74
Montreal Protocol 16, 686
MOS technology 75
MOSFET 25, 60, 62, 137

motherboard assembly 350
mounting equipment, SMCs
  for manual assembly 173
  multiple arm pick-and-place (in-line) machines 315
  parallel, simultaneous or multi mounting machines 316-321
  parallel-sequential machines 321
  pick-and-place (sequential) machines 309-315
  robots 324
  selection of equipment 326
  table-top 308
multichip module (MCM)
  a comparison with typical PBs 727
  definitions 156
  for small and powerful designs 155
  high density interconnect technology (HDI) 742
  high density, high speed circuits 156
  joining methods used 728
  major types of MCM 156
  multilayer MCM introduced by Kyocera 15
  the pinnacle of hybrid technology? 94
  use worldwide 157
multilayer boards
  clearance hole method of interconnection 338
  definition 336
  Hazeltine perfect the 'multiplanar' process 8
  PB-to-component area ratio 343
  PTH method of interconnection 339
  relative cost of a PB (schematic) 344
  replacement with double-sided boards 165
  the plated pillar method of layer interconnection 339
  use of buried via holes 339
multiplexers 83
multivibrator 83
Multiwire 9, 743
Mylar 44

nickel
  as a diffusion barrier 620
  as a transistor contact material 23
  as an impurity in solder 637
  dissolution in molten solder 551
  film for PTC resistors 108
  for gramophone record master pattern 19
  in nickel-chromium film resistors 37
  in thick film inks 97

nickel-cadmium cell 20
nickel-iron 11, 20, 143
  rate of solution in tin and tin-lead 672
  tin-nickel electrodeposit 14
  to prevent leaching of silver by solder 115
  undercoat for precious metal deposit 14, 17, 156
nitrogen or inert gas soldering 688
nitrogen soldering
  advantages 689
noise 35, 119, 166
  factors affecting noise immunity 166
  suppressors 119
numeric display 72

off-line sequencing of PIH components 240
oil 710
Onnes, K.
  superconductivity 3
operators
  involvement in manual assembly 173
  requirements for manual assembly 172
optical coupler 71
optical fibre 23, 204, 205
  optical fibre displays as assembly aids 204
optical pen 399
optical verification systems
  for automatic insertion machines 234
  in SMA inspection 303
optoelectronic devices
  LAS and LASCR 70
  light emitting diode (LED) 71
  liquid crystal displays 73
  optical coupler 71
  photodarlington 71
  photodiodes 69
  phototransistors 70
  solid state displays 71
  solid state laser 71
outer lead bonding (OLB) 737
oxygen
  as an impurity in solder 638
  for plasma desmearing polyimide materials 13

p-n junction 22, 56, 57, 59
packages
  BGA 151
  capacitors, conventional 40
  conventional integrated circuits 87
  CQFP 142

diodes, conventional 50
flatpacks 92
for MCM 156
for thin and thick film hybrids 97
LCCC 148
LDCC 150
MELF 106
optoelectronic devices 73
PFP 143
PGA 151
pin grid array 92
PLCC 144
PQFP 142
QFP 141
quad-in-line 89
resistors, conventional 33
single-in-line (SIL) 97
SOIC 139
SOJ 147
SOP 139
SOT-143 137
SOT-223 137
SOT-23 132, 135
SOT-89 137
SQFP 142
SSOIC 141
TO-116 251
TO-116 (DIP) tolerances for mounting 419
TO-236 137
TQFP 143
transistors, conventional 62
transistors, TO- series 62-64, 66-69
VSO 141
packaging
  axial components for assembly 215
  COB 155
  DIPS for assembly 221
  disposal of packaging materials 157
  hybrid circuits 97-100
  odd shaped components for automatic insertion 222
  radial leaded components for assembly 218
  SMCs by sticking to paper or plastic tape 159
  SMCs by tape-on-reel 160-163
  SMCs for delivery 157
  SMCs in bulk 157
  SMCs in magazines or cartridges 158
  SMCs in rail or tube magazines 158
  SMCs in trays or palettes 159
  TAB 152
  transistors in metal cans 66
packing density
  can be restricted by previously inserted components 362
  can be very high in plated-pillar MLBs 340
  improvements gained with polymer thick film 10
  restricted by orientation and insertion tooling 363
palladium
  a catalyst for many organic reactions 629
  as a metal glaze for film resistors 37
  as a solderable finish 628
  as palladium-silver for thick film conductors 97
  dissolution in molten solder 551
  in conductive adhesives 287
  palladium complexes produce best PTH adhesion 13
pallet 714
panel
  definition 332
  minimising laminate scrap 432
pantograph positioning in SMA 288, 289
pantograph positioning table 212, 213
paper 43, 47, 344, 642
Pareto 'Law' 386
pattern
  definition 332
PBs — see 'printed board (PB)'
personal computers
  social impact of electronic assemblies 1
  use of add-ons 351
  uses on the factory floor 292
PFP — plastic flat pack 143
PGA — pad grid array 151
phase diagram 605
phase-change cooling 535
phenolic resin discovered by Baekeland 10
phosphorus
  in solder 639
photodarlington 71
photodiodes 69
photographic 388, 390
photoplotting 398-400, 457, 458
photoresist 11
phototransistor 70
pick-up head
  see 'assembly, automatic, SMA, application head'
PIH assembly
  definition 332
  design rules 359-363

Subject Index

selection of a grid 413-421
plastic film capacitor 43
plated-through hole
  definition 333
  filling behaviour in contact with solder 593
  first used by Motorola 8
  hole-to-lead ratio 417
  measuring solderability 571
  see also 'PTH'
plating
  effect of conditions on conductance 3
  electroless, the beginning 8
  problems 629
  solderable finishes 627-629
plating resist 333
  definition 333
platinum
  dissolution in molten solder 551
  in ceramic chip capacitors 116
PLCC — plastic leaded chip carrier 144
point-to-point wiring 346, 347
polyester
  developed for flexible circuits 10
  impregnated fibreglass produced 1940/41 10
  metallised polyester capacitors (PIH) 43, 44
  metallised polyester capacitors (SMT) 120
  photoresist 11
polyimide 10, 11, 13, 287
populated PB
  definition 332
positioning table 179, 198, 212
power rating 34, 55
PQFP — plastic quad flat pack 142
preforming
  machines for axial leads 186, 190
  manual 186
  non-axial leaded components 190
  shapes for axial components 180-184
preheating 560, 696
primary side
  definition 333
printed board (PB)
  a 'surface mount' board 103
  classification of printed boards 334-340
  definition 332
  design 352-373
  essential elements of 328
  metal cored to reduce XY expansion 150
  solderability testing methods 570

the master drawing 393-395
thermal properties 480
printed board assembly
  component map 395
  definition of a PBA 332
  hole-to-lead ratios 417
  rework and repair 681-683
  see also 'design of printed board assemblies'
  thermal expansion mismatch 534
  types of PBA 103
printed board layout
  an example of good layout 413
  automatic film production 396-400
  CAD capability requirements 446-449
  CAD design checking 452
  examples of bad layout 409-413
  laser printers 460
  manual 384-393
  matrix plotters 459
  pen plotters 458
  restrictions imposed by soldering 359
  routing conductors using CAD 449-452
  use and selection of a grid 401-413
  using PIH (pin-in-hole) components 413-421
  using surface mounting components 421-432
printed circuit board
  basic elements 329
  thermal resistance 481
production master
  automatic film production 396-400
  definition 332
  multiple image 334
programmable logic device 87
  FPGA 87
  FPLA 87
  FPLS 87
programming
  automatic insertion machines 242-245
  automatic mounting machines 302
  for chain sequencing 229
PTH
  as a multilayer interconnection 339
  basic illustration 329
  can 'steal' solder 409
  causes of poor solder flow through the hole 599
  definition of a plated-through hole 333
  PB-to-component area ratio 343

relative cost of PB (schematic) 344
reliability relative to different joining methods 727
under SMDs 365
punching 12
pyrophosphate 13

QFP — flatpacks or quadpacks 141
quality assurance
 in microjoining 734
 in soldering 577-600
 QA involvement in board design 354, 361
 statistics in soldering 580-585
quality cost
 in soldering 541, 542

radial leaded component taping 218
radiation 3, 274-276, 284, 473, 686
rail and tube packaging PIH components 222
rail and tube packaging SMCs 158
RAM 23, 24, 85, 86
random access memory 85
rated continuous working voltage (RCWV) 35
reactive gas soldering 688
read only memory 84-86
rectifiers 51, 58, 67
reflow soldering
 'swimming' 431
 an aid for failure analysis 584
 attachment of small outline components 96
 can mean different things 687
 design requirements 428, 429
 surface mounting components 268
 vapour phase reflow introduced 14
reinforcement 11
resins
 desmearing 13
resist
 definition 333
resistance
 adjustment by trimming 96
 calculation 374-376
 contact resistance versus solderability 578
 distribution of contact resistance on copper 730
 of resistors 34-40
 resistor types 32
 thermal, of printed boards 481
 thermal, values for typical packages 509

zero electrical resistance 3
resistivity 3, 94, 287
 effect of reduction in temperature 3
resistor
 an early carbon film resistor patent 20
 characteristics 33
 chip 112
 construction of trimming resistors 130
 in single crystal silicon 94
 laser trimming 95
 manufacturing technology 35
 marking 37
 MELF 105-107
 surface mounting trimming resistors 129
 thin film nichrome 94
 variable 40
rework and repair
 link with testing 354
rework and repair of soldered joints 681-683
 savings resulting from automatic insertion 210
robots
 advantages and disadvantages for SMA 324
 for PIH assembly 257-259
 for soldering in electronics 723
ROM 24, 84-86
rosin 16, 285, 644
rotating heads and tables 233

Schottky 51, 133
SCR 67-70
screen printing
 as an aid to component assembly 200
 for SM adhesive application 281
 for solder paste application 285
 in production of thick film circuits 97
 until mid-sixties, mainstay of PB pattern production 11
screws 52
secondary side
 definition 333
semi-additive process
 definition 333
semiconductor diodes 50
 surface mounting 133
sequence assembly, manual 197
sequencing axial components 224
shear strength 624
shift register 77, 84
 description 84

Subject Index

shrinkage 234
signal diodes 51
silicon 20, 56, 60, 65-68, 74-77, 79, 80, 94
silicon controlled rectifiers 67
silicon-germanium 77
silver
  as a constituent of solder 633
  as a solderable finish 628
  as an impurity in solder 634-636
  as conductors in a proximity fuse 6
  as palladium-silver for thick film conductors 97
  dissolution in molten solder 551
  in conductive adhesives 287
  in conductive paint 13
  in silver-mica capacitors 44, 45
  in thick film hybrid inks 97
  in wet sintered anode tantalum capacitors 48
  overplated with gold 14
single-sided boards
  an advantage of SMT 165
  definition 335
  pad lift by shock or vibration 372
  PB-to-component area ration 343
  relative cost of PB (schematic) 344
slide projection
  an aid to assembly 202
SOIC — small outline integrated circuit synonym SOP 139
SOJ — small outline J-leaded package 147
solder
  balls 286, 648
  cracks in joints 650
  creep curves 655
  fatigue of solder joints 656-665
  high melting point solders 631
  impurities in solder 632-635
  in electronics 629-639
  low melting point solders 631
  mechanisms of joint failure 651-653
  specifications for solders 630
  stress relaxation 655
  tin pest 665, 666
  types of joint strength 653-656
solder balling 648
solder bumps 153, 733
solder globule method for solderability 569-572, 574
solder paste
  an unreliable adhesive for MELFs 107
  checking amount of paste deposited 588
  ease of contamination 177

flux vehicle 648, 649
  in surface mount assembly 269, 285-287
storage 650
solder resist
  definition 333
  for filling or bridging gaps between tracks 283
  for levelling under large packages 283
  possible effects on wave soldering 599
solder side
  definition 333
  projection of leads in PIH assembly 235
solder well
  more suitable for mass soldering PIH assemblies 174
solderability
  advantages of good solderability 561
  advantages of testing 561
  artificial ageing methods 572
  countermeasures for bad solderability 576
  maintaining good solderability 563
  meaning of good and bad 560
  of SM devices must be good 133
  testing methods 563-576
solderability testing methods for printed boards 570
soldering
  'expert' systems 598
  'spans the centuries' 26
  avoid glue on soldering areas 282
  backplane connectors 348
  blowholes created by preform shape 184
  defect definitions 594-598
  defects 590-598
  design for reliable solder joints 544
  ease of soldering various lead preforms 183
  economics of laser soldering 720
  economics of soldering 539-542
  education and training 579
  effect of impurities in solder 632-635
  effect of temperature and time 670
  fatigue of solder joints 656-665
  fluxes — see under 'fluxes'
  for SM soldering check CECC 00 802 113
  good and bad solderability 560
  hand soldering PBAs 680
  health and safety precautions 601-603

history 26, 27
indicators of wetting 555-558
inspection of soldered joints 586-590
inspection prior to soldering is good practice 259
isothermal soldering 740
joint cracking 650
laser solder joint 717, 718
lead type affects ease of soldering 105
major problem areas 542
materials layer build-up for reliable soldering 617
mechanisms of joint failure 651-653
metallurgy 549-554
preferred methods for SMAs 177
preproduction activities for reliable joints 547
removal of solder using a solder sucker 679
removal of solder using fluxed copper braid 683
restrictions imposed on layout 359
see also 'mass soldering'
see also 'soldering irons'
solderable coatings 618-629
solders for electronic purposes 629-639
step or sequential 632
storage of solderable materials 563
surface mount assemblies 366-371
temperature profiles 686
terms and definitions (see also under defects') 603-611
thermal considerations 558-560
troubleshooting chart for wave soldering 599
types of joint strength 653-656
use of statistics 580-585
vapour phase reflow introduced 14
with hot gas 290
soldering irons
criteria for selection 677
description of soldering iron tip 671
heat flow iron-to-workpiece 673
iron-plated tips 671
lowest allowable tip temperature 675
maximum allowable tip temperature 674
measuring tip temperature 675
rated wattage 676
shaped bits needed for SMCs 173
tip temperature and heat output 676
tip temperature/time diagrams 669
tip wear 673

use 668
solid state laser 71
solvents 696
SOP — small outline package synonym SOIC 139
SOT-143 137
SOT-223 137
SOT-23
  construction, dimensions and uses 136
  first small outline transistor package 132
SOT-89 137
spacing
  axial components for taping 215-217
  conductor spacing defined 330
  lead spacing of SOIC/VSO packages 140
spray fluxing 176, 694
spread test 569
spread test for solderability 569
SQFP — shrink quad flatpack 142
SSOIC — shrink small outline IC 141
stability
  of resistors 34-36
stage assembly
  manual assembly line 196, 197
standard cell device 86
statistical process control (SPC) 580-586
  a 'must' for high reliability boards 335
  a vital tool in soldering technology 580
step or sequential soldering 632
step-and-repeat
  an example 334
  definition 333
  to minimise laminate scrap 432
stitch bonding 66, 136
stitchwire 743
Streckfuss bench for component sequencing 199
structural 59
substrate
  advantages of alumina for heat dissipation 108
  for 'chip' resistors 114
subtractive process
  definition 334
sulphur 634, 638
  as an impurity in solder 638
super-VGA (super video graphics array) 465
superconductivity 3
supply packaging for SM assembly 157

surface mounting
  active components 132
  adhesive application by pin transfer 276
  adhesive application by screen printing 281
  adhesive application by syringe 280
  ancillary equipment 322-324
  board layout for SMC assembly 421-432
  commonly used abbreviations 267
  conductive adhesives 286
  criteria for adhesive dots 281
  curing adhesives 283
  definition 102
  design for SMC assembly 364-371
  for mounting equipment, see 'assembly, automatic, SMA'
  integrated assembly lines 324
  mounting components 287
  passive components 105
  problems with SMCs in manual assembly 176
  production cycle, components one side only 268
  production cycle, full mixprint board 272
  production cycle, mixprint with SMCs one side only 269
  production cycle, SMCs on both sides 269
  requirements for SMC adhesives 274
  selection of equipment 326
  solder pastes 285
  storage of boards after applying adhesive 283
surface tension 556

tantalum electrolytic capacitor
  'chip' capacitor 125
  surface mounting — construction and use 124
tantalum electrolytic capacitors
  conventional — construction 47, 48
tape automated bonding (TAB) 152-155, 735-738
tape-on-reel packaging SMCs 160-163, 297
taping
  leaded components for automatic assembly 215-220
  leaded components in sequences for assembly 224-229
TCR, 34, 36
terms and definitions
  printed circuits 329-334

soldering 603-611
soldering defects 594-598
testing
  an early contribution to the science 18
  fatigue properties of solder 662
  insertion verification on-line 236, 237
  off-line (stand-alone) component verification 228
  on-line component verification 227
  SMCs on-line 299-301
  solderability 561-576
  the soldering process 700-702
thermal considerations when soldering 558-560
  see also CECC 00 802 113
thermal design of electronic circuits
  'first look' temperature predictions 489
  accuracy of data 474
  coefficients of thermal expansion (typical) 657
  control of airflow in cabinets 530-533
  design decisions to be made 475
  effect of temperature on failure rate 477
  effects of printed board construction 478
  estimating effective thermal properties 485
  influence of components and cooling methods 470
  major function to ensure systems work 476
  more accurate temperature predictions 510
  phase-change cooling 535
  pitfalls of computer-supported methods 519
  relative effect of Al and Zn heatsinks 530
  temperature versus reliability and performance 474
  thermal expansion mismatch 534, 657
  thermal properties of PBs 480
  use of heat pipes 535
  use of liquid cooling 535
thermal inspection
  of solder joints 589
thermal properties of heatsinks 530
thermal properties of printed boards 480
thermistors
  chip 115
  temperature coefficient 115

thermocompression wire bonding 729
thermodes 733
thick film hybrids
　fabrication 97
　inks 97
thin film hybrids
　definitions 94
　production processes 94-96
through connection 334
　definition 334
throwing power 14
thyristors 67
tin
　as a solderable finish 618-620
　gold-tin solder preforms 148
　known in Mesopotamia 3200 years BC 26
　tin pest 665, 666
　use in microjoining 732
　whiskers 666
tin oxide
　as a conductive film 37
　in LED fabrication 73
　in solder dross 639
　in thick film resistors 97
　thermal conditions for reduction 646
tin pest 665, 666
tin-lead
　as a solderable finish 618-620
　use in coating SMC footprints 268
　use in microjoining 732
　volume resistivity of Sn60Pb40 287
tin-nickel 14
TO-116 251
TO-236 137
tooling hole
　definition 332
TQFP — thin quad flatpack 143
transistors
　characteristics 63
　Darlington pairs 62
　definition 58
　from small to ultra large scale integration 79
　in VLSI and ULSI circuits 24
　switching 25 billion times per second 25
　technology and manufacture 62-67
　the SOT-223 and other packages 137
　the SOT-23 package 136
　uses 58
　usual symbols 59
tray or palette packaging PIH components 222
tray or palette packaging SMCs 159

TRIAC 68, 69
trimming
　film resistors 36
　leads in automatic assembly 235
　leads in manual assembly 174-176
　thick film 'chip' resistors 113, 114
　with hardened steel templates 178
tunnel diode 22, 53
turret heads 303

ultrasonic (US) soldering 724
ultrasonic (US) welding 731
ultrasonic bonding 66
ultrasonic inspection
　of soldered joints 589
UNIX 468
US Department of Defense (DoD)
　withdrawal of MIL-STD-275 382

vapour phase soldering 712-715
varacter 49
variable capacitors 49
variable resistors 32, 33, 40
VCD inserters 241, 260
vertical mounting
　common preforming shapes 182-185
　preforming for automatic insertion 220
　selection of preform shape 182-185
　should be avoided if possible 343
VGA (video graphics array) 465
via holes
　buried vias 337
　definition 334
visual aids 200
visual inspection
　of automatically assembled PIH boards 260-262
　of boards assembled manually 206, 207
　of solder joints 586-588
voids 599
voltage dependent resistors 32
VSO — very small outline 141

water
　as a flux diluent 641, 645, 646
wave fluxing 693
wave shape 599
wave soldering
　a troubleshooting chart 599
　and controlled atmospheres 688-690
　Barnes, Elliott and Strauss patent the wave soldering machine 27
　bridging may occur with SO/VSO

packages 437
conveyor 691
different types of soldering wave 708
double wave recommended for SMAs 268
fixtures 691
fluxing 692
lands and layout 425-427, 431
machines developed for SMT 709
main stations of typical machine 690
oil in the wave 710
preheating 696
process control and machine maintenance 700-703
pulsed wave machines 710
remove excess lead length! 174
selecting a soldering machine 703
surface mount assemblies 366-371
the soldering machine 698-700, 705-711
very large boards need extra support 357
wear
  of finish on spring contacts 349
  of soldering iron tips 673
webbing 599
wedge bonding 735, 736
welding
  diffusion welding 729
  electrodes 733
  equipment for microjoining 731
  explosive welding 741
  impulse bonded wiring 742
  in microjoining 728-734
  resistance welding 729
  Stitchwire 743
  thermodes 733
  ultrasonic (US) welding 731
wettability 555
wetting
  causes of bad wetting 599
  definition 611
  definition of wetting defect 597
  effect of impurities on wetting 634
  indicators of solder wetting 555-558
  quality in terms of wetting angle 557
  variation of wetting time with temperature 593
wetting balance test 564, 574
wetting balance test, scanning method 565
whiskers 621, 666
white residues 599
wire bonding

description of methods 734, 735
flip-chip bonding easier and cheaper 152
similarity with other application methods 728
summary of types 66
wire wound resistors 36
wire wrapping
  backplanes 347, 348
  cold pressure diffusion welding 740
  dedicated automatic pin insertion 223
  developed in the 1940s 9
  pioneered by Bell Telephone Laboratories 21
workshop method of solderability testing 567

X-ray inspection 589
  of soldered joints 589

Zener diode 55, 133
zinc
  as a heatsink material 530
  as an impurity in solder 637
  cans for solid tantalum capacitors 48
  in metallised dielectric capacitors 44

— VOLUME 2 —
Contents

Section 1: Materials and Basic Methods of PB Manufacture

CHAPTER 1
Base Materials

CHAPTER 2
Machining

CHAPTER 3
Screen Printing

CHAPTER 4
Photoimaging

CHAPTER 5
Etch Resists and Etching

CHAPTER 6
Metal Deposition

CHAPTER 7
Other Materials and Processes

Section 2: Major Types of Printed Board

CHAPTER 8
Single- and Double-sided PBs

CHAPTER 9
Plated Through-hole PBs

CHAPTER 10
Multilayer PBs

CHAPTER 11
Flexible Circuits

CHAPTER 12
Polymer Thick-film PBs

CHAPTER 13
Moulded Interconnection Devices

CHAPTER 14
Discrete Wired PBs

Section 3: Cleaning and the Environment

CHAPTER 15
Cleaning and Contamination Control

CHAPTER 16
Water and Effluent Treatment

CHAPTER 17
Environmental Problems

CHAPTER 18
Occupational Health

Section 4: The Attainment of Quality

CHAPTER 19
Quality Assurance and Control

CHAPTER 20
Testing the Product

CHAPTER 21
Training and Education

Section 5: Markets and Finance

CHAPTER 22
Marketing PBs and Assemblies

CHAPTER 23
Product Costing

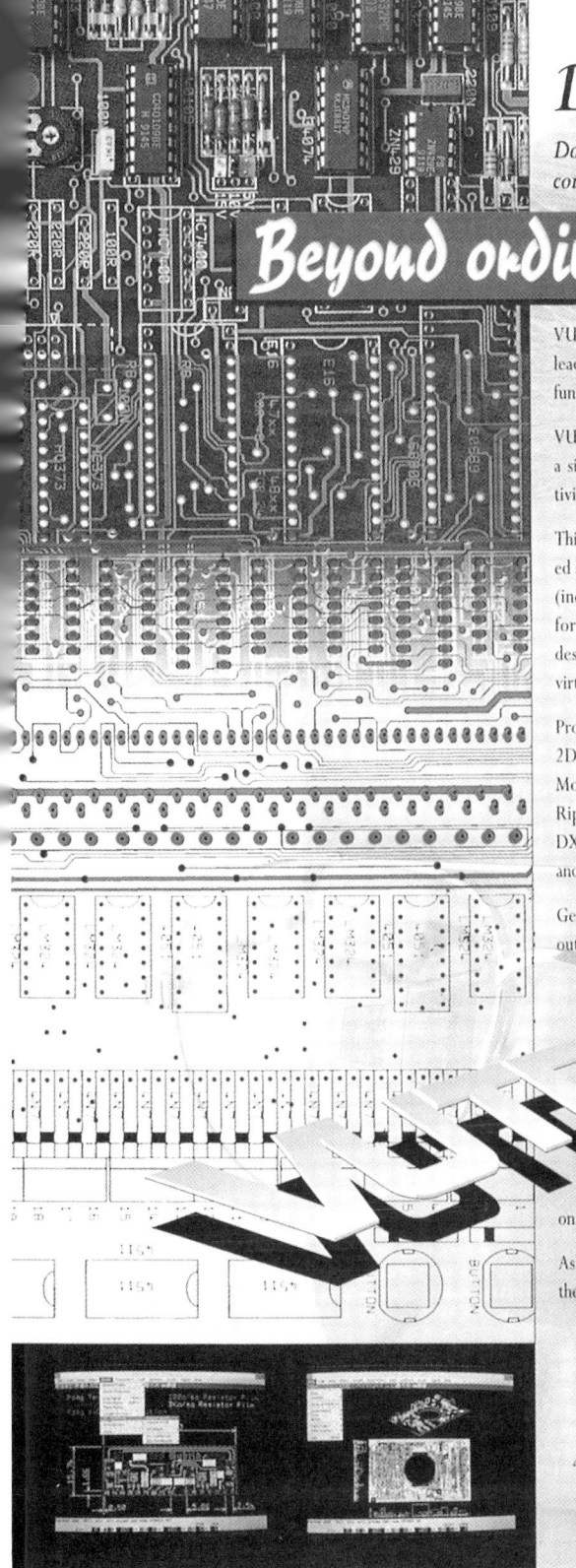

Do your circuits include analogue, RF and microwave design?

Do you need to incorporate mechanical design considerations on and around your boards?

Beyond ordinary PCB CAD

VUTRAX® always goes that one step beyond, to remain the leader in PCB design software for PCs...offering workstation functionality, but at lower cost.

VUTRAX® is modular, so you only buy what you need to create a single integrated design environment, for maximum productivity.

This is not a restricted netlist-driven system, but a fully integrated solution from schematic to manufacture on the same database (including mechanical sketching). So you can back-engineer, forward-engineer and customise a single item or complete design...yet still maintain integrity, to interface reliably with virtually all popular office and factory systems.

Proven facilities include: Schematic Capture, Design Validation, 2D Technical Drawing, SMT, On-line DRC, Bi-directional Modification, Groundplanes & Gluespot/Testpoint Analysis, Ripup & Shape-based Routing, Powerplanes, 3D, Gerber & DXF in/out, CNC, ATE, Pick & Place and Stock Control, Logic and Analogue Simulation.

Getting your job "right first time" is helped by true WYSIWYG output with the ability to mix graphics and text, scale all or part of the design and merge & modify all types of drawing in the same file (unlike other systems).

You also get the highest level of training and support direct from the authors, with modular upgrades to keep ahead of new technologies and benefit from continual development based on feedback from thousands of professional users.

Ask for an honest demo and compare our flexible approach with the rigid route that others impose to conceal their limitations!

COMPUTAMATION
SYSTEMS LIMITED

40 Lake Street, Leighton Buzzard, Bedfordshire LU7 8RX
Telephone 01525 378939 Facsimile 01525 850459

VUTRAX is a registered trademark of Computamation Systems Ltd.

Vantage Supply World Class Chemicals, Processes and Materials to the Electronics Industry

PRODUCTS SUPPLIED FROM VANTAGE ARE BEING USED BY LEADING PRINTED CIRCUIT MANUFACTURERS, COMPONENT MANUFACTURERS AND BOARD ASSEMBLERS. THIS **VANTAGE POINT** PROVIDES OUR CUSTOMERS WITH A UNIQUE INSIGHT INTO THIS INTERRELATED INDUSTRY.

VANTAGE PRODUCTS
- ▲ Tin/Lead Strippers
- ▲ Resist Strippers
- ▲ Copper Cleaners
- ▲ Fluxes and Oils
- ▲ Electronic Grade Solder Bar
- ▲ Enviroclean Range of Enviromentally Friendly Cleaners for Defluxing and removal of organics
- ▲ Waste Treatment Products

Dexter Adhesives Division
- ▲ Frekote mould release compounds

DuPont/Promosol
- ▲ Solderel® Solder Paste
- ▲ Promosol Fluxes
- ▲ Cleaning Products for Electronic Assembly

Pratta Inc.
- ▲ Unioils and Fluxes
- ▲ Evenflow Fluxes
- ▲ Etchants

Shikoku Chemicals
- △ Glicoat SMD - The Best Selling OSP

Shipley Chemicals Limited
- ▲ Electroless Copper Systems
- ▲ Crimson DP System
- ▲ Electroplating Solutions
- ▲ Black Oxide

3M
- ▲ Fluorad™ conformal coatings

Technic Inc.
- ▲ Electroplating processes and equipment for electronic components

Teledyne Electronic Technologies
- ▲ Unicote horizontal hot air levelling equipment
- ▲ DSA2000 Soldermask Spray equipment

VANTAGE CIRCUIT PRODUCTS LIMITED
Egerton Street, Farnworth, Bolton, BL4 7ER, United Kingdom
TELEPHONE 44 (0) 1204 861335 **FACSIMILE** 44 (0) 1204 861334

THICKNESS TESTING OF ELECTROPLATED AND RELATED COATINGS
by G. P. Ray

The purpose of this book is to describe the methods which can be used to measure the thickness of electrodeposited and 'electroless' (autocatalytic) coatings on both metallic and non-metallic substrates, conversion and hot-dipped coatings on metallic substrates, vitreous enamel coatings on metals, anodic coatings on aluminium and other metals and chemical replacement coatings on metallic substrates. Many of the methods, especially some of those based on non-destructive techniques, are equally applicable to organic coatings deposited on metallic substrates. The book attempts to give sufficient details of the test methods for them to be carried out without reference to any of the literature, apart from the instruction supplied by the instrument manufacturers for their particular instruments.

The first part of the book deals with 'Destructive Testing Methods' such as Microscopical examination of the cross-sections, Coulometric, Jet test and Strip and Weigh methods. The second part describes 'Non-destructive methods': Magnetic, Beta backscatter, X-ray Spectrometry, Eddy current and Breakdown voltage methods. Profilometric and Optical methods, which may be described as almost 'non-destructive' techniques, are included in the second part of the book. A brief list of some well-known instrument manufacturers has been included for guidance purposes. The book concludes with a list of current international (ISO), British (BS) and US (ASTM) Standards to assist both metal finishers and the users of their products.

For further information on this and other books, please contact:

Electrochemical Publications Limited
Asahi House, 10 Church Road, Port Erin, Isle of Man IM99 8HD, British Isles
Tel: +44 (0)1624 834941   Fax: +44 (0)1624 835400   E-mail: wela@enterprise.net

MANUFACTURING TECHNIQUES FOR SURFACE MOUNTED ASSEMBLIES
by R. J. Klein Wassink & M. M. F. Verguld   ISBN 0 901150 30 4

Pages—510 + xv; Tables—83; Figures—268; Size—23 x 16 cm. References—326.
ISBN 0 901150 30 4

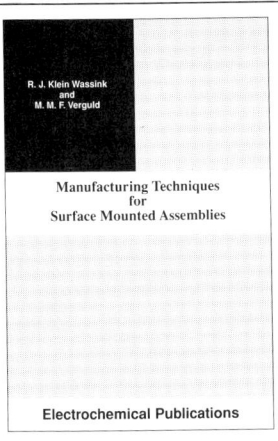

Electrochemical Publications

The book 'Manufacturing Techniques for Surface Mounted Assemblies' has been written for those who are responsible for and active in the development and manufacture of electronic assemblies, typically printed circuit boards carrying surface mounted devices. In view of the subjects treated, this book is indispensable to those employed in R&D and production, such as design engineers, production engineers and their managers. It is, however, not a handy 'do it yourself manual' for beginners.

This book intends to provide practical information, but related to background knowledge and understanding. The authors hope that the understanding provided will be of long-lasting significance to readers.

The book starts with a chapter identifying changes and trends in surface mounting. Components and boards, including flexible substrates, are treated as far as this is useful for the following chapters. A few examples of subjects are coplanarity of leads, multichip modules, ball grid arrays, accuracy and reference on the board, and board finishes. (The manufacturing techniques of printed circuit boards are not addressed.)

The manufacturing flow line for surface mounted assemblies (process integration) forms a central theme in the material discussed. Details of such a line are dealt with at length: the design of footprints on the board, the stencil printing of solder paste, the placement of components, reflow soldering, inspection of the product by visual methods and by using fully automatic techniques, plus repair of faulty connections. A full chapter is devoted to processes for microelectronics (chip on board, connection of foils, etc.), including wire bonding, TAB and adhesive bonding. Ample attention is given to laser soldering.

Considerable attention is paid to the principles of process control, with an elucidation of reflow soldering.

With regard to the solder metals to be used, information is given on compositions, cracking of joints and solderability assessment, etc., with special attention devoted to the matter of lead-free solders. With regard to fluxes and their environmental effects, methods of reducing the amount of flux are evaluated, with special attention to inert gas soldering, for both wave soldering and reflow soldering. The issue of cleaning (or not cleaning) is considered, in relation to the present environmental requirements.

Some 300 literature references are given at relevant places in the text, and a glossary of about 200 terms and acronyms used in the field of SMT is included.

- CHAPTER 1
  Trends in Surface Mount Technology
- CHAPTER 2
  Components and Boards
- CHAPTER 3
  Solder, Fluxes and Environment
- CHAPTER 4
  Technology Integration
- CHAPTER 5
  Guidelines for Design of Footprints
- CHAPTER 6
  Application of Solder Paste
- CHAPTER 7
  Placing of Components
- CHAPTER 8
  Soldering
- CHAPTER 9
  Inspection
- CHAPTER 10
  Automatic Inspection Techniques and Inspection Systems
- CHAPTER 11
  Processes for Microelectronics
- CHAPTER 12
  Process Control in Technologies (like Soldering)
- CHAPTER 13
  Repair of Fine-pitch Components and Very Small SMDs

For further information on this and other books, please contact:

Electrochemical Publications Limited
Asahi House, 10 Church Road, Port Erin, Isle of Man IM99 8HD, British Isles
Tel: +44 (0)1624 834941   Fax: +44 (0)1624 835400   E-mail: wela@enterprise.net

SOLDERING IN ELECTRONICS — Second Edition
A Comprehensive Treatise on Soldering Technology for Surface Mounting and Through-hole Techniques

by R. J. Klein Wassink ISBN 0 901150 24 X

The book 'Soldering in Electronics' has been written to cover soldering as a technique for mass production of electronic connection, although many items treated will be useful for other branches of soldering technology.

It is explained that reliable soldering can only be achieved if soldering is treated as a coherent system, all separate aspects of which are to be well prepared, often long before the actual moment of the soldering operation. This way of thinking has many implications for the other chapters, such as those on solderable coatings, solderability assessment, and joint design.

The book provides a combination of background knowledge and information for practice. In doing so it helps to clarify many points of basic discussion among soldering experts, such as wetting and dewetting, thermal influences, solderability and methods of its assessment, metallurgy of alloys and coatings, and process conditions. On the other hand it contains a wealth of detailed and practical information on soldering alloys, fluxes, coatings, soldering methods, pattern design, joint reliability and criteria for joint inspection. A separate chapter is devoted to the thermal aspects of soldering, a subject that is seldom discussed in literature despite its utmost importance for soldering.

In the second edition of this book (contrasting with the first edition) extensive attention is given to surface mounting technology, with subjects like component metallisations, dimension of footprints of leadless and multi-lead packages, application of solder paste and adhesive, adapted wave soldering and reflow soldering techniques, such as infra-red soldering, vapour phase soldering, resistance soldering and laser soldering.

Specific effects of surface mounting, such as misplacing of components, solder leaching, solder wicking, floating and drawbridging of components are dealt with in detail.

It is the combination of theory and practice which makes the book indispensable to a broad group of readers with a responsibility for electronics manufacturing, both in R & D and production, such as design engineers, process engineers, production engineers, quality engineers, metallurgists and chemists.

For further information on this and other books, please contact:

Electrochemical Publications Limited

Asahi House, 10 Church Road, Port Erin, Isle of Man IM99 8HD, British Isles
Tel: +44 (0)1624 834941   Fax: +44 (0)1624 835400

Setting Up In-house Surface Mount Technology
Practical Management and Technical Guidelines

by David Boswell   ISBN 0 901150 28 2

This book is believed to be the first to provide the necessary combination of management and technical guidelines for setting up and running an in-house surface mount capability. Aimed specifically at small and medium-sized companies, it contains a wealth of detail and data including high and low level management pitfalls, problems and solutions, typical capital investment and set-up costs for a range of budgets and technical guidance at all levels. Its advice is targeted at both beginners and more experienced teams.

The content is partitioned, the first half dealing with the vital changes in organisation, capacity and environment planning, product costing, product safety (hazard and risk analysis) and the management and workforce training needed for successful surface mount production. These factors are analysed for each departmental function in the company, including the chief executive, sales, accounts and site services, as well as the usual design, procurement/storage, manufacturing and quality teams.

The second half covers implementation in the circuit and printed board design stages in detail followed by realisation on the shop floor with reference to equipment selection and set-up for each process step, for materials specification and storage, for kitting, product assembly, inspection and test.

Having described the main set-up procedures, the book moves on to give guidance on the principal known failure mechanisms in assemblies and the related means for assessing product reliability. There are two chapters dealing with trouble-shooting for printed boards and surface mounted components and for each of the assembly processes in turn and a detailed description of the choice of tools for rework on a 'horses for courses' basis.

The book concludes with a series of management issues that include draft terms of reference for each department head and finally a brief project budget summary. The Bibliography and Definitions sections contain an extensive list of sources of help and training.

For further information on this and other books, please contact:

Electrochemical Publications Limited

Asahi House, 10 Church Road, Port Erin, Isle of Man IM99 8HD, British Isles
Tel: +44 (0)1624 834941   Fax: +44 (0)1624 835400

A Scientific Guide To Surface Mount Technology

by Colin Lea   ISBN 0 901150 22 3

The advantages of surface mounting electronic components, rather than insertion mounting leaded components, are a reduced board size and weight, an increased performance and reliability and a fully automated, rationalised production.

The purpose of this book is to give the reader a better scientific understanding of the physical mechanisms involved in the implementation of this novel technology. The aim is that, through a scientific perception, both the manufacturing efficiency and the product reliability will benefit from the revolutionary step of surface mounting, in offering the electronic assembly industry solutions to the new problems that are firmly based, technically and scientifically.

This book presents a full and rationalised coverage of all the major aspects of surface mounting technology, giving the underlying principles involved which in turn offer guidelines for design, assembly, inspection and test. The scientific underpinning of SMT is illustrated using existing components, substrates and soldering techniques, but in a format such that the guidelines can be readily extended for continued use as the technology inevitably progresses.

For further information on this and other books, please contact:

Electrochemical Publications Limited

Asahi House, 10 Church Road, Port Erin, Isle of Man IM99 8HD, British Isles
Tel: +44 (0)1624 834941   Fax: +44 (0)1624 835400

AFTER CFCS?
Options for Cleaning Electronics Assemblies

by C. Lea

Pages—395 + xix; Tables—60; Figures—133; References—263 ISBN 0 901150 25 8

In 1985 a hole was discovered in the Earth's stratospheric ozone layer. In 1987 the scientific evidence became incontrovertible, that the ozone hole was caused by emissions of man-made chemicals—the CFCs. In 1989 a schedule of cuts in production and consumption of CFCs came into force, under the banner of the Montreal Protocol. In 1990 the Protocol was seriously tightened by the London Revision. Before the year 2000 CFCs will be unavailable. By the year 2005 so will the very common solvent 1.1.1 trichloroethane.

High reliability electronics assemblies are cleaned to remove flux residues that prevent automatic electrical testing and also present a potential service reliability hazard. The great majority are cleaned using CFC-113. Many are cleaned using 1.1.1 trichloroethane. The electronics assembly industry is therefore faced with an urgent problem. It needs to find alternatives for these solvents.

This book describes the options available to the electronics assembly industry to address this problem. The scene is first set with an explanation of the mechanisms of ozone depletion and its threat to life on the planet Earth, followed by an account of the fascinating struggle towards an acceptance of worldwide regulatory control of such ubiquitous chemicals as CFCs. Then follows a brief look at how other industrial sectors are coping with the onset of life without CFCs.

All of the technological opportunities that have arisen for the electronics assembly industry, since the discrediting of CFC-113 as a solvent, are presented in detail. The literature has been carefully searched for available data. As can be seen from the contents list, (available on request) this book offers a comprehensive appraisal of the options. The information is given in a totally impartial manner, with the underlying principles of each technology used to illustrate its pros and cons. Suggestions are made as to how a choice finally can be made on both technical and financial grounds, optimised for a company's particular products and profile.

Reviews

"Who should buy—and use—this book? Those carrying responsibility for material, product and process development should have no hesitation in acquiring this book. It is this reviewer's experience, form the many question addressed to him regarding he cleaning issue, that such people frequently find themselves bewildered by the environmental regulations which pertain at the moment. Secondly, the growing army of people concerned with environmental matters in the area of electronics manufacturing will find this book of great help. Finally, the huge amount of general information on cleaning which this book includes means that it will be of use in many situations outside the field of electronics assembly—situations in which the environmental impact of cleaning is becoming an issue of increasing importance."
Soldering and Surface Mount Technology (UK)

"The strengths and weaknesses of this book? The main strength is a thorough, but comprehensible, treatment of the theoretical aspects of the vast subject. Other major positive features are its impartiality and its wide scope.

Do I recommend this book? Unconditionally, yes. This is a must for any technical bookshelf and extremely good value for money, as the wealth of information therein is unique.

As a single edition, it will undoubtedly be the major work for at least a few years. It is a well-written companion to the Electrochemical Publications works, worthy of the same success as Colin Lea's other book in the same series. It is produced with the same care, quality and attention to detail as its stable-mates."
Circuit World (UK)

"One of the most pressing considerations in the electronics industry at present is the problem of cleaning circuit assemblies, Colin Lea, who heads the soldering research facility at the National Physical Laboratory in Teddington, UK, is well known for his publications on the subject of cleaning and alternatives to CFCs. He is also the author of the excellent reference book 'A Scientific Guide to Surface Mount Technology'. With this pedigree, one would expect that a book on alternatives to CFCs would be an outstanding reference document. A am pleased to say that this new publication from Dr Lea does not disappoint.

All in all, this is an excellent work on what is a most important topic. Both the technical content and readability of the book are outstanding."
Hybrid Circuits (UK)

"... as would be expected of an author so well qualified, it is an excellent historical record over the events of the last few years. It is an authoritive review of the present position which should be read by all those having an interest in this special subject."
Microelectronics and Reliability (UK)

ELECTROCHEMICAL PUBLICATIONS LIMITED
Asahi House, 10 Church Road, Port Erin, Isle of Man IM99 8HD, British Isles.
Tel: +44 (0)1624 834941   Fax: +44 (0) 1624 835400   E-mail: wela@enterprise.net

CLEANING AND CONTAMINATION OF ELECTRONICS COMPONENTS AND ASSEMBLIES

by B. N. Ellis

Pages—365 + xxi; Tables—17; Figures—99; References—159; Size—23 x 15 cm
ISBN 0 901150 20 7

This book is a practical guide for all persons who are in any way concerned with cleaning or contamination control of either components or assembled circuits. It is also useful to those specifying components and, in particular, printed circuits. It equally covers the theoretical side to a sufficient extent to allow the average engineer or technician who is not a specialist in the field to understand the mechanisms involved in contamination and cleaning.

As a reference book, the text is divided into seven parts, logically divided into some thirty chapters illustrated by photographs, line drawings, graphs and tables. Each chapter has its own reference list. The introductory section comprises an historical background to cleaning in the electronics industry, a very complete chapter of definitions of all the terms employed and in the particular context, a short chapter on units employed, a theoretical treatise on the mechanics, physics and chemistry of cleaning (in simple terms) and one on the cost of cleaning.

The second part comprises come seven chapters cataloguing the diverse ways that contamination can occur in the electronics industry, whereas the third part describes, over three chapters, what effects contamination can have during the various manufacturing processes of components and assemblies and over the whole of their subsequent lives. Part 4 will be considered as being the most important by some production engineers because its four chapters describe all the currently used methods of cleaning and flux removal for the small, medium or large user with considerable detail on the products usually employed.

The fifth section deals with ionic contamination control. The first two chapters discuss respectively the American military specifications and the new British DEF standards. The third one gives a general view of the different instruments commercially available for measuring or detecting ionic contamination. The last chapter of this section gives an insight into some aspects of the theory of ionic contamination measurement and the solutions used for it. The next part treats the detection and measurement of contamination by other methods, with particular emphasis on non-ionic contaminants. Insulation resistance measurement is discussed in a separate chapter of this section. The last part, divided into three chapters, relates to the particular problems imposed by the use of surface mounted components and solder creams and pastes.

It is felt that this book will become a valuable reference work for the bookshelves of all companies involved in any aspect of electronics, particularly component, printed or hybrid circuit manufacturers or assemblers.

Reviews

"The publishers are to be complemented in adding to their range of reference books for our industry one which will become cleaning in process and who wish to know more of the reasons for the methods which are being recommended. It is the sort of book which you can pick up and read a chapter in isolation and feel that you have read an article of merit...A book well worth having on the reference bookshelf." Circuit World (UK)

ELECTROCHEMICAL PUBLICATIONS LIMITED
Asahi House, 10 Church Road, Port Erin, Isle of Man IM99 8HD, British Isles.
Tel: +44 (0)1624 834941   Fax: +44 (0) 1624 835400   E-mail: wela@enterprise.net